THE FRONTIERS COLLECTION

The books in this collection are devoted to challenging and open problems at the forefront of modern science and scholarship, including related philosophical debates. In contrast to typical research monographs, however, they strive to present their topics in a manner accessible also to scientifically literate non-specialists wishing to gain insight into the deeper implications and fascinating questions involved. Taken as a whole, the series reflects the need for a fundamental and interdisciplinary approach to modern science and research. Furthermore, it is intended to encourage active academics in all fields to ponder over important and perhaps controversial issues beyond their own speciality. Extending from quantum physics and relativity to entropy, consciousness, language and complex systems—the Frontiers Collection will inspire readers to push back the frontiers of their own knowledge.

More information about this series at http://www.springer.com/series/5342

Boris Murmann · Bernd Hoefflinger
Editors

NANO-CHIPS 2030

On-Chip AI for an Efficient Data-Driven World

Springer

Editors
Boris Murmann
Department of Electrical Engineering
Stanford University
Stanford, CA, USA

Bernd Hoefflinger
Sindelfingen, Baden-Württemberg, Germany

ISSN 1612-3018 ISSN 2197-6619 (electronic)
The Frontiers Collection
ISBN 978-3-030-18337-0 ISBN 978-3-030-18338-7 (eBook)
https://doi.org/10.1007/978-3-030-18338-7

This Springer imprint is published by the registered company Springer Nature Switzerland AG
The registered company address is: Gewerbestrasse 11, 6330 Cham, Switzerland

Contents

Editors and Contributors

About the Editors

Boris Murmann received his Ph.D. degree from the University of California, Berkeley in 2003 and serves as a professor of electrical engineering at Stanford University. His research interests are in mixed-signal integrated circuit design, with a focus on sensor interfaces, data converters, and custom circuits for embedded machine learning. He has served as an associate editor of the IEEE Journal of Solid-State Circuits, an AdCom member, and a distinguished lecturer of the IEEE Solid-State Circuits Society, as well as the data converter subcommittee chair and the technical program chair of the IEEE International Solid-State Circuits Conference (ISSCC). He is a fellow of the IEEE.

Bernd Hoefflinger became an assistant professor at Cornell University, Ithaca, NY, USA, after completing his Ph.D. at the Technical University of Munich, Germany. He was a co-founder of the MOS Division of Siemens in Munich and founded the electrical engineering department of the University of Dortmund, Germany, which houses the first Ion-Implanted BiCMOS production line. After serving as the head of the electrical engineering departments at the University of Minnesota and then at Purdue University in Indiana, he established the Institute of Microelectronics Stuttgart, Germany, as the first ISO 9000-certified research and manufacturing facility—a leader in ASICs, HDR vision, and e-beam-driven nanotechnology.

Contributors

Masoud Babaie received the Ph.D. (cum laude) degree from the Delft University of Technology, Delft, The Netherlands, in 2016. In 2006, he joined the Kavoshcom Research and Development Group, Tehran, where he was involved in designing wireless communication systems. From 2009 to 2011, he was a CTO of that company. From 2014 to 2015, he was a visiting scholar researcher with the Berkeley Wireless Research Center, Berkeley, CA, USA. In 2016, he joined the Delft University of Technology, where he is currently an assistant professor (tenured). He has co-authored one book, two chapters, 11 patents, and over 40 technical publications. His current research interests include RF/millimeter-wave integrated circuits and systems for wireless communications and cryogenic electronics for quantum computation. He has been a committee member of ISSCC Student Research Preview since 2017 and will join as a technical program committee of ESSCIRC in 2020. He was a co-recipient of the 2015–2016 IEEE Solid-State Circuits Society Pre-Doctoral Achievement Award and the 2019 IEEE ISSCC Best Demo Award. In 2019, he received the Veni Award from the Netherlands Organization for Scientific Research.

Harry S. Barowski studied physics at the University of Regensburg, where he graduated in low-temperature physics 1994. He finished his doctoral degree in 1997 on far and infrared spectroscopy on various types of cuprate-based high-temperature superconductors. After finishing his doctoral study, he started his career as logic designer of mainframe processors at IBM Research and Development in Boeblingen, Germany. Here he worked in logic design as logic design engineer professional of computational units for high-performance processors for P and Z systems and gaming consoles (Cell processor for Sony Playstation). He also was engaged in the field of formal logic verification and custom circuit design. He currently is focussed on custom array design and physical and logical design verification methodologies. As IBM master inventor, he drives technical innovations and new technologies. He contributed to several aspects of IBM system Q, such as design of classical hardware components to the IBM Q System One, superconducting chip design and was coaching the IBM Extreme Blue internship on implementation of the well-known HHL algorithm to solve a system of linear equations into Qiskit.

Andrew Bartolo received the B.S. and M.S. degrees in computer science from Stanford University, Stanford, CA, USA, where he is currently working toward the Ph.D. degree in computer science. His work focuses on hardware and software support for computation immersed in memory. His current research interests include the fields of architecture, compilers, digital design, and networking.

Luca Benini holds the chair of digital circuits and systems at ETHZ and is a full professor at the Università di Bologna. His research interests are in energy-efficient parallel computing systems, smart sensing micro-systems, and machine learning hardware. He has published more than 1000 peer-reviewed papers and five books. He is a fellow of the IEEE, of the ACM and a member of the Academia Europaea.

Alexander Bleitner received the B.Sc. and M.Sc. degrees in embedded systems engineering from the University of Freiburg, Freiburg, Germany, in 2013 and 2017, respectively, where he is currently pursuing the Ph.D. degree at the Fritz Huettinger Chair of Microelectronics, Department of Microsystems Engineering—IMTEK. From 2010 to 2013, he was an intern at Siemens Medical Solutions in Forchheim,

Germany. There, he developed electronic circuits and FPGA designs for medical imaging. Additionally, in 2015, he pursued an internship at the Bosch Research and Technology Center in Palo Alto, CA, USA. During this internship, he was working on image processing algorithms and integrated circuit designs for a low-power image sensor. Currently, his main research interests evolve around ultra-low-voltage and ultra-low-power circuits using Schmitt trigger structures in the sub-threshold region with a focus on digital standard cell and memory design.

Markus Brink as a physics graduate student in Paul McEuen's laboratory at Cornell University, investigated electronic properties of low-dimensional nanostructures using cryogenic scanning force microscopy combined with electronic transport measurements. He graduated from Cornell University in 2006 and started as postdoctoral associate in QLab with Michel Devoret at Yale University, where he built and characterized his first superconducting qubit circuits. In 2010, he joined IBM as a researcher at the T. J. Watson Research Center, where he has worked on scaling of classical (CMOS) and quantum hardware.

Carlo Gilardo received his B.S. (2015) and M.S. (2017) in engineering physics at Politecnico di Milano, Milan, Italy, and is currently pursuing his Ph.D. in electrical engineering at Stanford University, Stanford, CA. His current research interests include modeling and simulation of nanoscale carbon nanotube electronic devices and technology assessment and benchmarking.

Edoardo Charbon (SM'00–F'17) received the Diploma from ETH Zurich, the M.S. from the University of California at San Diego, and the Ph.D. from the University of California at Berkeley in 1988, 1991, and 1995, respectively, all in electrical engineering and EECS. He has consulted with numerous organizations, including Bosch, X-Fabs, Texas Instruments, Maxim, Sony, Agilent, and the Carlyle Group. He was with Cadence Design Systems from 1995 to 2000 and with Canesta from 2000 to 2002. Since 2002, he is a member of the faculty of EPFL, full professor since 2015. From 2008 to 2016, he was a full professor and chair of VLSI design with Delft University of Technology. He has been the driving force behind the creation of CMOS SPAD technology, which is mass produced since 2015 and the core of telemeters, proximity sensors, and medical diagnostics. His interests span from 3D vision, LiDAR, FLIM, FCS, NIROT to super-resolution microscopy, time-resolved Raman spectroscopy, and cryo-CMOS circuits and systems for quantum computing. He has authored or co-authored over 350 papers and two books, and he holds 21 patents. He has received several best paper awards and the prestigious award for best academic research team in Europe (London, 2019). He is a distinguished visiting scholar of the W. M. Keck Institute for Space at Caltech, a fellow of the Kavli Institute of Nanoscience Delft, and a fellow of the IEEE.

E. J. Chichilnisky is the John R. Adler Professor of Neurosurgery and Professor of Ophthalmology, at Stanford University, where he has worked since 2013. Previously, he worked at the Salk Institute for Biological Studies for 15 years. He received his B.A. in mathematics from Princeton University and his M.S. in mathematics and Ph.D. in neuroscience from Stanford University. His research has focused on understanding the spatiotemporal patterns of electrical activity in the retina that convey visual information to the brain, and their origins in retinal circuitry, using large-scale multi-electrode recordings. His ongoing work now focuses on using basic science knowledge along with electrical stimulation to develop a novel high-fidelity artificial retina for treating incurable blindness. He is the recipient of an Alfred P. Sloan Research Fellowship, a McKnight Scholar Award, a McKnight Technological Innovation in Neuroscience Award, and a Research to Prevent Blindness Stein Innovation Award.

Francesco Conti is a postdoctoral researcher at the IIS Laboratory, ETH Zurich, Switzerland, and the EEES Laboratory, University of Bologna, Italy, where he received his Ph.D. in 2016. His research focuses on enabling sophisticated AI capabilities on ultra-low-power computers, working on the full pipeline "from algorithm to silicon" to achieve that goal. He has co-authored more than 40 papers in international conferences and journals, and he has been the recipient of three Best Paper Awards and the 2018 HiPEAC Tech Transfer Award.

Wim Dehaene was born in Nijmegen, The Netherlands, in 1967. He received the M. Sc. degree in electrical and mechanical engineering in 1991 from the Katholieke Universiteit Leuven. In November 1996, he received the Ph.D degree at the Katholieke Universiteit Leuven. In the beginning of his career, he joined Alcatel Microelectronics, Belgium. There he was a senior project leader for the feasibility, design, and development of mixed mode systems on chip. The application domains were telephony, xDSL, and high-speed wireless LAN. In July 2002, he joined the staff of the ESAT-MICAS laboratory of the Katholieke Universiteit Leuven where he is now a full professor and head of the MICAS division. His research domain is the circuit-level design of digital circuits. The current focus is on ultra-low-power signal processing and memories in advanced CMOS technologies. Part of this research is performed in cooperation with IMEC, Belgium, where he is also a part-time principal scientist. He is a senior member of the IEEE. He was the technical program chair for ESSCIRC 2017. He is a member of the ESSCIRC/ESSDERC steering committee and the ESSCIRC technical program committee. He has also served for several years on the ISSCC program committee.

Albert Frisch studied quantum optics and atomic physics at the Institute of Experimental Physics at the University of Innsbruck where he graduated in 2014. His PhD thesis on Dipolar Quantum Gases of Erbium was awarded the thesis prize by the Institute of Quantum Optics and Quantum Information at the Austrian Academy of Sciences. In 2015, he worked at this institute as a postdoctoral research scientist before he joined IBM Research and Development in Böblingen, Germany. He started as a circuit design engineering professional working on IBM's high-performance processor series P and Z. He primarily focussed on array design and physical design verification. Further, he co-developed an automated build process for synthesized soft arrays. Since 2017, he was additionally engaged in the hardware and software development for IBM Q. He contributed on several layers of the system including the design of classical hardware components for the IBM Q

System One, the superconducting chip design, as well as the quantum software development kit Qiskit. He co-supervised an Extreme Blue internship out of which an efficient implementation of the well-known HHL quantum algorithm for solving linear systems of equations was developed and contributed to Qiskit. Today, he is working on ion trap quantum computers.

Jacob Goeppert was born in Lahr, Germany, in 1984. He received the Dipl.-Ing. (M.Sc.) degree in microsystems engineering from the University of Freiburg, Germany, in 2010. From 2010 to 2019, he was with the Fritz Huettinger Chair of Microelectronics, Department of Microsystems Engineering (IMTEK), University of Freiburg, Germany, where he worked in his Ph.D. degree. During this time, his research focus lay on of ultra-low-voltage, ultra-low-power digital circuits, and thermoelectrical energy-harvesting interfaces. Since 2019, he is with Hahn-Schickard, Villingen-Schwenningen, Germany, where he is working as a research engineer continuing his work on low-voltage digital circuits in addition to inference hardware for machine learning systems.

Thorsten Hehn received the Dipl.-Ing. degree in microsystems engineering from the University of Freiburg, Germany, in 2006 and the Dr.-Ing. degree in microsystems engineering from the University of Freiburg in 2014.

From December 2006 to September 2012, he was a research assistant with the Fritz Huettinger Chair of Microelectronics, Department of Microsystems Engineering (IMTEK), University of Freiburg. From December 2006 to December 2009, he was a fellow in the graduate school micro energy harvesting, funded by the German Research Foundation (DFG).

In October 2012, he joined Hahn-Schickard, Villingen-Schwenningen, Germany, as a research assistant in the group "Energy Autonomous Systems." As of February 2016, he is leading the group "Electronic Systems" which develops low-power embedded hardware and software for sensor systems. His research interests include low-power electronics, cyber-physical sensor systems, power management, energy harvesting, and stress measurement.

Daniel Hoffmann received his Dipl.-Ing. degree (M.Eng.) in mechanical engineering from the Technical University of Ilmenau, Germany, in 2002. From 2002 to 2006, he was a research scientist at the Tyndall National Institute in Cork, Ireland, and obtained his Ph.D. degree in microelectronics engineering from University College Cork, Ireland, in 2006.

In 2007, he joined the group "Energy Autonomous Systems" at Hahn-Schickard in Villingen-Schwenningen, Germany, where he worked in the field of kinetic energy harvesting systems as a postdoctoral research fellow.

From 2016 to 2020, he was the head of the group "Energy Autonomous Systems" at Hahn-Schickard. His main field of research was the design and optimization of kinetic energy harvesting devices with emphasis on frequency-tunable devices and rotational systems. He now joined the group "Inertial Sensor Systems" at Hahn-Schickard working on low-power and zero power microsystems.

Makoto Ikeda received the B.E., M.E. and Ph.D. degrees in electrical engineering from the University of Tokyo, Tokyo, Japan, in 1991, 1993 and 1996, respectively. He is a professor at the University of Tokyo. His research interests including hardware security, asynchronous circuits design, smart image sensor for 3D range finding and time-domain circuits for associate memories. He is a program vice-chair of International Solid-State Circuits Conference 2020 (ISSCC 2020). He was a technical program chair (2016–2017), a symposium chair and an executive committee member for the symposium on VLSI circuits, a technical program chair for Asian Solid-State Circuits Conference 2015 (A-SSCC 2015) and also a chair for IEEE SSCS Japan Chapter. He is an elected IEEE SSCS AdCom member for 2020–2022. He is a senior member of IEEE, IEICE Japan, and member of IPSJ and ACM.

Shiro Kamohara received the B. S. degree in physics from Keio University in 1986, M. E. degree from University of Tokyo Institute Technology in nuclear engineering in 1988 and Ph.D. from Tokyo metropolitan University in electrical and electronic engineering in 2008. He joined the Central Research Laboratory of Hitachi Ltd. in 1988. Since 1995, he was the member of Semiconductor & and Integrated Circuit Div. of Hitachi Ltd. He is a member of Renesas Technology Corp since 2003 and Renesas Electronics Corp since 2010. He is now the director of business unit for SOTB products. He was the visiting industrial fellow of the University of California at Berkeley in 1996.

Alireza Kaviani is a distinguished engineer at Xilinx Research Laboratories with a focus on the next-generation FPGA architectures and tools. He has more than 25 years of FPGA and ASIC industry experience in the areas of architecture, tools, IC design, and applications. He has authored more than 55 patents and publications in a number of areas, including clocking, asynchronous design, FPGA architecture, and CAD tools. He is a senior IEEE member and holds a Ph.D. degree from University of Toronto in electrical and computer engineering.

Katsuyuki Kimura received the M.S. in information and computer science from Keio University in 2002. His major is processor and LSI architecture. From 2002 to 2016, he worked on the hardware of media processors, multimedia CODECs and image recognition at Center for Research and Development, Toshiba Corporation. Since 2017, he has been working on the development of an automotive SoC at Electronic Devices and Storage Research and Development Center, Toshiba Electronic Devices and Storage Corporation.

Dr. Binh Le is an assistant professor of electrical engineering at San Jose State University and a researcher at Stanford University. His current research interests include non-volatile memory, carbon electronics, brain-inspired computing, brain–machine interface, and especially high-performance and energy-efficient analog/digital machine learning systems using emerging memory technologies. His recent publication on the topic was highlighted in the Nature Electronics journal, February 2019 issue. He received his B.S. degree in electrical engineering and computer sciences from the University of California, Berkeley, in 1994, his M.S and Ph.D. degrees in electrical engineering from Stanford University in 1999 and 2004, respectively. He has more than 20 years of experience working in the semiconductor industry where his last position was the director of Design Engineering at SanDisk Corporation. He is a senior member of the Institute of Electrical and Electronics Engineers (IEEE) and has authored or co-authored 59 US patents.

Zhichun Lei was born in Qian'an, China, in 1964. He received the B.S. degree in communication technology from Tianjin University, China, in 1986, the M.S. degree in communication and electronic system from Tianjin University in 1989 and the Ph.D. degree in electronic engineering and information technology from University of Dortmund, Germany, in 1998. From 1989 to 1993, he was a scientific assistant in Department of Electronic Engineering, Tianjin University, China. From 1994 to 1998, he was a scientific assistant at Chair for Communication Technology, Circuits and Systems Laboratory, University of Dortmund, Germany. From 1998 to 2001, he was a scientific assistant at Philips Research Laboratory, Aachen, Germany. From 2001 to 2011, he was an engineer at Sony European Technology Center in Stuttgart, Germany. Since 2011, he has been a professor at Institute of Measurement Engineering and Sensor Technology, University of Applied Sciences Ruhr West, Germany. Since 2014, he has held a second professorship at School of Microelectronics, Tianjin University, China. His research interests include high dynamic range and wide color gamut.

Haitong Li is a Ph.D. candidate in electrical engineering at Stanford University, supervised by Prof. H.-S. Philip Wong. He received M.S. in electrical engineering from Stanford University in 2017 and B.S. in microelectronics from Peking University, China, in 2015. His research theme is centered around energy-efficient machine learning hardware enabled by emerging nanotechnologies (e.g., 3D resistive memories), with over 30 publications and 800 citations to date. He received 2019 IEEE EDS Ph.D. Student Fellowship, 2016 IEEE EDS Masters Student Fellowship, Best Paper Award at 2016 SRC TechCon, and Best Paper nomination at 2016 Symposium of VLSI Technology.

Yiannos Manoli holds a B.A. degree (summa cum laude) in physics and mathematics, a M.S. degree in electrical engineering and computer science from the University of California, Berkeley, and the Dr.-Ing. degree in electrical engineering from the Gerhard Mercator University, Duisburg, Germany.

He was a research assistant at the University of Dortmund, Germany, before joining the Fraunhofer Institute of Microelectronic Circuits and Systems in Duisburg in 1985. There he established a design group for microsystem and microcontroller integrated circuits. In1996, he was appointed as a chair of microelectronics in the department of electrical engineering at the University of Saarland, Saarbrücken, Germany.

He joined the department of microsystems engineering (IMTEK) at the University of Freiburg, Germany, in 2001, where he holds the Fritz Huettinger Chair of Microelectronics. Since 2005, he additionally serves as a director of the Hahn-Schickard Institute in Villingen-Schwenningen, Germany.

He received Best Paper Awards from ESSCIRC 1988, 2009, and 2012, MWSCAS 2007, MSE 2007, and PowerMEMS 2006. For his creative and effective contributions to the teaching of microelectronics and the design of a web-based visualization of circuit functionality (Spicy VOLTsim, www.imtek.de/svs), he received various awards including the Excellence in Teaching Award of the University of Freiburg and the Teaching Award of the State of Baden-Württemberg, both in 2010.

Subhasish Mitra is a professor of electrical engineering and of computer science at Stanford University, where he co-leads the computation focus area of the Stanford SystemX Alliance and is a faculty member of the Wu Tsai Neurosciences Institute. He also holds the Carnot Chair of Excellence in NanoSystems at CEA-LETI in Grenoble, France. His research ranges across robust computing, NanoSystems, electronic design automation (EDA), and neurosciences. Results from his research group have been widely deployed by industry and have inspired significant development efforts by government and research organizations in multiple countries. Jointly with his students and collaborators, he demonstrated the first carbon nanotube computer and the first three-dimensional NanoSystem with computation immersed in data storage. These demonstrations received widespread recognitions: cover of NATURE, Research Highlight to the U.S. Congress by the NSF, and highlights by news organizations worldwide. In the field of robust computing, he and his students created key approaches for soft error resilience, circuit failure prediction, CASP on-line self-test and diagnostics, and QED design verification and system validation. His X-Compact test compression has proven essential to cost-effective manufacturing and high-quality testing of almost all electronic systems. X-Compact and its derivatives have been implemented in widely used commercial EDA tools. His honors include the ACM SIGDA / IEEE CEDA Newton Technical Impact Award in EDA (a test of time honor), the Semiconductor Research Corporation's Technical Excellence Award (for innovation that significantly enhances the semiconductor industry), the Intel Achievement Award (Intel's highest corporate honor), and the Presidential Early Career Award for Scientists and Engineers from the White House. He and his students have published many award-winning papers at major venues. He is a fellow of the ACM and the IEEE.

Dante Gabriel Muratore is an assistant professor of Microelectronics at the Delft University of Technology. He received the B.S. degree and the M.S. degree in electrical engineering from Politecnico of Turin in 2012 and 2013, respectively. He received the Ph.D. degree in microelectronics from University of Pavia in 2017. From 2015 to 2016, he was a visiting scholar at MTL laboratories at the Massachusetts Institute of Technology. From 2016 to 2020, he was a postdoctoral fellow at Stanford University. He is the recipient of the Wu Tsai Neurosciences Institute Interdisciplinary Scholar Award. His research focuses on hardware design for brain–machine interfaces, bioelectronics, sensor interfaces, and machine learning.

Clara Nieto Taladriz Moreno was born in Madrid, Spain, in 1994. She received the B.S. with honors in Ingeniería de Tecnologías y Servicios de Telecomunicación from the Univeridad Politécnica de Madrid (UPM), Spain, in 2016. She got her M.S. degree with honors in telecommunication engineering from the UPM in 2018, after completing the second year in the University of KU Leuven (KUL), Belgium. Since 2018, she has been a research assistant at the MICAS division of the department of electrical engineering (ESAT) at the KU Leuven. She has been working on the design of ultra-low-voltage energy-efficient digital circuits under the supervision of Prof. Wim Dehaene.

Kunle Olukotun is the Cadence Design Systems professor of electrical engineering and computer science at Stanford University. Olukotun is well known as a pioneer in multicore processor design and the leader of the Stanford Hydra chip multiprocessor (CMP) research project and the founder of Afara Websystems, which designed the Niagara processors. Olukotun currently directs the Stanford Pervasive Parallelism Laboratory (PPL), which seeks to proliferate the use of heterogeneous parallelism in all application areas using domain-specific languages (DSLs). Olukotun is an ACM fellow and IEEE fellow.

Zvi Or-Bach is the founder and CEO of MonolithIC 3D™ Inc. He is a world recognized expert in monolithic 3D technologies, past chairman of the 3D of IEEE S3S Conference, and is active as an invited speaker and tutorial instructor in the USA, Korea, and Japan. He is currently the chairman of the Board for Zeno Semiconductors and VisuMenu. Prior to founding and running MonolithIC 3D since 2009, he founded eASIC in 1999 and served as its CEO for six years. eASIC was funded by leading investors, such as Vinod Khosla and KPCB, in three successive rounds. Intel acquired eASIC in 2018. Under his leadership, eASIC won the prestigious EETimes' 2005 ACE Award for Ultimate Product of the year in the Logic and Programmable Logic category and the Innovator of the Year Award and was selected by EE Times to be part of the "Disruptors—The people, products and technologies that are changing the way we live, work and play." Earlier, he founded Chip Express in 1989 and served as the company's president and CEO for 10 years, bringing the company to $40M revenue. Chip Express was acquired by GigOptix which was later acquired by IDT. He received his B.Sc. degree (1975) cum laude in electrical engineering from the Technion-Israel Institute of Technology, and M.Sc. (1979) with distinction in computer science, from the Weizmann Institute, Israel. He holds over 251 issued patents.

Rebecca Park received her B.S. from Cornell University and is currently a Ph.D. candidate in electrical engineering at Stanford University, under the supervision of Professor H.-S. Philip Wong and co-advised by Professor Subhasish Mitra. Her current research interest is in the development of high-performance and energy–efficient nanoelectronics, in which she has focused on carbon nanotube-based FETs. She is a recipient of the Intel/SRCEA Masters Scholarship (2014–2016), the Intel/SRCEA Ph.D. Fellowship (2016–2019), and a finalist to participate in the Rising Stars Women in Engineering Workshop (2018). She has interned at IBM (summer 2017) working on carbon nanotube transistors and at Apple (summer 2018) as a flat panel display engineer.

Raghu Prabhakar is a senior principal engineer and one of the founding engineers at SambaNova Systems. His research interests include designing high-level programming models, compiler optimizations, and novel architectures with a focus on reconfigurable hardware. He holds a Ph.D. in computer science from Stanford University, where he was advised by Prof. Kunle Olukotun and Prof. Christos Kozyrakis. He earned a M.S. in computer science from UCLA. He is a member of IEEE and ACM.

Robert M. Radway is a Ph.D. student in electrical engineering at Stanford University. His research focuses on performance benefits, design, and thermal considerations for monolithic 3D systems (through the DARPA 3DSoC Program). He has worked on emerging non-volatile memories such as resistive RAM, aiming to increasing effective density (multiple bits-per-cell, 1T4R designs) and improving resilience through development of endurance management techniques. He completed his bachelor's and master's at the Massachusetts Institute of Technology in EECS. For his Master's thesis, he developed designs and processes for thermally efficient GaN HEMTs fabricated via wafer bonding.

Dennis Rich is currently pursuing a Ph.D. at Stanford University, advised by Prof. Subhasish Mitra. He received his dual B.S. in electrical engineering and engineering physics from the University of Illinois at Urbana-Champaign, investigating thin-film device fabrication advised by Prof. Can Bayram. His current research interests include addressing system-level design and fabrication challenges of emerging nanotechnologies and nanosystems. Previously, he worked at Northwestern University investigating carbon nanotube fabrication and at Silicon Laboratories. He is a recipient of the Goldwater Scholarship, the Bardeen Undergraduate Award, and the Robert C. MacClinchie Scholarship.

Peter Hans Roth received his Dipl.-Ing. degree in electrical engineering and his Dr.-Ing. degree from the Stuttgart University in 1979 and 1985, respectively. In 1985, he joined the IBM Germany Research and Development Laboratory, starting in the department of VLSI logic chip development. Since 1987, he has been leading the VLSI test and characterization team of the Boeblingen Laboratory. Later on, he was leading several development projects in the area of IBM's mainframe and power microprocessors. He also was heavily involved in the development of gaming products like the cell processor for the Sony Playstation. He owned also the hardware strategy for the IBM Germany Research and Development Laboratory. In early 2017, he formed a quantum computing team in the IBM Boeblingen Laboratory, and today he is still involved on several quantum engagements.

Ulrich Rueckert received the diploma degree (M.Sc.) with honor in computer science and the Dr.Ing. degree (Ph.D.) with honors in electrical engineering from the University of Dortmund, Germany, in 1984 and in 1989, respectively. He joined the Department of Electronic Components, University of Dortmund, in 1985, where he developed the first VLSI implementation of artificial neural networks in Europe. In February 1990, he accepted the position as a senior researcher at the Department of Electronic Components, University of Dortmund. From 1993 to 1995, he was an associate professor of microelectronics technology of the University of Hamburg-Harburg. From 1995 until 2009, he was a full professor of electrical engineering at the University of Paderborn. As a member of the Heinrich-Nixdorf Institute, he held the Chair "System and Circuit Technology". Since 2009, he has been at the Cognitive Interaction Technology Center of Excellence at the University of Bielefeld, heading the Research Group Cognitronics and Sensor Systems. Since 2001, he has been adjunct professor of the Faculty of Information Technology at the University of Technology, Brisbane, Australia. In 2006, he received the first Innovation Award of North-Rhine Westfalia (together with his colleague Prof. Noé). His main research interests are bio-inspired architectures for neural information processing and cognitive robotics. He has held many international positions, including the committee for the European Brain project.

Dr. Manuele Rusci is currently a postdoctoral researcher at the Energy-Efficient Embedded Systems Laboratory, University of Bologna, Italy. He received the Ph.D. degree in electronics from the same university in 2018. His main research interests include low-power embedded systems and AI-powered smart sensors, working at the intersection between machine learning and HW-SW system co-design.

Mohamed M. Sabry Aly is an assistant professor at Nanyang Technological University, Singapore. He received his Ph.D. degree in electrical and computer engineering from École Polytechnique Fédérale de Lausanne (EPFL), in 2013. He was a postdoctoral research fellow at Stanford University from 2014 to 2017. His current research interests include system-level design and optimization of computing systems enabled by emerging technologies. He is an active close collaborator with Stanford University and a founding member of the N3XT project at Stanford University. He was the recipient of the Swiss National Science Foundation Early Postdoctoral Mobility Fellowship in 2013.

Daniel A. Sanchez was born in Tepic, Mexico. He received his B.Sc. degree in electronic engineering from the Western Institute of Science and Technology (ITESO), Guadalajara, Mexico, in 2006, and the M.Sc. in microsystems engineering from the University of Freiburg, Germany, in 2012.

From 2006 to 2009, he worked in the Research and Development Center at Continental Automotive, Guadalajara, Mexico, designing instrument clusters for passenger cars. Between 2010 and 2011, he was with Corporate Technology, Siemens AG, Munich, Germany, working on piezoelectric harvesters. Between 2012 and 2017, he was with the Fritz Huettinger Chair of Microelectronics, University of Freiburg, Germany, working toward his Ph.D. degree, where he researched about efficient interface circuits for energy harvesting devices with special focus on integrated circuit design. Since 2017, he is with Hahn-Schickard in Freiburg, Germany, where he leads the microelectronics group.

He received second place in the statewide basic sciences challenge, Nayarit, Mexico, in 2002. In 2009, he was awarded with a DAAD-CONACYT scholarship and in 2016 with a Fritz Huettinger fellowship.

Fabio Sebastiano received the B.Sc. (cum laude) and M.Sc. (cum laude) degrees in electrical engineering from the University of Pisa, Italy, in 2003 and 2005, respectively, the M.Sc. degree (cum laude) from Sant'Anna school of Advanced Studies, Pisa, Italy, in 2006, and the Ph.D. degree from Delft University of Technology, The Netherlands, in 2011. From 2006 to 2013, he was with NXP Semiconductors Research in Eindhoven, The Netherlands, where he conducted research on fully integrated CMOS frequency references, deep-submicron temperature sensors, and area-efficient interfaces for magnetic sensors. In 2013, he joined Delft University of Technology, where he is currently an assistant professor. He has authored or co-authored one book, 11 patents, and over 60 technical publications. His main research interests are cryogenic electronics for quantum computing, quantum computing, sensor read-outs, and fully integrated frequency references. He has been a member of the "Emerging technologies" subcommittee of the technical program committee of the RFIC symposium. He was a co-recipient of the best student paper at ISCAS in 2008, the Best Paper Award at IWASI in 2017, and the Best IP Award at DATE in 2018. He is a distinguished lecturer of the Solid-State Circuit Society.

Daniel Schillinger was born in Freiburg im Breisgau, Germany, in 1985. At the Albert Ludwig University, Department of Microsystems Engineering–IMTEK, Freiburg im Breisgau, Germany, he received the bachelor's degree in microsystems engineering in 2009 as well as the master's degree in 2012. After that he became a Ph.D. candidate in the Fritz Huettinger Chair of Microelectronics.

Since 2017, he is additionally a part of the electronics systems design group, Hahn-Schickard, Villingen-Schwenningen, Germany, where he is working in the field of energy harvesting and RFID.

His current research interests are in efficient power processing circuits for vibration-based energy harvesters, where he is focusing on integrated circuit design as well as implementations with discrete components, and the power management for wireless and self-sustaining autonomous sensor systems. Currently, he is also involved in the development of long-lasting highly efficient LED drivers for lighting applications as well as in the development of a telemetric interface for a strain measurement chip in industrial applications.

He was a recipient of the German Research Foundation scholarship GRK 1322: Micro Energy Harvesting, in 2013.

Markus Strobel received his degree in electrical engineering (Dipl.-Ing.) from the University Stuttgart, Germany, and heads the Department Vision Sensors at the Institute for Microelectronics Stuttgart (IMS CHIPS). He has been with IMS CHIPS since 1997 and focusses on CMOS Imaging including the development of logarithmic high dynamic range CMOS (HDRC) image sensors, optical characterization, optical and electrical test environments as well as camera system integration for automotive, autonomous, industrial and custom-specific applications. Recent research topics cover high dynamic range imagers with linear characteristics as well as photonic or plasmonic nanostructures to build multispectral CMOS sensors.

Nobuyuki Sugii received the B.S., M.S. and Ph.D. degrees in applied chemistry from the University of Tokyo in 1986, 1988 and 1995, respectively. He joined the Central Research Laboratory, Hitachi, Ltd. in 1988. Since 1996, he has been working on the research and development of CMOS devices, including strained silicon and SOI. From 2010 to 2015, he served as a research group leader for the Low-Power Electronics Association and Project and developed ultralow-power thin-BOX FDSOI (named SOTB) process and design environment for ICs operating down to 0.4 V. From 2004 to 2015 and 2018–2020, he served as a visiting professor with the Tokyo Institute of Technology. He is currently with the Research and Development Group, Hitachi, Ltd., where he is researching on sensing devices and systems, and energy management systems fully utilizing renewables. He has been a technical committee member of the IEEE VLSI Technology (2012-2019), the International Symposium on Solid-State Devices and Materials (2014–2017) and the IEEE S3S Conference (2011-). He is a fellow of the Japan Society of Applied Physics.

Roel Uytterhoeven was born in Sint-Truiden, Belgium, in 1992. He received the M.S. degree in electrical engineering magna cum laude from KU Leuven, Belgium, in 2015. He is currently working as a research assistant with the ESAT-MICAS laboratories at KU Leuven. Here, he pursues a Ph.D. degree under the supervision of Prof. Dr. Ir. Wim Dehaene in the field of ultra-energy-efficient digital circuits through near/subthreshold supply voltage operation. For this research, he is being sponsored by the Research Foundation Flanders (FWO) with an SB-fellowship.

Bob Vanhoof was born in Leuven, Belgium, in 1994. In 2017, he received the M.Sc. degree in electrical engineering from the KU Leuven with a master's thesis about the implementation on FPGA of the IEEE802.11ad PHY. Currently, he is a research assistant at MICAS where he is working toward the Ph.D. degree. His research interest lies in the fields of ultra-low-energy design and SRAM design.

Marian Verhelst is an associate professor at the MICAS laboratories of the EE department of KU Leuven. Her research focuses on embedded machine learning, hardware accelerators, HW-algorithm co-design, and low-power edge processing. Before that, she received a Ph.D. from KU Leuven in 2008, was a visiting scholar at the BWRC of UC Berkeley in the summer of 2005, and worked as a research scientist at Intel Laboratories, Hillsboro, OR from 2008 to 2011. She is a member of the DATE and ISSCC executive committees, is TPC co-chair of AICAS2020 and tinyML2020, and TPC member DATE and ESSCIRC. She is an SSCS distinguished lecturer, was a member of the Young Academy of Belgium, an associate editor for TVLSI, TCAS-II, and JSSC, and a member of the STEM advisory committee to the Flemish Government. She currently holds a prestigious ERC Starting Grant from the European Union and was the laureate of the Royal Academy of Belgium in 2016.

Andrei Vladimirescu received the M.S. and Ph.D. degrees in EECS from the University of California, Berkeley, where he was a key contributor to the SPICE simulator, releasing the SPICE2G6 production-level SW in 1981. He pioneered electrical simulation on parallel computers with the CLASSIE simulator as part of his Ph.D. He is the author of "The SPICE Book" published by J. Wiley and Sons. For many years, he was R&D director leading the design and implementation of innovative software and hardware electronic design automation (EDA) products for Analog Devices Inc., Daisy Systems, Analog Design Tools, Valid Logic, and Cadence Design Systems. Currently, he is a professor involved in research at the University of California at Berkeley, Delft University of Technology, and the Institut Superieur d'Electronique de Paris, ISEP, as well as a consultant to industry. His research activities are in the areas of design, simulation, and modeling of CMOS circuits, new devices, and circuits for quantum computing. He is an IEEE life fellow.

Gordon Wetzstein is an assistant professor of electrical engineering and, by courtesy, of computer science at Stanford University. He is the leader of the Stanford Computational Imaging Laboratory and a faculty co-director of the Stanford Center for Image Systems Engineering. At the intersection of computer graphics, machine vision, optics, scientific computing, and applied vision science, his research has a wide range of applications in next-generation imaging, display, wearable computing, and microscopy systems. Prior to joining Stanford in 2014, he was a research scientist in the Camera Culture Group at MIT. He received a Ph.D. in computer science from the University of British Columbia in 2011 and graduated with honors from the Bauhaus in Weimar, Germany, before that. He is the recipient of an NSF CAREER Award, an Alfred P. Sloan Fellowship, an ACM SIGGRAPH Significant New Researcher Award, a Presidential Early Career Award for Scientists and Engineers (PECASE), an SPIE Early Career Achievement Award, a Terman Fellowship, an Okawa Research Grant, the Electronic Imaging Scientist of the Year 2017 Award, an Alain Fournier Ph.D. Dissertation Award, and a Laval Virtual Award as well as Best Paper and Demo Awards at ICCP 2011, 2014, and 2016 and at ICIP 2016.

H.-S. Philip Wong is the Willard R. and Inez Kerr Bell Professor in the School of Engineering. He joined Stanford University as a professor of electrical engineering in September 2004. From 1988 to 2004, he was with the IBM T.J. Watson Research Center. At IBM, he held various positions from research staff member to manager and senior manager. While he was a senior manager, he had the responsibility of shaping and executing IBM's strategy on nanoscale science and technology as well as exploratory silicon devices and semiconductor technologies. His research aims at translating discoveries in science into practical technologies. His works have contributed to advancements in nanoscale science and technology, semiconductor technology, solid-state devices, and electronic imaging. His present research covers a broad range of topics including carbon electronics, 2D-layered materials, wireless implantable biosensors, directed self-assembly, device modeling, brain-inspired computing, non-volatile memory, and monolithic 3D integration. He is a fellow of the IEEE. He served as the editor-in-chief of the IEEE Transactions on Nanotechnology (2005–2006), subcommittee chair of the ISSCC (2003–2004), general chair of the IEDM (2007), and is currently the chair of the IEEE Executive Committee of the Symposia of VLSI Technology and Circuits. He is the faculty director of the Stanford Non-Volatile Memory Technology Research Initiative (NMTRI) and is the founding faculty co-director of the Stanford SystemX Alliance—an industrial affiliate program focused on building systems.

Yutaka Yamada received the B.E. and M.E. degrees from Keio University, Yokohama, Japan, in 2003 and 2005, respectively. He joined Toshiba Corporation, Kawasaki, Japan, in 2005. He was involved in the research and development of reconfigurable processors and image processing accelerators. Since 2017, He is currently working as a research engineer in the Research and Development Center, Toshiba Electronic Devices and Storage Corporation, Kawasaki, Japan. His current research interests include the area of image recognition accelerators, image signal processors and system architecture for automotive SoCs. He is a member of IEEE and IEICE.

Ms. Xin Yu was born in Zhangjiakou, China, in 1995. She has received the B.S. degree in communication engineering from Hebei University of Technology, Tianjin, China, in 2017 and received the M.S. degree in information and communication engineering from Tianjin University, Tianjin, China, in 2020. From 2017 to 2020, she was a postgraduate at School of Microelectronics, Tianjin University, China. Her research interests include image and video acquisition and reproduction, image and video codec. As of March 2020, she will work as a software engineer at Huawei Technologies Co., Ltd.

Yaqi Zhang is a Ph.D. candidate in the electrical engineering department at Stanford University. She received a B.S. in electrical engineering from Duke University. Her research has been focusing on the architectural design of and compilation to reconfigurable hardware accelerators. Specifically, she has worked on an on-chip network design that improves the scalability of the accelerator. She also worked on compilation techniques that enhance programming abstraction and enhances the utilization and composability of reconfigurable architectures.

Chapter 1
The New Era of Nano-chips: Green and Intelligent

Boris Murmann and Bernd Hoefflinger

Since their invention in 1959 by Robert Noyce, silicon integrated circuits have followed a unique history of steep exponential progress. Moore's Law, which was articulated by Noyce's friend and partner Gordon Moore in 1964, drove the semiconductor industry into a widely agreed upon roadmap of doubling the number of transistors per chip every 18 months. Guided by the "International Technology Roadmap for Semiconductors (ITRS)," this strategy worked well until about 2010, and was driven by mass-produced memory chips and von-Neumann computing architectures ranging from microcontrollers to microcomputers and supercomputers. The dynamics that shaped this epoque and how it changed from bipolar to CMOS technology leadership was described in the 2012 edition of "CHIPS 2020—A Guide to the Future of Nanoelectronics" [1]. Here, it was also predicted that the ITRS would end in 2016 (at 10 nm) and that future progress would be driven by

- The need for entirely new levels of energy efficiency,
- Ultra- low voltage Fully Depleted Silicon-on-Insulator (SOI) CMOS,
- 3D Integration,
- Intelligent, neuromorphic architectures,
- Human-Visual-System (HVS)-inspired video.

Shortly after the ITRS program ended in 2015, "CHIPS 2020, Vol. 2—New Vistas in Nanoelectronics" was published [2] and delivered a broad range of contributions focusing on the above-listed topics. Since then, the continuing global wave

B. Murmann (✉)
Stanford University, Stanford, CA, USA
e-mail: murmann@stanford.edu

B. Hoefflinger
Sindelfingen, Germany
e-mail: bhoefflinger@t-online.de

B. Murmann and B. Hoefflinger (eds.), *NANO-CHIPS 2030*,
The Frontiers Collection, https://doi.org/10.1007/978-3-030-18338-7_1

toward innovative, sustainable, energy-efficient and intelligent nano-chip systems has inspired us to compile this book on a vision for 2030 and beyond. As shown in Table 1.1, we identified five major thrust areas that are covered by 28 chapter contributions from world-leading experts. Different from previous versions of this book, a larger fraction of the presented material is application focused, aiming to highlight the challenges that new applications will define for the semiconductor industry (see e.g., Chap. 25, "Augmented and Virtual Reality"). For the remainder of this introduction, we briefly discuss the positioning and interplay of these contributions within each thrust.

1.1 Robust and Energy-Efficient Silicon

The International Technology Roadmap for Semiconductors (ITRS) had been critically evaluated in 2012 in CHIPS 2020 [1] and in 2014 in CHIPS 2020 Vol. 2 [2], leading to the prediction that it would end in 2016 at the 10 nm node. And indeed, the ITRS program ended in 2015 with a forecast limit of 10 nm, and the birth of the "International Roadmap for Devices and Systems" (IRDS) [3]. The IRDS inherited rich know-how and data from the ITRS and is now being continuously updated under the umbrella of the IEEE. Important focus areas of the IRDS are highlighted in Chap. 2, including the unique and sustained importance of silicon, as well as the need for more 3D Integration. These topics were already at the core of [1, 2] and continue to be the main technology underpinning for this book. The ITRS ended mainly because of diminishing gains in speed and in energy efficiency of von-Neumann computer architectures. This perceived wall increased the interest in "Rebooting Computing," a program that is reviewed in Chap. 2. Rebooting Computing started in 2012 and focuses on long-term research such as quantum computing, which gets special attention in Chaps. 26 and 27. In addition, Chap. 16 provides an update on trends in more conventional supercomputing platforms.

The IEEE conference S3S ("Silicon-on Insulator, 3D Integration and Sub-Threshold MOS") also gets a special mention in Chap. 2, because it reflects the technology base of [1, 2], which is substantially expanded in the present book, particularly in Chaps. 4, 6–11, 13–15, 19, 20, and 23. The unique and fundamental importance of the Silico/Silicon-Dioxide system is emphasized again in Chap. 4 for its lasting significance over the coming decades. Its optimum incorporation for processing functions in complementary MOS (CMOS) is highlighted in Chap. 4. Its optimum downscaling for process complexity, stability, speed, voltage and energy is presented in Chap. 6 for the most advanced realization of fully depleted CMOS Silicon-on-Insulator (SOI). Such low-power technologies are propelling a variety of applications, such as energy-autonomous microcontrollers for the IoT (Internet of Things). A critical aspect here is robustness, as the underlying circuits are typically operated in subthreshold and at very low supply voltages. Chapter 7 takes a look at this problem in the context of robust and energy-optimal design using differential-transmission-gate logic. Finally, manufacturing at the advanced nodes continues to

Table 1.1 Overview of chapter contributions

Chapter		Robust and Efficient Silicon	Real-World Electronics	Neuromorphic Architectures	AI On-Chip and 3D Integration	Man-Machine Cooperation
2	IRDS—International Roadmap for Devices and Systems, Rebooting Computing, S3S	✓		✓	✓	
3	Real-World Electronics	✓	✓	✓	✓	✓
4	Silicon Complementary MOS in its 7th Decade	✓	✓		✓	
5	Nanolithography	✓				
6	The Future of Ultra-Low-Power SOTB CMOS	✓	✓			
7	Dealing with the Energy versus Performance Tradeoff in Future CMOS Digital Circuit Design	✓	✓		✓	
8	Monolithic 3D Integration—An Update	✓			✓	
9	Heterogeneous Monolithic 3D Nano-Systems: The N3XT Approach	✓			✓	
10	High-Speed 3D Memories Enabling AI Future	✓			✓	
11	3D for Efficient FPGA	✓	✓			
12	Digital Neural Networks	✓	✓	✓	✓	✓
13	Enabling Domain-specific Architectures with Programmable Devices	✓			✓	
14	Coarse-Grained Reconfigurable Architectures	✓			✓	
15	A 1000x Improvement of the Processor-Memory Gap	✓	✓		✓	
16	High-Performance Computing Trends	✓		✓	✓	
17	Analog-to-Information Conversion	✓	✓		✓	
18	Machine Learning at the Edge	✓	✓		✓	✓

(continued)

Table 1.1 (continued)

Chapter		Robust and Efficient Silicon	Real-World Electronics	Neuromorphic Architectures	AI On-Chip and 3D Integration	Man-Machine Cooperation
19	The Memory Challenge in Ultra-Low Power Deep Learning	✓	✓		✓	✓
20	Multi-Sensor, Intelligent Microsystems	✓	✓			✓
21	High-Dynamic-Range and Wide-Color-Gamut Video	✓	✓		✓	✓
22	Update on Brain-Inspired Systems	✓	✓	✓		✓
23	Energy-Autonomous Chip-Systems	✓	✓			✓
24	Artificial Retina: A Future Cellular-Resolution Brain-Machine Interface	✓	✓			✓
25	Augmented and Virtual Reality		✓		✓	✓
26	Cryogenic-CMOS for Quantum Computing	✓			✓	
27	Quantum Computing—Large-scale Quantum Systems based on Superconducting Qubits	✓			✓	
28	Man-Machine Cooperation and Cognitronics	✓	✓	✓	✓	✓
29	Efficient System-on-Chip (SOC) for Autonomous Driving with High Safety	✓	✓	✓	✓	✓

be challenging. The most expensive part, nanolithography, is handled by one of the Focus Teams in the IRDS structure, and it is reviewed in Chap. 5.

The CMOS technology base described in Chaps. 4, 6 and 7 is essential and virtually un-contested for providing continuous future growth, if combined with sustained efforts in 3D integration for sensing, memory and actuating, all aimed at realizing intelligent, energy-efficient architectures for real-world electronics.

1.2 Real-World Electronics

As electronic systems increase their interaction with the real world and strive to become significantly more power efficient, many of the traditional "brute force" data acquisition and signal processing approaches are being called into question. Chapter 3 of this book motivates this general trend and underlines the importance of log-domain perception, which is further underpinned in Chap. 21 on high dynamic range video. Chapter 17 advocates the concept of "Analog-to-Information" (A-to-I) conversion along similar lines as a replacement to conventional analog-to-digital conversion interfaces, which are bound to hit fundamental efficiency limits in the coming decade. An instantiation of A-to-I concepts is also found in Chap. 24, which details a massively parallel and data-compressive interface for cell mapping in the human retina. Finally, as many modern sensor interfaces to the real world take on the shape of large arrays, new ways to interface and integrate these with silicon must be found. Chapter 21 presents a cutting-edge example on a 3D-integrated photonic system for LIDAR and thereby builds bridges to Chap. 11 on 3D ASICS, as well as Chap. 29 on autonomous driving.

1.3 Neuromorphic Architectures and the Human Visual System

Reverse engineering and mimicking the brain has been an intriguing research direction in our long-standing quest on achieving the ultimate compute efficiency for intelligent systems. Recently, renewed interest in this topic has been fueled in part by large investments into the European Human Brain Project as well commercial activities such as Intel's Lohi development. Chapters 12 and 22 provide an overview of these activities and review the state of the art in brain-inspired architectures.

Ultimately, neuromorphic design is linked to our knowledge base in neuroscience, which nowadays is tightly coupled to progress in brain-machine interfaces and artificial intelligence research [4]. The dynamics between these fields are as exciting as ever and are beginning to inform the design of devices that would have been deemed science fiction not too long ago. An example covered in this book pertains to the next generation of artificial retina devices as described in Chap. 24. This application pushes our silicon capabilities to the limit and may enable the first high-fidelity prosthesis for restoring sight for age-related blindness. Another strong technology pull is expected to come from the processing needs for augmented and virtual reality (see Chap. 25), which will potentially redefine how we communicate, collaborate and learn. Chapter 28 expands on this trend with a more general discussion of man-machine collaboration and cognitronics and its technological needs.

1.4 AI on-Chip and 3D Integration

The past decade has brought renewed interest in deep neural networks as a cornerstone in our quest toward artificial intelligence (AI). While some of the core concepts behind these networks have been established decades ago, they are only now becoming mainstream with substantial application pull. The main factors behind this trend are the availability of immense amounts of data for training, as well as powerful computer hardware that can handle these data at high computational throughput. Figure 1.1 provides a simple, yet insightful cartoon that explains the success of deep learning. While older algorithms, often based on "hand-crafted" machine learning features, appeared to be superior with limited amounts of training data, deep learning approaches have shown unprecedented learning and classification abilities in today's environment with nearly unlimited data. Here, it is worth noting that the blue line in Fig. 1.1 is still sloping upward as of 2020, i.e. the algorithms continue to improve as we collect and use more training data.

A grand challenge that arises from the aforementioned trend is the insatiable demand for memory and computing power, which persists across the various implementation scales of deep learning (servers, gateway systems, edge computing units, and tiny embedded systems). This book contains a number of contributions that discuss the underlying challenges and opportunities. Chapters 13 and 14 look at domain specific and coarse grain reconfigurable architectures (CGRAs) as a means to provide the required compute power while retaining a high degree of programmability that is needed in light of ever-changing algorithms and network topologies. Chapters 18 and 19 zoom in on relevant aspects for low-power edge systems, where the industry is already actively engaged in the developing custom deep learning processors.

A common denominator across all implementation scales is the challenge of memory access and data movement. These are discussed in detail in Chaps. 8–11,

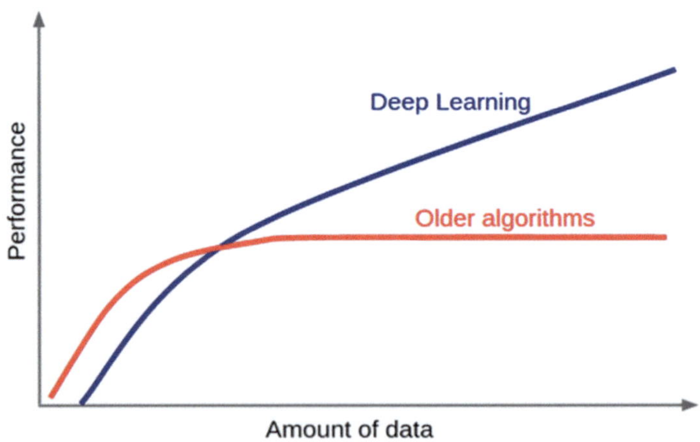

Fig. 1.1 The success story of deep learning. Adopted from Andrew Ng, Stanford University

15, 18 and 19. In conventional 2D chips, designers are currently trying to tackle the issue using various forms of in-memory computing (see e.g., Chap. 18). For the long term, however, there is a growing consensus that we must explore the third dimension to couple memory and compute more closely. Through Chaps. 8–11 and 15, this book provides a comprehensive overview of the various competing approaches to 3D integration from chip stacking to monolithic integration.

1.5 Man-Machine Cooperation and Safe Control

The nano-electronic realization of artificial intelligence towards 2030 and beyond is among the key topics of this book, as already discussed. In most application scenarios, these chip systems are part of a "machine," as for instance a navigator, a surgery support system, a prosthesis, a robot, a "carebot" or a vehicle. As all of these machines are trending toward increasingly autonomous actions, effective communication and cooperation with them becomes essential and critical. Cognitive actions and special features on both sides, humans and machines, as well as within their class, must be planned, interpreted and understood in real time. A special overview on this subject is presented in Chap. 28, while virtually all chapters contain contributions that are relevant to the construction of such complex and intelligent systems. A leading system-on-chip for autonomous driving at level 4, which entails avoiding collisions with other vehicles and pedestrians, is described in Chap. 29. A recurring theme here is to devise safe architectures that can autonomously adapt to failures and operate in an error-resilient manner and with robust performance within dynamically changing and uncertain environments.

To realize the ultimate vision of effective man-machine cooperation, order-of-magnitude improvements in all aspects within the process technology, circuit and system stack are needed. We hope that the pathfinding discussions in this book will help the community to drive the next decade of great opportunities and benefits from the application and continuing development of nano-chips.

References

1. B. Hoefflinger (ed.), in *CHIPS 2020—A Guide to the Future of Microelectronics* (Springer Science and Business Media, 2012). ISBN 978-3-642-22399-0
2. B. Hoefflinger (ed.), in *CHIPS 2020 Vol. 2—New Vistas in Nanoelectronics* (Springer Science and Business Media, 2016). ISBN 078-3-319-22092-5
3. International Roadmap for Devices and Systems. https://irds.ieee.org/
4. N. Savage, How AI and neuroscience drive each other forwards. Nature **571**, S15–S17 (2019)

Chapter 2
IRDS—International Roadmap for Devices and Systems, Rebooting Computing, S3S

Bernd Hoefflinger

2.1 International Roadmap for Devices and Systems (IRDS)

The International Technology Roadmap for Semiconductors (ITRS) had been founded by the Semiconductor Industry Association (SIA) in 1992. This unique, quantitative strategy of an industry was presented and analyzed in its 20th year in CHIPS 2020, Chap. 7 [1]. A critical review followed in CHIPS 2020 Vol. 2 [2]. At virtually the same time in 2015, the work of the ITRS groups was terminated. The hundreds of experts and thousands of trend documents were re-organized in a new program, managed by IEEE organizations [3]: The International Roadmap for Devices and Systems (IRDS). It is organized in 12 International Focus Teams (IFT's):

- Application Benchmarking
- Systems and Architectures
- Outside Systems Connectivity
- More Moore
- Lithograpy
- Factory Integration
- Yield
- Beyond CMOS
- Cryogenic Electronics and Quantum Information Processing
- Packaging Integration
- Metrology
- Environment, Safety, Health, and Sustainability.

Several of the key IFT's will be treated in the following sub-sections. Lithography is addressed in Chap. 5.

B. Hoefflinger (✉)
Sindelfingen, Baden-Württemberg, Germany
e-mail: bhoefflinger@t-online.de

B. Murmann and B. Hoefflinger (eds.), *NANO-CHIPS 2030*,
The Frontiers Collection, https://doi.org/10.1007/978-3-030-18338-7_2

The Executive Summary of 2018 shows a focus on gate-all-around (GAA) MOS transistors with vertical channels. This has had remarkable success in vertical NAND FLASH_RAM, where channel quality is not so critical. And it is of interest as an active vertical 3D interconnect, in particular between memory and logic. For a long-term 3D strategy 2030–2035, a multiple-transistor-layer topography is proposed, as shown in Fig. 2.1 [4]. This topography is basically attractive, particularly for logic because of its short-interconnect lengths, both laterally and vertically. An early version of this 3D integration was presented in 1985 for the high-density layout of the NMOS logic for a full-adder (Fig. 2.2) with 12 transistors in three transistor layers, requiring just 10 pitch-unit squares [5, 6]. This workshop in 1985 in Shujenji, Japan, remains as a historic highlight with its title: “Future Electron Devices: SOI and 3D Integration”. These focus areas have remained as top areas, and they make up two of the three in the S3S Program (Sect. 2.3). The technology-of-choice in 1985 for achieving high-quality vertical growth was selective silicon epitaxy with lateral overgrowth [6], even more attractive today with lateral overgrowth scaled down to ~25 nm, compared with the published 3D logic of 1992 built with 20 μm lateral overgrowth [7].

3D integration received significant coverage in CHIPS 2020 Vol. 2 [8], and it is emphasized further in Chaps. 8–11, 13 and 15 of this book. By contrast, the 2018 Executive Summary expects the dominance of 3D in VLSI logic in 2030 and later (Fig. 2.1).

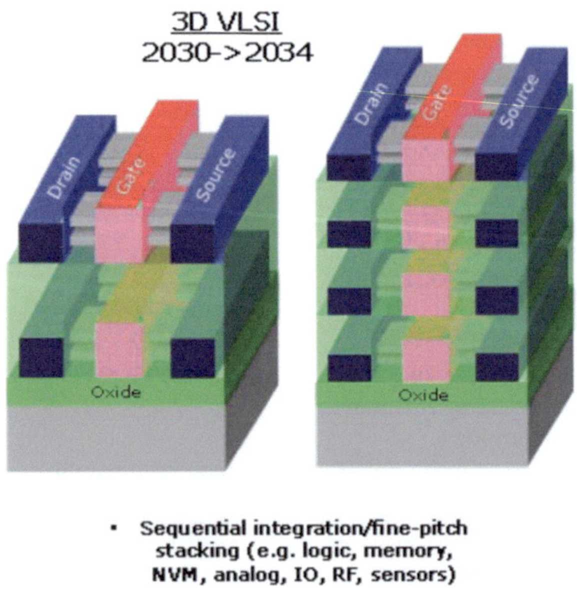

Fig. 2.1 3D integration in the IRDS executive summary for manufacturing >2030 [4]. © IEEE 2018

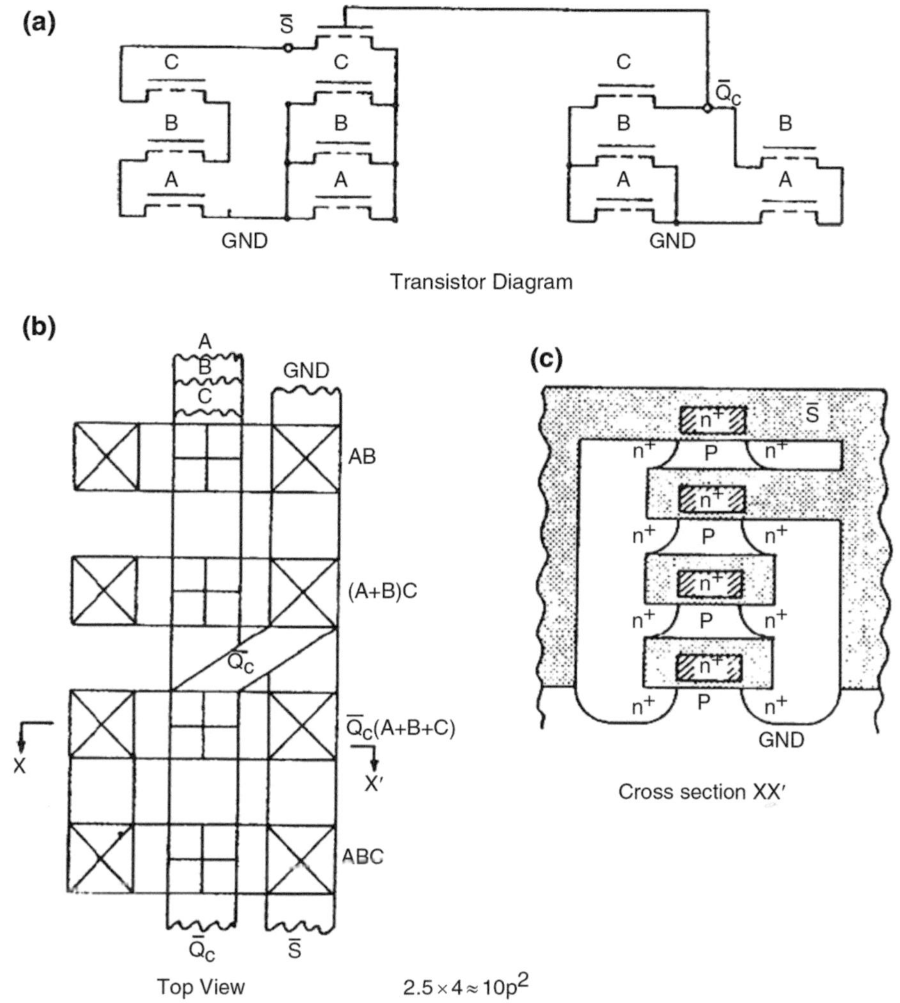

Fig. 2.2 1985 concept for the NMOS logic of a full-adder with 12 transistors on 10 pitch-unit squares, the equivalent of the footprint of 4 lateral transistors

2.1.1 More Moore

"More Moore" means a creative continuation of transistor- and on-chip-interconnect scaling. As a lesson from the ending of ITRS, the rate of changes has been adjusted, as is evident in Fig. 2.3.

The data in this figure is related to logic. HP: High-performance logic.

These lateral geometries are dramatic, conservative corrections. The physical gate-length limit of 12 nm confirms the arguments from 2012 [1] and from the review in [2]. The projected gate pitch, from 54 nm to 40 nm in 2034, reflects the new interest

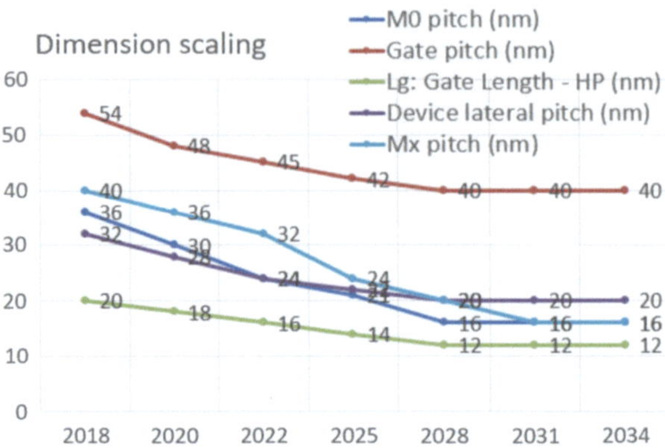

Fig. 2.3 IRDS 2018 projected scaling of key ground rules [12]. © IEEE 2019

in lateral gate-all-around (LGAA), vertical GAA transistors (VGAA), as well as multiple transistor layers. In all these topographies, the transistor bodies and their interconnects have significant space requirements. These dimensions deliver serious arguments for

- **10x investments into 3D integration,**

in order to achieve sustained progress in performance and energy efficiency. One example is sketched in Fig. 2.1, projected for 2030 manufacturing. This must happen earlier.

The performance estimates are concentrated on multi-core central-processor units (CPU) as an 80 mm^2 System-on-chip (SOC). Figure 2.4 shows the trend towards hundreds of floating-point units per chip. Integrated liquid cooling is assumed for maximum throughput of several Tera (10^{12}) floating-point operations per second (TFLOPS = TFLOP's/s), and the alternative mode would run with limited power density, as projected in Fig. 2.5.

Fig. 2.4 Number of floating-point processing cores on-chip [12]. © IEEE 2019

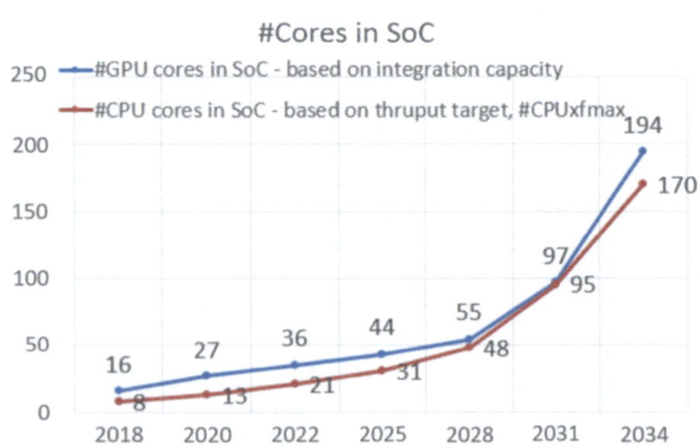

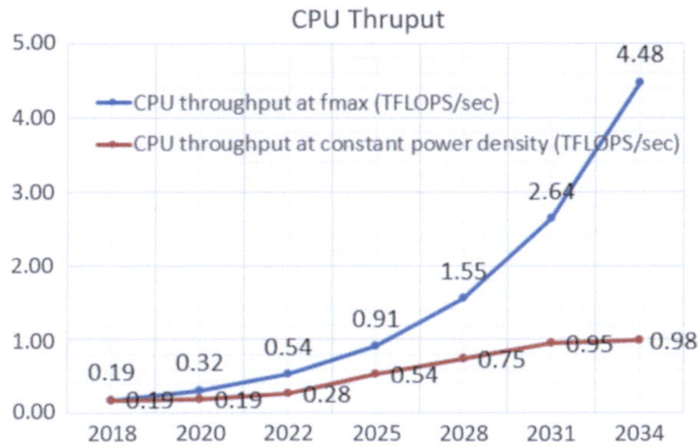

Fig. 2.5 Throughput (TFPO's/s) of multi-core CPU's in an SOC [12]. © IEEE 2019

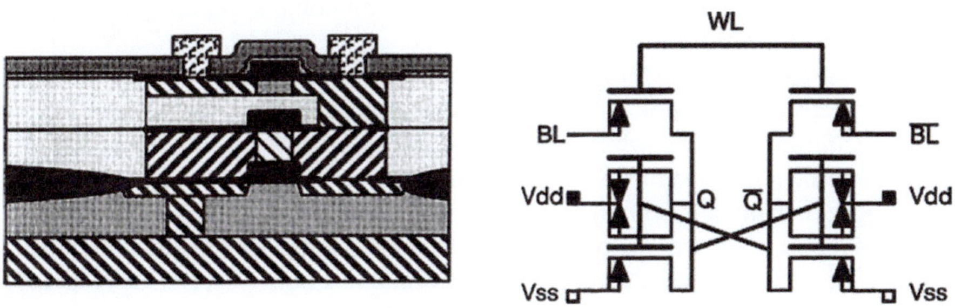

Fig. 2.6 Cross-section and transistor diagram of a 3D 6T CMOS SRAM cell with dual-gate PMOS. Implementation with selective epitaxy and lateral overgrowth [7]. © IEEE1992

The memory part of the More-Moore report lists all varieties of memory options, mostly with scaling parameters. Technology solutions, architectures and AI applications are treated extensively in Chaps. 8–11, 13, 15, and 19 of this book.

The Static RAM (SRAM) is listed as a major challenge in size, energy, and speed as the standard cache memory in direct co-operation with the processing units. The 6-transistor CMOS SRAM is the highest-speed, most robust, ultra-low-voltage write-read memory cell. The most efficient, high-quality 3D implementation was published in 1992 [7] and shown in Fig. 2.6. This exemplary memory cell received detailed treatment in [6], and it is central in Chap. 4 of this book.

Finally, the MM report points to the IFT "Beyond CMOS" for perspectives.

2.1.2 IRDS 2017 Report "Beyond CMOS (BC)"

This report [9] is an elaborate listing, with over one thousand references, of virtually all enhancements of and alternatives to CMOS for realizations of processing and memory. Inputs and outputs should be a voltage, a current or a charge. Inside, a wide spectrum of solid-state phenomena is considered:

III-V Compound Semiconductors
Tunneling (TFET)
2D layers like Graphene
Carbon Nano Tubes (CNT)
Ferroelectric
Thermal Phase Changes
Superconducting Electronics (SCE), Cryogenic Electronics
Magnetism
Spin
Quantum Effects
MEMS Switches.

Among these, cryogenic electronics and quantum comp computing are covered in Chaps. 26 and 27 of this book.

In storage tasks, beyond DRAM and Multi-Level, Vertical Flash NVRAM, there is a larger spectrum of technology alternatives for specific applications. Processing has some specific applications, where sensing and analog processing are particularly efficient like some IOT's, wearables and medical (Chap. 24). Digital processing remains as the biggest challenge, both in von-Neumann and in neural-network architectures. Here, the report compares many results focused on energy-per-operation and delay. Figure 2.7 shows these results for a 32-bit Arithmetic-Logic Unit (ALU) (Fig. 2.7).

The ultimate performance target is the lower left corner with a throughput figure-of-merit (FOM) of 100 TOPS/pJ. The 45-degree lines mean a constant throughput FOM. CMOS HP, High-Performance = speed-maximized, and Enhanced-CMOS with various Tunneling-FET technologies show the best results, like the thin TFET processing unit with an FOM of 1 TOPS/pJ. To calibrate these results, we can refer to CHIPS 2020 [6], where we identified a potential 16b multiplier with 600 MOPS (a delay of 1.6 ns) and an energy of 1fJ/operation with key innovations like

- A Leading-Ones-First (LOF) multiplier
- Ultra-Low-Voltage, Differential Transmission-Gate (ULVDTG) Logic,

which offer a 10x improvement in the throughput FOM, as treated in Chaps. 3 and 7. The Thin-TFET results emphasize the attention, which they received in CHIPS 2020, Vol. 2 [2].

The key digital alternatives to standard arithmetic-logic processing units are cellular neural networks (CNN), as shown in the IRDS overview Fig. 2.8. Their more-up-to-date performance overview is presented in Fig. 3.5, in Chap. 3 of this book, concentrated on real products with typically 1024 multipliers-accumulators.

As far as processing is concerned, the "Beyond CMOS" Report has the following assessment of requirements (quote) (Table 2.1).

These criteria make up a strong vote for "Enhanced CMOS", and the report concludes (quote):

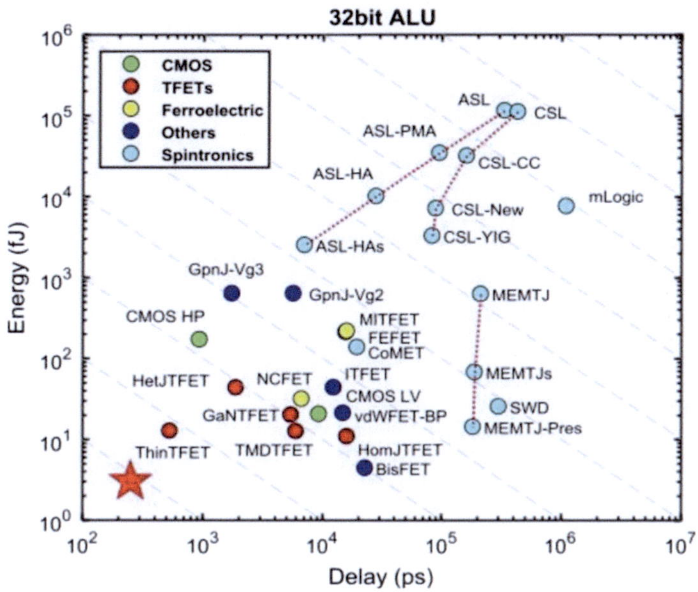

Fig. 2.7 Energy/operation (fJ) versus delay (ps) for 32b Arithmetic Logic Units (ALU's) [9]. © IEEE 2018

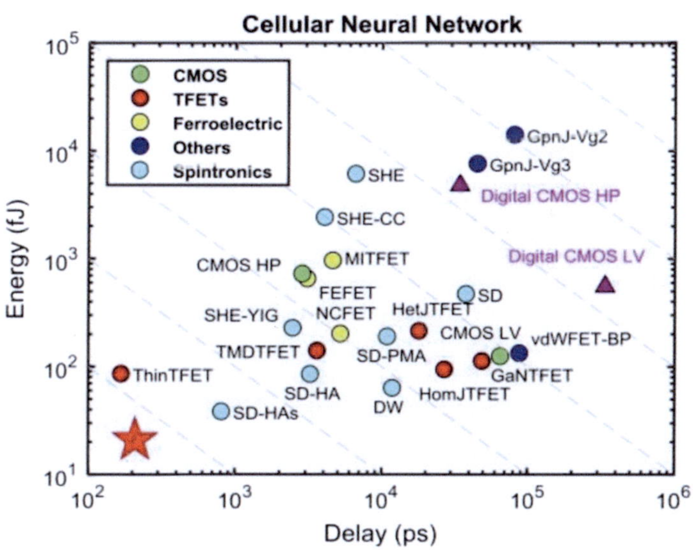

Fig. 2.8 Energy per CNN (MAC) operation (fJ) versus delay (ns) for cellular neural networks [9]. © IEEE 2018

Table 2.1 Requirements for "beyond CMOS" technologies, from [9]

• Inversion and flexibility (can form an infinite number of logic functions)
• Isolation (output does not affect input)
• Logic gain (output may drive more than one following gate and provides a high I_{on}/I_{off} ratio)
• Logical completeness (the device is capable of realizing any arbitrary logic function)
• Self restoring/stable (signal quality restored in each gate)
• Low cost manufacturability (acceptable process tolerance)
• Reliability (aging, wear-out, radiation immunity)
• Performance (transaction throughput improvement)

> Based on the current data and observations, it is clear that CMOS will remain the primary basis for IC chips for the coming years. While it is unlikely that any of the current emerging devices could entirely replace CMOS, several do seem to offer advantages such as ultra-low power or non-volatility….

These topics are central in all chapters of this boo. Enhancing CMOS gets an extra treatment in Chaps. 4 and 6.

2.2 Rebooting Computing

The annual International Conference for Rebooting Computing (ICRC) started in 2016. The highlights in November 2018 were [10]:

- Stochastic Computing
- Fault-Tolerant Computing with Interconnect Crosstalk
- Superconducting Optoelectronic Neuromorphic
- Large Fan-In Optical Logic Circuits
- Modular Multiplication with Fourier Optics
- Optical Parallel Multiplier Exploiting Approximate Logarithms
- Image Recognition with Resistive Coupled Vanadium-Dioxide Oscillators
- Molecular Quantum-Dot Cellular Automata
- Hardware-Software Co-Design for an Analog-Digital Accelerator.

These subjects show the longer-term research structure of Rebooting Computing with practical implications beyond 2030. At least one paper addresses the exploitation of logarithms for multiplication (see Chap. 3).

2.3 S3S—Silicon-on-Insulator, 3D, Sub-threshold MOS

S3S is a working group within IEEE, which started in 2014 with its own annual conference held in Monterey, California, and producing its own proceedings [11].

The three columns of S3S are a perfect match with key subjects in our books CHIPS 2020 and CHIPS 2020, Vol. 2, and they are central in the present book:

- Silicon-on-Insulator: Chapters 3,4,6,7 and 19,
- 3D: Chapters 3, 4, 8–11, 13, 15, 19 and 23,
- Sub-Threshold Operation: [6] and Chaps. 3, 4, 6, 7, 17, 24 and 30.

2.4 Conclusion

The IRDS is a major correction of the ITRS. It is fundamental in projecting a minimum physical gate length of 12 nm towards the late 20s, and it is realistic in lateral densities. Important details regarding the nm-meaning in the so-called "industry logic node" are listed in Chap. 5 on Nanolithography. Rebooting Computing concerns long-term research on alternative technologies. The S3S subjects Silicon-on-Insulator, 3D, and Subthreshold are treated extensively in this book.

References

1. B. Hoefflinger, The international technology roadmap for semiconductors, Chap. 7, in *CHIPS 2020—A Guide to the Future of Microelectronics* (Springer Science and Business Media, 2012). ISBN 978-3-642-22399-0
2. B. Hoefflinger, ITRS 2028, Chap. 7, in *CHIPS 2020 Vol. 2—New Vistas in Nanoelectronics* (Springer Science and Business Media, 2016). ISBN 078-3-319-22092-5
3. www:ieee.irds.org
4. IRDS—ES.pdf
5. B. Hoefflinger, Circuit considerations for future 3-dimensional integrated circuits, in *Proceedings of 2nd International Workshop on Future Electron Devices—SOI Technology and 3D Integration* Shujenzi, Japan (1985)
6. B. Hoefflinger, The future of 8 chip technologies, Chap. 3, in *CHIPS 2020—A Guide to the Future of Microelectronics* (Springer Science and Business Media, 2012). ISBN 978-3-642-22399-0
7. G. Roos, B. Hoefflinger, Complex 3D CMOS circuits based on a triple-decker cell. IEEE J. Solid-State Circ. **27**, 1067 (1992)
8. Z. Or-Bach, in *Monolithic 3D Integration, Chapter 3 in CHIPS 2020 Vol. 2* (Springer International Publishing, 2016). ISBN 978-3-319-22092-5
9. IRDS-BC.pdf
10. ieeetv.ieee.org/ieee-international-conference-on-rebooting-computing-2018
11. s3sconference.org
12. IRDS-MM.pdf

Chapter 3
Real-World Electronics

Bernd Hoefflinger

3.1 Introduction

Face-to-face with the challenges and opportunities of intelligent systems, electronic circuits should finally be driven by their real-wold relevance, after a century of numbers- and math-driven computing including the accident of linear CCD imaging (after 150 years of logarithmic quality photography and film).

The fundamentally logarithmic real world (Weber's Law) is leveraged perfectly in the logarithmic slide-rule, invented 1622 in Cambridge, which reduces multiplications into simple additions. Multipliers are very transistor- and energy-hungry, as well as time-consuming. It is incredible that world-wide multiplication starts with the irrelevant least-significant bits (LSB)-first, while human intelligence has looked, for thousands of years, for the leading numbers first on the Abacus, to immediately get the order-of-magnitude of a multiplication. Transistor-count, energy and speed can be improved by an order-of-magnitude each with leading-ones-first (LOF) multiplication, and we have demonstrated such circuits since the 1990s.

The most multiplier-hungry circuits are multi-layer perceptrons, the dominant form of digital neural networks. Every synapse multiplies its signal with a weight, and it is here that the LOF-first multiplier delivers the biggest gains.

The present explosion of digital neural networks presents major challenges for robust ultra-low-voltage, high-speed, low-energy, scalable digital circuits. We showed in the year 2000 that ultra-low-voltage, differential transmission-gate (LVDTG) logic is the most resilient and efficient logic. Wim Dehaene shows in Chap. 7, how LVDTG continues to hold this leadership.

B. Hoefflinger (✉)
Sindelfingen, Baden-Württemberg, Germany
e-mail: bhoefflinger@t-online.de

B. Murmann and B. Hoefflinger (eds.), *NANO-CHIPS 2030*,
The Frontiers Collection, https://doi.org/10.1007/978-3-030-18338-7_3

3.2 Efficient Electronic Processing of Real-World Information

The Morse communication, the telephone, radio and television have been an analog-electronics art for about 100 years into the 1950s, driven by vacuum tubes and finally by early transistors. Dealing with real-world issues, the quality of signals and results is determined by our perception of the real world. The technical and scientific evaluation, control and improvement of this analog world led to the development of analog computers.

A totally different world evolved with the human need for number crunching, mostly for trade and money. The support of mathematics has seen endless inventions of mathematical systems with mechanical accelerators. The Morse zero-one relay led the number crunchers to the binary digit, which, together with the silicon transistor, has enabled an unprecedented economic growth and data explosion for 60 years, based on a one-dimensional Micrometer- and then Nanometer-Roadmap.

This unparalleled growth was described and analyzed in CHIPS 2020, published in 2012 [1], with clear arguments, why this roadmap would come to halt in 2016. In the same year, CHIPS 2020, Vol. 2, was published [2], expanding the 2012 quest for orders-of-magnitude improvements of energy efficiency and intelligent processing to sustain the growth of an information- and communication-dependent world. These new priorities have picked up remarkably over the past five years, and they are central for the present book.

Under the inertia and the dominance of the digital number crunchers, it pays off to start with the fundamentals of the real world.

3.3 The Perception of the Real World: The Weber and Fechner Law

The 19th century saw the widest expansion of measuring, perceiving, analyzing, modelling and mathematically describing our real world. A very central finding was that our perception and measurement of real-world quantities is governed and limited by a logarithmic response. On a distance, a weight, a sound or a brightness of magnitude N, the just noticeable difference dN is a constant fraction of N, say a:

$$dN/N = a.$$

If our measured value is y(N), $y(N) = a \ln(N) + c$.

This logarithmic metric of our world has a very long history: Instead of linear scaling of money, the 1-2-5-10 scaling has been common as well as the 1-3-12 scaling of length and the very broad log2 scaling 1-2-4-8-16 and, on the big scale, the decimal system. In all practical cases of real-world quantities, the knowledge of its magnitude

(= the leading number) and its relative accuracy are of biggest interest. And the log or approximate-log system has had its biggest effect over thousands of years in the task of multiplying two quantities like the weight of something and its unit price. Inventive manual calculators like the "Abacus", were designed to get the product of the leading numbers immediately and then, step-by-step, improve the accuracy depending on demand. Logarithmic multiplication tables became a big must-have in the 19th century The biggest jump was achieved with the logarithmic slide rule, invented in 1622 in Cambridge, which has one bar with a high-quality log scale for the multiplicand and an identical sliding bar for the multiplicator, which allow the direct visual addition of the logarithms to read the value of the product.

The two most important real-world sensing quantities are sound and vision. When telephony became economic for efficient and robust digital coding, quasi-logarithmic conversion was invented, and it benefitted directly from logarithmic compression, more natural listening and better sensitivity at low voice levels. With the A-law or μ-law standards, the converters (encoders) convert a continuous analog audio signal with 12-bit dynamic range into 8-bit data [3].

The most significant and essential logarithmic sensing system is the human visual system (HVS). It converts light intensity (the photon current) with an instantaneous dynamic range of close to 1Mio./1 into a logarithmic response with just noticeable differences of 1% over six decades of brightness [4] (measured in cd/m^2 or lumen).

The HVS receives further attention in several chapters of this book. In Fig. 3.1, you also find the response curve of the High-Dynamic-Range CMOS (HDRC®) sensor with its instantaneous eye-like response curve over seven orders of magnitude, first published in 1993 [4–6]. The eye's real-world-logarithmic response has been adopted for centuries in the log scaling of aperture and brightness: f = 1.4, 2, 4, 8, 16, ... And it has been the core in the invention and in the improvement of chemical photo material to achieve a logarithmic response with a dynamic range of at least 4 orders of magnitude.

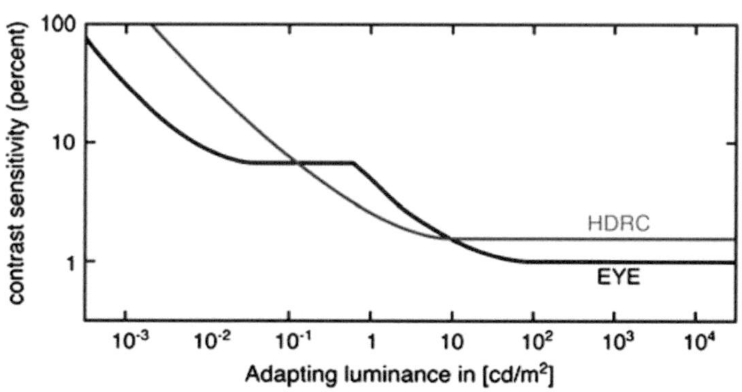

Fig. 3.1 Contrast sensitivity, the derivative of a photoreceptor signal: The HDRC® sensor [4–6] shows a natural response over seven orders of magnitude (© Springer 2006). Below 1 cd/m^2, the eye achieves sensitivity with long-term adaptation

The historic consequences of the linear response of the charge-coupled-device (CCD) imager in 1970 with a dynamic range of less than 3 orders-of-magnitude (60 db) have been dramatic: The loss of image quality with poor contrast resolution in shaded regions and quick white saturation in bright regions require multiple exposures to get a valid frame and, for faster response, parallel sub-pixels for different brightness regions. In any event, the seemingly cheap linear CMOS pixels need at least 3 exposures or 3 parallel sub-pixels to achieve the 7 orders-of-magnitude dynamic range of the HDRC® sensor or the human eye. Linear electronic vision not only needs at least 3-times more bits for a valid pixel information. Its response is also fundamentally alien to any image processing like contour detection. That is why logarithmic conversion of linear (or piece-wise linear) sensor data has received attention in logarithmic, perception- or HVS-inspired, efficient image processing [7]. A monograph on Logarithmic IMAGE processing appeared in 2016 [8], and a 2019 example of efficient log data compression was presented in [9], where log gradients make object detection independent of luminance effects, due to the fundamental that Log Lightness intensity = Log Luminance + Log Reflectance/Chrominance.

High-Dynamic-Range vision will receive further treatment in Sect. 3.6 and in Chap. 21.

Besides continuous analog signals, other types of signals merit efficient acquisition and processing for intelligent understanding and action:

Real-world pulses with their shapes, amplitudes, frequencies and densities contain essential information, and they are central in neural systems. Large gains in efficiency and intelligence are possible in this domain [10–13]. These signals have become an essential driving force for neuromorphic and brain-inspired processing [14–16].

3.4 Silicon Electronics for the Real World

Our mimicking of the real world—and our efforts to out-perform it—again and again have led to exploring alternatives to the electron and to silicon. Electrochemistry, Ionics, molecular electronics, photonics, and magnetism have received reviews again, fueled by the end of the nanometer roadmap. In addition to their fundamental compatibility problems with the real world, these alternatives would need another 50 years to achieve the world ranking and market of silicon electronics. Its maturity is considered by many to be a handicap for further significant and sustainable growth. Contrary to S-curve economics, this book shows that further quantum jumps and orders-of-magnitude improvements are on their way and realistic in the 2020s, since

- Intelligent Data have become more important than Bits,
- Neuromorphic Architectures have taken the lead from Von-Neumann Architectures,
- The resulting gains in energy efficiency and performance enable autonomous systems.

The unique features of silicon microelectronics, their development and their future were pursued in this chapter of [1] and in Chap. 2 of [2]. The fundamentals are listed here to check, if they are still un-contested:

- The silicon—silicon-dioxide (and nitride) system of semiconductor and insulator
- This system provides non-volatile data storage
- Practical temperature range −50 °C to +200 °C
- Complementary Transistors (unique history in Chap. 4)
- Photodiode and Solar Cell
- Electromechanical sensing and actuating
- Selective epitaxial growth and overgrowth (no lithography need)
- Power Devices
- Si Substrate for heterogeneous systems (photonics and chip carriers)
- Flexible Chips
- Robustness for 3D integration
- Poly-Si Large Arrays = Displays
- The Si-SiO_2 system has been the base for scaling and for the nanometer-roadmap, and its efficient and creative use enables further sustainable progress.

3.5 From Number-Crunching to Real-World Multiply-Accumulate

In Sect. 3.2, we pointed out, how efficient real-world multiplication of two numbers has occupied mankind, and that the Weber-Fechner-law characterizes real quantities with their relative (percent) accuracy. They are handled most effectively with the ABACUS in its leading-numbers-first multiply mode or by the slide-rule because of its log scale. The electronic analog computer, by nature, followed these accuracy laws. But school arithmetic and its computerized acceleration start with least-significant numbers (or bits) first, and the result is a multiplier with a complexity that grows with the product of the word-lengths and a result word-length equal to the sum of the two. This result is irrelevant in a real-world problem, where the resulting accuracy is only that of the less accurate factor. Thus conventional multipliers have become a tremendous waste of resources, energy, calculation-times and chip area.

Multipliers and multi-input accumulators have become a central problem in digital neural networks (DNN's). The early example of a DNN for a lane-keeping assistant of 1993 [17] with 5 inputs, one inner layer with 15 neurons, and the steering angle as output needed already 90 multipliers (Fig. 3.2), which motivated the development of an efficient and high-throughput real-world digital multiplier. The most efficient logic for this real-world task is the leading-ones-first (LOF) multiplier [18, 19] shown in Table 3.1.

In a practical DNN, b would represent the instantaneous data, and a would represent the weight, which changes slowly, in learning or repair, or not at all in certified operation.

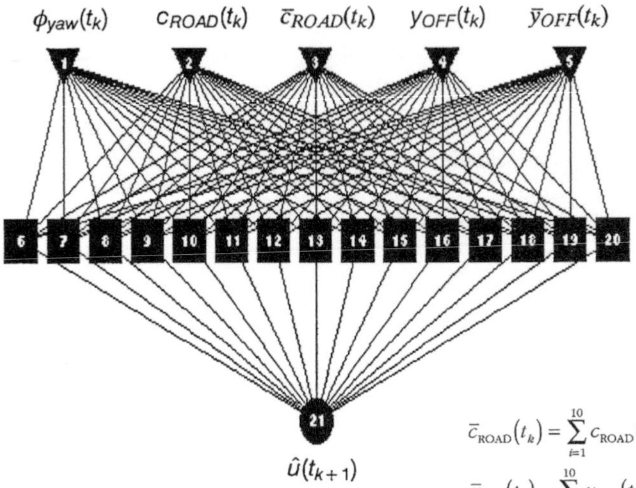

$$\bar{c}_{ROAD}(t_k) = \sum_{i=1}^{10} c_{ROAD}(t_{k-i}) * \left(1 - \frac{i}{10}\right)$$

$$\bar{y}_{OFF}(t_k) = \sum_{i=1}^{10} y_{OFF}(t_{k-i}) * \left(1 - \frac{i}{10}\right)$$

Fig. 3.2 Digital steering assistant with 21 neurons [17]. All lines mean synapses, multiplying instantaneous input with weight factors resulting from learning the task. © IEEE 1993

Table 3.1 Leading-Ones-First (LOF) Integer Multiplier with 6b Accuracy. The list of adder inputs for leading $a_j = 1$ and $b_k = 1$. The complexity is of the order $O(6^2/2)$, and the adder has a carry-look-ahead length of 6b, both independent of the data- or weight-word lengths

Integer index	j + k	−1	−2	−3	−4	−5
$a_j = 1, b_k = 1$	1	b_{k-1}	b_{k-2}	b_{k-3}	b_{k-4}	b_{k-5}
If $a_{j-1} = 1$[a]		(1)	(b_{k-1})	(b_{k-2})	(b_{k-3})	(b_{k-4})
If $a_{j-2} = 1$[a]			(1)	(b_{k-1})	(b_{k-2})	(b_{k-3})
If $a_{j-3} = 1$[a]				(1)	(b_{k-1})	(b_{k-2})
If $a_{j-4} = 1$[a]					(1)	(b_{k-1})
If $a_{j-5} = 1$[a]						(1)

[a]Otherwise the inputs from this line are 0

The CMOS transistor count and the energy of this LOF multiplier would be 6-times less than a Booth-Wallace multiplier for a 16 b × 16 b multiplication with 6 b accuracy (Table 3.2). And the speed would be three times higher. The straight integer processing is effective for the multi-input accumulators in DNN's.

- For real-world digital multipliers/accumulators, an order-of-magnitude improvement is possible in transistor count and energy with the LOF architecture.

The benefits of logarithmic computing were presented by the LOGNET results, based on ($\log_2$ 4 b) weights in a neural network [20]. (Log_2 4 b) weights were also used in a log computing neural net with highly effective 3D-stacked, low-latency 96 MB SRAM, inductively connected [21]. It should be pointed out that the attractive

Table 3.2 Transistor counts for standard multipliers and for the precision-oriented Leading-Ones-First (LOF) "reality" multipliers [18]

Word length n	8 b	16 b	24 b
Standard Booth-Wallace	1600	6400	14,400
LOF 6b precision (1.6%) 10 b precision (0.1%)	540	1030 2100	1620 2700

addition in the log multipliers still needs output decoding for the following accumulators, while the LOF multiplier uses and produces signals in the standard compatible form so that no encoding and decoding will be needed.

Further significant gains in energy, throughput and robustness are possible with ultra-low-voltage, sub-threshold differential transmission-gate logic, as published in [22–24] and well described in Chap. 7 of this book.

3.6 From 200 EV/Bit in One NVRAM Transistor to 30 Giga EV Per Long-Dist. Internet Bit

One essence of the energy-efficiency focus in CHIPS 2020, [1, 2], is that the remarkable progress in electron-volts (eV) per bit in a multilevel one-transistor memory cell is tough to realize off-chip, and the long-distance Internet bit continues to be very expensive energy-wise, even with photonics progress. 2018 estimates are shown in Table 3.3.

The CISCO forecast for mobile and total Internet traffic [25] continues to predict very large further growth, as shown in Fig. 3.3 for mobile traffic.

The mobile traffic reaches 20% of the total Internet traffic in 2020, and its share keeps growing. Furthermore, 80% of the mobile traffic is video with an annual growth of >65%, strongly driven by 5G, which will produce 10-times more traffic than a 4G phone [26]. Autonomous vehicles are video-driven, enhancing the video challenge. We quote in Chap. 16 that the Internet needs about 300 GW in 2020, heading towards >900 GW in 2030, which would then be 21% of the total global electric power. This

Table 3.3 Energy per bit in electron-volts (eV). 1 eV = 1.6×10^{-19} J (Ws)

Distance	Task	Electron volt (eV)
100 nm	1 bit SRAM cell	1.000
1 cm	Length on-chip	100.000
Brain synapse		60.000
10 cm	Circuit board	6M
1 m	Computer-rack	50M
1 km	Com cell	1G
1000 km	Cell-server	30G

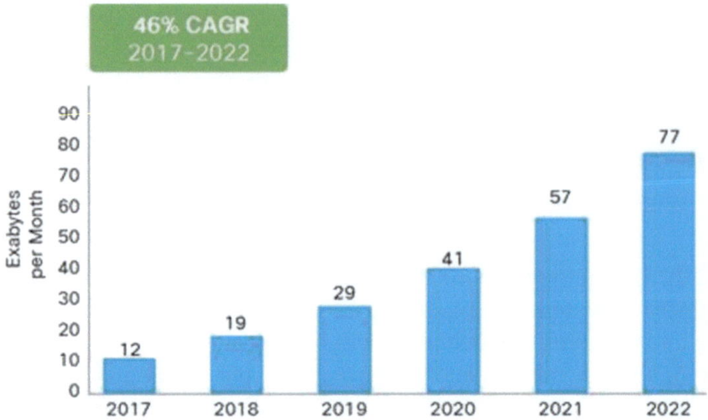

Fig. 3.3 CISCO forecast 2017 for mobile internet Traffic [25]

would mean an extra 100 GW every 3 years, the equivalent of 100 nuclear plants or 100 super wind farms with 200 wheels each or 5000 km^2. Considering the energy of the Internet bit in Table 3.3 and its limited scaling potential over longer distances, we have to reduce by orders of magnitude the numbers of bits per job or product, which we send into or request from the Internet. Given the overwhelming video challenge, we introduced, in Chap. 20 of [2], the energy per video frame as a figure-of-merit.

3.7 From Energy Per Operation to Energy Per Video Frame

Bits and operations are means to produce a result. In bits- and operations-hungry video, a quality video frame is such a result. That is, why we introduced energy/frame as a figure-of-merit in Chap. 20 of [2]. And we identified six innovations, which are most critical and which have the largest potential for orders-of-magnitude improvement, as illustrated in Fig. 3.4. These six special efforts are treated in the chapters of this book, and their progress since 2015 is rated here.

3.8 Efficient, High-Throughput Digital Neural Nets—a Giant Step for Real-World Electronic Intelligence

Heterogeneous Mega- to Giga-input information units are the central challenge for real-world perception and action. Again, vision is the dominant example where a multi-layer neural net needs Tera-(10^{12}) to Peta-(10^{15}) multiply-accumulate (MAC)

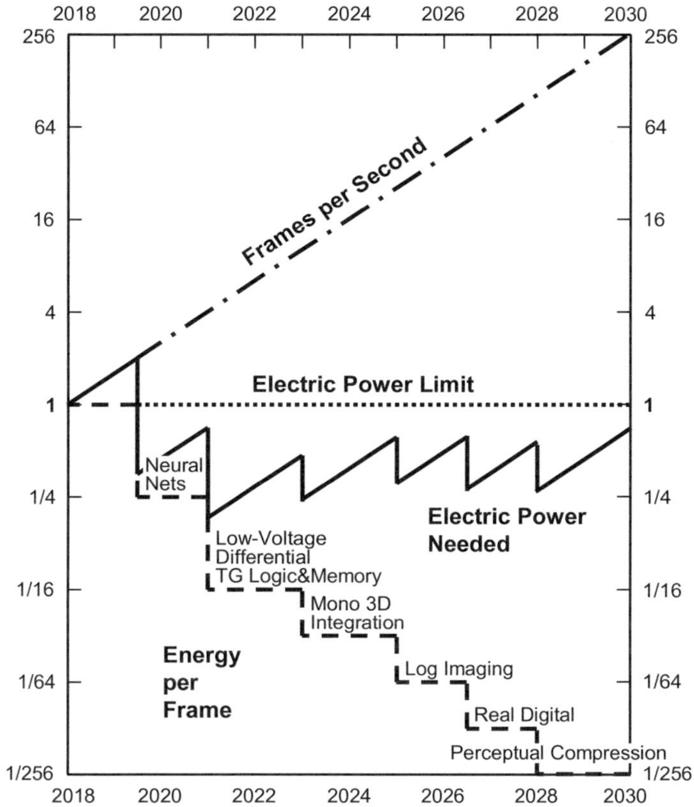

Fig. 3.4 Illustration (schematic) of potential improvements in energy per video frame with six special innovation efforts [29] (© Springer 2016)

operations per second to enable satisfactory perception and action. Technology nodes proceeding to 16 nm have enabled ultra-large-scale integration levels of thousands of processing units/chip and further to wafer-scale integration to realize these processing nets. Parallelism, an original means to achieve throughput, has become a natural architecture in DNN's plus the processing power of the depth of the network. Typical 2019 state-of-the-art performance data is listed in Table 3.4. A wide overview is given in Fig. 3.5.

Basic conclusions are that

- Arrays of 1024 MAC's have reached Throughputs of >1 TOPS at efficiency levels of <1 pJ per operation on 16 b word lengths.
- The word length enters with a quadratic effect into the energy efficiency because of the area needed for standard multipliers. Going from 8b to 16b means a 4-times drop in energy efficiency.
- As the figure-of-merit lines in Fig. 3.5 indicate, the area penalty still has a strong effect: Increasing the number of MAC's (= throughput) by three orders

Table 3.4 2020 MAC projection of 2012 [1] and 2019 State-of-the-Art 1000 MAC's DNN's

Source	Tech. (nm)	Volt (V)	MAC's	Word bit	Acc. bit	Throughput TOPS	Efficiency fJ/operation	FOM POPS/nJ
Song et al. [27]	8	0.5	1024	8 b		1.9	86 fJ	13.9
Yamada et al. [30] ch.29	16	0.4	1024	16 b	IEEE	1.6	1 pJ	1.6
Hoefflinger [1]	10	0.4	1	16 b	6 b	0.0007	1 fJ	0.7

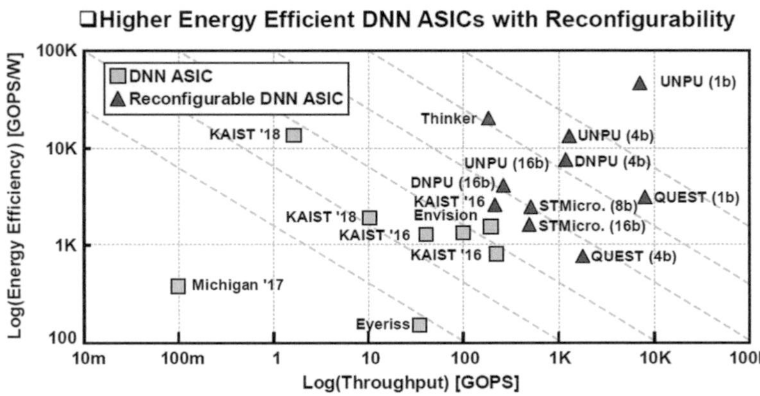

Fig. 3.5 Throughput and Energy Efficiency of Digital Neural Networks [16]. © IEEE 2019

of magnitude, raises the energy needed per operation by estimated two orders of magnitude.

In spite of or just because of the remarkable progress and development intensity since 2016, the state-of-the art provides strong arguments for the innovations emphasized in this chapter and central in other chapters of this book:

- Reduce word lengths in video with HVS-driven log image acquisition and processing.
- Use real-world LOF multipliers with 10x less transistors and area and 3x higher speed.
- Use ultra-low-voltage, sub-threshold, robust differential-transmission-gate CMOS design with highest efficiency and speed.
- Push 3D integration for drastically reducing signal paths.

Table 3.4 shows the projection for 2020 of a 16b LOF-multiplier needing just 1 fJ for a throughput of close to 1 GOPS (Sect. 3.6 in [1]). In chips like [27], the individual 8 b MAC has to run at ~1 fJ to enable 86 fJ in the 1024-MAC-system.

The implementation of all the innovations just listed, enables

Two orders-of-magnitude improvement in energy efficiency and 10-times higher throughput in the 2020s for 1024 MAC's Digital Neural Networks.

3.9 Conclusion

The progress in the design and realization of learning DNN accelerators, typically with 1024 MAC-type processors, has been so strong that neural-network-inspired architectures have taken over the lead in solving real-world problems from math-model-based, number-crunching von-Neumann computers.

This has been one big step: Real-world-inspired intelligent electronic processing. Other well-known fundamentals of real-world perception and their electronic Implementation have also been demonstrated in the 1990s: The logarithmic metric of real-world information is overwhelmingly alive in human vision. The log HDRC® CMOS sensor [4] surpasses the human eye in dynamic range and in robustness. The benefits of log imaging are manifold [4, 8], and three orders-of-magnitude improvements in energy-per-video frame can be identified. One further benefit of log imaging, which naturally creates the additive superposition of log luminance and log chrominance, is the high-dynamic-range (HDR) display, effectively used in the invention of the two-layer HDR display with the LED luminance panel and the LCD chrominance panel [28], the basis of DolbyVision™.

The other log invention to be re-vitalized in digital electronics, is the slide-rule, invented in 1622, best implemented in the binary integer leading-ones-first (LOF) multiplier.

Finally, more emphasis on 3D integration of the memory-processor system allows dramatic improvements.

References

1. B. Hoefflinger (ed.), in *CHIPS 2020—A Guide to the Future of Nanoelectronics* (Springer International). ISBN 978-3-642-23096-20127
2. B. Hoefflinger (ed.), in *CHIPS 2020, Vol.2—New Vistas in Nanoelectronics* (Springer, 2016). ISBN 978-3-319-22093-2
3. B. Hoefflinger, in Intelligent data versus big data, Chap. 12 in [2]
4. B. Hoefflinger (ed.), *High-Dynamic-Range (HDR) Vision* (Springer, Berlin, Heidelberg, 2007). ISBN-13 978-3-540-44432-9
5. B. Hoefflinger, U. Seger, M.E. Landgraf, U.S. Patent 5609204, filed 05–23, 1993, issued 03-04-1997
6. B. Hoefflinger, in HDR- and 3D-vision sensors, Chap. 13 in [2]
7. R.K. Mantiuk, K. Myszkowski, in Perception-inspired high dynamic range video coding and compression, Chap. 14 in [2]
8. M. Jourlin, in *Logarithmic Image Processing: Theory and Applications* (Elsevier, 2016)

9. A. Young et al., A data-compressive 1.5b/2.75b log-gradient QVGA image sensor with multi-scale readout for always-on object detection, in *ISSCC Digest of Technical Papers*, San Francisco (2019), pp. 98–100
10. S.V. Vandebroek, Three pillars enabling the internet of everything: smart everyday objects, information-centric networks, and automated real-time insights, in *IEEE International Solid-State Circuits Conference 2016, Technical Digest paper 1.2* (2016)
11. U. Rueckert, in Brain-inspired architectures for nanoelectronics, Chap. 18 in [2]
12. P. Cong, Neural interfaces for implantable medical devices. IEEE Solid-State Circ. Magazine Fall 48–56
13. M. Keller, B. Murmann, Y. Manoli, in Analog-digital interfaces—review and current trends, Chap. 4 in [2]
14. L. Spaanenburg, W.J. Jansen, in Networked neural systems, Chap. 16 in [2]
15. M. Verhelst, B. Moon, Embedded deep neural network processing. IEEE Solid-State Circ. Mag. Fall 55–65 (2017)
16. H.J. Yoo, Intelligence on silicon: from DNN accelerators to brain-mimicking AI SOC's, in *ISSCC 2019 Digest of Technical Papers* (2019), pp. 20–26
17. S. Neusser et al., Neurocontrol for lateral vehicle guidance. IEEE Micro **13**(1), 57–63 (1993)
18. B. Hoefflinger, in Digital multiplier for reality data, Sect. 12.3 in [2]
19. B. Hoefflinger, M. Selzer, F. Warkowski, Digital logarithmic CMOS multiplier for very-high-speed signal processing, in *IEEE 1991 Custom-Integrated-Circuits Conference (CICC), Digest* (1991), pp. 16.7.1–5
20. E.H. Lee, D. Miyashita et al., LOGNET: energy-efficient neural networks using logarithmic computation, in *IEEE International Conference on ASSP 2017* (2017),pp. 5900–5904
21. Takamaeda-Yamazaki S. et al., QUEST: A 7.49 TOPS multi-purpose log-quantized inference engine stacked on 95 mb SRAM using inductive coupling technology in 40 nm CMOS, in *2018 International Solid-State Circuits Conference, Digest of Technical Papers* (2018), pp. 216–218
22. N. Reynders, W. Dehaene, A 210 mV 5 MHz variation-resilient near-threshold JPEG encoder in 40 nm CMOS, in *2014 ISSCC Digest Digital Papers, paper 27.3, and private communication* (2014), pp. 457–458
23. N. Reynders, W. Dehaene, Variation-resilient building blocks for ultra-low-energy sub-threshold design. IEEE Trans. Circ. Syst.-II **59**(2), 898–902 (2012)
24. N. Reynders, W. Dehaene, in *Ultra-Low Voltage Design of Energy-Efficient Digital Circuits* (Springer, 2015). ISBN 978–3-318-16135-8
25. CISCO, in *The Zettabyte Era, Trends and Analysis* (CISCO Public, 2017)
26. S. Mattison, An overview of 5G requirements and future wireless networks. Solid-State Circ. Mag. Summer 53–60 (2018)
27. J. Song et al., An 11.5 TOPS/W 1.024-MAC butterfly structure dual-core sparsity-aware neural processing unit in 8 nm flagship mobile SoC, in *2019 International Solid-State Circuits Conference, Digest of Technical papers* (2019), pp. 130–131
28. H. Seetsen, in High-dynamic-range displays, Chap. 14 in [5]
29. Chapter 20 in [2]
30. T. Yamada et al., A 20.5 TOPS and 217.3 GOPS/mm^2 multi-core SOC with DNN accelerator and image signal processor complying with ISO 26262 for automotive applications, in *2019 International Solid-State Circuits Conference Digest of Technical Papers* (2019), pp. 132–133

Chapter 4
Silicon Complementary MOS into Its 7th Decade

Bernd Hoefflinger

4.1 The Complementary NMOS/PMOS Transistor Pair and the Quad

The magic behind the digital world is the binary on-off switch in computer science. The electronic engineers concentrated on the voltage control of an ideal inverter with a perfect ONE, a perfect ZERO, a transition with infinite voltage gain, offering a noise margin of 50% of the supply voltage, infinite current gain with both a high pull-up and pull-down current, for charging and discharging the following gates (Fig. 4.1).

The original patent directly shows the pair of complementary NMOS and PMOS transistors connected for the inverter function:

Input = #100 = V_I, Output = #101 = V_O.

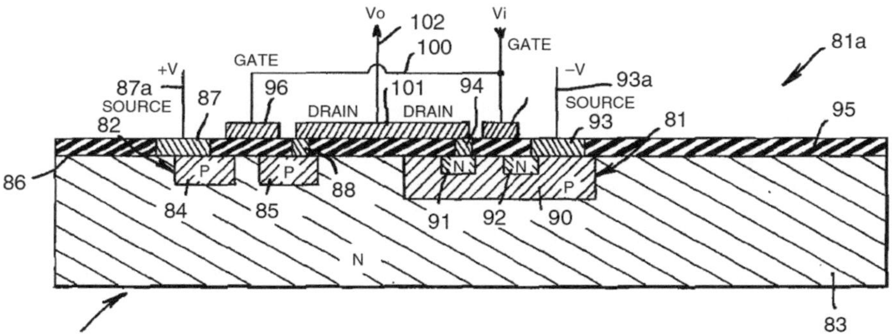

Fig. 4.1 The original patent of 1963 by Frank Wanlass for a planar integration of NMOS and PMOS transistors with junction isolation [12]. © USPTO

B. Hoefflinger (✉)
Sindelfingen, Baden-Württemberg, Germany
e-mail: bhoefflinger@t-online.de

B. Murmann and B. Hoefflinger (eds.), *NANO-CHIPS 2030*,
The Frontiers Collection, https://doi.org/10.1007/978-3-030-18338-7_4

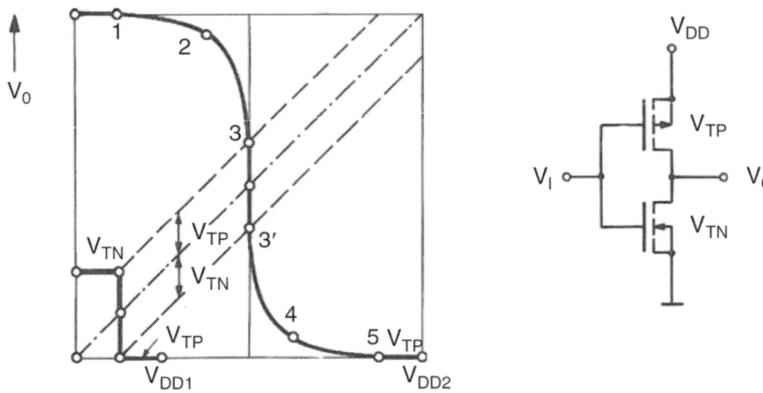

Fig. 4.2 Simplified transfer characteristics of a CMOS inverter [4] with transistor threshold voltages V_{TN} and V_{TP}, not considering the essence of the more-and-more important sub-threshold operation for the optimum energy efficiency. © Springer 2012

The classical modelling of the transfer characteristic of this inverter is shown in Fig. 4.2 with the threshold voltages V_{TN} and V_{TP} and a minimum operating voltage $V_{TN} + V_{TP}$ and a transition region with infinite voltage gain.

It is good to remember this ideal characteristic, because it was behind the invention of the CMOS inverter in 1963. CMOS technology advanced quickly into digital watches because of their low supply voltages and minimum currents of logic gates, outside their switching moments. Robust, readily computerized, effective design of fully complementary logic gates enabled a niche industry with the reputation of being expensive because of the number of processing steps, starting with 20 μm in the late 60s. Nevertheless, two specialties were promoted early [15]:

Silicon-on-Sapphire:

Perfect isolation of the PMOS and NMOS transistors, minimum parasitics because of the sapphire insulator as well as radiation hardness because of minimum transistor volumes [1]. The predecessor of today's SOI-CMOS [3] and Chap. 6.

The CMOS Static Random-Access Memory (SRAM):

The cross-coupled CMOS transistor pair is the most robust binary memory cell with perfect full-swing differential data levels, minimum standby power, maximum drive capability for lowest latency. The 6-transistor cell including the differential access transistors is shown in Fig. 4.3. The first 64 b chips were presented in 1968, and it has kept its performance lead ever since because of its scalability and low-voltage compatibility. The quad of 4 transistors is also the core of the differential output drivers in the ultra-low-voltage differential transmission-gate (LVGTG) logic (Sect. 4.3 and Chap. 7).

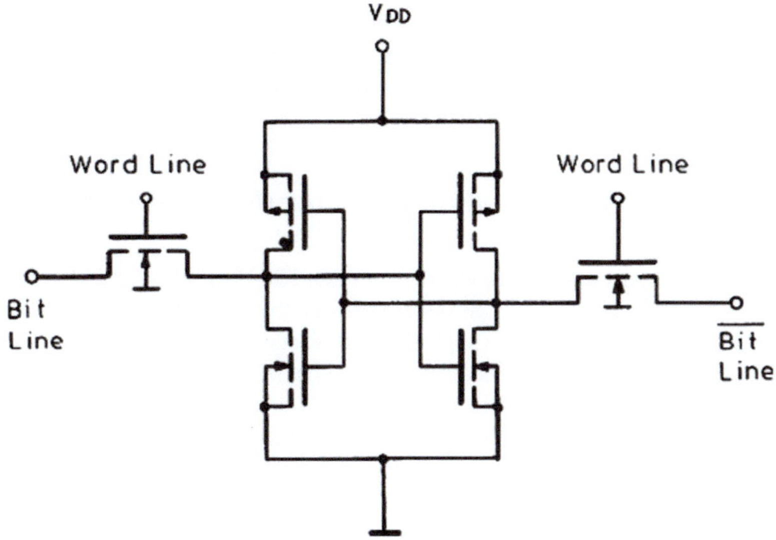

Fig. 4.3 6-transistor CMOS SRAM memory cell [4]. © Springer 2012

Because of the high transistor count of standard CMOS logic, it had a slow penetration against the leading NMOS technologies, until voltage down-scaling for transistor scaling and power reduction became a serious issue in the 80s, as shown in Fig. 4.4, where CMOS standard cells became convenient and effective for CAD.

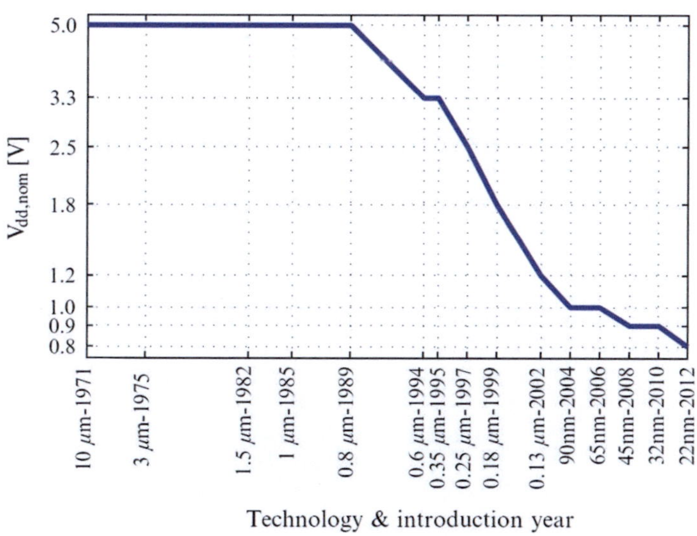

Fig. 4.4 MOS technology nodes and supply voltages [7]. © Springer 2016

For scaled-down CMOS standard-cells, the supply-voltage range 2018 has become 0.7–1.2 V with losses in circuit speed so that new, efficient circuit techniques have become a challenge, which is addressed comprehensively in Chap. 7.

4.2 Fully Depleted Silicon-on-Insulator (FD-SOI) CMOS

The cost of a silicon-on-sapphire wafer made this 1964 invention of SOI-CMOS an expensive specialty. The Si-SiO_2-Si system became the technology direction because its interfaces received sustained, sophisticated research and development since the early 70s, exemplified by the Silicon Interface Specialists Conference, the origin of today's S3S program [2]. One key for efficient Si-on SiO_2-on Si wafer production became the Smart-Cut process of 1995 [1], developed in Grenoble, France, which had already been a center of the sapphire era, and which is a leading SOI center today [3].

The ideal MOS transistor would have its Gate All Around (GAA) its channel. For Ultra-Large-Scale Integration (ULSI), this transistor topology has been realized in regular memory structures like Vertical NMOS NAND Flash-RAM. For general-purpose and complementary MOS ULSI circuits, with

- Optimum gate—hi-k oxide—channel quality,
- Minimum lateral parasitics,
- Minimum substrate leakage,
- Highest lateral density,
- Maximum frequency, equivalent to the ratio of transconductance (drain current over gate voltage), divided by the transistor capacitance,
- Minimum switching energy,

the fully oxide-isolated, thin fully-depleted-channel MOS transistor with buried-oxide (BOX) bias is the optimum MOS transistor for down-scaling and low-Voltage, high intrinsic-speed operation. This transistor type is the reference transistor in [4], including the highly critical variance of nm-size n- and p-channels with only a few doping atoms inside the channel for threshold control. The state-of-the-art of FD-SOI nano-circuits with typically 5 nm channel thickness and physical gate lengths of 10–20 nm has been covered in the tutorial [3] (Fig. 4.5).

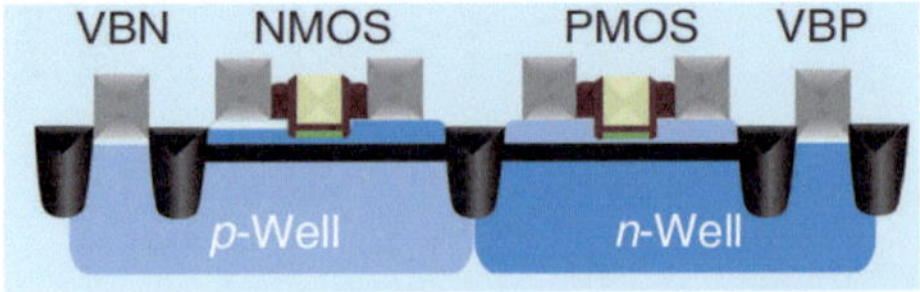

Fig. 4.5 Schematic cross-section of a FD-SOI CMOS technology with body-bias [3]. Black: oxide isolations. VBN: back-bias voltage for NMOS transistor, VBP: back-bias voltage for PMOS transistor. © IEEE 2018

The bandwidth capability of 22 nm FD-SOI transistors is shown in Fig. 4.6 with a maximum frequency of 330 GHz in a comparison with a 14 nm FinFET reaching 220 GHz, which, by construction, has a higher intrinsic capacitance in spite of a shorter channel length.

Furthermore, a comparison of a bulk CMOS microprocessor with an FD-SOI microprocessor in CHIPS 2020, Vol. 2, of 2016 [5] shows that the more ideal SOI transistors deliver a factor two in energy efficiency per operation together with a two-times higher frequency (Fig. 4.7). The future of this Japanese FD-SOI technology, CMOS on Thin Buried Oxide (SOTB), is treated further in Chap. 6 of this book.

The increase in energy below 0.35 V shows the limits of on-off current control in the sub-threshold operation of multi-input MOS gates, where gate voltage changes

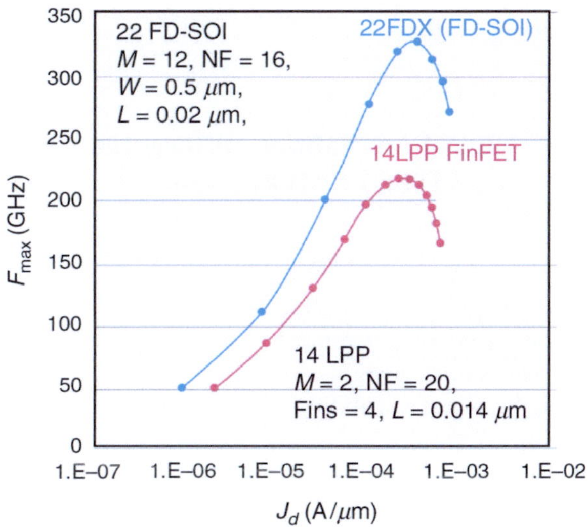

Fig. 4.6 The maximum frequency of a 22 nm FD-SOI transistor in comparison with a 14 nm FinFET [3, 13]. © IEEE 2018

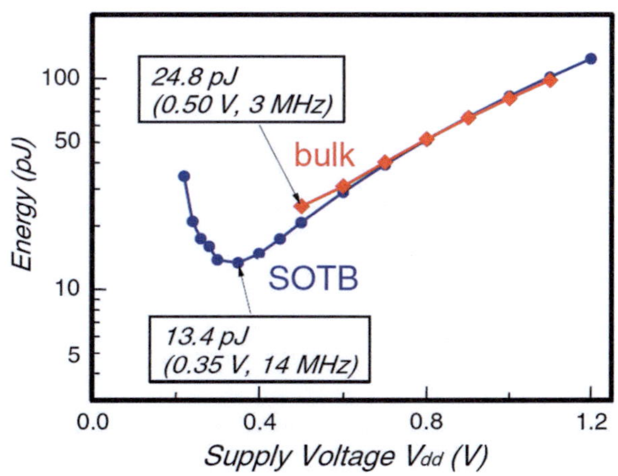

Fig. 4.7 The energy-per-operation of a microprocessor as a function of the supply voltage [5]

of typically 150 mV are needed to change the drain currents by a decade (Chap. 3 in [4]). This limit provides motivation for

- Other types of CMOS logic (the following section and Chap. 7),
- High-k gate insulators
- Lower temperatures (cooled CMOS)
- Tunneling FET's
- Enhance Si.

Other low-voltage limits of CMOS standard-cell logic are the variance of nano-transistors (Sect. 3.2.1 in [4]), zero noise margins, no rail-to-rail outputs, and, most seriously, drastic reductions in switching speeds of high fan-in gates. As a consequence, efficient and robust ultra-low-voltage CMOS design became an issue in the 80s with a major breakthrough published in 2000 [6], with an overview in the following section.

4.3 Ultra-Low-Voltage Differential Transmission—Gate (ULVDTG) CMOS Logic

The ULVDTG CMOS logic, [4, 7] and Chap. 7, has the specific features of

- Minimum robust supply voltage
- Rail-to-rail output voltages
- Highest noise margin
- Highest drive capability
- Highest speed
- Minimum energy per operation
- Best figure-of-merit: Ratio of Speed over (Energy/Operation).

An exemplary gate from the 2000 publication [6] is shown in Fig. 4.8.

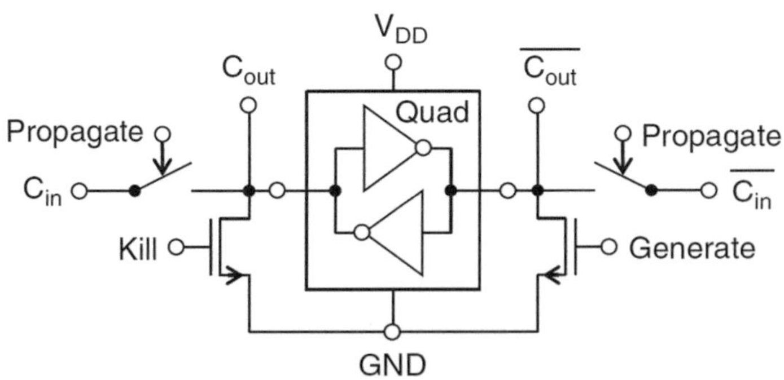

Fig. 4.8 A differential transmission gate logic element from a manchester-carry chain with differential inputs and outputs [6]. Quad = cross-coupled CMOS inverter pair. © IEEE 2000

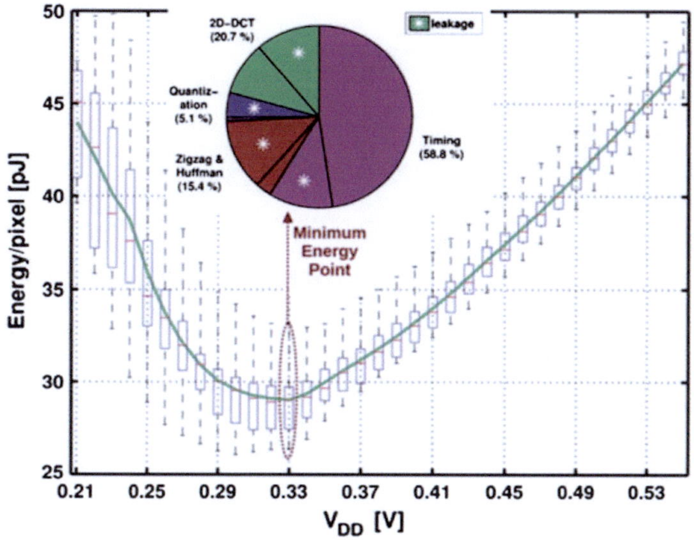

Fig. 4.9 The energy/pixel in a 16 b JPEG encoder with ULVDTG CMOS logic in 40 nm SOI technology [8]. Still the world's leading result, status 2019. © IEEE 2014

The "Quad" is the cross-coupled CMOS transistor pair, which is also the heart of the SRAM memory cell, for rail-to-rail output signals with maximum drive capability, independent of gate fan-in [4, 6, 7].

The most remarkable results for this logic were presented in 2014 for a JPEG coder [8].

In a 40 nm SOI technology, a minimum supply voltage of 210 mV, with minimum energy/pixel at 330 mV was achieved in a production-style test of 20 wafers (Fig. 4.9). Typical of the ULVDTG CMOS minimum transistor sizes and robust gate-output drive capabilities, the speed penalty at very low supply voltages is less serious than in standard-cell CMOS logic, where it is heavy [9].

A strategic overview of ULVDTG CMOS logic is presented in Chap. 7 of this book.

4.4 The CMOS SRAM Cell and 3D CMOS

The cross-coupled pair of complementary MOS transistors, which we called the "Quad" in CHIPS 2020 [4], the heart of the SRAM cell (Fig. 4.3) (and the output of the ULVDT Gate, Fig. 4.8), has been identified as a key benchmark item in the IRDS Report "More Moore" [10] (Fig. 4.10).

High-density SRAM's have been realized since 1980 with poly-silicon transistor layers on top of a high-quality NMOS base layer. The poly-Si PMOS transistors,

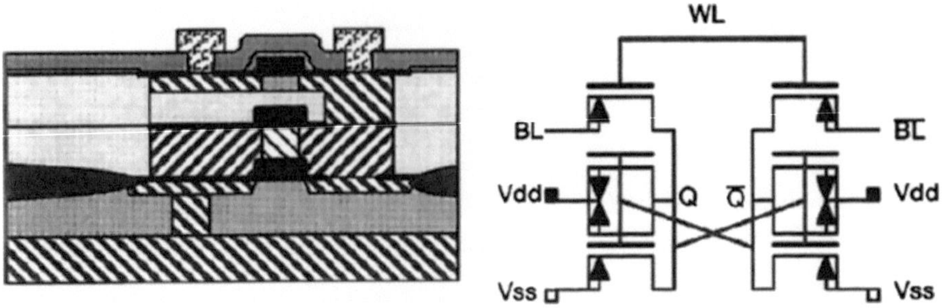

Fig. 4.10 Cross-section and transistor diagram of a 3D 6T CMOS SRAM cell with dual-gate PMOS. Implementation with selective epitaxy and lateral overgrowth [14]. © IEEE 1992

with their reduced conductance, still enabled the "active pull-up", and in a further layer, poly-Si NMOS transfer transistors played the cell-selection role.

The epi-grown, monolithic, high-quality cell has been projected to the 10 nm node for 2020 [11] with

- A footprint of 120 $F^2 = 12\ mm^2$/Gb
- Access time 0.6 ns
- Supply voltage 0.3 V (standby 0.1 V)
- Dynamic energy 7 eV/bit.

This energy of 7 eV/bit continues to be the lowest realistic energy for a memory cell with write- and read-capability. With a write- and read-voltage of 300 mV, sub-ns write and read, it is perfectly compatible with ULVDTG 300 mV logic for a local memory. This most energy-efficient combination of logic and memory would benefit significantly from the optimum 3D building block of four transistors, the "Quad", identified as a benchmark in in the IRDS IFT "More Moore" [10].

4.5 Conclusions

Nano-CMOS technology with fully depleted transistor channels and a body back-bias delivers robust low-voltage operation with the best energy efficiency and highest speed. Ultra- low voltage differential transmission-gate logic in 3D communication with local 3D CMOS SRAM, both at 300 mV, provide orders-of-magnitude improvements in intelligent operations/s/W. The transistor bandwidth >300 GHz enables transceiver integration.

References

1. Section 3.4 in [4]
2. s3sconference.org
3. R.Y. Nguyen, P. Flatress et al., A path to energy efficiency and reliability for IC's. IEEE Solid-State Circ. Mag. Fall **2018**, 24–33 (2018)
4. B. Hoefflinger, The future of 8 chip technologies. Chapter 3, in *CHIPS 2020—A Guide to the Future of Nanoelectronics* (Springer Science and Business Media, 2012). ISBN 978-3-642-22399-0
5. T. Masuhara, The future of low-power electronics, Chap. 2, in *CHIPS 2020, Vol. 2—New Vistas in Nanoelectronics* (Springer, 2016). ISBN 978-3-319-22093-2
6. R. Grube et al., 0.5 volt CMOS logic delivering 25 million 16 × 16 multiplications/s at 400 fJ on a 100 nm T-Gate SOI technology, in *IEEE Computer Elements Workshop*, Mesa, CO (2000)
7. N. Reynders, W. Dehaene, in *Ultra-Low Voltage Design of Energy-Efficient Digital Circuits* (Springe, 2015). ISBN 978-3-318-16135-8
8. N. Reynders, W. Dehaene, A 210 mV 5 MHz variation-resilient near-threshold JPEG encoder in 40 nm CMOS, in *2014 ISSCC Digest of Digital Papers*, paper 27.3 (2014), pp. 457–458
9. Section 3.6, in *CHIPS 2020 Vol. 2—New Vistas in Nanoelectronics* (Springer International Publishing, 2016)
10. 2018 IRDS-MM.pdf
11. Section 11.1 in [4]
12. F.M. Wanlass, Low stand-by power complementary field effect circuitry. US Patent No. 3356858, filed 18 June 1963. Issued 5 Dec 1967
13. G. Schaeffer, 5G wireless solutions for mobile and IOT products, in *Proceedings Linley Mobile and Wearables Conference* (2016)
14. G. Roos, B. Hoefflinger, Complex 3D CMOS circuits based on a triple-decker cell. IEEE J. Solid-State Circ. **27**, 1067 (1992)
15. B. Hoefflinger, New CMOS technologies, in *Solid-state devices 1980*, ed. by J.E. Carroll. Institute of Physics Conference Series, No. 57 (Institute of Physics Publishing, Bristol, 1981), pp. 85–139

Chapter 5
Nanolithography

Bernd Hoefflinger

5.1 IRDS Lithography Roadmap

One of the International Focus Teams (IFT's) of the International Roadmap for Devices and Systems (IRDS) [1] is the one for "Lithography" [2]. In its 2017 Report, it gives a summary on the presently valid critical dimensions for lithographic realization on wafers. A comparison of this data with the ITRS (International Technology Roadmap for Semiconductors) [3, 4], shows the major shift and corrections, which had to be made since 2009. They indicate also many of the challenges. Some of the aggressive items of the ITRS roadmap of 2009 are evident in Table 5.1, in particular the physical gate lengths below 10 nm. The arguments against this scaling were a fundamental issue in CHIPS 2020 [3, 4]. The corrections are now clearly visible in Table 5.2, representing the data valid in 2019, where physical gate lengths are 16 nm in 2021 and settling at 12 nm. The density of transistors and interconnects is represented by the half-pitch = (line-width + space)/2.

Regarding the half-pitches, the goals have been shifted by three to five years, and a settling is noticeable in the mid 20s for

Table 5.1 Short Overview of the ITRS 2009

Long-term years	2018	2020	2022	2024
Flash poly Si ½ pitch [nm]	12.6	10.0	8.0	6.3
MPU/ASIC first metal ½ pitch [nm]	15.0	11.9	9.5	7.5
MPU physical gate length [nm]	12.8	10.7	8.9	7.4

From CHIPS 2020 [3]. © Springer 2012 [1]

B. Hoefflinger (✉)
Sindelfingen, Baden-Württemberg, Germany
e-mail: bhoefflinger@t-online.de

B. Murmann and B. Hoefflinger (eds.), *NANO-CHIPS 2030*,
The Frontiers Collection, https://doi.org/10.1007/978-3-030-18338-7_5

Table 5.2 Short overview of the 2017 LITHOGRAPHY REPORT of the IRDS [2]

Year of production	2021	2024	2027	2030
Logic industry node labeling (nm)	"5 nm"	"3 nm"	"2.1 nm"	"1.5 nm"
DRAM minimum half-pitch (nm)	17.0	14.0	11.0	8.4
2D flash half-pitch (nm)	15	15	15	15
Flash 3D channel half-pitch (nm)	80	<80	<80	<80
MPU/ASIC metal half-pitch (nm)	12	10	7.0	7.0
Physical gate length HP logic (nm)	16	14	12	12
Contact CD after etch (nm)	12	10.5	7.0	7.0

- Min. physical gate length 12 nm,
- Min. metal half-pitch 7 nm (not contacted),
- Min. contact CD 7 nm.

The most confusing new quantity is the *Logic Industry Node in nm.*

Its specifications are shown in Fig. 5.1.

The "7 nm" Logic Node, for production in 2019, correlates with a min. physical gate length of 16 nm and with a high-performance (HP)-logic metal half-pitch of 14 nm. So product announcements need careful reading of which nano-meters are meant. The lithography options in Fig. 5.1 are

- 193 nm QP: Deep UV Immersion Quadruple Projection,
- NIL: Nano-Imprint Lithography,
- EUV SP: Extreme UV (13.5 nm) single-projection,
- EUV DP: EUV Double projection,
- DSA: Directed Self-Assembly.

EUV has been introduced into volume manufacturing, after reported 2Mio. processed wafers [5] until then, in 2018 by one supplier (ASML, ZEISS) and three customers, serving the "7 nm" industry node with 18 delivered systems and 40 planned for 2019.

5.2 EUV Lithography

Extreme ultra-violet (13.5 nm) lithography has had a unique, dramatic R&D and investment history with several critical reviews, among them one in [6], and with a comprehensive coverage in "EUV Lithography 2nd Edition" [5], published in 2018. A schematic graph of an EUV exposure system with reflective-mirror optics is presented in Fig. 5.2 [5].

The source needs a high-power (>100 W) CO_2 laser. It is focused on Tin (Sn) droplets, falling at a rate of 50,000/s, producing a plasma, which sends 13.5 nm radiation to the illumination. The 4x mask has a numerical aperture of 0.0835, enabling

Semiconductor Product node	*Minimum half pitch (nm)*	Production Year Minimum lithographically defined half pitch for MPU and DRAM Possible Option
"10nm" Logic Node 18nm DRAM	18	193nm QP 193nm QP
"7nm" Logic Node	14	193nm QP EUV SP NIL
"5nm" logic Node "3nm" Logic Node 14nm DRAM	12 10.5 14	193nm QP EUV DP NIL DSA
"2.1nm" Logic Node and below < 10nm DRAM	7 9.2, 7.7	EUV DP High NA EUV DSA plus EUV

Fig. 5.1 Industry product nodes, typical half-pitches and lithography options [2]. © IEEE 2017

an aperture of 0.33 at the wafer-level. The system has many serious challenges, covered in the >600 pages of [5]. We select two state-of-the-art results for the optics to illustrate requirements:

The precision of the mirrors is illustrated in Fig. 5.3 [7]. The mirror diameters have reached >1 m in 2018, and their surface root-means-square (rms) non-uniformities have advanced from 0.3 nmin 2012 to <0.1 nm in 2018. The EUV reflecting mask

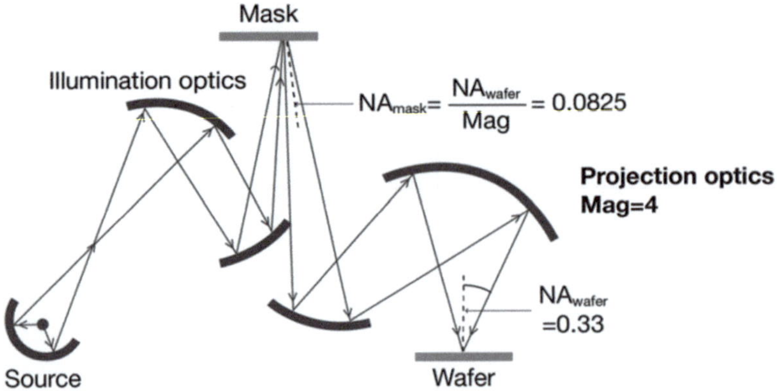

Fig. 5.2 Schematic of an EUV lithography system [5]. © SPIE 2018

Optics	MET	ADT	3100	3300 / 3400
Photos show relative mirror size		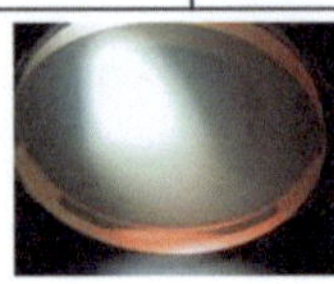		
Figure [pm rms] → aberrations	350	250	140	< 75
MSFR [pm rms] → flare	250	200	130	< 80
HSFR [pm rms] → light loss	300	250	150	< 100

Fig. 5.3 EUV mirror uniformity, generations from 2012 to 2018 [7]. © SPIE 2018

is precision-etched at 4x magnification into a multilayer $MoSi_2/Si$ surface of 50 bi-layers with an individual thickness variation of 0.03 nm (see Fig. 5.4), delivering a peak reflectance of 70%.

EUV lithography manufacturing systems for the “5 nm” Industry Node” have been under construction since 2018 for manufacturing of minimum structures of 12 nm in 2021.

Besides the exciting optical system, EUV has several other essential subjects, treated in [5]:

- Metrology
- Mask Making
- Pellicles
- Contamination

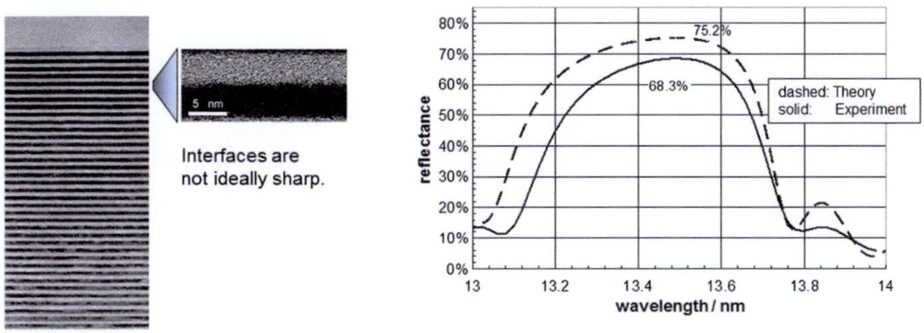

Fig. 5.4 Reflectance of a multi-$MoSi_2$/Si bi-layer mask [7]. © SPIE 2018

- Photoresist
- EUV Scanners.

All of these subjects make this lithography the core of selected manufacturers for aggressively down-scaled, high-volume and highly valued chip products.

5.3 Conclusion

The IRDS Lithography roadmap of 2017 has introduced an ambitious, but more realistic roadmap for minimum physical gate lengths of 12 nm and minimum contact holes of 7 nm in 2030. In any event, 2018 saw the introduction of volume manufacturing with EUV lithography for the leading *industry* "7 nm" node with a minimum half-pith of 14 nm by three leading chip makers. These leaders, together with the EUV providers ASML + ZEISS, will implement the roadmap. The major drive towards 3D integration (Chaps. 8–11, 13 and 15) will introduce new patterning and alignment techniques as well as self-assembly.

References

1. www.ieee.irds.org
2. 2017 IRDS—LI.pdf
3. B. Hoefflinger (ed.), The international technology roadmap for semiconductors, Chap. 7, in *CHIPS 2020—A Guide to the Future of Microelectronics* (Springer Science and Business Media, 2012). ISBN 978-3-642-22
4. B. Hoefflinger (ed.), ITRS 2028, Chap. 7, in *CHIPS 2020, Vol. 2, New Vistas in Nanoelectronics* (Springer, 2016). ISBN 978-3-319-22093-2
5. V. Bakshi (ed.), in *EUV Lithography*, 2nd edn. (SPIE Press, (ePub), 2018). ISBN 9781510616806
6. B. Lin, Nanolithography, Chap. 8, in *CHIPS 2020—A Guide to the Future of Microelectronics* (Springer Science and Business Media, 2012). ISBN 978-3-642-22399-0
7. S. Migura, W. Kaiser et al., in Optical systems for EUVL, Chap. 5 of ref. 5

Chapter 6
The Future of Ultra-Low Power SOTB CMOS Technology and Applications

Nobuyuki Sugii, Shiro Kamohara and Makoto Ikeda

6.1 Ultra-Low-Power CMOS in IoT Front-End Devices

The numbers of Internet-of-Things (IoT)[1] connected devices are reported to be about 15 billion in 2015 and continuously increasing with a high CAGR (compound annual growth rate) of ~10% or more [2–4] as shown in Fig. 6.1. The major functions of the IoT frond-end devices are sensing data and actuating something from/to the real physical world in the cyber-physical systems [5]. The actuating part may employ various types of devices such as displays and mechanical assemblies. The sensing part is usually comprised of sensors, analog front-end circuits, analog-to-digital convertors, edge processors, and data-transmission circuits. The data transmission often uses wireless communication to avoid any wiring, and for the same reason, the sensing part is better working without outer power supply, by using batteries or energy harvesters (EH). Since the wireless communication consumes relatively much power, in the operation with a standalone (battery or EH) power source, decreasing the data transmission rate is an effective way to reduce the power consumption of each IoT connected device. The use of the edge processing with ultra-low-power consumption has thus drawn much attention in recent years as well as decreasing latency for the data communication like the 5G technology.

[1]The term "Internet of Things" was firstly used by Kevin Ashton in 1999 [1], while the term has become widely known in the early 2010s.

N. Sugii (✉)
Hitachi, Ltd., Tokyo, Japan
e-mail: nobuyuki.sugii.df@hitachi.com; n-sugii@ieee.org

S. Kamohara
Renesas Electronics Corp., Tokyo, Japan

M. Ikeda
The University of Tokyo, Tokyo, Japan

B. Murmann and B. Hoefflinger (eds.), *NANO-CHIPS 2030*,
The Frontiers Collection, https://doi.org/10.1007/978-3-030-18338-7_6

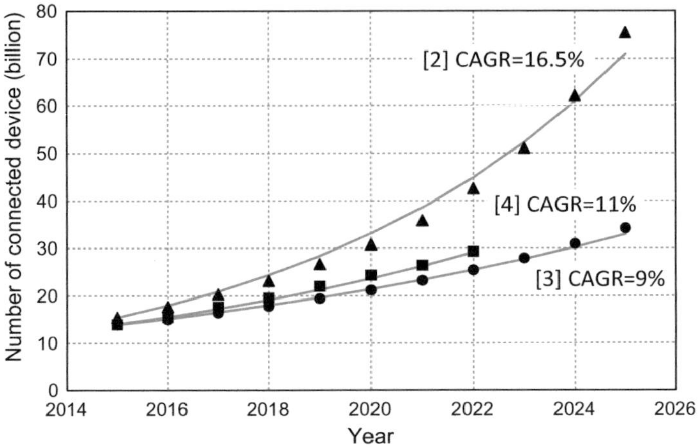

Fig. 6.1 Annual growth forecast of number of connected devices [2–4]

Continuous progress of semiconductor devices, especially, CMOS integrated circuits (ICs) with an aggressive miniaturization, has enabled both the performance improvement and decreasing the power consumption. It is well known, however, in the recent generations, the increase in the performance-per-power efficiency has been slowed down as the article named "The Free Lunch Is Over" [6] clearly depicted. As seen in Fig. 6.2 on the power and power-performance efficiency of top-class supercomputers [7], the maximum performance is dominated by available power: about 20 MW, this maximum power level has been unchanged in recent years. This situation (power-limited performance) is common from a large-scale supercomputer system

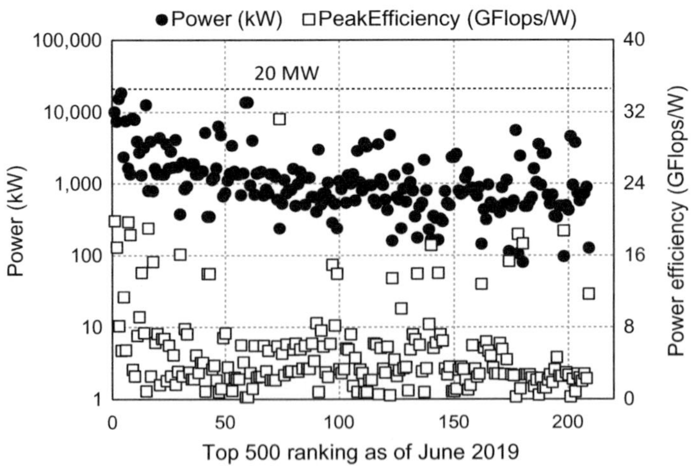

Fig. 6.2 Power consumption and power efficiency of top 500 supercomputers [7]

to a tiny IoT sensor-node device. In the large system, the performance is dominated by the available power from the power grid and also by the cooling capability, and in the standalone wireless sensor node, the performance is determined simply by the battery life or the maximum generating power of EH. The only way to increase the performance for each system, regardless of its size or capacity, is thus to improve the performance-per-power efficiency.

6.2 Available Power for a Standalone Sensor Node and Power Requirement for a Micro Controller Unit

In the battery-operated system, in general, life of the battery (or interval of charging cycle for the rechargeable battery) is determined by the average current.[2] Figure 6.3 depicts the life of popular small-size batteries as a function of average current. Considering the battery life (or charging interval) of about one year, the average current of about 100 μA will be required. For the devices powered by EH, the situation is similar. The available power of popular EH sources is on the order of 100 μW [8]. Note that the photovoltaic cells are most powerful among the EH sources, however, the available power by indoor light stays on the same level.

In order to decrease the average current consumption to the above mentioned level, it is useful to reduce the working duration of wireless communication since the

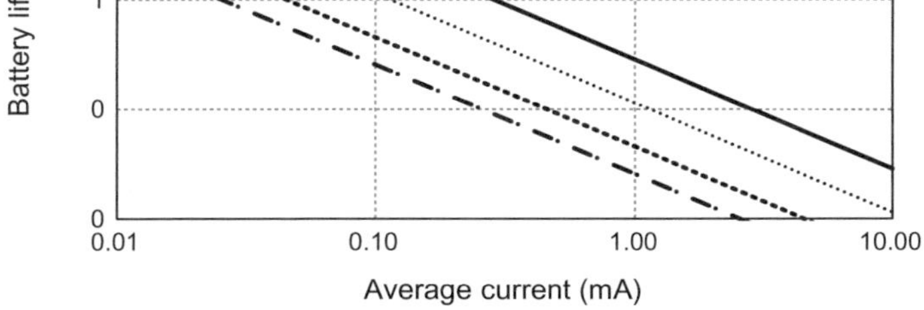

Fig. 6.3 Battery life as a function of average current

[2]In the general system with a series regulator connected to a power source, the power due to the difference between supply and operating voltages is consumed in the series regulator, and thus the total power consumption is proportional to the current consumption. On the other hand, in the system with a power management using a dc-dc converter, the battery life is determined by the average power consumption of the system.

power consumption of wireless communication is in general tenth or hundredth of mW level. For most of the sensing nodes, the required data are not continuous, and thus intermittent operation of sensing and data transmission is a realistic solution. Reducing the power for the data processing at the sensing node (edge) is important as well, and this is the main topic of this chapter. In the standalone sensor node, a micro controller-unit (MCU) is usually used.

Let us consider the required power consumption level of an MCU for this purpose. The power efficiency metrics for MCU are in general active current per clock frequency (μA/MHz) and standby current (μA). Considering the target power consumption level of 100 μW to 1 mW, operation voltage at or less than 1 V, and clock frequency of 10 MHz or more, that is, the typical operation conditions of an MCU for the IoT sensor-node application, the active current level of 10–100 μA/MHz is required. Since the power consumption of the sensing node does not drop to zero due to the leakage current even for the intermittent operation, the standby current of MCU should be considerably low. Although most of MCUs have multiple sleep modes to reduce the standby current, it is essential to reduce the leakage of CMOS circuits themselves, and thus the low-leakage CMOS device technology is important. In the next section, the factors to reduce both the active and standby powers (currents) of the CMOS unit circuit are briefly reviewed.

6.3 Energy Efficient CMOS Operation

The power consumed by the CMOS circuit has two components, that is, active (dynamic, or switching) and leakage (or static) power. The power consumption of CMOS inverters, as a representative of CMOS circuits, can be expressed as

$$P = n\left(\alpha C_{load} V_{dd}^2 f + I_{leakage} V_{dd}\right), \tag{6.1}$$

where n is the number of transistors, α is an activity factor (including time averaged active ratio of transistors), C_{load} is the load capacitance, V_{dd} is the operation (or supply) voltage, f is the clock frequency, and I_{leakage} is the leakage current. Note that the power due to the short-circuit current $\left(I_{sc} \propto (V_{dd}/2 - V_{th})^2\right)$ is omitted for simplicity. The energy consumed by a logic (switching) operation is more important than the power because it directly reflects the efficiency of the information processing. Divided by αf, the energy per a single switching operation (energy per cycle) is thus written as,

$$E = n\left(C_{load} V_{dd}^2 + I_{leakage} V_{dd} / \alpha f\right). \tag{6.2}$$

The first and second terms correspond to active and leakage energies, respectively. In the same technology node (feature size of CMOS integrated circuits), C_{load} is constant and thus decreasing the V_{dd} is effective for decreasing the active energy. On the other hand, to minimize the leakage energy, the operation frequency should be taken into account. Let us consider the operation at the maximum operating frequency here that is determined by the propagation delay time t_{pd} of CMOS circuits,

$$t_{pd} \propto C_{load} V_{dd} / (V_{dd} - V_{th})^m, \quad (6.3)$$

where V_{th} is the threshold voltage assuming a symmetrical CMOS operation: $V_{\mathrm{th}} = V_{\mathrm{thp}} = -V_{\mathrm{thn}}$ where V_{thp} and V_{thn} are threshold voltages of p- and n-type MOSFETs, respectively, and m is a factor taking the velocity saturation into account. For the recent CMOS technologies, the m value is about 1.2.

The energy E in (6.2) is thus determined by V_{dd}, V_{th}, and f where C_{load} and α are assumed to be constant. Although f can be arbitrarily set, in order to minimize E, f should be maximized within the range satisfying (6.3). As an example, E as a function of V_{dd} and V_{th} behaves as shown in Fig. 6.4 [9]. As indicated by the energy contours and a dashed line, the optimal combination of V_{dd} and V_{th} is determined according to the required frequency. The absolute minimum energy point (MEP) is also shown in the graph, however, this point lies at near the condition where V_{dd} is slightly less than V_{th} (so called sub-threshold operation), and its frequency is very slow. A practically useful approach is thus to follow the dashed line to increase f at the expense of the increase in E. This means that V_{th} should be controlled together with V_{dd}. The only way to control V_{th} within the conventional CMOS circuit operation scheme is applying back bias.

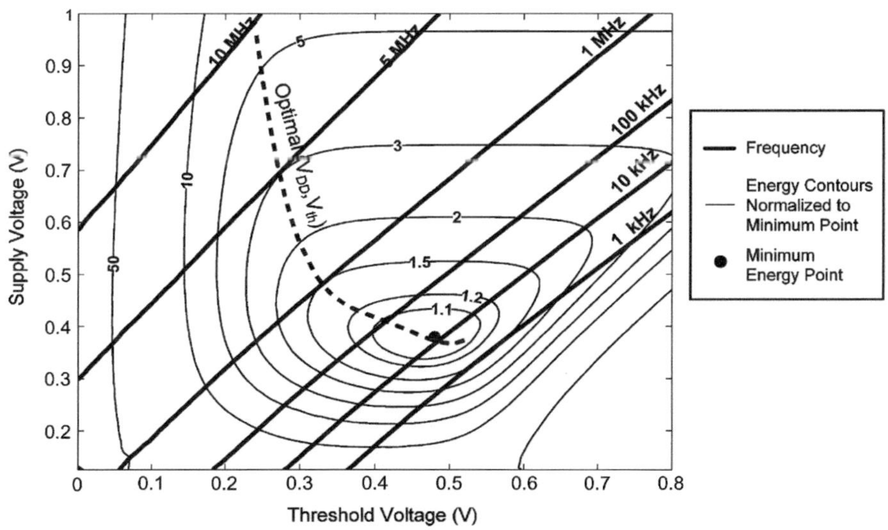

Fig. 6.4 Minimum energy point and energy contours © 2005 IEEE [9]

6.4 Suitable CMOS Technology Node for IoT Front-End Devices

The miniaturization of the CMOS technology has enabled to decrease C_{load}, and together with decreasing V_{dd}, the active energy has significantly reduced. The leakage energy, however, has been rather increased for the recent highly scaled generations due to tunnel leakage around the gate electrode, short-channel-effect related subthreshold leakage, and so on. By taking the required frequency and the acceptable leakage-current level into account, one can determine the optimum CMOS technology node and its technology flavor such as general purpose or low standby power. There are many reports on the energy minimizing in various technology generations, for example, following the conventional scaling model, the energy at MEP decreases with decreasing the technology node from 65 to 22 nm [10], whereas in the low-energy dedicated design (subthreshold logic), the energy at MEP hits the bottom at 90 nm [11]. The change of the MEP behavior with different technology nodes is schematically shown in Fig. 6.5 as another example. In this calculation, the typical 65 and 28 nm processes of relatively high performance and relatively low leakage processes, respectively, are assumed. The typical parameters are: relative C_{load} of 1 and 0.25, V_{th} of 0.25 and 0.15 V, and subthreshold swing of 75 and 85 mV/decade, respectively. For the performance-dedicated applications, it is preferred to use more advanced process, 28 nm in this figure, and the energy at MEP is higher and also the voltage of MEP is higher (at 0.6 V). On the other hand, for the low-energy dedicated applications, the voltage of MEP is about 0.4 V or less. The minimum energy can decrease whereas the clock frequency is not so high. For the IoT front-end device applications, the low-energy dedicated option will be preferred.

In addition to the above energy-performance trade-off, the production cost is another factor to choose the adequate CMOS technology. In the past generations,

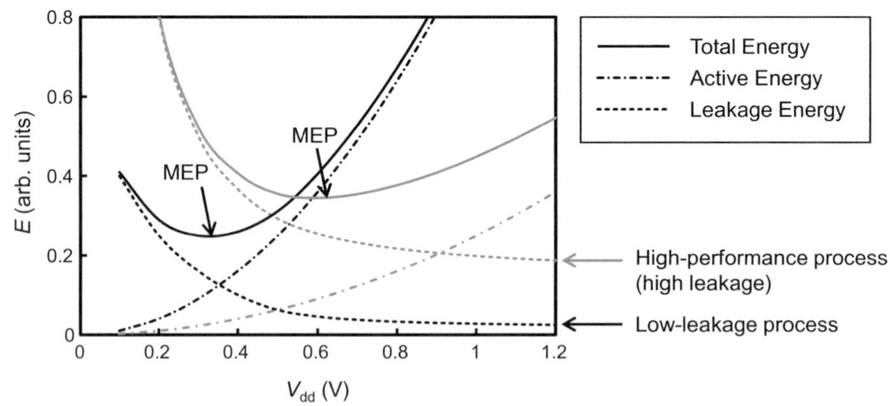

Fig. 6.5 Comparison of minimum energy points for high-performance and low-leakage processes

the most advanced process, that is, most miniaturized process, was the most cost-effective option. In the recent few generations, the situation has been different. There is a report that the lowest cost for an IC die with a high-volume production is achieved in the 65 nm process [12]. Although the situation may change by the maturity of the advanced CMOS processes and circuit design technologies, staying on at 65–40 nm nodes is currently the preferred option for the IoT front-end device applications.

6.5 The Variability Issue that Hinders Low-Voltage Operation

As discussed in the previous section, decreasing the operation voltage V_{dd} to that of the MEP condition is an important approach for the energy efficient CMOS ICs. The CMOS scaling, however, has brought another important issue: statistical characteristic variability of transistors. In the ultra-large-scale ICs with the most advanced process, the number of transistors exceeds billions. Decreasing the characteristic variability of transistors is thus crucial problem to operate the ICs without any functional errors, especially for low-voltage operation since the voltage margin should be minimized. There are many reports on the lowest operating voltage taking the variability into account. The lowest voltages of both logic and static random-access memory (SRAM) circuits have rather increased in the recent CMOS technologies [10]. Among many types of CMOS transistor characteristics, variability of V_{th} is most important. It is well-known that the V_{th} variability is defined as [13]

$$\sigma V_{th} = A_{VT}/\sqrt{LW}, \tag{6.4}$$

and in the conventional bulk MOS transistor, it can be written as

$$\sigma V_{th} \propto t_{ox} N_{imp}^{1/4}/\sqrt{LW}, \tag{6.5}$$

where σV_{th} is the standard deviation of V_{th}, A_{VT} is the Pelgrom coefficient, L is the channel length, W is the channel width, t_{ox} is the gate-oxide thickness, and N_{imp} is the impurity density of the channel region.

If we follow the ideal scaling rule [14], t_{ox}, L, and W decrease and N_{imp} increases at a constant rate by generation. This device scaling strategy inherently increases σV_{th} slightly by generation [15]. Moreover, in the recent generations, t_{ox} has not been scaled down sufficiently due to the increase of gate leakage current. This further increases the σV_{th}. It is thus very difficult to decrease the V_{th} variability with the conventional bulk MOS transistors. Using the transistor structures with a fully-depleted (FD) channel is a possible solution because, in these structures, the transistor characteristic can be controlled without increasing the channel impurity density N_{imp}, and thus σV_{th} can be significantly decreased as indicated by (6.5).

6.6 SOTB Technology

It is a long history to commercialize the FD transistors. There were several important proposals regarding the transistor structures of FD or a reduced impurity-density channel in the late 1980s–early 1990s: for example, the intrinsic-channel (epitaxially grown channel) structure of the bulk MOSFETs [16], the planar double-gate structure [17], and the DELTA structure [18]. Note that the DELTA structure is the original name that is now well-known as FinFET with three-dimensional channels. As a family of the planar FD transistors using an SOI (silicon on insulator) wafer with a very thin BOX (buried oxide) layer, and adding new features of a high V_{th} controllability and a high compatibility with the existing bulk CMOS technology, the SOTB (Silicon on Thin Buried Oxide) transistor was proposed in 2004 [19]. The schematic cross section of the SOTB structure is shown in Fig. 6.6.

The advantages of SOTB transistors are listed as follows:

1. Excellent short-channel-effect (SCE) immunity due to better electrostatic control enabled by thin SOI and BOX layers and a underlying ground plane (GP).
2. Small V_{th} variability and low sensitivity to SOI-thickness variation due to a low N_{imp} SOI channel.
3. Flexible V_{th} control by N_{imp} and depth profile of the GP region.
4. Back-gate-bias control by applying voltages to the GP regions of p- and n-type transistors through the V_{bp} and V_{bn} terminals. Deep n-well region secures separation between the two GP regions if proper back-gate bias voltages are applied.
5. A hybrid bulk transistor can be integrated on the same wafer by removing the SOI and BOX layers. Patterning the gate electrode and shallow-trench isolation of both SOTB and hybrid bulk transistors can be done because step-height difference between the SOTB and hybrid bulk regions is small due to thin SOI and BOX layers.

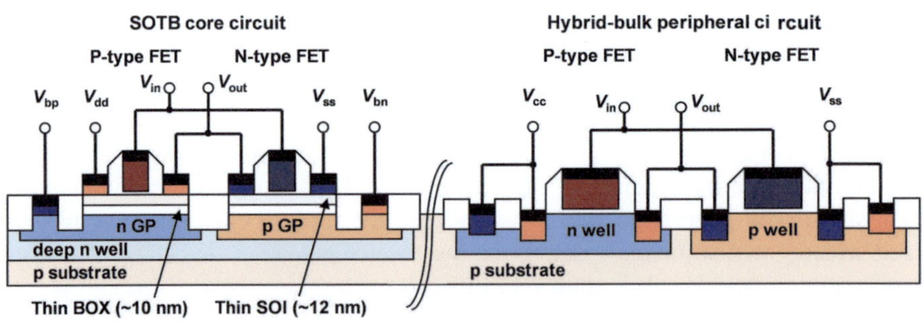

Fig. 6.6 Schematic cross section of SOTB transistors. Hybrid bulk transistors are shown. SOTB transistors are used in low-voltage (< ~1.5 V) logic and analog circuits including SRAMs. Bulk transistors are used in peripheral, ESD-protection, high-voltage analog and power circuits, on-chip, flash memory, and reuse of legacy circuits

6. The planar layout of transistors and logic cells is the same as that of the existing bulk technology.
7. High soft-error (single-event-upset) immunity against a high-energy-particle irradiation such as alpha particles and neutrons due to the thin active (channel) layer separated from the substrate by the BOX layer.

Proper V_{th} control is important to solve the performance and power trade-offs as described in Sect. 6.3. In the SOTB technology, the V_{th}s of different flavors such as those suitable for ultra-low-voltage (V_{dd} down to 0.4 V) or ultra-low leakage (off-current down to pA/μm level) operations are controlled by selecting proper high-k gate-stack materials and changing the impurity density of the GP region [20].

Back-gate-bias controllability is an important point of the SOTB transistor design. In the SOTB structure, the GP layers act as back-gate electrodes. To achieve high back-gate bias controllability, it is important to thin down the BOX-layer as well as decreasing the depletion-layer thickness in the GP layer under the whole range of the bias voltages because the depletion layer also acts as a dielectric layer between the channel and the back gate. In the typical SOTB transistor design, for example, the back-gate-bias coefficient (γ)[3] is about 0.16 for the design with 10-nm BOX thickness and nearly uniform impurity-density profile of 1×10^{18} cm^{-3} in the GP region just below the BOX layer [15].

The range of back-gate-bias voltage is limited by the leakage current between the two GP layers through the deep n-well. The voltage difference, $V_{bp} - V_{bn}$ (see Fig. 6.6), depends on back-bias voltage (V_{bb}) and operation voltage (V_{dd}), where $V_{bp} = V_{dd} - V_{bb}$ and $V_{bn} = V_{bb}$. In the reverse back biasing to increase V_{th} ($V_{bb} < 0$), V_{bp} is positive, and V_{bn} is negative. In such a case, the junction between the nGP and pGP is reversely biased. In the forward biasing condition, $V_{bn} > V_{bp}$, that is, $V_{bb} > V_{dd}/2$, the junction is positively biased. The maximum applicable positive back-bias voltage is thus limited to the condition $V_{bn} - V_{bp} < 0.5$ (built in potential of pn junction); that is, $V_{bb} < 0.25 + V_{dd}/2$. A significant amount of leakage current flows from the pGP to the nGP when this condition is not satisfied. To apply higher forward V_{bb}, The flip-well structure (the conduction types of the GP layers in Fig. 6.6 are swapped each other) was proposed [21]. The forward V_{bb} significantly increases the maximum clock frequency of circuits, but also increase the static leakage current. The flip-well technology is thus suitable for high-performance applications. In the SOTB technology in this section, the normal conduction types of the GP layer are preferred since the static leakage reduction by the reverse back biasing is important for the IoT device applications.

[3] γ is defined as $\partial V_{th}/\partial V_{bb}$. at around $V_{bb} = 0$.

6.7 Reduction of V_{th} Variation and Ultra-Low-Voltage SRAM Operation

The V_{th} variation of SOTB transistors was demonstrated to be about half of the bulk transistors of the same size both for p- and n-types [22]. Recent results [23] with the effect of back biasing are shown in Fig. 6.7. It should be noted that the V_{th} variation under reverse back biasing at $V_{bb} = -2$ V, that is effective for the static leakage reduction of a few orders of magnitude, is the same as that at $V_{bb} = 0$ V.

It is known that the low-voltage operation of SRAM is more difficult than the general CMOS logic circuits as indicated in [10]. It is thus important to investigate the lowest operation voltage (V_{min}) of SRAM to verify the effect of the characteristic variability reduction. For the SOTB SRAM of 0.54 μm^2 area of the conventional 6-transistor layout, the V_{min} of 0.37 V was reported [22]. It was demonstrated that this V_{min} can be achieved by controlling V_{bb} regardless of temperature variation from −30° to 80 °C. Figure 6.8 shows the V_{min} of SOTB SRAMs with different V_{th} flavors (high speed or low leakage). Lower V_{min} even at higher V_{th} than bulk SRAM was demonstrated.

It should be noted that the SOTB SRAM can store the data at very small leakage current level (cell leakage about 1.2 pA [22]) by applying a proper reverse V_{bb}. Taking advantage of this feature, the SOTB SRAM can be used as a *pseudo-nonvolatile* memory in the specific applications.

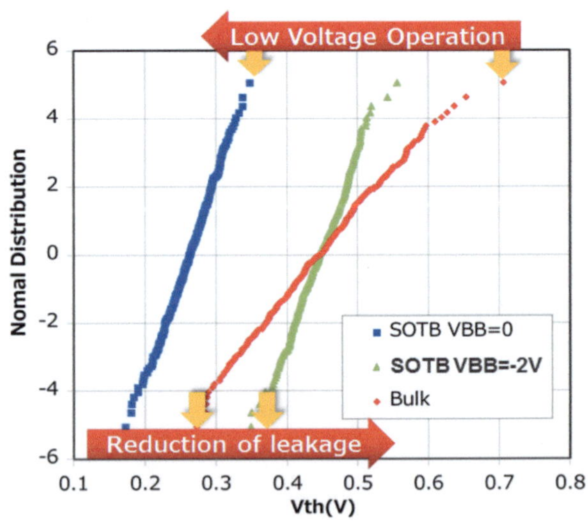

Fig. 6.7 V_{th} distribution of one-million n-type SOTB transistors compared with bulk transistors of the same size © 2017 IEEE [23]

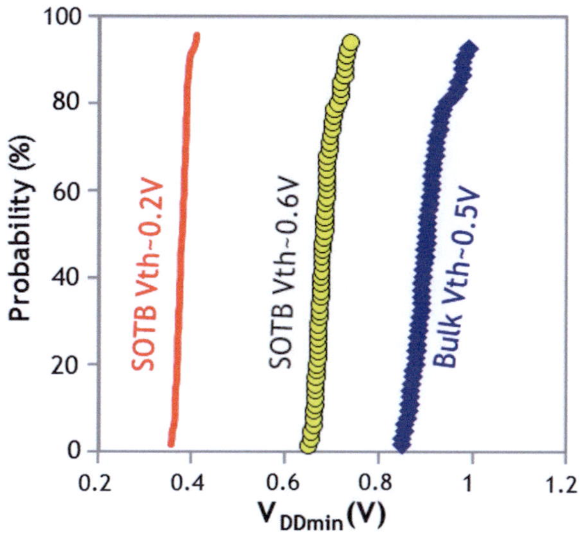

Fig. 6.8 $V_{\min}$ (V_{DDmin}) of SOTB SRAMs compared with bulk SRAMs © 2017 IEEE [23]

6.8 Circuit Design Environment and Open Shuttle Activity

The design flow of the SOTB ICs can be built based on that for the bulk CMOS technology of the same technology node. The electronic design automation (EDA) tools and their file formats are completely the same as the existing ones from the register transfer level (RTL) to the layout (graphic database system: GDS). The circuits including both the SOTB and the hybrid bulk transistors can be designed at a time.

The design (mask) layer, layout rules and their verification files (including the antenna effect) should be revised or added to match the characteristics of SOTB. There are a few additional points to be specially considered related to the back-gate biasing. The location and distance of the back-gate-bias voltage taps are important design points for compromising the back-gate voltage stability and the integration density. These are generally embedded in the layout rule file and the standard-cell layouts. In some applications using different back-bias voltages in fine grained back-bias domains (that will be shown in Sect. 6.11), the spacing between the deep-n-well islands is preferably decreased, and it is also a trade off among the size, the leakage current, and the range of the back-bias voltage [24].

The compact model of transistors is indispensable for the circuit design with high accuracy of both timing and power estimation. Currently available transistor models for bulk transistors cannot be used accurately for the SOTB technology because the transistor characteristics under various back-bias voltages cannot be reproduced. New SPICE (Simulation Program with Integrated Circuit Emphasis) models for the SOTB and related thin-BOX FDSOI transistors with back biasing, namely,

HiSIM-SOTB [25] and BSIM-IMG [26], have thus been developed. Both models are based on a surface-potential expression and they represent well the behavior of transistor characteristics with varying back-bias voltages.

Detailed studies on circuit design such as body-biasing schemes, delay variability reduction, signal voltage design on ultra-low-voltage macros, energy minimization have been reported [27–31]. Moreover, various types of circuit designs have been implemented using the SOTB shuttle service operated by the University of Tokyo in collaboration with Renesas Electronics Corp. from FY 2015 [32]. About 12 chip designs per shuttle run were fabricated in FY 2018. This shuttle is not restricted to academia, but can be used for the commercial proto typing. Most of the circuit design examples that will be mentioned in the later sections are demonstrated using this shuttle service.

6.9 MCU with Back-Gate Bias Control

A low-energy-consumption central processing unit (CPU) core for the MCU application was demonstrated using the 65-nm SOTB process [33]. The CPU core consists of an in-order 5-stage pipeline, and 4 blocks of 32 kword × 9 data memory. The scale of integration for this 32-bit CPU core is 50.1 kgate logic and 144 kB SRAM arrays and the area is 2.1 mm^2. As shown in Fig. 6.9, the core is functional down to $V_{dd} = 0.22$ V at 1 MHz clock frequency, whereas the same core fabricated by the conventional bulk process operates down to 0.5 V. The MEP of the SOTB core is 13.4 pJ/cycle at $V_{dd} = 0.35$ V as shown in Fig. 6.10, which corresponds to 38 μA/MHz. This is a good number for the IoT application chip (see Sect. 6.2). The optimization of the energy is done by controlling the back-gate bias V_{bb}. The sleep current is only 0.14 μA at V_{dd} and V_{bb} of 0.35 and −2.5 V, respectively. Considering the intermittent operation, the average current consumptions for the activity (ratio

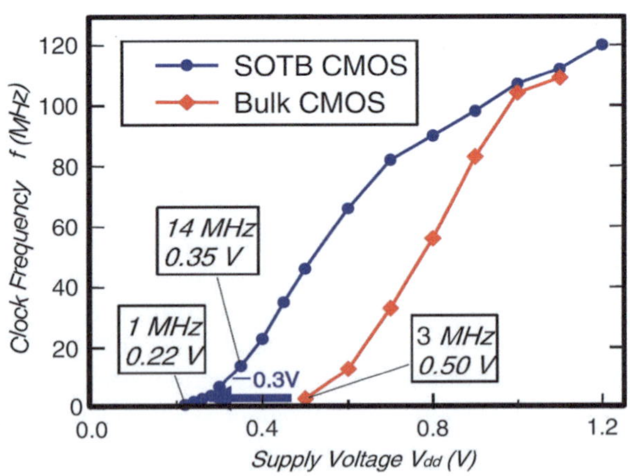

Fig. 6.9 Maximum operating frequency of 32-bit CPU core fabricated by SOTB and bulk technologies [33] © 2014 IEEE (same figure as Fig. 2.10 (a) in CHIPS 2020 Vol. 2)

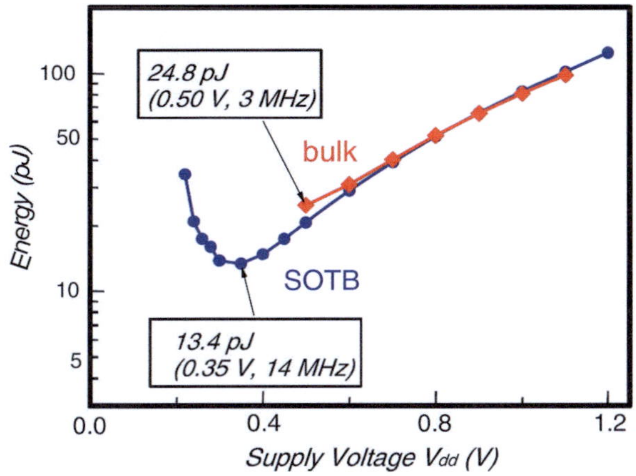

Fig. 6.10 Energy per cycle of 32-bit CPU core fabricated by SOTB and bulk technologies [33] © 2014 IEEE (same figure as Fig. 2.10 (b) in CHIPS 2020 Vol. 2)

of wakeup time) of 0.1, 1, and 10% are 0.52, 3.94, and 38.1 μA, respectively, and these are suited for both the battery- or EH-powered operations. Note that the current consumption of the V_{bb} generator circuit should be taken into account for this type of operation, because the generator should work throughout in the standby state. It was reported that the current consumption can be less than 1 μA [34], and thus this current consumption level is negligibly small. These data, low energy and sleep current, proves that the SOTB technology is a suitable for the energy-efficient MCU in the IoT applications.

The advanced MCU chip design equipped with an on-chip V_{bb} generator and various peripheral circuits was demonstrated [23]. Assuming the application with the EH power source and rf communication, the process and operating conditions of SOTB is slightly modified (with higher V_{th}) from those of [33]. The scale of integration for this MCU chip is 64 kgate 32-bit CPU logic and 64 kB SRAM. The maximum operation frequency can be controlled by V_{dd} and V_{bb} as shown in Fig. 6.11. At V_{dd} and V_{bb} of 0.75 and 0 V, respectively, the maximum frequency is 75 MHz and the active current is 37 μA/MHz. The leakage currents at $V_{bb} = 0$ and -1.5 V are 4.3 μA and 45 nA, respectively. In this design, due to higher V_{dd} and frequency than the design of the previous paragraph [33], the energy per cycle might be higher than MEP, nevertheless, it provides a practically useful option (the active current is the same level), as far as the available supply voltage matches the required V_{dd} (for example, using the dry cell of 0.8 V end voltage). In the above MCU designs, the nonvolatile memory macro, usually used to store the program code, are not implemented. In these chips, however, the code can be fetched from the SRAM and it can be stored taking advantage of a very small leakage current by applying the reverse V_{bb}.

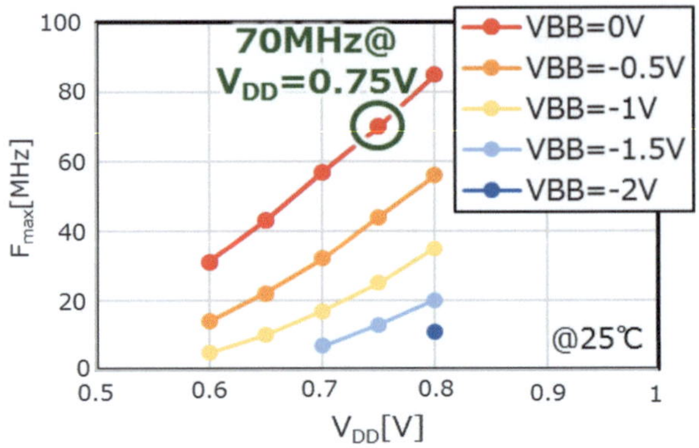

Fig. 6.11 Maximum clock frequency of MCU chip as a function of V_{dd} and V_{bb} © 2017 IEEE [23]

Table 6.1 Comparison of energy and current consumption of various 32-bit CPU cores

Technology	65 nm SOTB		32 nm bulk		180 nm bulk	
SRAM (kByte)	128	64	16	16	3	3
E (pJ/cycle)	13.4	27.8	170	347	28.9	37.4
f_{clock} (MHz)	14	70	60	500	0.073	1
V_{dd} (V)	0.35	0.75	0.45	0.8	0.4	0.5
Active current (μA/MHz)	38.3	37.01	377.8	433.8	72.3	74.8
Standby current (μA)	0.14	0.045	9330	25,000	0.00025	0.00092
REF	[33]	[23]	[35]		[36]	

The energy and current consumption of the CPU cores are compared in Table 6.1. It is remarkable that both the active and standby currents for the SOTB CPUs are small.

6.10 MCU with Embedded Memory

In many applications using MCUs, the embedded nonvolatile memory is useful for storing the program code, parameters and the various data, from the sensors for example. Taking advantage of the hybrid bulk integration capability of the SOTB technology, the conventional embedded flash-memory macro can be integrated with the SOTB MCU core. The integration of a two-transistor type metal-oxide-nitride-oxide-silicon (MONOS) flash memory macro was demonstrated [37]. A new sense amplifier and a data transmission circuit were designed to utilize the SOTB's low-energy and low-voltage capability. The memory operates at 64 MHz, and its read energy and current are 0.22 pJ/bit and 6.32 μA/MHz (32 bit bus).

There are various types of embedded memory candidates, among them, the code memory using the atom switch of lower energy than the conventional flash was

Table 6.2 Comparison of various MPUs with embedded flash memories

Technology	65 nm SOTB		130 nm bulk	180 nm bulk
Memory type	MONOS	Atom switch	FeRAM[b]	ReRAM
ROM capacity (kByte)	1500	16	16	64
Read energy (pJ/bit)	0.22	0.14	–	–
CPU (bit)	32	32	16	8
V_{dd} (V)	0.75	0.39	1.8	1.8
f_{clock} (MHz)	64	25	24	10
Total energy (pJ)	35.8	18.3	147.6	378
CPU energy (pJ)	27.8	13.8	–	–
Active current (μA/MHz)	43.32	46.8	82	210
Standby power (μW)	0.034	0.63	4.8	0.11
Standby current (μA)	0.045	1.57	2.67	0.06
REF	[23, 37][a]	[38]	[40]	[41]

[a]Current consumption and energy were simply added to the data of [23, 37] by the author and not the reported values
[b]FeRAM: Ferroelectric RAM

demonstrated [38]. The atom switch is a family of the resistive random-access memory (ReRAM) utilizing the polymer electrolyte and metal (copper and ruthenium) electrodes. The advantages are a low writing voltage as low as 2 V and a high on-off ratio. The 32-bit MCU test chip with the atom-switch code memory was fabricated. The chip can operate at 25 MHz at $V_{dd} = 0.39$ V. The energies per cycle for memory and total (memory and logic) are 4.48 pJ (0.14 pJ/bit) and 18.26 pJ, respectively. The latter corresponds to the active current of 46.82 μA/MHz. Moreover, the nonvolatile programmable-logic circuits can be embedded with the atom-switch technology on the SOTB CMOS platform [39]. This circuit acts as an off-loader to improve the total energy efficiency (the same processing with less clock cycle) compared with the CPU-only circuits. Table 6.2 and Fig. 6.12 compare the performances of various MPU chips with different technologies and types of the embedded flash memory.

Finally in this section, the features and properties of the first commercial MCU chip of the SOTB technology are briefly described [42]. The CPU core is the Cortex M0$^+$ (32 bit, two-stage pipeline) with a 1.5 MByte flash memory and 256 kByte SRAM. It operates up to 64 MHz, and the active and the standby currents are 20 μA/MHz and 200 nA, respectively. The energy performance seems to be improved from [23]. Various peripheral IP (intellectual property) cores are also embedded in the chip: analog-digital converters (ADCs), digital-analog converters (DACs), a temperature sensor, timers, serial interfaces, display interfaces, and security functions, as shown in Fig. 6.13. The unique feature of the chip is the embedded EH controller. Various types of harvesters and an energy-storage capacitor can be controlled by this chip. Due to the outstanding low-power performance, this chip seems to be a very suitable option to be used in the IoT front-end devices.

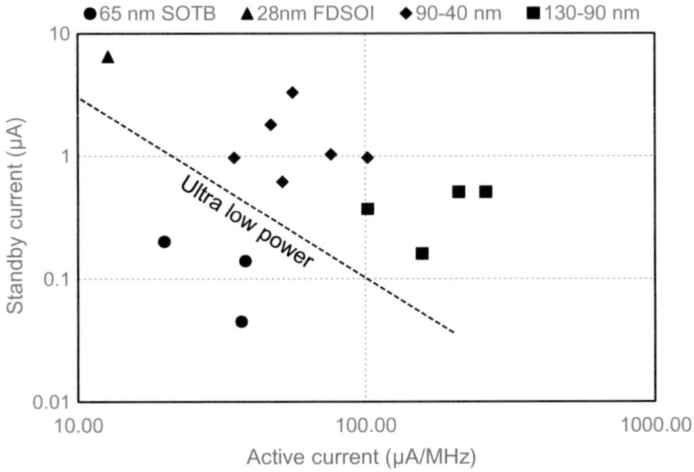

Fig. 6.12 Benchmark of active and standby currents for various MCUs [23, 32, 42]

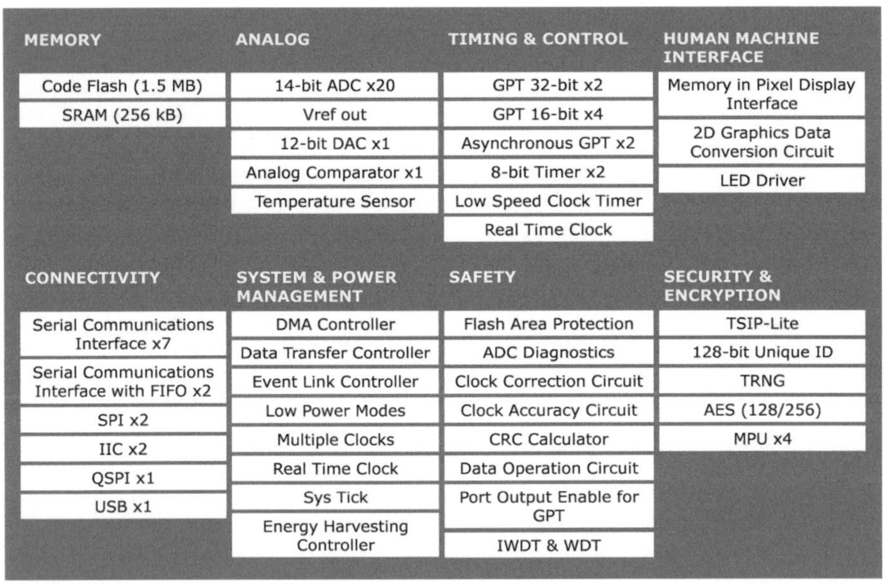

Fig. 6.13 Block diagram of commercial MCU chip on SOTB technology [42] (SRAM: Static Random-Access Memory, ADC: Analog-Digital Converter, Vref: Reference Voltage, DAC: Digital-Analog Converter, GPT: General PWM Timer, PWM: Pulse Width Modulation, LED: Light Emitting Diode, FIFO: First-In First-Out, SPI: Serial Peripheral Interface, IIC: Inter-Integrated Circuit, QSPI: Quad SPI, USB: Universal Serial Bus, DMA: Direct Memory Access, CRC: Cyclic Redundancy Check, IWDT: Independent Watchdog Timer, WDT: Watchdog Timer, TSIP: Trusted Secure IP, TRNG: True Random Number Generator, AES: Advanced Encryption Standard, MPU: Memory Protection Unit)

6.11 Reconfigurable Circuits

In this section, the circuits are described, where the back-bias control has a strong effect for the optimization of performance and power (especially, static power). The reconfigurable circuits, such as the field-programmable gate array (FPGA), are widely used. It is well known that the flexibility and the power-performance efficiency are a trade-off relationship. For example, the hard-wired logic circuits such as the application-specific integrated circuit (ASIC) are overwhelmingly efficient compared to the software-defined circuits such as the microprocessor. However, there is no flexibility of changing the function of the circuits. Moreover, high required number of production for the custom ICs like ASIC is another obstacle for the small-volume products. The FPGA is a good compromise for this tradeoff and it is thus widely used. To optimize the power efficiency in the reconfigurable circuits, however, there is a problem to solve. In the design of the hard wired logic circuits, the designer can select the technology options, that is, V_{th} flavors, in each specific part of the circuits. In general, the critical paths are found through the timing analysis, and the low-V_{th} transistors are used only in these critical paths. By selecting proper V_{th} options, the performance and power of the circuits can be optimized. In the reconfigurable circuits, however, the speed requirement in each processing element (PE) is not determined at the time of the circuit design. In the conventional FPGA, therefore, all the PEs need to set to have the highest speed: low V_{th}. Since not all the PEs need to work with full activity in most of applications, there is a huge power loss in the conventional FPGA.

The independent back-biasing in each PE is thus a strong way to reduce the power consumption of the reconfigurable circuits. The important insight of the back-biasing for these circuits is that only the performance of the PEs in the bottle-neck process is needed to speed-up, and at the same time, the other PEs are better to a slow-down (with reverse back biasing) to reduce leakage power while securing the total performance (clock frequency).

The significant improvement of the power efficiency for FPGAs was demonstrated with independent back biasing for each PE in the FPGA, named Flex-Power FPGA [43] using the 65 nm SOTB process. The schematic architecture of the Flex-Power FPGA is shown in Fig. 6.14. Each PE has a body bias (back bias) selector connected to the body bias voltage lines for p- and n-type SOTBs (V_{bp} and V_{bn}). By using the specially designed mapping tool for the Flex-Power FPGA, the circuit is mapped on the look-up table of the FPGA. At the same time, the critical paths are found and the proper body-bias-selector information is also mapped. As an example, the result for the 32-bit binary counter is shown in Figs. 6.15 and 6.16. The counter operates from 14 to 72 MHz at V_{dd} from 0.5 to 1.2 V, respectively. It should be noted that the frequency does not change with the reverse back-bias voltages (VRBB). This is because the above mapping software sets the reverse-bias flags only for the non-critical paths. The static power can be reduced by the reverse bias by from 59 to 80% for V_{dd} of 1.2 and 0.5 V, respectively, as shown in Fig. 6.14. The detailed analyses on performance and power of the Flex-Power FPGA are described in [44].

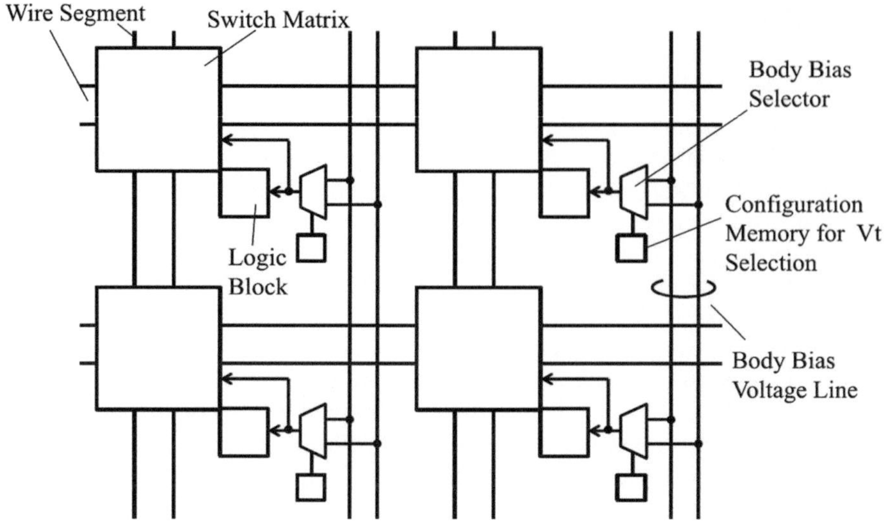

Fig. 6.14 Schematic FPGA architecture with independent back-biasing © 2016 IEICE [43]

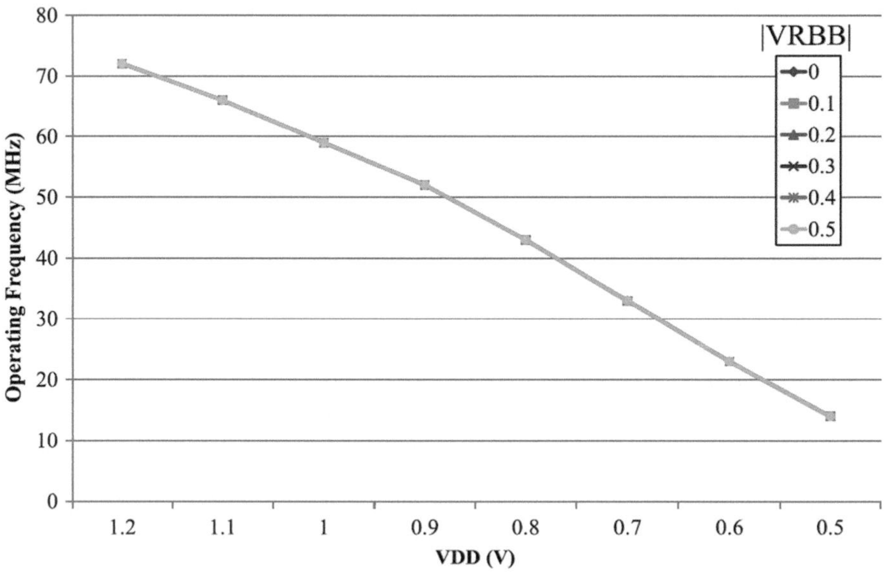

Fig. 6.15 Operation frequency of Flex-Power FPGA with different back-bias voltages © 2016 IEICE [43]

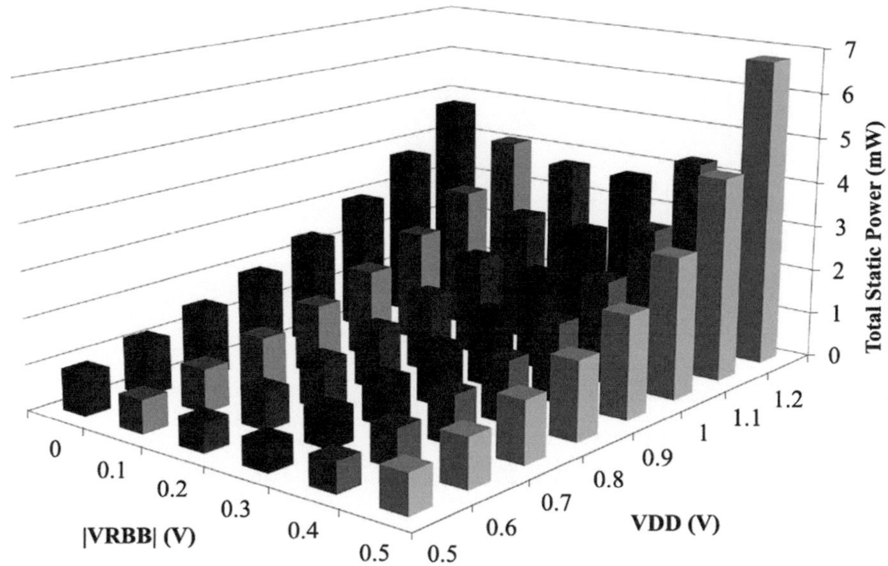

Fig. 6.16 Static power reduction by back biasing in Flex-Power FPGA © 2016 IEICE [43]

Another significant power saving, regarding the reconfigurable circuits with the back-biasing, was demonstrated on the reconfigurable accelerator circuits named cool mega array (CMA) [45]. There are various types of the reconfigurable circuits: FPGA, dynamic reconfigurable processor array (DRPA), etc., with different time scales of the reconfiguration action. The CMA is designed as an off-loading processor of various image or sensing data dedicated for the low-power battery operating applications by reducing the power from those of the existing DRPAs (but without dynamic reconfigurability). The block diagram of CMA is shown in Fig. 6.17. It has a large PE array without memory elements for mapping the data flow of the application program, and has a small programmable micro controller for the data management. Results for the typical image processing (alpha blender, sepia filter, and gray-scale filter) are shown in Fig. 6.18. The maximum performance of 743 MOPS/mW, which corresponds to 1.35 pJ per operation cycle, is achieved at $V_{dd} = 0.5$ V with the optimized back-bias voltage application. Note that the curve in this graph is similar to the behavior of energy per cycle versus V_{dd} as shown in Figs. 6.5 and 6.10. The image processing on an evaluation board was demonstrated using lemon batteries [45] or indoor solar cells.

In the back-bias operation of these reconfigurable circuits, the granularity of the back-bias domains is an important design point. Considering the effect of the back-biasing, it is ideal that all the domains should be independently controlled, however, this has a high area penalty. The optimization of the domain division size is investigated for CMA [46]. The sizes are selected from 1×1 to 4×4, where their area penalties varied from 12 to 1%. Figure 6.19 shows the power reduction ratio

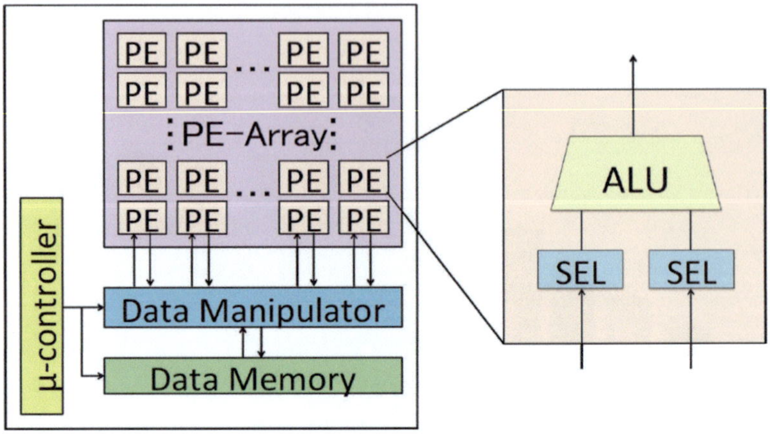

Fig. 6.17 Block diagram of a CMA © 2015 IEEE [45] (ALU: arithmetic and logic unit, SEL: switching element)

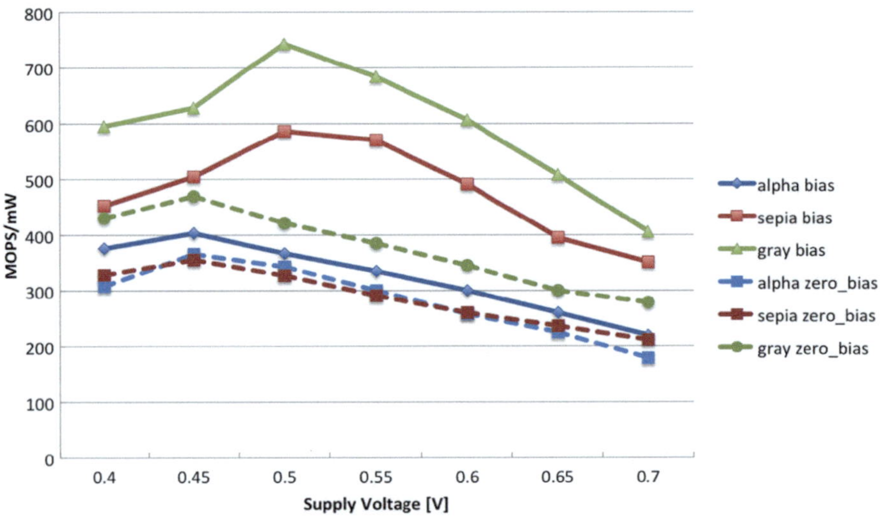

Fig. 6.18 Performance of the CMA for various image processing with and without back-biasing © 2015 IEEE [45]

compared to the case that all the PEs operate under zero back-bias voltage for various image processing algorithms. The back-bias voltages are optimized for each algorithm and each power domain. Although the results are slightly different for the algorithms, where the usage of PEs is different, they clearly depict that there are optimum domain sizes for different algorithms with both low power and small area penalty.

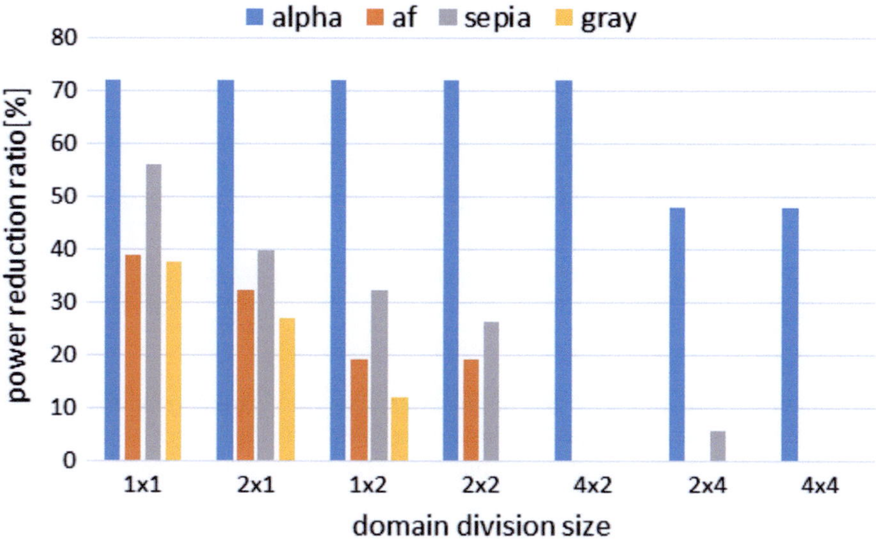

Fig. 6.19 Power reduction ratios under optimal back bias compared to zero back bias for CMA of various domain division sizes © 2016 IEEE [46] (alpha: 8-bit alpha blender and af: 24-bit RGB alpha blender)

6.12 Data Processing Circuits

Low-power data processors for data query, pattern matching, database operation, signal processing, etc. are important building blocks in the IoT edge processing. As well as the parallel operation by general-purpose computing on graphics processing units (GPGPU) and FPGA, dedicated data processing units are useful in terms of higher energy efficiency. In this section, the data processing circuits based on the content-addressable memory (CAM) and the coordinate-rotation digital computer (CORDIC) algorithm are described.

The CAM-based pattern matching system for two-dimensional image search is implemented on the SOTB technology [47]. The system consists of a CAM block, a shift circuit, multiplexers, an AND logic, and a finite-state machine (FSM) controller. The CAM memory block is designed by using the two-port SRAM macro of the 65-nm SOTB technology library. Back-bias flexibly controls the active performance under the operation state, and a reverse bias of -1.2 V reduces the leakage current down to 2 μA (0.2 mA with zero bias) under the standby state. Table 6.3 compares the performance of the system with that of the bulk 65-nm process. Significant increase of energy efficiency (more than $\times 5$) with comparable search-time performance is achieved by the SOTB technology.

The bitmap indexing is a kind of database index that is used for improving the speed of database retrieval, and is useful for various data analytics. The bitmap index creator (BIC) chip with high energy efficiency was demonstrated [49]. The block

Table 6.3 Performance comparison of CAM-based pattern matching systems

Technology	65-nm SOTB	65-nm bulk
V_{dd}	0.4 V	1.2 V
System size	256 word × 8 bit	128 word × 512 bit
Area	1.6 mm^2	1.62 mm^2
Search time	520 ns (256 patterns)	283 ns (128 patterns)
	260 ns (128 patterns, est.)	
Power	0.59 mW (12 pJ/search)	3.39 mW
Reference	[47]	[48]

diagram of the BIC core is shown in Fig. 6.20. This core is used to index N records by M given keys. The record R1 is fed into the CAM with all M keys. If R1 contains some keys, bit flags turns on (one by one for all the M keys) at the specific positions of M × N bit matrix that is finally stored in the BI memory. The chip fabricated by the 65-nm SOTB technology operates at 41 MHz (at $V_{dd} = 1.2$ V) and 10 MHz (at $V_{dd} = 0.4$ V) where energy consumptions are 163 and 19 pJ/cycle, respectively. Remarkably small standby power of 2.64 nW (0.31 pW/bit) is achieved at $V_{dd} = 0.4$ V with reverse back bias of −2 V.

An adaptive CORDIC-based FFT (fast Fourier transformation) macro was implemented on the 65-nm SOTB technology [50]. By utilizing both forward and reverse back biasing, the active energy performance and the leakage can be optimized. The clock frequency is 43 MHz at $V_{dd} = 1.0$ V with zero back bias where the energy is 10.27 pJ/cycle. The energy can be decreased to about 3 pJ/cycle by decreasing V_{dd} down to 0.5 V and controlling V_{bb} to satisfy the required delay. Table 6.4 compares

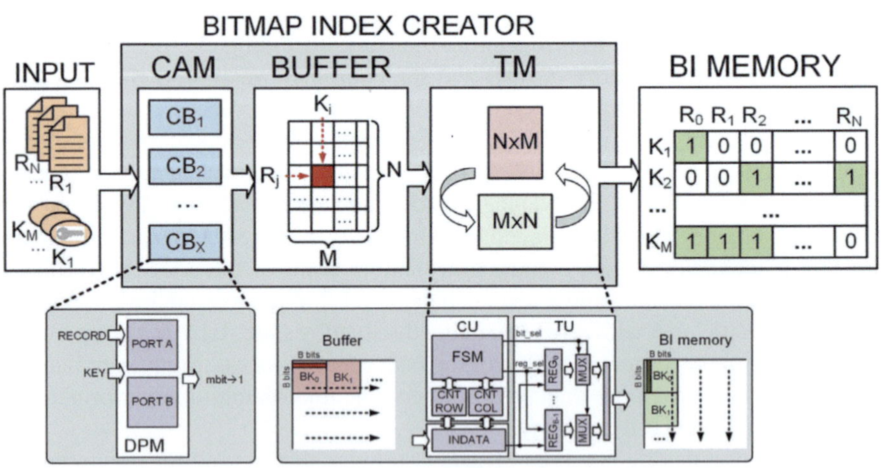

Fig. 6.20 Block diagram of the BIC core (TM: transpose matrix, CU: control unit, TU: transpose unit) © 2019 Elsevier [49]

Table 6.4 Performance comparison of FFT macros [50]

Technology	Bulk[a]	Bulk[a]	SOTB
Architecture	Look-up Table	Look-up Table	Adaptive CORDIC
Area (μm^2)	203,013	211,379	86,721
Delay (ns)	5.70	6.06	23.25
V_{dd} (V)	1.1	1.1	0.75
Power (mW)	173.63	194.92	1.03
Energy (pJ/cycle)	1736	1949	10.3
Reference	[51]	[51]	[50]

[a]The data scaled to match the 65-nm bulk technology

performances of the FFT macros. A remarkable reduction in energy is demonstrated by both the Adaptive CORDIC architecture and the SOTB technology.

6.13 Security Circuits

It is widely accepted that the IoT devices should be robust in terms of security against any attack via the network or outside of the device physically. There are various studies on the circuits regarding the security. This section describes typical circuits such as encryption and physically unclonable function (PUF) of ultra-low-power consumption suited for the IoT devices.

The advanced encryption standard (AES) is widely used as an encryption method [52]. Area penalty, encryption speed, and low power are main issues on the AES encryption macros. The AES encryption circuits with a simple clock-gating technique were implemented by using the 65 nm SOTB process [53, 54]. The performances of the 8-bit AES encryption circuits are compared in Table 6.5. Significant energy reduction is achieved by the 65 nm SOTB technology while keeping the frequency relatively high.

Table 6.5 Comparison of performances of 8-bit AES encryption circuits

Technology	Number of gates (kgates)	V_{dd} (V)	Frequency (MHz)	Energy (μW/MHz)	Reference
65 nm SOTB	2.6	0.55	130.9	0.40	[53]
130 nm bulk	3.2	1.2	130.0	30	[55]
22 nm tri-gate	2.0	0.9	1133	11.8	[56]
65 nm bulk	(0.012 mm^2)[a]	0.5	11.0	1.33	[57]

[a]The number of gates might be the same as [53] because the area is similar

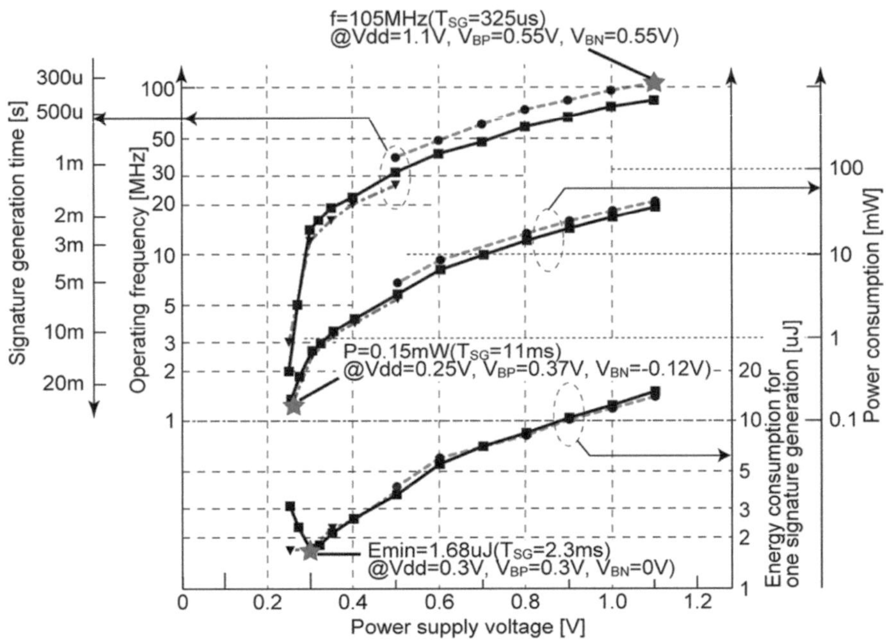

Fig. 6.21 Signature generation time, operating frequency, power and energy consumptions for ECC circuits implemented on 65-nm SOTB technology © 2016 IEEE [58]

The generation circuits of elliptic-curve cryptography (ECC), with smaller key size than the conventional RSA that is widely used for digital signatures, were developed as a suitable candidate for the small IoT devices [58, 59]. By the improvement of the signature generation architecture and the optimization of V_{dd} and V_{bb} utilizing the 65-nm SOTB technology, smaller energy and faster signature generation time (T_{sig}) is demonstrated. Figure 6.21 plots signature generation time, operating frequency, power consumption, and energy consumption per one-signature generation as a function of V_{dd} for the ECC circuits [58]. The minimum energy is 1.68 μJ at $V_{dd} = 0.3$ V and $T_{sig} = 2.3$ ms. On the process with higher V_{dd} flavor [59], the signature generation speed increased about 10 times higher while the energy twice. The performances of the ECC circuits with Galois field of 256 bits are compared in Table 6.6. Among the circuits of the state-of-the-art technologies, the ECC circuits with the SOTB process are advantageous for both the generation time and energy.

The physically unclonable functions (PUFs) can be used for IC authentication like a fingerprint preventing from counterfeit. Among various types of PUFs, the PUF using the silicon technology, in general, generates the individual identification data extracting from the characteristic variability of each chip, such as the power-on initial value of SRAMs or the delays of gates. The low-power PUF macro is implemented on the 65-nm SOTB technology [63]. The circuit consists of two chains of selectors generating a delay variation and a flip flop acting as an arbiter, as shown in Fig. 6.22. A concern arises on implementing the PUF on SOTB, that is, the SOTB's small

Table 6.6 Performance comparison of ECC circuits with a Galois field of 256 bits

Technology	Number of gates (kgates)	V_{dd} (V)	Frequency (MHz)	T_{sig} (μs)	Energy (μJ)	Reference
65 nm SOTB[a]	1575	0.75	98.0	76.0	9.32	[59]
65 nm SOTB[a]	1575	0.45	76.0	210	3.28	[59]
65 nm SOTB	2493	1.1	105	325	13.9	[58]
65 nm SOTB	2493	0.3	14	2300	1.68	[58]
90 nm bulk	NA[b]	NA	157.2	320	NA	[60]
90 nm bulk	168	1.2	256	1890	80.0	[61]
90 nm bulk	168[c]	1.2	256	740	20.0	[61]
90 nm bulk	342[c]	1.2	214	0.29	57.0	[62]

[a]SOTB technology for low-standby-power application (with higher V_{th} and nominal V_{dd})
[b]Implemented on Stratix II FPGA
[c]Galois field of 160 bits

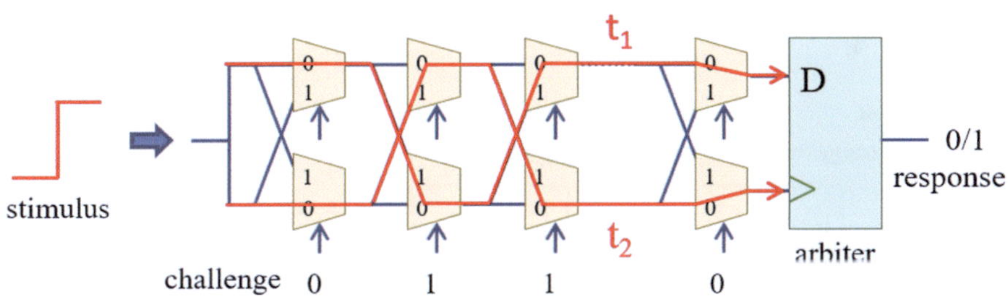

Fig. 6.22 Block diagram of PUF circuit © 2017 IEEE [63]

variability can deteriorate the uniqueness of the PUF. The result shows that the identification error rate is rather high in the voltage range as the conventional bulk-CMOS, however, by applying the reverse back-bias or decreasing the V_{dd}, the error-rate decreases due to increasing the delay variability. This means that the SOTB PUF can be used under the condition of lower voltage and lower power consumption than the conventional bulk PUF. The USB stick sized PUF module is also implemented by using this technology [64].

6.14 Analog and Rf Circuits

In this section, various analog and rf circuit implementations are described.

ADCs (analog-digital converters) are indispensable parts in MCUs, and successive-approximation-register (SAR) type or Δ-Σ type ADCs are frequently used. A very low power Δ-Σ modulator circuit was demonstrated [65]. Figure 6.23 shows the block diagram. By adequately controlling the back-bias voltages, the mid rail is tuned to half V_{dd} with the symmetrical operation of inverters that drive the switched capacitors, and this enables very low V_{dd} operation. The modulator operates at $V_{dd} = 0.5$ V and achieves 910 nW power consumption (0.07 μW/MHz) and the conversion figure of merit (FoM) of 46 fJ/conversion.

A voltage-controlled oscillator (VCO) with back-bias control was implemented on the 65-nm SOTB technology [66]. As shown in Fig. 6.24, the VCO consists of a ring oscillator. Figure 6.25 shows oscillation frequency and current consumption. They are controlled by the back-bias voltage, where $V_{c,dif} = V_{bp} - V_{bn}$ and $V_{c,com} = (V_{bp} + V_{bn})/2 = V_{dd}/2$. The oscillator operates at $V_{dd} = 0.55$ V with the tuning range from 377 to 556 MHzand achieves FoM = −158 dBc/Hz. This FoM value is the best among the CMOS ring-type VCO operating less than 1.0 V.

An ultra-low-power rf receiver and transmitter for wireless sensor node are described. The on-off-keying (OOK) modulation is a simple modulation scheme and suitable for low-power applications. A 312–315 MHz receiver circuit was designed

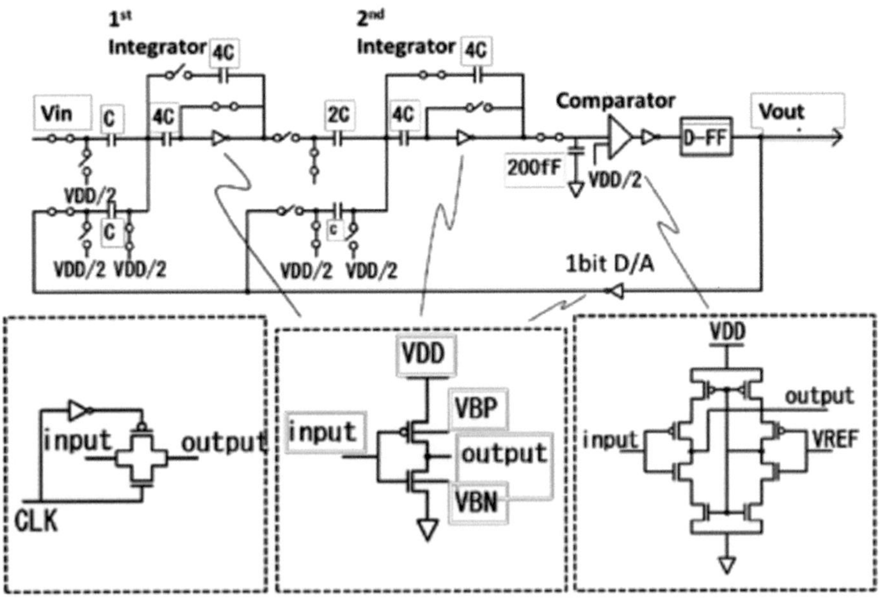

Fig. 6.23 Block diagram of the Δ-Σ modulator. Two integrators, 1 bit DAC, and a comparator are composed of back-gate controlled inverters without differential amplifiers © 2017 IEEE [65]

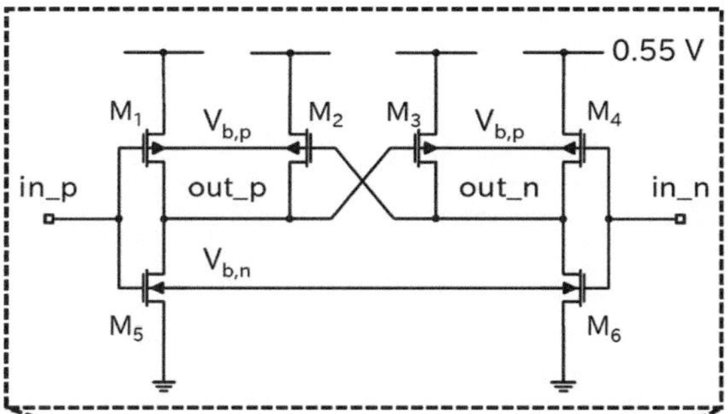

	Width [μm]	Length [nm]
M_1,M_4	1 x 12	65
M_2,M_3	1 x 8	65
M_5,M_6	0.5 x 20	65

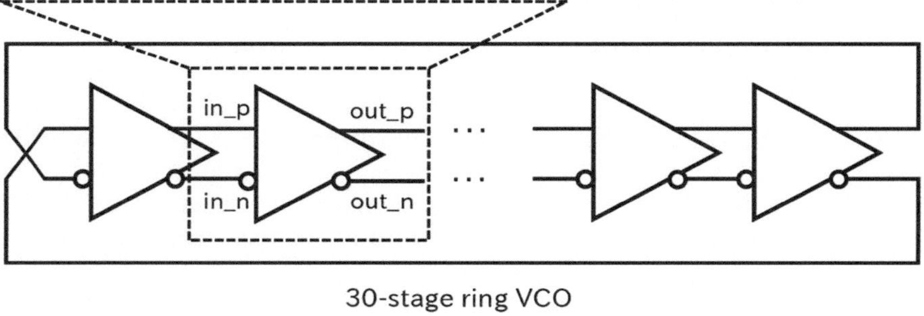

Fig. 6.24 Block diagram of a ring-type VCO © 2017 IEEE [66]

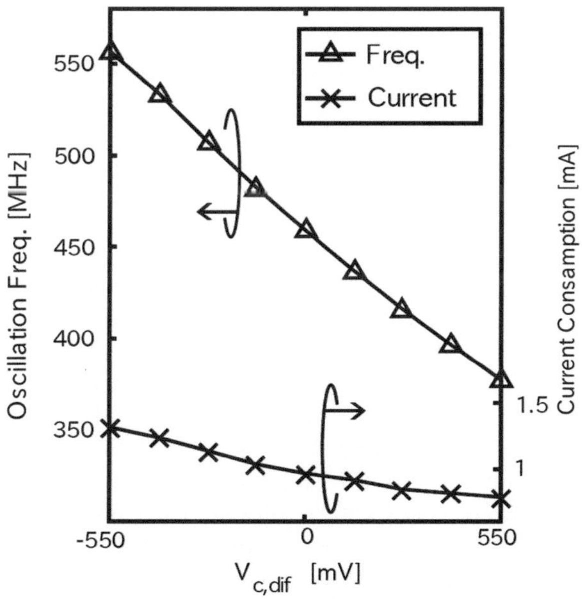

Fig. 6.25 Oscillation frequency and current consumption controlled by back-bias voltage © 2017 IEEE [66]

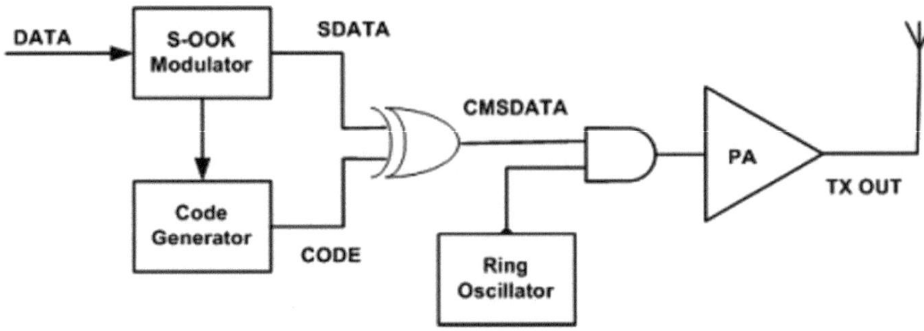

Fig. 6.26 Block diagram of a CMS-OOK transmitter © 2018 IEICE [68]

on the 65 nm SOTB technology [67]. Post-layout simulation showed −58.5 dBm sensitivity with 1.36 and 8.39 μW power consumption corresponding to 10 kbps and 100 kbps data rate, respectively. The code-modulated synchronized (CMS)-OOK modulation transmitter with a normally-off intermittent operation scheme is implemented using the 65-nm SOTB technology to significantly reduce the power consumption of the transmitter [68]. The digital part is implemented on FPGA. By employing the CMS scheme as shown in Fig. 6.26, a ring-oscillator type internal carrier oscillator with relatively high jitter can be used. This enables to turn the carrier generator on quickly (reducing the on duration of rf transmission) and to reduce the power consumption in the intermittent operation. Also, the peak output power can be reduced by diffusing the carrier frequency with the back-bias of triangular waveform. A signal modulation via back-bias terminals is a unique feature of SOTB for analog application. As a result, −62 dBm/MHz peak power spectrum density at 15 MHz bandwidth is achieved. The chip consumes 83 μW in average according to 83 nJ/bit at 1 kbps data transmission. (The analog part of the power amplifier operates at 1.0 V, and 0.75 V for the rest of the part.)

The dynamic threshold MOSFET (DTMOS) operation can be done with the SOTB technology by applying the same signal as the front gates to the back gates. The rf energy-harvesting circuit is implemented by using the SOTB DTMOS [69]. This harvester consists of three-stage cross-couple rectifiers as shown in Fig. 6.27 connected in series. The rf signals collected from an antenna are fed into V_{IN} terminals and the rectifier outputs a dc voltage from the V_{DC} terminal. The nodes (N1, N2, P1, and P2) are boosted by additional two floating nodes (not shown) of the similar structure as in Fig. 6.27 to improve the rectifying operation in a small input power range. The experimental result shows that the output dc voltage exceeds 1000 mV at input 954-MHz power of −9 dBm. With the 18-cm dipole antenna collecting rf in the laboratory environment, the output voltage is 130 mV.

The low-frequency noise characteristics of SOTB have been extensively studied [70]. Figure 6.28 shows distribution of the drain-current normalized current noise intensity for bulk and SOTB MOSFETs. Due to the low impurity density of the channel region, the variation in the noise characteristics is smaller than in bulk CMOS. Although the median value of noise is higher than in the bulk due to an additional

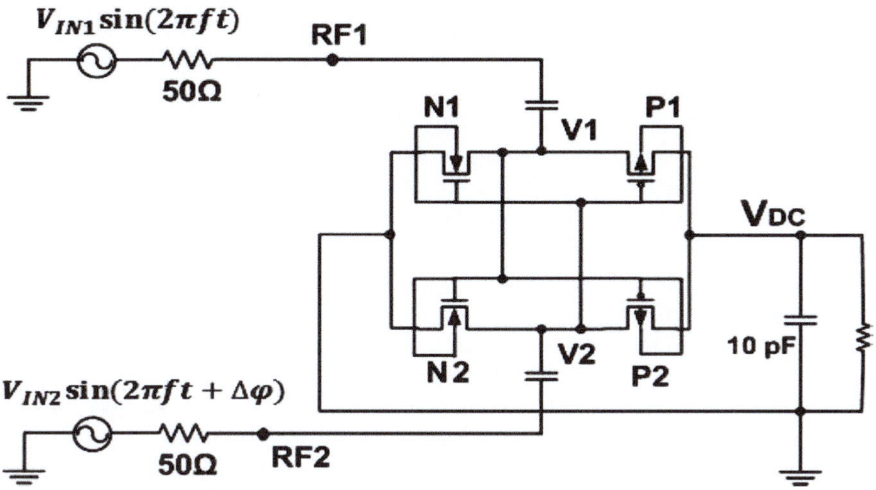

Fig. 6.27 Circuit schematic of cross-couple rectifier © 2019 IEEE [69]

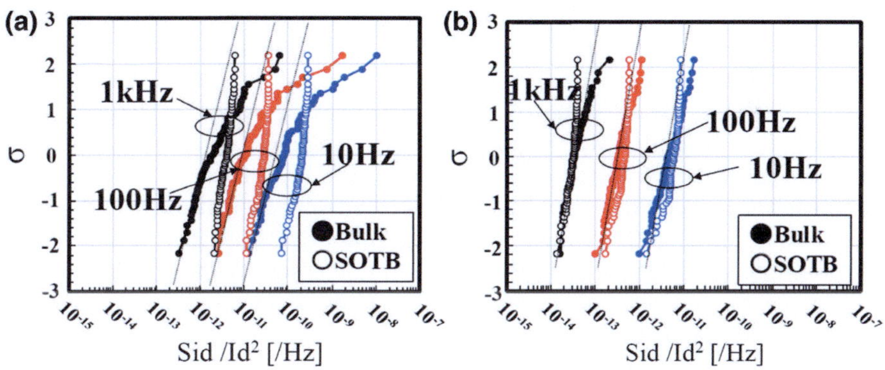

Fig. 6.28 Cumulative frequency distribution of drain-current normalized current noise intensity (S_{id}/I_d^2) in bulk and SOTB MOSFET. **a** Weak inversion state and **b** strong inversion state © 2018 JSAP [70]

interface between channel and the BOX layer, considering the variability tail, the noise characteristics of the SOTB is better.

6.15 Soft-Error Reliability

There are various reliability issues on silicon CMOS. In the FDSOI structure like SOTB, there are additional reliability issues such as, the bias temperature instability related to different electric field in the SOTB transistor from that of the bulk CMOS

[71, 72], the antenna effect (plasma damage during the fabrication process) [73, 74], and the gate-oxide reliability of the hybrid bulk CMOS fabricated on the exposed surface by removing the SOI and BOX layers [75]. In addition to these transistor process related issues, the soft error, especially, the single event upset (SEU) of SRAMs and logic circuits is a serious reliability problem for ICs. The SOI CMOS transistors have inherently a higher soft-error immunity than the bulk CMOS transistors because of its structure with the BOX insulating layer that prevents most of the charges generated by the ion incidence from flowing to the channel. In this section, the soft error of SRAMs, logic circuits, and combined effects for the chip-level soft-error immunity are described below.

The SEU caused by alpha and neutron irradiation on the SOTB SRAM is thoroughly studied in comparison with the bulk SRAM of the same footprint [76]. The SOTB SRAM can operate at low voltage down to ~0.4 V [22], however, this can increase the risk of a soft error versus the conventional bulk SRAM operating at higher V_{dd} such as 1.0 V. The measurement results for both alpha and neutron irradiation show that the soft-error immunity of the SOTB device is superior to that of the bulk SRAM. In the SRAMs that require high reliability, the error correction code (ECC) is implemented. If multiple memory cells in a row are attacked at a time by a single particle incidence, however, there is some possibility that the ECC cannot completely work. The multiple cell upset (MCU: not the micro controller unit in this section) is thus a significant point to be considered for the SRAM reliability. As shown in Fig. 6.29, the MCU rate (FIT: failure in time) for the SOTB SRAM is lower than that of the bulk SRAM. Complete dielectric separation between transistors by both the shallow trench isolation (STI) and the BOX layer in the FDSOI transistor contributes to reduce the risk of MCU. This result suggest that the SOTB SRAM is more robust even at 0.4 V compared to the bulk SRAM at 1.0 V. Moreover, the soft-error rate under the reverse back-bias condition is significantly reduced.

Recently, the soft error due to muon irradiation draws much attention, especially for the SRAMs fabricated by the highly scaled process. The muon soft-error rate

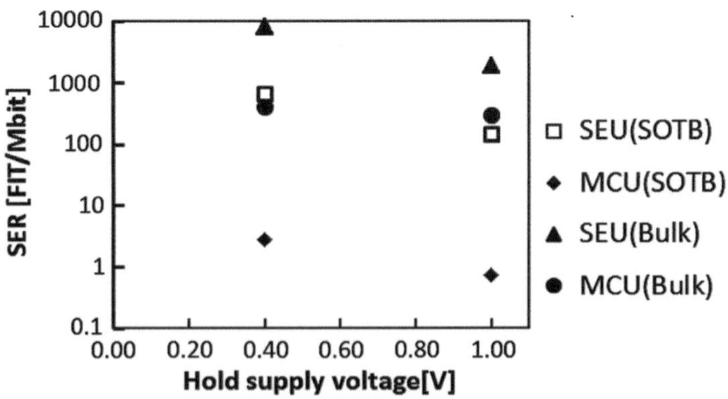

Fig. 6.29 Measured neutron-induced SEU and MCU as a function of supply voltage © 2015 IEEE [76]

for both the bulk and the SOTB SRAMs was studied [77]. The experimental results reveal that the effect of muons is not significant compared to neutron effects for the 65-nm technologies and the SOTB is less sensitive to the muon irradiation than the bulk.

A new type of soft error was found on the SOTB SRAM [78]. In contrast to the above mentioned superior results on the SOTB's soft error immunity, a 100-fold increase is observed under the reverse back-bias compared to the zero back-bias. A remarkable phenomenon is that the multiple-bit error occurs along the bit line direction. In this direction, the p-well (p GP in Fig. 6.6) is common in an array of the SOTB SRAM. As schematically depicted in Fig. 6.30, electrons generated by the incident ions can modulate the potential of the p-well (p GP) layer, and this effect is significant if this layer is in the reverse bias state. Although this phenomenon is not a favorable characteristic in terms of the low-power circuit operation that tends to use the reverse biasing, the modeling of this soft error [79] can contribute to optimize the triple-well structure and the BOX thickness, and its reliability will be improved further.

The soft error caused in the logic circuits can seriously affect the operation, because there is generally no way of salvation like the ECC for SRAMs, other than using the redundant circuits with majority logic. Especially, it is known that the flip-flop (FF) circuit is relatively weak among various logic circuits. The experimental results for alpha and neutron irradiation were reported [80]. Figure 6.31 shows the neutron results for D-type FF as a function of back-bias voltage. It is remarkable that the soft-error immunity of SOTB D-FF is about 20 times better than the bulk D-FF, and the immunity of the SOTB D-FF becomes stronger with reverse back-bias whereas that of the bulk D-FF slightly increases.

There are various FF structures for radiation hardening such as the dual interlocked storage cell (DICE) latch [81]. In the FDSOI structures, with the same reason as the

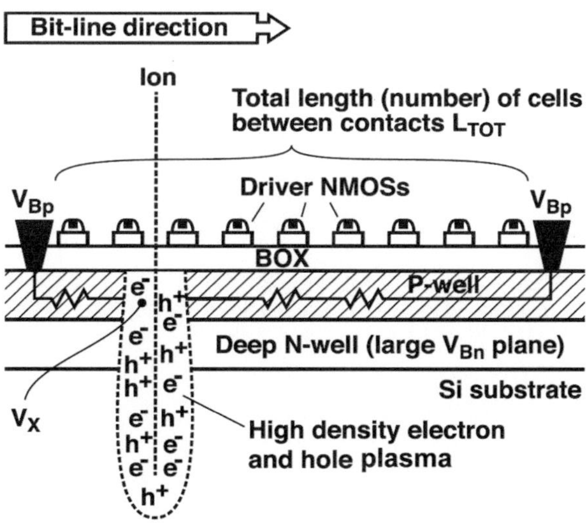

Fig. 6.30 Schematic illustration to explain multiple cell upset through a p-well layer underneath the BOX layer © 2018 IEEE [78]

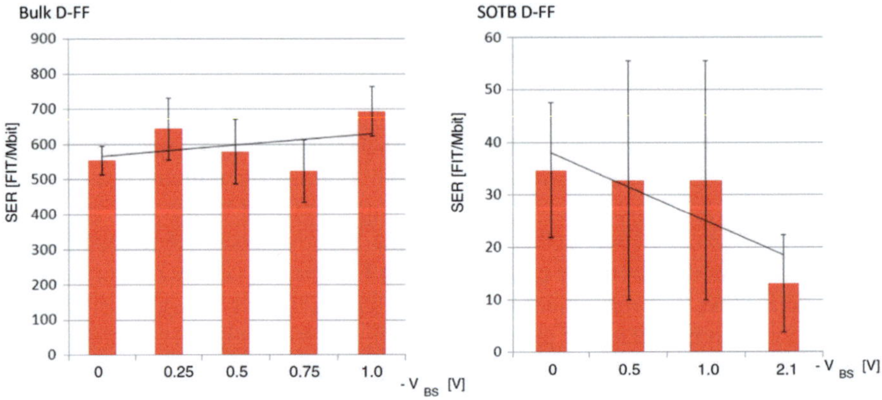

Fig. 6.31 Soft-error rate by neutron irradiation for bulk and SOTB D-FFs © 2014 IEEE [80]

robustness over multiple cell upset in SRAMs, the impact of the single event in one transistor to the adjacent transistor is weaker for the SOTB transistor than the bulk one. A series connection of two transistors is thus effective way to improve the soft error immunity like the stacked inverter structure [82]. There are trade-off relationships between the soft-error immunity of the circuit and its size and delay because the soft error immune circuit tends to require additional transistors. The study to solve these trade-offs was reported on the SOTB circuits with various circuit topologies [83–85]. Figures 6.32 and 6.33 show the circuit schematics of the conventional transmission-gate FF (TGFF) and the feedback recovery FF (FRFF),

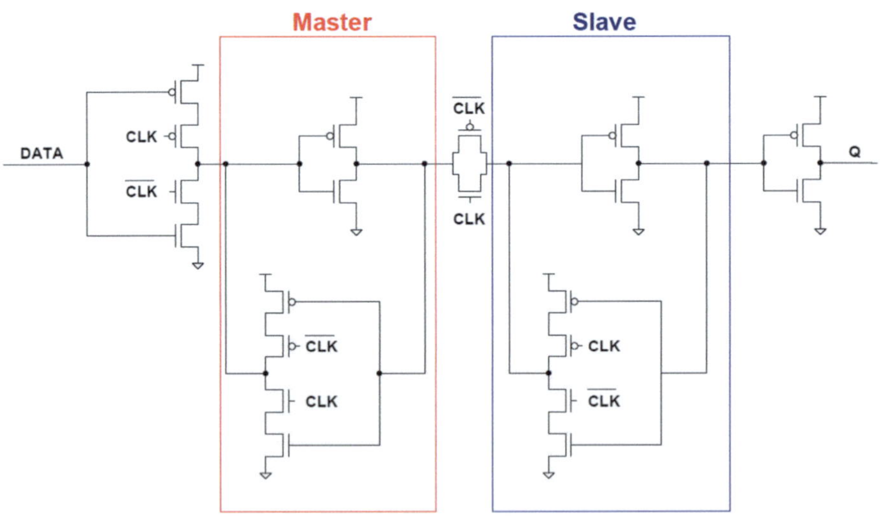

Fig. 6.32 Circuit schematic of conventional transmission-gate FF (TGFF) ©2019 Prof. Kobayashi [85]

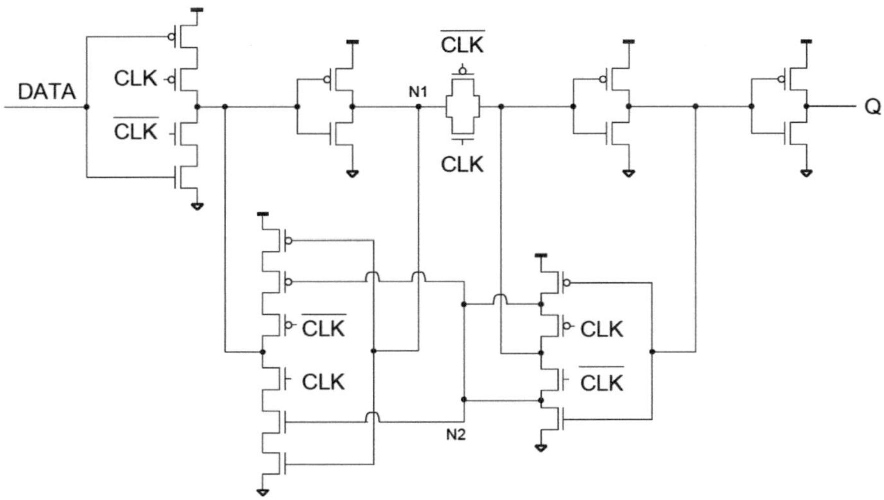

Fig. 6.33 Circuit schematic of feedback recovery FF (FRFF) ©2019 Prof. Kobayashi [85]

respectively. The latter one is considered to be a superior structure in terms of the above trade-offs among the SOI FFs, and its feature is additional feedback lines indicated by N2 with only two additional transistors. The area, delay, and power of the latter increase only by 6%, 6%, and 3%, respectively, from the former (conventional TGFF), and the average soft-error rate by neutrons is 1/3 of that for the TGFF. The average cross section over heavy ions (Ar and Kr) is also 1/2 of the TGFF.

By combining the results of SRAM and FF, the chip-level soft-error rate (SER) was estimated [86]. Assuming two types of typical processor chips: a high-performance processor of 6×6 mm^2 size with 50% SRAM area and an embedded (open RISC) processor of 1×1 mm^2 size with 91% SRAM area, the chip-level SERs for the bulk and SOTB chips operated at 0.5 and 1.0 V were calculated. Most of (>95%) the errors occur in the SRAM area when ECC is turned off. By applying ECC, the error-rates of the SOTB and bulk chips were drastically reduce by two orders and one order of magnitude, respectively. The smaller risk of MCU for SOTB enhances the effect of ECC. The results with ECC are shown in Fig. 6.34. Significant decrease in the chip-level SER for SOTB was demonstrated. By applying ECC, the majority of errors occur in the FF area. Note that the data of the conventional D-FF are used in this estimation. By using highly immune FF structures as described in the previous paragraph, the chip-level SER is anticipated to be improved further.

6.16 Summary of SOTB Chip Implementation

The various examples of the SOTB chip implementation described in this chapter are summarized in this section.

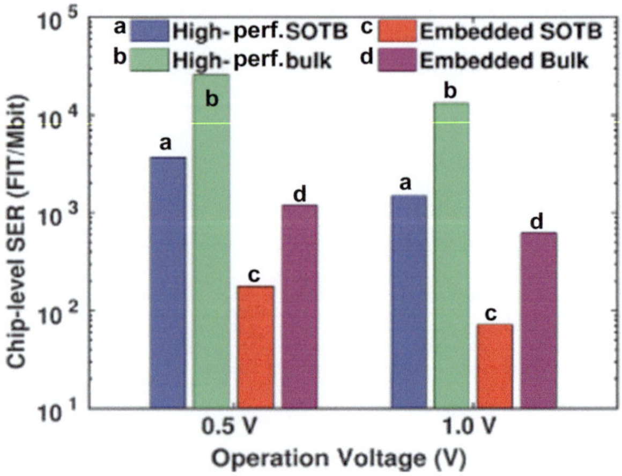

Fig. 6.34 Comparison of chip-level SER with ECC © 2019 IEEE [86]

- Low-voltage SRAM of minimum V_{dd} down to 0.37 V in 6.6:
  - Reverse V_{bb} (back bias) enables to store the data with very small leakage current.
- MCU (microcontroller unit) in Sects. 6.9 and 6.10:
  - MEP (minimum energy point) operation at V_{dd} around 0.4 V with both small active and leakage currents.
  - MCU chip with embedded flash memory.
  - Commercial MCU chip with various IPs and embedded EH (energy harvesting) controller.
- Reconfigurable circuits in Sect. 6.11:
  - FPGA (field-programmable gate array) with drastically reduced leakage current due to independent V_{bb} control on each processing element.
  - Reconfigurable accelerator circuit, CMA (cool mega array), with optimized V_{bb} and optimized domain size.
- Data processing circuits in Sect. 6.12:
  - CAM (content-addressable memory) for pattern-matching systems and database operation.
  - FFT (fast Fourier transformation) macro using the coordinate rotation digital computer (CORDIC) algorithm.
- Security circuits in Sect. 6.13
  - AES encryption circuits with enhanced performance and energy efficiency.
  - PUF (physically unclonable function) circuits, using small variability transistors.

- Analog and rf circuits in Sect. 6.14
  - Δ-Σ modulator with high conversion figure of merit.
  - VCO (voltage-controlled oscillator) with V_{bb} control.
  - OOK (on-off keying) receiver and transmitter for the IoT node.
  - RF energy harvester by using the SOTB as a dynamic threshold MOSFET.
  - Small noise variability of the SOTB transistors.
- Soft-error immune SRAM and logic circuits in Sect. 6.15
  - SRAMs with significantly reduced single-event as well as multiple-cell upsets.
  - Reduced soft-error rate for FF (flip-flop) circuits and circuit topologies to obtain further robustness.

6.17 Future Perspective

A drastic decrease of connectivity cost for IoT devices and a popularization of prototyping tools such as 3D printers with various easy-to-use 3D-CAD tools and tiny development boards with microcontrollers like Arduino and Raspberry Pi have accelerated the democratization of manufacturing, and they have opened a door of the makers movement [87, 88] with the open-source hardware. This will significantly accelerate the production of a wide variety of applications bridging the cyber and physical worlds through sensing, processing, networking, and actuating. Note that the opensource hardware is not restricted to the education and the hobby. For example, the industry-grade Raspberry Pi is already a strong candidate to be used in various control devices in the industry because of both low hardware and development costs.

In this context, the ultra-low-power electronic devices including microcontrollers and various accelerating engines will be more important in the future. With increasing the number of IoT connecting devices, the required specifications of ICs will be upgraded to satisfy the needs of increasing the performance of the edge processing. Considering the limited power for most of the IoT devices, improving the energy efficiency is still important as described in the first section. Let us quote the insightful words by Mark Horowitz, "*Unfortunately, many of the magic bullets for decreasing energy without affecting performance have already been found and exploited. While there are no quick fixes, power growth must be addressed by application specific system level optimization, increasing use of specialized functional units and parallelism, and more adaptive control.*" [89]. The authors consider that the highly optimized combination of dedicated functional logic engines, reconfigurable processors, and central processing units (microcontrollers), all with the adaptive control, will be a gold solution. The adaptive control function and low-voltage operation capability of the SOTB technology should contribute to each processing part working with the best energy efficiency. Some indications are believed to have been shown in Sects. 6.11–6.13.

On the logic engines and microcontrollers, the important factor is that the hardware should be released with an easy development environment to be a defacto standard. In the Arduino family, for example, the integrated development environment (IDE) is ultra easy to use, and this leads to a positive feedback of increasing its users, growing user communities, and further improving its environment. It is known that developing the system working on FPGA is not easy because it usually needs to use the hardware description language such as VHDL or Verilog HDL. The Arduino family, however, already released the development environment integrated with the Arduino IDE [90]. This will accelerate the users to take advantage of the higher-performance hardware. Therefore, highly sophisticated design environments also contribute to the highly energy-efficient logic engines to become popular.

Another important trend to be considered is novel computing architectures such as neuromorphic computing and quantum computing. Let us go back to "The Free Lunch Is Over" [6], it was shown that the increase of the clock frequency has already slowed down (currently, maximum frequency as high as 5 GHz). Furthermore, considering the MEP in Sects. 6.3 and 6.4, rather lower frequencies are preferred to improve the energy efficiency in the current logic-circuit framework with CMOS transistors. It is known that the human brain processing speed is, however, about 60 Hz [91] and the structure with massive and reconfigurable wiring might be another significant difference. A combination of the neuromorphic computing architecture and the 3D integration technology with moderate clock frequencies thus can be a new paradigm of high-efficiency computing. The quantum computing can be another new paradigm. The inherent parallel computing architecture enables high-performance computing with slower clock frequencies.[4] In the quantum computing with superconducting qubits, the cryogenic interface with the conventional electronic devices is important [93]. The FDSOI transistors can work at cryogenic temperatures (with a proper design) [94] and operation with minimum heat dissipation by the MEP operation will be an important design issue. These novel computing schemes, mean new scenarios, in which the SOTB technology can contribute to energy efficient computing.

Acknowledgements The part of the work, especially on developing the SOTB technology by the Low-power Electronics Association and Project (LEAP), is supported by the Ministry of Economy, Trade and Industry (METI) and the New Energy and Industrial Technology Development Organization (NEDO). Part of the chip fabrication by the universities is done under a support of VLSI Design and Education Center (VDEC) in collaboration with Renesas Electronics Corporation, Cadence Corporation, Synopsys Corporation and Mentor Graphics Corporation.

References

1. www.rfidjournal.com/article/view/4986
2. https://www.statista.com/statistics/471264/iot-number-of-connected-devices-worldwide/

[4] In quantum computing, the key index of progress is not the size, voltage, nor clock frequency. The quantum decoherence can be the alternative [92].

3. https://iot-analytics.com/state-of-the-iot-update-q1-q2-2018-number-of-iot-devices-now-7b/
4. https://www.ericsson.com/en/mobility-report/internet-of-things-forecast
5. E.A. Lee, Cyber physical systems: design challenges, Technical Report No. UCB/EECS-2008-8, University of California, Berkeley, 23 Jan 2008. http://www2.eecs.berkeley.edu/Pubs/TechRpts/2008/EECS-2008-8.pdf
6. H. Sutter, The free lunch is over. http://www.gotw.ca/publications/concurrency-ddj.htm (Initial version: http://www.drdobbs.com/web-development/a-fundamental-turn-toward-concurrency-in/184405990?queryText=Free%2BLunch)
7. https://www.top500.org/green500/
8. Y. Lee, D. Blaauw, D. Sylvester, Ultralow power circuit design for wireless sensor nodes for structural health monitoring. Proc. IEEE **104**(8), 1529–1546 (2016). https://doi.org/10.1109/JPROC.2016.2547946
9. A. Wang, A. Chandrakasan, A 180-mV subthreshold FFT processor using a minimum energy design methodology. IEEE J. Solid-State Circuits **40**(1), 310–319 (2005). https://doi.org/10.1109/JSSC.2004.837945
10. A.P. Chandrakasan et al., Technologies for ultradynamic voltage scaling. Proc. IEEE **98**(2), 191–214 (2010). https://doi.org/10.1109/JPROC.2009.2033621
11. D. Bol, D. Kamel, D. Flandre, J.-D. Legat, Nanometer MOSFET effects on the minimum-energy point of 45 nm subthreshold logic, in *ISLPED'09*, San Francisco, California, USA, 19–21 Aug 2009. https://doi.org/10.1145/1594233.1594237
12. R. Aitken, V. Chandra, J. Myers, B. Sandhu, L. Shifren, G. Yeric, Device and technology implications of the Internet of Things, in *2014 Symposium on VLSI Technology (VLSI-Technology): Digest of Technical Papers*, Honolulu, HI (2014), pp. 1–4. https://doi.org/10.1109/VLSIT.2014.6894339
13. M.J.M. Pelgrom, A.C.J. Duinmaijer, A.P.G. Welbers, Matching properties of MOS transistors. IEEE J. Solid-State Circuits **24**(5), 1433–1439 (1989). https://doi.org/10.1109/JSSC.1989.572629
14. R.H. Dennard, F.H. Gaensslen, V.L. Rideout, E. Bassous, A.R. LeBlanc, Design of ion-implanted MOSFET's with very small physical dimensions. IEEE J. Solid-State Circuits **9**(5), 256–268 (1974). https://doi.org/10.1109/JSSC.1974.1050511
15. N. Sugii, R. Tsuchiya, T. Ishigaki, Y. Morita, H. Yoshimoto, S. Kimura, Local V_{th} variability and scalability in silicon-on-thin-BOX (SOTB) CMOS with small random-dopant fluctuation. IEEE Trans. Electron Devices **57**(4), 835–845 (2010). https://doi.org/10.1109/TED.2010.2040664
16. M. Aoki et al., 0.1 mu m CMOS devices using low-impurity-channel transistors (LICT), in *International Technical Digest on Electron Devices*, San Francisco, CA, USA (1990), pp. 939–941. https://doi.org/10.1109/iedm.1990.237087
17. M. Fukuma, Limitations on MOS ULSIs, in *Symposium on VLSI Technology*, May 1988, pp. 7, 8
18. D. Hisamoto, T. Kaga, Y. Kawamoto, E. Takeda, A fully depleted lean-channel transistor (DELTA)-a novel vertical ultra thin SOI MOSFET, in *International Technical Digest on Electron Devices Meeting*, Washington, DC, USA (1989), pp. 833–836. https://doi.org/10.1109/iedm.1989.74182
19. R. Tsuchiya, M. Horiuchi, S. Kimura, M. Yamaoka, T. Kawahara, S. Maegawa, T. Ipposhi, Y. Ohji, H. Matsuoka, in *Electron Devices Meeting, 2004. IEDM Technical Digest. IEEE International*, 13–15 Dec 2004, pp. 631, 634. https://doi.org/10.1109/IEDM.2004.1419245
20. Y. Yamamoto, H. Makiyama, T. Tsunomura, T. Iwamatsu, H. Oda, N. Sugii, Y. Yamaguchi, T. Mizutani, T. Hiramoto, Poly/high-k/SiON gate stack and novel profile engineering dedicated for ultralow-voltage silicon-on-thin-BOX (SOTB) CMOS operation, in *2012 Symposium on VLSI Technology (VLSIT)*, pp. 109, 110, 12–14 June 2012. https://doi.org/10.1109/vlsit.2012.6242485
21. O. Weber et al., Work-function engineering in gate first technology for multi-V_T dual-gate FDSOI CMOS on UTBOX, in *2010 International Electron Devices Meeting*, San Francisco, CA (2010), pp. 3.4.1–3.4.4. https://doi.org/10.1109/iedm.2010.5703289

22. Y. Yamamoto, H. Makiyama, H. Shinohara, T. Iwamatsu, H. Oda, S. Kamohara, N. Sugii, Y. Yamaguchi, T. Mizutani, T. Hiramoto, Ultralow-voltage operation of silicon-on-thin-BOX (SOTB) 2Mbit SRAM down to 0.37 V utilizing adaptive back bias, in *2013 Symposium on VLSI Technology*, Kyoto (2013), pp. T212–T213. http://ieeexplore.ieee.org/stamp/stamp.jsp?tp=&arnumber=6576627&isnumber=6576594. See also: T. Masuhara, The future of low-power electronics, Chap. 2, in *CHIPS 2020*, vol. 2 (Springer International Publishing, 2016)
23. T. Hasegawa, Y. Yamamoto, H. Makiyama, H. Shinkawata, S. Kamohara, Y. Yamaguchi, SOTB (Silicon on Thin Buried Oxide): more than Moore technology for IoT and automotive, *2017 IEEE International Conference on IC Design and Technology (ICICDT)*, Austin, TX (2017), pp. 1–4. https://doi.org/10.1109/icicdt.2017.7993512
24. Y. Ogasahara, T. Sekigawa, M. Hioki, T. Nakagawa, T. Tsutsumi, H. Koike, Reduction of overhead in adaptive body bias technology due to triple-well structure based on measurement and simulation, in *Proceedings of the 2015 International Conference on Microelectronic Test Structures*, Tempe, AZ (2015), pp. 207–211. https://doi.org/10.1109/icmts.2015.7106154
25. M. Miura-Mattausch, U. Feldmann, Y. Fukunaga, M. Miyake, H. Kikuchihara, F. Ueno, H.J. Mattausch, T. Nakagawa, N. Sugii, Compact modeling of SOI MOSFETs with ultrathin silicon and BOX layers. IEEE Trans. Electron Devices **61**(2), 255–265 (2014). https://doi.org/10.1109/TED.2013.2286206
26. S. Khandelwal, Y. Singh Chauhan, D.D. Lu, S. Venugopalan, M.A.U. Karim, A.B. Sachid, B.-Y. Nguyen, O. Rozeau, O. Faynot, A.M. Niknejad, C.C. Hu, BSIM-IMG: a compact model for ultrathin-body SOI MOSFETs with back-gate control. IEEE Trans. Electron Devices **59**(8), 2019–2026 (2012). https://doi.org/10.1109/ted.2012.2198065
27. H. Okuhara, A. Ben Ahmed, J.M. Kühn, H. Amano, Asymmetric body bias control with low-power FD-SOI technologies: modeling and power optimization. IEEE Trans. Very Large Scale Integr. (VLSI) Syst. **26**(7), 1254–1267 (2018). https://doi.org/10.1109/tvlsi.2018.2812893
28. H. Okuhara, A.B. Ahmed, H. Amano, Digitally assisted on-chip body bias tuning scheme for ultra low-power VLSI systems. IEEE Trans. Circuits Syst. I Regul. Pap. **65**(10), 3241–3254 (2018). https://doi.org/10.1109/TCSI.2018.2811504
29. A.K.M.M. Islam, R. Shimizu, H. Onodera, Effect of logic depth and switching speed on random telegraph noise induced delay fluctuation, in *2019 IEEE 32nd International Conference on Microelectronic Test Structures (ICMTS)*, Kita-Kyushu City, Fukuoka, Japan (2019), pp. 166–170. https://doi.org/10.1109/icmts.2019.8730976
30. K. Usami, S. Kogure, Y. Yoshida, R. Magasaki, H. Amano, Level-shifter free approach for multi-Vdd SOTB employing adaptive Vt modulation for pMOSFET, in *2017 IEEE SOI-3D-Subthreshold Microelectronics Technology Unified Conference (S3S)*, Burlingame, CA (2017), pp. 1–3. https://doi.org/10.1109/S3S.2017.8309226
31. S. Nakamura, J. Kawasaki, Y. Kumagai, K. Usami, Measurement of the minimum energy point in Silicon on Thin-BOX(SOTB) and bulk MOSFET, in *EUROSOI-ULIS 2015: 2015 Joint International EUROSOI Workshop and International Conference on Ultimate Integration on Silicon*, Bologna (2015), pp. 193–196. https://doi.org/10.1109/ulis.2015.7063746
32. http://www.vdec.u-tokyo.ac.jp/English/index.html; http://www.vdec.u-tokyo.ac.jp/English/VDECReport18E.pdf
33. K. Ishibashi et al., A Perpetuum Mobile 32bit CPU with 13.4 pJ/cycle, 0.14 μA sleep current using Reverse Body Bias Assisted 65 nm SOTB CMOS technology, in *2014 IEEE COOL Chips XVII*, Yokohama (2014), pp. 1–3. https://doi.org/10.1109/coolchips.2014.6842954. See also: T. Masuhara, The future of low-power electronics, Chap. 2, in *CHIPS 2020*, vol. 2 (Springer International Publishing, 2016)
34. H. Nagatomi, N. Sugii, S. Kamohara, K. Ishibashi, A 361nA thermal run-away immune VBB generator using dynamic substrate controlled charge pump for ultra low sleep current logic on 65 nm SOTB, in *2014 SOI-3D-Subthreshold Microelectronics Technology Unified Conference (S3S)*, Millbrae, CA (2014), pp. 1–2. https://doi.org/10.1109/S3S.2014.7028184
35. S. Jain et al., A 280 mV-to-1.2 V wide-operating-range IA-32 processor in 32 nm CMOS, in *2012 IEEE International Solid-State Circuits Conference*, San Francisco, CA (2012), pp. 66–68. https://doi.org/10.1109/ISSCC.2012.6176932

36. G. Chen et al., Millimeter-scale nearly perpetual sensor system with stacked battery and solar cells, in *2010 IEEE International Solid-State Circuits Conference—(ISSCC)*, San Francisco, CA (2010), pp. 288–289. https://doi.org/10.1109/ISSCC.2010.5433921
37. K. Matsubara et al., A 65 nm silicon-on-thin-Box (SOTB) embedded 2T-MONOS flash achieving 0.22 pJ/bit read energy with 64 MHz access for IoT applications, in *2019 Symposium on VLSI Circuits*, Kyoto, Japan (2019), pp. C202–C203. https://doi.org/10.23919/VLSIC.2019.8778078
38. T. Sakamoto et al., A silicon-on-thin-buried-oxide CMOS microcontroller with embedded atom-switch ROM. IEEE Micro **35**(6), 13–23 (2015). https://doi.org/10.1109/mm.2015.142. See also T. Masuhara, The future of low-power electronics, Chap. 2, in *CHIPS 2020*, vol. 2 (Springer International Publishing, 2016)
39. Y. Tsuji et al., Sub-μW standby power, <18 μW/DMIPS@25 MHz MCU with embedded atom-switch programmable logic and ROM, in *2015 Symposium on VLSI Circuits (VLSI Circuits)*, Kyoto (2015), pp. T86–T87. https://doi.org/10.1109/VLSIC.2015.7231363
40. M. Zwerg et al., An 82 μA/MHz microcontroller with embedded FeRAM for energy-harvesting applications, in *2011 IEEE International Solid-State Circuits Conference*, San Francisco, CA (2011), pp. 334–336. https://doi.org/10.1109/ISSCC.2011.5746342
41. https://industrial.panasonic.com/content/data/SC/ds/ds4/MN101L05__E.pdf
42. https://www.eenewseurope.com/design-center/ultra-low-power-microcontrollers-enabling-energy-harvesting-applications-0
43. M. Hioki, H. Koike, Low overhead design of power reconfigurable FPGA with fine-grained body biasing on 65-nm SOTB CMOS technology. IEICE Trans. Inf. Syst. **E99-D**(12), 3082–3089 (2016). https://doi.org/10.1587/transinf.2016edp7129
44. T. Katashita, M. Hioki, Y. Hori, H. Koike, Development of an evaluation platform and performance experimentation of flex power FPGA device. IEICE Trans. Inf. Syst. **E101-D**(2), 303–313 (2018). https://doi.org/10.1587/transinf.2017rcp0003
45. K. Masuyama, Y. Fujita, H. Okuhara, H. Amano, 7MOPS/lemon-battery image processing demonstration with an ultra-low power reconfigurable accelerator CMA-SOTB-2, in *2015 25th International Conference on Field Programmable Logic and Applications (FPL)*, London (2015), p. 1. https://doi.org/10.1109/fpl.2015.7293964
46. Y. Matsushita, H. Okuhara, K. Masuyama, Y. Fujita, R. Kawano, H. Amano, Body bias grain size exploration for a coarse grained reconfigurable accelerator, in *2016 26th International Conference on Field Programmable Logic and Applications (FPL)*, Lausanne (2016), pp. 1–4. https://doi.org/10.1109/fpl.2016.7577346
47. D.-H. Le, N. Sugii, S. Kamohara, Hong-Thu Nguyen, K. Ishibashi, C.-K. Pham, A 400 mV 0.59 mW low-power CAM-based pattern matching system on 65 nm SOTB process, in *TENCON 2015—2015 IEEE Region 10 Conference*, Macao (2015), pp. 1–2. https://doi.org/10.1109/tencon.2015.7372913
48. H.J. Mattausch, M. Yasuda, A. Kawabata, W. Imafuku, T. Koide, A 381 fs/bit, 51.7 nW/bit nearest hamming-distance search circuit in 65 nm CMOS, in *2011 Symposium on VLSI Circuits—Digest of Technical Papers*, Honolulu, HI (2011), pp. 192–193. http://ieeexplore.ieee.org/stamp/stamp.jsp?tp=&arnumber=5986100&isnumber=5985982
49. X.-T. Nguyen, T.-T. Hoang, H.-T. Nguyen, K. Inoue, C.-K. Pham, A 1.2-V 162.9 pJ/cycle bitmap index creation core with 0.31-pW/bit standby power on 65-nm SOTB. Microprocess. Microsyst. **69**, 112–117 (2019). https://doi.org/10.1016/j.micpro.2019.05.008
50. T. Hoang, X. Nguyen, D. Le, C. Pham, Low-power floating-point adaptive-CORDIC-based FFT twiddle factor on 65-nm silicon-on-thin-BOX (SOTB) with back-gate bias. IEEE Trans. Circuits Syst. II Express Briefs **66**, 1723–1727 (2019). https://doi.org/10.1109/tcsii.2019.2928138
51. J.H. Min, S.-W. Kim, E.E. Swartzlander, A floating-point fused FFT butterfly arithmetic unit with merged multiple-constant multipliers, in *Conference Record of the 45th Asilomar Conference on Signals, Systems and Computers*, Pacific Grove, USA, Nov 2011, pp. 520–524
52. Institute of Standards and Technology (NIST), Advanced encryption standard (AES), *Federal Information Processing Standards (NIST FIPS)*, vol. 197, Nov 2001. https://doi.org/10.6028/nist.fips.197, https://www.nist.gov/publications/advanced-encryption-standard-aes

53. V.P. Hoang, V.L. Dao, C.K. Pham, Design of ultra-low power AES encryption cores with silicon demonstration in SOTB CMOS process. Electron. Lett. **53**(23), 1512–1514 (2017). https://doi.org/10.1049/el.2017.2151
54. V. Hoang, V. Nguyen, A. Nguyen, C. Pham, A low power AES-GCM authenticated encryption core in 65 nm SOTB CMOS process, in *2017 IEEE 60th International Midwest Symposium on Circuits and Systems (MWSCAS)*, Boston, MA (2017), pp. 112–115. https://doi.org/10.1109/MWSCAS.2017.8052873
55. P. Hamalainen, T. Alho, M. Hannikainen T.D. Hamalainen, Design and implementation of low-area and low-power AES encryption hardware core, in *9th EUROMICRO Conference on Digital System Design (DSD'06)*, Dubrovnik (2006), pp. 577–583. https://doi.org/10.1109/dsd.2006.40
56. S. Mathew et al., 340 mV–1.1 V, 289 Gbps/W, 2090-gate NanoAES hardware accelerator with area-optimized encrypt/decrypt $GF(2^4)^2$ polynomials in 22 nm tri-gate CMOS. IEEE J. Solid-State Circuits **50**(4), 1048–1058 (2015). https://doi.org/10.1109/JSSC.2014.2384039
57. W. Zhao, Y. Ha, M. Alioto, AES architectures for minimum-energy operation and silicon demonstration in 65 nm with lowest energy per encryption, in *2015 IEEE International Symposium on Circuits and Systems (ISCAS)*, Lisbon (2015), pp. 2349–2352. https://doi.org/10.1109/ISCAS.2015.7169155
58. M. Tamura, M. Ikeda, 1.68 μJ/signature-generation 256-bit ECDSA over GF(p) signature generator for IoT devices, in *2016 IEEE Asian Solid-State Circuits Conference (A-SSCC)*, Toyama (2016), pp. 341–344. https://doi.org/10.1109/ASSCC.2016.7844205
59. S. Sugiyama, H. Awano, M. Ikeda, 31.3 μs/signature-generation 256-bit $_p$ ECDSA cryptoprocessor, in *2018 IEEE Asian Solid-State Circuits Conference (A-SSCC)*, Tainan (2018), pp. 153–156. https://doi.org/10.1109/ASSCC.2018.8579287
60. N. Guillermin, A high speed coprocessor for elliptic curve scalar multiplications over $_p$, in *Cryptographic Hardware and Embedded Systems, CHES 2010. CHES 2010*, ed. by S. Mangard, F.X. Standaert. Lecture Notes in Computer Science, vol. 6225 (Springer, Berlin, 2010). https://doi.org/10.1007/978-3-642-15031-9_4
61. J. Lee, J. Hsiao, H. Chang, C. Lee, An efficient DPA countermeasure with randomized montgomery operations for DF-ECC processor. IEEE Trans. Circuits Syst. II Express Briefs **59**(5), 287–291 (2012). https://doi.org/10.1109/TCSII.2012.2190857
62. J. Lee, S. Chung, H. Chang, C. Lee, Processor with side-channel attack resistance, in *2013 IEEE International Solid-State Circuits Conference Digest of Technical Papers*, San Francisco, CA (2013), pp. 50–51. https://doi.org/10.1109/ISSCC.2013.6487632
63. Y. Hori, T. Katashita, Y. Ogasahara, A 65-nm SOTB implementation of a physically unclonable function and its performance improvement by body bias control, in *2017 IEEE SOI-3D-Subthreshold Microelectronics Technology Unified Conference (S3S)*, Burlingame, CA (2017), pp. 1–3. https://doi.org/10.1109/S3S.2017.8309209
64. T. Katashita, Y. Hori, Y. Ogasahara, Prototype of USB stick-sized PUF module for authentication and key generation, in *2017 IEEE 6th Global Conference on Consumer Electronics (GCCE)*, Nagoya (2017), pp. 1–2. https://doi.org/10.1109/gcce.2017.8229225
65. K. Ishibashi, J. Kikuchi, N. Sugii, A 910nW delta sigma modulator using 65 nm SOTB technology for mixed signal IC of IoT applications, in *2017 IEEE International Conference on IC Design and Technology (ICICDT)*, Austin, TX (2017), pp. 1–3. https://doi.org/10.1109/icicdt.2017.7993514
66. T. Yoshio, T. Kihara, T. Yoshimura, A 0.55 V back-gate controlled ring VCO for ADCs in 65 nm SOTB CMOS, in *2017 IEEE Asia Pacific Microwave Conference (APMC)*, Kuala Lumpar (2017), pp. 946–948. https://doi.org/10.1109/APMC.2017.8251606
67. M.-T. Hoang, N. Sugii, K. Ishibashi, A 1.36 μW 312–315 MHz synchronized-OOK receiver for wireless sensor networks using 65 nm SOTB CMOS technology. Solid-State Electron. **117**, 161–169 (2016). https://doi.org/10.1016/j.sse.2015.11.016
68. V.-T. Nguyen, R. Ishikawa, K. Ishibashi, 83nJ/bit transmitter using code-modulated synchronized-OOK on 65 nm SOTB for normally-off wireless sensor networks. IEICE Trans. Electron. **E101.C**(7), 472–479 (2018). https://doi.org/10.1587/transele.e101.c.472

69. T.-L. Nguyen, S. Takahashi, Y. Sato, K. Ishibashi, RF energy harvesting using cross-couple rectifier DTMOS on SOTB with phase effect of paired RF inputs, in *16th International Conference on Electrical Engineering/Electronics, Computer, Telecommunications and Information Technology (ECTI-CON 2019)*, Pattaya Beach, Thailand, July 2019. https://doi.org/10.1109/ECTI-CON47248.2019.8955229
70. K. Maekawa, H. Makiyama, Y. Yamamoto, T. Hasegawa, S. Okanishi, K. Sonoda, H. Shinkawata, T. Yamashita, S. Kamohara, Y. Yamaguchi, Comprehensive analysis of low-frequency noise variability components in bulk and fully depleted silicon-on-insulator metal–oxide–semiconductor field-effect transistor. Jpn. J. Appl. Phys. **57**(4S), 04FD19 (2018). https://doi.org/10.7567/jjap.57.04fd19
71. T. Ishigaki, R. Tsuchiya, Y. Morita, N. Sugii, S. Kimura, Effects of device structure and back biasing on HCI and NBTI in silicon-on-thin-BOX (SOTB) CMOSFET. IEEE Trans. Electron Devices **58**(4), 1197–1204 (2011). https://doi.org/10.1109/TED.2011.2107520
72. R. Kishida, T. Asuke, J. Furuta, K. Kobayashil, Extracting BTI-induced degradation without temporal factors by using BTI-sensitive and BTI-insensitive ring oscillators, in *2019 IEEE 32nd International Conference on Microelectronic Test Structures (ICMTS)*, Kita-Kyushu City, Fukuoka, Japan (2019), pp. 24–27. https://doi.org/10.1109/icmts.2019.8730967
73. R. Kishida, K. Kobayashi, Correlations between plasma induced damage and negative bias temperature instability in 65 nm bulk and thin-BOX FDSOI processes, *2016 IEEE SOI-3D-Subthreshold Microelectronics Technology Unified Conference (S3S)*, Burlingame, CA (2016), pp. 1–3. https://doi.org/10.1109/S3S.2016.7804371
74. Y. Yamamoto et al., The study of plasma induced damage on 65-nm Silicon on thin BOX transistor. IEEE J. Electron Devices Soc. **7**, 825–828 (2019). https://doi.org/10.1109/JEDS.2019.2917893
75. T. Ishigaki, R. Tsuchiya, Y. Morita, N. Sugii, S. Kimura, T. Iwamatsu, T. Ipposhi, Y. Inoue, T. Hiramoto, Wide-range threshold voltage controllable silicon on thin buried oxide integrated with bulk complementary metal oxide semiconductor featuring fully silicided NiSi gate electrode. Jpn. J. Appl. Phys. **47**(4), 2585–2588 (2008). https://doi.org/10.1143/jjap.47.2585
76. S. Hirokawa, R. Harada, M. Hashimoto, T. Onoye, Characterizing alpha- and neutron-induced SEU and MCU on SOTB and bulk 0.4-V SRAMs. IEEE Trans. Nucl. Sci. **62**(2), 420–427 (2015). https://doi.org/10.1109/TNS.2015.2403265
77. S. Manabe, Y. Watanabe, W. Liao, M. Hashimoto, S. Abe, Estimation of muon-induced SEU rates for 65-nm bulk and UTBB-SOI SRAMs. IEEE Trans. Nucl. Sci. **66**(7), 1398–1403 (2019). https://doi.org/10.1109/TNS.2019.2916191
78. D. Kobayashi et al., Heavy-ion soft errors in back-biased thin-BOX SOI SRAMs: hundredfold sensitivity due to line-type multicell upsets. IEEE Trans. Nucl. Sci. **65**(1), 523–532 (2018). https://doi.org/10.1109/TNS.2017.2774805
79. C. Chung, D. Kobayashi, K. Hirose, Resistance-based modeling for soft errors in SOI SRAMs caused by radiation-induced potential perturbation under the BOX. IEEE Trans. Device Mater. Reliab. **18**(4), 574–582 (2018). https://doi.org/10.1109/TDMR.2018.2873220
80. K. Zhang, Y. Manzawa, K. Kobayashi, Impact of body bias on soft error tolerance of bulk and Silicon on Thin BOX structure in 65-nm process, in *2014 IEEE International Reliability Physics Symposium*, Waikoloa, HI (2014), pp. SE.2.1–SE.2.4. https://doi.org/10.1109/irps.2014.6861174
81. T. Calin, M. Nicolaidis, R. Velazco, Upset hardened memory design for submicron CMOS technology. IEEE Trans. Nucl. Sci. **43**(6), 2874–2878 (1996). https://doi.org/10.1109/23.556880
82. A. Makihara et al., SEE in a 0.15/spl mu/m fully depleted CMOS/SOI commercial process. IEEE Trans. Nucl. Sci. **51**(6), 3621–3625 (2004). https://doi.org/10.1109/TNS.2004.839155
83. J. Yamaguchi, J. Furuta, K. Kobayashi, A radiation-hardened non-redundant flip-flop, stacked leveling critical charge flip-flop in a 65 nm thin BOX FD-SOI process, in *2015 15th European Conference on Radiation and Its Effects on Components and Systems (RADECS)*, Moscow (2015), pp. 1–4. https://doi.org/10.1109/radecs.2015.7365581

84. K. Yamada, M. Ebara, K. Kojima, Y. Tsukita, J. Furuta, K. Kobayashi, Radiation-hardened structure to reduce sensitive range of a stacked structure for FDSOI. IEEE Trans. Nucl. Sci. **66**(7), 1418–1426 (2019). https://doi.org/10.1109/TNS.2019.2908722
85. M. Ebara, K. Yamada, K. Kojima, Y. Tsukita, J. Furuta, K. Kobayashi, Evaluation of soft-error tolerance by neutrons and heavy ions on flip flops with guard gates in a 65 nm thin BOX FDSOI process, in *2019 19th European Conference on Radiation and Its Effects on Components and Systems (RADECS)*, Montpellier (2019), pp. 1–5
86. W. Liao, M. Hashimoto, Analyzing impacts of SRAM, FF and combinational circuit on chip-level neutron-induced soft error rate. IEICE Trans. Electron. **E102.C**(4), 296–302 (2019). https://doi.org/10.1587/transele.2018cdp0004
87. Dale Dougherty, The maker movement. Innovations Technol. Governance Globalization **7**(3), 11–14 (2012). https://doi.org/10.1162/INOV_a_00135
88. C. Anderson, *Makers: The New Industrial Revolution* (Crown Business, New York, 2012). ISBN: 978-0307720955
89. M. Horowitz, E. Alon, D. Patil, S. Naffziger, R. Kumar, K. Bernstein, Scaling, power, and the future of CMOS, in *IEEE International Electron Devices Meeting, 2005. IEDM Technical Digest.*, Washington, DC (2005), pp. 7–15. https://doi.org/10.1109/iedm.2005.1609253
90. https://www.arduino.cc/en/Guide/MKRVidor4000
91. F.M. del Prado Martín, The thermodynamics of human reaction times (2009). arXiv:0908.3170, https://arxiv.org/abs/0908.3170
92. M.H. Devoret, R.J. Schoelkopf, Superconducting circuits for quantum information: an outlook. Science **339**(6124), 1169–1174 (2013). https://doi.org/10.1126/science.1231930
93. E. Charbon et al., Cryo-CMOS for quantum computing, in *2016 IEEE International Electron Devices Meeting (IEDM)*, San Francisco, CA (2016), pp. 13.5.1–13.5.4. https://doi.org/10.1109/iedm.2016.7838410
94. A. Beckers, F. Jazaeri, H. Bohuslavskyi, L. Hutin, S. De Franceschi, C. Enz, Design-oriented modeling of 28 nm FDSOI CMOS technology down to 4.2 K for quantum computing, in *2018 Joint International EUROSOI Workshop and International Conference on Ultimate Integration on Silicon (EUROSOI-ULIS)*, Granada (2018), pp. 1–4. https://doi.org/10.1109/ulis.2018.8354742

Chapter 7
Dealing with the Energy Versus Performance Tradeoff in Future CMOS Digital Circuit Design

Wim Dehaene, Roel Uytterhoeven, Clara Nieto Taladriz Moreno and Bob Vanhoof

List of Symbols

V_{DD}	Supply voltage
V_T	Threshold voltage
B	Current factor in MOSFET model
I_0	Reference current in leakage model
I_{on}	On current of a transistor
I_{off}	Off current of a transistor (~leakage)
t_d	Delay time of a switching operation
C	General switching capacitance
kT/q	Thermal voltage where k is the Boltzmann constant, T the absolute temperature and q the charge of an electron
SNM_{hold}	Static Noise Margin of the SRAM memory cell in hold mode
SNM_{tran}	Transient Noise Margin of the SRAM memory cell during read
SNM_{read}	Static Noise Margin of the SRAM memory cell during read

7.1 Introduction

Classic IC technology scaling laws show that transistor performance and dynamic energy consumption improve with reduced dimensions and scaling voltages. Ideal scaling is called constant electric field scaling or Dennard scaling [1]. Thus, for a few decades, technology evolution has brought "more for less". For a given functionality less area and less energy per operation was required with each new technology generation. This led us to the smartphone, GPS, (almost) self-driving car, Internet of

W. Dehaene (✉) · R. Uytterhoeven · C. Nieto Taladriz Moreno · B. Vanhoof
KU Leuven, Leuven, Belgium
e-mail: wim.dehaene@kuleuven.be

B. Murmann and B. Hoefflinger (eds.), *NANO-CHIPS 2030*,
The Frontiers Collection, https://doi.org/10.1007/978-3-030-18338-7_7

Things, era we live today. However, advanced scaling puts an end to the fairy tale: increased leakage and increased technological variability complicates the "more for less" paradigm. Today, and in the foreseeable future, we must deal with the energy performance trade off in a much more active and deliberate way. For digital design, this means that for physical design the circuit level is back from never fully gone. Careful library design, optimal choice of supply voltages, plural and advanced architectural techniques to deal with timing variability impose themselves. In this chapter an overview of this scenery is given. We start by describing the energy versus performance trade off, also introducing active energy reduction. Section 7.3 deals with leakage. In Sect. 7.4, ultimate supply voltage reduction, leading to near threshold logic is discussed. The following section addresses timing margin reduction with in situ timing detection. This is required to deal with the enhanced variability of advanced CMOS technology. Section 7.6 gives a brief introduction on how to deal with energy versus performance in SRAM design.

7.2 Setting the Scene: Energy, Performance, Supply Voltage, Threshold Voltages

The performance of a digital gate is governed by the time it takes to charge and discharge the parasitic capacitance on its output node. The delay is given in first order by the ratio of the charge to the available on current:

$$t_d = \frac{aCV_{dd}}{I_{on}} \tag{7.1}$$

The model conveniently used in this context for the current is the Sakurai-Newton model [2]:

$$I_{on} = \beta(V_{dd} - V_T)^{\alpha} \quad \text{with } 1 \leq \alpha \leq 2 \tag{7.2}$$

Combining both equations shows that a reduction in V_{dd} also implies a reduced threshold voltage V_T, otherwise no current would be left:

$$t_d = \frac{V_{dd}}{\beta(V_{dd} - V_T)^{\alpha}} \tag{7.3}$$

When reaching the 130 nm node, scaling has led to a no longer negligible amount of leakage current given by the equation below:

$$I_{leak} = I_0 e^{\frac{-V_T}{nkT/q}} \tag{7.4}$$

The energy and power consumed when operating a digital circuit is also related to switching and leakage. The dynamic, switching energy per operation is given by:

$$E_{dyn/op} = CV_{dd}^2 \tag{7.5}$$

The static, leakage energy per operation is given by:

$$E_{stat/op} = V_{dd} I_{leak} t_d \tag{7.6}$$

For completeness, it should be mentioned that there is a third kind of energy consumption: short circuit energy. This is caused by the fact that during switching there is a short period of time during which both the pull up and the pull-down network of a standard CMOS gate is on, typically when the input signal is around half the power supply voltage. However, due to the ever-increasing switching speed, and the more aggressive scaling on V_{dd} than on V_T, the short circuit energy can be neglected in modern digital circuits.

It was already stated that according to (7.3) a decrease in V_{dd} should be followed by a decrease in V_T to keep the delay performance constant. At first, a decrease in V_{dd} looks advantageous because in that case the active energy and the leakage energy are also reduced. However, the delay should also be considered. If V_T is left untouched, the delay will increase leading to increased leakage energy per operation according to (7.6). If V_T is decreased as well, to keep delay performance constant, the leakage energy will also augment due to an exponential increase in leakage current according to (7.4). This is a clear indication that unbridled voltage reduction is not a meaningful energy optimisation strategy. V_{dd} reduction is mainly a means to combat dynamic energy consumption. It comes at the cost of increased leakage energy. Therefore, a minimum energy point (MEP) exists. This is shown in Fig. 7.1. The supply voltage at

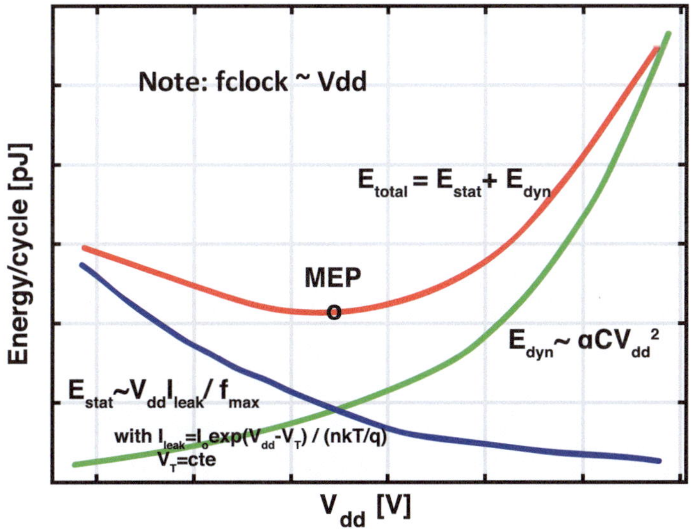

Fig. 7.1 Theoretical graph showing active versus dynamic energy trade off and the existence of a MEP

which this MEP occurs is dependent on the circuit activity. For an increase in activity, the MEP moves to lower voltages, again demonstrating that low voltage operation is a countermeasure for dynamic energy consumption.

It can thus be concluded that the higher the activity in a circuit the lower the MEP-voltage will be. The MEP voltage of a standalone MAC unit will thus be lower than that of a complete data path. A data path in its turn will have a lower MEP-voltage than a complete microprocessor. Most memory circuits have a very low activity. Therefore, it is a very relevant question whether low voltage operation, a low V_{min} as it is called in classic SRAM specifications e.g., is a good idea after all. This will be discussed in Sect. 7.6 of this chapter.

It should be noted that in the above discussion leakage was only considered when the circuit is active. This implies that the circuit is power gated when it is not in use. If state retention is needed this is not possible. In that case, either non-volatile registers [3] must be used or the state must first be saved to a non-volatile memory. Using non-volatile storage introduces an energy overhead. Thus, it limits the duty cycle at which the circuit can be power gated. This is not further discussed here.

In the by now iconic paper [4] the authors showed that increased parallelism in a DSP architecture can reduce the dynamic energy per operation if the power supply is also reduced. This is explained in Fig. 7.2. The paper cited above only mentions dynamic energy as it dates from the pre-leakage era. Leakage again introduces a

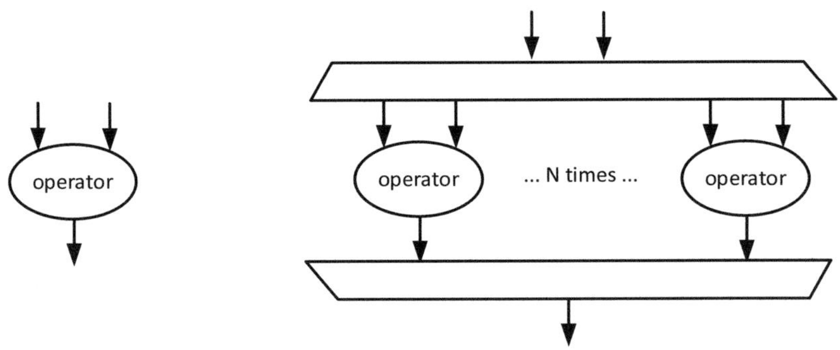

Reference: no parallellism

$$E_{dyn/op\_ref} = aCV_{dd\_ref}^{2}$$

$$E_{leak/op\_ref} = V_{dd\_ref} I_{leak} t_{d\_ref}$$

assumption:

$$t_d \sim \frac{1}{V_{dd}}, \text{ so } V_{dd} \text{ can be N times lower}$$

With N times parallelism

$$E_{dyn/op} = a(N + overhead)C\left(\frac{V_{dd\_ref}}{N}\right)^2 \approx \frac{E_{dyn/op\_ref}}{N}$$

$$E_{leak/op} = (N + overhead)\left(\frac{V_{dd\_ref}}{N}\right) Ileak(N.t_{d\_ref}) = N\, E_{leak/op\_ref}$$

Fig. 7.2 Increased parallelism reduces dynamic energy per operation if the power supply is also lowered but increases leakage energy

dynamic versus static energy trade off, this time in the architectural choice for parallelism. Increased parallelism also means increased leakage. This is also explained in Fig. 7.2.

7.3 Leakage Reduction Techniques

From the discussion in the previous paragraph, it is clear that reducing the power supply voltage also requires a strategy for leakage reduction. Reducing the static energy will shift the minimal energy point to a lower supply voltage, especially when lower threshold voltages are introduced to maintain speed performance. It should be noted here that supply voltage reduction also reduces static power. Yet voltage reduction leads to additional delay. Integrating the power over this delay thus leads to an increased static energy beyond a certain, optimal, decrease in supply voltage. Therefore, additional leakage power reduction techniques are needed to shift the minimum energy point to a lower power supply voltage.

Several options exist to combat leakage current. A first class of techniques uses the backgate terminal of the transistors to adjust the threshold voltages. This technique becomes less effective when scaling continues as the sensitivity of the threshold voltage to the backgate voltage reduces. The technique comes in two variants. Reverse body bias is used to reduce leakage compared to a nominal point at the cost of speed. The second variant is forward body bias. Here the source and drain junction diodes are more forward biased. This leads to an increase in leakage but also in speed performance compared to the reference point. Forward body bias is even harder to control than reverse body bias especially for elevated temperatures [5]. For all these reasons, backgate biasing is not a very popular technique when bulk CMOS technology is used. This might change with the advent of ultra-thin box, fully depleted SOI technologies (UTB-FDSOI) [6]. In these technologies, good modulation of the threshold voltage in both directions is possible leading to a large, tuneable range of speed performance for the same design. See [6–9], for examples. More research is required to determine whether FDSOI technologies are really a game changer for energy efficient digital circuit design. That includes also economical and strategical aspects that come with the technology choice.

A technique that is more suited is the use of multiple libraries during logic synthesis. In that case, each cell is implemented twice: once in a slow, low leakage version and once in a fast but leakier version. The synthesis tool will only use the fast, leaky cells on the critical paths as required for speed performance. Introduction of a second library shifts the MEP from 440 to 370 mV for an ARM cortex M0 processor as shown in [10]. To create both libraries two viable options exist. The first option is to make use of the different threshold voltages that are available in modern technologies. Each cell gets two variants in that case: one with a low V_T and another one with a higher V_T. The problem with this technique lies in the large difference in speed between the two V_T variants. This is shown in Fig. 7.3. This implies that the slower cells are not often used during synthesis and consequently the reduction in

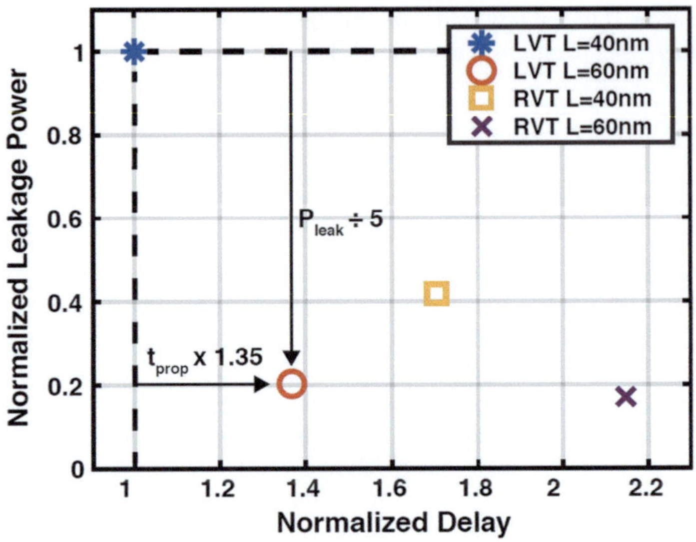

Fig. 7.3 Normalized leakage power and propagation delay of an nMOS stacked invertor with 40 and 60 nm length a, low VT (LVT) and regular VT (RVT) [10]

leakage is limited. A more effective technique is the use of different lengths for the transistors, also shown in Fig. 7.3. An increase of the length from 40 to 60 nm gives rise to a small but effective increase in V_T. The consequence is a leakage current divided by 5 at the cost of a delay increase of only 35%. The area impact of this technique is almost negligible, as the area of a standard cell is no longer dominated by the length of the transistors. This technique was used to create the libraries used in [10].

The invertor used in Fig. 7.3 is actually a variant with two stacked nMOS transistors, see Fig. 7.4. This type is used in technologies where the difference in current between a pMOS and an nMOS in weak inversion is almost an order of magnitude.

Fig. 7.4 Stacked nMOS invertor

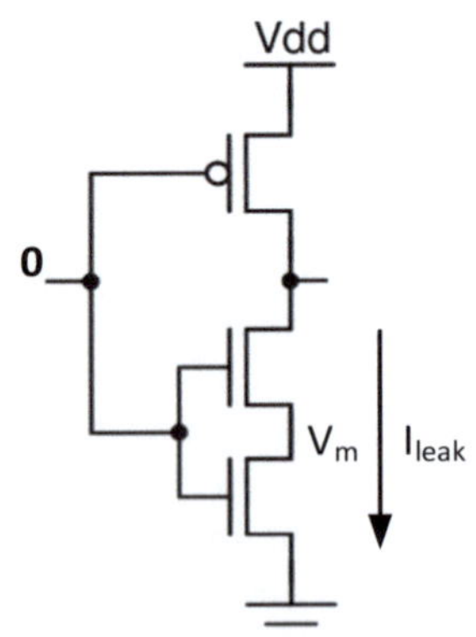

$$I_{leak} = I_0 e^{\frac{-V_m - V_T}{nkT/q}}$$

$$= I_{leak\_unstacked} e^{\frac{-Vm}{nkT/q}}$$

$$\approx \frac{I_{leak\_unstacked}}{5..8}$$

This is the case in 40 nm, general purpose, technology. The effect of stacking is also explained in Fig. 7.4. The build-up of voltage on the intermediary node reversely biases the upper transistor in the stack. On top of that, the effect is enhanced by the bulk effect on the upper transistor and the reduced drain induced barrier lowering (DIBL) on both transistors.

7.4 Near-Threshold Logic for Low Power DSP

In the previous sections, it has become clear that, provided the leakage current can be reduced, aggressive downscaling of the power supply voltage makes sense. This statement is even more true for applications that require low speed performance like IoT nodes or processors for medical implants. An example of this is given in [11]. In that paper, a processor for medical signals is described, running at 1 MHz with a power supply voltage of 400 mV. The processor has been demonstrated with several medical signal processing algorithms such as EEG or ECG.

When the power supply voltage is reduced below ca. 400 mV, the transistors will mostly work in weak inversion. In that case, the relation between drain current and gate-source voltage is exponential:

$$I_{Drain} = I_0 e^{\frac{V_{GS}-V_T}{nkT/q}} \tag{7.7}$$

Logic operating in this regime is called sub- or near-threshold logic in literature. The difference between sub- and near-threshold is not important. It mainly depends on the definition of the threshold voltage, which is usually taken in modern technologies as the gate-source voltage at which the current has an arbitrary, small value, e.g. 500 nA. In this text, near-threshold (NT) logic is used.

To design NT circuits, careful selection of the technology is required. Most modern technologies have variants that are labelled "Low Power (LP)". With this label, the vendor usually means optimized for low leakage, typically at the cost of some speed performance. This implies that the weak inversion current of the devices in such a technology will be low. For NT design, this has a direct impact on the available on current in the exponential regime. This means that the speed performance of NT circuits in an LP technology is usually limited to the sub 1 MHz range. This is only feasible for a limited amount of niche applications. The circuits that are discussed here target a speed performance range from 1 up to 50 MHz. 20–30 MHz is considered the sweet spot.

The first challenge in designing NT logic is of course to guarantee nominal functionality of the gates at the envisaged low voltages. However, our previous analysis has shown that a MEP exists. It does not make sense to reduce the supply voltage beyond that point. Even worse, when logic gates are (over)sized or special techniques [12] are used to work at extreme low voltages, in the order of 100 mV or lower, they

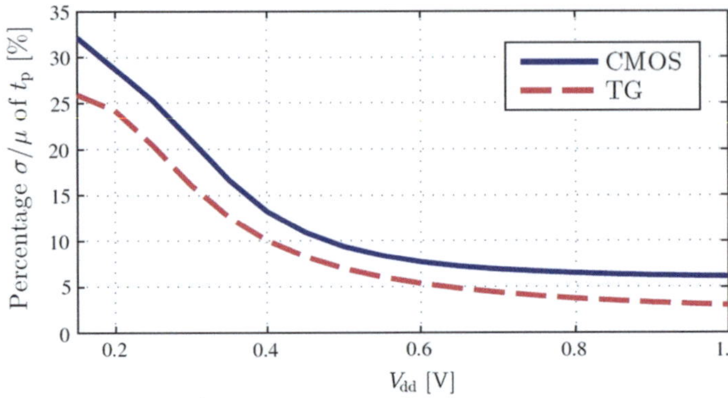

Fig. 7.5 Delay variation of TG based NOR gate compared to a standard CMOS NOR gate relative to the mean delay [13]

become suboptimal for operation at the MEP. For optimized designs, the MEP is often situated between 300 and 400 mV, but this depends on logic family and circuit architecture.

The second main challenge is how to deal with the increased sensitivity to variations of the design when operated in the NT region. In [13] it was demonstrated in detail that transmission gate (TG) logic is a favourable countermeasure. Transmission gates consist of a pMOS and nMOS switched in parallel to each other. Both transistors are active during the better part of digital transition. This implies that TG logic is less sensitive to variations because the variation on its on current is the combined variation of both n and pMOS. In other words, the on current of the TG is only compromised when both transistors are compromised. This is statistically less likely than for a single transistor. This is more quantitatively illustrated in Fig. 7.5.

TG logic requires for each signal also the complementary signal at its input. This makes it possible to design differential logic circuits without too much area overhead: only the number of local wires is increased. Furthermore, many local inverters can be saved because every gate now also produces complementary outputs. Differential logic is also more resilient against variations. However, differential logic synthesis is not available in a typical digital design flow. In [8] a solution for this was proposed. First, a standard cell library with differential logic cells is created and characterised. This differential library is accompanied with a pseudo single ended library. It contains a single ended version of each cell in the differential library, but the timing given for these single ended cells is the worst-case transition for the corresponding differential cell. Timing driven synthesis is now performed with the pseudo single ended library. After synthesis, the resulting netlist is transformed into a differential netlist by replacing the cells with their differential counterpart and introducing the necessary complementary signals. Because the worst-case timing of the differential was already taken into account, the timing closure problem for the differential netlist remains feasible. It stands to reason that the creation of the pseudo single ended library and the transformation of the single ended netlist into a differential one is automated. This is done with python scripts that enhance the flow with the necessary manipulations. The same script also removes superfluous invertors from the pseudo single ended netlist. As all signals need to be complementary anyway in the differential netlist, not every TG based logic cell needs local inverters to generate complementary signals.

To make this discussion more concrete we will discuss a few of the designs that we performed with NT, TG logic. We started from a relatively small multiply-accumulate block (MAC). Based on the learnings from this design, we designed a computation intensive data path for a JPEG encoder. The final designs were a couple of ARM cortex M0 processors. Ordered like that the designs start from high activity to low activity. This also means that the urge for leakage reduction is most prominent in the ARM cores.

The MAC blocks were designed both in 90 nm and in 40 nm general purpose technologies. Their properties are summarized in Fig. 7.6. This clearly shows that, depending on the required performance, technology scaling is not necessarily favourable for energy efficient design. Actually, the 90 nm MAC outperforms the

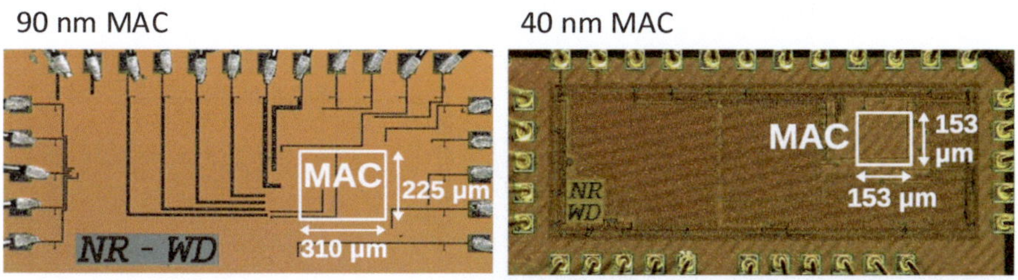

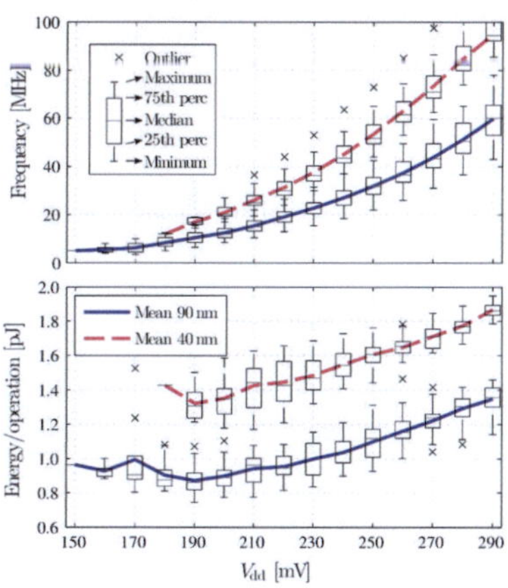

CMOS technology	90nm	40nm	Δ
# measured dies	34	20	
Active area [µm²]	225x310	153x 153	-66%
$V_{dd,min}$ [mV]	150	180	+20%
Clock freq. [MHz]			
@ $V_{dd,min}$	5.0	12.0	
@ 190mV	10.48	17.06	+63%
@ 250mV	31.88	53.48	+68%
σ/µ @ 190-290mV	16.77%	11.93%	-29%
Energy/oper. [pJ]			
@ $V_{dd,min}$	0.97	1.43	
@ 190mV	0.87	1.32	+51%
@ 250mV	1.10	1.61	+46%
σ/µ @ 190-290mV	7.94%	6.10%	-23%

Fig. 7.6 Summary of the ultra-low energy MAC designs [13]

40 nm block in terms of energy delay product (EDP). However, the 40 nm MAC has a better speed performance than the 90 nm Block. Needless to say that the area of the 40 nm block is also smaller.

As a next step, the data path of JPEG encoder was designed. The required memory was designed as a serial in, parallel out shift register. This is suboptimal in terms of energy consumption. This choice was made because in the corresponding research set up there was no room for SRAM design. The pipeline of the data path is latch based with a two phase, non-overlapping clock. This proved to be power hungry but allowed for time borrowing between the pipeline stages, thus enhancing the variability robustness of the design. The properties of the ultra-low power JPEG encoder are summarized in Fig. 7.7. These designs clearly show that low speed, energy efficient digital signal processing is feasible with NT TG logic. State of the art comparison is difficult for such a data path but to our opinion, the presented JPEG encoder outperforms the encoder published in [14].

The designs until now were mainly crafted at the circuit level. For the JPEG encoder, a data path generator tool was used [15]. However, for NT TG based design to be industrially viable, NT TG logic must be incorporated in the standard cell based digital design flow . This was realised for the ARM Cortex M0 cores. The

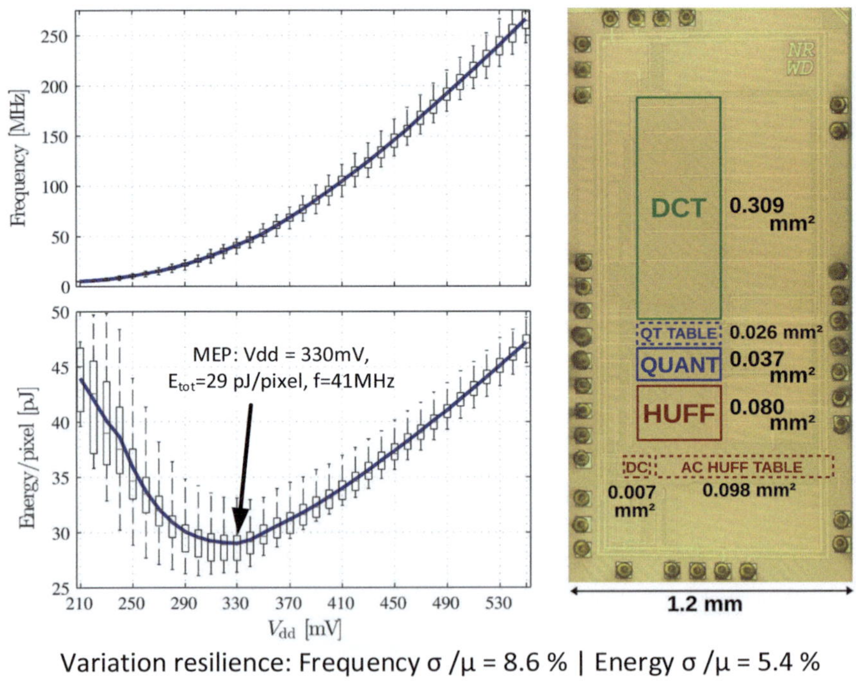

Fig. 7.7 Summary of the ultra-low energy JPEG encoder [13]

enhancements to the digital design flow are dealing with the differential nature of the logic as already discussed above. The properties of the ARM cortex M0 cores are summarized in Fig. 7.8. These designs outperform the state of the art at the time of this writing.

Summarizing this section, it can be stated that NT logic is a viable, feasible option for ultra-low energy digital signal processing. Design and demonstration of several blocks operating in the near-threshold logic proof this. The used logic family is based on differential transmission gates. It remains subject to future research how this will evolve in more advanced technology nodes. Do we still need stacking with finfet transistors in 16 nm and beyond? Will the improved subthreshold swing live up to its promises or will enhanced variability ruin the picture? Required design margins in NT operation remain painfully large even in established nodes like 40 nm or 28 nm. In situ timing detection can mitigate that to some extent.

7.5 In Situ Timing Detection to Deal with Variability and Margins

Almost all digital systems are designed to operate in a synchronous way. This basically means that all events in the system are related to a clock signal. This is the most feasible and probably only efficient way to guarantee timing correctness. However, the timing of a signal path is influenced by all kind of variations the circuit suffers from. These variations can be divided into global and local variations. Global variations affect all transistors on a die equally and thus consider the variations from one die to another, i.e. inter-die variations. Typical sources of global variations are, voltage drops, temperature fluctuations, aging effects and inter-die process variations. Their resulting effects on the transistor circuit behaviour are generally modelled with corners in which worst-case deviations are considered. Local variations affect each transistor on a die individually and thus consider so-called intra-die effects. The main source behind local variations is random mismatch governed by Pelgrom's law. Their impact is modelled through normal or Gaussian statistical distributions accounting for the randomness of these variations.

In digital circuits, designers have to make sure that the correct data is captured by the sequential elements (i.e. flip-flops or latches) at each rising (falling) edge of the clock signal. Therefore, they have to employ two types of timing checks; one for setup timing and one for hold timing as depicted in Fig. 7.9. The former checks that data arrives at the sequential element before the rising (falling) edge of the clock. This means setup timing mainly deals with late arriving data signals on critical paths of the circuit. The latter checks that data is captured in the sequential element before it is overwritten by new data of the next clock edge. Hold timing thus deals with fast propagating paths between registers with significant clock skew. Note that violation of any timing check in the system can lead to unexpected behaviour. Therefore, timing violations cannot be tolerated in most systems.

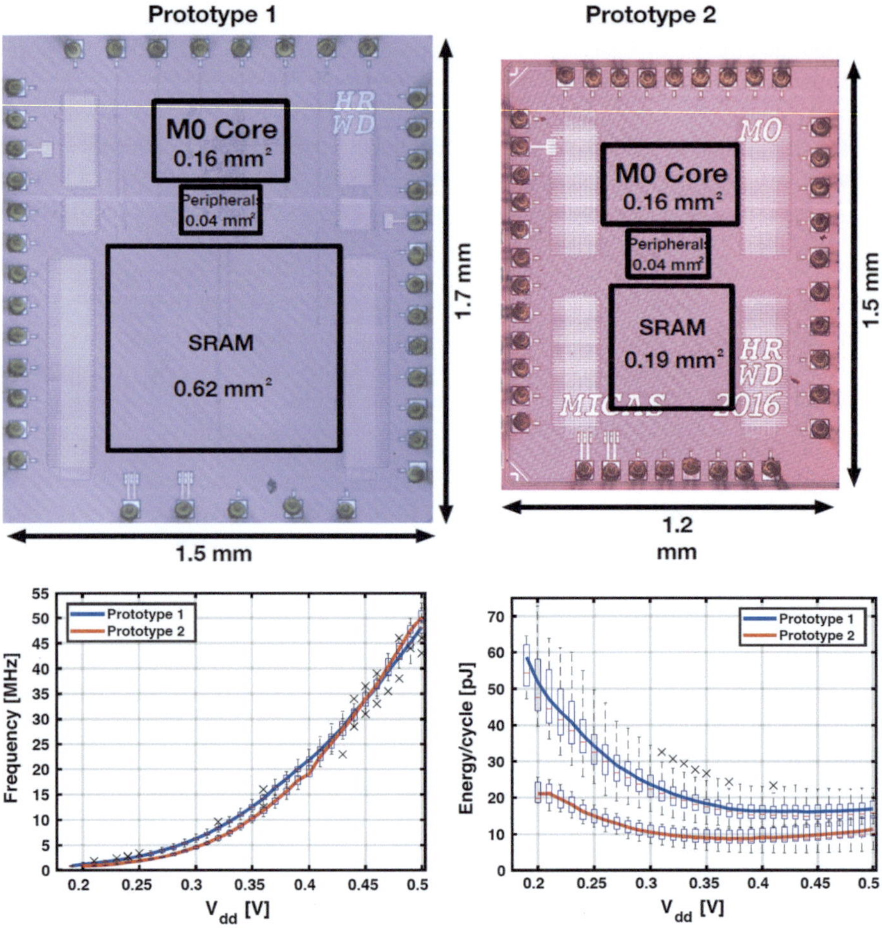

	Prototype 1	Protoype 2
System	ARM Cortex-M0	ARM Cortex-M0
Peripherals	Debugger, UART, GPIO	Debugger, UART, GPIO
Memory	2x 128KB SRAM block 0.9V $V_{dd,nominal}$	1x 64KB SRAM block 0.9V $V_{dd,nominal}$
Library	ULV Differential 40 nm gate length optimized for $V_{dd} = 250$ mV	ULV Differential 40 nm/60 nm gate length optimized for $V_{dd} = 300$ mV
Self-invert technique	No	Yes
Slow-slow sign-off	0.4 MHz, 250 mV	0.7 MHz, 300 mV
Power domains	3 Core, Peripherals, Memory	2 Core logic, Memory

Fig. 7.8 Summary of the ARM cortex M0 designs [10]

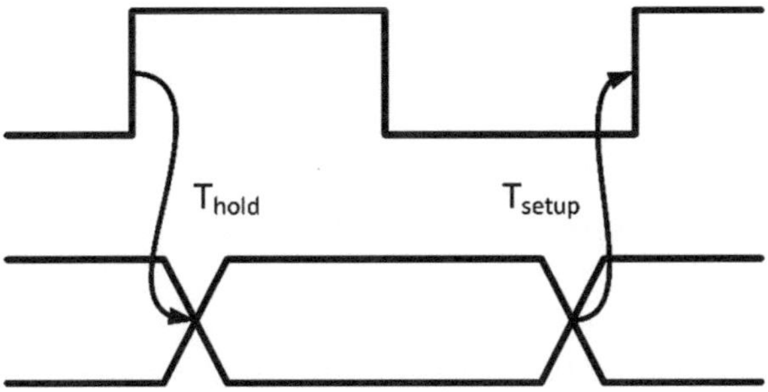

Fig. 7.9 Illustration of setup and hold timing

The uncertainty introduced by the aforementioned variations has a strong impact on the timing checks. These checks have to guarantee correct behaviour under all conditions, which means that they typically have to account for the worst-case over all variations. This leads to significantly larger design/safety margins that enforce an over designed system that does not operate at its maximum capabilities since the likelihood of a worst-case sample is rather slim. Hence, the uncertainty from variations degrades both performance and energy-efficiency for all but some samples.

As stated in the previous section, using ultra-low supply voltages increases the circuit sensitivity to variations resulting in large design margins. This is caused by the transition from a quadratic current relation in strong inversion to an exponential relation in weak inversion given by (7.7). The exponential function enhances the variations of the parameters that it encloses, which includes V_T. In advanced nm scale technologies V_T suffers strongly from local variations. As a result, the relative impact of local variations increases strongly at low voltages. Hence the large design margins. This leads to significant additional overhead and thus losses in performance, energy, area, and cost that overthrow the advantages of near-threshold designs. In other words, the in super-threshold logic widely used worst-case guard band timing mechanism becomes extremely inefficient for NT circuits due to large design margins.

7.5.1 Design Margin Reduction Techniques

Over the past decade, two categories of techniques that reclaim design margins have emerged in research. Both of them track the critical paths and apply dynamic voltage and/or frequency scaling (DVFS) to optimize energy-efficiency. On the one hand, replica or canary techniques try to track the behaviour of the critical paths as close as possible and allow diminishing margins against global variations. On the other hand, in situ timing monitoring techniques go in search of the point-of-first failure (PoFF) and by doing so, bypass the margins taken to deal with both global and local variations. Both techniques have been applied numerous times in super-threshold designs, but only few near-threshold implementations exist. This is in contrast with the fact that low voltage circuits can benefit the most for these techniques and leverage them to their full potential. Below, both techniques are discussed in detail.

7.5.2 Replica Monitoring or Canary Circuit

A replica of the critical paths is integrated on the same die and its performance is continuously monitored. The replica path shares process corner, global voltage and global temperature with the actual critical paths, becoming the reference to predict actual circuit performance. To provide an always-correct monitoring scheme, some margins are added to the replica to ensure that it fails before any of the actual critical paths. The performance can then be tuned using DVFS.

A typical method to implement such a replica path is with a ring oscillator. The number of inverter stages is chosen so that the oscillation period matches as close as possible with the critical timing of the circuit's critical paths. Often, the output of the ring oscillator is directly used as a clock signal for the system. This avoids having to tune the system's clock period towards the replica period. Rather it immediately generates a clock that tracks changes in supply voltage, temperature, aging and other global variation effects.

To improve the matching between the replica and the actual critical paths, the exact timing of the replica can often be fine-tuned during testing of the sample. This tuning allows removing some of the margin that would otherwise be required to guarantee that the replica is slower than all critical paths over all samples. However, even with tunability, some margin remains as the matching between replica and critical path will drift with PVT variations.

Furthermore, the replica can track neither intra-die variations, nor local fluctuations in temperature, supply voltage and aging. This results in an additional mismatch between the replica and the actual critical path. As this mismatch is created by local variations, no amount of tuning can alleviate it and safety margins remain required to resolve it.

The replica technique has been integrated in super-threshold designs with success because of the dominant impact of global variations. However, it lacks effectiveness in NT threshold designs, where the local mismatch becomes equally or even more important than global effects. See [16] for an example of a processor using replica monitoring.

7.5.3 In Situ Monitoring

In contrast with the previous mechanism, this strategy monitors each critical path to eliminate the local on-chip variation uncertainty. They key idea relies on the fact that the data path takes a finite amount of time to complete its operations. On this premise, the flip-flops of each critical path are equipped with an extra sequential

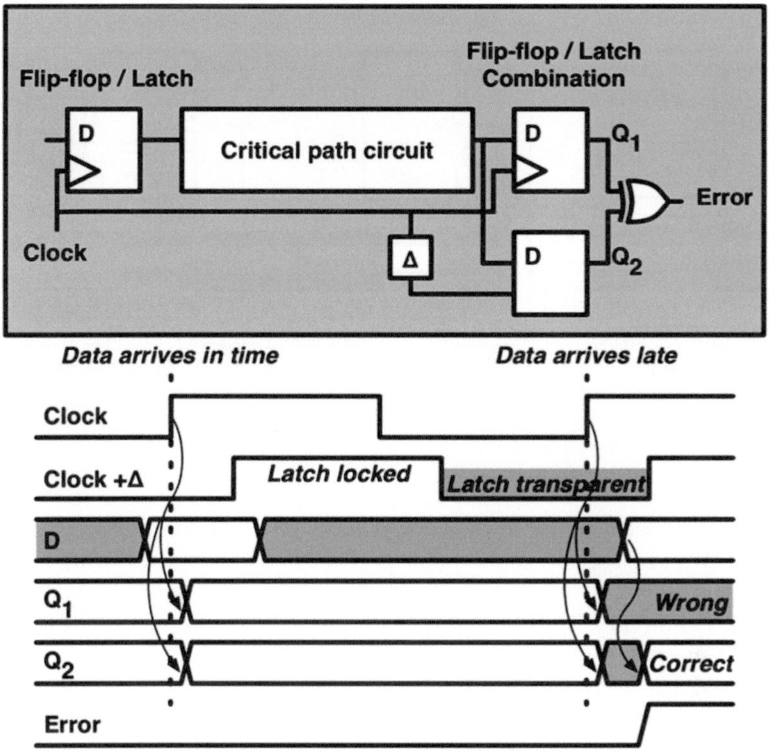

Fig. 7.10 Double sampling principle overview and timing diagram [10]

element (flip-flop, latch) to perform the additional sample, as depicted in Fig. 7.10. Any difference between the outputs of Q1 and Q2 means that the path made a last-minute change, indicating that the circuit operates at or close to the point of first failure (PoFF). Such a difference is flagged as a timing-error and is communicated to the error-processor.

The error processor continuously monitors error rates and adjusts frequency and/or voltage accordingly. This enables the tracking of the PoFF over all possible PVT variations. This is particularly convenient in conditions where it is hard to predict critical paths and their performance due to high variation sensitivity, as it is the case in NT operation. Furthermore, thanks to the locality of the error detection, the tracking can overcome margins against local variations making this technique even more interesting in NT designs.

Based on the double sampling with the two sequential elements from Fig. 7.10, two strategies can be considered.

- Error prediction (Fig. 7.11a): the additional sequential element samples the data first, while the original element samples the data a time Δt later. Thus, Δt works as a margin to predict the PoFF for a die and to tune system parameters to operate

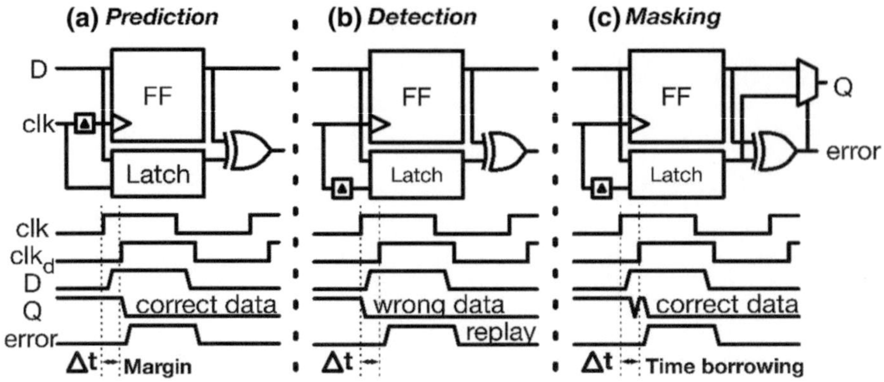

Fig. 7.11 Schematic overview of **a** error prediction, **b** detection and **c** masking with timing diagram implemented with double sampling [10]

near this predicted point. In the comparison, the prediction assumes the output sample Q1 to be correct, so this value is propagated and no errors are introduced in the data path.

- Error detection (Fig. 7.11b): the original sequential element samples first i.e. at the rising edge of the system's clock. The extra elements samples Δt later, capturing potentially late arriving data. In case of an error, a data correction strategy is required to feed the Q2 output value into the pipeline instead of the Q1 value. The inherent benefit is that no margins are introduced. Only actual incorrect data is flagged.
  The main drawback lies in the endorsement of strict and large hold constraints on all monitored paths. These constraints ensure that the value of a monitored path remains fixed after the rising edge of the clock so that it can be resampled Δt later. Without such a constraint, data from a new clock cycle could arrive so fast that it looks like late arriving data from the previous cycle triggering a false error. Depending on the system architecture and the chosen Δt, meeting the hold constraints can require a significant amount of delay cells and thus energy overhead.

In case of error detection, a recovery mechanism must be applied. Current correction strategies avail the fact that with the previous detection technique the correct data is readily available:

- Replay: the error signal selects the correct data through a MUX in case of error. This way, the flip-flop reinserts and feeds forward the correct data the next clock cycle. However, it requires stalling the entire pipeline for a single cycle. By consequence, the error signal needs to notify the entire pipeline of this stall before the next clock edge.
- Masking (Fig. 7.11c): is the straightforward mechanism. The available correct value is tunneled to the output immediately, despite arriving after the clock edge. The reinsertion of the data can be explicit via a MUX after the main sequential

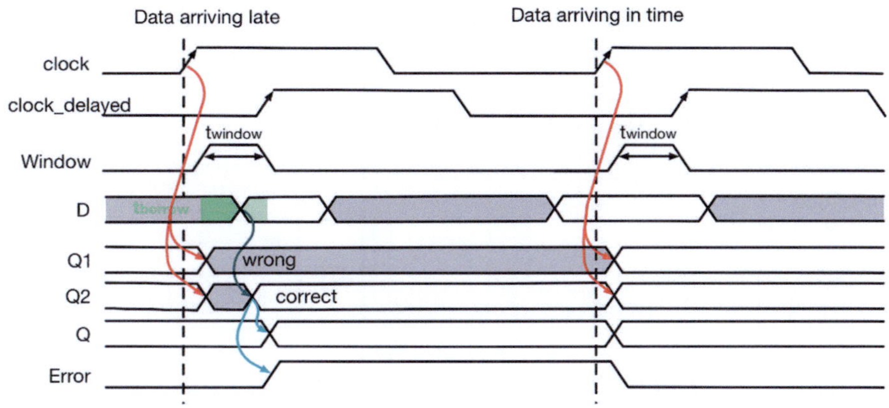

Fig. 7.12 Timing diagram of the timing error masking operation

element or implicit via time borrowing to reduce timing penalty, as depicted in Fig. 7.12. It requires using modified flip-flops or latches so that the Δt delay of the master clock edge defines a transparency window instead of a hard boundary for capturing the data. Therefore, data arriving after the slave clock rising edge but before the master clock rising edge can instantly propagate to the output. The window inherently allows borrowing time from subsequent pipeline stages to correct the data and propagate the correct value. Therefore, it is crucial to guarantee that the following stage has enough slack to perform a normal operation despite the stolen time.

A common practice lies in the combination of both detection and correction techniques, known as an error and correction (EDAC) system. Different combinations are possible, depending on the target circuit and its requirements. A real implementation of an EDAC system is discussed next.

7.5.4 EDAC System Example

The solution proposed in [8] presents an example of an in situ error detection and correction technique implemented in an ARM Cortex M0 microcontroller system in 40 nm CMOS. The EDAC system combines error detection and time borrowing correction. The architecture of the monitors is shown in Fig. 7.13. It combines a timing control block, a soft-edge flip-flop (SEFF), a transition detector (TD), and an error latch to detect and inherently correct data, which arrives late.

The SEFF is the main element as it holds the master and slave latch. The timing control is responsible for delaying the master clock Δt timing units regarding the slave clock. By doing so, it creates and specifies the size of the transparency window. The contribution of the window is twofold. First, it enables double sampling to detect

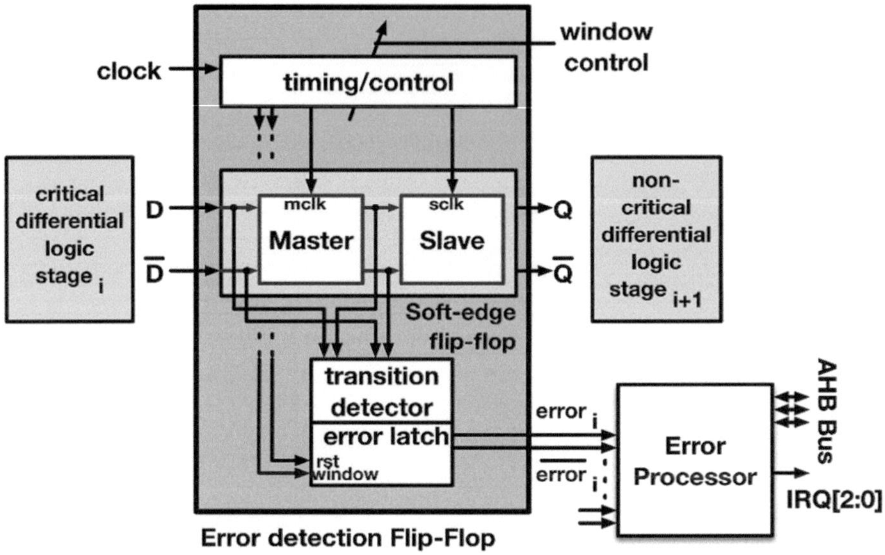

Fig. 7.13 Diagram of the proposed EDAC architecture [10]

late data arrival that otherwise would result in faulty operation. Secondly, it inherently allows borrowing time from subsequent pipeline stage to correct the error.

The transition detector is responsible for flagging the late arriving data. The TD compares data before and after the master latch. Since the latter is transparent at this moment, incoming data transitions result in a detectable delay. When both samples differ, the TD triggers a set dominant error latch, rising the associated error signal. The error processor evaluates the incoming error signal so an autonomous dynamic voltage scaling (DVS) loop can run.

While the additional logic is necessary to flag a timing error event, the timing error correction is inherent to the system. Because the SEFF allows data to propagate after the clock edge (during the transparency window), normally wrong data is propagated correctly because of time borrowing (Fig. 7.12). This allows operation at or close to the PoFF.

This technique was evaluated in several dies to average results. The results, depicted in Fig. 7.14, show a core energy consumption of 11–18 pJ/cycle for a frequency range of 5–30 MHz, and timing error detection is realized down to 290 mV and 5 MHz. The MEP is achieved at 7.5 MHz, 11.12 pJ/cycle and 310 mV.

The slow-slow corner static timing analysis sign-off points were 1 MHz at 350 mV and 5 MHz at 500 mV and the energy consumption are 75.80 pJ/cycle and 51.72 pJ/cycle respectively. When running both at 5 MHz the EDAC energy consumption is 75% lower.

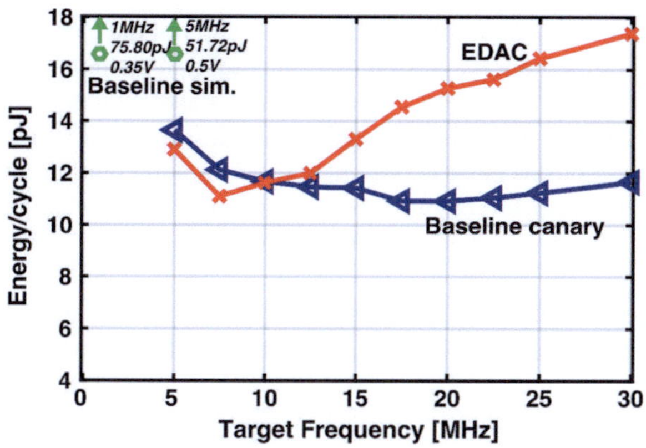

Fig. 7.14 Measurement of the PoFF curve for a wide frequency range, showing required energy consumption for the achieved target frequency [10]

The replica technique achieves better energy performance at high target frequencies (higher V_{dd}). At the lowest target frequencies (lowest V_{dd}) intra-die variation results in a high margin. Here, the EDAC approach allows working close to the ideal baseline design without margin.

7.5.5 Digital Flow Integration

Tools that automate the implementation process from an RTL level architectural description to an actual physical layout are indispensable. They map the architecture to the available logic cells, they ensure that the design meets the desired timing, they do a DRC clean placement and routing and they allow simulating the design along all steps of the implementation flow. Making today's complex architectures without this tool flow is simply impossible. Therefore, any novel technique must somehow fit into this flow.

Conventionally, the flow uses a worst-case based approach to deal with both global and local variations. This neglects the averaging effect that is present with local variations (i.e. Pelgrom's Law) over different cells. As a result, the tools provide an extremely conservative amount of design margin to account for these local variations. Yet, in super-threshold designs, this does not infer a large overhead, as global variations are the dominant source of design margins. However, in near-threshold designs this approach leads to excessive design margins as they suffer more from local variations. To overcome this, some tools offer statistical timing calculations on top of the typical worst-case approach. This allows putting in place design margins based on a targeted yield number and results in smaller and more realistic design margins for NT designs.

In their current state, the implementation tools do not have dedicated commands or procedures to implement the design margin reduction techniques discussed in the previous sections. Yet, integration of the aforementioned techniques into these tools is of vital importance to make them useful in an industrial context. Depending on the technique, several options exist to automate their implementation. In general, these options rely on the insertion of black boxes, the engineering change of order (eco) capabilities of the tool and the scripting language that allows interfacing with the tool.

For the design of the replica, the simplest option would be an implementation as analog macro. This gives the designer full control over the tuning and matching of the replica path to the circuit's critical timing. The timing itself is available in the tools timing reports. Once designed the macro, it can be simply added to the rest of the design as a black box. The downside of this approach is that the designer must repeat the manual matching procedure each time the critical timing changes. To avoid this, a more complex approach could leverage the tools timing optimization capabilities to implement a dummy path with a desired set of cells and the same timing constraint as the actual data path. The disadvantages of this more complex approach are a less fine control over the replica path (e.g. no custom layout) and the additional scripting.

In situ timing detection requires the insertion of individual sequential cells at specific locations. In order to be practical this process should be automated, meaning the automatic selection of the system's paths that require monitoring. This could be achieved by post-processing the tools timing reports. Second, the engineering change of order (ECO) flow of the tools could be used to place and connect the cells. Using the ECO flow ensures that little to no modification will be made to the existing design. Finally, the part that remains a challenge is providing a valid system wide simulation and testing of such an implementation. The creation of EDAC enabled design tools is subject to research at the moment of this writing.

7.5.6 Future Work

As of today, several error detection and correction (EDAC) tools have emerged in research. These approaches manifest and evidence the viability of its integration in NT logic, showing energy savings close to the 30% [8]. Nevertheless, several research questions still need an answer before these techniques can be applied on a wide scale in commercial systems. The most important remaining questions are briefly discussed below.

How to select the minimal set of paths to monitor so that detection is still guaranteed but overhead is minimized? This is a trade-off between the ability to guarantee detection of all possible errors and the overhead introduced by the error-detection circuits.

How to determine the size of the timing window Δt? A large window provides better detection but also requires larger hold constraints. This also links back to the previous question, as smaller windows tend to require a larger insertion rate to maintain good error visibility.

How does the activity, i.e. which paths are actually used by the program that runs on the processor, influence the visibility of timing errors? Will we always be able to see the PoFF or is it possible to write a program that 'circumvents' all monitored paths?

Finally, using time borrowing as a correction methodology further increases the complexity of the timing closure as the late arrival of one path has a ripple effect towards the available slack of another path in the next clock cycle. This could lead to unobservable errors or loops in which correction becomes impossible.

7.6 Multiple Supply Voltages for Energy Efficient SRAMs

Addressing the low energy design of SRAM is of a different nature than the design of low energy logic. Decreasing the power supply voltage is mainly an answer to dynamic energy, mostly helped by the quadratic dependency on that voltage. In the meantime, it must be avoided that leakage becomes the dominant contribution. However, in low power memory design in general and SRAM design in particular, leakage power is the dominant contribution. Therefore, another energy reduction strategy is more appropriate. Before diving into this strategy, it should be noted that the SRAM memories under consideration here only have moderate performance specifications. Read and write access times should be well below 1 ns. Using faster memories in a low energy system is a contradiction. 1 GHz clock frequency is already rather high for a low energy system after all.

SRAM memory design is split in two parts: cell matrix design and periphery design. The periphery includes all devices to access the memory cells for read or write, for example the address decoder and the sense amplifiers. The periphery design can rely on the same techniques as low energy logic design. These are summarized in Table 7.1.

Table 7.1 Overview of energy reduction techniques giving their influence on dynamic and/or static energy

Technique	Active energy	Leakage
Voltage scaling	✔	✔
Transmission gate logic		✔
Pipelining and parallelism	✔	
Low swing signaling	✔	
Power gating		✔
Vt modulation		✔
Transistor stacking		✔

The problem of cell matrix design is again twofold. The cell as such must be designed but also the architecture of the matrix is critical. The general idea behind the SRAM cell design is as follows. First, select a relatively high threshold voltage for the SRAM cells. This reduces the leakage power of the cell but compromises its speed performance. The speed performance of an SRAM cell depends on the read current. This current depends on $(V_{dd} - V_T)^{\alpha}$. Consequently, a supply voltage needs to be chosen that is sufficiently above the chosen threshold voltage to enable the required speed performance. If the resulting supply voltage is higher than what is technologically feasible, either the threshold voltage choice must be revisited, or the architecture of the memory could be reconsidered in order to reduce the bit line capacitance. Of course, this reasoning is insufficient to complete the design. Still circuit and architectural design choices will influence the energy consumption of the memory. In principle, the leakage reduction techniques of Table 7.1 are also applicable to SRAM circuits except for power gating. The use of transmission gates or transistor stacking can be very effective but leads to larger cells than the classical 6T cell. Commercially this is mostly avoided for area cost reasons.

7.6.1 Cell Design

Design of the cell comes with an extra set of constraints. First, the area of the cell is extremely critical. For SRAM cells, organized in a regular matrix, the lithography is pushed to its limits to minimize the area. The circuit topology choices are limited due to area constraints. Using transmission gates or upsizing transistors to improve matching is therefore not feasible. Also increasing transistor count beyond the classic 6T cell (see Fig. 7.15a) is not very popular in commercial SRAM for area reasons.

The stability of an SRAM cell is characterised by its static noise margin in hold (SNM_{hold}) as introduced by Seevinck et al. [17]. The typical butterfly curves (see

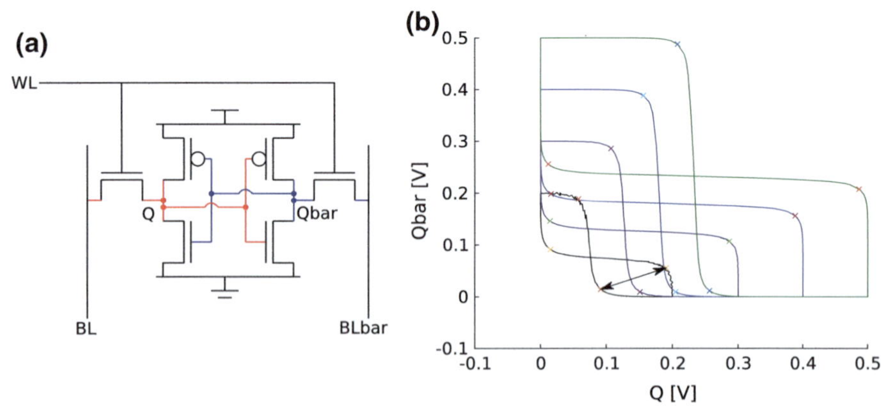

Fig. 7.15 **a** Schematic of a classic 6T SRAM cell and **b** the butterfly curves showing the SNM_{hold}

Fig. 7.15b) are used for this purpose. When the power supply of the cell is scaled, the eye opening of the butterfly curve reduces, showing that the stability of the SRAM reduces. This is quantified by inserting a square into the eye. The size of the square is the SNM_{hold}. In practice, the SNM_{hold} is the maximum allowed offset voltage caused by variation between the cross-coupled invertor pair for the cell to remain stable. A minimum SNM_{hold} of e.g. 50 mV is required for robust data retention. In addition, the read upset problem limits voltage reduction in SRAM cells. When the pass transistors of the cell are enabled, the internal node that is low is pulled up. The large charge on the parasitic capacitor of the precharged bit lines can thus upset the internal nodes since the cross-coupled inverter pair cannot sink that charge instantly. If the voltage rise on node Q becomes too large, the cell may flip and its content is destroyed. The voltage rise of the internal node can be incorporated in the noise margin. This leads to a static read noise margin SNM_{read}, which is smaller than SNM_{hold}. However, reading is dynamic behaviour especially for short bit lines with smaller bit line capacitance. Consequently, SNM_{read} is a too conservative metric in that case. Therefore, a transient noise margin, SNM_{tran}, is defined [18]. $SNMt_{ran}$ is the maximum offset voltage between the two cross-coupled inverters for which the cell content is not destroyed during transient read operations. Similar to SNM_{hold}, SNM_{tran} also decreases with decreased supply voltage. In practice, SNM_{tran} thus also limits voltage scaling.

For writing, write buffers must be able to overpower the drive strength of the cross-coupled inverter pair, to ensure the bit cell holds the correct state. Here, a trade-off rises: The inverter pair must be stable enough to survive reading and weak enough to enable correct writing. Another way of modulating the readability and writability is to look at the strength of the pass transistors: Increasing its strength makes them more suitable for writing, since the overpowering current is larger. Reducing its strength makes the voltage rise of the internal nodes less severe, increasing readability.

Increasing the strength of a transistor in a RAM cell can only be achieved by boosting its gate-source voltage. Upsizing of the transistor is not really an option for area reasons. For the inverter pair, this requires a boost of the supply voltage or a lowering of the ground voltage. Increasing the pass transistor strength requires a boost of the word line voltage or a reduction of the bit line voltage. Moving these voltages in the opposite directions leads a decrease in transistor strength. Table 7.2 provides a summary. This again shows that supply reduction for SRAM severely has its limits, since SRAM is more prone to readability issues.

Table 7.2 Voltage boosting techniques summary

Readability increase	
Inverter pair strength increase	Supply boost, ground reduction
Pass transistor strength reduction	WL reduction, BL reduction
Writability increase	
Inverter pair strength reduction	supply reduction, ground boost
Pass transistor strength increase	WL boost, negative BL

7.6.2 Architectural Techniques

The bit lines in SRAM are inherently long and depend on the size of the memory. Thus, bit lines have a large parasitic capacitance, which results in long discharge times during read. However, bit lines can be divided into short local bit lines (LBL), connecting only a limited number of cells and global bit lines (GBL), which transfer the data of the LBL to the side of the memory, where the interface with other blocks is located. This is shown in Fig. 7.16. See [18] for an example of an SRAM using this hierarchy.

Now that division between LBL and GBL is present, further optimization is possible by using low-swing signalling on the GBL in order to save active energy. Then, a read buffer translates the full-swing LBL information to a low-swing GBL. A write buffer senses the low swing GBL information and drives the LBLs to the appropriate state. Since the buffer circuits are shared for all cells on the LBL, a higher leakage can be tolerated and can hence be implemented with lower V_T transistors, speeding up the memory.

Similar to BL division, a division on word lines is possible as well. Splitting up word lines in local word lines (LWL) and global word lines (GWL) lowers the fan-out on the WL, reducing the capacitive load. An additional advantage occurs when the LWL size is equal to the word size: the half-select read upset and dynamic power

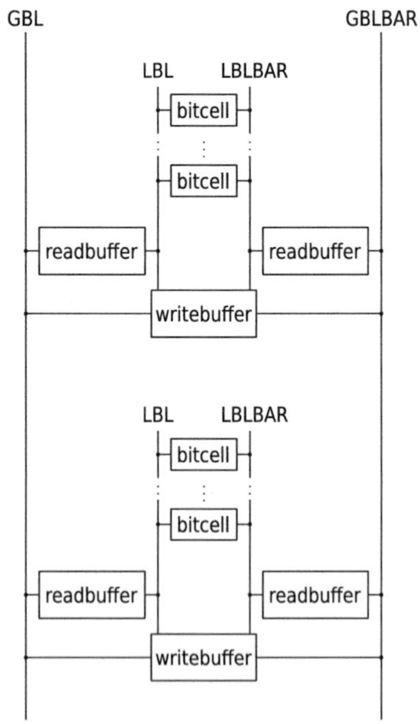

Fig. 7.16 Hierarchical bit line architecture

consumption of non-accessed words is avoided. However, WL division comes at a cost. The last decoder stage is pushed into the memory array, resulting in an area and leakage power increase.

When designing SRAM using advanced architectures with low-swing techniques, the leakage of the assist circuits needs to be monitored very carefully. Some of the assist circuits can be power gated when not in use. However, for low speed memories with a very low activity factor, it might be more beneficial to tolerate the higher dynamic power consumption of the signalling itself, since the leakage of the assist circuits would dominate.

An increased memory supply voltage, compared to the logic, can lead to an energy efficient SRAM, but level shifting between the logic supply domain and the memory-core supply domain is required. Careful design of the level shifters is mandatory. The challenge in level shifter design is to realize a large shift in voltage range both up and down while maintaining energy efficiency. This is best taken care of at the memory side. In that case, the digital design flow is agnostic to this level shift because the memory looks like low voltage on the interface towards the digital designer. From that point of view, only an additional, relatively high, supply voltage must be provided to the memory.

7.6.3 Summary

Power supply reduction in SRAM is an even more subtle story than for logic. The dominance of leakage energy in the matrix combined with reduced read and hold stability for lower voltage supplies dictates a higher power supply. The increased active energy that comes with this can be tackled with advanced architectures and low swing signalling. Low swing signalling is also advantageous for speed performance. An energy efficient system will thus end up with two power supply voltages: a low, near-threshold supply voltage for the logic and a higher voltage for the memory cells. The fact that this moves cell operation away from the near-threshold regime will definitely increase the variability resilience.

7.7 Conclusion

In this chapter, we have shown that designing for energy efficiency holds multiple challenges. A careful approach is needed to deal with the energy versus performance trade off. Simply relying on technology scaling is not possible long before scaling ends. Just reducing the power supply while ensuring functionality is not good enough. It will lead to suboptimal designs in terms of trading leakage versus active energy. Given the different relative importance of leakage and dynamic energy in logic and memory, different approaches are needed for both types of circuits. For logic, drastic supply reduction is appropriate, but not beyond the point where leakage starts to dominate. There is a minimum energy point! For memory, leakage is conspicuously more dominant. Therefore, it is wise to start from higher threshold voltages. This implies an increased power supply voltage due to performance requirements. Active energy in memory must be tackled at the architectural level.

Having dealt with leakage, variability is even a larger threat for energy efficient design at low supply voltage because the sensitivity to technology variations. If this is not handled, the required design margins will thus increase to the point where the energy benefits of low voltage design are overruled by the energy loss due to enhanced design margins. Current research focusses on more advanced ways of dealing with the margins. This can e.g. be realised by extending the circuits with in situ timing detection. This way, timing errors are avoided before they can happen and/or eventual timing are corrected so that any influence on circuit performance is avoided. Several strategies are under study at the time of this writing, but more research is required to turn in situ timing into a standard, automated, design strategy for digital circuits.

We have shown that further reduction of the energy consumption of modern digital electronic systems is still possible. It is also mandatory if we want to realise our dreams in an ever more connected and intelligent world in a sustainable way.

References

1. R. Dennard et al., Design of ion-implanted MOSFETs with very small physical dimensions. IEEE J. Solid-State Circ. **SC-9**(5), 256–268 (1974)
2. T. Sakurai, R. Newton, Alpha-power law MOSFET model and its applications to CMOS inverter delay and other formulas. IEEE J. Solid-State Circ. **25**(2), 584–594 (1990)
3. M. Natsui et al., An FPGA-accelerated fully nonvolatile microcontroller unit for sensor-node applications in 40 nm CMOS/MTJ-hybrid technology achieving 47.14 μW operation at 200 MHz, in *International Solid-State Circuits Conference (ISSCC)* (2019)
4. A. Chandrakasan et al., Low-power CMOS digital design. IEEE J. Solid-State Circ. **27**(4), 473–484 (1992)
5. S. Narendra et al., Forward body bias for microprocessors in 130-nm technology generation and beyond. IEEE J. Solid-State Circ. **38**(5), 696–701 (2003)
6. E. Beigné, FDSOI circuit design for high energy efficiency: wide operating range and ULP applications—a 7-year experience, in *IEEE 44th European Solid-State Circuits Conference (ESSCIRC)* (2018), pp. 216–216
7. R. UytterHoeven, W. Dehaene, A sub 10 pJ/cycle over a 2 to 200 MHz performance range RISC-V microprocessor in 28 nm FDSOI, in *IEEE 44th European Solid-State Circuits Conference (ESSCIRC)* (2018), pp. 236–239
8. H. Reyserhove, N. Reynders, W. Dehaene, Ultra-low voltage datapath blocks in 28 nm UTBB FD-SOI, in *IEEE Asian Solid-State Circuits Conference (A-SSCC)* (2014), pp. 49–52
9. A. Quelen, G. Pillonnet, A 2.5 μW 0.0067 mm^2 automatic back-biasing compensation unit achieving 50% leakage reduction in FDSOI 28 nm over 0.35-to-1V VDD range, in *2018 IEEE International Solid-State Circuits Conference (ISSCC)*, San Francisco, USA (2018)
10. H. Reyserhove, W. Dehaene, *Efficient Design of Variation-Resilient Ultra-low Energy Digital Processors* (Springer, Switzerland, 2019)
11. M. Ashouei et al., A voltage-scalable biomedical signal processor running ECG using 13 pJ/cycle at 1 MHz and 0.4 V, in *ISSCC* (2011)
12. N. Lotze, Y. Manoli, 62 mV 0.13 μm CMOS standard-cell-based design technique using Schmitt-trigger logic. IEEE J. Solid-State Circ. **47**(1), 47–60 (2012)

13. N. Reynders, W. Dehaene, *Ultra-Low-Voltage Design of Energy-Efficient Digital Circuits* (Springer, Leuven, 2015)
14. Y. Pu, J. Pineda de Gyvez, H. Corporaal, Y. Ha, An ultra-low-energy multi-standard JPEG co-processor in 65 nm CMOS with sub/near threshold supply voltage. IEEE J. Solid-State Circ. **45**(3), 668–680 (2010)
15. O. Weiss, M. Gansen, T. Noll, A flexible datapath generator for physical oriented design, in *Proceedings of the 27th European Solid-State Circuits Conference* (2001), pp. 393–396
16. T. Kuroda et al., A 0.9-V, 150-MHz, 10-mW, 4 mm^2, 2-D discrete cosine transform core processor with variable threshold-voltage (VT) scheme. IEEE J. Solid-State Circ. **31**(11), 1770–1779 (1996)
17. E. Seevinck et al., Static-noise margin analysis of MOS SRAM cells. IEEE J. Solid-State Circ. **22**(5), 748–754 (1987)
18. S. Cosemans et al., A 3.6 pJ/access 480 MHz, 128 kb on-chip SRAM with 850 MHz boost mode in 90 nm CMOS with tunable sense amplifiers. IEEE J. Solid-State Circ. **44**(7), 2065–2077 (2009)

Chapter 8
Monolithic 3D Integration—An Update

Zvi Or-Bach

8.1 Precise Bonder Enables Monolithic 3D Integration

The 3D IC space is considered to have two main branches—Through Silicon Via—"TSV" and Monolithic 3D. Some call the first branch '3D Parallel' and the second branch '3D Sequential.' The key differentiating aspect is the vertical connectivity density or pitch as is illustrated in Fig. 8.1, taken from a recent article [1] entitled "CoolCube™: More than a True 3D VLSI Alternative to Scaling."

Now that advanced precision bonders such as EVG-GEMINI® FB XT [2] and TEL-Synapse™ Si [3] are at the 50 nm (3σ) alignment precision range, a 3D Parallel integration flow could enable 50 nm like vertical pitch, which represents the

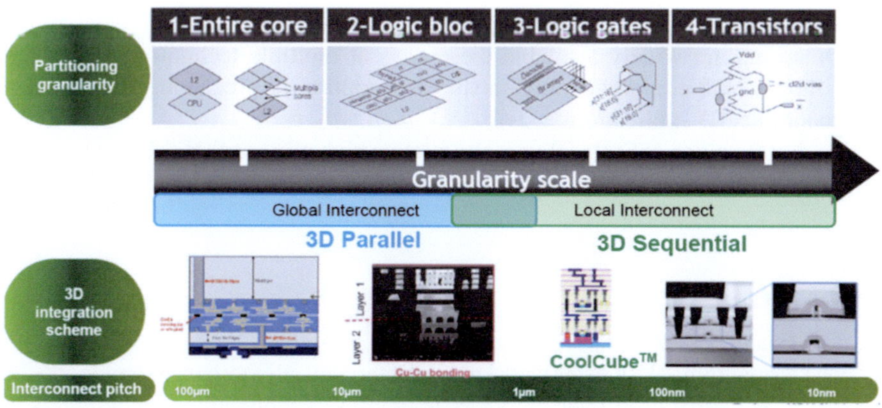

Fig. 8.1 Two 3D VLSE complementary approaches by CEA-Leti

Z. Or-Bach (✉)
MonolithIC 3D Inc., 3555 Woodford Drive, San Jose, CA 95124, USA
e-mail: Zvi@MonolithIC3D.com

B. Murmann and B. Hoefflinger (eds.), *NANO-CHIPS 2030*,
The Frontiers Collection, https://doi.org/10.1007/978-3-030-18338-7_8

Monolithic (or Sequential) 3D level of vertical connectivity. Such bonding precision could be assisted and enhanced by other technologies such as "Smart Alignment" [4, Chap. 3.3.2], Staggering [5, 6, Fig. 21A–C] and "Electronic Alignment" [7, Figs. 1–3].

8.2 Thinning the Transferred Layer—"Cut-Layer"

A second enabling technology for monolithic 3D using precision bonders is wafer thinning technology, especially for applications of more than two levels. To achieve high-density vertical connectivity, one needs to have a through silicon via with a diameter far less than 1 μm, compared to the >5 μm diameter of the common TSV technologies. For small diameter through silicon vias, the silicon layer needs to be very thin, as the aspect ratio for etching and filling such a via needs to be less than 1–10. In common TSV technologies, the transfer wafer is first thinned by backgrinding to a thickness of about 50 μm. It was found that thinning below 50 μm makes handling of the wafer unpractical—hence the >5 μm via diameter of common TSV technologies.

However, for the monolithic 3D application the thinning would take place after the transferred wafer has been bonded to the target wafer, thus achieving mechanical stability from the target wafer. In many applications, the desired thinning could be to 50 nm or even less. Such aggressive thinning would require a built-in control to avoid over thinning. Currently, without a built-in control, manufacturers avoid thinning below 10 μm. We can call such a built-in control a 'Cut-Layer.' One such built-in control is the BOX (Buried Oxide) of SOI wafer as was invented by IBM [8] and been used for many years by MIT Lincoln Lab [9] (Fig. 8.2).

SOI wafers are widely available these days at multiple technology nodes and wafer fabs, which could encourage a smooth adoption of precision wafer bonders for monolithic 3D applications.

One disadvantage of SOI wafers is the relative high price of SOI substrates. A few innovative alternatives for the BOX as a 'cut-layer' are presented in the following.

8.2.1 SiGe

Silicon Germanium—"SiGe" is well-known material in silicon-based semiconductor devices. It has been used over the years for multiple applications including as an alternative channel material or as a way to form stress. It is a well characterized material which can be epitaxially grown. Additionally, there are well-known etch process both wet and dry to allow a selective etch of SiGe versus Silicon. The use of SiGe as an etch stop layer for 3D using layer transfer has been proposed many years

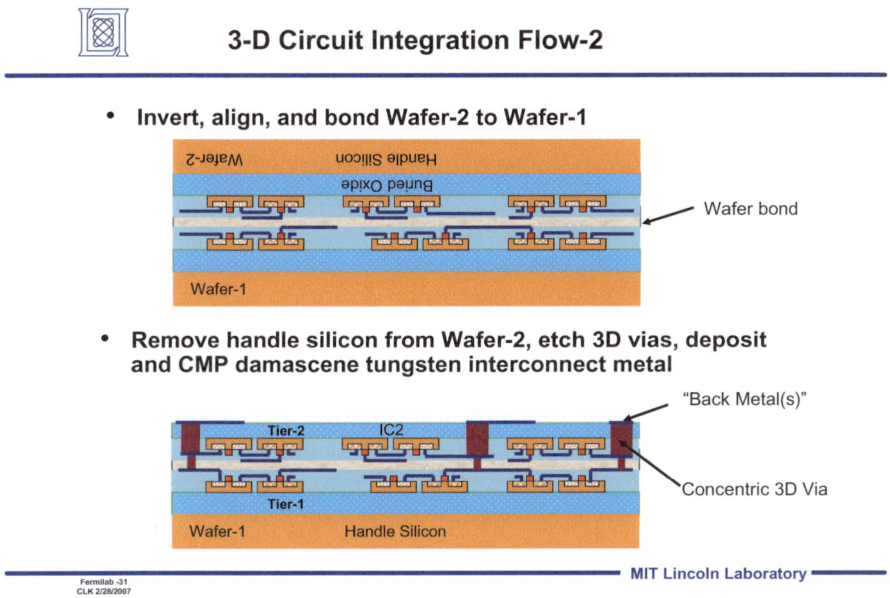

Fig. 8.2 Slide illustrating the use of SOI wafers, having the BOX as a 'Cut-Layer'

ego [10]. Recently, SiGe has been used for next generation device Nanowire/Nanosheet for which SiGe could be selectively dry etched in a multilayer structure to allow a gate-all-around structure to be formed (Fig. 8.3).

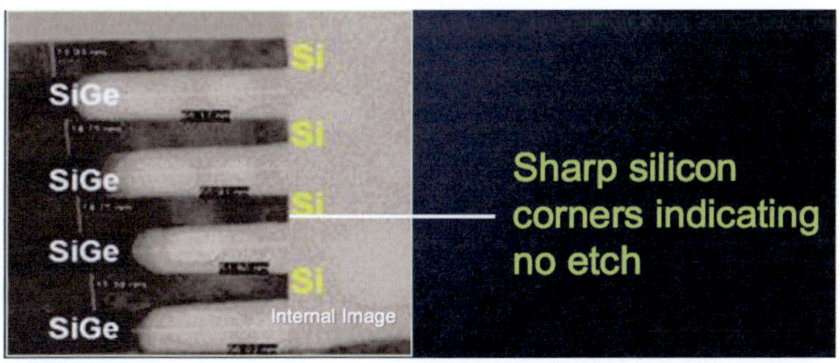

Fig. 8.3 SEM cross-section showing excellent etch of SiGe within alternating Si/SiGe layers, as will be needed for gate-all-around (GAA) horizontal nanowire (NW) transistor formation. *Source* Applied Materials

8.2.2 Doped Layer

Doped silicon, such as a deep N+ well or deep P+ well, could be used as 'cut-layer'. Preparing such a substrate could be prepared by the substrate provider or the foundry. The use of such a 'cut-layer' is simple with a wet etch or anodizing wet etch for which good selectivity is available to use the 'cut-layer' as an etch stop.

8.2.3 Ion-Cut

While Ion-Cut is the preferred choice for fabricating SOI substrates, it is not attractive for "3D Parallel" due to the damaging aspect of the Ion Implant as it passes through active transistors. Performing the H+ implant prior to the transistor formation is not effective due to the high temperature (>600 °C) activation process and other high temperature processes associated with the transistor formation.

8.2.4 Substrate Re-use

Additional savings could be achieved if instead of grind and etch back all the way to the 'cut-layer', a real cut could be used to achieve reuse of the substrate. Such a "cut" with a re-useable substrate could be accomplished by the use of a Modified ELTRAN® [11] process, the use of SiGe with a dry under-etch [12], or under-cut special etc. converting the buried SiGe to tear-able porous layer [5]. These processes are not standard with the current industry and accordingly might be adopted much later, if at all (Fig. 8.4).

8.2.5 Early Adoption

The 3D NAND vendor YMTC (Yangtze Memory Technologies Co.) is one of the early adopters of leveraging precision bonders for monolithic 3D volume applications. Many of the 3D NAND vendors are placing the periphery circuit at the periphery of the 3D memory array. Micron and Intel were the first to place the memory control circuit under the array calling it CuA (CMOS under Array) achieving about 90% array efficiency compared to about 70% for periphery next to the array. At the 2018 Flash Memory Summit YMTC introduce its Xtacking technology which places the peripheral circuitry on top of the memory array instead of underneath it. YMTC uses face-to-face wafer bonding as illustrated in Fig. 8.5.

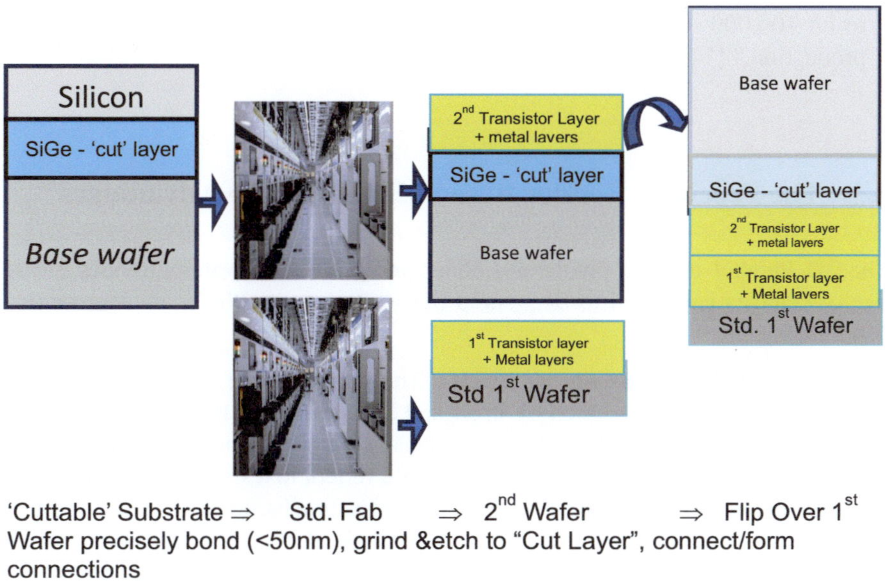

Fig. 8.4 Monolithic 3D integration flow using precise wafer bonder and wafers with built-in SiGe "cut-layer"

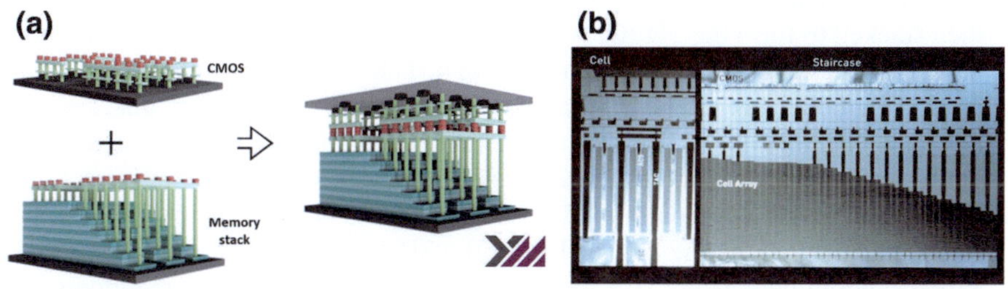

Fig. 8.5 **a** The Xtacking flow. **b** SEM of Xtacking device

"Under Xtacking addressing and I/O circuits are made on a separate wafer *(180 nm)* to the vertically stacked NAND cells and then bonded to them face-to-face through millions of **vertical** vias at the wafer-scale to complete the memory."

- "YMTC pushed its "pitches at several microns" down to about 100 nm for use in 3D NAND." [13]
- "YMTC has started delivering samples of its 64-layer 3D NAND chip with volume production likely to kick off in the third quarter of 2019, … Xtacking architecture is already adopted in the company's 64-layer 3D NAND engineering samples. Xtacking enables YMTC's 64-layer 3D NAND to be competitive with the available 96-layer 3D NAND solutions, … company expects its monthly production capacity

to hit 100,000 wafers after moving 64-layer 3D NAND technology to volume production." [14]

8.3 The Precision Bonder Based Monolithic 3D Advantages

The 3D Parallel using a precision bonder and 'Cuttable' wafer provides attractive advantages compared to Sequential 3D while keeping the equivalent vertical connectivity.

- **Standard Fab process for all levels**—The nature of the parallel flow is that each level is being processed by itself and accordingly its thermal budget is not impacting the other levels in the stack. This is extremely important advantage as the present IC fabrication complexity forces a vendor to resist any process change.
- **Heterogeneous Integration**—These days fabrication facilities are being designed and constructed to support a specific type/class of products such as a specific technology node, a specific type of circuit—logic, memory, analog, power, RF, …, a specific type of substrate—Bulk Silicon, SOI, …. In parallel 3D mix and match of different types of wafers in the stack provides an unparalleled advantage. Some specific applications will be covered in Chaps. 11 and 15.
- **Time to Market**—In parallel 3D, all levels could be fabricated in parallel and then stacked to form the 3D IC. With today's complex processing advanced node parallel processing could take 3 months. For a 3D IC with four levels the sequential processing could take more than a year which might introduce an unacceptable time to market challenge. While in 3D Parallel the fabrication of even a ten-level 3D IC stack could be done in less than five months.
- **Per Level Testing**—For parallel 3D, each level could be tested before being added to the stack to reduce the risk of losing the 3D IC because of a defective level. While random defects are still likely and should be managed by redundancy or other techniques, a total level failure could be managed.

In short: Precision wafer bonders with a 'Cuttable' wafer provide a very attractive technology for Monolithic 3D integration and could enable a broad industry adoption of monolithic 3D IC technology.

8.4 Update on Sequential Monolithic 3D

The research activity for Sequential Monolithic 3D is ongoing by the world leading semiconductor labs with good device demonstration as reported in IEDM 2018.

8.4.1 CEA Leti

CEA Leti reported breakthrough progress [15] with their CoolCube™ program. Implementing six major process changes "to limit the thermal budget of top tier processing to low temperature (LT) (i.e. TTOP = 500 °C) in order to ensure the stability of the bottom devices." The technology provides top level transistor performance compatible with the standard base level. Such thermal budget allows the use of tungsten for local interconnect in between top level and base level but would not allow copper or aluminum type interconnect. Additional progress reported by CEA Leti is the use of modified Smart Cut™ for the upper level silicon substrate. So instead of bonding SOI wafer and then grind and etch it, CEA Leti could use the modified Smart Cut™ process. So, wafers go through ion implant, then are bonded, cleaved and annealed within the allowed thermal budget of 500 °C. Since the transferred layer has no active device ion-cut could be used with the proper annealing steps and other process adjustment reducing the overall cost of forming the upper layer substrate.

8.4.2 imec

The imec report [16] titled "First Demonstration of 3D stacked FinFETs at a 45 nm fin pitch and 110 nm gate pitch technology on 300 mm wafers." is a relatively new entry to the monolithic 3D space. Like prior CoolCube™ work, imec use SOI wafer bonding with grind and etch back to form the upper level silicon substrate, and imec thermal budget is similar too (T < 525 °C), yet imec chose to use junction-less transistors for the top level to comply with the thermal budget challenge, yet reporting compatible performance with the base level transistors.

8.4.3 National Nano Device Laboratories (NDL)

National Nano Device Laboratories has been developing Sequential Monolithic 3D technologies utilizing laser re-crystallization for the upper level devices. Their recent progress, with support of additional partners [17], was: "location-controlled-grain technique is presented for fabricating BEOL monolithic 3D FinFET ICs over SiO_2. The grain-boundary free Si FinFETs thus fabricated exhibit steep sub-threshold swing (<70 mV/dec), high driving currents (n-type: 363 μA/μm and p-type: 385 μA/μm), and high Ion/Ioff (>10^6)".

8.5 Update on 3D Heterogeneous Integration

The old International Technology Roadmap for Semiconductors (ITRS) has ceased, acknowledging the sunsetting of Moore's law and ITRS issued in 2016 its final roadmap. A new initiative for a more generalized semiconductor road-mapping was started through the IEEE's Rebooting Computing initiative, called the International Roadmap for Devices and Systems (IRDS). One part of this new IRDS roadmap under IEEE has been the Heterogeneous Integration Roadmap that recently released its 2019 Edition [18]. This new report references the opportunities with 3D integration associated with heterogeneous integration similar to those covered in Chap. 15.

Additionally, many foundries have embarked on an effort to add wafer stacking technologies, and specifically hybrid bonding, to their offering. GlobalFoundries recently issued a press release about their collaboration with ARM to demonstrate High-Density 3D Stack Test Chips for High Performance Compute Applications, stating "the companies validated a 3D Design-for-Test (DFT) methodology, using GF's hybrid wafer-to-wafer bonding that can enable up to 1 million 3D connections per mm^2, extending the ability to scale 12 nm designs long into the future" [19]. This followed TSMC announcing a similar type of collaboration [20]. The TSMC program, called SoIC for 3D Integration, is a part of TSMC's advanced packaging options as presented in Fig. 8.6.

Technology	2.5D	3D-IC	SoIC
Structure cross-section	SoC1, SoC2, μbump, BEOL, Interposer	SoC1, μbump, SoC2	SoC1, SoIC Bond, SoC2
Interconnect	μbump + BEOL	μbump	SoIC bond
Chip Distance	**~100 μm**	**~30 μm**	**0**
Bong-pad Pitch	**36μm (1.0X)**	**36μm (1.0X)**	**9μm (0.25X)**
Speed	0.01X	1.0X	**11.9X**
Bandwidth Density	0.01X	1.0X	**191.0X**
Power Efficiency (Energy/bit)	22.9X	1.0X	**0.05X**

Fig. 8.6 Comparison of multi-die integration technologies. 2.5D and 3D-IC use backend equipment, SoICs frontend (wafer fab) technology. In SoIC, there is virtually no distance between integrated chips. It achieves a very small bond-pad pitch of 9 μm for good scalability. *Source* TSMC

These new advances are most effective for two-wafer face-to-face integration, as it avoids the challenges associated with TSV for many applications. While limited, they are an important step toward adoption of the 3D concept presented in this chapter.

References

1. J.-E. Michallet, CoolCube™: more than a true 3D VLSI alternative to scaling, 3d incites, 27 Mar 2019
2. EV Group, EV Group accelerates 3D-IC packaging roadmap with breakthrough wafer bonding technology, Press release 3 Jul 2018
3. TEL, Wafer Bonder/Debonder Synapse™ Series. https://www.tel.com/product/synapse.html
4. Z. Or-Bach, *Chips 2020 Vol. 2*, chap. 3.3.2 (Springer, Switzerland, 2016)
5. WO 2018/071143, PCT application
6. Z. Or-Bach, A 1,000x improvement in computer systems by bridging the processor-memory gap, in *2017 IEEE SOI-3D-Subthreshold Microelectronics Technology Unified Conference (S3S)* (IEEE, 2017)
7. WO 2019/060798, PCT application publication
8. U. S. Patent 6,821,826
9. V. Suntharalingam et al., Megapixel CMOS image sensor fabricated in three-dimensional integrated circuit technology, in *ISSCC. 2005 IEEE International Digest of Technical Papers. Solid-State Circuits Conference, 2005* (IEEE, 2005)
10. U. S. Patent 6,521,041
11. Z. Or-Bach et al., Modified ELTRAN®—a game changer for Monolithic 3D, in *2015 IEEE SOI-3D-Subthreshold Microelectronics Technology Unified Conference (S3S)* (IEEE, 2015)
12. WO 2017/053329, PCT application
13. R. Merritt, The latest in NAND. EE Times, 9 Aug 2018
14. DocMemory, Yangtze Memory to quickly migrate from 64 layers to 128 layers on 3D NAND. Simmtester.com. 15 Nov 2018
15. L. Brunet et al., Breakthroughs in 3D Sequential technology, in *2018 IEEE International Electron Devices Meeting (IEDM)* (IEEE, 2018)
16. A. Vandooren et al., First demonstration of 3D stacked FinFETs at a 45 nm fin pitch and 110 nm gate pitch technology on 300 mm wafers, in *2018 IEEE International Electron Devices Meeting (IEDM)* (IEEE, 2018)
17. C.-C. Yang et al., Location-controlled-grain technique for Monolithic 3D BEOL FinFET circuits, in *2018 IEEE International Electron Devices Meeting (IEDM)* (IEEE, 2018)
18. https://eps.ieee.org/technology/heterogeneous-integration-roadmap.html (2019)
19. GLOBALFOUNDRIES and Arm demonstrate high-density 3D stack test chip for high performance compute applications, Press release 07 Aug 2019
20. A. Patterson, TSMC, Arm show 3DIC made of chiplets. EE Times, 29 Sept 2019
21. Korczynski, Ed. Applied Materials releases selective etch tool, 13 (2016)

Chapter 9
Heterogeneous 3D Nano-systems: The N3XT Approach?

Dennis Rich, Andrew Bartolo, Carlo Gilardo, Binh Le, Haitong Li, Rebecca Park, Robert M. Radway, Mohamed M. Sabry Aly, H.-S. Philip Wong and Subhasish Mitra

9.1 The N3XT Architecture for Abundant-Data Applications

The future of computing is in crisis. Progress in abundant-data applications—including those with massive memory footprints (such as deep learning, brain-inspired computing, graph analytics, and natural language processing)—is demanding more of hardware than ever before. At the same time, implementations of these applications on Si CMOS continue to encounter the memory wall, i.e., the time and energy required to move data between memory and the relevant compute units is becoming very significant [1]. Slowing progress in component technologies compounds the problem as Dennard scaling of transistors meets fundamental limits. While 2D miniaturization (Moore's Law) continues, it, too, may soon hit fundamental limits [2]. Thus, business as usual cannot address this crisis. Instead, new kinds of architectures enabled by their underlying nanotechnologies must lead the way. N3XT (Nano-Engineered Computing Systems Technology) architectures [1, 3] embrace this paradigm (Fig. 9.1).

Traditional 2D system architectures generally consist of separate computing and memory chips connected by sparse chip-to-chip interconnects, resulting in expensive

D. Rich · C. Gilardo · H. Li · R. Park · R. M. Radway · H.-S. P. Wong
EE, Stanford University, Gates Building, Room 334, 353 Serra Mall, Stanford, CA 94305, USA

A. Bartolo · S. Mitra (✉)
CS, Stanford University, Gates Building, Room 334, 353 Serra Mall, Stanford, CA 94305, USA
e-mail: subh@stanford.edu

B. Le
EE, San Jose State University, Charles W. Davidson College of Engineering Building, Room 365, 1 Washington Square, San Jose, CA 95192, USA

M. M. Sabry Aly
SCSE, Nanyang Technological University, 50 Nanyang Avenue, N4-02c-92, Singapore 639798, Singapore

B. Murmann and B. Hoefflinger (eds.), *NANO-CHIPS 2030*,
The Frontiers Collection, https://doi.org/10.1007/978-3-030-18338-7_9

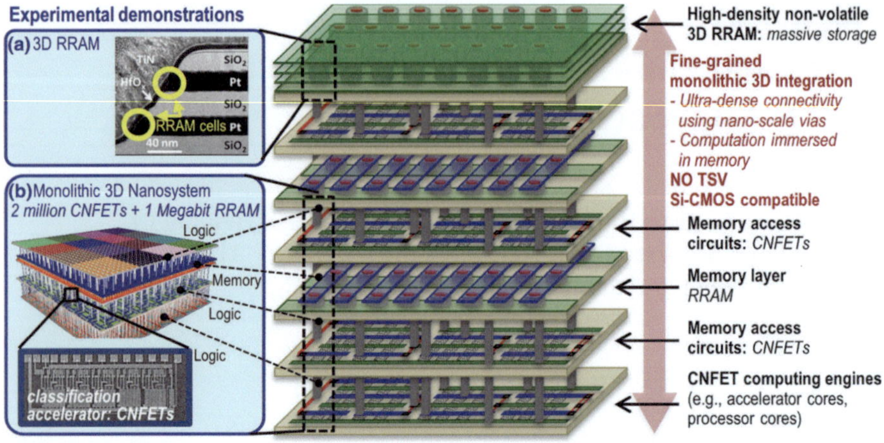

Fig. 9.1 Overview of N3XT architectures (taken from [3]). ©2015 IEEE

(in energy and time) off-chip memory accesses. Some system architectures recognize this problem and mitigate it with denser connections through 2.5D integration and chip stacking using through-silicon vias (TSVs). However, the via density (about $0.2\,\mu m^{-1}$ [4, 5]) doesn't provide the bandwidth required for many memory-intensive workloads [1, 6].

N3XT architectures finely integrate interleaved thin layers of logic and memory in 3D with ultra-dense interlayer vias (ILVs), which gives rise to its hallmark feature: ultra-dense inter-layer connectivity. This advantage grants chip architects an extremely wide, parallel interface for moving data between layers resulting in computation immersed in memory (in addition to the fact that the number of logic and memory elements is no longer constrained by the chip footprint). One use case for this wide interface is connecting compute logic with memory; however, other use cases exist for a wide range of application domains. For instance, a system using ILVs to shuttle data from an upper layer of sensors into memory and compute layers underneath was demonstrated in [7].

It should be noted that N3XT technologies do not preclude the use of the 2.5D technologies discussed above; for instance, a "multi-N3XT" system comprised of multiple N3XT chips could be interconnected together on an interposer.

9.2 Realizing the N3XT Architecture

N3XT can be implemented using many combinations of emerging technologies: in this chapter, we focus on a specific N3XT implementation (described in [1], outlined in Fig. 9.2) using Carbon Nanotube NFET (CNFET) transistors, Spin-Transfer Torque Magnetoresistive RAM (STT-MRAM) and Resistive RAM (RRAM)

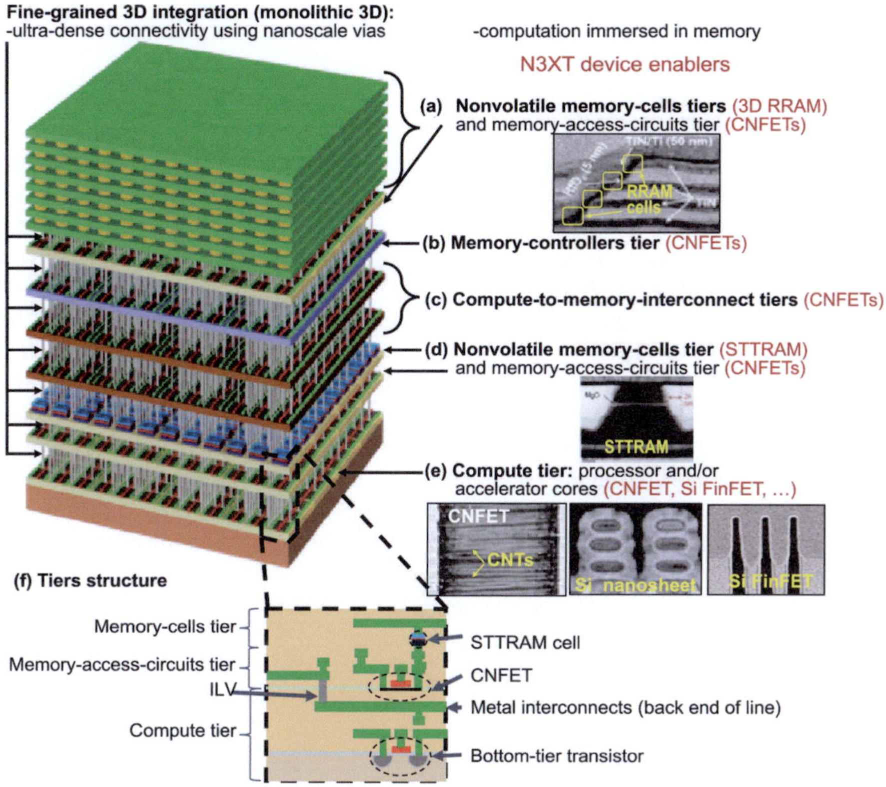

Fig. 9.2 Envisioned architecture for first-generation N3XT [1]. ©2019 IEEE

memories. These component technologies, fabricated at temperatures under 400 °C, together enable monolithic (ultra-dense) 3D integration. Thus, the ILVs can be implemented using back-end-of-line (BEOL) metal vias (that are already present today in conventional integrated circuits for connecting interconnect layers). The component technologies, as well as other potential candidates for fulfilling those roles in the N3XT architecture, will be discussed in later sections. The first generation of N3XT architectures, as shown in Fig. 9.2, uses CNFET logic for computation in lower layers and for implementing memory access mechanisms on the upper layers. Computations occur on the bottom-most layer (near the heatsink) for thermal reasons. Upper layers consist of STT-MRAM and RRAM memory arrays (and corresponding logic circuits to access those memories). A later section addresses possible methods of addressing the thermal challenge such that computations can be implemented using logic on the upper layers (with interleaved memory layers).

9.2.1 Benefits of N3XT

The high level of integration offered by N3XT makes it particularly attractive for abundant-data applications (i.e., applications with large working sets that must be shuttled back and forth to memory) resulting in significant benefits in application-level energy and execution time. Workloads spanning graph analytics, conventional machine learning, and deep learning all achieve large execution time and energy benefits when running on N3XT architectures as compared to traditional 2D silicon baseline architectures (Fig. 9.3).

Simulations are performed at a system level for a 2D baseline architecture and the first-generation N3XT architecture. Both a general-purpose CPU and a DNN accelerator are designed with the same memory size and configuration, and technology node. One part of the benefits, of course, results from more efficient component technologies (CNFETs and RRAM/STT-MRAM). However, more important in many cases is the increased memory-compute bandwidth from ultra-dense (monolithic) interconnects, which allows N3XT to overcome the memory wall. For example, in the graph analytics simulation, the 2D processor core spent 95.1% of its execution time on memory access, while first-generation N3XT spent just 2.1% of the 2D core's execution time on the same. In compute-bound applications, the first-generation N3XT architecture achieves 13–16× EDP improvements since this latter advantage is negated.

	CPU-BASED			DNN-ACCELERATOR-BASED	
	Graph analytics	**Conventional machine learning**	**Deep learning**	**ResNet (CNN)**	**Language Model (LSTM)**
Execution time	22x	18x	17x	13.6x	32x
Energy consumption	39x	37x	40x	11.7x	61.6x
EDP	858x	666x	680x	159x	1,971x

Fig. 9.3 Benefits of first-generation N3XT (Fig. 9.2) against a 2D baseline system with breakdowns for classes of abundant-data applications on a CPU-based architecture and a Deep Neural Network (DNN) accelerator. Details of how these benefits are achieved are discussed in [1]

Thermal analysis also shows approximately the same temperature between 2D and first-generation N3XT for both architectures (61–63 °C for the CPU architecture and 35–36 °C for the DNN).

Since the main hallmark of N3XT is ultra-dense 3D connectivity, any compatible component technology could be used in place of the assumptions made in the simulations for Fig. 9.3. Logic alternatives for the upper tiers can be any low-temperature compatible FETs: CNFETS, 2-D materials, thin-film transistors [8], or alternative fabrication methods for traditional silicon [9]. Furthermore, there is a wide suite of compatible memory technologies explored in Sect. 9.3.

The rest of this chapter will discuss the various components of N3XT systems, their challenges, and recent work to overcome these challenges.

9.3 Memory Technologies for Monolithic 3D Integration

The N3XT approach enables Nano-Systems with compute ***immersed*** in memory. These systems, as explained in the previous section, can give substantial system-level energy and execution time benefits across a wide range of abundant-data applications. However, this architecture requires highly capable memory technologies to achieve such benefits. Monolithically integrated 3D systems require that the memory technology is low temperature fabricable (<400 °C). While this integration technique greatly reduces memory access time and energy, the memory itself should also have low latency and low-energy access to achieve maximal benefits (e.g., avoid the pitfalls of Amdahl's Law). The enormously increased interconnectivity of N3XT also means that we can access substantially more memory at higher bandwidths than previously achievable. Memory must therefore be extremely dense to provide the capacity to solve abundant data problems that benefit from such bandwidths. To achieve this density, simple lithographic scaling will be insufficient—vertical integration schemes with multiple layers of memory cells are required to provide increased density. Memory technologies that are difficult to stack and scale up vertically (e.g., magnetic RAM, due to fabrication challenges) therefore have limited benefits. Embedded Flash (eFlash) was used for low-cost, on-chip data storage, but is not scalable both in terms of density and energy beyond 40-nm technology nodes. Finally, memory non-volatility (e.g. data retention without external power) is desired from various perspectives: storing and processing abundant data might otherwise mean abundant static/refresh power ***unless*** the memory is non-volatile. In particular, for embedded systems and applications non-volatility is critical to achieve systems with useable battery life. Moreover, on-chip non-volatile memory provides the ability to power gate the system in a temporally fine-grained fashion [10], providing substantial energy savings (and system battery life) over traditional off-chip NVMs.

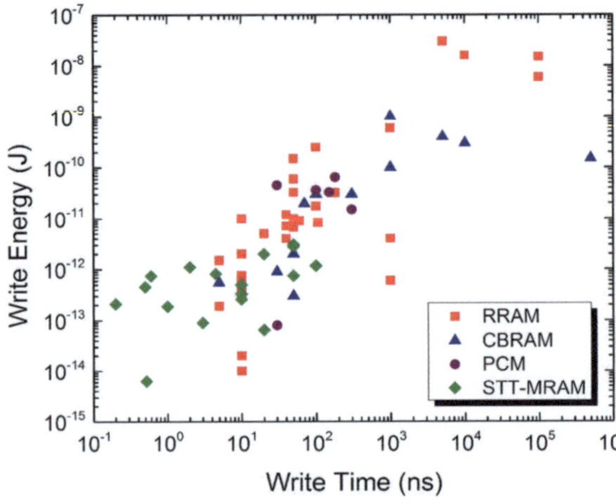

Fig. 9.4 Write energy and speed trends of various emerging non-volatile memory technologies that can be integrated on chip. Data from [12]

9.3.1 Emerging Technologies to Pave the Way

Several non-volatile memory (NVM) technologies are emerging to serve as high-capacity on-chip memories that overcome the limitations of conventional DRAM and Flash [11], while being compatible with the N3XT paradigm. Promising candidates include resistive RAM (RRAM), magnetic RAM (including spin-torque-transfer (STT), spin-orbit-torque (SOT), and voltage-controlled MRAM), ferroelectric RAM (FeRAM), and phase change memory (PCM). These memory technologies have different material systems and device structures than conventional silicon devices. Because of different switching mechanisms coupled with a variety of material systems, a wide spectrum of energy and speed characteristics have been demonstrated at device level for RRAM, PCM, and STT-MRAM, as summarized in Fig. 9.4. NVMs have asymmetric read and write properties, where read operations are typically faster and consume less energy than write operations.

Next, we will discuss the status of NVM technologies with an emphasis on their integration capabilities and reliability.

9.3.2 Summary of Emerging Memory Technologies

Metal-oxide RRAM. A typical RRAM device consists of a metal-oxide switching layer (e.g., HfO_x, TaO_x, TiO_x, AlO_x by atomic layer deposition) sandwiched by top and bottom metal electrodes, forming a two-terminal metal-insulator-metal (MIM) structure. As a CMOS-compatible NVM device, RRAM can be directly fabricated in the back-end-of-line (BEOL) process without impacting the front-end-of-line (FEOL) portion of a silicon chip. Leading foundries have successfully demonstrated RRAM macros integrated with Si CMOS logic process [13, 14]. Unlike charge-based volatile memories, information is encoded by high ('0') and low ('1') resistance states in RRAM and is held during the whole retention time (typically 10 years under 85 °C). In addition to low energy (<pJ/bit), RRAM is promising for high-capacity on-chip data storage due to good scalability and large-scale monolithic integration capability. RRAM enables high cell density ($<12\ F^2$), with a compact 1T-1R structure with a transistor as the selection device, or a crossbar structure with integrated two-terminal selectors. Sub-20-nm and sub-10-nm RRAM devices have been reported with good memory performance and reliability [15–17]. Full RRAM memory chips have been reported, with capacities ranging from Mbit scale to Gbit scale [12]. Notable high-capacity demonstrations include a 16-Gbit chip [18] and a 32-Gbit chip [19].

Monolithic 3D integration can further enable higher density by stacking memory layers and scaling in the third dimension. There are two viable 3D architectures for RRAM: 3D vertical structure [20] and stacked cross-point structure [19]. 3D vertical RRAM (VRRAM), similar to vertical-channel 3D NAND Flash, has multi-layer RRAM cells sandwiched between horizontal plane electrodes and vertical pillar electrodes. Vertical pillars and multi-layer RRAM cells can be individually accessed and modulated by select transistors underneath, as demonstrated by experiments on four-layer 3D VRRAM integrated with select transistors [21, 22]. The key design considerations targeting ultra-high density ($>$Gbit/mm^2) are the driving capabilities of select transistors for reliable write operations [22] and the non-linearity of bit cells for reliable read operations.

STT-MRAM. A STT-MRAM cell typically consists of a magnetic tunnel junction and a select transistor, as a 1T-1MTJ structure. STT-MRAM is being pursued as a fast (~ns) and low-energy (<pJ/bit) cache-like memory. The scalability of STT-MRAM depends on the MTJ and the select transistor. MTJ diameters have been shown to be down-scalable to 11 nm [23]. However, the overall cell size is typically bounded by the select transistor, which needs to be sized properly to provide enough write current (10s of μA) for stable subsequent read operations (correlated with retention time as well). For chip prototypes, a 65 nm silicon-CMOS chip with STT-MRAM cache has been reported. The last-level cache (LLC) was 4 Mbit with 3.3-ns read speed [24]. A 7-Mbit embedded STT-MRAM chip has been recently integrated with a 22-nm FinFET process [25]. The effective memory density is mainly limited by two factors: planar scaling limitation (the need for high write current) and vertical scaling limitation. Vertical integration capabilities of STT-MRAM haven't yet been explored and reported, potentially due to complex material stacks used in today's STT-MRAM technologies.

PCM. PCM materials and device technologies have been extensively studied for decades. PCM cells can be made in 1T-1R structures similar to planar RRAM cells. As PCM programming relies on current-induced Joule heating, the major scaling consideration is write current (typically 10s to 100s μA), which is provided by the select transistor. Scaling below 10 nm is feasible for PCM cells for low-current operations [12]. Gbit-scale PCM chips have been reported for standalone memory applications [26, 27].

FeRAM. Traditional FeRAM with a PZT-based ferroelectric capacitor has been hard to scale (in terms of thickness) and suffers from high voltage and latency. Recently, ferroelectricity in ultra-thin HfO_2 layers leads to a more scalable, single-transistor FeRAM (or FeFET) technology. While it is still in early phase of explorations, integration with 22-nm FDSOI CMOS logic with measurements on a 32-Mbit prototype array has been demonstrated [28].

3D NVM. Stacking memory layers and scaling in the third dimension make 3D NVM a key technology enabler for realizing N3XT architectures that requires high memory capacity and bandwidth. There are two viable 3D NVM architectures: 3D vertical structure for RRAM [17] and stacked cross-point structure for both PCM and RRAM [16]. 3D vertical RRAM (VRRAM), like vertical-channel 3D NAND Flash, cost-effectively achieves the full potential of vertical scaling. Vertical pillars and multi-layer RRAM cells can be individually accessed and modulated by select transistors underneath, as demonstrated by experiments on four-layer 3D VRRAM integrated with select transistors [18, 19]. The key design considerations targeting ultra-high density (>Gbit/mm^2) are the driving capability of select transistors for reliable write operations [19] and the non-linearity of bit cells for reliable read operations.

9.3.3 Challenges of Emerging Memory Technologies

Endurance. STT-MRAM excels at high endurance due to magnetic switching. At the same time, the energy barrier for magnetic switching also leads to an important tradeoff between retention time and write energy. Hence, retention statistics at an array level becomes a key reliability optimization goal for STT-MRAM. RRAM and PCM studies primarily focus on the endurance aspect instead. 10^{12} endurance cycles have been reported for RRAM at cell level [29, 30]. The improvement over typical endurance cycles of RRAM (10^6–10^9) is attributed to interface and device stack optimization. However, few endurance studies are done at array or system level; these studies are key to understanding the impact of cell-to-cell variability and its impact on overall system lifetime. Recently, array-level and system-level endurance results have been reported, including array-level statistics for up to 10^7 cycles [31]. While substantially better than Flash endurance (10^4–10^5), such endurance levels at first appear insufficient to operate as a general-use RAM (vs. purely as a ROM for model parameters or instructions). As will be discussed in Sect. 9.3.4 these challenges are ripe for cross-layer solutions. Understanding abundant data applications and their write patterns can allow for better endurance resilience techniques to mitigate endurance-related failures.

Variability. In addition to endurance, cell-to-cell variability is an oft-cited challenge of RRAM, PCM, and STT-MRAM [11, 32–38]. Due to the material-dependent switching (e.g. of magnetic field, material phase, or ionic density) these technologies can be especially sensitive to manufacturing variation and defects across the memory array, which yield variations in the resistive values of different cells for each state [39]. These variations must be understood to develop effective circuit-level techniques to manage variability and enable effective use of the memory technology. Moreover, understanding the sensitivity of abundant data applications to bit errors in these technologies will allow for effective cross-layer optimization of the algorithm, memory controller and cell design.

9.3.4 Cross-Layer Solutions for Emerging Memory Technologies Across Device, Circuit, Architecture and Application Layers

RRAM variations, originating from device-to-device non-uniformity and the stochastic nature of resistive switching process, need to be properly managed in memory circuits. This becomes even more critical for exploiting RRAM's capability of programming and storing multiple bits per cell, where resistance distributions need to fall under distinct ranges to represent multiple levels that can be correctly read out. While there has been some work on multiple bits-per-cell RRAM mostly at the cell level [40–43], much of this work did not take a systematic approach to programming in light of cell-to cell variations. Recently, 3-bit-per-cell at a full array level has been reported on 4 kbit 1T-1R RRAM arrays [44]. Leveraging the knowledge of array-level statistical distributions for RRAM resistance values, sigma-based allocation and bitline voltage allocation techniques are developed and coordinated to enable successful sensing of 3 bits with iterative write-verify programming as shown in Fig. 9.5. Multi-bit per cell effectively enables higher bit density and overall memory capacity.

A similar approach was used in [10] where 2.3-bits-per-cell (e.g. 5 levels) were achieved at the full system level, operating on 4-kBytes of 1T-1R RRAM. Even more recently, a 1T-4R 1 Mbit RRAM array was demonstrated (where 4 RRAM cells are driven by a single select transistor) with 2-bits-per-cell demonstrated on a 1 kbit sub-array [45]; a single select transistor accesses up to a byte of data. This work used a novel gradual SET/RESET scheme to allow for precise control of each of the RRAM cells in the 1T4R structure (to account for disturb induced during neighboring same-transistor cell writes). The gradual SET/RESET scheme enabled such substantial control over the cell resistance that 128 levels (7-bits-per-cell) were demonstrated on a single-cell. By combining the refined control achievable through the methods in [45] and the variation-aware range definition schemes in [10, 44] multiple bits-per-cell capabilities can be achieved with low bit-error rates (BERs) across large arrays.

A critical insight in the multiple bits-per-cell work has been in how such capacity is used. With emerging memories, the technologies bit-error rate can be quite high, especially when pushing the boundary with what is achievable in terms of write

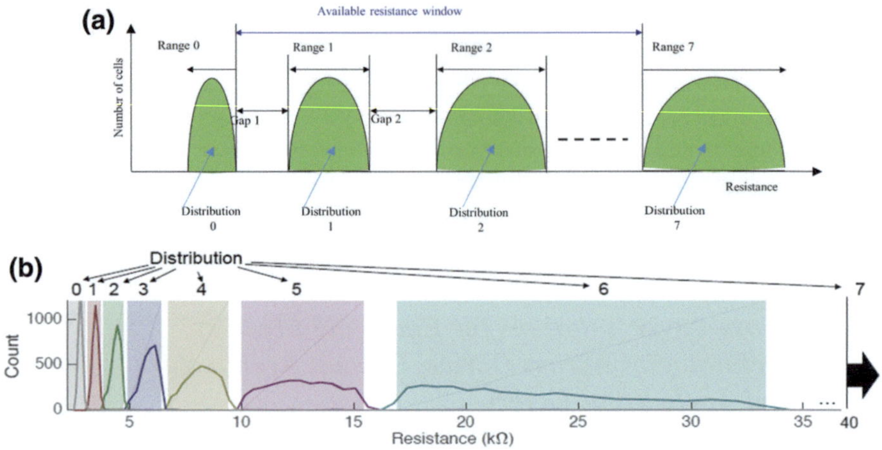

Fig. 9.5 **a** Sigma-based allocation concept, where wider ranges are allocated for higher resistance levels. **b** Measurement result on 4-kbit 1T-1R RRAM array [44]. ©2019 IEEE

speed, energy, and storage density. For many abundant data applications, bit-error is tolerable [10, 45]. Thus, increased bits-per-cell storage at the cost of increased bit-error rates can be tolerated. Wu et al. [10] demonstrate a 2.3× increase in a deep neural network model accuracy that can be achieved by using increased bits-per-cell storage, even though bit-error rates are quite high (2.24%). Hsieh et al. [45] show that even with bit-error rates of 1.56% for 2-bits-per-cell 1T-4R RRAM, the model inference accuracy is expected to be within 0.01% of ideal (no BER). Critically, for [45] a unique encoding scheme was used in weight storage to minimize error—as shown in Fig. 9.6, by co-optimizing the workload, substantial errors in the multi-bit storage can be tolerated and the benefits of increased storage capacity (larger models with improved inference accuracy) can be realized.

Similar cross-layer optimizations have been developed to manage the endurance challenges of RRAM, e.g., ENDURER—an endurance resilience technique utilizing address remapping and write redistributions with small SRAM buffers (Fig. 9.7). The SRAM write-back buffer reduces wear by filtering frequently written addresses in memory, while the random address remapping simultaneously distributes wear to all words in the memory at the word level. Such a combination is proven to bound (with high probability) the number of writes to any word in the memory [1]. Simulation results [31] across a wide range of deep neural network applications indicate that such a method can guarantee years of operating resiliency. Moreover, flash-inspired endurance resilience methods are insufficient, providing less than a year of operating lifetime [31]. Hardware results running continuous machine learning inference on an RRAM-based microcontroller [10] demonstrated 10-year lifetime through measurement. Finally, the unique physics and characteristics, such as inherent stochasticity, analog programmability, and ultra-dense (monolithic) connectivity, can be collectively exploited with circuit, architecture, and application

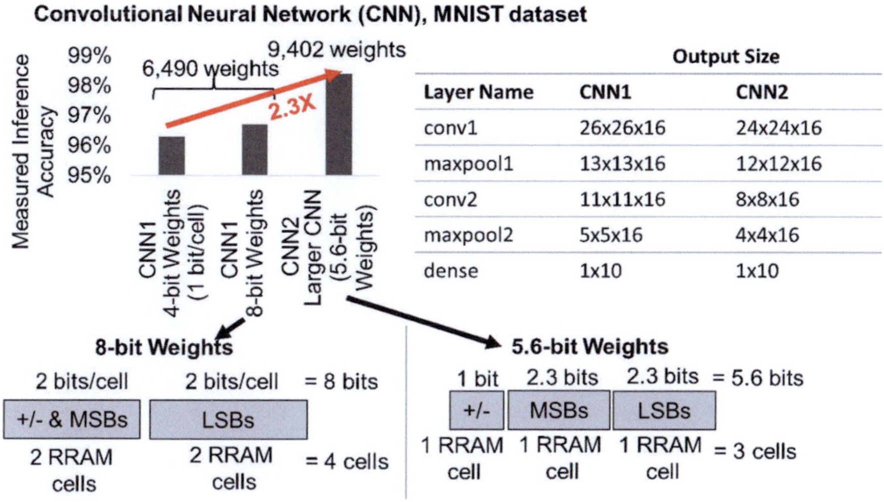

	Output Size	
Layer Name	CNN1	CNN2
conv1	26x26x16	24x24x16
maxpool1	13x13x16	12x12x16
conv2	11x11x16	8x8x16
maxpool2	5x5x16	4x4x16
dense	1x10	1x10

Fig. 9.6 Co-optimizing both network weight encoding and network size achieved the biggest increase in inference accuracy while maintaining model size within the same number of RRAM cells [10]. ©2019 IEEE

layers to produce computation results natively inside memories [46]. Together with the aforementioned cross-layer solutions, when operations are properly orchestrated, the massive memories on chip can be further exploited with less data movement and "denser" functionalities for additional energy and area efficiency benefits.

9.4 CNFETs as Logic for Monolithic 3D Integration

The CNFET (Fig. 9.8a) is an essential component to enabling monolithic 3D integration. Although the CNT growth itself requires high temperature (865 °C), the transfer of CNTs to any layer or substrate separates the temperature requirement for the material growth and the transistor fabrication, allowing CNFETs to be processed at very low temperatures (<200 °C) [49].

CNFETs are projected to offer an order of magnitude improvement compared to SiFETs in the Energy Delay Product (EDP) at the processor scale, as shown in Fig. 9.8b [48]. One key advantage of CNFETs comes from CNTs' inherently high mobility (>2,500 cm^2/V s) [50] and injection velocity (4.1×10^7 cm/s) [51] even at very thin T_{BODY} (1–2 nm). This results in a superior drive strength that enables CNFETs to reach higher effective current at a reduced supply voltage, compared to SiFET [52, 53]. The CNFET short electrostatic scale length, resulting from the very small T_{BODY}, allows to scale the gate length (L_G) of the devices preserving steep sub-threshold slope [52] and hence low leakage current. This, coupled with the low parasitic capacitances of a planar geometry, enables even further benefits, given the

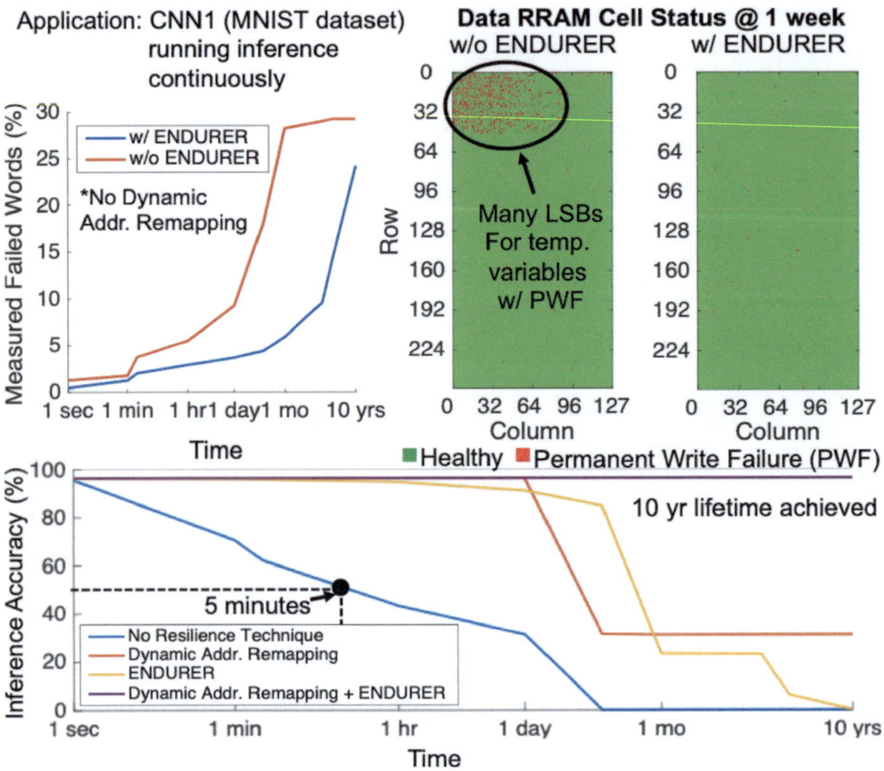

Fig. 9.7 Measurement results for ENDURER. Without ENDURER, RRAM writes are centralized to few cells, which fail quickly (top), rapidly degrading neural network inference performance (bottom). With ENDURER, writes are distributed to the whole array, reducing failures and improving system lifetime to 10 years [10]. ©2019 IEEE

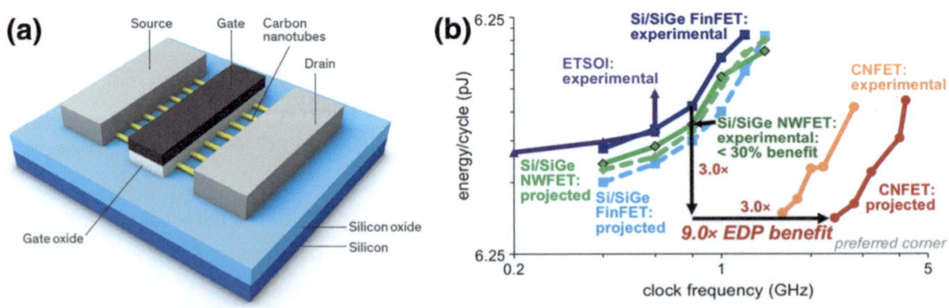

Fig. 9.8 **a** Schematic of a CNFET [47]. CNTs are used as channel material in place of silicon. **b** Projected CNFET offers 9.0× EDP benefit versus experimental Si/SiGe FinFET for the same I_{OFF} density (100 nA/μm) and power density (~65 W/cm^2); experimental Si/SiGe NWFET offers <30% EDP benefit [48]

reduction of the total circuit capacitance. CNFETs can hence operate at a 3.0× higher clock frequency dissipating 3.0× less energy per clock cycle, giving a total of 9.0× EDP benefit at the processor level.

Despite major projected benefits, CNFETs have been characterized by device imperfections and variations, such as mispositioned CNTs, metallic CNTs (m-CNTs), and variability in the CNT count, that impose major hurdles to practical implementations of CNFET-based VLSI logic circuits. In the following section, we will review the process and design techniques that have been developed to overcome these challenges, making CNFETs a viable alternative to replace and outperform SiFETs.

9.4.1 Misaligned and Mispositioned CNTs [54]

One of the main challenges in CNT manufacturing is the accurate placement and positioning of all the CNTs at VLSI scale. In particular, both misaligned CNTs (CNTs that deviate from the crystal orientation of the single crystal quartz substrate during the growth process) and mispositioned CNTs (CNTs that lie outside the gate region) may cause incorrect logic functionality. Figure 9.9 shows an example of a misaligned CNT causing incorrect functionality. To address the challenges of imperfections of CNT synthesis, algorithms to determine vulnerability to—and to implement CNFET logic circuits immune to—CNT imperfections have been developed. CNFET logic circuits have been designed and experimentally demonstrated to implement correct logic function even in the presence of large numbers of misaligned and mispositioned CNTs without any die-specific customization.

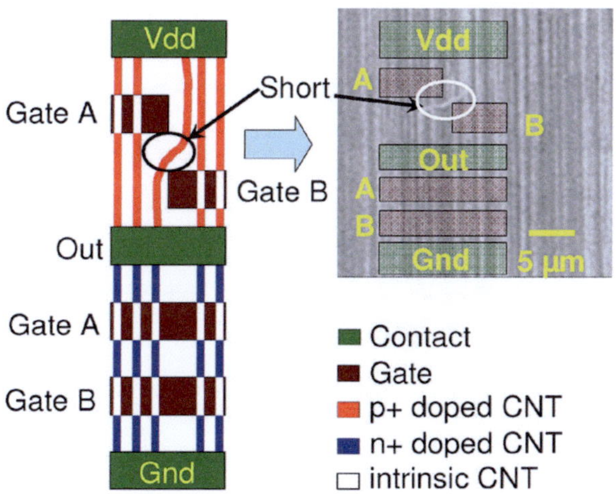

Fig. 9.9 Misaligned CNT causing short in NAND logic gate [55]. ©2008 IEEE

9.4.2 Density Enhancement of CNTs for Faster Speed

Producing wafer-scale high CNT density is crucial as it improves the speed of the CNFET, which corresponds to the current-drive per unit layout width (I_{ON}). To improve upon the transfer procedure [49], a multiple transfer technique called Controlled IDC Density Enhancement by Repeated transfers (CIDER) has been developed. Density above 100 CNTs/μm has been achieved, yielding CNFETs with high current-drive (>100 μA/μm at 400 nm channel length and 1 V V_{DS}) and high on-off ratio (>5,000) [56]. CIDER can be repeatedly implemented to increase the CNT density to any arbitrary density value required for a given application.

9.4.3 Removal of Metallic CNTs

Current CNT growth process yields approximately 1/3 metallic CNTs (m-CNTs) and 2/3 semiconducting CNTs (s-CNTs) [57], and there are no known CNT synthesis techniques that exclusively grow s-CNTs as of today. The presence of m-CNTs results in high CNFET off-state leakage current (I_{OFF}), leading to degraded noise margins and incorrect functionalities. A VLSI-compatible m-CNT Removal (VMR) technique [58] is performed by applying a high voltage across inter-digitated VMR electrodes, while applying a back-gate bias to turn off the s-CNTs (Fig. 9.10). As the s-CNTs are turned off, large current can only flow through and electrically break down the m-CNTs. Improving upon VMR, a Scalable m-CNT Removal (SMR) [59] technique has been demonstrated with high selectivity (≥99.99% of m-CNT removal with ≤1% of s-CNT removal) and high scalability (applicable to any arbitrary CPP).

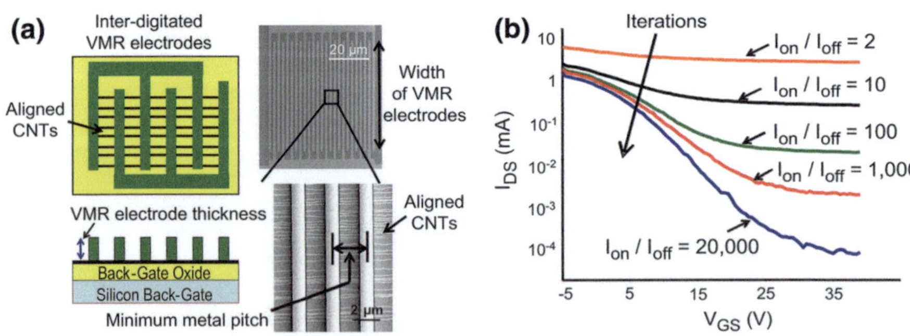

Fig. 9.10 **a** Schematic (left) and SEM images (right) of VMR electrodes with CNTs. **b** Increase in I_{ON}/I_{OFF} after several iterations of VMR [58]. ©2009 IEEE

9.4.4 DREAM: Designing Resiliency Against m-CNTs

The requirements on s-CNT purity to ensure low leakage and acceptable noise immunity in VLSI digital circuits can be further reduced by a factor of 10,000 leveraging DREAM, a wafer-scale and VLSI compatible design technique that does not involve any additional process steps [60]. At the logical synthesis, the DREAM technique maps the original circuit to another one with the same logical functionality but avoiding the logic gates pairs in which the noise margin is below a certain acceptance threshold. Fully functional circuits can thus be manufactured with only a 99.99% s-CNT purity requirement (a purity level already available today) and with only minimal decrease in energy efficiency and increase in overall area of the realized circuit.

9.4.5 System-Level Experimental Demonstrations

In addition to the imperfection-immune system-design techniques, significant progress has been made in device-level challenges. These include solution-based CNT purification [61–63], hysteresis reduction [64], control of doping [65–68], contact resistance studies [69–71], investigation of threshold voltage variation [72, 73], etc.

Due to the combined efforts from system designers and device engineers, CNT technology is the first and only nanotechnology to demonstrate large-scale applications and enable novel architectures. Recently, a 16-bit microprocessor, RV16X-NANO built entirely with CNFETs has been demonstrated [60]. RV16X-NANO comprises more than ten thousand CMOS CNFETs and leverages unique process and design techniques that are able to overcome the defects and variability issues inherent to CNFET technology. Most importantly, the microprocessor is fully compatible with commercial silicon CMOS manufacturing and is designed with standard EDA tools. CNFETs have also been successfully integrated to realize 1 kbit 6 transistors static random-access memory (SRAM) arrays [74] that leverage the CNFET low temperature fabrication process to enable small area SRAM, e.g., placing the cell on top of a logic layer. New design techniques, exploiting emerging memories, such as RRAM, have been developed to realize fully functional analog circuits [75], where the constraints on s-CNT purity are more stringent and new techniques are hence necessary. Lastly, monolithic 3D integration of CNFET logic devices and memory has already been demonstrated in practice many times. One such demonstration is the 3D nanosystems shown in Fig. 9.11. It consists of four monolithically integrated vertical layers, connected through dense vertical inter-connects. From top to bottom: CNFET sensors and logic (including more than one million CNFET inverters, which operate as gas sensors), 1 Mbit RRAM, CNFET logic (the CNFET row decoders and CNFET classification accelerator), and SiFET logic [7]. Another demonstration is a brain-inspired hyperdimensional computing nanosystem entirely realized with

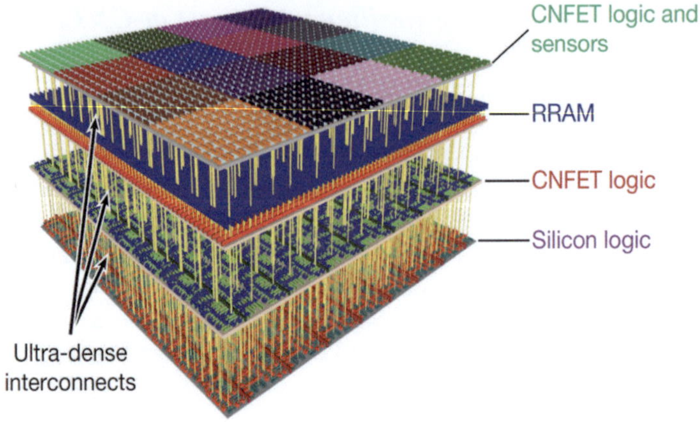

Fig. 9.11 Illustration of a nanosystem: 4 vertically stacked layers of logic, memory and sensors, comprising more than two million CNFETs and more than one million memory cells and fully compatible with silicon technology [7]. ©2017 Springer Nature

RRAM and CNFETs [76], able to perform language recognition with an accuracy of 98%.

9.5 The Thermal Challenge

Heat dissipation remains a bottleneck even for high-performance single-layer CPUs [77], which produce power on the order of 100 W/cm^2. In a 3D integrated stack, this footprint power density scales with the number of layers. Furthermore, each layer is separated from bottom-mounted heatsinks by the preceding layers, adding vertical thermal resistance that pushes junction temperatures still higher.

Monolithic 3D architectures face an additional thermal challenge in their unique form factor which impedes the horizontal heat dissipation [78] critical for quashing hotspots that decrease device lifetimes. Furthermore, layers of SiO_2 thinner than 100 nm have thermal resistivity twice that of bulk layers [79].

Current monolithic 3D implementations alleviate this problem by stacking power-intensive computation layers at the bottom of the device, leaving higher layers for low-power memory [1]. Although this decreases vertical thermal resistance to the chip floor for each computation layer, it results in increased latency between data in memory and the computation layers that need it. A more efficient structure intersperses layers of processing and memory, despite the thermal cost [80]. While CNFETs and RRAM reduce the total power dissipation (being more efficient technologies than Si), the thermal limits of such structures are still unknown.

Several avenues remain ripe for exploration to widen the thermal bottleneck. One approach is to increase horizontal thermal conductivity, reducing hotspots, and vertical conductivity, allowing external heatsinks to better cool every layer in the stack.

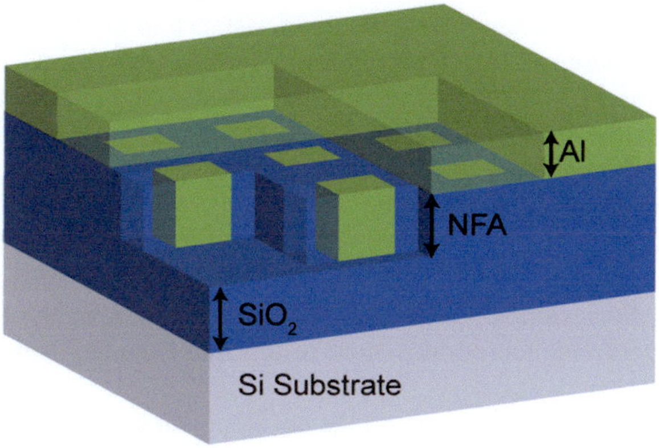

Fig. 9.12 A sample test structure to probe aggregate interlayer via conductivity [81]. ©2017 Taylor & Francis

For example, the uniquely high interlayer via densities promised by monolithic 3D ICs could achieve additional vertical conductivity. Park et al. [81] show this experimentally, finding that denser, smaller vias with high aspect ratio increase vertical conductivity even when metal density is held constant (Fig. 9.12). Continued efforts to better understand the thermal effects of vias and interconnects on the system level [82, 83] will further designers' ability to and achieve optimal thermal structures.

Modifying the materials that make up the system could also result in increased thermal conductivity. For instance, conventional oxide-based interlayer dielectrics (ILDs) that separate layers of compute and memory have low thermal conductivity [84]. Instead, they are designed for low dielectric response, preserving interconnect bandwidth and reducing overall parasitic capacitance. They must also resist exposure to water and heat (up to 400 °C), which are common process components, as well as exhibit a high Young's modulus for mechanical stability. Any candidate for a replacement ILD must maintain these advantages as well as increase thermal conductivity. Dielectric response and thermal conductivity are known to be strongly correlated [85], but creative solutions such as heterogeneous material construction, porous layers, or 2D materials [86] could decouple these parameters.

More ambitiously, active cooling that scales with height, and not just footprint like mounted heatsinks, could have a place in future monolithic 3D devices. Interlayer microfluidic channels meet this requirement: each added layer adds space for more channels. Furthermore, encapsulated phase-change materials offer potential for dense thermal energy storage [87], further improving the thermal capacity of each channel. Unfortunately, current microfluidic channels, even under ideal conditions, can only reach pitches of 4 μm [88]. This is an order of magnitude larger than monolithic 3D layers, posing two problems. First, interlayer vias would need to traverse additional distance to be routed properly around the channels, increasing aspect ratios and reducing their density. These routing issues could effectively eliminate the bandwidth gains from monolithic 3D integration. Second, the material that surrounds the microchannels would add resistance in the thermal path to either a channel or the bottom-mounted heatsink.

Testing any of these methods in transient simulation remains a challenge for monolithic 3D as well. They require high resolution (to capture the effect of dense via structures and component-level activity), large area (to generalize to the system level), and speed (for designers to iteratively test and adjust their layouts). For speedy, large-area simulation, analytical models have been devised, as in [89, 90]. More accuracy can be gained, however, by incorporating higher-precision models into these analytics. For example, Wei et al. [82] perform FEA analysis on individual interconnect layers and incorporates those results into the analytical model given by Kemper et al. [91].

There has been significant focus on the development and use of thermal resistor networks for modeling monolithic 3D systems and cooling solutions for the same. These networks seem to provide a good compromise between resolution, area, and speed. HotSpot, one such network solver, has added easy modeling of large vias [92], although monolithic 3D ILVs are too small to accurately describe and model in its framework. A more recent network solver, 3D-ICE, was designed to model the thermal effects of microfluidic channels on 3D systems [93]. Generation and analysis of thermal resistor networks that consider individual interconnects, vias, and components remains an open problem.

Further investigation into both passive and active cooling methods will allow monolithic 3D technology to achieve tolerance for more layers of computation and the ability to intersperse them with memory. These improvements, along with scaling to smaller nodes, would increase interconnectivity and lead to monolithic 3D devices with unprecedented efficiency.

9.6 Conclusion

Deep learning, brain-inspired computing, and other abundant-data applications require radically new NanoSystems built using next-generation component technologies. The N3XT approach leverages these emerging logic and memory technologies for their individual system benefits as well as their amenability to ultra-dense 3D integration. One first-generation N3XT architecture overcomes the 'memory wall' bottleneck with increased memory capacity and dramatically improved memory-compute bandwidth stemming from monolithic 3D integration. This implementation could yield system-level energy × execution time benefits of 1000× over 2D Si CMOS [1].

In addition to specific technologies used in our first-generation N3XT architecture (CNTs, STT-MRAM, RRAM, monolithic 3D), other technologies for realizing N3XT must also be explored. There are many avenues open to achieving still larger benefits in future generations of N3XT. So far, N3XT simulated benefits have been demonstrated using existing software designed for existing 2D systems, absent any N3XT-specific optimizations. Even greater benefits could be realized by co-optimizing the entire application-software-hardware stack for N3XT (e.g., using domain-specific languages, and appropriate workload partitioning, data placement, and scheduling [94]). Yield, reliability, and cost aspects of N3XT are addressed through technology-, circuit-, architecture- and application-level techniques [95–97].

As two-dimensional miniaturization reaches fundamental limits, realizing N3XT becomes even more critical with fully interspersed compute and memory (beyond the first-generation N3XT architecture where the bulk of computation occurs in lower tiers, close to the heatsink, and logic on the upper layers is mostly used for supporting memory accesses). Mitigation of the corresponding thermal effects requires architecture-driven thermal solutions co-optimized across technology, architecture, and software layers.

Meanwhile, N3XT technologies have already been demonstrated in commercial fabrication facilities [98]. Although further work is necessary to demonstrate the simulated 1000× benefits on an experimental system, intermediate hardware prototypes demonstrate the practicality and potential of N3XT.

References

1. M.M.S. Aly, T.F. Wu, A. Bartolo, Y.H. Malviya, W. Hwang, G. Hills, I. Markov, M. Wootters, M.M. Shulaker, H.-P. Wong, S. Mitra, The N3XT approach to energy-efficient abundant-data computing. Proc. IEEE **107**, 19–48 (2019). https://doi.org/10.1109/JPROC.2018.2882603
2. P. Wong, What will the next node offer us? in *Hot Chips* (2019). https://www.hotchips.org/hc31-keynotes-available-to-all/
3. M.M.S. Aly, M. Gao, G. Hills, C. Lee, G. Pitner, M.M. Shulaker, T.F. Wu, M. Asheghi, J. Bokor, F. Franchetti, K.E. Goodson, C. Kozyrakis, I. Markov, K. Olukotun, L. Pileggi, E. Pop, J. Rabaey, C. Ré, H.-P. Wong, S. Mitra, Energy-efficient abundant-data computing: the N3XT 1,000x. Computer **48**, 24–33 (2015). https://doi.org/10.1109/MC.2015.376
4. S. Van Huylenbroeck, M. Stucchi, Y. Li, J. Slabbekoorn, N. Tutunjyan, S. Sardo, N. Jourdan, L. Bogaerts, F. Beirnaert, G. Beyer et al., Small pitch, high aspect ratio via-last TSV module, in *2016 IEEE 66th Electronic Components and Technology Conference (ECTC)* (IEEE, 2016), pp. 43–49
5. S.-W. Kim, M. Detalle, L. Peng, P. Nolmans, N. Heylen, D. Velenis, A. Miller, G. Beyer, E. Beyne, Ultra-fine pitch 3D integration using face-to-face hybrid wafer bonding combined with a via-middle through-silicon-via process, in *2016 IEEE 66th Electronic Components and Technology Conference (ECTC)* (IEEE, 2016), pp. 1179–1185
6. P. Batude, M. Vinet, B. Previtali, C. Tabone, C. Xu, J. Mazurier, O. Weber, F. Andrieu, L. Tosti, L. Brevard, B. Sklenard, P. Coudrain, S. Bobba, H. Ben Jamaa, P.-E. Gaillardon, A. Pouydebasque, O. Thomas, C. Le Royer, J.-M. Hartmann, L. Sanchez, L. Baud, V. Carron, L. Clavelier, G. De Micheli, S. Deleonibus, O. Faynot, T. Poiroux, Advances, challenges and opportunities in 3D CMOS sequential integration, in *2011 International Electron Devices Meeting* (2011), pp. 7.3.1–7.3.4. https://doi.org/10.1109/IEDM.2011.6131506
7. M.M. Shulaker, G. Hills, R.S. Park, R.T. Howe, K. Saraswat, H.-S.P. Wong, S. Mitra, Three-dimensional integration of nanotechnologies for computing and data storage on a single chip. Nature **547**, 74–78 (2017). https://doi.org/10.1038/nature22994
8. T. Naito, T. Ishida, T. Onoduka, M. Nishigoori, T. Nakayama, Y. Ueno, Y. Ishimoto, A. Suzuki, W. Chung, R. Madurawe, S. Wu, S. Ikeda, H. Oyamatsu, World's first monolithic 3D-FPGA

with TFT SRAM over 90 nm 9 layer Cu CMOS, in *2010 Symposium on VLSI Technology* (2010), pp. 219–220. https://doi.org/10.1109/VLSIT.2010.5556234
9. L. Brunet, P. Batude, C. Fenouillet-Beranger, P. Besombes, L. Hortemel, F. Ponthenier, B. Previtali, C. Tabone, A. Royer, C. Agraffeil, C. Euvrard-Colnat, A. Seignard, C. Morales, F. Fournel, L. Benaissa, T. Signamarcheix, P. Besson, M. Jourdan, R. Kachtouli, V. Benevent, J.-M. Hartmann, C. Comboroure, N. Allouti, N. Posseme, C. Vizioz, C. Arvet, S. Barnola, S. Kerdiles, L. Baud, L. Pasini, C.-M.V. Lu, F. Deprat, A. Toffoli, G. Romano, C. Guedj, V. Delaye, F. Boeuf, O. Faynot, M. Vinet, First demonstration of a CMOS over CMOS 3D VLSI CoolCube™ integration on 300 mm wafers, in *2016 IEEE Symposium on VLSI Technology* (2016), pp. 1–2. https://doi.org/10.1109/VLSIT.2016.7573428
10. T.F. Wu, B.Q. Le, R. Radway, A. Bartolo, W. Hwang, S. Jeong, H. Li, P. Tandon, E. Vianello, P. Vivet, E. Nowak, M.K. Wootters, H.-P. Wong, M.M.S. Aly, E. Beigne, S. Mitra, 14.3 A 43pJ/cycle non-volatile microcontroller with 4.7 μs shutdown/wake-up integrating 2.3-bit/cell resistive RAM and resilience techniques, in *2019 IEEE International Solid-State Circuits Conference (ISSCC)* (2019), pp. 226–228. https://doi.org/10.1109/ISSCC.2019.8662402
11. H.-S.P. Wong, S. Salahuddin, Memory leads the way to better computing. Nat. Nanotechnol. **10**, 191 (2015)
12. H. Wong, C. Ahn, J. Cao, H. Chen, S. Fong, Z. Jiang, C. Neumann, S. Qin, J. Sohn, Y. Wu, X. Zheng, Stanford memory trends (2019). https://nano.stanford.edu/stanford-memory-trends/. Accessed 9 Sept 2019
13. C.-C. Chou, Z.-J. Lin, P.-L. Tseng, C.-F. Li, C.-Y. Chang, W.-C. Chen, Y.-D. Chih, T.-Y.J. Chang, An N40 256Kx 44 embedded RRAM macro with SL-precharge SA and low-voltage current limiter to improve read and write performance, in *2018 IEEE International Solid-State Circuits Conference (ISSCC)* (IEEE, 2018), pp. 478–480
14. P. Jain, U. Arslan, M. Sekhar, B.C. Lin, L. Wei, T. Sahu, J. Alzate-vinasco, A. Vangapaty, M. Meterelliyoz, N. Strutt et al., 13.2 A 3.6 Mb 10.1 Mb/mm^2 embedded non-volatile ReRAM macro in 22 nm FinFET technology with adaptive forming/set/reset schemes yielding down to 0.5 V with sensing time of 5 ns at 0.7 V, in *2019 IEEE International Solid-State Circuits Conference (ISSCC)* (IEEE, 2019), pp. 212–214
15. Y. Wu, H. Yi, Z. Zhang, Z. Jiang, J. Sohn, S. Wong, H.-S.P. Wong, First demonstration of RRAM patterned by block copolymer self-assembly, in *2013 IEEE International Electron Devices Meeting* (IEEE, 2013), pp. 20–28
16. B. Govoreanu, G. Kar, Y. Chen, V. Paraschiv, S. Kubicek, A. Fantini, I. Radu, L. Goux, S. Clima, R. Degraeve et al., 10×10 nm^2 Hf/HfO$_x$ crossbar resistive RAM with excellent performance, reliability and low-energy operation, in *2011 International Electron Devices Meeting* (IEEE, 2011), pp. 31–36
17. K.-S. Li, C. Ho, M.-T. Lee, M.-C. Chen, C.-L. Hsu, J. Lu, C. Lin, C. Chen, B. Wu, Y. Hou et al., Utilizing sub-5 nm sidewall electrode technology for atomic-scale resistive memory fabrication, in *2014 Symposium on VLSI Technology (VLSI-Technology): Digest of Technical Papers* (IEEE, 2014), pp. 1–2
18. R. Fackenthal, M. Kitagawa, W. Otsuka, K. Prall, D. Mills, K. Tsutsui, J. Javanifard, K. Tedrow, T. Tsushima, Y. Shibahara et al., 19.7 A 16 Gb ReRAM with 200 MB/s write and 1 GB/s read in 27 nm technology, in *2014 IEEE International Solid-State Circuits Conference Digest of Technical Papers (ISSCC)* (IEEE, 2014), pp. 338–339
19. T. Liu, T.H. Yan, R. Scheuerlein, Y. Chen, J.K. Lee, G. Balakrishnan, G. Yee, H. Zhang, A. Yap, J. Ouyang, T. Sasaki, S. Addepalli, A. Al-Shamma, C. Chen, M. Gupta, G. Hilton, S. Joshi, A. Kathuria, V. Lai, D. Masiwal, M. Matsumoto, A. Nigam, A. Pai, J. Pakhale, C.H. Siau, X. Wu, R. Yin, L. Peng, J.Y. Kang, S. Huynh, H. Wang, N. Nagel, Y. Tanaka, M. Higashitani, T. Minvielle, C. Gorla, T. Tsukamoto, T. Yamaguchi, M. Okajima, T. Okamura, S. Takase, T. Hara, H. Inoue, L. Fasoli, M. Mofidi, R. Shrivastava, K. Quader, A 130.7mm^2 2-layer 32 Gb ReRAM memory device in 24 nm technology, in *2013 IEEE International Solid-State Circuits Conference Digest of Technical Papers* (2013), pp. 210–211. https://doi.org/10.1109/ISSCC.2013.6487703

20. H.-Y. Chen, S. Yu, B. Gao, P. Huang, J. Kang, H.-S.P. Wong, HfO_x based vertical resistive random access memory for cost-effective 3D cross-point architecture without cell selector, in *2012 International Electron Devices Meeting* (IEEE, 2012), pp. 20–27
21. F.-K. Hsueh, C.-H. Shen, J.-M. Shieh, K.-S. Li, H.-C. Chen, W.-H. Huang, H.-H. Wang, C.-C. Yang, T.-Y. Hsieh, C.-H. Lin et al., First fully functionalized monolithic 3D + IoT chip with 0.5 V light-electricity power management, 6.8 GHz wireless-communication VCO, and 4-layer vertical ReRAM, in *2016 IEEE International Electron Devices Meeting (IEDM)* (IEEE, 2016), pp. 2–3
22. H. Li, K.-S. Li, C.-H. Lin, J.-L. Hsu, W.-C. Chiu, M.-C. Chen, T.-T. Wu, J. Sohn, S.B. Eryilmaz, J.-M. Shieh et al., Four-layer 3D vertical RRAM integrated with FinFET as a versatile computing unit for brain-inspired cognitive information processing, in *2016 IEEE Symposium on VLSI Technology* (IEEE, 2016), pp. 1–2
23. J.J. Nowak, R.P. Robertazzi, J.Z. Sun, G. Hu, J.-H. Park, J. Lee, A.J. Annunziata, G.P. Lauer, R. Kothandaraman, E.J. O'Sullivan, others, Dependence of voltage and size on write error rates in spin-transfer torque magnetic random-access memory. IEEE Magn. Lett. **7**, 1–4 (2016)
24. H. Noguchi, K. Ikegami, S. Takaya, E. Arima, K. Kushida, A. Kawasumi, H. Hara, K. Abe, N. Shimomura, J. Ito et al., 7.2 4 MB STT-MRAM-based cache with memory-access-aware power optimization and write-verify-write/read-modify-write scheme, in *2016 IEEE International Solid-State Circuits Conference (ISSCC)* (IEEE, 2016), pp. 132–133
25. L. Wei, J.G. Alzate, U. Arslan, J. Brockman, N. Das, K. Fischer, T. Ghani, O. Golonzka, P. Hentges, R. Jahan et al., 13.3 A 7 Mb STT-MRAM in 22FFL FinFET technology with 4 ns read sensing time at 0.9 V using write-verify-write scheme and offset-cancellation sensing technique, in *2019 IEEE International Solid-State Circuits Conference (ISSCC)* (IEEE, 2019), pp. 214–216
26. C. Villa, D. Mills, G. Barkley, H. Giduturi, S. Schippers, D. Vimercati, A 45 nm 1 Gb 1.8 V phase-change memory, in *2010 IEEE International Solid-State Circuits Conference (ISSCC)* (IEEE, 2010), pp. 270–271
27. Y. Choi, I. Song, M.-H. Park, H. Chung, S. Chang, B. Cho, J. Kim, Y. Oh, D. Kwon, J. Sunwoo et al., A 20 nm 1.8 V 8 Gb PRAM with 40 MB/s program bandwidth, in *2012 IEEE International Solid-State Circuits Conference* (IEEE, 2012), pp. 46–48
28. S. Dünkel, M. Trentzsch, R. Richter, P. Moll, C. Fuchs, O. Gehring, M. Majer, S. Wittek, B. Müller, T. Melde, H. Mulaosmanovic, S. Slesazeck, S. Müller, J. Ocker, M. Noack, D. Löhr, P. Polakowski, J. Müller, T. Mikolajick, J. Höntschel, B. Rice, J. Pellerin, S. Beyer, A FeFET based super-low-power ultra-fast embedded NVM technology for 22 nm FDSOI and beyond, in *2017 IEEE International Electron Devices Meeting (IEDM)* (2017), pp. 19.7.1–19.7.4. https://doi.org/10.1109/IEDM.2017.8268425
29. Y.-B. Kim, S.R. Lee, D. Lee, C.B. Lee, M. Chang, J.H. Hur, M.-J. Lee, G.-S. Park, C.J. Kim, U.-I. Chung et al., Bi-layered RRAM with unlimited endurance and extremely uniform switching, in *2011 Symposium on VLSI Technology-Digest of Technical Papers* (IEEE, 2011), pp. 52–53
30. C.-W. Hsu, I.-T. Wang, C.-L. Lo, M.-C. Chiang, W.-Y. Jang, C.-H. Lin, T.-H. Hou, Self-rectifying bipolar TaO_x/TiO_2 RRAM with superior endurance over 10^{12} cycles for 3D high-density storage-class memory, in *2013 Symposium on VLSI Technology* (IEEE, 2013), pp. T166–T167
31. A. Grossi, E. Vianello, M.M. Sabry, M. Barlas, L. Grenouillet, J. Coignus, E. Beigne, T. Wu, B.Q. Le, M.K. Wootters, C. Zambelli, E. Nowak, S. Mitra, Resistive RAM endurance: array-level characterization and correction techniques targeting deep learning applications. IEEE Trans. Electron Device **66**, 1281–1288 (2019). https://doi.org/10.1109/TED.2019.2894387
32. J. Park, S. Kim, J. Baek, D. Seo, J. Chun, K. Kwon, Analysis of resistance variations and variance-aware read circuit for cross-point ReRAM, in *2013 5th IEEE International Memory Workshop* (2013), pp. 112–115. https://doi.org/10.1109/IMW.2013.6582111
33. M. Chang, S. Sheu, K. Lin, C. Wu, C. Kuo, P. Chiu, Y. Yang, Y. Chen, H. Lee, C. Lien, F.T. Chen, K. Su, T. Ku, M. Kao, M. Tsai, A high-speed 7.2-ns read-write random access 4-Mb embedded resistive RAM (ReRAM) macro using process-variation-tolerant current-mode read

schemes. IEEE J. Solid-State Circuits **48**, 878–891 (2013). https://doi.org/10.1109/JSSC.2012.2230515
34. S. Sheu, K. Cheng, M. Chang, P. Chiang, W. Lin, H. Lee, P. Chen, Y. Chen, T. Wu, F.T. Chen, K. Su, M. Kao, M. Tsai, Fast-write resistive RAM (RRAM) for embedded applications. IEEE Des. Test Comput. **28**, 64–71 (2011). https://doi.org/10.1109/MDT.2010.96
35. A. Chen, M. Lin, Variability of resistive switching memories and its impact on crossbar array performance, in *2011 International Reliability Physics Symposium* (2011), pp. MY.7.1–MY.7.4. https://doi.org/10.1109/IRPS.2011.5784590
36. S. Hamdioui, P. Pouyan, H. Li, Y. Wang, A. Raychowdhur, I. Yoon, Test and reliability of emerging non-volatile memories, in *2017 IEEE 26th Asian Test Symposium (ATS)* (2017), pp. 175–183. https://doi.org/10.1109/ATS.2017.42
37. P. Pouyan, E. Amat, S. Hamdioui, A. Rubio, RRAM variability and its mitigation schemes, in *2016 26th International Workshop on Power and Timing Modeling, Optimization and Simulation (PATMOS)* (2016), pp. 141–146. https://doi.org/10.1109/PATMOS.2016.7833679
38. A. Fantini, L. Goux, R. Degraeve, D.J. Wouters, N. Raghavan, G. Kar, A. Belmonte, Y. Chen, B. Govoreanu, M. Jurczak, Intrinsic switching variability in HfO_2 RRAM, in *2013 5th IEEE International Memory Workshop* (2013), pp. 30–33. https://doi.org/10.1109/IMW.2013.6582090
39. A. Grossi, E. Nowak, C. Zambelli, C. Pellissier, S. Bernasconi, G. Cibrario, K.E. Hajjam, R. Crochemore, J.F. Nodin, P. Olivo, L. Perniola, Fundamental variability limits of filament-based RRAM, in *2016 IEEE International Electron Devices Meeting (IEDM)* (2016), pp. 4.7.1–4.7.4. https://doi.org/10.1109/IEDM.2016.7838348
40. W. Chien, M. Lee, F.-M. Lee, Y. Lin, H.-L. Lung, K. Hsieh, C.-Y. Lu, Multi-level 40 nm WOX resistive memory with excellent reliability, in *2011 International Electron Devices Meeting* (2011), pp. 31.5.1–31.5.4. https://doi.org/10.1109/IEDM.2011.6131651
41. A. Prakash, J. Park, J. Song, J. Woo, E. Cha, H. Hwang, Demonstration of low power 3-bit multilevel cell characteristics in a TaO_x-based RRAM by stack engineering. IEEE Electron Device Lett. **36**, 32–34 (2015). https://doi.org/10.1109/LED.2014.2375200
42. S. Stathopoulos, A. Khiat, M. Trapatseli, S. Cortese, A. Serb, I. Valov, T. Prodromakis, Multibit memory operation of metal-oxide bi-layer memristors. Sci. Rep. **7**, 1–7 (2017). https://doi.org/10.1038/s41598-017-17785-1
43. S. Sheu, P. Chiang, W. Lin, H. Lee, P. Chen, Y. Chen, T. Wu, F.T. Chen, K. Su, M. Kao, K. Cheng, M. Tsai, A 5 ns fast write multi-level non-volatile 1 K bits RRAM memory with advance write scheme, in *2009 Symposium on VLSI Circuits* (2009), pp. 82–83
44. B.Q. Le, A. Grossi, E. Vianello, T. Wu, G. Lama, E. Beigne, H.-S.P. Wong, S. Mitra, Resistive RAM with multiple bits per cell: array-level demonstration of 3 bits per cell. IEEE Trans. Electron Device **66**, 641–646 (2018)
45. E.R. Hsieh et al., High-density multiple bits-per-cell 1T4R RRAM array with gradual SET/RESET and its effectiveness for deep learning, in *IEEE International Electron Devices Meeting (IEDM)* (2019)
46. D. Ielmini, H.-S.P. Wong, In-memory computing with resistive switching devices. Nat. Electron. **1**, 333–343 (2018). https://doi.org/10.1038/s41928-018-0092-2
47. M. Shulaker, H.-P. Wong, S. Mitra, Computing with carbon nanotubes. IEEE Spectr. **53**, 26–52 (2016). https://doi.org/10.1109/MSPEC.2016.7498155
48. G. Hills, M.G. Bardon, G. Doornbos, D. Yakimets, P. Schuddinck, R. Baert, D. Jang, L. Mattii, S.M.Y. Sherazi, D. Rodopoulos, R. Ritzenthaler, C. Lee, A.V. Thean, I. Radu, A. Spessot, P. Debacker, F. Catthoor, P. Raghavan, M.M. Shulaker, H.-P. Wong, S. Mitra, Understanding energy efficiency benefits of carbon nanotube field-effect transistors for digital VLSI. IEEE Trans. Nanotechnol. **17**, 1259–1269 (2018). https://doi.org/10.1109/TNANO.2018.2871841
49. N. Patil, A. Lin, E.R. Myers, K. Ryu, A. Badmaev, C. Zhou, H.-P. Wong, S. Mitra, Wafer-scale growth and transfer of aligned single-walled carbon nanotubes. IEEE Trans. Nanotechnol. **8**, 498–504 (2009). https://doi.org/10.1109/TNANO.2009.2016562
50. X. Zhou, J.-Y. Park, S. Huang, J. Liu, P.L. McEuen, Band structure, phonon scattering, and the performance limit of single-walled carbon nanotube transistors. Phys. Rev. Lett. **95**, 146805 (2005). https://doi.org/10.1103/PhysRevLett.95.146805

51. C. Lee, E. Pop, A.D. Franklin, W. Haensch, H.-P. Wong, A compact virtual-source model for carbon nanotube FETs in the sub-10-nm regime—Part I: Intrinsic elements. IEEE Trans. Electron Device **62**, 3061–3069 (2015). https://doi.org/10.1109/TED.2015.2457453
52. C. Qiu, Z. Zhang, M. Xiao, Y. Yang, D. Zhong, L.-M. Peng, Scaling carbon nanotube complementary transistors to 5-nm gate lengths. Science **355**, 271–276 (2017). https://doi.org/10.1126/science.aaj1628
53. A.D. Franklin, M. Luisier, S.-J. Han, G. Tulevski, C.M. Breslin, L. Gignac, M.S. Lundstrom, W. Haensch, Sub-10 nm carbon nanotube transistor. Nano Lett. **12**, 758–762 (2012). https://doi.org/10.1021/nl203701g
54. N. Patil, A. Lin, J. Zhang, H.-P. Wong, S. Mitra, Digital VLSI logic technology using carbon nanotube FETs: frequently asked questions, in *2009 46th ACM/IEEE Design Automation Conference* (2009), pp. 304–309. https://doi.org/10.1145/1629911.1629995
55. N. Patil, J. Deng, A. Lin, H.-P. Wong, S. Mitra, Design methods for misaligned and mispositioned carbon-nanotube immune circuits. IEEE Trans. Comput. Aided Des. Integr. Circuits Syst. **27**, 1725–1736 (2008). https://doi.org/10.1109/TCAD.2008.2003278
56. M.M. Shulaker, G. Pitner, G. Hills, M. Giachino, H.-P. Wong, S. Mitra, High-performance carbon nanotube field-effect transistors, in *2014 IEEE International Electron Devices Meeting* (2014), pp. 33.6.1–33.6.4. https://doi.org/10.1109/IEDM.2014.7047164
57. M.S. Dresselhaus, R.E. Smalley, G. Dresselhaus, P. Avouris, Carbon Nanotubes: Synthesis, Structure, Properties, and Applications (Springer, Berlin, Heidelberg, 2001). https://books.google.com/books?id=dkvDhZJnafgC
58. N. Patil, A. Lin, J. Zhang, H. Wei, K. Anderson, H.-P. Wong, S. Mitra, VMR: VLSI-compatible metallic carbon nanotube removal for imperfection-immune cascaded multi-stage digital logic circuits using carbon nanotube FETs, in *2009 IEEE International Electron Devices Meeting (IEDM)* (2009), pp. 1–4. https://doi.org/10.1109/IEDM.2009.5424295
59. M.M. Shulaker, G. Hills, T.F. Wu, Z. Bao, H.-P. Wong, S. Mitra, Efficient metallic carbon nanotube removal for highly-scaled technologies, in *2015 IEEE International Electron Devices Meeting (IEDM)* (2015), pp. 32.4.1–32.4.4. https://doi.org/10.1109/IEDM.2015.7409815
60. G. Hills, C. Lau, A. Wright, S. Fuller, M.D. Bishop, T. Srimani, P. Kanhaiya, R. Ho, A. Amer, Y. Stein, D. Murphy, Arvind, A. Chandrakasan, M.M. Shulaker, Modern microprocessor built from complementary carbon nanotube transistors. Nature **572**, 595–602 (2019). https://doi.org/10.1038/s41586-019-1493-8
61. T. Lei, X. Chen, G. Pitner, H.-S.P. Wong, Z. Bao, Removable and recyclable conjugated polymers for highly selective and high-yield dispersion and release of low-cost carbon nanotubes. J. Am. Chem. Soc. **138** (2016). https://doi.org/10.1021/jacs.5b12797
62. G. Tulevski, A. Franklin, A. Afzali, High purity isolation and quantification of semiconducting carbon nanotubes via column chromatography. ACS Nano **7** (2013). https://doi.org/10.1021/nn400053k
63. Q. Cao, J. Tersoff, D. Farmer, Y. Zhu, S.-J. Han, Carbon nanotube transistors scaled to a 40-nanometer footprint. Science **356**, 1369–1372 (2017). https://doi.org/10.1126/science.aan2476
64. R.S. Park, G. Hills, J. Sohn, S. Mitra, M.M. Shulaker, H.-S.P. Wong, Hysteresis-free carbon nanotube field-effect transistors. ACS Nano **11**, 4785–4791 (2017). https://doi.org/10.1021/acsnano.7b01164
65. L. Suriyasena Liyanage, X. Xu, G. Pitner, Z. Bao, H.-S. Philip Wong, VLSI-compatible carbon nanotube doping technique with low work-function metal oxides. Nano Lett. **14** (2014). https://doi.org/10.1021/nl404654j
66. C. Lau, T. Srimani, M.D. Bishop, G. Hills, M.M. Shulaker, Tunable n-type doping of carbon nanotubes through engineered atomic layer deposition HfO_X films. ACS Nano **12** (2018). https://doi.org/10.1021/acsnano.8b04208
67. D. Shahrjerdi, A. Franklin, S. Oida, J. Ott, G. Tulevski, W. Haensch, High-performance air-stable n-type carbon nanotube transistors with erbium contacts. ACS Nano **7** (2013). https://doi.org/10.1021/nn403935v
68. J. Tang, D. Farmer, S. Bangsaruntip, K.-C. Chiu, B. Kumar, S.-J. Han, Contact engineering and channel doping for robust carbon nanotube NFETs, in *2017 International Symposium*

on VLSI Technology, Systems and Application (VLSI-TSA) (2017), pp. 1–2. https://doi.org/10.1109/VLSI-TSA.2017.7942478
69. A. Franklin, Z. Chen, Length scaling of carbon nanotube transistors. Nat. Nanotechnol. **5**, 858–862 (2010). https://doi.org/10.1038/nnano.2010.220
70. A.D. Franklin, W. Haensch, Defining and overcoming the contact resistance challenge in scaled carbon nanotube transistors, in *72nd Device Research Conference* (2014), pp. 191–192. https://doi.org/10.1109/DRC.2014.6872362
71. G. Pitner, G. Hills, J. Pablo Llinas, K.-M. Persson, R.S. Park, J. Bokor, S. Mitra, H.-S.P. Wong, Low-temperature side-contact to carbon nanotube transistors: resistance distributions down to 10 nm contact length. Nano Lett. **19** (2019). https://doi.org/10.1021/acs.nanolett.8b04370
72. A. Franklin, G. Tulevski, S.-J. Han, D. Shahrjerdi, Q. Cao, H.-Y. Chen, H.-S. Philip Wong, W. Haensch, Variability in carbon nanotube transistors: improving device-to-device consistency. ACS Nano **6**, 1109–1115 (2012). https://doi.org/10.1021/nn203516z
73. Q. Cao, S.-J. Han, A. Penumatcha, M. Frank, G. Tulevski, J. Tersoff, W.E. Haensch, The origins and characteristics of the threshold voltage variability of quasi-ballistic single-walled carbon nanotube field-effect transistors. ACS Nano **9** (2015). https://doi.org/10.1021/nn506839p
74. P.S. Kanhaiya, C. Lau, G. Hills, M. Bishop, M.M. Shulaker, 1 Kbit 6T SRAM arrays in carbon nanotube FET CMOS, in *2019 Symposium on VLSI Technology* (2019), pp. T54–T55. https://doi.org/10.23919/VLSIT.2019.8776563
75. A.G. Amer, R. Ho, G. Hills, A.P. Chandrakasan, M.M. Shulaker, 29.8 SHARC: self-healing analog with RRAM and CNFETs, in *2019 IEEE International Solid-State Circuits Conference (ISSCC)* (2019), pp. 470–472. https://doi.org/10.1109/ISSCC.2019.8662377
76. T.F. Wu, H. Li, P. Huang, A. Rahimi, J.M. Rabaey, H.-P. Wong, M.M. Shulaker, S. Mitra, Brain-inspired computing exploiting carbon nanotube FETs and resistive RAM: hyperdimensional computing case study, in *2018 IEEE International Solid-State Circuits Conference (ISSCC)* (2018), pp. 492–494. https://doi.org/10.1109/ISSCC.2018.8310399
77. M.S. Bakir, C. King, D. Sekar, H. Thacker, B. Dang, G. Huang, A. Naeemi, J.D. Meindl, 3D heterogeneous integrated systems: liquid cooling, power delivery, and implementation, in *2008 IEEE Custom Integrated Circuits Conference* (2008), pp. 663–670. https://doi.org/10.1109/CICC.2008.4672173
78. P. Shukla, A.K. Coskun, V.F. Pavlidis, E. Salman, An overview of thermal challenges and opportunities for monolithic 3D ICs, in *Proceedings of the 2019 on Great Lakes Symposium on VLSI* (ACM, New York, NY, USA, 2019), pp. 439–444. https://doi.org/10.1145/3299874.3319485
79. Y.S. Ju, K.E. Goodson, Phonon scattering in silicon films with thickness of order 100 nm. Appl. Phys. Lett. **74**, 3005–3007 (1999). https://doi.org/10.1063/1.123994
80. T.-Y. Chiang, S.J. Souri, C.O. Chui, K.C. Saraswat, Thermal analysis of heterogeneous 3D ICs with various integration scenarios, in *International Electron Devices Meeting. Technical Digest* (Cat. No. 01CH37224) (2001), pp. 31.2.1–31.2.4. https://doi.org/10.1109/IEDM.2001.979599
81. W. Park, A. Sood, J. Park, M. Asheghi, R. Sinclair, K.E. Goodson, Enhanced thermal conduction through nanostructured interfaces. Nanoscale Microscale Thermophys. Eng. **21**, 134–144 (2017). https://doi.org/10.1080/15567265.2017.1296910
82. H. Wei, T.F. Wu, D. Sekar, B. Cronquist, R.F. Pease, S. Mitra, Cooling three-dimensional integrated circuits using power delivery networks, in *2012 International Electron Devices Meeting* (2012), pp. 14.2.1–14.2.4. https://doi.org/10.1109/IEDM.2012.6479040
83. T.-Y. Chiang, K. Banerjee, K.C. Saraswat, Effect of via separation and low-k dielectric materials on the thermal characteristics of Cu interconnects, in *International Electron Devices Meeting 2000. Technical Digest*. IEDM (Cat. No. 00CH37138) (2000), pp. 261–264. https://doi.org/10.1109/IEDM.2000.904306
84. K. Banerjee, A. Amerasekera, G. Dixit, C. Hu, The effect of interconnect scaling and low-k dielectric on the thermal characteristics of the IC metal, in *International Electron Devices Meeting. Technical Digest* (1996), pp. 65–68. https://doi.org/10.1109/IEDM.1996.553123

85. M. Baklanov, M. Green, K. Maex, *Dielectric Films for Advanced Microelectronics* (Wiley, New York, 2007). https://doi.org/10.1002/9780470017944
86. S. Subrina, D. Kotchetkov, A.A. Balandin, Heat removal in silicon-on-insulator integrated circuits with graphene lateral heat spreaders. IEEE Electron Device Lett. **30**, 1281–1283 (2009). https://doi.org/10.1109/LED.2009.2034116
87. M. Fuensanta, U. Paiphansiri, M.D. Romero-Sánchez, C. Guillem, Á.M. López-Buendía, K. Landfester, Thermal properties of a novel nanoencapsulated phase change material for thermal energy storage. Thermochim. Acta **565**, 95–101 (2013). https://doi.org/10.1016/j.tca.2013.04.028
88. Z. Cao, L. Yobas, Gel-free electrophoresis of DNA and proteins on chips featuring a 70 nm capillary-well motif. ACS Nano **9**, 427–435 (2015). https://doi.org/10.1021/nn505605e
89. C. Santos, P. Vivet, G. Matter, N. Peltier, S. Kaiser, R. Reis, Thermal modeling methodology for efficient system-level thermal analysis, in *Proceedings of the IEEE 2014 Custom Integrated Circuits Conference* (2014), pp. 1–4. https://doi.org/10.1109/CICC.2014.6946045
90. S.K. Samal, S. Panth, K. Samadi, M. Saedi, Y. Du, S.K. Lim, Fast and accurate thermal modeling and optimization for monolithic 3D ICs, in *2014 51st ACM/EDAC/IEEE Design Automation Conference (DAC)* (2014), pp. 1–6. https://doi.org/10.1145/2593069.2593140
91. T. Kemper, Y. Zhang, Z. Bian, A. Shakouri, Ultrafast temperature profile calculation in IC chips. ArXiv:0709.1850 [Cond-Mat] (2007). http://arxiv.org/abs/0709.1850. Accessed 29 Jul 2019
92. J. Meng, K. Kawakami, A.K. Coskun, Optimizing energy efficiency of 3-D multicore systems with stacked DRAM under power and thermal constraints, in *DAC Design Automation Conference 2012* (2012), pp. 648–655
93. A. Sridhar, A. Vincenzi, M. Ruggiero, T. Brunschwiler, D. Atienza, 3D-ICE: fast compact transient thermal modeling for 3D ICs with inter-tier liquid cooling, in *2010 IEEE/ACM International Conference on Computer-Aided Design (ICCAD)* (2010), pp. 463–470. https://doi.org/10.1109/ICCAD.2010.5653749
94. H. Kasture, X. Ji, N. El-Sayed, N. Beckmann, X. Ma, D. Sanchez, POSTER: improving datacenter efficiency through partitioning-aware scheduling, in *2017 26th International Conference on Parallel Architectures and Compilation Techniques (PACT)* (2017), pp. 134–135. https://doi.org/10.1109/PACT.2017.43
95. G. Hills, D. Bankman, B. Moons, L. Yang, J. Hillard, A. Kahng, R. Park, M. Verhelst, B. Murmann, M.M. Shulaker, H.-S.P. Wong, S. Mitra, TRIG: hardware accelerator for inference-based applications and experimental demonstration using carbon nanotube FETs, in *2018 55th ACM/ESDA/IEEE Design Automation Conference (DAC)* (2018), pp. 1–10. https://doi.org/10.1109/DAC.2018.8465852
96. Y. Li, E. Cheng, S. Makar, S. Mitra, Self-repair of uncore components in robust system-on-chips: an OpenSPARC T2 case study, in *2013 IEEE International Test Conference (ITC)* (2013), pp. 1–10. https://doi.org/10.1109/TEST.2013.6651907
97. H. Cho, L. Leem, S. Mitra, ERSA: error resilient system architecture for probabilistic applications. IEEE Trans. Comput. Aided Des. Integr. Circuits Syst. **31**, 546–558 (2012). https://doi.org/10.1109/TCAD.2011.2179038
98. First 3D nanotube and RRAM ICs come out of foundry. IEEE Spectrum: Technology, Engineering, and Science News (n.d.). https://spectrum.ieee.org/nanoclast/semiconductors/devices/first-3d-nanotube-and-rram-ics-come-out-of-foundry. Accessed 28 Oct 2019

Chapter 10
High-Speed 3D Memories Enabling the AI Future

Zvi Or-Bach

10.1 Stacked Capacitor DRAM

For the last two decades, Stacked Capacitor DRAM has been the technology of choice for high speed (<100 ns), high endurance (>10^{12}), and low cost (<$0.5/Gb) memory. Thus far, no alternative technology has been positioned to challenge DRAM. Figure 10.1 was presented by John Hennessy in multiple events during 2018 stating: "For many years we were achieving increases of about 50 percent a year that is going up slightly faster than Moore's law. Then we began a period of slowdown and if you look at what's happened in the last seven years, this technology we were used to seeing increased the number of megabits per chip more than doubling every two years but is now going up at about 10% a year and it's going to take about seven years to double now." Capacitor based DRAM technology needs a minimum size capacitor to keep enough charge so that the refresh rate would be kept, while scaling with reduced size make it harder to keep the charge leakage under control. It is now clear that capacitor-based DRAM scaling has leveled off.

During the last decade, it was observed that the need for DRAM in computing systems is limited, while the need for storage has kept growing. Accordingly, industry analysts were expecting the NAND market will become far larger than the DRAM market by now. But the re-birth of AI technology has reversed that trend in the recent years and DRAM demand has seen a rapid growth resulting in dramatic price increases for DRAM devices (Fig. 10.2).

The diminishing effectiveness of conventional scaling, at a time of accelerating AI-driven use, presents a tough challenge for the industry. However, at nearly the same time, the NAND industry was facing a scaling challenge. But then the industry was able to change course and adopt 3D scaling (Fig. 10.3).

Z. Or-Bach (✉)
MonolithIC 3D Inc., 3555 Woodford Drive, San Jose, CA 95124, USA
e-mail: Zvi@MonolithIC3D.com

B. Murmann and B. Hoefflinger (eds.), *NANO-CHIPS 2030*,
The Frontiers Collection, https://doi.org/10.1007/978-3-030-18338-7_10

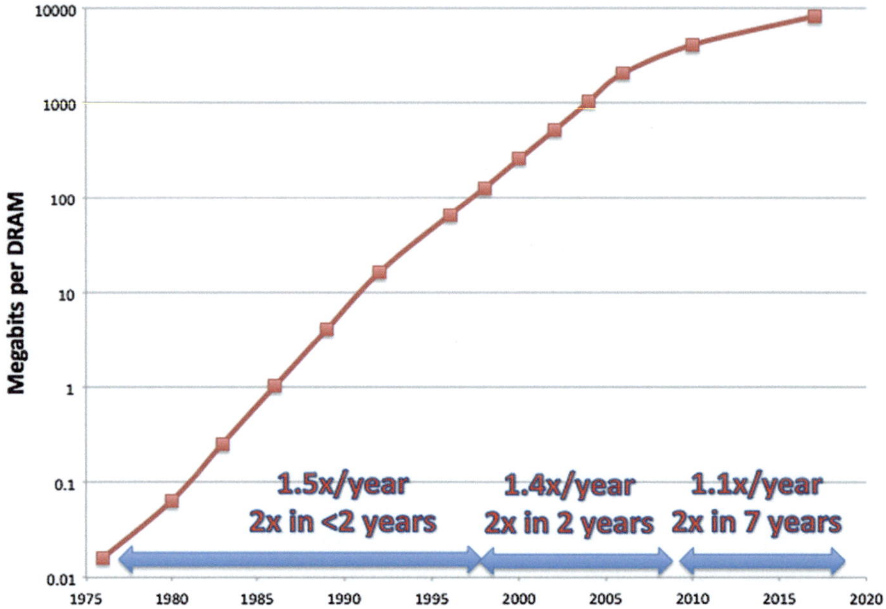

Fig. 10.1 Moore's law for DRAM—J. Hennessy 2018

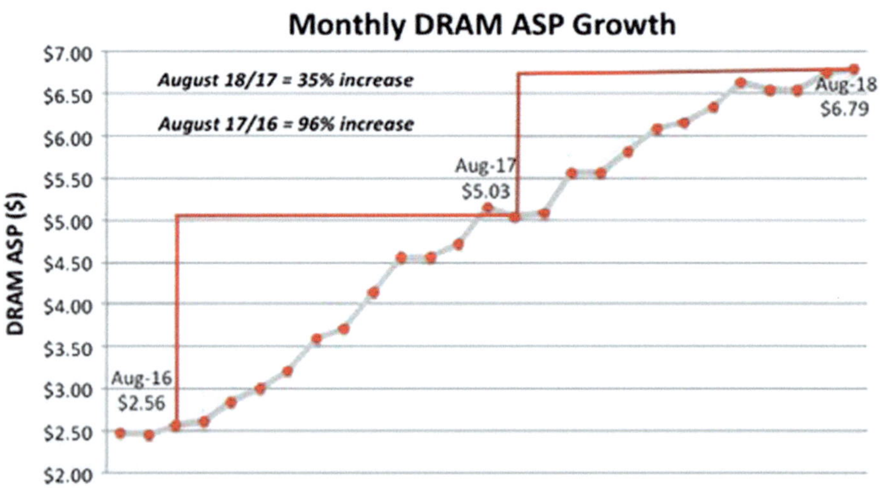

Fig. 10.2 Recent years DRAM device price appreciation

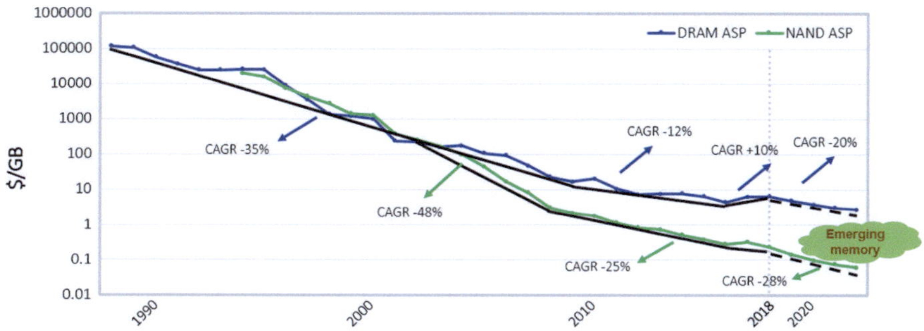

Fig. 10.3 DRAM versus flash ASP (average selling price). *Source* IDC

Early 3D NAND products used 24 layers in the 3D stack, and then the industry released 32, 64, 72, and recently 96 layers to production. This 3D-stack driven roadmap suggests continuing the 3D scaling towards a few hundred layers, thus keeping the scaling of NAND memory products to increase the memory capacity with the corresponding reduction in cost per bit.

Capacitor based DRAM would not allow such 3D scaling and no alternative has been proposed so far to do so for DRAM.

10.2 Alternative Memory Technologies

Over the past decades, a significant R&D effort has been devoted to developing alternative memory technologies. The leading alternative technologies are based on: Phase Change Materials (PCM), Resistive Memory (R-RAM) or Magnetic Memory (M-RAM). These alternative memory technologies have many variations and derivatives with other name branding as well. None of these alternative memory technologies seems to challenge the mainstream technologies—DRAM and NAND. And none of these technologies has been considered as a potential alternative to DRAM.

10.2.1 PCM—3D XPoint

Intel and Micron collaborated in releasing to the market a product named Optane™ as a Storage Class Memory (SCM) to bridge the growing gap between DRAM and 3D NAND. Also, it is considered a 3D memory as it is designed as a cross-point architecture and would not fit the low-cost 3D Scaling in which many memory layers are processed together following the same lithography step. 3D XPoint is not considered as a potential DRAM alternative due to access speed and endurance limitations.

10.2.2 R-RAM

R-RAM has been a very popular memory candidate with many alternative material and configurations. So far it seems that most of the effort is to position R-RAM as an attractive alternative for embedded non-volatile memory. R-RAM is not been proposed as a DRAM alternative mostly due to endurance limitations.

10.2.3 M-RAM

M-RAM has recently made good progress and is now being offered as a qualified non-volatile embedded memory by multiple vendors including TSMC, Samsung, and Intel. M-RAM has not been proposed as a DRAM alternative mostly due to the much larger memory cell size and concerns with the challenge of scaling to smaller technology nodes.

10.2.4 F-RAM

Ferro-Electric memory (F-RAM) is an established high-speed non-volatile specialty memory technology currently offered by Fujitsu and Cypress. It was considered a low-density memory due to the prohibitive thickness of the special Ferro-Electric materials. Recently, it was discovered that doped hafnium oxide (HfO_2) exhibits ferro-electric properties and could enable a high-density F-RAM [1, 2]. The technology was proposed to support capacitor-based DRAM or single transistor memory cells. So far, the endurance of single transistor FRAM memory cell has been at about 10^6 cycle, which is too low to be considered as a DRAM alternative.

10.3 Charge-Trap DRAM

Charge-Trap is the dominating technology for 3D NAND which is considered a slow Non-Volatile memory technology. In CT 3D NAND, the charge is trapped in a nitride layer of about 5–8 nm thick. A high-quality tunneling oxide is placed as a barrier between the trapping layer and the channel to keep the charge trapped for about 10 years. In a landmark paper by Wann and Hu in 1995 [3] it was presented that thinning the tunneling oxide to about 1 nm provides a memory with performance attractive for DRAM applications from all aspects (endurance, access time, size). So, by accepting the concept of refreshing the CT NAND memory cells, a Charge-Trap memory of thin tunneling oxide will give up a 10-year retention for few seconds of retention, but in return possess fast write times in the tens of ns and an endurance higher than 10^{12} cycles. A similar concept was published by IBM [4]. This work was confirmed and improved on by work such as Fujitsu's [5] and covered in patents filed by Macronix [6] and Micron [7]. It seems that the thin tunneling concept was proposed at a time floating gate rather than charge-trap was the Non-Volatile industry's technology of choice. Moreover, at that time DRAM scaling was in-step with the rest of the industry and a thin tunneling charge-trap did not offer enough of an advantage to be pursued by the memory industry.

In the Flash market for storage applications, NAND architecture became the industry choice as it provides a significantly higher density (lower cost) than a NOR architecture. As illustrated in Fig. 10.4 a NAND architecture with only two diffusion contacts could provide access to a long NAND string, thus reducing the effective size of a memory cell to $4F^2$ [8]. In the NOR architecture, the one diffusion contact per cell increases the cell size to $8F^2$, thus a higher memory cost. The NOR architecture does provide direct access to the selected cell which result in much faster read access time, consequently making it attractive for applications such as program code storage. An alternative architecture shown in Fig. 10.4 as AND architecture provides direct access with a better density than conventional NOR. This architecture often is also called NOR and could be attractive for 3D random access memory structures.

The success of the NAND industry with 3D NAND scaling could now be followed by adopting Charge-Trap for DRAM and changing the memory architecture from a NAND to a NOR (AND) architecture. Such a 3D architecture has been first proposed by Macronix [9], later by MonolithIC 3D Inc. with single crystal channel option [10], and then by Eli Harari [11] and his new company Sunrise Memory Corp (Eli Harari was the founder of SanDisk and won the National Medal of Technology and Innovation from President Barack Obama for his innovations and contributions to flash memory storage solutions). These proposals could be grouped into those with horizontal bit-line orientation and vertical bit-line orientation. In the following, the details of 3D NOR with a vertical bit-line orientation are presented. An important advantage of these structures is the similarity to the common 3D NAND 'Punch and Plug' process and accordingly the advantage in sharing the industry accumulated know-how and manufacturing infrastructure.

10.4 Charge-Trap 3D NOR (AND)

Just like in 3D NAND, the foundation fabric is a multi-layer fabric such as oxide layers with poly-silicon in-between. The number of poly-silicon layers is a linear relation to the number of memory cells in the 3D memory structure. And just as in 3D NAND the memory process is done for the full multi-layer fabric affecting all the levels together—hence 3D scaling (Fig. 10.5).

Figure 10.6 illustrates a side cut-view of the structure overlaying the structure transistor schematic. It represents an aggressive 3D NOR (AND) structure in which the bit-lines (B0–B4), in blue, serves as Source and Drain to cells on their right side and on their left side. These bit lines could be formed by filling the punch holes with

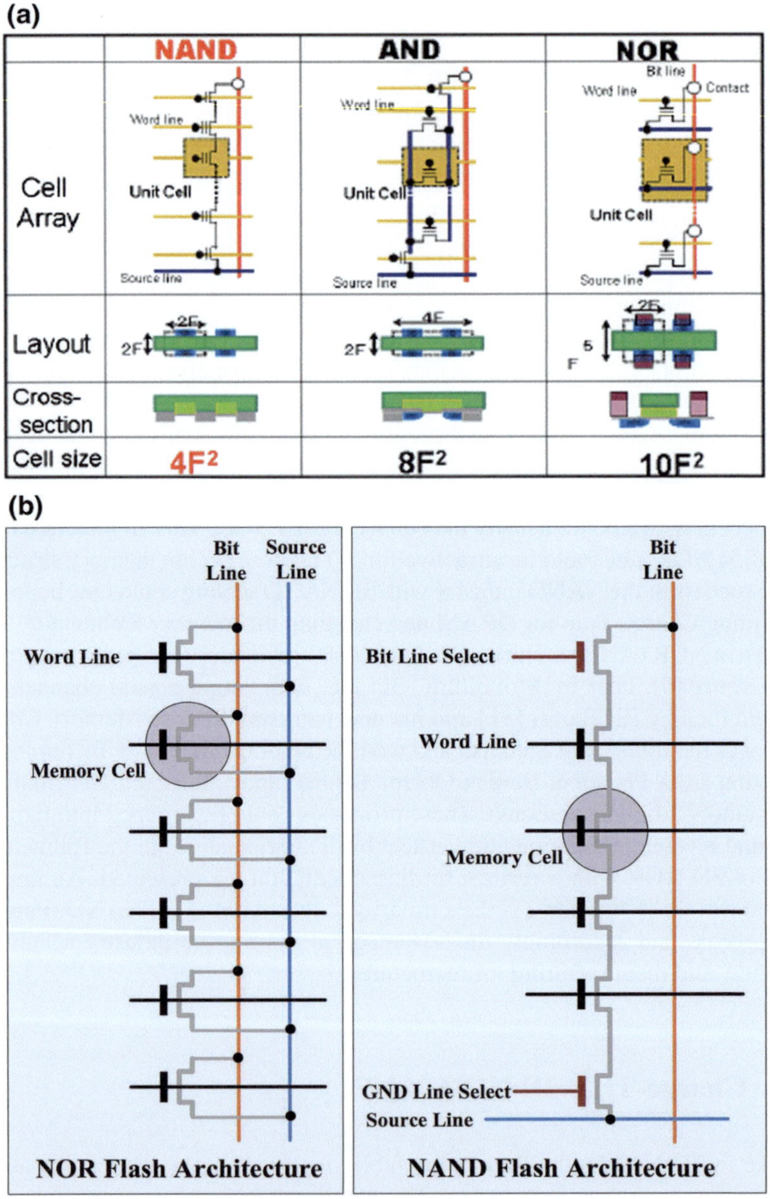

Fig. 10.4 **a** NOR versus NAND flash architecture. **b** NOR versus NAND flash architecture

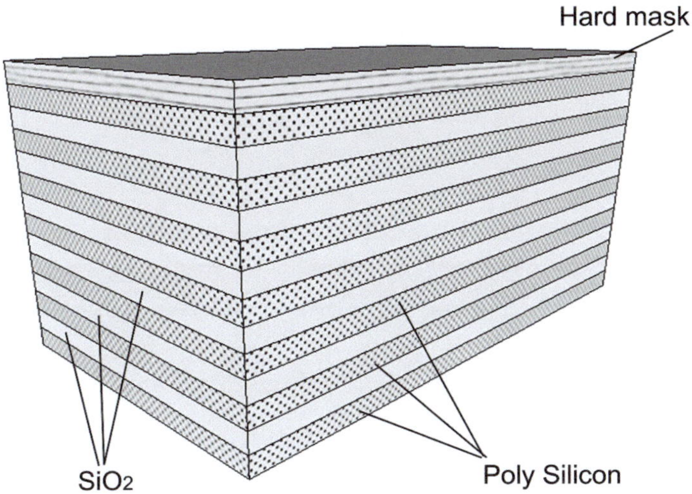

Fig. 10.5 Multilayer fabric as foundation for 3D NAND and 3D NOR

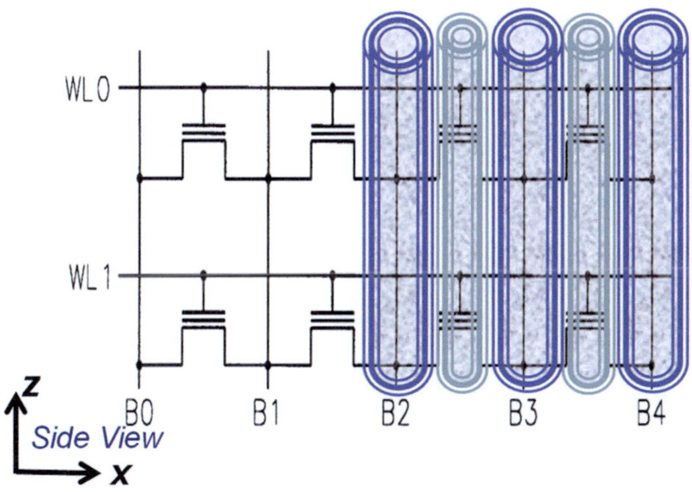

Fig. 10.6 Transistor schematic overlaid by punch and fill source/drain (bit-lines) in the odd holes and channel in the even holes

N+ silicon, or through a combination of N+ layers on the holes' walls and core of metal or even just metal for a Schottky-based structure. The channel holes in between are filled with un-doped polysilicon.

The structure looks like a 3D NAND with N+ holes punched between channel holes. Figure 10.7 illustrates a top view of the structure.

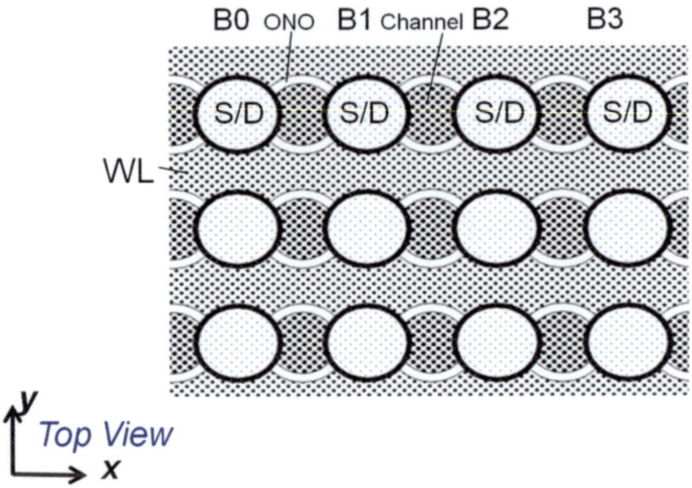

Fig. 10.7 Top-view, source/drain (bit-lines) in the odd holes and channel in the even holes

Figure 10.8 illustrates an alternative for the 3D NOR structure in which no additional holes are 'punched' for the channel but rather forming the channels by use of etch and deposition through the S/D holes.

Additional details for the 3D NOR structures and alternative process flows to form them could be found in the referenced patents and applications [9–12].

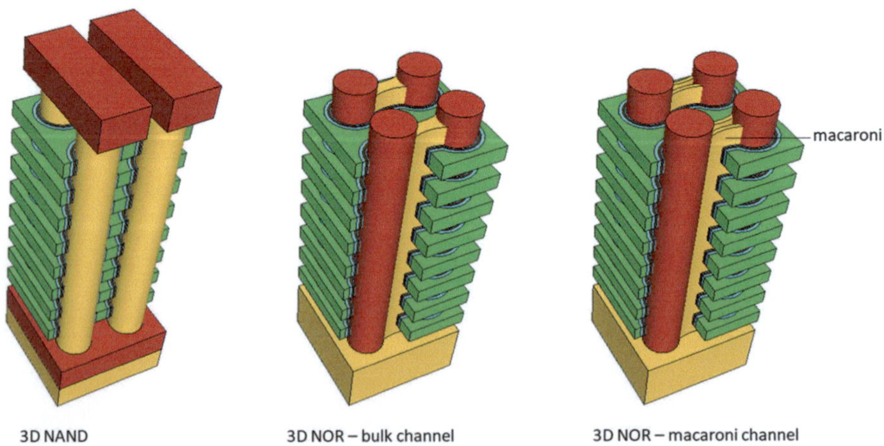

Fig. 10.8 Some alternatives for 3D memory structures

10.5 Schottky Barrier and Dopant Segregated Schottky Barrier ("DSSB")

In flash devices there are a few writing mechanisms that are frequently used. One is Fowler–Nordheim (FN) tunneling commonly used in NAND flash devices and another is Hot Carrier Injection (HCI), also called Hot Electron, often used in NOR flash devices. Flash cell writing using FN tunneling is orders of magnitudes more efficient than HCI as in FN most of the current is the tunneling current while in HCI only a small fraction of the current through the channel is actually the hot carriers being driven over the quantum barrier thus to be trapped.

In a paper [13] titled "Performance breakthrough in NOR flash memory with dopant-segregated Schottky-barrier (DSSB) SONOS devices" a few orders of magnitude improvements were reported by the use of Schottky Barrier devices, as is illustrated in Fig. 10.9 (Fig. 3 of the paper [13]).

This improvement in hot carrier write time and efficiency was reported in other papers including devices without dopant segregation, and devices utilizing poly silicon channels [14–16]. Using metalized Source/Drain lines in the 3D NOR device improves the bit-line conductivity and thus enhances the device P/E efficiency and speed.

Comparing such a 3D NOR technology to Stacked Capacitor DRAM suggests many advantages such as: higher density, 3D scaling, lower power, reduced rate of refresh, non-destructive read. Yet Charge-Trap 3D NOR is expected to have much longer erase time. Proper design of a 3D NOR device could support a full segment erase scheme, which combined with proper system design and support software, could compensate for the erase time deficiency.

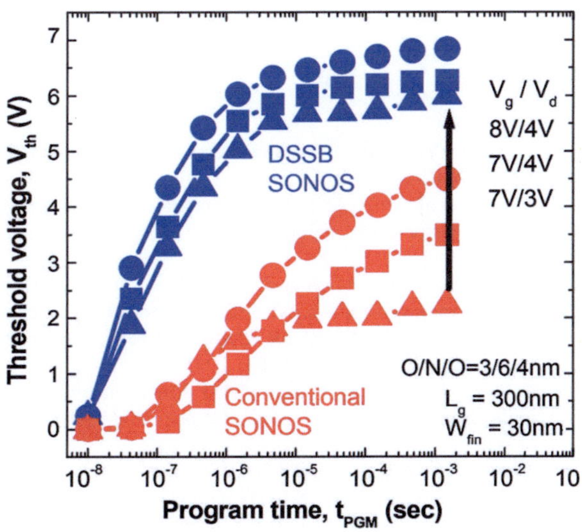

Fig. 10.9 Program/erase characteristics for NOR flash memory cell (double gate), DSSB and conventional SONOS devices

10.6 Periphery Under Cell ("PUC") or Over Cell ("POC")

To further enhance the 3D NOR structure to support DRAM applications, it useful to have the memory control circuits, often called periphery circuits, either under the memory array or on top of it. Some of the 3D NAND products in the market use periphery under cell, also called CMOS under Array ('CUA'), currently being produced by Micron and Intel. And as discussed in Chap. 8, YMTC use Xtaking to form the periphery over the memory array. For DRAM applications the 3D NOR structure could utilize these ideas further to break the array to hundreds or even thousands of small arrays, each with its own control circuits, to keep the memory control lines short and accordingly support very high-speed access.

10.7 Further Applications

3D NOR high speed memory could be an attractive architecture to many memory applications such as Storage Class Memory (SCM) and AI applications such as Neuromorphic Computing [17, 18]. The NOR (AND) architecture provides direct access to the selected cell, the 3D structure allows high density packing and reduce costs with 3D scaling. Supporting it, with periphery under the cell or over the cell, further enables high speed access and partitioning the memory into small arrays helps keep the memory access lines short. In summary, 3D integration technology is already a part of memory scaling and it is positioned to support the full range of memory applications required to keep advancing device integration to drive the AI era.

References

1. S. Mueller et al., Incipient ferroelectricity in Al-doped HfO_2 thin films. Adv. Funct. Mater. **22**(11), 2412–2417 (2012)
2. J. Müller et al., Ferroelectric hafnium oxide: a CMOS-compatible and highly scalable approach to future ferroelectric memories, in *2013 IEEE International Electron Devices Meeting* (IEEE, 2013)
3. H.C. Wann, C. Hu, High-endurance ultra-thin tunnel oxide in MONOS device structure for dynamic memory application. IEEE Electron Device Lett. **16**(11), 491–493 (1995)
4. H.I. Hanafi et al., A scalable low power vertical memory, in *Proceedings of International Electron Devices Meeting* (IEEE, 1995)
5. K. Tsunoda et al., Ultra-high speed direct tunneling memory (DTM) for embedded RAM applications, in *Digest of Technical Papers. 2004 Symposium on VLSI Technology, 2004* (IEEE, 2004)
6. Patents: US 7,848,148, US 8,705,278
7. Patents: US 6,249,460, US 6,639,835, US 6,730,960
8. J. Cooke, Flash Memory 101: An Introduction to NAND Flash. EE Times, 20 Mar 2006
9. Patents: US 8,203,187, US 8,426,294

10. Patents: US 10,014,318 and PCT application WO 2017/053329
11. Patents: US 9,842,651, US 9,892,800, US 9,911,497
12. Patent application: WO/2018/144957
13. S.-J. Choi et al., Performance breakthrough in NOR flash memory with dopant-segregated Schottky-barrier (DSSB) SONOS devices, in *2009 Symposium on VLSI Technology* (IEEE, 2009)
14. S.-J. Choi et al., A novel TFT with a laterally engineered bandgap for of 3D logic and flash memory, in *2010 Symposium on VLSI Technology* (IEEE, 2010)
15. C.-H. Shih et al., Source-side injection Schottky barrier flash memory cells. Semicond. Sci. Technol. **24**(2), 025013 (2009)
16. C.-H. Shih et al., Schottky barrier silicon nanowire SONOS memory with ultralow programming and erasing voltages. IEEE Electron Device Lett. **32**(11), 1477–1479 (2011)
17. Y. Noh et al., Synaptic devices based on 3-D AND flash memory architecture for neuromorphic computing, in *IEEE International Memory Workshop (IMW)* (2019)
18. H.-T. Lue et al., A novel 3D AND-type NVM architecture capable of high-density, low-power in-memory sum-of-product computation for artificial intelligence application, in *2018 IEEE Symposium on VLSI Technology* (IEEE, 2018)

Chapter 11
3D for Efficient FPGA

Zvi Or-Bach

11.1 Historical Prospective

Logic devices have amounted to about two thirds of the IC industry for many years. In logic devices, there has always been a tradeoff between the costs of developing the logic device in time and money, versus the cost of the end product in terms of performance, power, and cost ("PPC") as illustrated in Fig. 11.1.

In a fundamental work at the Berkeley Wireless Research Center and followed work at many other technology centers [1–3] this tradeoff has been characterized over two decades of designs and benchmarks (Fig. 11.2).

At the early days of the FPGA market, two programming technologies were competing—SRAM based Look Up Table (LUT), and Anti-Fuse. LUT eventually won because it allows easy technology scaling and unlimited reprogramming iterations. Yet, due to the severe PPC penalties of FPGA technology [4], the adoption of the FPGA technology remains limited (Fig. 11.3).

Adapting 3D technology to FPGA design could be cost-effective and might greatly reduce those PPC penalties.

11.2 Early Work on 3D FPGA

Early work on 3D FPGA considered that forming the SRAM of the LUT on top of the FPGA logic would be technologically possible and far less demanding than forming two levels of logic one on top of the other. Tier Logic collaborated with Toshiba [5] to build SRAM using Thin Film Transistors (TFT) for the FPGA LUT on top of the rest of the FPGA circuit. It believed it could have reduced the FPGA device area by

Z. Or-Bach (✉)
MonolithIC 3D Inc., 3555 Woodford Dr., San Jose, CA 95124, USA
e-mail: Zvi@MonolithIC3D.com

B. Murmann and B. Hoefflinger (eds.), *NANO-CHIPS 2030*,
The Frontiers Collection, https://doi.org/10.1007/978-3-030-18338-7_11

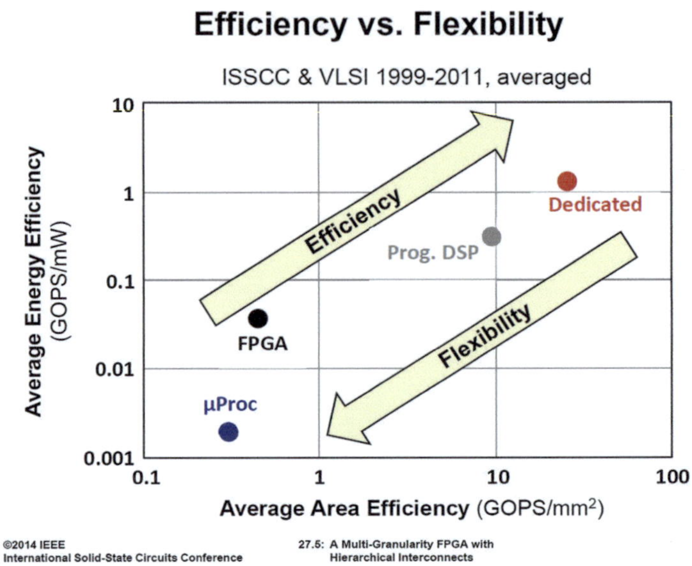

Fig. 11.1 Logic device tradeoff

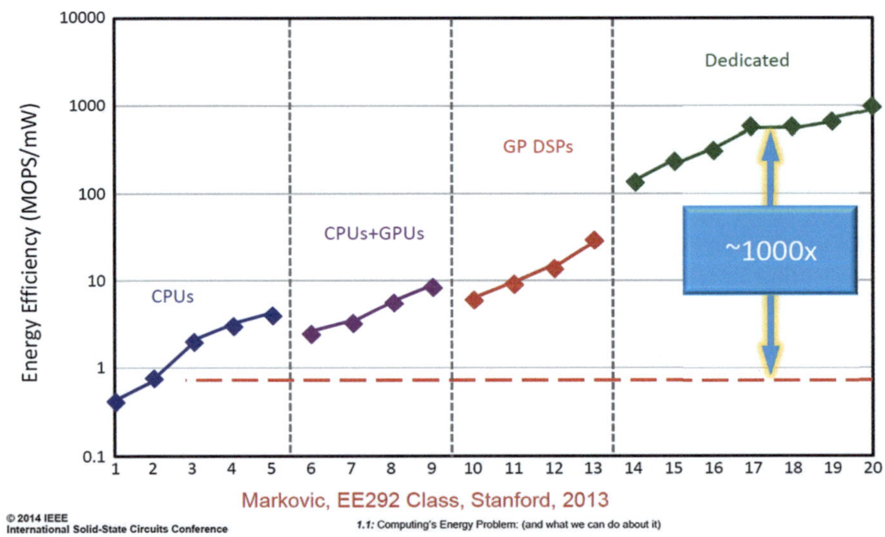

Fig. 11.2 Characterization of logic device tradeoff

about 20%, yet the effort failed, and the project was shut down. A similar concept using RRAM [6] on top of the logic instead of TFT reported potential 40% reduction compared to 2D FPGA but was not pursued commercially.

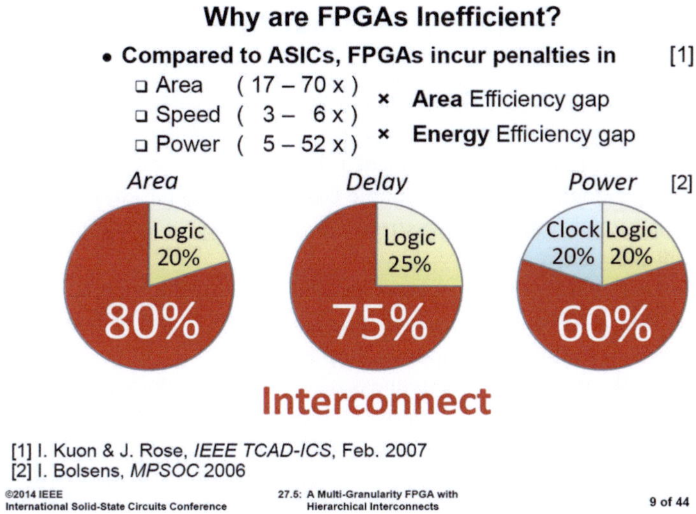

Fig. 11.3 The FPGA penalties

CEA Leti has been developing sequential monolithic 3D calling it CoolCube™. As a benchmark, they evaluated [7] applying their technology for FPGA putting logic over memory with the expectation to achieve 55% area reduction compared to 2D FPGA [9].

11.3 3D for Multi-configurations

Tabula, a recently failed start-up, had developed a unique type of FPGA—a real time reconfigurable FPGA. The concept tries to leverage FPGA reconfigurability through storing multiple configurations on-chip and swapping them as needed. It effectively attempted to compensate for the limited area efficiency of the FPGA by reusing the same chip's real estate for multiple purposes on the fly. The company even called its product a 3D FPGA, time being the 3rd dimension. Tabula had raised about $200M but eventually went out of business. An interesting concept that could be added to Tabula structure has been suggested [8] to leverage monolithic 3D technology for multi-stack to hold the multi configuration of the FPGA. Having more than one configuration of a device stack in 3D could allow switching between device configurations within just a few clock cycles and would not increase the device footprint.

11.4 3D for FPGA-ASIC Dual Mode Concept

An interesting alternative to FPGA was developed by eASIC [10], recently acquired by Intel. The original concept pioneered by eASIC was that the key deficiency of FPGA is its Programmable Interconnect ("PIC") rather than logic. Consequently, eASIC's early product used programmable LUT-4 (SRAM based) with mask-defined via interconnection. Figure 11.4 illustrates the advantage of via defined interconnect versus PIC at the 45 nm node.

It should be noted that PIC requires sharing some of the base silicon fabric and consumes additional routing resources by going down from the interconnect levels (metal layers 3–6) to the base silicon and up again.

Figure 11.5 illustrates the effectiveness of via-defined interconnect logic. It could potentially provide logic that has only a factor of 2–4 area penalty versus ASICs, with a power-speed penalty of 2–3.

Leveraging monolithic 3D technology could enable effective replacement of eASIC's via with electrically programmable anti-fuse, thus enabling FPGA devices with better than 10× improvement to PPC.

3D heterogeneous integration could help overcome some of the known limitations of anti-fuse technology. First, it allows using a standard fab and process for the base FPGA fabric. Second, it allows saving on the anti-fuse high voltage programming circuits overhead by moving them to an upper level.

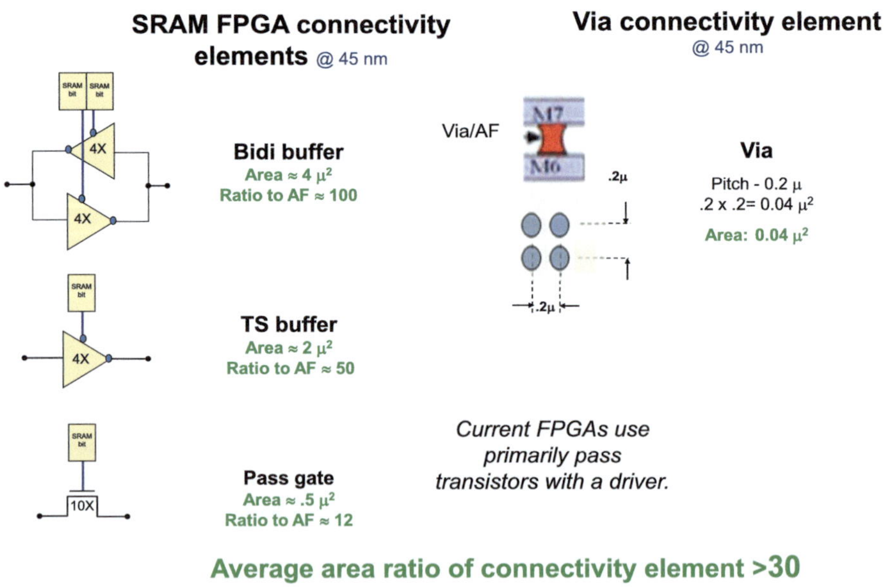

Fig. 11.4 Programmable interconnect versus masked defined interconnect

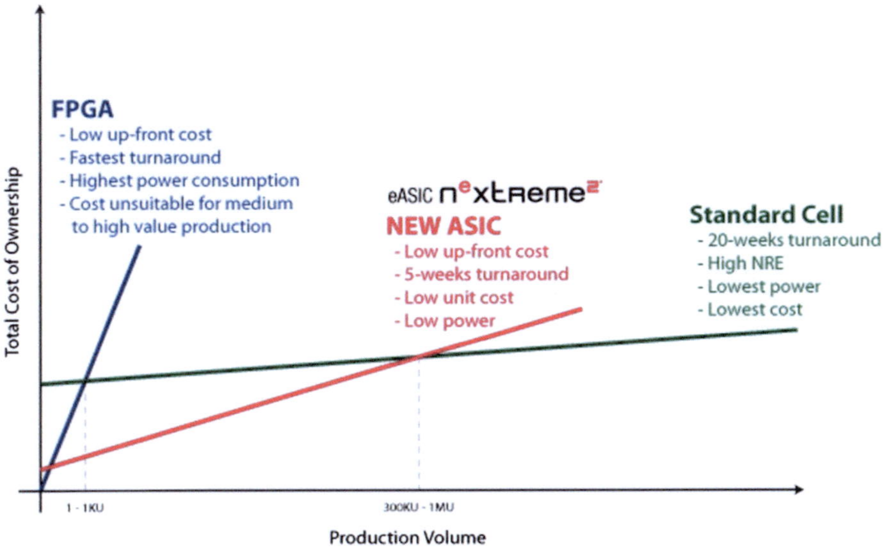

Fig. 11.5 eASIC versus FPGA and versus ASIC. *Source* eASIC web site

Replacing via-defined interconnect fabric with programmable anti-fuse interconnect fabric could be done with relatively low overhead (<20%) as is illustrated by Fig. 11.6.

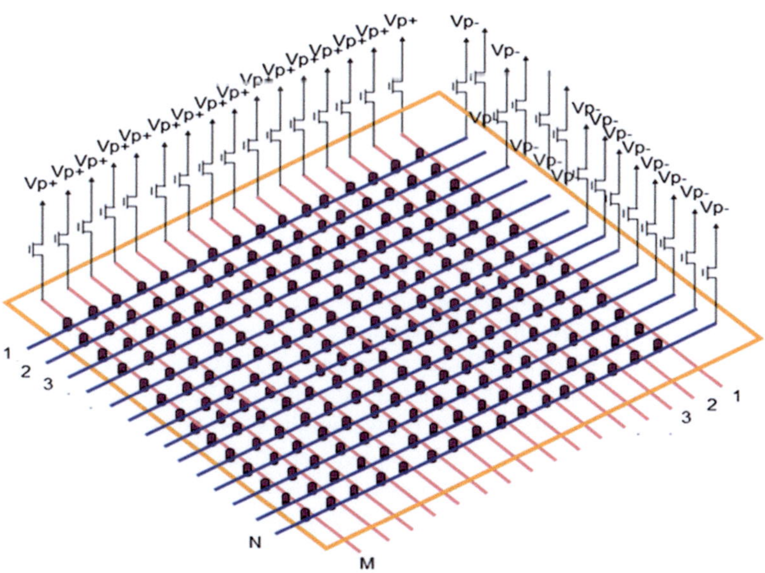

Fig. 11.6 Anti-fuse M × N fully populated crossbar interconnect structure

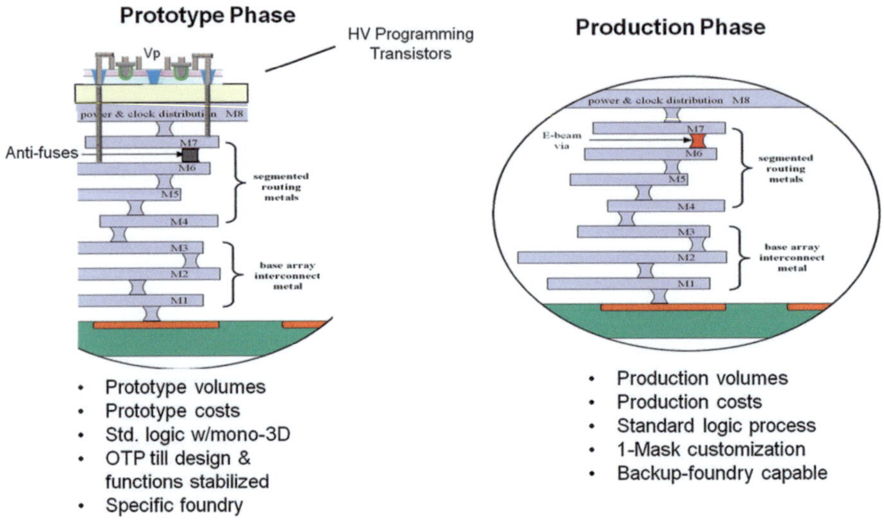

Fig. 11.7 Dual mode: FPGA for prototype and low volume, and mask-defined via for low cost

An additional advantage in which 3D heterogeneous integration could be applied is supporting dual mode of the custom logic: using field programmable device for prototypes and low volume, and form a low-cost compatible volume replacement device, in which the anti-fuses are replaced by a mask-defined via layer (Fig. 11.7).

Removing the anti-fuse and programming circuitry could reduce costs of the high-volume part for the relatively low cost of a single via mask.

11.5 Utilizing 3D Memory Fabric for FPGA Fabric

The breakthrough which was introduced with 3D NAND technology was the introduction of a new form of scaling—3D Scaling. In 3D scaling technology, more device transistors (or memory cells) are being produced for about the same manufacturing effort by having more layers in the substrate starting wafer. In Chap. 10 we presented a variation called 3D NOR which could be used to replace Stacked Capacitor DRAM technology. Here, a technology concept is presented to leverage 3D scaling for FPGA fabric. The technology has also been detailed in MonolithIC 3D, Inc. patent applications [11, 12]. The first structure [11] is leveraging 3D NOR memory fabric having a single crystal channel and vertically oriented word-lines for FPGA fabric. The second structure [12] leverages 3D NOR memory fabric having poly-crystalline channel and horizontally oriented word-lines for FPGA fabric. The following description is based on the first structure. First, a generic structure is constructed using shared lithography and processing, which later on could be programmed to function as an FPGA.

11.5.1 The Fabric

A key concept leveraging 3D NOR memory structure for FPGA application is using a flash memory for programmable logic applications [13–15] (Fig. 11.8).

A variation of the 3D NOR structure presented in Chap. 10 could include first epitaxial growth of multilayer SiGe over silicon for single crystal channel, or conventional multilayer deposition of polysilicon over oxide as common for 3D NAND. Then, etching the structure, forming rims and valleys takes place (Fig. 11.9).

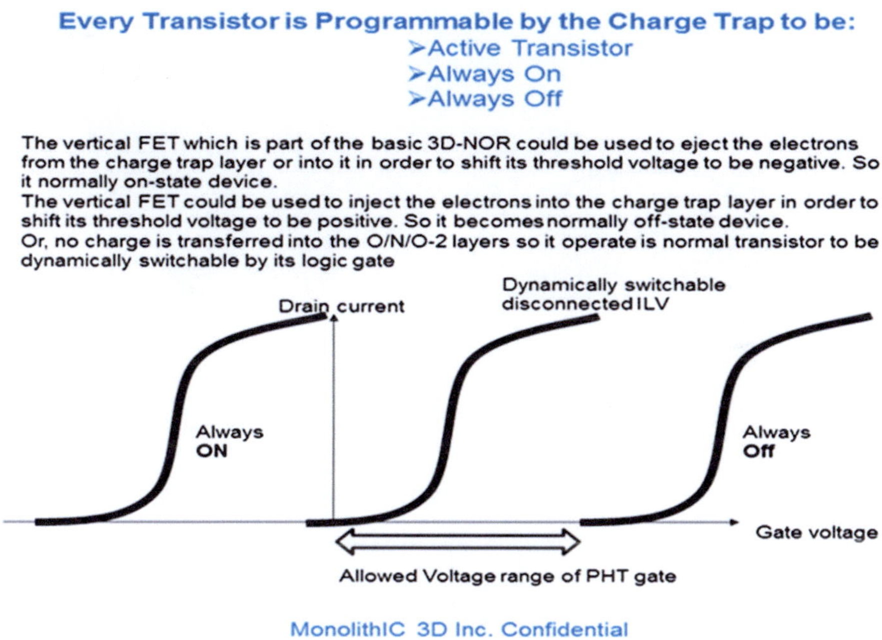

Fig. 11.8 Flash cell is a programmable logic function

Fig. 11.9 Multilayer substrate after etching forming ridges and valleys

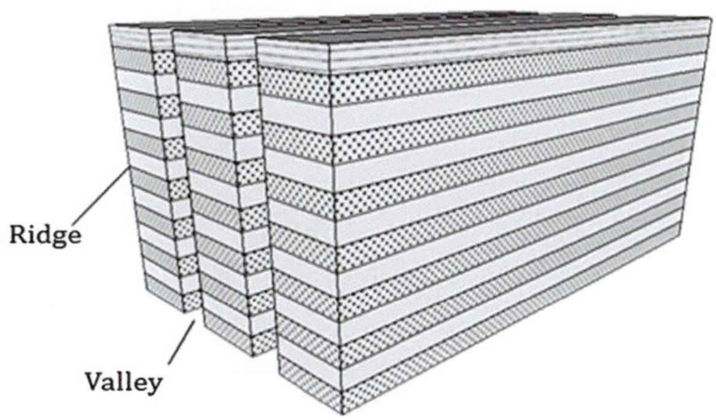

Next, depositing Oxide-Nitride-Oxide (O/N/O) makes the structure ready for charge trap memory function. Next, forming gates and a staircase makes the structure illustrated in Fig. 11.10.

The transistor schematic of one ridge is illustrated in Fig. 11.11.

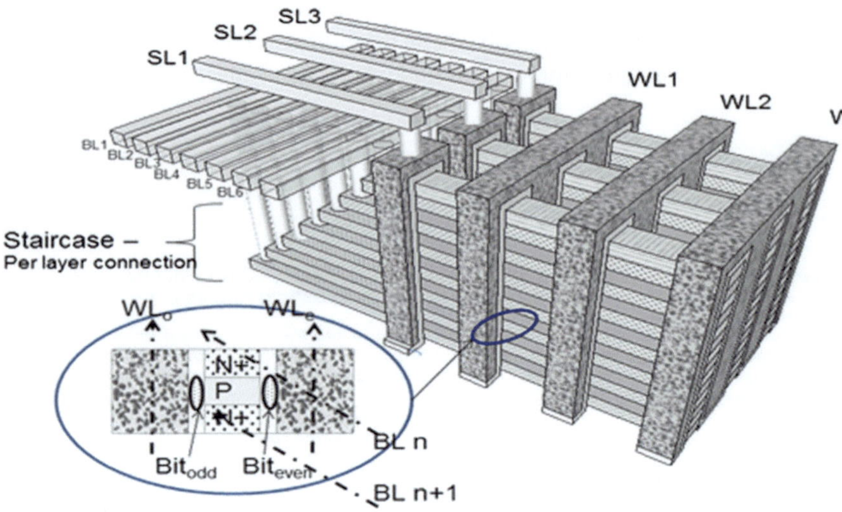

Fig. 11.10 Adding O/N/O, gates, and staircase access

Fig. 11.11 Transistor schematic along a ridge

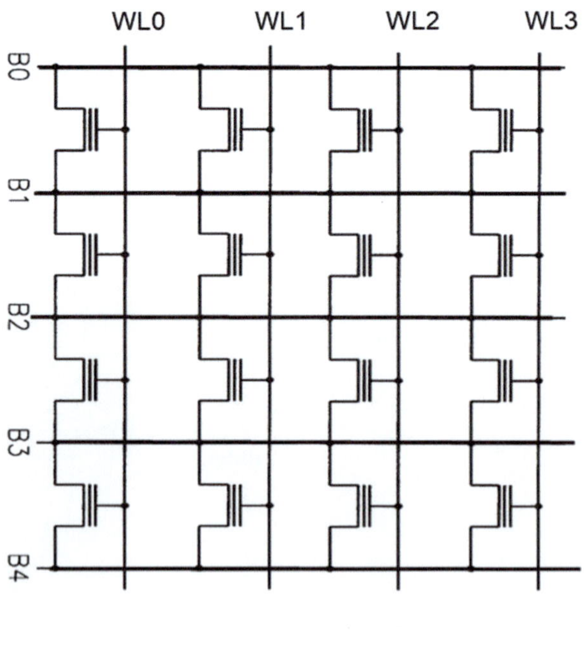

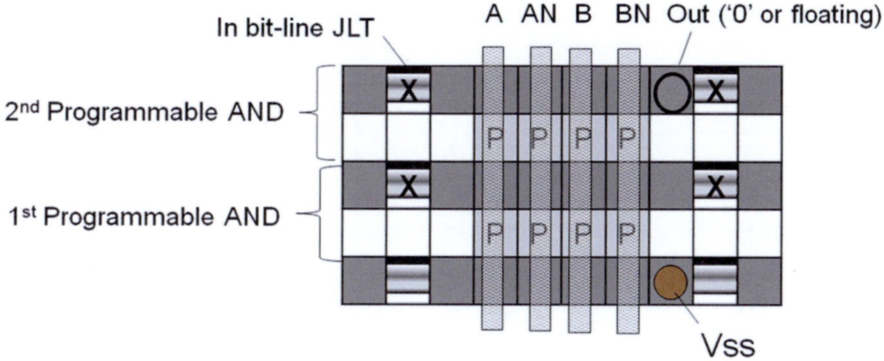

Fig. 11.12 LUT-2 could be formed in section of a 3D NOR structure

11.5.2 Programmable LUT-n Memory

The above structure could be used to form logic functions such as Look-Up-Table and programmable interconnect for FPGA applications. Figure 11.12 illustrates a LUT-2 formed in two layers of such a ridge.

The LUT-2 gates (A, AN, B, BN) are the WL0–WL3 (Fig. 11.11). The X represents an additional variation in which an in the bit-line junction-less-transistors ("JLT") is being formed. The details for such in bit-line JLT processing are detailed in PCT application WO 2017/053329. Such in bit-line JLT enable horizontal segmentation of the 3D NOR structure. The truth table of this LUT-2 structure is presented in Fig. 11.13 (Fig. 11.14).

The 3D NOR structure is a 3D matrix of n-type transistors. Accordingly, the logic functions formed in it utilize only n-type transistors. A transferred layer on top could be used to add full CMOS circuitry to complement the n-only programmable logic underneath. Logic circuits that utilize mainly n-type transistors had been proposed in the past [16]. One approach to reconstruct full swing signals from n-type only circuits is to use two complementing logic functions. Figure 11.15a, b illustrates the use of complementing LUT and LUT-N with top CMOS circuit to reconstruct full swing logic output.

For higher performance, a differential amplifier circuit could be used instead of the logic half-latch.

11.5.3 Programmable Interconnect in Memory

Differential logic could be extended to differential signaling throughout the FPGA. It could help reduce power and improve speed but, far more importantly, it allows using the 3D NOR fabric for programmable routing. Differential interconnects offer lower voltage swings with better noise immunity resulting in lower power. For years,

IN									OUT			
First AND				Second AND				a=	0	1	0	1
A	AN	B	BN	A	AN	B	BN	b=	0	0	1	1
T	T	T	T	T	T	T	T		0	0	0	0
X	T	X	T	T	T	T	T		0	0	0	1
T	X	X	T	T	T	T	T		0	0	1	0
X	X	X	T	T	T	T	T		0	0	1	1
X	T	T	X	T	T	T	T		0	1	0	0
X	T	X	X	T	T	T	T		0	1	0	1
X	T	T	X	T	X	X	T		0	1	1	0
X	T	T	X	X	X	X	T		0	1	1	1
T	X	T	X	T	T	T	T		1	0	0	0
T	X	T	X	X	T	X	T		1	0	0	1
T	X	X	X	T	T	T	T		1	0	1	0
T	X	X	X	X	X	X	T		1	0	1	1
X	X	T	X	T	T	T	T		1	1	0	0
X	X	T	X	X	T	X	X		1	1	0	1
X	X	T	X	T	X	X	X		1	1	1	0
			X	X	X	X	X		1	1	1	1

Fig. 11.13 Truth table of the programmable memory for LUT-2 function

interconnect delay has increased with scaling, while gate delay has decreased as has been illustrated in Fig. 15.2a, b. Yet, the interconnect effect on chip power had been managed by chip operating voltage scaling known as Dennard scaling (Fig. 11.16).

The end of Dennard Scaling made power the limiting factor. The constant charge and discharge of the interconnect capacitance now dominates chip power and performance (Fig. 11.17).

Yet, the industry has not adapted differential interconnect because it requires double the routing resources and additional support circuits. However, as power becomes a dominant problem, perhaps it is time for differential interconnects to take center role in new chip architectures.

3D scaling for configurable logic using shared litho and shared processing opens an iterating opportunity for new type of interconnect technology. In 3D scaling, many layers are processing together, allowing the effective processing of many layers of interconnect together as a generic 3D matrix, and later program them for specific interconnect functions.

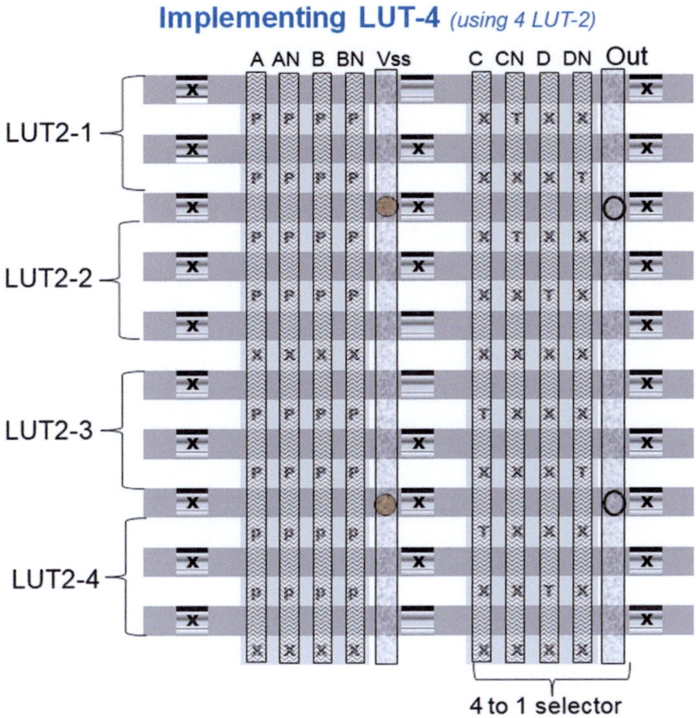

Fig. 11.14 LUT-4 could be formed in section of a 3D NOR ridge structure, having four LUT-2 vertically stacked within a ridge and adjacent 4 to 1 selector

For example, in a 3D fabric of 32 levels the top 10 could be used for the LUT-4 as is illustrated in Fig. 11.14 and the bottom 22 could be used for interconnect. The unused bit-lines of these 22 layers could function as horizontal ("X" direction) segments of the interconnect fabric. Vertical segment could be formed by depositing vertical ("Z" direction) conductive segments in-between the word-lines the structure—see Figs. 11.11 and 11.18a, b.

The programmable connectivity structure could use RRAM technology or anti-fuse (One Time Programmable—"OTP") technology. The connectivity segments in the horizontal direction vertical to the bit-line ("Y" direction), could add in using technology concept know as word-line replacement in 3D NAND (Fig. 11.19).

The support circuit on top could support the differential interconnect just like the differential logic.

The FPGA in memory fabric enables the formation of a multilayer (96–128) memory, such as 3D NOR, with the top 32 layers used for programmable logic while the rest for memory. Recently, logic in memory has become a popular concept as it fits very well many AI type applications. The 3D NOR with built-in FPGA could fit very well in this emerging space.

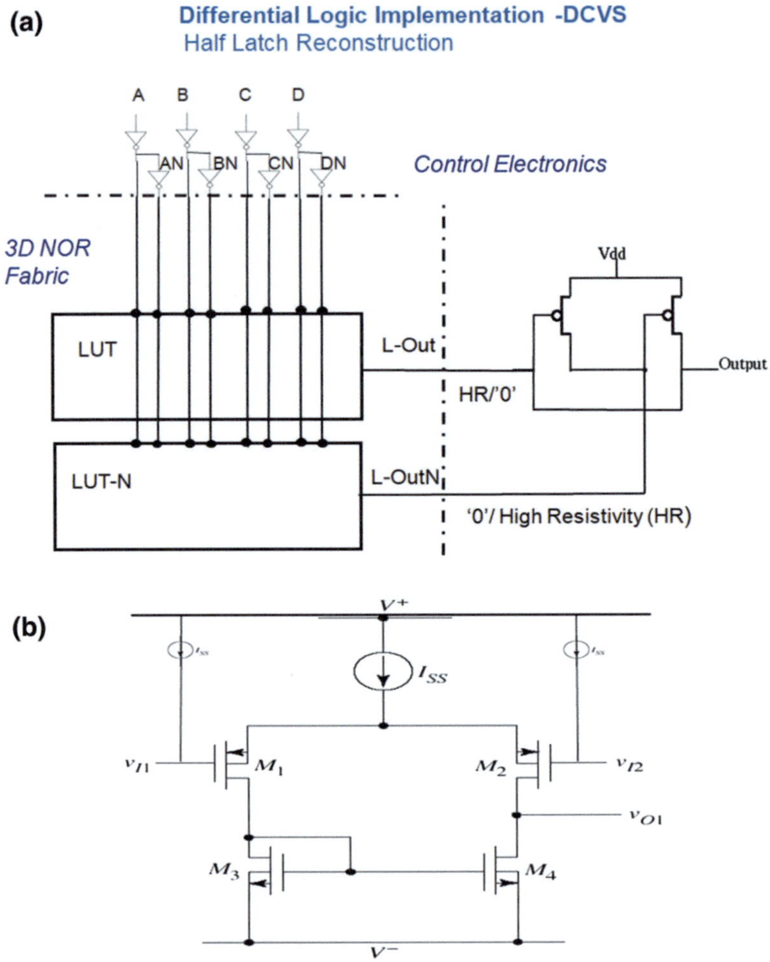

Fig. 11.15 **a** Two complementing LUT-4 with top lower control and reconstruction. **b** Optional differential amplifier top level reconstruction circuit

As a standalone FPGA product, 3D-NOR base FPGA could compete well with mask-defined standard cell designs. The LUT-4 footprint could be about (10 $\times$ 100 nm) $\times$ (2 $\times$ 100 nm) $=$ 0.2 μm^2 which represents a logic density of about 70 MGate/mm^2. The forecast for standard cells at the 7 nm node is about 20 MGate/mm.

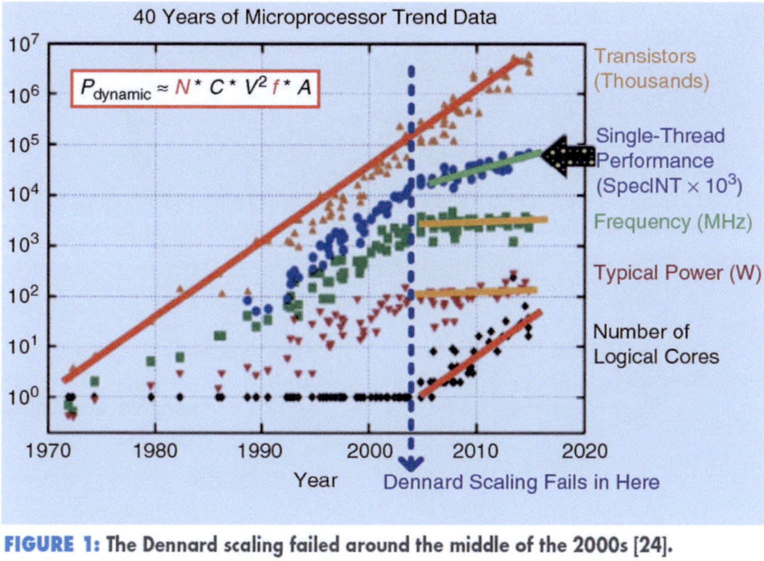

Fig. 11.16 End of Dennard scaling [17]

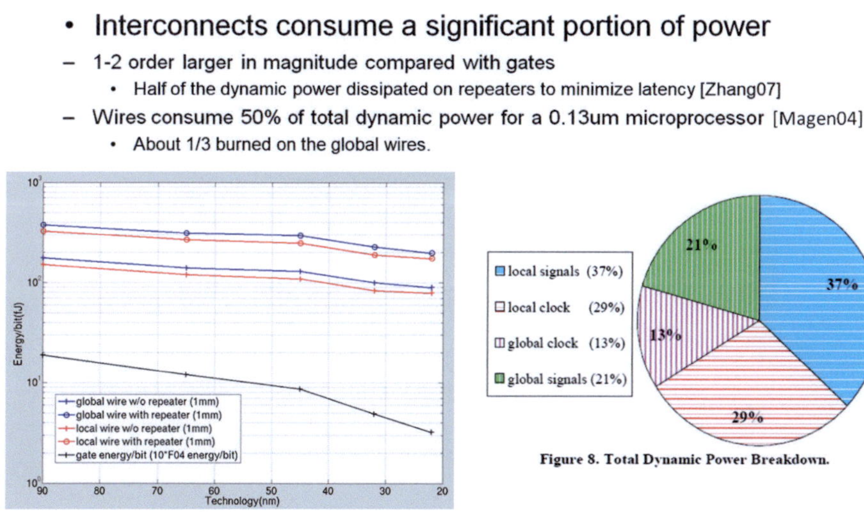

Fig. 11.17 Interconnect chip power [18]

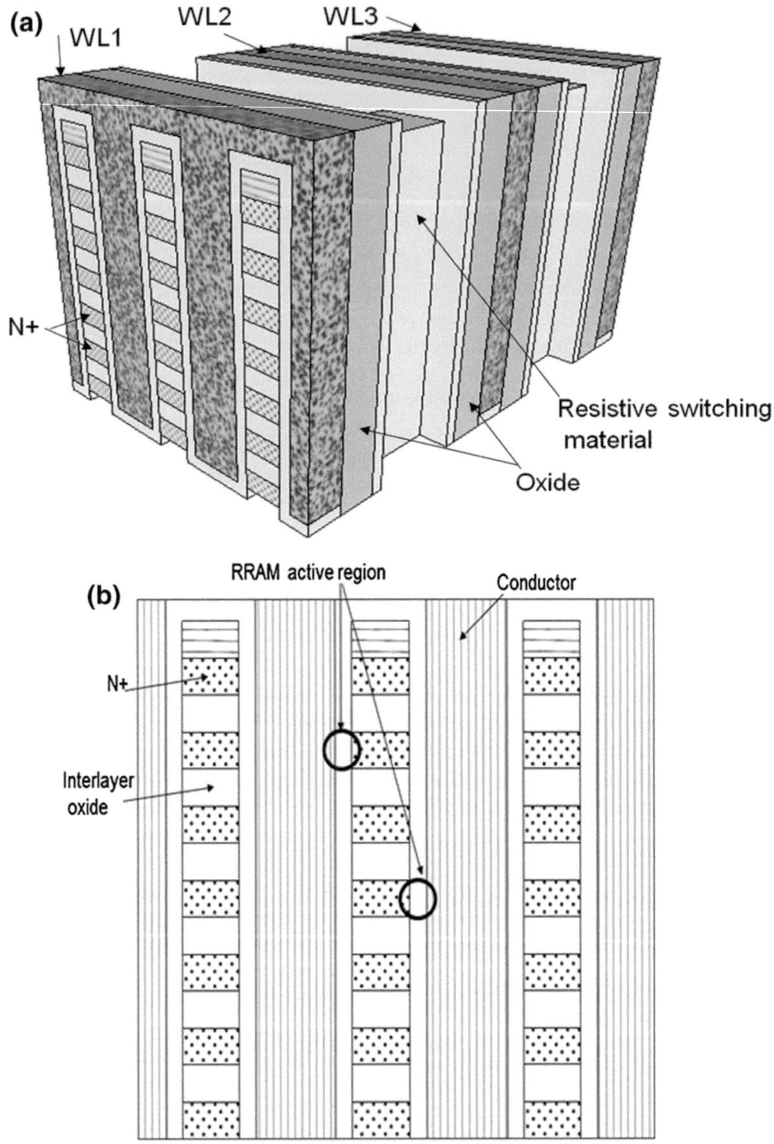

Fig. 11.18 **a** Preparing the structure for Z segments, **b** Z segments with anti-fusses

11.6 Summary

A few alternative concepts have been presented for use of 3D integration in FPGA applications. These alternatives offer different uses of 3D technologies resulting in

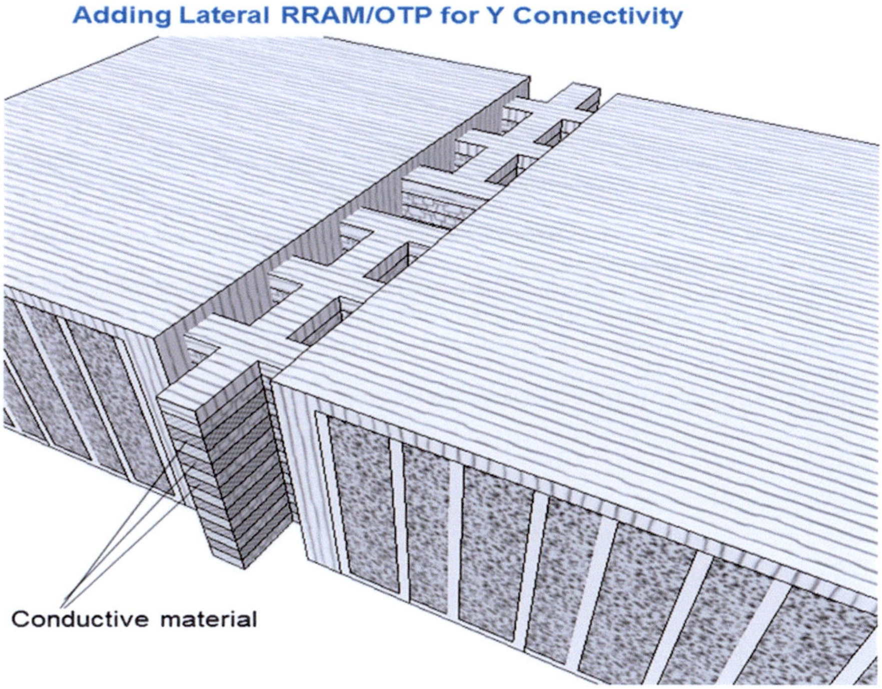

Fig. 11.19 3D structure with programmable logic and X-Y-Z programmable connectivity

different PPC, spanning the spectrum from 2× better FPGA, to about 0.4× of ASIC PPC, and to the 3D NOR FPGA, while having better PPC than ASICs.

References

1. N. Zhang, B. Brodersen, The cost of flexibility in systems on a chip design for signal processing applications. University of California, Berkeley, Tech. Rep. (2002)
2. B. Brodersen, Plenary Session, IEEE S3S 2013
3. M. Horowitz, 1.1 computing's energy problem (and what we can do about it), in *2014 IEEE International Solid-State Circuits Conference Digest of Technical Papers (ISSCC)* (IEEE, 2014)
4. F.-L. Yuan et al., A multi-granularity FPGA with hierarchical interconnects for efficient and flexible mobile computing. IEEE J. Solid-State Circuits **50**(1), 137–149 (2014)
5. T. Naito et al., World's first monolithic 3D-FPGA with TFT SRAM over 90 nm 9-layer Cu CMOS, in *2010 Symposium on VLSI Technology* (IEEE, 2010)
6. Y.Y. Liauw et al., Nonvolatile 3D-FPGA with monolithically stacked RRAM-based configuration memory, in *2012 IEEE International Solid-State Circuits Conference* (IEEE, 2012)
7. A. Mihal, S. Teig, A constraint satisfaction approach for programmable logic detailed placement, in *International Conference on Theory and Applications of Satisfiability Testing* (Springer, Berlin, Heidelberg, 2013)
8. US Patent 8,912,820

9. O. Turkyilmaz et al., 3D FPGA using high-density interconnect Monolithic Integration, in *2014 Design, Automation & Test in Europe Conference & Exhibition (DATE)* (IEEE, 2014)
10. US Patents: 6,642,744, 6,476,493, 6,819,136
11. Patent Application WO/2017/053329
12. Patent Application WO/2018/144957
13. C. Hu, Interconnect devices for field programmable gate array, in *1992 International Technical Digest on Electron Devices Meeting* (IEEE, 1992)
14. US Patent 5,633,518
15. T. Speers et al., 0.25 μm FLASH memory based FPGA for space applications. System 10000, 100000 (1999): 1000000
16. D. Somasekhar, K. Roy, Differential current switch logic: a low power DCVS logic family. IEEE J. Solid-State Circuits **31**(7), 981–991 (1996)
17. Liming Xiu, Time Moore: exploiting Moore's law from the perspective of time. IEEE Solid-State Circuits Mag. **11**, 39–55 (2019)
18. http://cseweb.ucsd.edu/~kuan/talk/interconnect0824.ppt

Chapter 12
Digital Neural Network Accelerators

Ulrich Rueckert

12.1 Introduction

Compared to other machine learning algorithms, Deep Neural Networks (DNNs) have achieved exciting accuracy improvements over the past decade. Hence, DNNs become a standard Artificial Neural Network (ANN) model today. Its underlying models and algorithms are still evolving, and hardware is trying to catch up with new architectures to accelerate the learning and inference phase of DNNs. The majority of learning is done on Graphics Processor Units (GPUs) in floating point on large server systems. However, further acceleration of the learning phase is needed which is the topic of another chapter in this book. In this chapter we focus on accelerators for the inference phase of already trained DNNs.

A DNN is composed of multiple convolutional layers, intermediate data operations (e.g. nonlinearity, pooling, normalization), and fully connected layers at the end of the processing chain. For example, Fig. 12.1 shows the VGG-16 DNN architecture [1]. It is a pre-trained model with 13 convolution and 3 fully connected layers (two with 4096 nodes, the output layer with 1000 nodes and softmax activation, model size about 528 MiB, about 138 M parameters). All convolutions use 3×3 filters and max pooling operations with 2×2 receptive fields. Basic operations in the inference phase are matrix-matrix- and vector-matrix-operations. Hence, multiply-accumulate (MAC) operations have by far the highest share in computation. For VGG-16 we come up with about 15 G (10^9) MAC, 20 M (10^6) compare, 29 M activation, and 1 K (10^3) for addition, division, exponential operations each [2]. Because of the large memory requirements the data transfer from memory to processing units and back is more costly in respect to time and energy than the computational cost. Hence, the reduction of the data transfer is the key to improving the resource-efficiency of DNN accelerators. For an introduction to efficient processing of DNNs see [3].

U. Rueckert (✉)
Universität Bielefeld, Bielefeld, Germany
e-mail: rueckert@cit-ec.uni-bielefeld.de

B. Murmann and B. Hoefflinger (eds.), *NANO-CHIPS 2030*,
The Frontiers Collection, https://doi.org/10.1007/978-3-030-18338-7_12

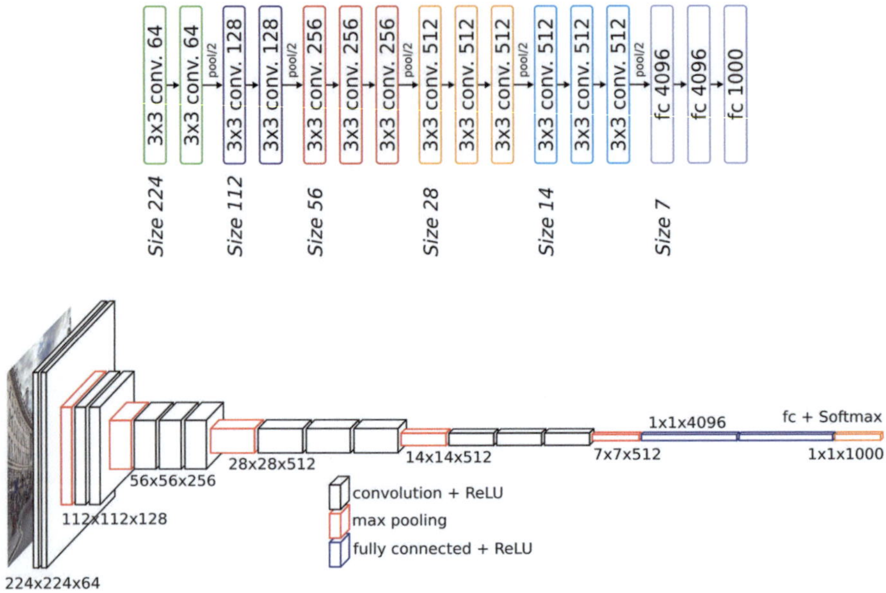

Fig. 12.1 Layer-architecture of the VGG-16 DNN [5]

VGG-16 showed good results (about 70% top-1 and 90% top-5 accuracy [1]) on the ImageNet dataset [4]. Executing the 1000-class ImageNet task requires about 31 GOPs/cl [10^9 operations per classification, 32-bit floating-point (FP32)] for VGG-16 [2]. "Deeper" networks, e.g. the SE-ResNet with 152 layers (winner in 2017, Fig. 12.2), achieve better results at considerably higher computational costs. In this chapter, VGG-16 serves as a representative example for comparing different hardware implementation approaches. It has a comprehensive structure and all characteristic aspects of DNN inference acceleration can be studied based on VGG-16.

In order to make DNNs more "hardware-friendly" approximations are applied. For DNN inference, approximation contributes to increases in throughput in three ways: increased parallelism, memory transfer reductions and workload reductions. Approximation algorithms can be classified into two broad categories: quantisation and weight reduction. Quantisation methods reduce the precision of weights, activations (neuron outputs) or both, while weight reduction removes redundant parameters through pruning and structural simplification leading to reductions in numbers of activations per network as well [6, 7]. When memory bound, the arithmetic performance of a platform does not scale with any increase in parallelism. When compute bound, all available processing resources are saturated.

As tasks increase in complexity, inference architectures become deeper (more layers) and more computationally expensive, and so methods for hardware oriented approximation have become a hot topic [6, 7]. The development of algorithms for

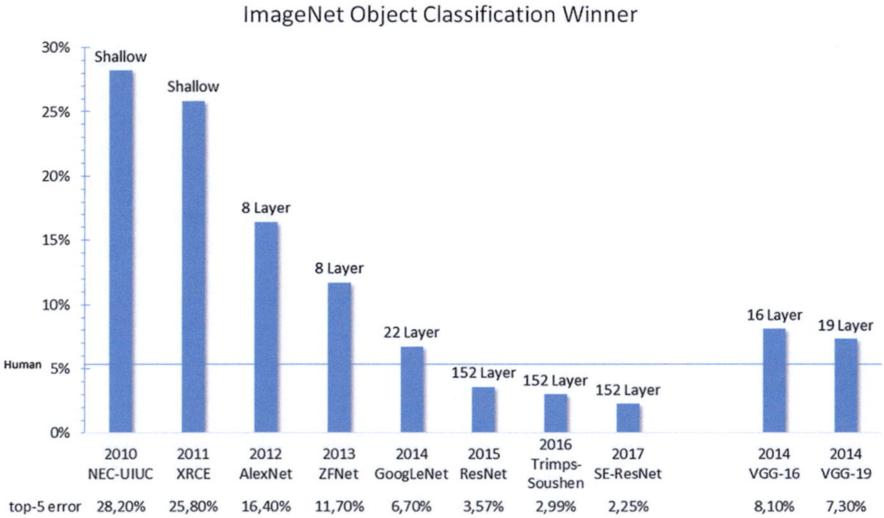

Fig. 12.2 DNN implementations winning the ILSVRC challenge based on the ImageNet dataset with 1000 object classes, 1.2 million training images (224 × 224), and 50,000 validation images [4]

reducing the computational and storage costs of DNN inference is therefore essential for resource-efficient processing of DNNs. Common evaluation criteria of DNN performance are:

Throughput:	classifications produced per second (cl/s) (classification rate);
Latency:	end-to-end processing time for one classification, in seconds (s);
Energy efficiency:	throughput obtained per unit power, expressed in cl/J;
Compression ratio:	the network's weight storage requirement vs. that of a baseline [0 < cr < 1];
Testing accuracy:	proportion of correct classifications over testing dataset [0 < cr < 1];

Other criteria are robustness, parameter tuning time, and design flexibility. Top-n accuracy, reported as percentages, captures the proportion of testing data for which any of the n highest-probability predictions match the correct result. Where comparisons are drawn against baselines, these are uncompressed implementations of the same network, trained and tested using identical datasets, with all data in IEEE-754 single-precision floating-point format (FP32) [4].

The increasing availability of parallel standard hardware such as Field-Programmable Gate Arrays (FPGAs), GPUs, and Multi-Core Processors (MCPs) offer new scopes and challenges in respect to resource-efficient implementation and real-time applications of DNNs. Because these devices are inexpensive and available, we can take the first step in accelerating DNNs with such standard devices. DNNs are inherently parallel, and hence it is obvious that MCPs are an attractive implementation platform for them. To improve the resource-efficiency, application specific hardware implementations are trying to take over the lead. However, as benchmarking of DNN accelerators is still in its infancy, there is no clear consensus about the right balance of computing power, memory capacity, and internal as well as external communication bandwidth for DNN accelerators. In the following, general aspects of DNN inference acceleration will be summarized and selected hardware implementations compared. Wherever performance data are available, we base our comparison on the VGG-16 network (batch size 1 = inference time for one image).

12.2 Graphics Processing Units

Graphics Processing Units are suited for single-instruction and multiple-data (SIMD) parallel processing. A GPU is a specialized integrated circuit designed to rapidly process floating-point-intensive calculations, related to graphics and rendering at interactive frame rates. The rapid evolution of GPU architectures from a configurable graphics processor to a programmable massively parallel co-processor makes them an attractive computing platform for graphics as well as other high performance computing domains having substantial inherent parallelism such as DNNs. The demand for faster and higher definition graphics continues to drive the development of increasingly parallel GPUs with more than 1000 processing cores and larger embedded memory at a power consumption of several watts. At the same time, GPU architectures will be extended to further increase the range of other applications such as DNNs. Specialized programming systems for GPUs evolved (e.g., CUDA [8] and OpenCL [9]) enabling the development of highly scalable parallel programs that can run across tens of thousands of concurrent threads and hundreds of processor cores. However, even with these programming systems, the design of efficient parallel algorithms on GPUs for other applications than graphics is not straightforward. Re-structuring of the algorithms is required in order to achieve high performance on GPUs. Furthermore, it is difficult to feed the GPUs fast enough with data to keep them busy. Nevertheless, an increasing number of papers on this topic shows that GPUs are currently the predominant implementation platform for simulating large DNNs [10]. The GPU´s SIMD architecture turned out to be a decent fit for DNN workloads. Hence, almost all GPU manufactures are active in developing ANN accelerators for data centres and scaled-down versions for edge devices as well as smart sensors.

In data server environments, high-end devices are employed in order to maximize throughput at the penalty of substantial power consumption. A representative example is the NVIDIA Tesla V100 accelerator based on the NVIDIA Volta GV100 GPU (Fig. 12.3). The GV100 GPU employs the Volta architecture and is fabricated in a 12 nm production process at TSMC [11]. With a die size of 815 mm^2 and a transistor count of 21.1 billion (25.9 million/mm^2) it features 5120 shading units, 320 texture mapping units and 640 tensor cores which help improve the speed of machine learning applications. Tensor cores are specialized execution units designed specifically for performing the tensor/matrix operations that are the core compute function used

Fig. 12.3 NVIDIA Tesla V100 SXM2 Module (**a**) with Volta GV100 GPU (**b**) [11]

in DNNs. The module comes with 16 GB GPU memory, 900 GB/s memory bandwidth, and a maximum of 300 W power consumption (1.3 GHz base clock, 1.5 GHz boost clock). Peak computation rates are 7.8 TFLOPS (Tera (10^{12}) Floating point Operations per Second) of double precision floating-point (FP64) performance, 15.7 TFLOPS of single precision (FP32) performance, and 125 Tensor TFLOPS (FP16) [11]. Running VGG-16 on V100 yields from 821 (batch size 1) up to 2067 (batch size 128) cl/s [12].

NVIDIAs Turing GPUs include a new version of the Tensor core design that has been enhanced with INT8 and INT4 precision modes for inference workloads that can tolerate quantization and don't require FP16 precision. The TU104 graphics processor [13] includes 320 such Tensor cores, and is built on the 12 nm TSMC process with a die area of 545 mm^2 and 13.6 billion transistors (25 million/mm^2). The GPU is operating at a frequency of 585 MHz, which can be boosted up to 1.6 GHz. NVIDIA has placed 16 GB GDDR6 memory on the Tesla T4 graphics card. The card measures 168 mm in length, and features a single-slot cooling solution for a 70 W power consumption maximum [13]. Running VGG-16 on Tesla T4 with 726 (batch size 1) up to 1956 cl/s (batch size 128) is a bit slower compared to Tesla V100, but more power efficient: 10 (batch size 1) to 28 (batch size 128) cl/s/W instead of 4 (batch size 1) to 10 (batch size 128) cl/s/W [12].

NVIDIA T4 data servers deliver more than 10,000 TOPS (Trillion Operations per Second) for real-time speech recognition and other real-time AI tasks. NVIDIA also targets low-latency edge AI (Artificial Intelligence) with the scalable EGX platform [14], an accelerated computing platform that enables to perceive, understand and act in real time on continuous streaming data. NVIDIA EGX was created to meet the growing demand for instantaneous, high-throughput AI at the edge—where data is created—with guaranteed response times, while reducing the amount of data that must be sent to the cloud. EGX starts with the tiny NVIDIA Jetson Nano™, which in a few watts can provide 0.5 TOPS for tasks such as image classification, and it spans all the way to a full rack of NVIDIA T4 servers. NVIDIA supports programmers with its TensorRT™ platform for high-performance deep learning inference. It includes a deep learning inference optimizer delivering low latency and high-throughput for DNN inference applications. TensorRT™ is built on CUDA, NVIDIA's parallel programming model, and optimizes neural network models trained in all major frameworks. Reduced precision inference significantly reduces application latency, which

Fig. 12.4 Chip micrograph of AMDs Vega GPU (7 nm CMOS, 1.6 GHz) [16]

is a requirement for many real-time services in embedded applications. For example, optimizing VGG-16 with TensorRT™ for Jetson TX2 yield about 26 (FP16) or 13 cl/s (FP32) resulting in 3.4 or 1.7 cl/s/W, respectively (7.5 W) [15].

Currently, GPUs are the dominant hardware platforms for DNN learning and inference. Other GPU manufacturer (e.g. AMD, ARM, INTEL, Qualcomm) are offering powerful chips and programming frameworks for mapping DNNs on their GPUs as well. For example, Fig. 12.4 shows the AMD's first 7 nm Vega GPU design improving performance per watt over previous generation products. It offers ultra-fast double precision performance with up to 7.4 TOPS (FP64) on the AMD Radeon Instinct™ MI60 Compute GPU. Optimized DNN operations with mixed FP16, FP32 and INT8 data representation support efficient learning and inference of DNNs. Two Infinity Fabric™ Links per GPU for high speed directly connected GPU clusters deliver up to 92 GB/s peer-to-peer bandwidth. Programmers are assisted by the ROCm open ecosystem that includes optimized libraries supporting frameworks like TensorFlow, PyTorch and Caffe 2 [16]. As benchmarking of inference accelerators is still in its infancy, a fair comparison of available GPUs is hard. Nevertheless, any of them can get the job efficiently done.

GPUs offer powerful and scalable solutions for DNN acceleration. For uncompressed DNN models, layer operations are mapped with the help of frameworks to dense floating-point or integer matrix multiplications, which can be processed efficiently in parallel by GPUs. However, GPUs may perform poorly when operating on sparse data and compressed DNNs via fine-grained weight reduction [6, 7]. Hence, there is still room for architectural improvements and alternative solutions.

12.3 Field-Programmable Gate Arrays

Field-Programmable Gate Arrays have a modular and regular architecture containing mainly programmable logic blocks, embedded memory, and a hierarchy of reconfigurable interconnects for wiring the logic blocks. Furthermore, they may contain digital signal-processing blocks and embedded processor cores. After manufacturing, they can be configured before and during runtime by the customer. Today, system-on-chip designs with a complexity of about several billion logic gates and several megabytes of internal SRAM (Static Random Access Memory) can be mapped on state-of-the-art FPGAs. Clock rates approach the GHz range boosting the chip-computational power in the order of GOPS (billion operations per second) at a power consumption of several watts. Hence, FPGAs offer an interesting alternative for parallel implementation of DNNs providing a high degree of flexibility and a minimal time to market. The time for the development of an FPGA or application specific integrated circuit (ASIC) design is comparable. A big advantage of FPGAs is that no time for fabrication is needed. A new design can be tested directly after synthesis for which efficient CAD tools are available. A disadvantage of FPGAs is the slower speed, bigger area, and higher power consumption compared to ASICs. Compared to software implementations, FPGAs offer a higher and a more specialized degree of parallelization. Vendors of FPGAs, which have long been used to accelerate signal processing algorithms, are refining their products to suit DNN acceleration.

The implementation of DNNs on FPGAs makes it possible to realize powerful designs that are optimized for dedicated algorithms [18]. Another great advantage is the feature of reconfigurability that enables the change to a more efficient algorithm whenever possible. Using a lower precision allows to set up an optimized architecture that can be faster, smaller, or more energy-efficient than a high-precision architecture. For fine-tuning of DNNs, the FPGA can be reconfigured to implement high-precision elements. Additionally, the implemented algorithms can be adapted to the network size that is required for a certain problem. Thus, always the most suitable algorithms and architectures can be used. Furthermore, dynamic (or runtime) reconfiguration enables to change the implementation on the FPGA during runtime [19]. Dynamic reconfiguration is used to execute different algorithms on the same resources. Thus, limited hardware resources can be used to implement a wide range of different algorithms. In DNN simulations, we are often interested in providing as much computing power as possible to the simulation of the algorithm. But pre- and post-processing of the input and output data often also requires quite a lot of calculations. In this case, dynamic reconfiguration offers the opportunity to implement special pre-processing algorithms in the beginning, switch to the DNN simulation and in the end reconfigure the system for post-processing [20].

There are different approaches to implement DNNs on FPGAs, either the network itself is implemented on the FPGA or a DNN processing engine is developed for the FPGA onto which the target network is mapped at run-time. The advantage of the first approach is that it is possible to fully optimize the network for the target FPGA and achieve the best possible performance and energy efficiency. However, at the same time, this removes most forms of flexibility, as the design only works for one specific network and any changes to the network or the integration of a new network will result in several weeks of changing the design or require a complete redevelopment of the design. On the other hand, the DNN processing engine approach allows for any DNN model to be accelerated on the FPGA, as only the most common and performance critical layers are calculated on the FPGA and everything else on the CPU or on a different accelerator. Because the network is not calculated on the FPGA entirely, there will be at least some communication overhead which reduces the overall performance. Additionally, as this is a rather generic approach, the performance will be below that of a fully optimized implementation of the target network.

In many cases the DNN processing engine is the most suitable approach as it provides a high degree of flexibility. However, in situations where energy efficiency is of importance or when the FPGA needs to operate stand-alone, i.e., without a dedicated CPU, manually implementing the network on the FPGA will be the better choice.

One example of a hand optimized FPGA implementation, based on a Hardware Description Language (HDL), is presented in [22]. The authors developed an FPGA design of the VGG-16 network using a binary neuronal network (BNN). BNNs store their weights as either 0 or 1, which, even though a BNN requires a modification of the original VGG-16 architecture, significantly reduces the amount of weight storage required for the network. The developed design achieved a performance of 40.8 TOPS which equals to about 115 cl/s on an Intel Arria 10 GX1150 FPGA. Energy efficiency reaches 849.38 GOPS/W.

Other examples that fully implement the network on FPGA like [23, 24] aimed to design templates for High Level Synthesis (HLS). Convolutional layers usually have the highest computational cost in DNNs. In [23] Winograd and Fast Fourier

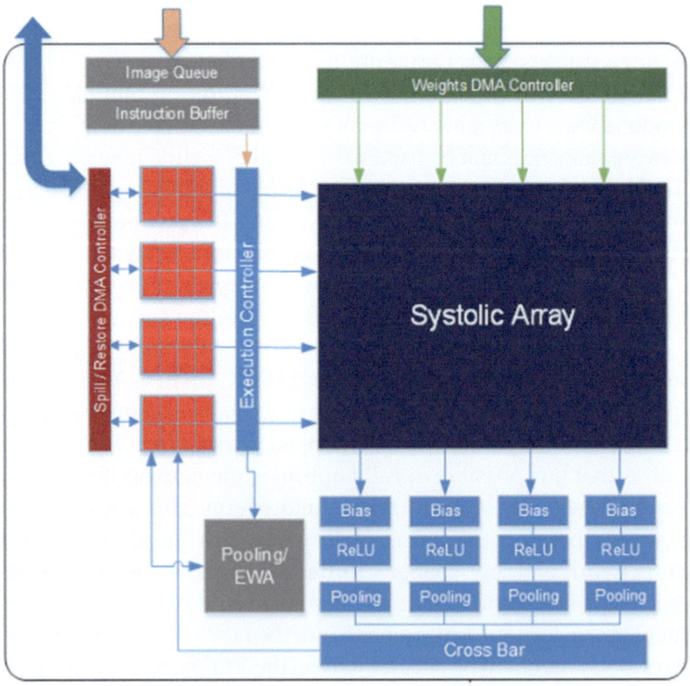

Fig. 12.5 Xilinx generic xDNN engine architecture for DNN inference [21]

Transformation (FFT) are implemented in processing elements (PEs) separately. Reuse of feature map data, pipelining and parallelization are applied as well. The authors are stating a performance of 2.5 TOPS for VGG-16 on the Xilinx ZC706 platform. In [24] also the Winograd optimization is applied to convolutional layers. Multiplications are simplified through the use of simple addition, subtraction or bit-shift operations, where possible. The implemented PEs are working in a systolic array and are organized in parallel working units. For VGG-16 a throughput of 3.8 TOPS on the Xilinx VCU118 is stated. A C3S network (3D CNN) is also implemented with the same HLS templates and achieves a 5$\times$ energy efficiency gain, compared to a GTX1080 GPU [24].

Convenient frameworks for high-level design support are an essential requirement for using FPGAs. High-level implementation tools, including Intel's OpenCL Software Development Kit and Xilinx Vivado High-Level Synthesis, and Python-to-netlist neural network frameworks, such as DNN Weaver [25], make the DNN hardware design process for both FPGAs and ASICs faster and simpler. The OpenVINO toolkit developed by Intel [26] allows DNNs to be accelerated using MCPs, GPUs, FPGAs, and VPUs (Vector Processing Units), meaning that it is possible to determine and use the best possible accelerator for a given DNN or even for specific layers of that DNN. Out of the box, OpenVINO supports a large amount of different networks for execution on FPGAs, such as AlexNet, GoogleNet, VGG-16, ResNet, Yolo and many more. Additional networks can be manually implemented using Caffe, MXNet, TensorFlow, Kaldi or ONNX models. While Intel provides FPGA implementations for most of the commonly used layers, some layers are not available on FPGAs either because they are rarely used special layers, layers that do not significantly affect the performance or are new/custom developed layers. For those layers it is possible to either specify a fall-back implementation, e.g., on CPU/GPU, or to provide a custom implementation.

Xilinx develops two frameworks for DNN processing on FPGAs. One is the generic xDNN accelerator (Fig. 12.5) in the ML-Suite [21], which can in general be compared to Intels OpenVINO toolkit. The other is the FINN Framework from Xilinx Research Labs [27], which targets DNN inference specifically for quantized neural networks. Such frameworks allows DNN architects to migrate their designs to custom hardware with relative ease. Reconfigurability enables rapid design iteration, making FPGAs ideal prototyping and deployment devices for future DNNs developments.

12.4 Application-Specific Hardware

Application-specific integrated circuits (ASICs) have the highest potential for major improvements in resource-efficient performance for DNN inference. Many various special-purpose hardware implementations for DNN inference have been proposed and the number of proposals is still increasing. Advances in technology have successively increased the ability to emulate neural networks with speed and accuracy.

Practically every processor vendor has specialized custom hardware for DNN acceleration. For digital ASICs, efficient software tools for a fast, reliable and implementation are available. Digital circuits can use standard technologies with the highest density in devices down to the lowest available structure sizes. Their time-consuming and resource-demanding fabrication processes, however, make it hard for them to keep up with the fast development of DNN algorithms.

One of the first custom ASICs for accelerating the inference phase of DNNs is Google's Tensor Processing Unit (TPU) [30]. The TPU was designed as a co-processor on a standard PCIe bus, so that it can be plugged into a server like a GPU card (Fig. 12.6). The TPU chip is programmed in the TensorFlow framework to drive many important applications in Google data centres, including image recognition, language translation, search, and game playing. A first generation TPU chip is capable of performing 92 TOPS. The die size in 28 nm CMOS is below 330 mm^2 and includes 28 MiB on-chip memory (29% of chip area), mainly for buffering neuron activations. The main logic block is the matrix multiply unit (24% chip area) with 256×256 8 Bit MAC operators. Clock speed is 700 MHz leading to 40 W measured power consumption when busy (28 W idle) [30]. As this first generation TPU was limited by memory bandwidth, the second generation design has an increased bandwidth of 600 GB/s. The second-generation TPUs can also calculate in floating point making them useful for both training and inference of DNNs. A third-generation TPU eight times as powerful as the second-generation TPUs is in use today (up to 100 Peta FLOPS) [30]. With its new Edge TPU Google offers an ASIC designed to offer DNN inference (INT8) for edge computing. The chip is much smaller and consumes far less power compared to the server TPUs. It is capable of performing 4 TOPS using 2 W resulting in about 130 cl/s for the VGG-16 DNN [31].

Besides application in data centres, DNN accelerators hold much promise for edge computing. Embedded Machine Learning (ML) at the edge can be applied to almost every electronic appliance, from production lines (industry 4.0) over house hold devices (smart home) to hand-held devices (smartphones). Applications that require resource-efficient implementations with respect to latency, power, and cost. Mobile devices are more and more equipped with sensors and embedded data processing (e.g. face detection, voice and gesture recognition, activity tracking, ...). For data processing ML methods as DNNs take over the lead and can be found in almost all new smartphones today. Hence, any smartphone vendor includes ML accelerators in their mobile SoC devices nowadays; e.g., Qualcomm in Snapdragon [33], HiSilicon in Kirin [34], Samsung in Exynos [35], or MediaTek in Helio [36]. A detailed analysis of DNN accelerators in smartphone SoCs can be found on the regularly updated official project website maintained by Andrey Ignatov (ETH Zurich, Computer Vision Lab, [37]).

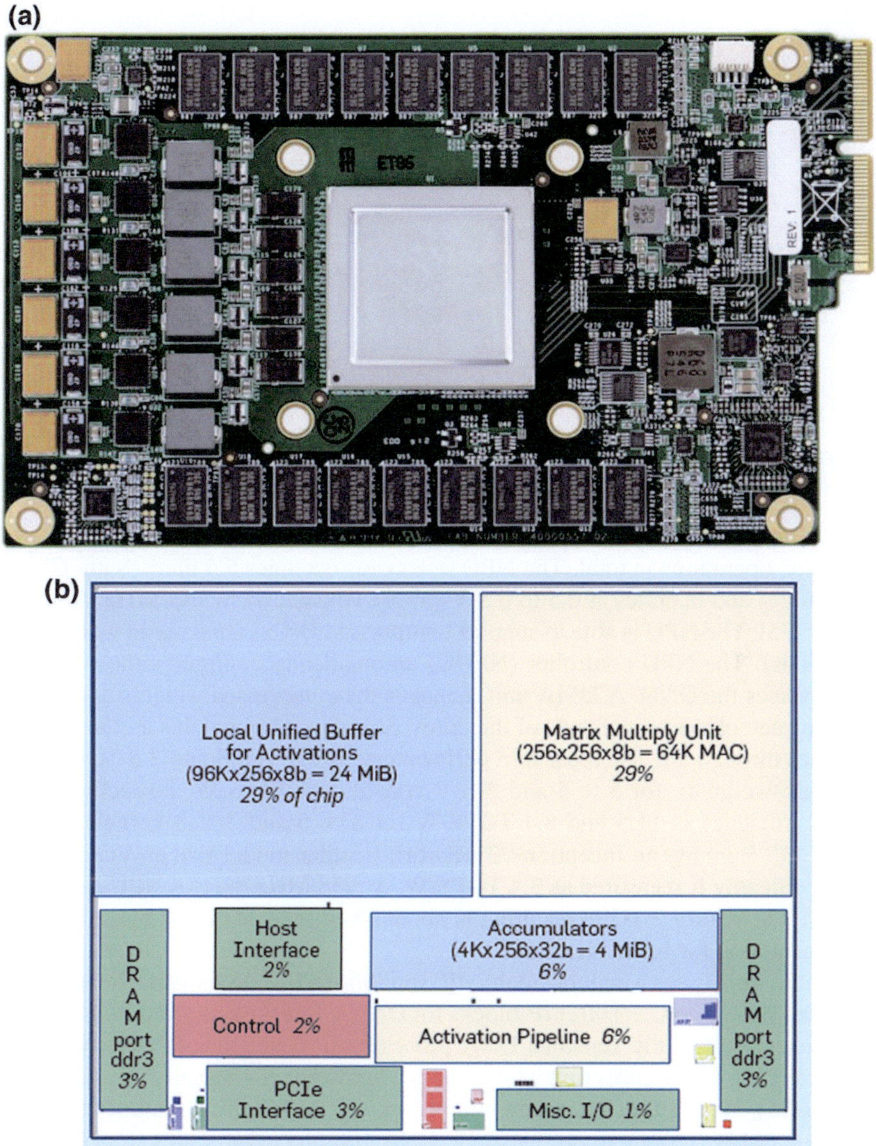

Fig. 12.6 **a** 1st generation TPU printed circuit board [32]; **b** floor plan of the TPU die [30]

For example, the Neural Processing Unit (NPU) from Samsung for their Exynos SoCs [35] features an energy-efficient butterfly-structure dual-core accelerator offering 1024 MAC operations (INT8) and three-fold parallelism in computing DNNs. The NPUs are optimized for DNN inference. The overall architecture of one NPU core is shown in Fig. 12.7. Each core has 16 arrays of dual-MAC units, performing

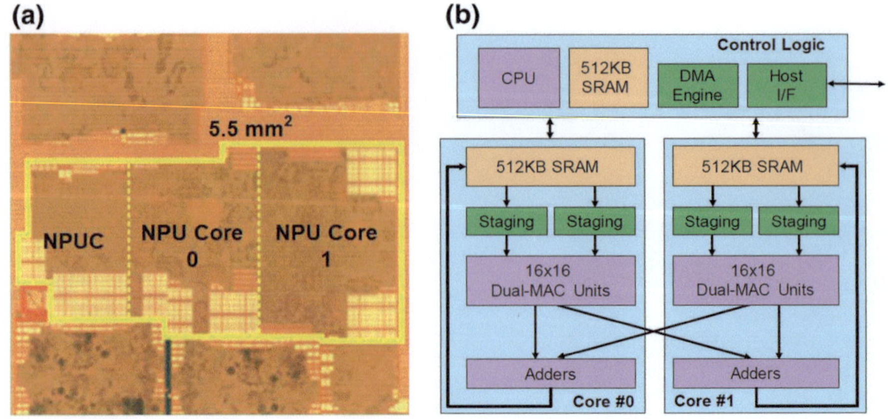

Fig. 12.7 Chip micrograph [35] (**a**) and overall architecture (**b**) of Samsung deep learning accelerator for Exynos chips [38]

512 MAC operations in total. The NPU in Exynos occupies 5.5 mm^2 (8 nm CMOS technology) and operates at 0.5 to 0.8 V supply voltage, 67 to 933-MHz clock frequency [35]. The NPU is able to support compressed DNNs (sparsity in weights and activations). The NPU controller (NPUC) automatically configures the two cores and traverses the DNN. A DMA unit manages the compressed weights and feature maps in each of the scratchpads of the cores. Skipping zero weights and activations increases throughput. The measured performance is 6.9 TOPS and 3.5 TOPS (with 75% zero-weights) for 5×5 and 3×3 convolutional kernels, respectively. The energy efficiency is 11.5 and 8.4 TOPS/W for 5×5 and 3×3 kernels, respectively [33]. Running an Inception-v3 network (similar model size as VGG-16) the energy efficieny is measured as 3.4 TOPS/W. At 933 MHz the two NPUs add up to nearly 1 TMAC/s or 2 TOPS resulting in about 65 cl/s and 43 cl/s/W of the VGG-16 implementation for ImageNet data.

Besides processor vendors, many IP vendors (ARM, Synopsis, Imagination, Cadence, VeriSilicon, ...) offer IP-blocks for DNN inference acceleration. For example, Cadence offers the Tensilica DNA processor IP (Intellectual Property) for AI inference [39]. The architecture incorporates a hardware engine and a Tensilica DSP (Vision C5). Heart of the hardware engine is a 4 K MAC array configuration with up to 3.4 TMAC/s/W in 16 nm. A single DNA 100 processor scales from 0.5 to 12 TMACs (INT8). Multiple processors can be stacked to achieve hundreds of TMACs. A Tensilica Neural Network Compiler maps a trained ANN into executable and optimized code. A SystemC model is provided for cycle-accurate system simulations [39]. Based on the estimated 3.4 TMAC/s/W a VGG-16 implementation on a DNN 100 may achieve about 110 cl/s/W.

Many startups come up with special architectures for DNN acceleration as well. They range from processor-in-memory computing (Mythic, Syntiant, Gyrfalcon) to processor-near-memory (Hailo): from programmable logic (Flex Logix) to RISC-V cores (Esperanto, GreenWaves); and from the tiny (Eta Compute) to the hyper-scale (Cerebas, Graphcore). Most of them aim for ML at the edge. For example, the Goya™ HL-1000 chip is an inference chip being developed by startup Habana Labs [40]. The scalable Goya platform architecture has been designed from the ground up for deep learning inference workloads. It comprises a fully programmable Tensor Processing Core (TPC™) along with its associated development tools, libraries and compiler. The platform is capable of massive data crunching with low latency and high accuracy. The TPC™ was designed to support deep learning workloads. It is a VLIW SIMD vector processor with ISA and hardware that was tailored to serve deep learning workloads efficiently. The HL-1000 chip uses a cluster of eight TPC™ cores and further dedicated hardware. The TPC™ natively supports several mixed-precision data types (FP32, INT32, INT16, INT8, UINT32, UINT16, UINT8). The performance achieved on VGG-16 inference is 1447 cl/s (batch size 1) with 1.1 ms latency [40]. More detailed Information on the architecture or about power consumption are currently not unavailable.

The Graphcore wafer-scale approach from Cerebas is another start-up example at the extreme end of the large spectrum of approaches [41]. The company claim to have built the largest chip ever with 1.2 trillion transistors on a 46,225 mm^2 silicon wafer (TSMC 16 nm process, Fig. 12.8). It contains 400,000 ML optimized cores, 18 GB on-chip memory, and 9 PetaByte/s memory bandwidth. The programmable cores with local memory are optimized for ML primitives and equipped with high-bandwidth and low latency connections. DNN approximations are incorporated, such as fine grained sparsity. The 2D mesh topology is a fully configurable fabric with hardware supported communication. The entire wafer operates on a single DNN and supports learning. Common ML-frameworks (e.g. Tensorflow, PyTorch) can be

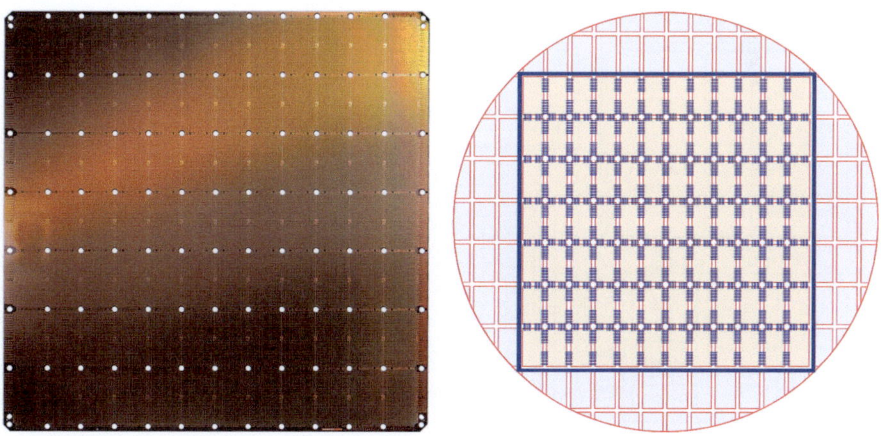

Fig. 12.8 Graphcore wafer from Cerebas: 400,000 ML-Cores on 46,225 mm^2 [41, 42]

used for programming the wafer engine, Cerebas tools map, place, and route the network layers onto the wafer. Redundancy for cores and links can be incorporated to replace defective elements [41]. The company announced that the system is running customers workload, but more detailed information on the architecture and performance data are not published yet.

Last but not least, academia is very active in the DNN chip landscape as well. Examples are the Eyeriss architecture from MIT [43], ENVISON from KU Leuven [44], STICKER-T from Tsinghua [45], or DNPU from KAIST [46]. The architectures have in common a 2d-array of special processing elements and controllers for an efficient data flow from and to the memory. For example, the ENVISION chip from KU Leuven is equipped with 2D- (for convolutions) and 1D-SIMD arrays (for ReLU, max-pooling), and a scalar unit (Fig. 12.9). An on-chip memory (DM) consists of 64×2 kB single-port SRAM macros which can be read or written in parallel [44]. The processor has a 16 bit SIMD instruction set extended with custom instructions. The chip is divided into three power- and body-bias domains to enable granular dynamic voltage scaling. Implemented in a 28 nm FDSOI technology on 1.87 mm^2, the chip runs at 200 MHz at 1 V and room temperature. Energy-efficiency is further improved by modulating the body bias in an FDSOI technology. This permits tuning of the dynamic versus leakage power balance while considering the computational precision. Efficiency is 2 TOPS/W on average for VGG-16 (about 13 cl/s/W) and up to 10 TOPS/W peak (about 64 cl/s/W).

In conclusion, the DNN accelerator development is progressing fast with a steady stream of new architectures coming up. At first, the acceleration of the data flow had the highest priority. A range of customized blocks of large parallel arrays multiply-add units for efficient and flexible computation of the many convolutions and fully connected layers were proposed. As DNNs got larger, the circuit designers realized

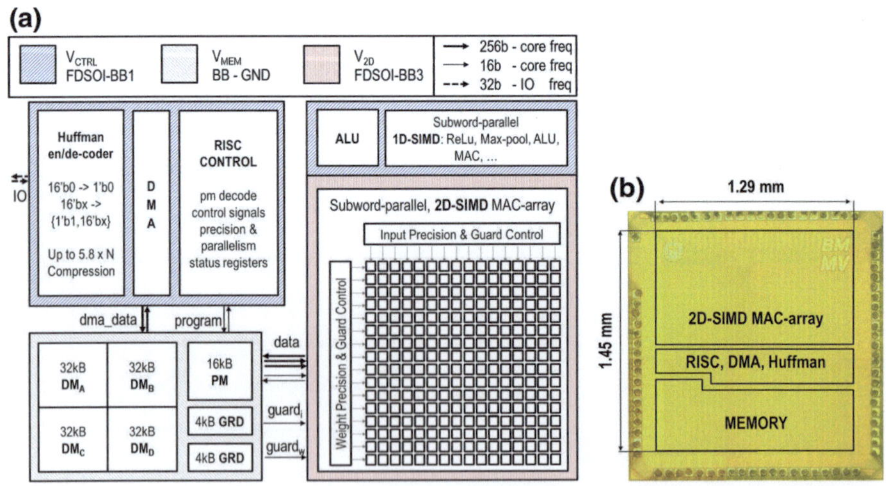

Fig. 12.9 Top level architecture of the ENVISON DNN accelerator (**a**) and chip micrograph (**b**) [44]

that memory access and data movement are more critical than arithmetic. Additional circuitry like buffers, transpose logic, nonlinearity logic must be employed to keep the MAC units busy while utilizing the memory bandwidth efficiently. With even larger networks DNN approximation and compression is used in order to match the application requirements for throughput, latency and energy efficiency. Counterintuitively with the growth of model size and complexity, it has been shown that out of the millions of parameters used in common DNN architectures, many of these can be removed with insignificant reductions in accuracy, leading to a much lower memory footprint for storing the model as well as less computations and significantly lower energy usage. This process is referred to as pruning and can be applied to both connection weights and neurons and there are various methods proposed in the literature [6, 7]. Another popular method to increase the efficiency reduces the numerical precision of weights and activations. This method is called quantization. Typically, single precision floating point numbers are used to represent weights and activations. However, networks with ternary weights (+1, 0, and −1) show in some cases only low performance loss when compared to the floating-point counterparts. In addition to pruning and quantization, there are many other methods that can be used to make DNN accelerators more efficient. However, combining these, in some cases contradictive methods, in an optimal way, as well as optimising a model to meet system requirements, is still an unsolved problem. Hence, system designers should jointly consider hardware/software issues for finding optimal compromises, so called pareto-optimal architectures for efficient and flexible implementation of DNN accelerators. The "AI chip landscape" [47] is not settled yet.

12.5 Benchmarking

There is a tremendous surge of innovation in DNN hardware, making it a very challenging task choosing the best hardware offering for a given application. Hardware vendors have yet no incentive to provide unbiased comparison and benchmarking. Hence, there is a high demand for benchmarking, the objective performance measuring based on specific indicators, of such systems. Though benchmarking of DNN accelerators is still in its infancy, there are first approaches to fill this gap. Baidu and Google together with researchers from several universities launched the MLPerf benchmark suite in May 2018 in part to create a fair way to measure the chips expected from "dozens and dozens" of startups. The MLPerf approach is supported by over 50 companies and researchers from 7 universities [48]. Hence, it has a chance to become for AI accelerators what SPEC benchmark is for CPUs. The suite itself consists of two major subparts: training and inference benchmarks. At time of writing, the suite is still in beta stadium (v0.6) and inference results are available only for a previous version of the suite. Training results can be found on their webpage [48]. Main metric of all benchmarks is the wallclock time to reach a pre-defined goal using a pre-defined network model (e.g. a certain accuracy on a machine learning task). The set of benchmarks comes together with a set of rules for submitting results, and most importantly, results have to be submitted together with the source code. This allows to reproduce the results to some extent if the respective hardware is at your disposal.

For smartphones based on Google's Android operating system and Mobile SoCs in general there is the AI-Benchmark [49]. Since 2019, the suite is also capable of benchmarking CPUs, GPUs and TPUs based on Tensorflow, allowing to compare workstation hardware with mobile hardware. The suite is focusing on inference and consists of 21 tests distributed over 11 benchmark sections. The overall rules are not that strict compared to the MLPerf approach. However, in this case benchmarking is strongly coupled to the implementation of the networks, relying on frameworks and drivers to efficiently map individual networks to the hardware. Results in the form of a ranking can be found online [50]. A similar approach is applied by the EEMBC MLMark benchmark library [51]. Instead of using the measured wallclock time as a benchmark metric, MLMark measures throughput, latency and accuracy targeting requirements of embedded applications. Currently, the suite consists of three models only: MobileNet [52], MobileNet-SSD and ResNet-50. The results are visible to registered members only.

One major point, when it comes to embedding DNN accelerators, is not only the wallclock time per inference or latency, which is constrained by e.g. real-time requirements of your application, but also the resource-efficiency. Mobile applications, like autonomous driving, robot control or everyday tasks on a smartphone, require resource-efficient implementations for DNN inference. Despite these demands, this benchmark measure is not served by any of the discussed benchmark approaches. The cost of employing a system of multiple GPUs/TPUs is also a very restricting factor in training deep networks, which is also not considered by any of the suites.

DNNs are a field with rapid development. This complicates representative and up-to-date benchmarking of hardware accelerators, which is reflected in the state of all machine learning benchmark suites. The performance of a DNN platform depends on many aspects, such as computational accuracy or tool assistance. A special architecture may perform well on a DNN of type A, but worse on another of type B. At present, a fair comparison is almost impossible. Only few chips have been fully described and benchmarked (e.g. Google's TPU) but the pipeline of new implementations is full.

In Fig. 12.10 the performance values of the introduced DNN accelerators (Tables 12.1, 12.2, 12.3) implementing VGG-16 have been merged. The expected clustering from low end edge devices over FPGA implementations to high end ASICs and GPUs can roughly be seen. The effect of the batch size is clearly visible [e.g. V100: 821 cl/s (batch size 1) up to 2845 cl/s (batch size 128)] as well. However, such figures have to be considered with caution. First of all, most of the data have been taken directly from the publications and are not a result of an objective benchmark measurement. In most cases it is unclear how these data are obtained. Especially, power data are in most cases missing. Second, though the comparison is based on a fixed DNN model (VGG-16) and data set (ImageNet) there are many additional architectural and technological attributes influencing the performance data. Obviously, the numeric precision for weights and activations plays an important role (e.g.

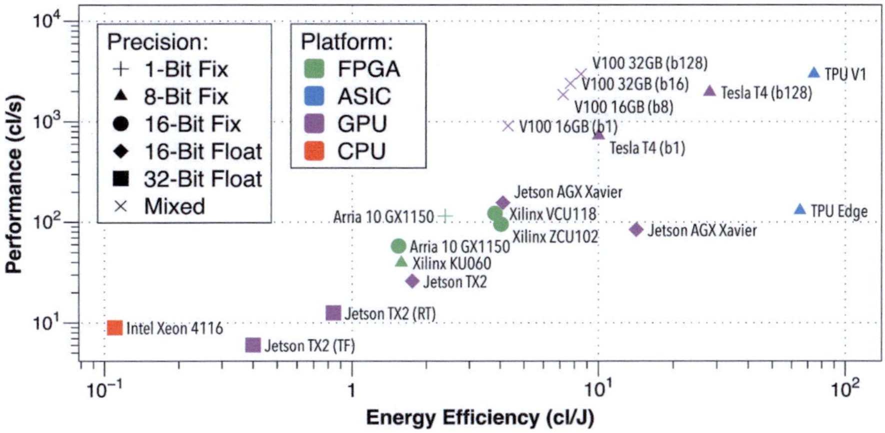

Fig. 12.10 Comparison of VGG-16 implementation results of selected accelerators

Table 12.1 VGG-16 inference performance data (batch size 1) of selected CPUs and GPUs (batch size 1, * estimated)

Ref.	Type	(GOPS/s)	(GOPS/s/W)	cl/s	cl/s/W	Power (W)	Precision
[17]	Intel Xeon	215	2,5*	9	0,11*	85	FP32
[12]	GV100	125,800	419*	821	24	300	FP16
[12]	TU104	178,000	774*	726	10	230	INT8
[15]	Jetson TX2	810	54*	26*	1,75*	15	FP16
[15]	Jetson TX2	391	26*	12,6*	0,84*	15	FP32

Table 12.2 VGG-16 inference performance data (batch size 1) of selected FPGA accelerators (all in 20 nm CMOS, * estimated)

Ref.	FPGA	GOPS/s	GOPS/s/W	cl/s	cl/s/W	Power (W)	Precision
[22]	Arria 10 GX1150	40,770	849,38	114,8	2,39	48	1-bit fix
[23]	Xilinx ZCU102	2940.7	124,6	95,05*	*4,03	23,6	INT16
[24]	Xilinx VCU118	3772	117,88	121,91*	3,81*	32	INT16
[28]	Arria 10 GX1150	1790	47,78	57,85*	*1,54	37,5	INT16
[29]	Xilinx KU060	1171.7	46,87	39,53*	1,58	25	INT8

Table 12.3 VGG-16 inference performance data (batch size 1) of ASIC DNN accelerators (* estimated)

Ref.	ASIC	(GOPS/s)	(GOPS/s/W)	cl/s	(cl/s/W)	Power (W)	Precision
[29]	TPU V1	92,000	2300	2968*	74.2*	40	INT8
[30]	TPU Edge	4000	2000	130*	65*	2	INT8
[33]	Exynos	2000	1333	65*	43*	1.5	INT8
[35]	DNA 100	3400	3400	110*	110*	1	INT8
[40]	ENVISON	2500	2000	16*	13*	0.8	INT6

Jetson TX2 about 13 cl/s for FP32 and 26 cl/s for FP16) as well as the data flow management. The size and the utilization of the implemented MAC arrays are essential as well. Utilization in this context is the percentage of the raw compute capabilities of the system that can be effectively used for a real workload (DNN model). Utilization varies with the DNN model, but utilization figures are rarely published. A high peak performance of an accelerator is no guarantee for a high inference performance. Last but not least, the framework used for DNN implementation has a high impact on inference performance (e.g. Jetson TX2 FP32 6 cl/s using TensorFlow and 13 cl/s using TensorRT).

12.6 Outlook

Parallel standard hardware like multi-core MPCs, GPUs, or FPGAs are cost effective, available, and benefit from market-driven development improvements in the future. They have the highest flexibility and are manufactured in standard technologies with highest device densities. They set the base-line with respect to cost and performance for DNN implementation. ASICs have the highest potential for major improvements in resource-efficient performance for DNNs. Currently, product developers and users have few real choices for hardware supporting efficient DNN implementation. Almost all major IT and chip companies are aggressively entering the market. However, the increased competition doesn't necessarily mean better choices, as customers still don't have the means to evaluate these different chipset platforms for the optimal integration with their AI-driven system and application demands.

Due to their highly regular and modular structure, inherent fault-tolerance, and learning ability, ANNs offer an attractive alternative for ultra-large-scale integration and the development of resource-efficient systems with minimal total energy consumption combined with a small size and fault-tolerant behaviour. Among the many different ANN models discussed in literature (see e.g. "The Neural network Zoo" [53]) DNNs serve in this chapter as a representative example architecture. Despite the impressive development of nanoelectronics during the last decades, there is still no clear consensus on how to exploit this technological potential for massively-parallel ANN implementations. Hence, it is currently quite difficult to determine the best way to perform DNN calculations for any given application. This is one reason for the huge variety of approaches to DNN hardware implementation known today.

DNNs look promising but have many variations, and the algorithms are still in development, so it is not clear how they may influence hardware development in the future. Implementations are still incomplete and immature. There is a lack of standardization, e.g. for model data formats, file formats to transfer models and data sets between frameworks, or interfaces to build engineering tools that work together. A first step in this direction is the specification of the Brain Floating Point (BFLOAT16) half-precision data format for DNN learning [54]. Its dynamic range is the same as that of FP32, making conversion between both straightforward, and training results are almost the same as with FP32. Industry-wide adoption of BFLOAT is expected.

Another challenge lies in mastering the design complexity and achieving economic viability for integrated systems with more than a billion devices per square centimetre. This requires system concepts that both exhaust the possibilities of future technologies and reduce the design- as well as the test-complexity. These arguments were already a strong motivation for ANN hardware in the 80s [55]. Flexibility is another important factor as researchers are coming up with ANN concepts all the time. While DNNs are the dominant model especially for image processing today, other types of ANN models are more suited to other applications, such as speech recognition or controlling tasks. Hence, today´s accelerators may be too specialized to accelerate future ANN models. The challenge is to find the right balance of flexibility, performance, and price for as many applications as possible. This should go hand in hand with efficient software frameworks for developers and users of ANN hardware in this rapidly evolving sector.

In conclusion, as the increase in processor speed slows down alternative architectures get a second chance today. The hunt for the right architecture is just beginning. As in the late 80s [56], within the second neural network hype, analogue computing, wafer-scale integration, 3D-integration, in-memory computing, massively parallelism, and even optical approaches are being explored again. Radically new ideas for circuit designs or system architectures are not in sight. At present, know-how from digital signal processing and data flow management from massively parallel computing architectures are combined in different ways. An obvious approach is to bring the memory closer to the arithmetic devices to mitigate the memory bottleneck and reduce power consumption. In-Memory-Computing (IMC) exploiting dense 2D memory arrays and matrix-vector multiplication offer an interesting alternative approach to achieve high throughput with low power requirements for DNN accelerators [57–59]. However, model sizes of today's DNNs are generally too large to fit into on-die memory resources. On-die memory can be used to mitigate the memory bandwidth problem, but deciding what stays on-die versus off-die requires careful memory management to achieve high performance. Even more computational power may be obtained by emerging technologies like quantum computing, molecular electronics, or novel nano-scale devices (memristors, spintronics, nanotubes (CMOL)), but these technologies will not be available on broad basis in the next decade. Today, we are still early in the efficient use of nanoelectronics, and we are keenly awaiting the technology we can use tomorrow.

References

1. K. Simonyan, A. Zisserman, Very deep convolutional networks for large-scale image recognition. arXiv:1409.1556v6 (2015)
2. D. Gschwend, ZynqNet: an FPGA-accelerated embedded convolutional neural network. Masters Thesis, ETH Zürich (2016)
3. V. Sze et al., Efficient processing of deep neural networks: a tutorial and survey. arXiv:1703.09039 (2017)
4. O. Russakovsky et al., ImageNet large scale visual recognition challenge. arXiv:1409.0575v3 (2015)
5. T.L.I. Sugata, C.KJ. Yang, Leaf App: leaf recognition with deep convolutional neural networks. In: IOP Conf. Series: Materials Science and Engineering, vol. 273 (2017)
6. E. Wang et al., Deep neural network approximation for custom hardware: where we've been, where we're going. ACM Comput. Surv. **52**(2), Article 40 (2019)
7. Y. Cheng et al., Model compression and acceleration for deep neural networks: the principles, progress, and challenges. IEEE Signal Process. Mag. **35**, 1 (2018)
8. M. Gerland et al., Parallel computing experiences with CUDA. IEEE Micro **28**(4), 13–27 (2008)
9. J.E. Stone et al., OpenCL: a parallel programming standard for heterogeneous computing systems. Comput. Sci. Eng. **12**(3), 66–72 (2010)
10. K.S. Oh, K. Jung, GPU implementation of neural networks. Pattern Recognit. **37**(6), 1311–1314 (2004)
11. NVIDIA, TESLA V100 GPU Architecture, Whitepaper (2017)
12. https://developer.nvidia.com/deep-learning-performance-training-inference (retrieved 31.10.2019)
13. NVIDIA, Turing GPU Architecture, Whitepaper, (2018)
14. https://www.nvidia.com/en-us/data-center/products/egx-edge-computing/ (retrieved 31.10.2019)
15. https://github.com/NVIDIA-AI-IOT/tf_to_trt_image_classification (retrieved 31.10.2019)
16. https://www.amd.com/en/technologies/vega7nm (retrieved 31.10.2019)

17. M. Almeida et al., EmBench: quantifying performance variations of deep neural networks across modern commodity devices. arXiv:1905.07346v1 (2019)
18. A.R. Omondi, J.C. Rajapakse (eds.), FPGA Implementations of Neural Networks (Springer, Berlin, 2005)
19. M. Koester et al., Design optimizations for tiled partially reconfigurable systems. IEEE Trans. Very Large Scale Integr. Syst. **19**(6), 1048–1061 (2010)
20. M. Porrmann, U. Witkowski, U. Rückert, Implementation of self-organizing feature maps in reconfigurable hardware, in *FPGA Implementations of Neural Networks*, ed. by A.R. Omondi, J.C. Rajapakse (Springer, Berlin, 2005), pp. 253–276
21. https://github.com/Xilinx/ml-suite/blob/master/docs/ml-suite-overview.md (retrieved 31.10.2019)
22. D.J.M. Moss et al., High performance binary neural networks on the Xeon + FPGA™ platform, in *27th International Conference on Field Programmable Logic and Applications (FPL), Ghent* (2017), pp. 1–4
23. Y. Liang, L. Lu, Q. Xiao, S. Yan, Evaluating fast algorithms for convolutional neural networks on FPGAs. IEEE Trans. Comput.-Aided Des. Integrat. Circuits Syst. https://doi.org/10.1109/tcad.2019.2897701 (2018)
24. J. Shen, Y. Huang, M. Wen, C. Zhang, Towards an efficient deep pipelined template-based architecture for accelerating the entire 2D and 3D CNNs on FPGA. IEEE Trans. Comput.-Aided Des. Integrat. Circuits Syst. https://doi.org/10.1109/tcad.2019.2912894 (2018)
25. H. Sharma et al., From high-level deep neural models to FPGAs, in *IEEE/ACM International Symposium on Microarchitecture* (2016)
26. https://docs.openvinotoolkit.org (retrieved 31.10.2019)
27. M. Blot et al.: FINN-R: an end-to-end deep-learning framework for fast exploration of quantized neural networks. arXiv:1809.04570v1 (2018)
28. J. Zhang, J. Li, Improving the performance of OpenCL-based FPGA accelerator for convolutional neural network, in *Proceedings of the 2017 ACM/SIGDA International Symposium on Field-Programmable Gate Arrays (FPGA '17)*. ACM, New York, NY, USA, 25–34 (2017)
29. C. Zhang et al., Caffeine: towards uniformed representation and acceleration for deep convolutional neural networks. IEEE Trans. Comput. Aided Des. Integrat. Circuits Syst. **38**(11), 2072–2085 (2019)
30. P. Norman et al., A domain-specific architecture for deep neural networks. Commun. ACM **61**, 9 (2018)
31. A. Reuther et al., Survey and benchmarking of machine learning accelerators. arXiv:1908.11348v1 (2019)
32. https://images.anandtech.com/doci/12195/google-tpu-board-2.png (retrieved 31.10.2019)
33. https://www.qualcomm.com/snapdragon (retrieved 31.10.2019)
34. http://www.hisilicon.com/en/Solutions/Kirin (retrieved 31.10.2019)
35. J. Song et al., An 11.5TOPS/W 1024-MAC butterfly structure dual-core sparsity-aware neural processing unit in 8 nm flagship mobile SoC, in *Proceeding of the IEEE International Solid-State Circuits Conference* (2019), pp. 130–132
36. https://i.mediatek.com/P60 (retrieved 31.10.2019)
37. http://ai-benchmark.com (retrieved 31.10.2019)
38. L. Gwennap, EXYNOS 9820 has Samsung AI Engine, Microprocessor Report (11 March 2019)
39. https://ip.cadence.com/ai, (retrieved 31.10.2019)
40. Habana Labs, Goya™ Inference Platform White Paper (2019)
41. https://www.graphcore.ai (retrieved 31.10.2019)
42. M. Demler, CEREBRAS BREAKS THE RETICLE BARRIER: Wafer-Scale Engineering Enables Integration of 1.2 Trillion Transistors, Microprocessor Report (2 Sept 2019)

43. Y-H. Chen et al., Eyeriss: an energy-efficient reconfigurable accelerator for deep convolutional neural networks. IEEE J. Solid-State Circuits **52**(1), 127–138 (2016)
44. B. Moons et al., ENVISION: a 0.26-to-10TOPS/W subword-parallel dynamic-voltage-accuracy-frequency-scalable convolutional neural network processor in 28 nm FDSOI, in *Proceeding of the IEEE International Solid-State Circuits Conference* (2017), pp. 146–148
45. J. Yue et al., A 65 nm 0.39-to-140.3TOPS/W 1-to-12b unified neural-network processor using block-circulant-enabled transpose-domain acceleration with 8.1 $\times$ higher TOPS/mm^2 and 6T HBST-TRAM-based 2D data-reuse architecture, in *Proceeding of the IEEE International Solid-State Circuits Conference* (2019), pp. 138–140
46. D. Shin et al., DNPU: An 8.1TOPS/W reconfigurable CNN-RNN processor for general-purpose deep neural networks, in *Proceeding of the IEEE International Solid-State Circuits Conference* (2017), pp. 140–142
47. https://basicmi.github.io/AI-Chip/ (retrieved 31.10.2019)
48. MLPerf: https://mlperf.org/ (retrieved 31.10.2019)
49. A. Ignatov et al., AI benchmark: all about deep learning on smartphones in 2019. ariv.191006663v1 (2019)
50. http://ai-benchmark.com/ (retrieved 31.10.2019)
51. https://www.eembc.org/mlmark/ (retrieved 31.10.2019)
52. A.G. Howard et al., Mobilenets: efficient convolutional neural networks for mobile vision applications. arXiv preprint arXiv:1704.04861 (2017)
53. https://www.asimovinstitute.org (retrieved 31.10.2019)
54. D. Kalamkar et al., A study of BFLOAT16 for deep learning training. arXiv:1905.12322v3 (2019)
55. U. Ramacher, U. Rückert (eds.), *VLSI Design of Neural Networks* (Kluwer Academic, Boston, 1991)
56. P. Inne, in *Digital Connectionist Hardware: Current Problems and Future Challenges, Biological and Artificial Computation: From Neuroscience to Technology*. Lecture Notes in Computer Science, vol. 1240 (Springer, Berlin, 1997), pp. 688–713
57. N. Verma et al., In-memory computing: advances and prospects. IEEE Solid-State Circuits Mag. **11**(3), 43–55 (2019)
58. C. Eckert et al., Neural cache: Bit-serial in-cache acceleration of deep neural networks. IEEE Micro **2019**, 11–19 (2019)
59. X. Si et al.: A twin-8T SRAM computation-in-memory macro for multiple-bit CNN-based machine learning, in *Proceeding of the IEEE International Solid-State Circuits Conference* (2019), pp. 396–398

Chapter 13
Enabling Domain-Specific Architectures with Programmable Devices

Alireza Kaviani

13.1 Introduction and Background

Advances in process technology and Moore's law have enabled programmable devices to grow more than four orders of magnitude in capacity. The performance of these devices has skyrocketed by a factor of 100, while cost and energy per operation have decreased by more than a factor of 1000 (see Fig. 13.1). Field Programmable Gate Arrays (FPGAs) were introduced in mid-80s, and later proved to be the dominant form of the programmable devices [1]. Today's FPGA market is more than $5 Billion and still growing three and a half decades after introduction. These advances have been fueled by process technology scaling, but the FPGA success story is also about architecture and software choices made in the industry.

The first wave of success for FPGAs came from replacing custom logic designs and Application-Specific Integrated Circuits (ASICs). In the 1980s, ASIC companies introduced the built-to-order custom integrated circuit to the electronics market with a powerful product. ASIC companies began fiercely competing to sell on the market; the winning attributes were low cost, high capacity and speed. At the time, FPGAs compared poorly on those measures, but they thrived by the virtue of programmability. At those early days transistors were of high value and FPGAs were disregarded and deemed the worst wastage of transistors. Additional transistors were utilized for accommodating field programmability by allowing the user to implement the designs on the manufactured devices off the shelf. Time has proven that such programmability and availability at the time of market need is highly valuable, which led to growth of FPGAs through the first wave.

A. Kaviani (✉)
Distinguished Engineer, Xilinx Research Labs, 2100 Logic Drive, San Jose, CA 95124, USA
e-mail: alireza.kaviani@xilinx.com

B. Murmann and B. Hoefflinger (eds.), *NANO-CHIPS 2030*,
The Frontiers Collection, https://doi.org/10.1007/978-3-030-18338-7_13

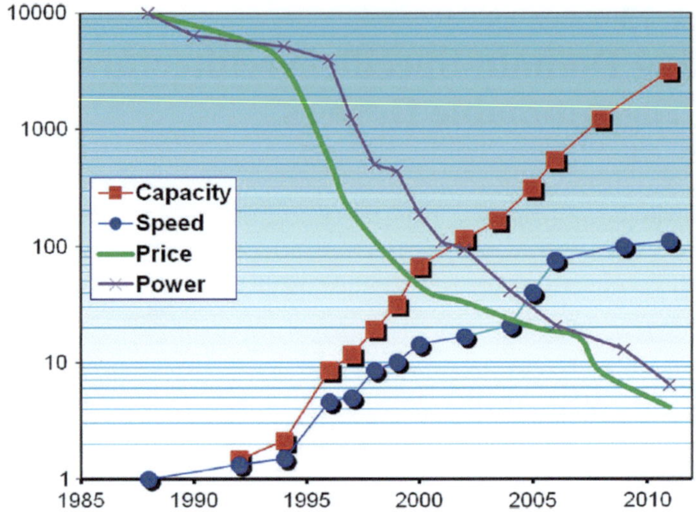

Fig. 13.1 Xilinx FPGA evolution since mid-80s (from [1], ©IEEE 2015). Capacity is logic cell count. Speed is same-function performance in programmable fabric. Price and Power are per logic cell and scaled by 10,000

After the first wave of replacing the custom logic, FPGAs were common components of digital systems. Moore's Law helped the capacity of FPGAs grow beyond a collection of LUTs, flip-flops, I/O and programmable routing. They included multipliers, RAM blocks, multiple microprocessors, and high speed transceivers. This enabled FPGAs to penetrate a huge market in the data communications industry. The FPGA business grew not from general ASIC replacement, but from adoption by the communications infrastructure. Companies such as Cisco Systems used FPGAs to make custom data paths for steering huge volumes of internet and packetized voice traffic through their switches and routers [2]. New network routing architectures and algorithms could quickly be implemented in FPGAs and updated in the field. Sales to the communications industry segment grew rapidly to well over half the total FPGA business during this 2nd wave of growth. The increasing cost and complexity of silicon manufacturing eliminated "casual" ASIC users. ASICs expanded by adding programmability in the form of application specific standard product (ASSP) and system-on-chip (SoC) devices. An SoC combines a collection of fixed function blocks along with a microprocessor subsystem. The function blocks are typically chosen for a specific application domain, such as image processing or networking. The SoC gave a structure to the hardware solution, and programming the microprocessors was easier than designing hardware. Leveraging the FPGA advantages, programmable ASSP devices served a broader market, amortizing their development costs more broadly. Companies building ASSP SoCs became fabless semiconductor vendors in their own right, able to meet sales targets required by high development costs.

The FPGA industry is currently at its early stage of riding a third wave, which is serving the computing market. Programmable devices have kept their key advantages of the first two waves for customized logic and communication and preparing for the compute and acceleration opportunities in the data centers. Confirming this trend, Intel acquired the 2nd largest FPGA company in 2015 at approximately $16.7 billion. The combination with FPGA technology is expected to enable new classes of products that meet customer needs in the data center and Internet of Things (IoT) market segments. In the words of Brian Krzanich, the CEO of Intel in 2015, "With this acquisition, we will harness the power of Moore's Law to make the next generation of solutions not just better, but able to do more. Whether to enable new growth in the network, large cloud data centers or IoT segments, our customers expect better performance at lower costs."

A fundamental early insight in the programmable logic business was that Moore's Law would eventually propel FPGA capability to cover ASIC requirements. Today, transistors are abundant, and their number is no longer a cost driver in the "FPGA versus ASIC" decision. Many ASIC customers use older process technology, lowering their NRE cost, but reducing the per-chip cost advantage. Instead, performance, time-to-market, power consumption, I/O features and other capabilities are the key factors. Solving transistor-level design problems such as testing, signal integrity, crosstalk, I/O design and clock distribution along with eliminating the up-front masking charges helped the FPGAs grow and have a prominent footprint in the semiconductor industry. Advances in process technology have enabled FPGAs to grow in capacity and implement large heterogenous systems in a device. The emerging devices are highly adaptable—making them the candidate of choice for a wide range of emerging domains from compute to networking. In the next section we will have a deeper dive to introduce various aspects of these devices that will be introduced to the market in the next few years with a special interest to address the compute domain.

13.2 Highly Integrated Emerging Programmable Devices

Xilinx is introducing the latest FPGAs in 7 nm process technology; a new heterogeneous compute family, called the Adaptive Compute Acceleration Platform (ACAP). In addition to the next generation Programmable Logic (PL), this monolithic platform includes vector and scalar processing elements tightly coupled together with a high-bandwidth network-on-chip (NoC), which provides memory-mapped access to all three processing element types. This tightly coupled hybrid architecture, is called Versal™ and is conceptually depicted in Fig. 13.2. It allows more dramatic customization and performance increase than any previous programmable device. This is an architecture solution for the computing and communication needs of modern applications. The scaler ARM processors and platform management controller occupy the lower left region of the chip. The adjacency of the Processor Subsystem (PS) to Gigabit Transceivers (GTs), memory controllers, and the NoC enables those

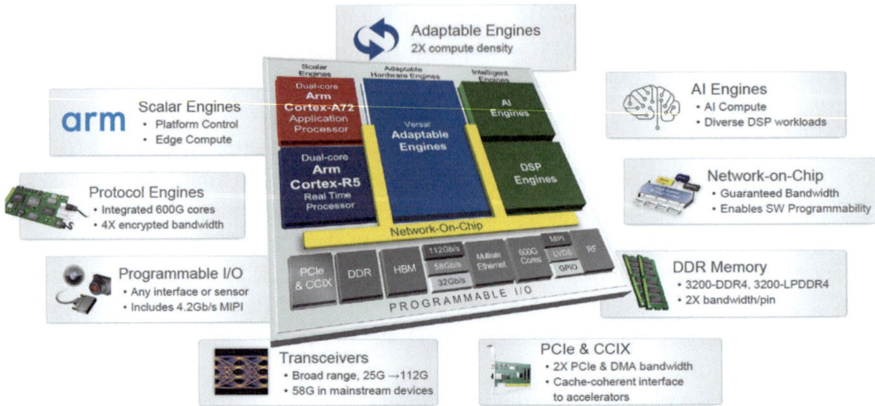

Fig. 13.2 Xilinx ACAP devices will include a number of heterogenous blocks to enable the new wave of customized compute

blocks to be used together without any of the fabric being programmed. GTs can occupy the left and right edges of the fabric regions. High speed IOs also run along the bottom edge of the die, which include hardened memory controllers to interface with off-chip memory such as DDR and HBM. Across the top of this example Versal architecture based floorplan is an array of AI Engines designed to accelerate math intensive functions for applications including machine learning and wireless. Finally, a hardened network-on-chip (NoC) augments the traditional fabric interconnect and enables a new class of high speed, system level communication between the various heterogeneous features, including the PS, DDR, AI Engines and FPGA fabric (in blue). In this section, we provide more detailed information on each heterogenous block, providing an overall understanding for the upcoming FPGAs in the next few years [3].

13.2.1 Programmable Fabric

Traditionally, the core architecture of an FPGA consists of an array of Configurable Logic Blocks (CLBs) and an interconnect with programmable switches, as simplified in Fig. 13.3a. This fabric, which is the core differentiation of FPGAs with other semiconductors, has benefited the most from Moore's law. In this subsection, we introduce the latest programmable fabric, highlighting the vast evolution compared to early FPGAs. In Fig. 13.3b, a representative device floorplan for upcoming Xilinx Versal architecture is depicted. The fabric portion of the simplified floorplan in Fig. 13.3b (in blue) is conceptually similar to a traditional FPGA; it includes resources such as LUTs, flip-flops, and a rich interconnect to connect them. Every

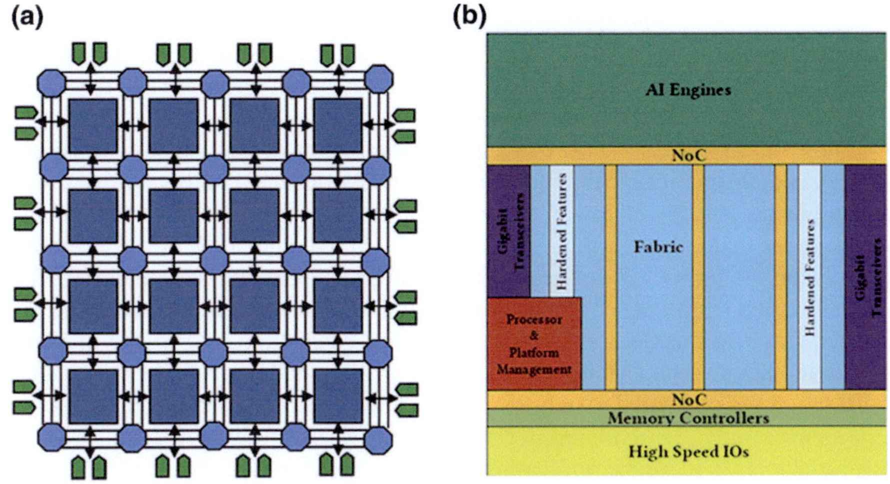

Fig. 13.3 **a** Conceptual 4 × 4 array of PL with three wiring tracks and switches at the intersection circles ([1], ©IEEE 2015). **b** Xilinx Versal representative device floorplan ([4], ©ACM 2019)

CLB contains 32 look-up tables (LUTs) and 64 flip-flops. The LUTs can be configured as either one 6-input LUT with one output, or as two 5-input LUTs with separate outputs but common inputs. Each LUT can optionally be registered in a flip-flop [3].

CLBs in early FPGAs contained a single LUT and register with 3 or 4 inputs. The CLB in the Versal architecture contains more than 60 times the amount of logic and registers in comparison with early FPGAs. By enlarging the CLB to include more logical elements, a significant fraction of local nets is subsumed internally, thereby reducing global track and wiring demand. A dedicated local interconnect structure resides within each CLB to support more versatile intra-CLB connectivity as shown in Fig. 13.4a. This is a clear architectural response to the technology scaling dynamics. Wire distances shrunk with scaling, but the cross-sectional area shrunk quadratically, resulting in a net increase in resistance for each generation. Despite the physical distance shrink and transistor delay speed up, total delay would have

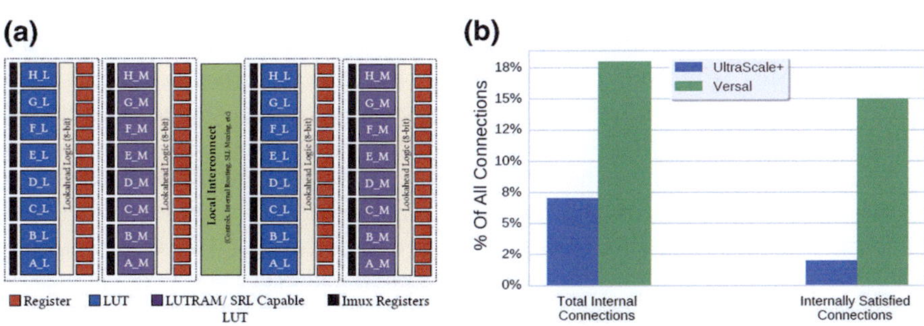

Fig. 13.4 **a** Versal CLB, **b** Internal routing structure in course-grained CLB ([4], ©ACM 2019)

increased with more advanced process nodes. Hence, the designers were forced to use thicker metal with lower resistance to reduce wire delays. As technology scales, metal resources became more expensive and architectural changes such as coarser CLBs were a necessity for more efficiency.

Empirical experiments show that a significant fraction of nets have very localized sources and destinations. Since local routes are shorter and can be squeezed with tighter pitches onto fewer, lower level metal layers, the implementation cost of local routes is substantially less than global routes. On average, 18% of all pin to pin connections are intra-CLB connections, in contrast to 7% within the smaller CLB in previous 16 nm Ultrascale™ architecture. Figure 13.4b denotes this as "Total Internal Connections." In practice, roughly 83% of those connections are routed in Versal due to limitations such as tools. This is noted as "Internally Satisfied Connections" in the figure and compared to only 28% of UltraScale theoretical connections. As the figure shows, only 2% of all nets in UltraScale are successfully routed within a CLB compared to 15% in Versal, increasing internal net routing by a factor of almost 8X while only modestly increasing the cost of the CLB.

In addition to the LUTs and flip-flops, the CLB contains dedicated circuitry such as arithmetic carry logic and multiplexers to create wider logic functions. Internals of the CLB, such as wide function muxes, carry chain, and internal connectivity are designed to increase total device capacity by reducing area per utilized logic function. Within each CLB, 16 LUTs can be configured as 64-bit RAM, 32-bit shift registers (SRL32), or two SRL16s. For every group of 64 flip-flops, there are four clocks signals, four set/reset signals, and 16 clock enables. There are dedicated local interconnect paths for connecting LUTs together without having to exit and re-enter a CLB. This enables a flexible carry logic structure that allows a carry chain to start at any bit in the chain [3].

Another interesting new feature for the fabric is called the Imux Register Interface (IRI), which aims to provide an easier implementation of high-performance designs. These are flexible registers on the input side of all blocks that can optionally be bypassed. Such architectural features will enable time borrowing or additional pipelining to improve the performance of designs. Adding additional pipeline registers to interconnect, with an approach where registers exist on every interconnect resource, is presented in Intel Statix10, claiming that overall performance would not be affected significantly when registers are not used [5]. The authors in [3], however, state that Imux Registers are a more cost-effective solution for increasing design speed while requiring less design adaptation, compared to "registers-everywhere approach" in [5]. Both these approaches are architectural decisions that lend themselves to emerging design that are highly pipelined.

Today's fabric portion of the device also contains hardened DSP and memory blocks, all arranged in a columnar topology. These generic hard blocks will enable the fabric to be customized for a large number of applications. This comes down to more than 20 MB of customizable on-chip memory and near 2000 DSP engines for the higher end devices. The largest Versal PL will contain close to 900K LUTs; more details of the CLB and the number of memory and DSP blocks can be found in [6].

13.2.2 Hardened Domain-Specific Features

Features in PL are often ubiquitous enough to be used for all domains, but the same trend that started in the last decade to support communication market will continue for other domains. ACAP will hardens all the necessary platform management functions and separates them from the FPGA core logic. The processor and platform management controller occupy the lower left region of the chip as s depicted in Fig. 13.3b. The example floorplan shows hardened scalar processor systems (PS), memory controllers and GTs. Versal architecture comprises a framework that enables swapping new domain-specific blocks that are market-driven and not always necessary. For example, some devices may have A-to-D converters replacing GTs. A variety of smaller domain hard IP blocks such as forward error correction, MAC blocks, Interlaken, or PCIE can occupy slots within the fabric array. In this respect, the Versal architecture enables a platform that continues the trend towards enabling families of domain specific devices.

Xilinx FPGAs have had columnar IOs over the last decade. There are several advantages to columnar IOs, including tight integration with the fabric and area efficiency. However, IO cells don't tend to shrink with Moore's Law and the cost of using long metal wires has increased. As a result, the interconnect delays and clock skew incurred by metal increase while crossing over large IO columns. The timing and spatial discontinuities in fabric have led to additional complexity for software tools mapping designs across those boundaries. Moreover, IO package trace breakouts from the die interior can be challenging and performance limiting with additional IOs required. Therefore, the implementation of perimeter IOs enables higher performance IOs and less fabric disruption. These high speed IOs at the bottom of Fig. 13.3b and their adjacent hardened memory controllers are often used for domains that require significant bandwidth for external memory access.

The Platform Management Controller (PMC) is another hard block that brings the platform to life and keeps it safe and secure. It boots and configures all of the blocks in the ACAP in milliseconds. All security, safety and reliability features are managed through this block. It cryptographically protects images for both hardware and software, while safely providing enhanced diagnostics, system monitoring and anti-tamper. All debug and system monitoring happen through PMC and high-speed chip-wide debug. A number of the application domains for FPGAs have been relying on partial reconfiguration of the blocks. Versal architecture offers an 8x configuration speed boost over previous generations by increasing the internal configuration bus width by 4x and a faster configuration clock. Leveraging the same configuration path speedups and rearranging CLB flop data into minimal number of frames enables up to 300X readback enhancements. Faster readback helps faster and more efficient debug of the designs.

13.2.3 Hardened Data Movement on Chip

Future cloud and high-performance computing (HPC) will be data-centric and leadership in data movement is critical to success. The consensus is that workloads in data centers will become more data-intensive and they need to manage three orders of magnitude of new data from 5G. New interconnect technologies to connect and communicate the data on and off the chip will be essential to accommodate the need for lower latency, while maintaining energy efficiency. FPGAs have been very successful in providing users with a bit level configurable interconnect. FPGA capacity has grown rapidly, and emerging applications comprise a large number of compute modules. The communication among these modules and external memory will cause routing congestion in fabric interconnect. This problem is more pronounced with process scaling since the technology is not improving wire resistance. System performance at high frequencies will require efficient global data movement across the chip from/to an external memory. Therefore, it makes sense to organize data movement into wide standardized bussed interfaces. A general technique to reduce interconnect burden is sharing the resources and Network-on-Chip (NoC) is a systematic method for sharing wires. The higher speed for the data movement makes possible the higher sharing level for valuable wire resources. ASICs and SoCs addressed a similar problem of moving many high bandwidth data streams by adding hardened NoCs. In packet switched NoCs, the same physical resource is used to route communication between multiple ports, thus increasing area efficiency.

For FPGAs, researchers have similarly proposed various techniques to improve on the efficiency of bit level interconnect. These include requiring users to reason at the word level rather than at bit level [7, 8], to implementing NoCs as hardened interconnect resources on the FPGA [9]. In the Versal architecture, a hardened NoC is a hardened layer of interconnect augmenting the traditional FPGA interconnect. Adding hard blocks for domains such as storage or compute is not new for FPGAs, but hardening data movement is a first in the industry. The traditional soft FPGA interconnect continues to provide bit level flexibility, but the NoC can absorb a significant portion of the interconnect demand. This separates system level communication implementation from compute portion. Consider the concrete case of a compute IP requiring access to some memory controller. In order to close timing at high frequencies (required to support high bandwidths), the compute would have to be placed close to the memory controller. Alternately, the physical implementation tools would have to be smart enough to insert on-demand pipelining. On the other hand, with NoC, it is possible for the compute to be implemented anywhere on the FPGA. All it needs to do is hook up to the nearest NoC port for communication to occur at a guaranteed bandwidth. This eases the timing closure for a large variety of the designs.

Mesh is a common topology for NoCs, but this is neither necessary nor useful in the FPGA case. Figure 13.3 shows a view of how the NoC integrates with the rest of the device. There are multiple Vertical NoC (VNoC) columns in the fabric and each master or slave clients simply connects to the nearest one. The figure also shows two more Horizontal NoC (HNoC) rows at the top and bottom of the floorplan. Adding more horizontal connections would not significantly improve access to the NoC, but significantly disrupt the fabric connectivity. Columnar integration with the fabric is natural in the context of FPGAs, because VNoCs will be added similar to any other columnar compute block within an FPGA. HNoCs are sized to have more physical channels than the VNoCs. This provides enough horizontal bandwidth for fabric clients attached to a particular VNoC to access memory controllers at all horizontal locations in the device—a key feature enabling a uniform view of memory across the entire device for all clients [10]. Versal NoC is a packet switched network that implements a deterministic routing flow with wormhole switches. It supports multiple Virtual Channels (VCs) to help avoid deadlock and head-of-line blocking. It also supports multiple Quality-of-Service (QoS) classes, the details of which are described in [10]. The Versal NoC is not a replacement for fabric interconnect; it provides a persistent interconnect that implements switching and routing functions that would previously have consumed fabric resources.

One key driver of the NoC requirements is to effectively manage access to external memory through DDR channels. The NoC bandwidth and resources scale both in terms of the device memory bandwidth and fabric size. The number of fabric ports on each VNoC scales with the height of the device and the number of VNoC columns scales with device memory bandwidth. This enables the NoC to support the entire memory bandwidth and at the same time allow for enough fabric access to consume it. Each horizontal and vertical line represents a full-duplex link of 128 bits wide and operating at 1 GHz. The upper bound of throughput of each unidirectional physical link is over 16 GB/s in each direction. Each VNoC contains two physical lanes, which sums up to 64 GB/s bidirectional bandwidth. HNoCs will have either 2 or 4 physical links depending on device size, which provides up to 128 GB/s horizontal bandwidth. The NoC provides unified, physically addressed access to all hard and soft components on the device. The NoC has programmable routing tables that must be initially programmed at boot time.

SoCs and ASICs have been using NoCs for many years. The requirement for such devices is different from those of a programmable device. In a programmable device NoC topology, bandwidth and QoS requirements depend on a mixture of fixed and programmable functions whose behavior varies substantially based on the application being mapped. This requires a high degree of programmability from the NoC. The Versal NoC architecture has to permit all possible point to point communication. Each egress port must be reachable from every ingress port. In a traditional NoC based system, one could have multiple instances of NoCs optimized for different needs. Within a programmable NoC platform, the compilers have to manage all the flows within the constraints of the hardened NoC architecture. This requires some level of over provisioning of the NoC resources and a high degree of programmability. For example, in the Versal NoC we provision for more VCs (8) and QoS classes (3) than would be required for typical applications. The entire topology of the NoC also needs to be designed using repeatable blocks. This permits easy integration and design of a family of devices with different communication and compute needs using the same blocks.

13.2.4 AI Engines

We mentioned in Sect. 13.1 that programmable devices will ride a prominent wave of compute-intensive applications such as 5G cellular and machine learning. 5G requires between five to 10 times higher compute density when compared with prior generations. The emergence of machine learning in many products also dramatically increases the compute-density requirements. Xilinx products started addressing computationally intense applications, by adding hardened multipliers developed with the Virtex®-II series of FPGAs in 2001. Today, there are over 12000 DSP slices in current devices—an increase of 3 orders of magnitude in compute resources over last 2 decades. The ACAP devices include a new type of programmable compute engine, called AI engine, as shown in the top of Fig. 13.3b. AI Engines are an array of VLIW SIMD processors that deliver up to 8X silicon compute density at 50% the power consumption of traditional programmable logic solutions [11]. AI Engines have been optimized for signal processing, meeting both the throughput and compute requirements to deliver the high bandwidth and accelerated speed required for wireless connectivity. AI Engine arrays offer a leap into computational applications. They can also be viewed as a commercial realization of Coarse Grained Reconfigurable Arrays (CGRAs). Chapter 14 provides a broader academic perspective on CGRAs and their advantages with a more in-depth look in some of the architectural and compilation aspects. In the remaining portion of this subsection we focus on describing the AI Engine architecture.

Figure 13.5 shows a 9 × 9 array of AI Engine tiles with detailed accounting of the resources in each tile. Engine core includes 16 KB instruction memory, 32 KB of RAM, 32b RISC scalar processor, and both 512b fixed-point and floating-point SIMD vector processor. AI Engines are interconnected using a combination of dedicated AXI bus routing and direct connection to neighboring engine tiles. For data movement, dedicated DMA engines and locks connect directly to dedicated AXI bus connectivity, data movement, and synchronization. The vector processors are composed of both integer and floating-point units. Operands of 8-bit, 16-bit, 32-bit, and single-precision floating point are supported. Two key architectural features

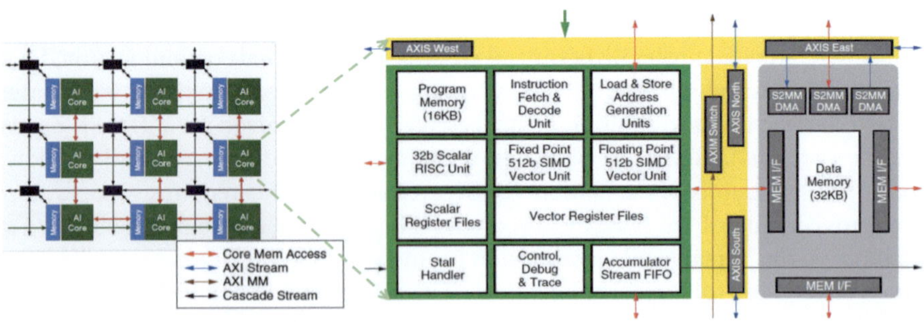

Fig. 13.5 AI Engines ([11], ©Xilinx 2019)

ensure deterministic timing: (1) Dedicated instruction and data memories and (2) Dedicated connectivity paired with DMA engines for scheduled data movement. The simplest form of inter-tile data movement is via the shared memory between immediate neighboring Tiles. This implies up to 128 KB addressable shared memories with neighbors. However, when the tiles are further away, then the AI Engine tile needs to use the AXI-Streaming dataflow. AXI-Streaming connectivity is predefined and programmed by the AI Engine complier tools based on the data flow graph. These streaming interfaces can also be used to interface directly to the PL and the NoC.

The architecture is modular and scalable; some of the devices will contain up to 400 of these tiles. One of the highest value propositions of this CGRA is the connectivity with adjacent fabric. Figure 13.6 illustrates the connectivity between the AI Engine array and the programmable logic. AXI-Streaming connectivity exists on each side of the AI Engine array interface, and extends connectivity into the programmable logic and separately into the network on chip (NoC). Leveraging NoC connectivity, AI engines communicates to the external memory. Processor subsystem (or scalar processors) on the device also manage configuration, debug and tracing of AI engines through HNoC. AI Engines are programmed using a C/C++ paradigm familiar to many programmers as will be explained in the following sections. AI Engines are integrated with Xilinx's Adaptable and Scalar Engines (PL and PS)

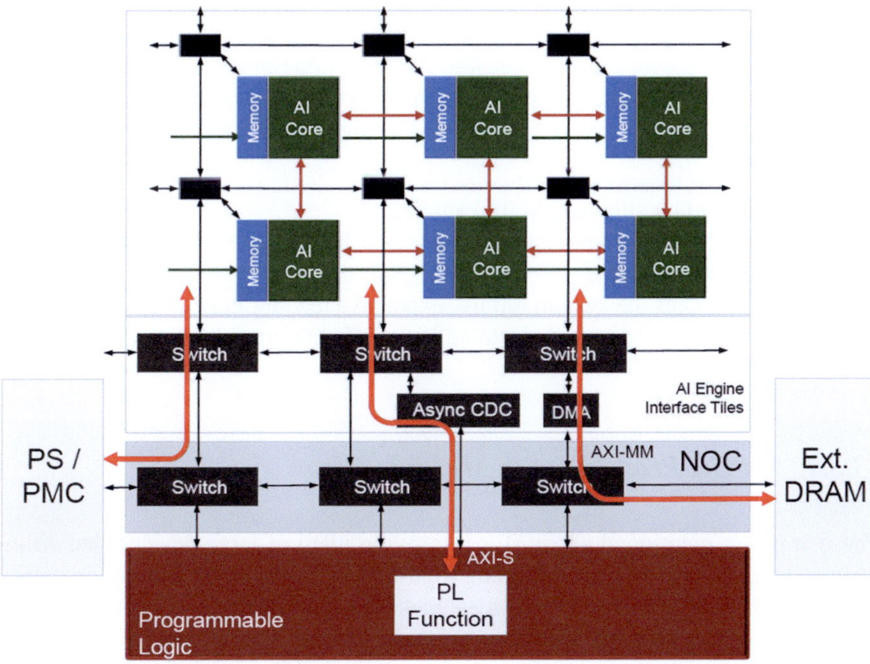

Fig. 13.6 AI Engines and connectivity with fabric

to provide a highly flexibly and capable overall solution. The key difference of an AI Engine array with traditional multicore computing engine is the dedicated not-blocking deterministic interconnect. Xilinx has provided results indicating 10X higher compute for ML inference, 5X higher 5G wireless bandwidth, and 40% less power compared to an earlier 16 nm FPGA devices [11].

13.3 Disaggregation Trend for Cost and Market Agility

Silicon transistors and wires are not providing much area or speed benefits due to the slowdown of Moore's Law, and the power per chip area is increasing (reflecting the end of Dennard scaling). Advances in process technology have enabled FPGAs to grow in capacity and implement large heterogenous systems in a monolithic device. The emerging ACAP devices explained in the previous section are highly adaptable—making them the candidate of choice for a wide range of emerging domains from compute to networking. However, performance or power improvements are no longer readily available from process technology and it is no longer trivial to build cost-efficient devices. A prominent trend to respond to rising fabrication cost is disaggregation of architecture components, since only parts of the system on the chip require expensive leading-edge process nodes. Disaggregation of monolithic systems means implementing the required connectivity needs at the wafer or package level. Disruptive technologies such as wafer level connectivity or advanced packaging are expected to evolve over the next decade. ACAP unique position is that it includes many heterogenous blocks. This provides an opportunity for cost reduction by selective disaggregation per domain of interest. The significant drivers for this trend are claimed to be:

- Improving yield by silicon split into smaller dies.
- It's the only way to get enough memory or a heterogenous technology such as photonics into the system.
- Only some parts of the system require expensive leading-edge nodes.
- It's a way to use the same silicon to address different configurations/markets.

13.3.1 FPGA Products with Multiple Dice

The first driver mentioned above (for improving yield of large devices) led Xilinx to develop a new approach for building high-capacity FPGAs for emulation market in early 2010s [12]. The new solution enables high-bandwidth connectivity between multiple dice by providing high density package connectivity. Combining several large dice in a single device is the only way to exceed the capacity and bandwidth offered by the largest monolithic devices. Figure 13.7 shows a side view of four large

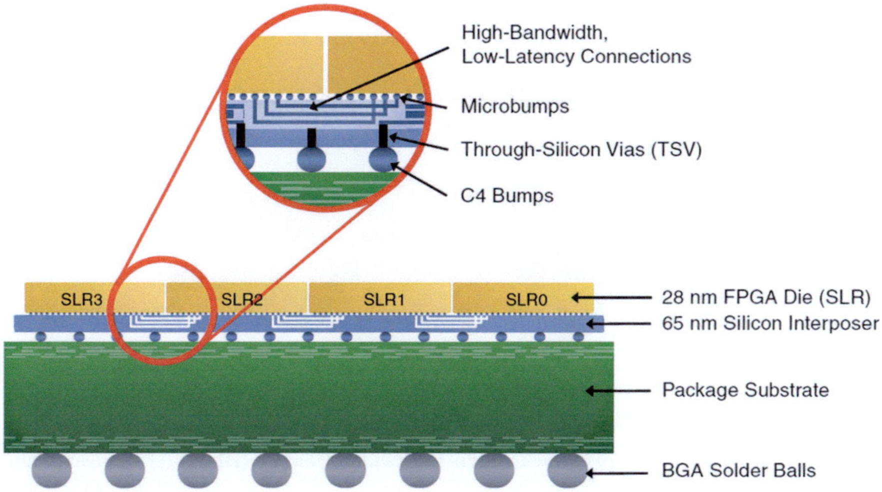

Fig. 13.7 Virtex®-7 2000T FPGA enabled by advanced packaging technology

FPGA dice along with a passive interposer that provides tens of thousands of die-to-die connections in the same package, responding to cost pressures of monolithic integration. The key to this enabling technology was combining Through-Silicon Via (TSV) and micro bump technology. The passive silicon interposer was a low-risk, high-yield 65 nm process that provided four layers of metallization for building the tens of thousands of traces that connect the logic regions of multiple FPGA die. C4 solder bumps connect interposer stack-up on a package substrate using flip-chip assembly techniques. This technology provided multi-Terabit-per-second die-to-die bandwidth through more than 10,000 device-scale connections—enough for the most complex multi-die FPGA product at 28 nm process technology (XC7V2000T).

Later on, the same technology was used for the integration of different types of die. Virtex-7 H870T FPGA, announced in 2012, ties together three homogeneous dice and a separate 28G transceiver chiplet via the silicon interposer. This was the world's first heterogeneous FPGA architecture—an FPGA consisting of heterogeneous die placed side-by-side to operate as one integrated device. While this product didn't have the market success of 2000T device for a number of reasons beyond the scope of this chapter, it was an important technology turning point for many more devices with heterogenous integration at present and future.

FPGA High Bandwidth Memory (HBM) devices, introduced in 2017, integrated 16 nm UltraScale+ FPGA fabric with HBM controller and memory stacks from Xilinx supply partners [13]. The HBM is integrated using a similar interposer-based stacking technology explained above and is depicted in Fig. 13.8. Such heterogenous integration enables more than 20X external memory bandwidth on the same device compared to that of PCB. Low power and high bandwidth memory access are essential requirements for emerging compute and data center domains. The AXI Interface in the HBM memory controller needs to be hardened to accommodate the aggregate

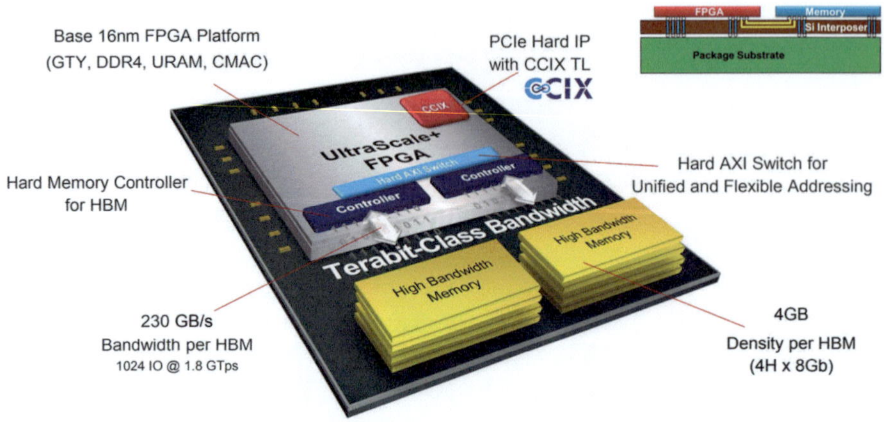

Fig. 13.8 Vitex Ultrascale+ FPGA with High Bandwidth Memory (HBM) ([13], ©Hot Chips, 2017)

high bandwidth between the local programmable routing and the HBM module. This structure significantly increases the user's AXI interface bandwidth, allowing for up to 3.7 Tb/s operation [14]. Recently, Xilinx introduces the Virtex® UltraScale+™ VU19P, the world's largest FPGA with the highest logic density and I/O count on a single device ever built, addressing new emulation market. The devices boast more than 4M LUTs in the same package, which would not have been possible without disaggregating the whole system into multiple dice.

Versal ACAP with highly integrated monolithic features was introduced in the previous section. However, the new fabric has a unique feature that can be leveraged for adding connectivity using silicon interposer technology. Current generation of interposer technology for HBM devices or VU19p only use the wires on interposer at the edge of die and in the vertical direction. In this case micro bumps are distributed in channels along the edge by displacing the CLBs. Versal ACAP fabric architecture embeds a number of these micro bumps in each CLB allowing them to be distributed evenly across the die. This enables the architecture to utilize more wiring on the interposer and in both directions. In this new routing architecture, interposer wires serve two purposes: (1) inter-chiplet connectivity, and (2) additional regular intra-chiplet long range routing. These wires on the interposer are 30% faster for the same distance and ideal for long reach die connectivity. This enables ultra-large ACAP devices with multiple active silicon dice stacked on a passive interposer and with ample routing wires on interposer that may be introduced to the market in the next few years. This architecture will reduce delays and routing congestion at the die boundaries and will consequently ease the software burden of partitioning the design to be mapped to multiple dice. The key enabler for this form of chiplet connectivity is 4X CLB granularity that was explained earlier. Further details and quantitative benefits can be found in [4].

Intel's recent 10-nm Intel® Agilex™ FPGAs are also built using a disaggregated chiplet architecture, which integrates heterogeneous technology elements in a System-in-Package (SiP). Leveraging a packaging technology, called Embedded Multi-Die Interconnect Bridge (EMIB), Intel uses the chiplet approach to combine a traditional FPGA die with purpose-built semiconductor die, creating devices that are uniquely optimized for target applications. EMIB silicon bridges are positioned as an alternative to 2.5D packages using silicon interposers. They often provide a similar connectivity density as interposers but take less area on the interposer. The attractiveness of EMIB is that silicon is used only in the areas where two dies connect. Since the main cost of such advanced packaging is assembly, it is not clear if any of these two methods have superiority, and hence both approaches are expected to stay around for the time being (Fig. 13.9). Intel is also using this technology to add advanced analog functions such as 112 Gbps PAM-4 transceivers to the programmable device, as shown in Fig. 13.10 [15]. Xilinx provides a similar GT functionality with a key differentiation that monolithic integration is used to add analog high-speed functionality in contrast with Intel disaggregation strategy. This ideally exemplifies how the old trend of monolithic die integration will continuously be considered and evaluated against disaggregated package integration in the next decade. The merit of each solution will depend on a number of factors including expertise in the company and agility to market, as will be discussed further in the next subsection.

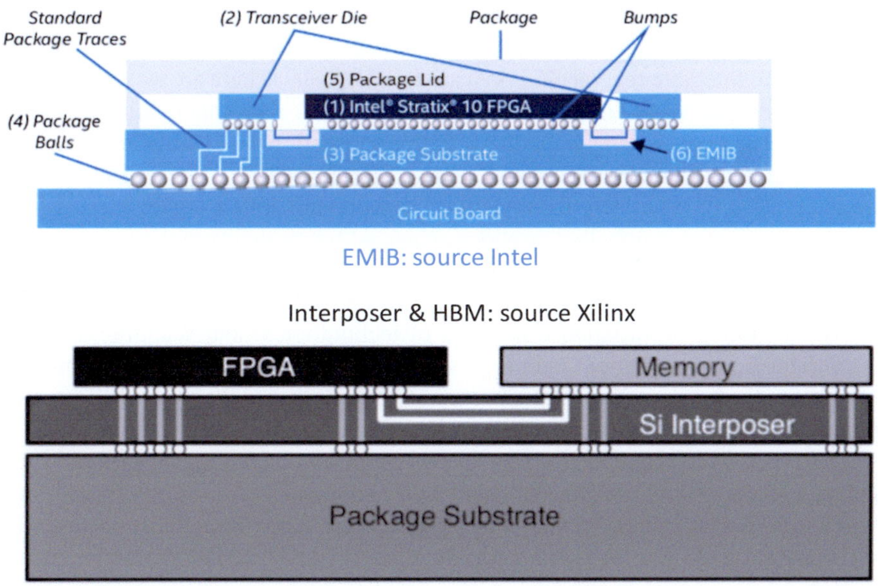

Fig. 13.9 EMIB or interposer 2.5D advanced package connectivity

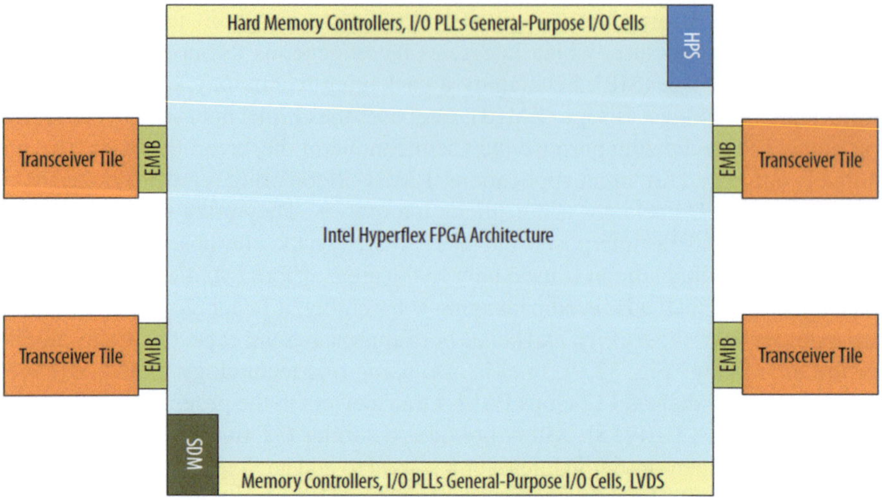

Fig. 13.10 Intel AgileX FPGA with EMIB package connectivity ([15], ©Intel, 2019)

13.3.2 Upcoming Heterogenous Integration Trends and Programmable Devices

Integration in the package is not new; OEMs have used Multi-Chip Modules (MCMs) in systems to integrate several chips in a module for years. There are two new dynamics, however, that raise the importance of SiP devices: rising fabrication costs and advancements in packaging technology. In this subsection we identify some of these trends in the context of programmable logic. Heterogeneous Integration refers to the integration of separately manufactured components into a higher-level System in Package (SiP) assembly that provides enhanced functionality and improved operating characteristics in the aggregate. There are many examples of Heterogenous Integration through SiP today as explained in the previous subsection. Heterogeneous Integration is initiating a new era of technological and scientific advances to continue and complement the progression of Moore's Law Scaling into the distant future. Packaging—from system packaging to device packaging—will form the vanguard to this enormous advance.

There is a wide range of heterogeneous integration technologies for both serial and parallel connectivity within the chiplets. We anticipate standards evolving around both Ultra-Short Reach (USR) serial and parallel (e.g. HBM-like) die-to-die interfacing. In addition to PHY layer standards, there are higher level data protocols, such as AMBA AXI, essential to any application. The key enabling metrics include energy efficiency and aggregate throughput delivery for data movement between chiplets. Figure 13.11 shows the energy efficiency of existing and emerging solutions, approximated in oval areas. There is a two orders of magnitude power gap between today's

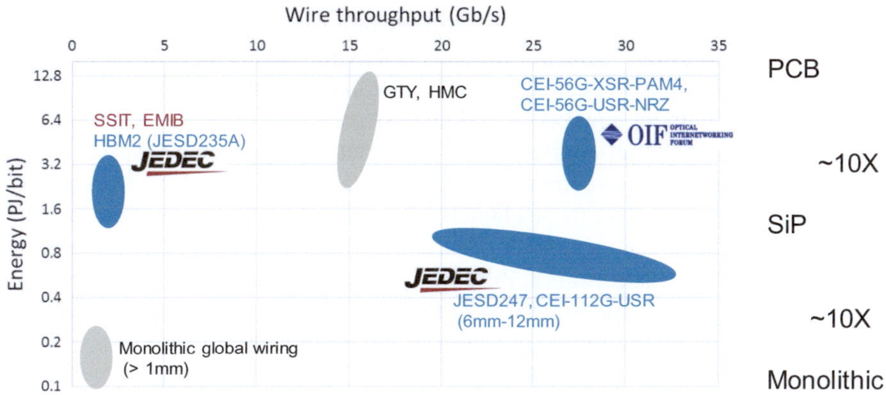

Fig. 13.11 SiP connectivity landscape for energy efficiency

existing PCB solutions (such as Xilinx GTY or HMC) and monolithic implementations of data movement, as noted by gray ovals. The blue oval areas in the figure show the emerging packaging connectivity based on recent published work. Some of these emerging SiP solutions are reducing the power gap by more than one order of magnitude. On the other hand, monolithic implementation of global data movement between heterogeneous blocks will require additional overhead such as shims or synchronization modules, leading to the energy increase. This provides an opportunity for reaching near-monolithic energy efficiency, especially with leveraging domain optimizations at the software level. Another take-away from Fig. 13.11 is that we need to be on the right side of the chart in terms of wire throughput in order to use MCM packaging technology with coarse wire pitch. In contrast, interposer fine micro bump pitch (such as that of HBM or EMIB) enables larger counts of interface wires to run at lower frequency as shown on the left side of the figure.

Aggregating multiple existing serial interfaces to amortize PLL power reduces energy consumption. Removing the Clock Data Recovery (CDR) blocks, often used in serial interfaces in lieu of a source synchronous data transmission, offers another degree of potential power reduction. This is a reasonable decision since distance within a package between the chiplets is short. Assuming a bump pitch of 130–150 μm for the substrate, we can fit around 44–59 bumps in 1 mm^2 of silicon area. By reserving about 10 bumps for power and source synchronous clocking, we can deliver 1 Tb/s bandwidth if each wire carries data at a rate above 20 Gb/s. This approach translates to differential Gigabit Transceivers (GTs) running in the range of 40–56 Gb/s. Fortunately, GT blocks are emerging to offer these ranges. Published literature introduces a USR IP in 28 nm, demonstrating less than 1 pJ/bit using a test chip [16]. This USR interfacing approach uses the CNRZ-5 coding layer that is added on top of aggregating bunch of GTs together. They claim that the coding may contribute to 2X of the power reduction. There is also a recent JESD247 standard, which is based on this IP [17]. A group of other low power GT efforts circle around extensions of OIF standard, which is mostly focused on optical and photonic connectivity. There are test chips published that claim a bit more than 2 pJ/bit for such interfacing without any special coding [18].

Both parallel and serial interfacing are viable technology trends that will enable inter-die connectivity in the coming years. The key enabling factor will be an ecosystem or market place to have those chiplets, which would be after standardization of those interfaces. There are a number of initiatives such as DARPA CHIPS program and Open Compute Platform (OCP) OSDA efforts to push in this new direction. FPGAs can provide an important role in this new paradigm. Moore's law has significantly reduced the traditional overhead of programmability in FPGAs, which is attributed to using LUTs and interconnect. Modern FPGAs include a number of heterogeneous blocks such as processors, memory, and high speed I/O, as explained in previous section. The new programming overhead will be the unused blocks on highly integrated FPGAs. A programmable fabric chiplet along with specific domain chiplets in an SiP enables a wide range of applications, expanding the fast time to market and customization benefits of FPGAs to this new paradigm. The future package connectivity was classified by power targets in a recent keynote at Hot Interconnect 2019. For 2.5 D technologies, he envisioned 1 pJ/bit for organic substrate (which is achievable today) and 0.3 pJ/bit for interposer or EMIB connectivity. Moreover, he estimated 0.15 pJ/bit for an SiP connectivity to which he referred as 3D [19]. This is getting close to the power for long distance wires within a chip with monolithic integration. Hierarchical Integration Roadmap [20] anticipates 3D interconnect with micro bump density of less than 10um to be available in 10–15 years. FPGAs will significantly benefit from such technology as regular repeatable patterns in fabric can leverage such dense connectivity.

13.4 Software Implication and Trends

Discussing programmable devices would not be complete without understanding of the software design flow. Traditional design process for FPGAs involves transforming the design from a preferred design entry to a configuration bitstream that can be downloaded into the device. This process consists of a sequence of major steps:

(1) synthesizing the design into the fundamental architecture blocks such as LUTs and flip-flops,
(2) place and route those blocks under the given timing and area constraints, and
(3) generating a configuration bitstream to program the device.

The goal of this section is not an in-depth discussion of these tools. Instead, we highlight a few meta-level trends in the recent years with a look into the future. FPGA devices started capturing market share by replacing ASICs as explained in Sect. 13.1. Therefore, the CAD tools started being EDA-like with one significant difference: reduced cost. FPGA companies started building their own CAD tools for configuring devices and offered it to the customers at a highly subsidized price. This was in contrast with the ASIC tools that were from 3rd party and often at higher cost.

FPGA capacity and complexity grew rapidly and as a response, the design entry abstraction was raised to mitigate productivity. This trend, which is shown in Fig. 13.12, occurred with a combination of organic growth and acquisition of third-party tool providers. The figure shows how the design entry abstraction is raised from schematic design entry in 1990s to RTL design entry in the last decade. Today, it is possible to use high-level programming languages such as C and Python as a method of design entry for FPGAs. The most recent Xilinx announcement is a unified software platform, called Vitis [21], that enables the development of embedded software and accelerated applications on heterogeneous Xilinx platforms including FPGAs, SoCs, and Versal ACAPs. Vitis enables integration with high-level frameworks and development in C, C++, or Python and is available free of charge.

The key trend that is prominent for programmable devices is going in the direction of catering to software programmers. The programming aspect of these devices uniquely positions them somewhere between ASIC hardware platforms (with spatial design) and CPUs (with temporal programming). Software programming models have closely tracked the evolution of processor architecture, evolving from a focus on single-core, central memory machines towards multi-core, domain-specific accelerators. As a result, the FPGA tools need to move towards what software developers expect, with improving productivity. One clear architectural attempt in this direction is adding the new CGRA-like components as explained in subsection 13.2.4. These AI engines can be programmed using C/C++ similar to other software programming platforms. AI Engine simulation can be functional or cycle accurate using

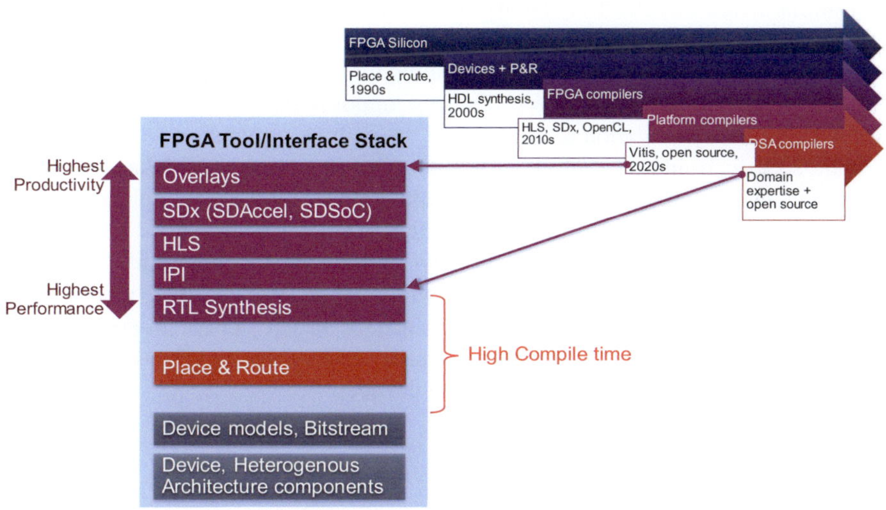

Fig. 13.12 Raising the abstraction of design entry for programmable devices

an x86-based simulation environment. For system-level simulation, a System-C virtual platform is available that supports both AI engines and traditional arm-based processor (scalar engines) integrated on the chip.

In addition to adding new components that are software friendly, it is expected that software for the fabric moves in the same direction. Today's barrier of entry to new markets such as HPC and compute is not the hardware limitations of programmable devices, but the software productivity and efficiency with new methods of design entry. Software developers expect better user experience such as faster compile time for the backend (see Fig. 13.12) and higher flexibility for building their own flows. The tools for programmable devices are likely to address these issues by two approaches: domain specific overlays and open source. Post-bitstream, programmable domain-specific soft overlays and pre-implemented shells enable fast compilation to the fabric. FPGA designers familiar with hardware will design these overlays; this enables domain programmers to leverage customized memory and interconnect architectures without the need to be an FPGA design expert. Overlays offer user programmability within a given domain. However, scaling this concept to more domains requires new tools and an active ecosystem of new domain experts.

An efficient way to enable an ecosystem of FPGA domain compilers is tapping into open source dynamics in software community. The open source movement relies on free software to stimulate innovation and progress. Software development has become significantly more complex than hardware design and open source is analogous to Moore's law (of technology scaling) for software. FPGA tools are also expected to move towards open source in the next decade. For example, RapidWright, an open source platform, was introduced recently to provide a gateway to Xilinx's back-end implementation tools [22]. The goal is to raise the implementation abstraction similar to the way the design entry abstractions are raised, while maintaining the full potential of advanced FPGA silicon. Such framework can help building domain-specific backend tools in two ways: (1) creating highly optimized overlay and shells, and (2) domain-specific compilers. The best opportunity for domain design tasks lies with domain application architects and the path to automation would require a domain-specific front-end compiler. This compiler may be an LLVM data flow graph parser that can automatically identify domain operators with high replication. This concept is depicted in Fig. 13.13 by application examples in domains 1 and 2. Open domain data flow and HLS compilers may be built by the community or will be available as more of free Vitis framework will be open. We anticipate great interest to maximize existing FPGA silicon performance for the age of domain-specific compute. RapidWright or similar open source frameworks are likely to be the enabler for a significant part of that journey in the next decade.

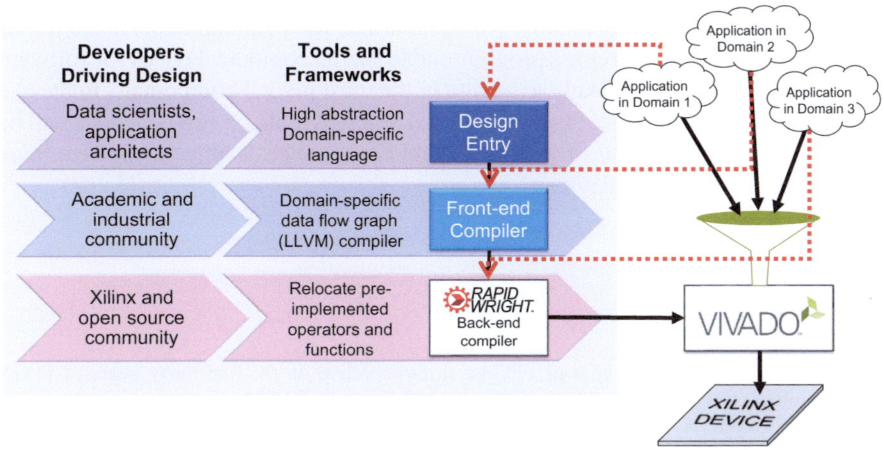

Fig. 13.13 Leveraging the community and open source towards domain-specific compilers

13.5 Concluding Remarks

Field programmable devices (FPGAs) are integrated circuits designed to be configured by the customer after manufacturing. Initially they contained an array of programmable logic blocks and a hierarchy of reconfigurable interconnects that allow the blocks to communicate. Reconfiguration and hardware customization were the key differentiating attributes of FPGAs that facilitated the market expansion by replacing custom logic ASICs. Riding Moore's law, FPGAs grew more than four orders of magnitude in capacity and captured a significant portion of communication domain in addition to ASIC replacement. Today, the FPGA market is more than $5 billion and still growing by penetrating in domains such as machine learning and datacenter networking.

We highlighted some of the key features that enable programmable devices to enter the compute domain as accelerators. We summarized how Xilinx will address current semiconductor technological, economical, and scalability challenges with the new 7 nm ACAP compute platform. The Versal architecture tightly integrates programmable fabric, CPUs, and software-programmable acceleration engines into a single device that enables higher levels of software abstraction, enabling more rapid development of hardware accelerators that solve next generation problems. Such high-level integration is a direct result of process technology advancements; these new complex products introduced to the market in the next few years will be a testament to the success of such a trend.

The slow-down of Moore's law and fabrication cost pressures will also set a disaggregation trend for semiconductor industry. Programmable devices with multiple dice in the package are already introduced to the market for yield improvement or heterogenous integration. We believe this approach will continue and we provided some guidelines and insight for how programmable devices will add value to SiP devices of the next decade. Finally, we summarized the automation software trends and the steps required to prepare programmable spatial compute devices for software programmers. FPGAs are likely to be one of the most pivotal components in the age of domain specific compute. Open source and efficient tools for these devices will be developed to enable software-friendly customer experience despite the high complex functionality of the latest devices.

References

1. Steve Trimberger, Three ages of FPGAs: a retrospective on the first thirty years of FPGA technology, in *Proceedings of the IEEE*, vol. 103, issue 3 (2015 Mar)
2. J.W. Lockwood, N. Naufel, J.S. Turner, D.E. Taylor, Reprogrammable network packet processing on the field programmable port extender (FPX), in *Proceedings of ISFPGA 2001*, ACM, pp. 87–93
3. Xilinx white paper, Versal: The First Adaptive Compute Acceleration Platform (ACAP), WP505 (v1.0.1) (23 Sept 2019)
4. B. Gaide et. al., Xilinx Adaptive Compute Acceleration Platform: Versal™ Architecture, FPGA '2019 (24–26 Feb 2019)
5. D. Lewis et al., The Stratix™ 10 Highly Pipelined FPGA Architecture (2016)
6. Xilinx advance product specification, Versal Architecture and Product Data Sheet: Overview, DS950 (v1.2) (3 July 2019)
7. N. Napre, J. Gray, Hoplite: building austere overlay NoCs for FPGAs, in *International Conference on Field-Programmable Logic and Applications* (2015)
8. P. Maidee, A. Kaviani, K. Zeng, LinkBlaze: efficient global data movement for FPGAs, in *IEEE reconfig* (2017)
9. M.S. Abdelfattah, V. Betz, LYNX: CAD for FPGA-based network-on-chip, in *International Conference on Field-Programmable Logic and Applications* (2016)
10. I. Swarbrick et.al., Network-on-chip programmable platform in Versal™ ACAP architecture, in *ACM FPGA '2019* (Feb 2019)
11. Xilinx white paper, "Xilinx AI Engines and Their Applications," WP506 (v1.0.2) (3 Oct 2018)
12. Xilinx white paper, "Xilinx Stacked Silicon Interconnect Technology Delivers Breakthrough FPGA Capacity, Bandwidth, and Power Efficiency," WP380 (v1.2) (11 Dec 2012)
13. G. Singh et. al., Xilinx 16 nm datacenter device family within-package HBM and CCIX interconnect, in *2017 Hot Chips*
14. M. Wissolik et al., Virtex UltraScale+ HBM FPGA: a revolutionary increase in memory performance, WP485 (v1.1) (15 July 2019)
15. Intel Product announcement, Intel® Agilex™ FPGA Advanced Information Brief. www.intel.com/content/www/us/en/products/programmable/fpga/agilex.html
16. A. Shokrollahi et al., 10.1 A pin-efficient 20.83 Gb/s/wire 0.94pJ/bit forwarded clock CNRZ 5-coded SerDes up to 12 mm for MCM packages in 28 nm CMOS, in *2016 IEEE International Solid State Circuits Conference (ISSCC), San Francisco, CA, USA* (2016)
17. JEDEC, Multi-wire Multi-level I/O Standard (June 2016). http://www.jedec.org/standards-documents/results/jesd247
18. M. Erett et al., A 126mW 56 Gb/s NRZ wireline transceiver for synchronous short-reach applications in 16 nm FinFET, in IEEE ISSCC2018
19. U. Cummings, CTO of DCG Connectivity Group, Intel, "From Microns to Miles—The Broad Spectrum of Intel's Interconnect Technology Strategy", Hot interconnect (2019)
20. Heterogeneous Integration Roadmap, "Interconnects for 2D and 3D architectures," HIR version 1.0, chapter 22, eps.ieee.org/hir (2019)

21. Vitis Unified Software Platform, https://www.xilinx.com/products/design-tools/vitis/vitis-platform.html, Xilinx
22. C. Lavin et al., RapidWright: enabling custom crafted implementations for FPGAs, in *IEEE FCCM* (2018)

Chapter 14
Coarse-Grained Reconfigurable Architectures

Raghu Prabhakar, Yaqi Zhang and Kunle Olukotun

14.1 Introduction

Rapid algorithmic and technological innovations in fields such as genome sequencing, data analytics, machine learning, and software-defined networking have placed greater compute and memory demands on the underlying computing systems. At the same time, technology scaling challenges with the slowdown of Moore's law and the end of Dennard scaling has made it increasingly difficult to scale processor performance in an area and energy-efficient manner. Consequently, the computer architecture community has ushered in the era of specialized accelerators [1–4]. Accelerators implement customized data and control paths to suit a domain of applications, thereby avoiding many of the overheads of flexibility in general-purpose processors. However, specialization in the form of dedicated ASICs is expensive due to the high NRE costs for design and fabrication, as well as the high deployment and iteration times. Furthermore, applications and algorithms evolve at a rapid pace; for example, the number of machine learning articles posted on arXiv.org has grown faster than Moore's law in the past decade [5]. An ASIC designed to accelerate a specific set of algorithms can immediately be rendered obsolete when ASIC design time is considered along with the rate of algorithmic innovation [6]. This makes ASIC accelerators impractical for all but the most common and unchanging applications. Achieving a balance between specialization and flexibility in the underlying system is thus a critical task.

Flexibility in architecture can be achieved in multiple ways. General-purpose CPUs and GPGPUs achieve flexibility by implementing a well-defined Instruction

R. Prabhakar (✉)
SambaNova Systems Inc, Palo Alto, CA, USA
e-mail: raghu.prabhakar@sambanova.ai

R. Prabhakar · Y. Zhang · K. Olukotun
Stanford University, Stanford, CA, USA

B. Murmann and B. Hoefflinger (eds.), *NANO-CHIPS 2030*,
The Frontiers Collection, https://doi.org/10.1007/978-3-030-18338-7_14

Set Architecture (ISA). Applications can be executed on such architectures by compiling them to a sequence of instructions in the ISA, which are then executed using one or more threads. ISA-based processors and thread-based execution models are ubiquitous today. However, instruction pipelines incur a nontrivial amount of hardware area and power overheads [7]; events such as instruction fetch, decode, and register file access account for about 40% of the data path energy on the CPU [8], and about 30% [9] of the total dynamic power on the GPU. Furthermore, studies have shown that using a reconfigurable data path in place of a conventional instruction pipeline in a GPU reduces energy consumption by about 57% [10]. Techniques like SIMD execution [11] amortize the overheads to some extent by performing more useful work per instruction and can achieve energy efficiency improvements of 4% to 1.9× [12]. However, applications often contain parallelism at multiple levels of nesting [13]. ISAs typically offer limited support to exploit such nested parallelism even with SIMD, as costly synchronization mechanisms in software would be necessary to orchestrate execution. Architectures that allow a finer degree of customization can better exploit nested parallelism without incurring the overhead of instructions.

On the other hand, reconfigurable architectures like Field Programmable Gate Arrays (FPGAs) achieve energy efficiency by providing statically reconfigurable compute elements and on-chip memories in a programmable interconnect that can be configured to implement customized data paths. In FPGAs, these custom data paths are configurable at the bit level, allowing users to prototype arbitrary digital logic and take advantage of architectural support for arbitrary precision computation. However, FPGAs have long suffered from programming inefficiencies due to low-level programming models and long compile times. Furthermore, architectural inefficiencies due to bit-level reconfigurability in computation and interconnect resources result in significant area and power overheads. For example, over 60% of the chip area and power in an FPGA is spent in the programmable interconnect [14–17]. In contrast, a study on an AMD GPU reports that up to 14% of the dynamic power is consumed in the interconnect [18].

To mitigate programming and architectural inefficiencies, architects from industry and academia alike have attempted to raise the hardware abstraction level by introducing coarser-grained building blocks such as ALUs, register files, and memory controllers. These architectures are referred to as Coarse-Grained Reconfigurable Architectures (CGRAs). More generally, CGRAs are characterized by reconfigurable compute and memory elements in a programmable interconnection network. CGRAs have been shown to achieve higher performance and energy efficiency compared to conventional instruction-based architectures by avoiding instruction overheads with reconfigurable data and control paths. CGRAs also avoid the hardware and programming overheads incurred by fine-grained alternatives such as FPGAs by providing dense compute resources, power efficiency, and clock frequencies up to an order of magnitude higher than FPGAs. Furthermore, although symmetry is not always a strict design goal, most CGRAs tend to be inherently symmetrical, with repeated patterns of reused components. From a practical standpoint, such symmetry lowers design and verification complexity. Symmetry also simplifies the hardware-software interface and increases flexibility, which allows the developing of aggressive optimizing compilers and higher-level programming models. Several surveys [19–21] provide a broad overview of prior work on CGRAs.

Harnessing the full potential of CGRAs requires co-designing the architectural primitives with the programming model and compilation flow. Choosing the right granularity for a CGRA requires navigating a large multidimensional design space and depends on common patterns of computation for which the CGRA is being built for. This choice, in turn, impacts the programming model, and the complexity and efficiency of the compiler.

This chapter reviews some fundamental concepts in CGRAs. A canonical way to quantitatively reason about CGRA granularity is presented in terms of compute, memory, and interconnect. An automatic compilation flow is then described that maps applications, which are described as arbitrary hierarchies of loop controllers, into parallelized, nested pipelines on the target CGRA. The chapter concludes with a case study using the Plasticine CGRA [22].

14.2 Key Elements of CGRAs: Compute, Memory, and Interconnect

CGRAs can differ widely based on the granularity of reconfigurable elements that compose them. This choice of granularity presents a fundamental architectural tradeoff between flexibility, programmability, and efficiency. Consider the following example, where sixteen functional units (FUs) are used to compose two different CGRA architectures.

Figure 14.1 shows an example of two CGRA designs. Figure 14.1a organizes the sixteen functional units as a 4 × 4 mesh, connected by 20 interconnect switches. In this CGRA, each FU consumes inputs and produces outputs directly to the interconnect switches. Figure 14.1b organizes the same FUs in a hierarchy, where four FUs

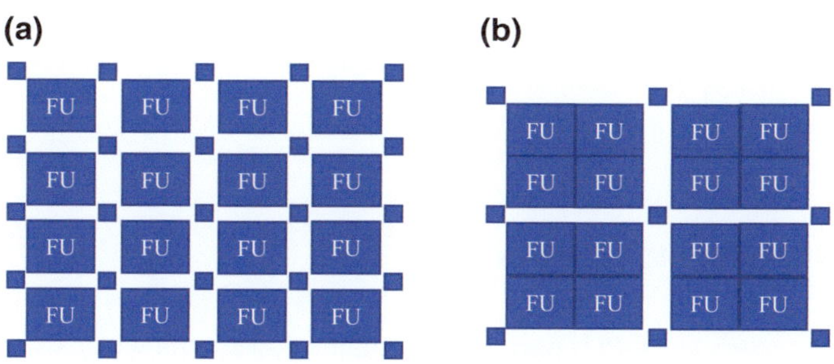

Fig. 14.1 Two example CGRA designs with 16 functional units. **a** shows a 4 × 4 topology, where each FU connects to a mesh interconnect directly. **b** shows a 2 × 2 topology, where 4 FUs are grouped together into clusters to implement a 2-stage SIMD pipeline, with 2 SIMD lanes per pipeline stage

are combined to create a two-stage SIMD pipeline with two lanes per pipeline stage. In this design, only the first two FUs can consume inputs from the interconnect, and only the last two FUs produce outputs into the interconnect. The four FU clusters are organized as a 2 × 2 mesh, connected together by 6 interconnect switches.

The contrived designs in Fig. 14.1 illustrate a fundamental tradeoff between flexibility, efficiency, and programmability, and underscores the importance of granularity selection. The design in Fig. 14.1a is more flexible than the design in Fig. 14.1b to map arbitrary data flow graphs, as each operation can be independently mapped to an FU in Fig. 14.1a. Figure 14.1b requires making the choice of mapping an operation to an FU within a cluster as opposed to across clusters, and manage data dependencies accordingly. On the other hand, Fig. 14.1b has greater compute density, which would translate to greater performance-per-area than Fig. 14.1a, if the FUs are utilizable.

Primitive elements that compose a CGRA can be grouped into the following three categories:

1. **Compute**: Primitive resources used to perform compute operations such as multipliers, ALUs, DSP blocks, etc. are referred to as "compute primitives". The compute primitives in a CGRA are grouped hierarchically into larger blocks called "compute units" (CUs). Each CU consists of one or more compute primitives organized into pipeline stages and SIMD lanes. In the example above, Fig. 14.1a has 16 CUs, where each CU has just a single compute primitive. Figure 14.1b has 4 CUs, where each CU has 4 compute primitives organized as a 2-stage SIMD pipeline. Additional supporting pipeline resources such as register files are considered to be part of compute resources.
2. **Memory**: In addition to compute resources, many CGRAs also contain on-chip memory resources such as caches, FIFOs, and/or programmer-managed scratchpad memories. Memory is characterized in terms of capacity and bandwidth and is influenced by the organization of compute resources.
3. **Interconnect**: The topology, the programmable switch fabric connecting compute and memory resources, and the routing policies is collectively referred to as the interconnect. A CGRA may have one or more interconnects between its resources to carry different types of data. In addition, each interconnect may have different routing characteristics, such as statically routed vs. circuit-switched versus packet-switched routing, bus width, number of virtual channels, and buffer depths.

14.3 Compiling to CGRAs

Modern processors and GPUs pack more compute power with wider vector instructions and more cores. However, efficiently utilizing these resources is often a challenge, as applications often contain parallelizable tasks interleaved with non-parallelizable tasks. Furthermore, communication and synchronization overheads in multi-threaded programming models increase with more parallelism and creates a performance plateau. In contrast, spatial architectures such as CGRAs present an alternative solution to improve application throughput. Hardware resources can be allocated proportional to the amount of compute at various stages in the program. Data and task parallelism in applications is exploited by spatially mapping the application as hierarchical pipelines. Intermediate results in the pipeline are allocated on-chip to avoid main memory access. Recent studies have shown up to 30–50× speedup over a general-purpose GPU (GPGPU) on single-batch Recurrent Neural Network Serving (RNN) [23, 24] using a spatial architecture due to better utilization of compute resources.

The pipelined execution model raises new requirements on on-chip interconnect and memory bandwidth. Multithreaded execution models rely on the processor's ability to interleave independent operations from one or more instruction streams such that long latency operations such as memory accesses are interleaved with computation. With pipelined execution models, several CUs can simultaneously produce and consume intermediate results every clock cycle. Managing these parallel data streams requires managing on-chip interconnect and memory resources carefully, such that bottlenecks are avoided. Specifically, partitioning and allocating data structures in a program to on-chip resources to match the bandwidth requirements of all parallel CUs is critical to sustain compute throughput.

A common technique to improve memory bandwidth is to modify the data layout such that all parallel accesses hit different memory banks. A compiler can statically analyze various access patterns in the program and remap address space automatically, which is a technique called static banking. To perform such analysis, a compiler must have a global view of all access to a data structure. Such information can be acquired from analyzing a high-level program or domain-specific language. Instead of modeling the memory as a global address space like on CPUs, the high-level language should capture access patterns of all parallel CUs to individual data structures with disjoint address space. These individual memories can be mapped onto distributed memory resources without synchronization. In addition to static banking, the memory needs to be buffered such that pipelined readers and writers can access different copies of the data from different iterations. Finally, the imbalanced data path in the program needs to be retimed with sufficient buffers to maintain full pipelining throughput.

While hierarchy in CGRAs improves the scalability and compute density of the architecture, it also introduces fragmentation in mapping, which can potentially underutilize hardware resources. The compiler needs to decompose computation and memories into smaller fragments that satisfy hardware constraints. When memory is distributed, the compiler also needs to allocate synchronization logic to preserve coherent view of the memory. The coherence protocol for global address space can introduce large overhead in both performance and energy. A combined software-hardware codesign approach can dramatically reduce such cost by allocating synchronization logic per data structure between all parallel readers and writers. Figure 14.2a shows how a large compute graph can be partitioned to multiple subgraphs, where each subgraph satisfies a set of hardware constraints, such as operation types, number of operations and I/Os, etc. Figure 14.2b shows one logical memory is

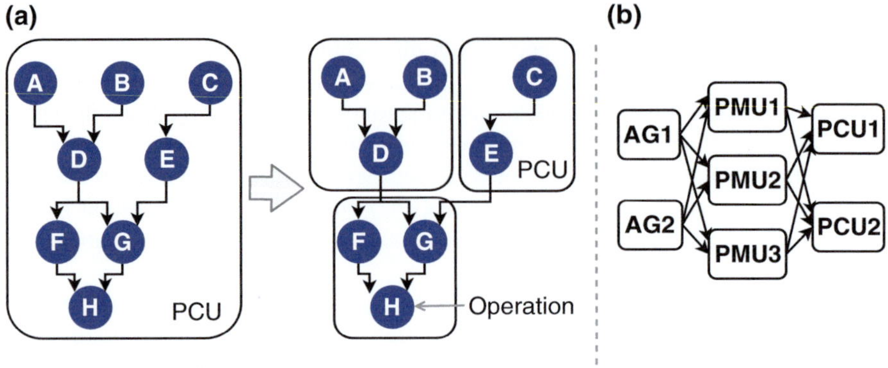

Fig. 14.2 Partitioning of compute and memory

mapped to multiple physical memory blocks; each contains one or more banks. The compiler preschedules the address generation (AG) to an independent CU, broadcasting requests to all memories, and responses are pipelined to all consumers. With static banking, it is guaranteed that only one request will hit each bank per cycle. In cases where banks accessed by each AG can be statically resolved, the compiler can optimize away all unnecessary data paths between memory and compute to a partial crossbar or one-to-one communication. The high fan-in in the crossbar can be partitioned to a tree across CUs to scale in network bandwidth with large parallelization. This allows the performance to scale linearly with parallelization in spatial architectures, at the cost of an exponential increase in interconnect resource consumption.

Once the program graph is decomposed in subgraphs that can fit in hardware, a place and route (PaR) tool can map the application onto the network array. This process is similar to the FPGA PaR process with lesser complexity. A prior study [25] has shown that compiler can improve PaR quality with static program analysis in a tightly integrated system.

14.4 Case Study: The Plasticine CGRA

Plasticine is a CGRA from Stanford University [22] consisting of reconfigurable Pattern Compute Units (PCUs) and Pattern Memory Units (PMUs), which we refer to collectively simply as "CUs". Figure 14.3 shows the chip-level architecture. CUs communicate with three kinds of interconnect: word-level scalar, multi-word-level vector, and bit-level control interconnects. Plasticine's array of CUs interfaces with DRAM through multiple DDR channels. Each channel has an associated coalescing unit that arbitrates between multiple address streams and consists of buffers to support multiple outstanding memory requests and address coalescing to minimize DRAM accesses. Each Plasticine component is used to map specific parts of applications:

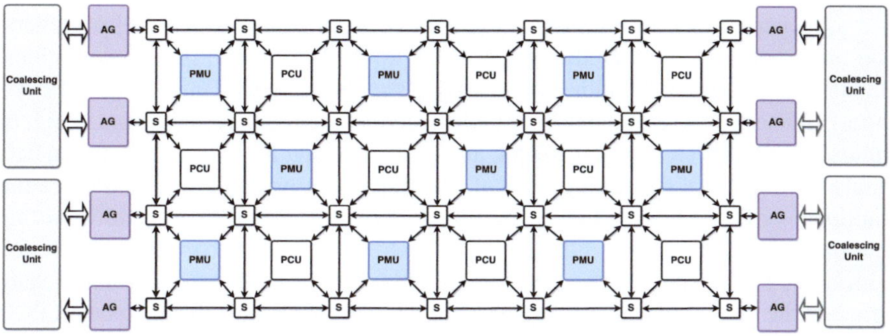

Fig. 14.3 Plasticine chip-level architecture (actual organization 16 × 8) [22]. All three networks have the same structure. PCU Pattern Compute Unit, PMU Pattern Memory Unit, AG Address Generator, S Switch Box. ©ACM 2017

local address calculation is done in PMUs, DRAM address computation happens in the DRAM address generation (AG) units, and the remaining data computation happens in PCUs.

14.4.1 Pattern Compute Unit (PCU)

The PCU is designed to exploit the fine-grained data and pipeline parallelism in a single, innermost parallel pattern in an application. Figure 14.4 shows the architecture of a PCU.

The PCU data path is organized as a multi-stage, reconfigurable SIMD pipeline. Each stage consists of several functional units (FUs) operating in SIMD fashion,

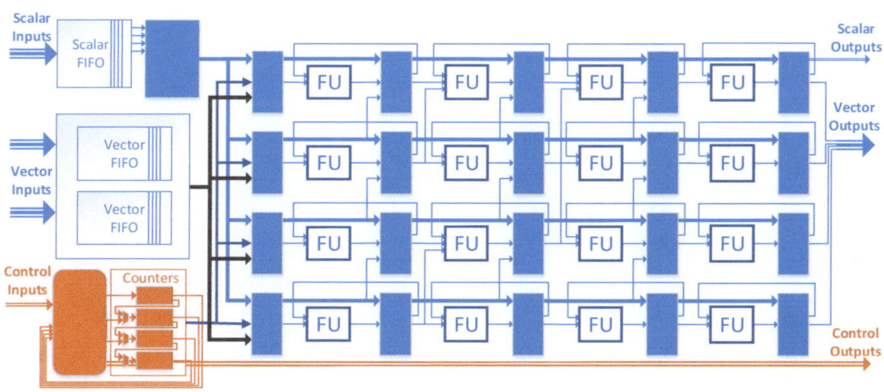

Fig. 14.4 Pattern Compute Unit (PCU) architecture [22]. We show only 4 stages and 4 SIMD lanes, and omit some control signals. ©ACM 2017

and associated pipeline registers (PR). FUs perform 32-bit word-level arithmetic and binary operations, including support for floating-point and integer operations. Reductions and sliding window operations are supported across lanes using dedicated intra-PCU networks. PCUs interface with the global interconnect using three kinds of inputs and outputs (IO): scalar, vector, and control. Scalar IO is used to communicate single words of data, such as the result of reductions. Vector IO allows communicating multiple words, such as reading and writing to scratchpads in PMUs and transmitting intermediate data between multiple PCUs. Control IO is used to coordinate execution with other PCUs and PMUs. A reconfigurable control block and a counter chain generate the necessary control signals and loop iterators, respectively, to begin PCU execution.

14.4.2 Pattern Memory Unit (PMU)

PMUs contain the on-chip memory as programmer managed scratchpads. Figure 14.5 shows the architecture of a PMU.

Scratchpads are built with multiple SRAM banks matching the number of PCU lanes. Address decoding logic around the scratchpad can be configured to operate in several banking modes to support various access patterns. Strided banking mode supports linear access patterns often found on dense data structures. FIFO mode supports streaming accesses. Line buffer mode captures access patterns resembling a sliding window. Duplication mode duplicates contents across all memory banks to support parallel indirect reads. In addition to banking, the scratchpad address space can be partitioned to implement generalized double buffering, or N-buffering, to support hierarchical pipelines. Each PMU contains a reconfigurable scalar data path intended for address calculation, which better utilizes PCU resources. A programmable counter chain and control block triggers PMU execution similar to the PCU.

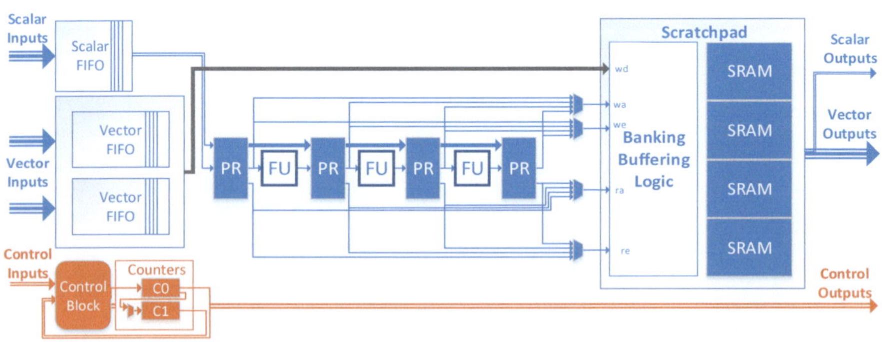

Fig. 14.5 Pattern Memory Unit (PMU) architecture [22]: configurable scratchpad, address calculation data path, and control. ©ACM 2017

14.4.3 Pipelined Switches

Plasticine is designed with three interconnects of different granularities; a bus-level vector network, a word-level scalar network, and a bit-level control network. All network switches are statically configured, and switches include pipeline registers to avoid long wire delays. Scalar and control switches share a reconfigurable control block and counters to efficiently map outer pipeline logic and increase PCU utilization and reduce routing hotspots.

14.4.4 Static-Dynamic Hybrid Interconnect

In an extended network study [25], a dynamic network is introduced in addition to the pipelined static network in Plasticine. The dynamic network only has a single vector network with support to partially clock gate buffers when transmitting scalar data. To optimize for multicast communication, a common communication pattern in spatial architectures, the router only duplicates packets where routes branch off for different destinations. The router contains a parameterizable number of virtual channel (VC) buffers to prevent deadlock. A static PaR tool decides which network the logical flows from the program get mapped to, and routes links for both static and dynamic network. The router dynamically looks up the statically assigned routes with packet headers. Unlike a static network that provides dedicated bandwidth to a logical flow, the dynamic network can share physical link resources across multiple flows.

14.4.5 Off-Chip Memory Access

Memory requests to external DRAM are generated in specialized reconfigurable scalar data paths called address generators (AG). A coalescing unit arbitrates between multiple AGs sharing a single DRAM channel. AGs can generate memory commands that are either dense or sparse. Dense requests are converted to multiple DRAM burst requests in the coalescing unit, while sparse requests engage the scatter-gather engine within the coalescing unit to minimize issuing DRAM requests to the same burst.

14.4.6 CU Granularity Selection

Section 14.2 outlined some key concepts around CGRA element granularity and identified the impact of granularity on the tradeoff between flexibility, efficiency, and programmability. This section uses the Plasticine CGRA as an example and discusses various sensitivity analysis experiments performed to select the granularity of each PCU and PMU.

Empirical selection of CGRA granularity requires representative benchmarks that stress the key patterns of computation. For Plasticine, a variety of compute-bound and memory-bound benchmarks from various domains were chosen to study the organization of various compute primitives within a PCU. Specifically, the study focused on area overheads of a PCU for various numbers of stages, registers, scalar and vector inputs and outputs to the PCU. Figure 14.6 heatmap shows the result of the experiments performed.

To drive the above study, benchmark-normalized area overhead is used as a cost metric for useful PCU area. First, the area of a single PCU is modeled as the sum of the area of its control box, FUs, pipeline registers, input FIFOs, and output crossbars. A sweep is performed for each parameter of interest. For each proposed value, a sweep is performed on the remaining parameter space to find the minimum possible PCU Area (AreaPCU). This area is then normalized based on their minimum (MinPCU) and report the overhead of each possible parameter value.

Figure 14.6a and b show that 6 stages and 6 pipeline registers per stage achieves the best area/op for the selected benchmarks. Four scalar and vector inputs and outputs

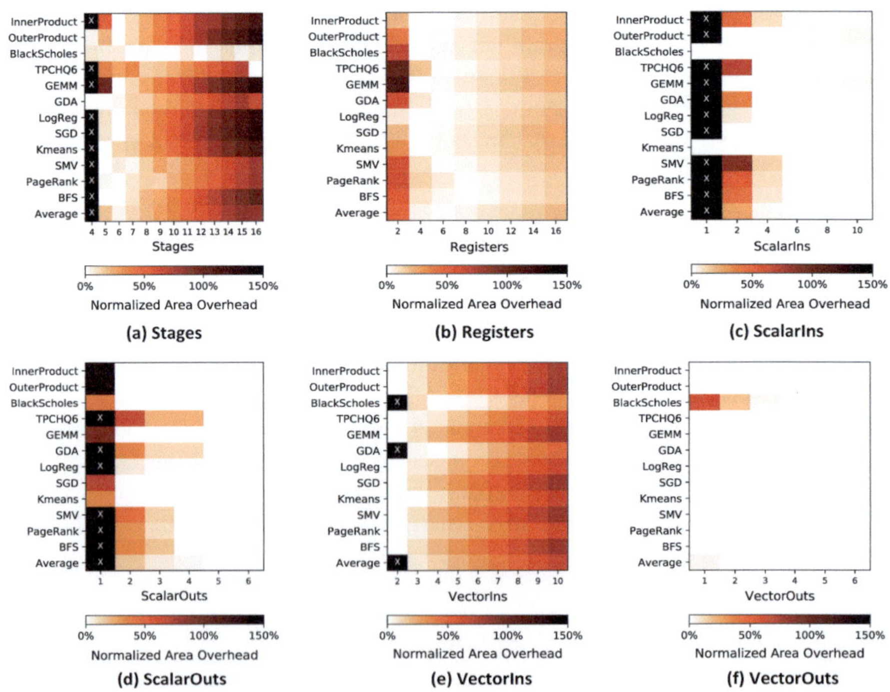

Fig. 14.6 Heatmap [22] (lighter is better) of PCU area overhead while sweeping various CU parameters for a subset of benchmarks from a variety of domains. 'X's mark invalid parameters for a given benchmark. ©ACM 2017

Table 14.1 Final CU granularities in Plasticine

	Component	Range	Final value
PCU	Lanes	4, 8, 16, 32	16
	Stages	1–16	6
	Registers/stage	2–16	6
	Scalar inputs	1–16	6
	Scalar outputs	1–6	5
	Vector inputs	1–10	3
	Vector outputs	1–6	3
PMU	Bank size	4, 8, 16, 32,	16 KB
	Scratchpad	64 KB	16
	Banks	= PCU lanes	256 KB
	Total scratchpad	Bank size *	4
	Stages	banks	6
	Registers/stage	1–16	4
	Scalar inputs	2–16	0
	Scalar outputs	1–16	3
	Vector inputs	0–6	1
	Vector outputs	1–10	
		1–6	
Plasticine	PCUs	–	64
	PMUs	–	64

are similarly shown to have the best area efficiency. Based on these experiments, the final CU granularity chosen for Plasticine is summarized in Table 14.1.

Similar to compute and memory granularity selection, interconnect parameters are chosen empirically by performing a parameter sweep to characterize the area, energy, and power overheads of interconnect switches. The graphs in Fig. 14.7 show the results of this study.

Figure 14.7 shows the characterization of area and energy for various parameters in the static-dynamic hybrid network. Figure 14.7d–f present the energy necessary to transmit a single bit through a switch or router. Figure 14.7a demonstrates the roughly quadratic scaling of switch area with the number of links between adjacent switches. Figure 14.8 shows the power consumption of one design point of switch and router with varying testbench duty cycles. Figure 14.7 shows that the router design point with *lower* throughput and power than the switch design point in Fig. 14.8 consumes *more* energy to transmit the same amount of data. This is due to heavy buffering and logics for VC and switch allocations required in the router. Figure 14.7 further suggests the switch consumes a significant amount of power when inactive. The zero-load dynamic power attributes to high fan-out clock tree in the switch. In summary, the study suggests that scaling network bandwidth is more efficient with a static network, and network throughput is critical for system performance on a pipelined spatial architecture. However, overprovisioning in the static network can be expensive in power because the switch cannot be effectively clock-gated when they are not used. Therefore, using a low-bandwidth dynamic network for infrequent traffic and a high-bandwidth static network for bandwidth-sensitive traffic can achieve the best system-level efficiency. The study further shows that the PaR converges to less overall

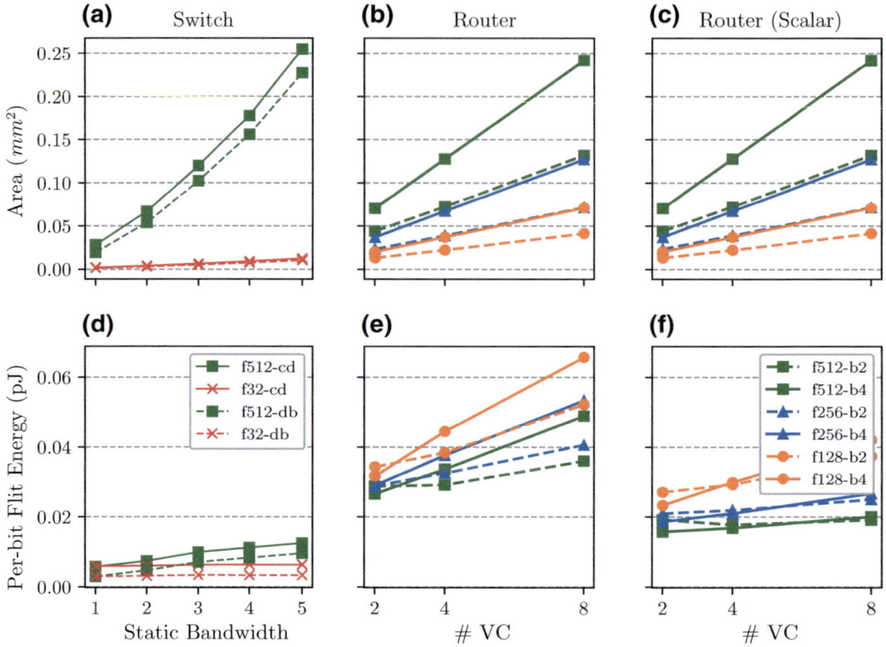

Fig. 14.7 Switch and router power and area characterization [25] (f# flit-width, cd credit-based flow control, db double-buffered flow-control, b# buffer depth, static bandwidth # links between adjacent switches). ©ACM 2019

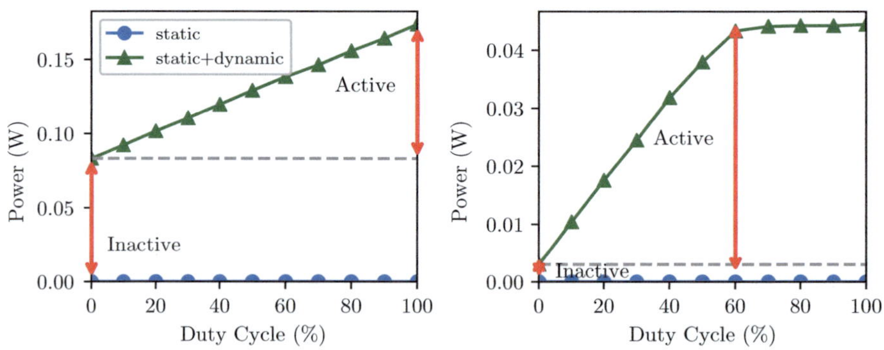

Fig. 14.8 Switch and router power with varying testbench duty cycle [25]. The switch corresponds to design point f512-db with static bandwidth equals to 2. The router corresponds to f512-b4 with 4VCs. ©ACM 2019

routing distance when using the dynamic network as an escape path, which reduces data movement and improves network energy efficiency. The end-to-end evaluation with a variety of benchmarks shows that a hybrid network matches the performance and area of a pure static network with slightly higher bandwidth while providing 1.8× improvement in network energy efficiency.

14.4.7 Compiling to Plasticine

The programming language for Plasticine is Spatial [26], a hardware-centric DSL expresses applications with nested loops and parallel patterns. Figure 14.9 describes the compilation flow at a high level, along with various analyses and transformations performed.

Spatial exposes several built-in memory types such as DRAM, SRAM, FIFO, along with explicit on/off-chip memory transfer operations to allow users to have explicit control over the memory hierarchy. Data structures are declared explicitly using these types, where memory variables indicate non-overlapping regions in the address space. Users describe applications with untimed, unparalleled loops, branch statements, and FSMs. The language exposes loop parallelization factors, loop scheduling, and tiling factors as parameters in applications. The compiler automatically banks and buffers the memories when the user chooses to parallelize and pipeline loop nests in the application.

To map applications described in Spatial to Plasticine, the Plasticine compiler converts the controller hierarchy in Spatial to a distributed control and data flow graph (CDFG) that gets lowered to hardware configuration in a series of transformations. Figure 14.10 shows an example of the key steps involved.

Figure 14.10a shows a controller hierarchy in Spatial, which is transformed automatically into a virtual CU data-flow graph shown in Fig. 14.10c. To do so, the compiler assigns each innermost controller, which corresponds to a basic block in the program, into a virtual CU. The data-flow graph inside the basic block is mapped across stages of the SIMD pipeline, and a parallelized innermost loop is vectorized

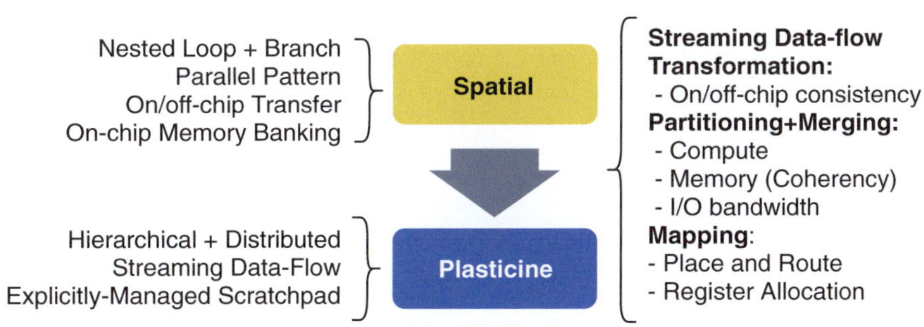

Fig. 14.9 Compilation flow to target plasticine

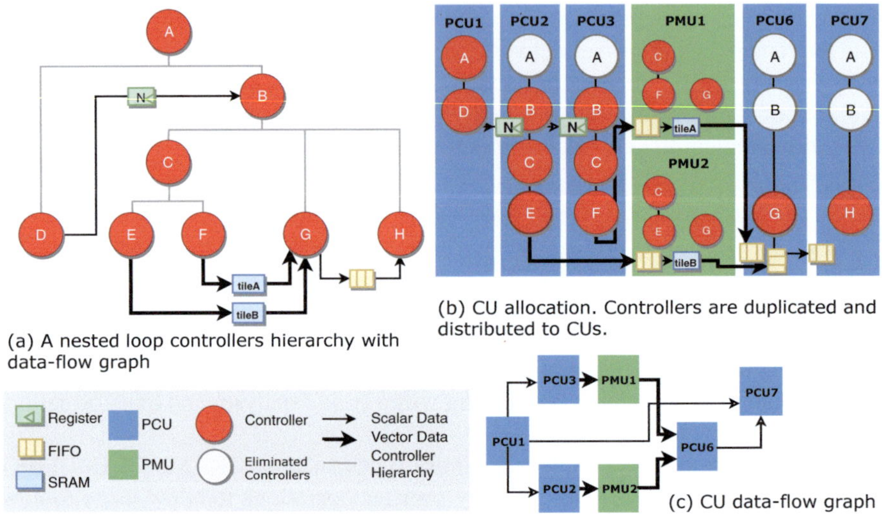

Fig. 14.10 Mapping hierarchical control-flow to distributed execution

across SIMD lanes. The compiler duplicates the outer controllers for each innermost controller into their assigned CUs, as shown in Fig. 14.10b. These controllers are mapped to the control block of the CU and used to synchronize incoming and outing streams between CUs. Outer controllers that are not used to synchronize streams can be eliminated from the graph. To maintain memory consistency in a data flow fashion, the compiler introduces additional control tokens between multiple accessors of a memory to synchronize their access orders. After allocating the CU data flow, the compiler partitions the program graph based on a Plasticine specification that limits compute and memory resources in each physical CU type. After the program is partitioned, a PaR tool maps the program graph onto the network array. The PaR also allocates VCs to each link in the program graph to avoid deadlock in the dynamic network.

Evaluation over several benchmarks [23, 25, 22] has shown that Plasticine provides up to a 95× improvement in performance and up to 77× improvement in performance/watt over a Stratix V FPGA, a geometric mean speedup of 30× compared to the Tesla V100 GPU, and 2× compared with Microsoft's BrainWave due to better utilization of compute and memory resources.

14.5 Related Work

Several researchers [19–21, 27–31] and industry practitioners [32, 33] alike have explored various flavors of coarse-grained building blocks to build reconfigurable architectures. Several surveys [19–21] provide a broad overview of prior work on CGRAs. We discuss a few relevant bodies of work under the following categories:

14.5.1 CGRAs with Reconfigurable Scratchpads

ADRES [34], DySER [27], Garp [28], and Tartan [29] closely couple a reconfigurable fabric with a CPU. These architectures access main memory through the cache hierarchy shared with the host CPU. ADRES and DySER tightly integrate the reconfigurable fabric into the execution stage of the processor pipeline, and hence depend on the processor's load/store unit for memory accesses. ADRES consists of a network of functional units, reconfigurable elements with register files, and a shared multi-ported register file. DySER is a reconfigurable array with a statically configured interconnect designed to execute innermost loop bodies in a pipelined fashion. Garp consists of a MIPS CPU core and an FPGA-like coprocessor. Piperench [30] consists of a pipelined sequence of "stripes" of functional units (FUs). A word-level crossbar separates each stripe. Each FU has an associated register file that holds temporary results. Tartan [29] consists of a RISC core and an asynchronous, coarse-grained reconfigurable fabric (RF). The RF architecture is hierarchical with a dynamic interconnect at the topmost level, and a static interconnect in the inner level. The architecture of the innermost RF core is modeled after Piperench [30].

14.5.2 Architectures with Reconfigurable Data Paths

TRIPS [31] is a tiled architecture where execution proceeds dynamically in a dataflow fashion, while instructions are statically issued within a block. TRIPS does not have a static interconnect, but contains two dynamic interconnect ion networks [35]: an operand network (OPN) to route operands between tiles, and an on-chip network (OCN) to communicate with cache banks. The Raw microprocessor [36] is a tiled architecture where each tile consists of a single-issue in-order processor, a floating-point unit, a data cache, and a software-managed instruction cache. Tiles communicate with their nearest neighbors using pipelined, word-level static and dynamic networks. Plasticine does not incur the overheads of dynamic networks and general-purpose processors mentioned above. Using hardware managed caches in place of reconfigurable scratchpads reduces power and area efficiency in favor of generality.

14.5.3 Dense Data Paths and Hierarchical Pipelines

RaPiD [37] is a one-dimensional array of ALUs, registers, and memories with hardware support for static and dynamic control. A subsequent research project called

Mosaic [38] includes a static hybrid interconnect along with hardware support to switch between multiple interconnect configurations. HRL [39] combines coarse-grained and fine-grained logic blocks with a hybrid static interconnect. While a centralized scratchpad enables some on-chip buffering, the architecture is primarily designed for memory-intensive applications with little locality and nested parallelism. Triggered instructions [40] is an architecture consisting of coarse-grained processing elements (PEs) of ALUs and registers in a static interconnect. Each PE contains a scheduler and a predicate register to implement dataflow execution using triggers and guarded actions. The control flow mechanism used in Plasticine has some similarities with Triggered instructions. Wavescalar [41] is another tiled dynamic dataflow architecture with four levels of hierarchy, connected by dynamic interconnects that vary in topology and bandwidth at each level. While execution is dataflow driven, the data path is not reconfigurable, and broadcast and dynamic interconnects are used for communication. Coarse-grained parallelism can be exploited using multi-threaded support and barriers to achieve synchronization. However, the lack of a distributed scratchpad means that parallel memory accesses are serialized at the memory interface.

14.5.4 Statically Scheduled Interconnects

Some architectures allow interconnect configurations to change periodically based on a statically determined schedule, to allow for greater interconnect link utilization compared to a fully static network [38, 42, 43]. Such interconnects typically require the compiler to provide a valid static schedule using modulo scheduling. While this approach is effective for inner loops with predictable latencies and fixed Initiation Interval (II), variable latency operations and hierarchical loop nests complicates the compiler by creating scheduling complexities to arrive at a single module schedule. HyCube [44] has a similar statically scheduled network with the added ability to bypass intermediate switches in the same cycle. This approach allows operands to travel multiple hops in a single cycle, but creates long wires and combinational paths, which adversely affects the clock period and scalability.

14.6 Summary and Conclusions

Coarse-Grained Reconfigurable Architectures are a class of architectures characterized by reconfigurable compute and memory elements in a programmable interconnection network. CGRAs have been shown to achieve higher performance and energy efficiency compared to conventional instruction-based architectures by avoiding instruction overheads with reconfigurable data and control paths. CGRAs also avoid the hardware and programming overheads of fine-grained alternatives such as FPGAs by raising the hardware abstraction.

Harnessing the full potential of CGRAs requires co-designing the architectural primitives with the programming model and compilation flow. Choosing the right granularity for a CGRA requires navigating a large multidimensional design space and depends on common patterns of computation for which the CGRA is being built for. This choice, in turn, impacts the programming model, and the complexity and efficiency of the compiler.

Using the Plasticine CGRA as a case study, a canonical way to quantitatively reason about CGRA granularity is presented in terms of compute, memory, and interconnect. Compute is characterized in terms of the number of ALUs in a compute element, and their organization into pipeline stages and lanes. Memory is characterized in terms of capacity and bandwidth. Interconnect is characterized in terms of topology, bus width, and routing flexibility with static and dynamic routing.

An automatic compilation flow is described that maps applications, which are described as arbitrary hierarchies of loop controllers, into parallelized, nested pipelines on the target CGRA. Compiling an arbitrary loop nest on to a distributed architecture like a CGRA requires lowering the program from a monolithic loop-centric representation into a control and data flow graph (CDFG) with distributed data and control flow.

In the wake of technology scaling challenges and the ever-increasing appetite for greater compute, CGRAs show a promising path to design the next generation of chip architectures. Given that a large fraction of modern ASIC development costs goes towards software development [45], co-designing the hardware architecture with the programming model can provide a huge practical advantage; applications can be written using high-level constructs and compiled to the desired architecture several generations ahead of the actual hardware. This approach also provides necessary feedback between hardware and software early on in the design process, which helps making more informed hardware and software design decisions.

References

1. Y. Chen, T. Luo, S. Liu, S. Zhang, L. He, J. Wang, L. Li, T. Chen, Z. Xu, N. Sun, O. Temam, Dadiannao: a machine-learning supercomputer, in *2014 47th Annual IEEE/ACM International Symposium on Microarchitecture* (Dec 2014), pp. 609–622
2. L. Wu, A. Lottarini, T.K. Paine, M.A. Kim, K.A. Ross, Q100: the architecture and design of a database processing unit, in *Proceedings of the 19th International Conference on Architectural Support for Programming Languages and Operating Systems, ASPLOS '14, New York, NY, USA*. (ACM, 2014), pp. 255–268
3. Y. Chen, T. Krishna, J. Emer, V. Sze, Eyeriss: an energy-efficient reconfigurable accelerator for deep convolutional neural networks, in *2016 IEEE International Solid-State Circuits Conference (ISSCC)* (IEEE, 2016), pp. 262–263
4. S. Han, X. Liu, H. Mao, J. Pu, A. Pedram, M.A. Horowitz, W.J. Dally, Eie: efficient inference engine on compressed deep neural network. arXiv preprint arXiv:1602.01528 (2016)
5. J. Dean, D. Patterson, C. Young, A new golden age in computer architecture: empowering the machine-learning revolution. IEEE Micro **38**(2), 21–29 (2018)
6. K. Olukotun, Designing computer systems for software 2.0, isca 2018 keynote. http://iscaconf.org/isca2018/docs/Kunle-ISCA-Keynote-2018.pdf (2018)

7. M. Horowitz, 1.1 Computing's energy problem (and what we can do about it), in *IEEE International Solid-State Circuits Conference Digest of Technical Papers (ISSCC)* (Feb 2014), pp. 10–14
8. R. Hameed, W. Qadeer, M. Wachs, O. Azizi, A. Solomatnikov, B.C. Lee, S. Richardson, C. Kozyrakis, M. Horowitz, Understanding sources of inefficiency in general-purpose chips, in *Proceedings of the 37th Annual International Symposium on Computer Architecture, ISCA '10, New York, NY, USA* (ACM, 2010), pp. 37–47
9. J. Leng, T. Hetherington, A.E. Tantawy, S. Gilani, N.S. Kim, T.M. Aamodt, V.J. Reddi, Gpuwattch: enabling energy optimizations in gpgpus, in *Proceedings of the 40th Annual International Symposium on Computer Architecture, ISCA '13, New York, NY, USA* (2013, ACM), pp. 487–498
10. D. Voitsechov, Y. Etsion, Single-graph multiple flows: energy efficient design alternative for gpgpus, in *Proceeding of the 41st Annual International Symposium on Computer Architecture, ISCA '14, Piscataway, NJ, USA* (IEEE Press, 2014), pp. 205–216
11. C.J. Hughes, Single-instruction multiple-data execution, in *Synthesis Lectures on Computer Architecture* (2015), pp. 1–121
12. K. Czechowski, V.W. Lee, E. Grochowski, R. Ronen, R. Singhal, R. Vuduc, P. Dube, Improving the energy efficiency of big cores, in *Proceeding of the 41st Annual International Symposium on Computer Architecture, ISCA '14, Piscataway, NJ, USA* (IEEE Press, 2014), pp. 493–504
13. H.J. Lee, K.J. Brown, A.K. Sujeeth, T. Rompf, K. Olukotun, Locality-aware mapping of nested parallel patterns on gpus, in *Proceedings of the 47th Annual IEEE/ACM International Symposium on Microarchitecture, IEEE Micro* (2014)
14. I. Kuon, R. Tessier, J. Rose, Fpga architecture: survey and challenges. Found. Trends Electron. Des. Autom. **2**(2), 135–253 (2008)
15. B.H. Calhoun, J.F. Ryan, S. Khanna, M. Putic, J. Lach, Flexible circuits and architectures for ultralow power. Proc. IEEE **98**(2), 267–282 (2010)
16. I. Bolsens, Programming modern fpgas, international forum on embedded multiprocessor soc, keynote. http://www.xilinx.com/univ/mpsoc2006keynote.pdf (2006)
17. K.K.W. Poon, S.J.E. Wilton, A. Yan, A detailed power model for field-programmable gate arrays. ACM Trans. Des. Autom. Electron. Syst. **10**(2), 279–302 (2005)
18. V. Adhinarayanan, I. Paul, J.L. Greathouse, W. Huang, A. Pattnaik, W. Feng, Measuring and modeling on-chip interconnect power on real hardware, in *2016 IEEE International Symposium on Workload Characterization (IISWC)* (Sept 2016), pp. 1–11
19. R. Tessier, K. Pocek, A. DeHon, Reconfigurable computing architectures. Proc. IEEE **103**(3), 332–354 (2015)
20. T.J. Todman, G.A. Constantinides, S.J.E. Wilton, O. Mencer, W. Luk, P.Y.K. Cheung, Reconfigurable computing: architectures and design methods. IEE Proc. Comput. Digit. Tech. **152**(2), 193–207 (2005)
21. R. Hartenstein, A decade of reconfigurable computing: a visionary retrospective, in *Proceedings of the Conference on Design, Automation and Test in Europe, DATE '01, Piscataway, NJ, USA* (IEEE Press, 2001), pp. 642–649
22. R. Prabhakar, Y. Zhang, D. Koeplinger, M. Feldman, T. Zhao, S. Hadjis, A. Pedram, C. Kozyrakis, K. Olukotun, Plasticine: a reconfigurable architecture for parallel patterns, in *Proceedings of the 44th Annual International Symposium on Computer Architecture* (ACM, 2017), 389–402
23. T. Zhao, Y. Zhang, K. Olukotun, Serving recurrent neural networks efficiently with a spatial accelerator, in *Proceedings of the 2nd SysML Conference (SysML 2019). Palo Alto, CA, USA*
24. J. Fowers, K. Ovtcharov, M. Papamichael, T. Massengill, M. Liu, D. Lo, S. Alkalay, M. Haselman, L. Adams, M. Ghandi, S. Heil, P. Patel, A. Sapek, G. Weisz, L. Woods, S. Lanka, S.K. Reinhardt, A.M. Caulfield, E.S. Chung, D. Burger, A configurable cloud-scale DNN processor for real-time AI, in *45th ACM/IEEE Annual International Symposium on Computer Architecture, ISCA 2018, Los Angeles, CA, USA*
25. Y. Zhang, A. Rucker, M. Vilim, R. Prabhakar, W. Hwang, K. Olukotun, Scalable interconnects for reconfigurable spatial architectures, in *Proceedings of the 46th International Symposium on Computer Architecture (ISCA 2019). Phoenix, AZ, USA*

26. D. Koeplinger, M. Feldman, R. Prabhakar, Y. Zhang, S. Hadjis, R. Fiszel, T. Zhao, L. Nardi, A. Pedram, C. Kozyrakis, K. Olukotun, Spatial: a language and compiler for application accelerators, in *Proceedings of the 39th ACM SIGPLAN Conference on Programming Language Design and Implementation (PLDI 2018)* (ACM, New York, NY, USA)
27. V. Govindaraju, C. Ho, T. Nowatzki, J. Chhugani, N. Satish, K. Sankaralingam, C. Kim, Dyser: unifying functionality and parallelism specialization for energy-efficient computing. IEEE Micro **32**(5), 38–51 (2012)
28. T.J. Callahan, J.R. Hauser, J. Wawrzynek, The garp architecture and c compiler. Computer **33**(4), 62–69 (2000)
29. M. Mishra, T.J. Callahan, T. Chelcea, G. Venkataramani, S.C. Goldstein, M. Budiu, Tartan: evaluating spatial computation for whole program execution, in *Proceedings of the 12th International Conference on Architectural Support for Programming Languages and Operating Systems, ASP-LOS XII, New York, NY, USA* (ACM, 2006), pp. 163–174
30. S.C. Goldstein, H. Schmit, M. Moe, M. Budiu, S. Cadambi, R.R. Taylor, R. Laufer, Piperench: a co/processor for streaming multimedia acceleration, in *Proceedings of the 26th Annual International Symposium on Computer Architecture, ISCA '99, Washington, DC, USA* (IEEE Computer Society, 1999), pp. 28–39
31. K. Sankaralingam, R. Nagarajan, H. Liu, C. Kim, J. Huh, D. Burger, S.W. Keckler, C.R. Moore, *Exploiting ilp, tlp, and dlp with the polymorphous trips architecture*, in *Proceedings of the 30th Annual International Symposium on Computer Architecture, ISCA '03, New York, NY, USA* (ACM, 2003), pp. 422–433
32. Wave Computing Launches Machine Learning Appliance. https://www.top500.org/news/wave-computing-launches-machine-learning-appliance/
33. J. Noguera, C. Dick, V. Kathail, G. Singh, K. Vissers, R. Wittig, 2018. Xilinx Project Everest: 'HW/SW Programmable Engine' (Hot Chips 30). http://www.hotchips.org/hc30/2conf/2.03_Xilinx_Juanjo_XilinxSWPEHotChips20180819.pdf
34. B. Mei, S. Vernalde, D. Verkest, H. De Man, R. Lauwereins, *ADRES: An Architecture with Tightly Coupled VLIW Processor and Coarse-Grained Reconfigurable Matrix* (Springer, Berlin, 2003), pp. 61–70
35. P. Gratz, C. Kim, K. Sankaralingam, H. Hanson, P. Shivakumar, S.W. Keckler, D. Burger, On-chip interconnection networks of the trips chip. IEEE Micro **27**(5), 41–50 (2007)
36. M.B. Taylor, J. Kim, J. Miller, D. Wentzlaff, F. Ghodrat, B. Greenwald, H. Hoffman, P. Johnson, J.W. Lee, W. Lee, A. Ma, A. Saraf, M. Seneski, N. Shnidman, V. Strumpen, M. Frank, S. Amarasinghe, A. Agarwal, The raw microprocessor: a computational fabric for software circuits and general-purpose programs. IEEE Micro **22**(2), 25–35 (2002)
37. D.C. Cronquist, C. Fisher, M. Figueroa, P. Franklin, C. Ebeling, Architecture design of reconfigurable pipelined datapaths, in *Proceedings. 20th Anniversary Conference on Advanced Research in VLSI, 1999* (Mar 1999), pp. 23–40
38. B. Van Essen, A. Wood, A. Carroll, S. Friedman, R. Panda, B. Ylvisaker, C. Ebeling, S. Hauck, Static versus scheduled interconnect in coarse-grained reconfigurable arrays, in *2009 International Conference on Field Programmable Logic and Applications* (Aug 2009), pp. 268–275
39. M. Gao, C. Kozyrakis, Hrl: Efficient and flexible reconfigurable logic for near-data processing, in *2016 IEEE International Symposium on High Performance Computer Architecture (HPCA)* (March 2016), pp. 126–137
40. A. Parashar, M. Pellauer, M. Adler, B. Ahsan, N. Crago, D. Lustig, V. Pavlov, A. Zhai, M. Gambhir, A. Jaleel, R. Allmon, R. Rayess, S. Maresh, J. Emer, Triggered instructions: a control paradigm for spatially-programmed architectures, in *Proceedings of the 40th Annual International Symposium on Computer Architecture, ISCA '13, New York, NY, USA* (ACM, 2013), pp. 142–153
41. S. Swanson, A. Schwerin, M. Mercaldi, A. Petersen, A. Putnam, K. Michelson, M. Oskin, S.J. Eggers, The wavescalar architecture. ACM Trans. Comput. Syst. **25**(2), 4:1–4:54 (May 2007)
42. G. Dimitroulakos, M.D. Galanis, C.E. Goutis, Exploring the design space of an optimized compiler approach for mesh-like coarse-grained reconfigurable architectures, in *Parallel and Distributed Processing Symposium, 2006. IPDPS 2006. 20th International* (IEEE, 2006), pp. 10

43. C. Nicol, A coarse grain reconfigurable array (cgra) for statically scheduled data flow computing
44. M. Karunaratne, A. Kulkarni Mohite, T. Mitra, L.S. Peh, Hycube: a cgra with reconfigurable single-cycle multi-hop interconnect, in *Proceedings of the 54th Annual Design Automation Conference 2017, DAC '17, New York, NY, USA* ACM, 2017), pp. 45:1– 45:6
45. H. Jones, Strategies in optimizing market positions for semiconductor vendors based on ip leverage, ibs white paper. http://www.ibs-inc.net (2014)

Chapter 15
A 1000× Improvement of the Processor-Memory Gap

Zvi Or-Bach

15.1 Historical Prospective

Over more than 50 years, the Integrated Circuit (IC) industry has grown from nothing to over $500 B/year. The driving force was the ability to scale down, known as Moore's Law, where with each new node the number of integrated elements doubles at about the same overall cost and with better speed and lower power. In the deep sub-micron regime such scaling has come at an exponentially higher development and infrastructure cost, usually consisting of many $B. From over 50 IC companies pursuing scaling just 20 years ago, we now have merely three committed to the 7 nm node. Additionally, these handful of companies are integrating just few flavors of logic circuits. Memory circuits are being produced separately by special fabs dedicated to memory. These are DRAM fabs, which at advanced nodes are currently produced by only three vendors, and storage fabs such as 3D NAND. The full system is typically achieved by integrating logic and memory using a Printed Circuits Board (PCB) or 2.5D (chip-on-substrate) packaging. The overall system performance is limited by the off-chip interconnection that lags way behind IC interconnection (Figs. 15.1 and 15.2).

While on-chip interconnects have improved faster than off chip interconnects, they are still far worse than the transistor performance improvement with scaling. And the performance gap between logic gate delay and the on-chip interconnect delay is getting exponentially worse with scaling.

The combination of these effects has been the source of what was called by John L. Hennessy and David A. Patterson the "Memory Wall" [1] or the Processor-Memory Gap. This performance gap has grown by about 50% per year. Figure 15.3a and b shed some more light on this gap.

Z. Or-Bach (✉)
MonolithIC 3D Inc., 3555 Woodford Dr, San Jose, CA 95124, USA
e-mail: Zvi@MonolithIC3D.com; or_bach@yahoo.com

B. Murmann and B. Hoefflinger (eds.), *NANO-CHIPS 2030*,
The Frontiers Collection, https://doi.org/10.1007/978-3-030-18338-7_15

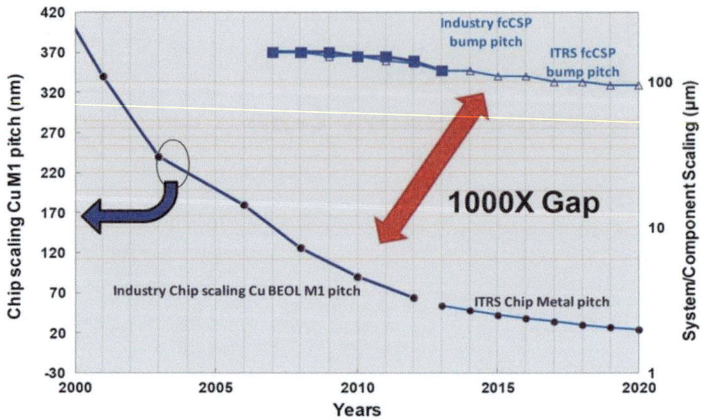

Fig. 15.1 Gap between on-chip interconnect and off-chip interconnect. *Source* VLSI 2013, Dr. Jack Sun, CTO of TSMC

In a report named "Why we need Exascale and why we won't get there by 2020" [3] the problem with the wires has been nicely articulated (see Fig. 15.4).

3D integration leveraging the concepts presented in Chaps. 8 and 10 could help overcoming the memory wall and the tyranny of interconnects to enhance computer systems by orders of magnitude.

The use of Monolithic 3D integration for 1000× improvement in computer performance has been reported [4–6], work on it is now supported in DARPA's 3DSoC program and is also detailed in Chap. 9 of this book (Fig. 15.5).

15.2 Precise Wafer Bonding to Overcome the Memory Wall

The advantage of 3D integration using precise wafer bonders, as detailed in Chap. 8, is the ability to keep using existing wafer processing fabs and processes while allowing 3D heterogeneous integration. Such 3D heterogeneous integration enables overcoming the "Memory Wall" just as suggested in the work by Stanford [4–6].

In a following work [7] the concept of 3D integration has been further advanced to enable first aggregating memory layers, such as conventional DRAM, to create a 3D array of memory with enough capacity and then integrating it with logic to complete the 1000× improved computing system. This concept has been designed to keep the 3D integration as simple as Place-Bond-Thin ("cut")-and Place again. Such simplified 3D integration can leverage Hybrid Bonding [8–11] in which the boding process allows for oxide to oxide and metal to metal bonding, thus achieving both mechanical bonding of the two wafers and formation of electrical connections between the landing pads of the bottom wafer and the connection pins of the top wafers. This could be further enhanced using a technology called "Fusion Hybrid

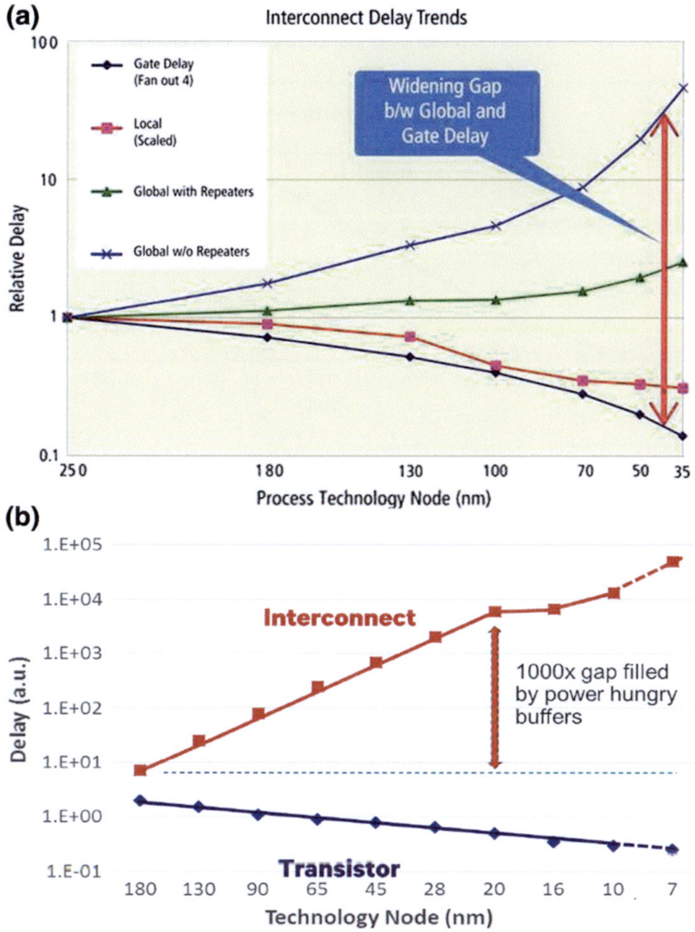

Fig. 15.2 **a** Gap between on chip interconnect and logic gate delay. **b** Another look at the gap between on-chip interconnect and logic gate delay. *Source* ITRS

Bonding" [12] which would work well for precise wafer bonding as discussed in Chap. 8, possibly including a "check and correct" step. It starts with both wafers precisely placed one on top of the other. The wafers are then lightly bonded at about room temperature. The bond surface might be pre-treated such as with plasma to enable nearly contact bonding. Once the initial bond has been established and alignment has been verified, an elevated temperature (100–200 °C) is used to finalize the bonding, achieving a permanent strong bonding between the two wafers.

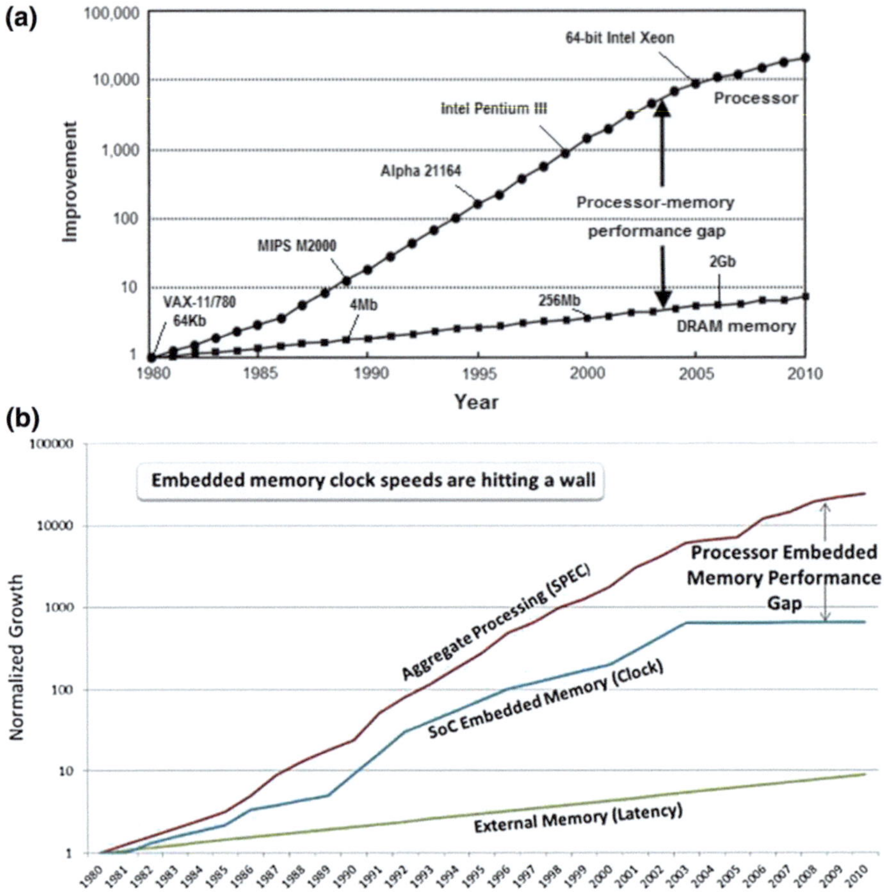

Fig. 15.3 **a** Yearly improvement of processor and DRAM memory speeds over three decades (*Source* [2]). **b** Embedded memory performance gap (*Source* semiwiki.com)

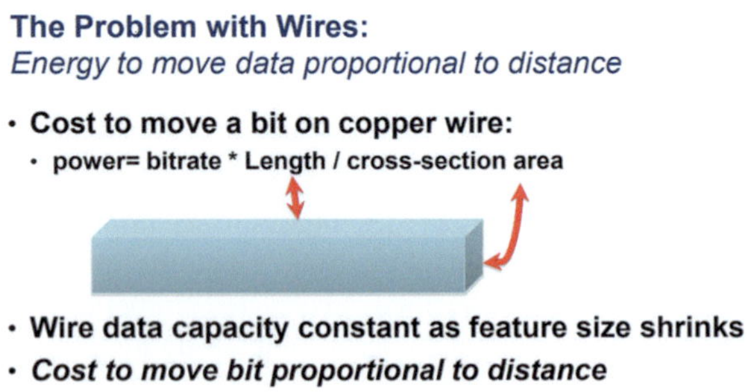

Fig. 15.4 The problem with wires

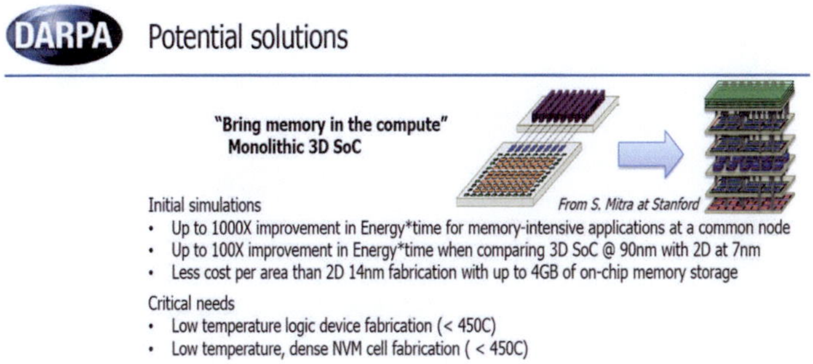

Fig. 15.5 1000× improvement in energy × time by Monolithic 3D SoC

15.3 The Memory Stack

As presented in Chap. 8, it is desirable to have a 'cut layer' built in the base wafer used for the memory stack. Such could be a SiGe layer or an oxide layer or other etch-selective layer. For DRAM wafers the use of the N+ deep well which is common for DRAM wafers could be a convenient option. The use of SOI wafers is also attractive as it allows the use of advanced fab lines such as the GlobalFoundries or Samsung. An additional advantage in the use of SOI, such as GlobalFoundries' 22FDX process, is having a substrate contact as part of the PDK to provide for back-bias. Such substrate contacts could be used as part of the 'nano-TSV,' also called through-layer-via, as illustrated in Figs. 15.6 and 15.7. Vertical pillars are formed with stacking of nano-TSVs.

Use of a 'cuttable' wafer enables a controlled removal of the substrate, after its flipping and bonding, by grinding and etching, using the BOX (the 'cut-layer') as an etch stop. Accordingly, the 'nano-TSV' is made similar to inter-metal via of the corresponding process, which allows about 10,000× higher vertical connectivity [~$(5\mu/50n)^2$]. It should be noted that 'nano-TSV' process needs to be all the way to the cut layer, so it could be easily turn into pin or landing pad after flipping, bonding, and cut process, as is illustrated in Fig. 15.7a, b.

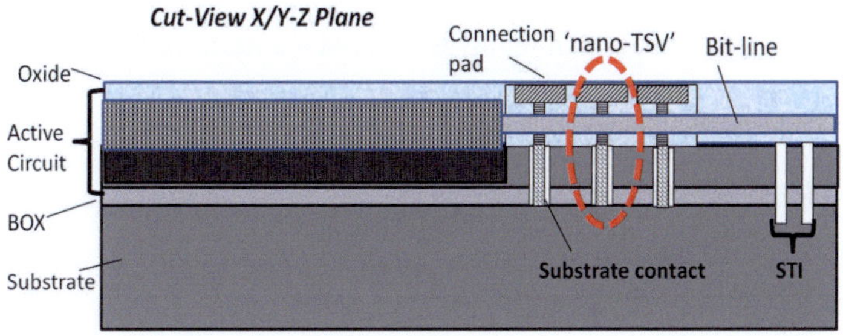

Fig. 15.6 Bit-cell array on SOI wafers with vertical pillar

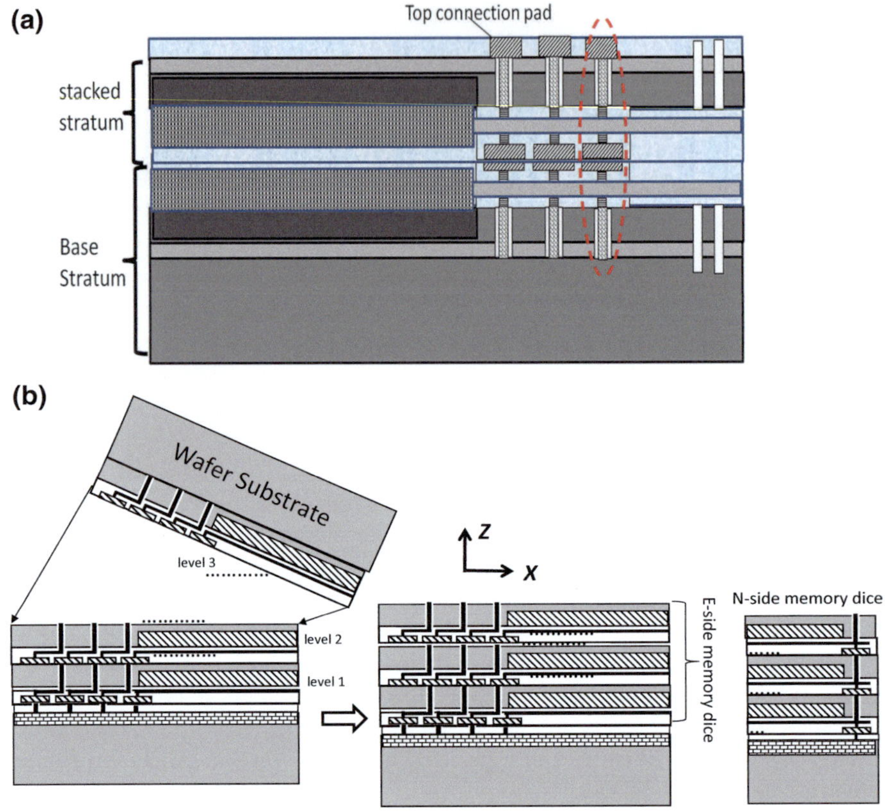

Fig. 15.7 **a** Two memory strata, vertical pillar marked. **b** Illustrating formation flow of three memory strata

The other element enabling fine grain vertical connectivity relates to the stacking misalignment. Until recently, bonder misalignment was on the order of 1 μm, which severely impacted the effective vertical connectivity. To combat that MonolithIC 3D has developed an innovative alignment technique called 'Smart Alignment' [13, 14].

As detailed in Chap. 8 herein, precise bonders are now capable of better than 50 nm (3σ) alignment precision, which removes some of the need for Smart Alignment.

15.4 The Architecture

The suggested computer architecture includes the following strata: Bit-cell array, Memory control, processor, and I/O. Figure 15.5a, b illustrate two optional configurations.

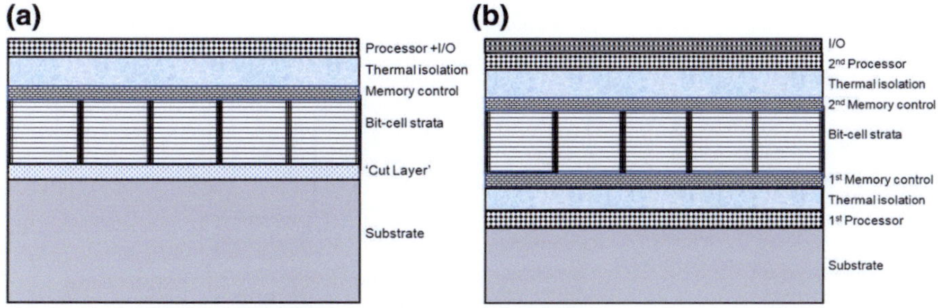

Fig. 15.8 **a** Single side configuration. **b** Dual side configuration

The configuration of Fig. 15.8a is built on a 'cuttable' substrate allowing the use of the illustrated structure as a transferable structure for further 3D integration. The bit-cell memory stack is built by stacking memory strata as will be detailed later. A memory control stratum provides the peripheral circuits for each of the memory units using per unit vertical pillars of global bit-lines and global word-lines. These pillars are formed by stacks of nano-TSVs as illustrated in Figs. 15.6, 15.7. The memory control is interfaced to the processor stratum through a thermal isolation layer designed to isolate the heat generated at the processor stratum from the memory stack underneath it. The processor stratum could include the 3D SoC I/O circuits, or the I/O could occupy its own stratum. Figure 15.8b illustrates an alternative 3D SoC. The base wafer could be any 2D wafer including the most advanced process node for the first processor stratum. Through thermal isolation layer it is connected with the first memory control stratum, which provides bottom peripheral circuits to the memory strata. The memory strata include feed-throughs to allow the bottom side and the top side (2nd memory control) to synchronize their memory access. Overlaying the memory strata is the 2nd memory control stratum, connecting with the 2nd processor stratum built on a 'cuttable' wafer, such as a standard foundry SOI wafer. An I/O stratum overlays the structure, thus providing system connections to the external devices. Such an I/O stratum could be built on a design-rule relaxed SOI process, such as RF-SOI, and could include a wireless communication channel or be built on a wafer supporting optical communication channels.

15.5 Details of the Memory Stack

The memory stack is built by stacking wafers structured as units of bit-cell array [13].

Figure 15.9 illustrates a small 3 × 3 region of an array of units forming the bit-cell array stratum. The unit size is about 200 μm × 200 μm while the connectivity lane between units, intended for inter-stratum connections, is about 1 μm wide (the drawing is not to scale). Each unit is a mini array of tightly packed bit-cells. The

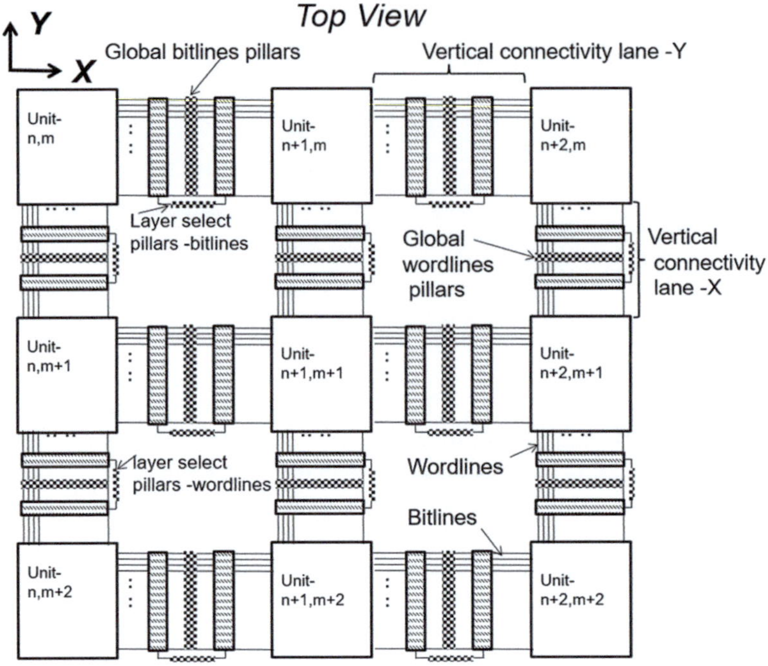

Fig. 15.9 Exemplary 3 × 3 units region of the bit-cell array stratum

bit-lines and the perpendicularly-oriented word-lines allow control of the individual bit-cells within a unit. These memory control lines extend across units, yet as part of a connectivity lane they have a connectivity control, called layer select, as illustrated in Fig. 15.10a, b. The local bit-line of line j (L-BLj) will be connected to the corresponding global bit-line j pillar (G-BLj) through a select transistor controlled by layer select i (SLi). The connection lane between units, carry the corresponding layer select per unit per control line, controlling the connection, to the global pillar of that control line (bit-lines, word-lines).

Fig. 15.10 **a** Bit-line layer select. **b** Word-line layer select

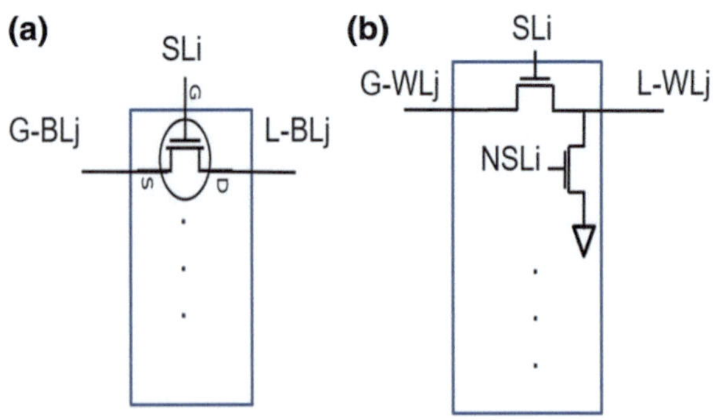

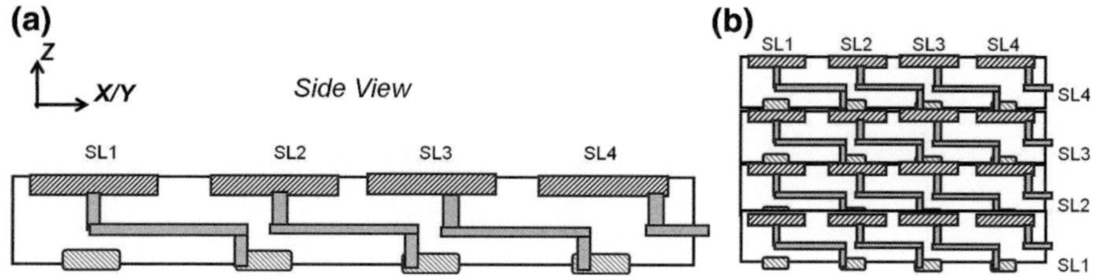

Fig. 15.11 **a** Side shift structure per stratum. **b** 4 strata with ripple

The overlaying (or underlying) memory control stratum will provide decode control to each of the bit-line and word-line pillars to be selectively connected to one of the underlying bit-cell units. Dedicated layer-select pillars provide the signals selecting the target stratum within the memory stack. A technique to route an individual layer select signal from the memory control stratum to each of the memory strata uses a 'shift to the right' concept as is illustrated in Fig. 15.11a, b. Figure 15.11a illustrates a side view of the per-stratum side shift structure and Fig. 15.11b illustrates 4 strata stack allowing top layer memory control access to ripple down the per-stratum select.

Alternative techniques for forming layer select could be the use of vertical shift registers or per layer decoding circuits. These techniques require small active circuits in the memory strata.

The technical concepts described here support high yield and low cost 3D heterogeneous integration by keeping the process at the stacking fab simple, while using existing high yielding complex semiconductor standard processes for all the individual strata forming the 3D computing structure. The use of face-to-face hybrid fusion bonding achieving high yield and dense vertical connectivity as part of the stacking process is combined with a very simple 3D system integration. Furthermore, the alternative presented here of using FDSOI substrate contacts, the 'nano-TSV', could be formed as part of the standard foundry process. In such a flow the stacking fab job is quite simple:

1. Flip and bond (preferably using a precise bonder with <100 nm misalignment)
2. Remove the 'top' layer substrate by grind and etch, using the BOX as an etch stop
3. Build landing pads from the now exposed substrate contacts
4. Repeat with subsequent strata.

The following describes an alternative approach using a precise bonder.

In advanced memories the bit-line and the word-line pitch can be 80 nm or even smaller. With 50 nm bonding misalignment the vertical connectivity pad could to be about 200 nm × 200 nm. It is suggested staggering the vertical pillars to accommodate the misalignment, while keeping a high memory control line pitch as illustrated for a small connection lane region in Fig. 15.12a–c. The vertical connections between the 'Pad' and the 'Pin' forms the 'nano-TSV', and through the stacking process forming the Global bit-line and word-line.

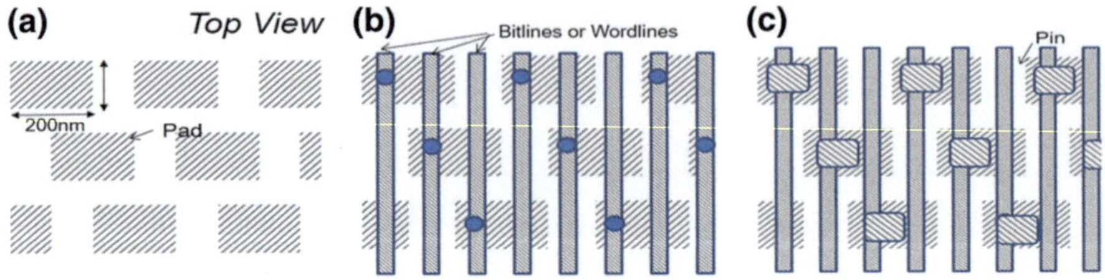

Fig. 15.12 **a** Staggering 3 rows, b via to control lines, **c** via to pins

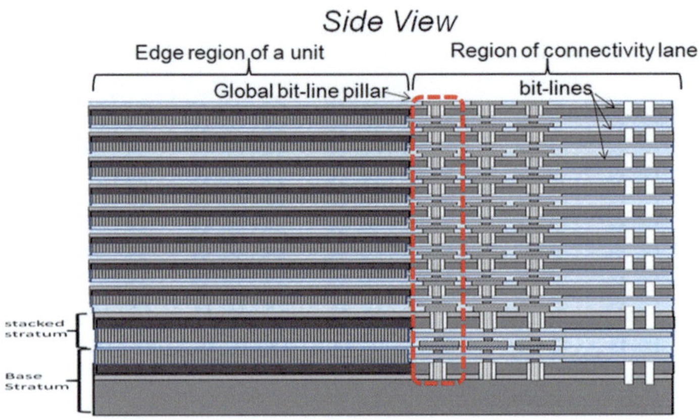

Fig. 15.13 Side view of memory strata formed by successive stacking

Figure 15.13 illustrates a side view of a region of memory stack having 10 strata. Such would allow for generic memory strata, which could be sized for the actually desired memory capacity by choice of the number of strata in the stack. Using the layer select concept of Fig. 15.11a, b requires up-front accommodation for the maximum number of strata in the memory stack.

Figure 15.14 illustrates a 3D view of a 3D memory stack with a row decoder being fed from the logic level using a scheme similar to the one illustrated in Fig. 15.11a/b.

An alternative layer-select could be formed by a vertical shift register which would remove the need for up-front maximum stack-size decision and the associated need for per-layer pads. A combination of these techniques or adding per-stratum layer decoding functions could be integrated in some versions.

An additional advantage of the presented technical approach is the option to mix memory technologies, provided they all obey the same word-line and bit-line pitch and are designed to the same unit and connectivity lane sizes. Accordingly, the memory stack could include a top stratum of high speed memory while the bottom stratum could be low power, high density memories. The memory control could include multiple control strata dedicated to the specific memory in the memory

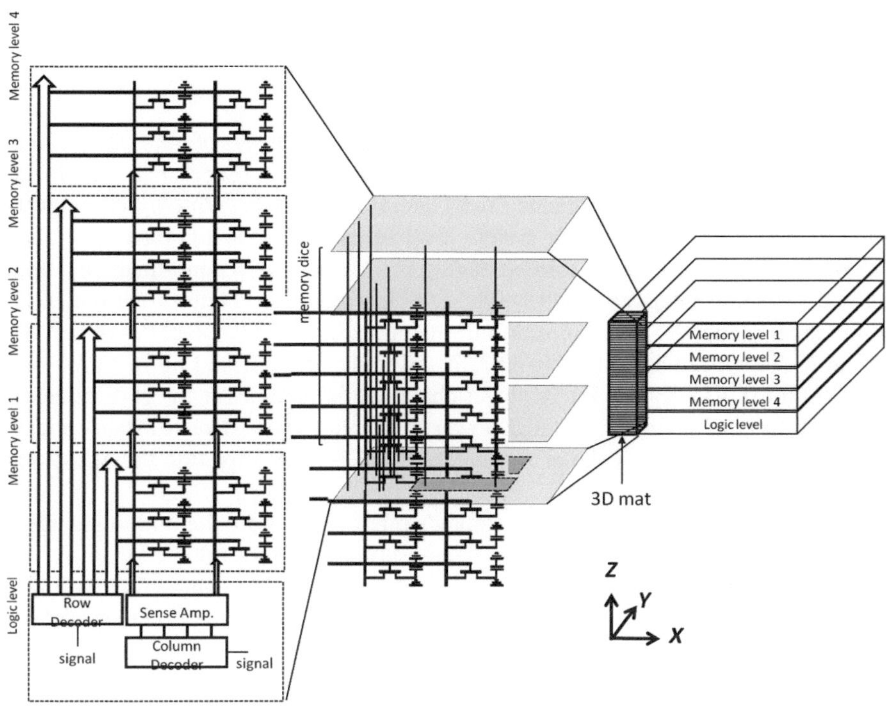

Fig. 15.14 3D illustration of memory strata formed by successive stacking

stack. Additionally, a parallel high speed, data transfer between strata in the stack can be facilitated using the proposed architecture.

The memory stack design also includes pass-through pillars, which allow transferring signals through it such as to allow synchronization of the memory control strata for the case in which one is under the memory strata and another is overlying it. The pass-through pillars could be used also for I/O when a processor stratum is placed underneath the memory strata as the base wafer, while the I/O stratum is placed at the top of the SoC stack. Thermal vias could be included to help thermal management.

Additional power delivery pillars can be included in the memory stack both for supplying memory power needs and to deliver power through the memory stack to strata underneath it.

An important advantage of this proposed architecture is the ability to form a per-unit redundancy. By having a redundancy stratum and proper circuitry in the memory controller, the layer select decoding circuit could include a mapping table to skip a 'bad' unit stratum and replace it with a unit in the redundancy stratum. Having thousands of units per die allows repair even in memory strata with tens of defects. This concept could also be used for field repair, providing a valuable advantage of this architecture.

15.6 3D Heterogeneous Integration Enables Electromagnetic Waves Interconnects

A modular 3D IC system, as suggested here, that utilizes arrays of units each with its unit 3D memory cell block, memory control circuit block, processing logic block, and I/O block, needs good in-plane (X-Y) lateral interconnect with high throughput and low power consumption for system level functionality. While the out-of-plane (Z) vertical interconnects are formed having vertical vias with nano-meter and up to micron sizes and relatively short heights, the interconnect length in the horizontal in-plane direction (X-Y) remains at millimeter sizes, from die level (3–16 mm, for X and Y sides), reticle level (20–30 mm), to multi reticles, and up to wafer sizes (60–300 mm). Clearly the interconnect challenge is greater for the X-Y interconnect and the propagation delay and power dissipation using low-resistance metals such as copper and low-k dielectric material will end up impeding the 3D system performance.

As presented in Figs. 15.1–15.2b, today's interconnects are the limiting factor of computing electronics. The simple voltage representation of a logic signal is very sensitive to the interconnect RC. The most effective path to overcome this fundamental physical limitation is to shift from voltage logic representation to modulated electromagnetic (EM) wave of signal representation [15, 16] (Fig. 15.15).

The spectrum of the EM wave could be selected to fit the average target distance and the access to the appropriate technology. 3D heterogeneous structures could open the door to EM interconnects by adding strata of RF or Optical drivers, receivers, modulators and waveguides. In conventional 2D devices the cost of new nodes development and infrastructure drove vendors to focus their development to the most critical functions of logic and SRAM. Accordingly, any design targeting advanced manufacturing nodes must exclude anything other than what leading fabs include in their technology offering, which would be logic gates, SRAM and some I/O and basic support for analog function. The implication is that in advanced nodes RF or optical functions are not available and X-Y interconnects would be limited to RC Repeaters. Adapting 3D heterogeneous integration enables adding strata that could be built in other types of fab, such as RF-SOI lines, enabling the use of them

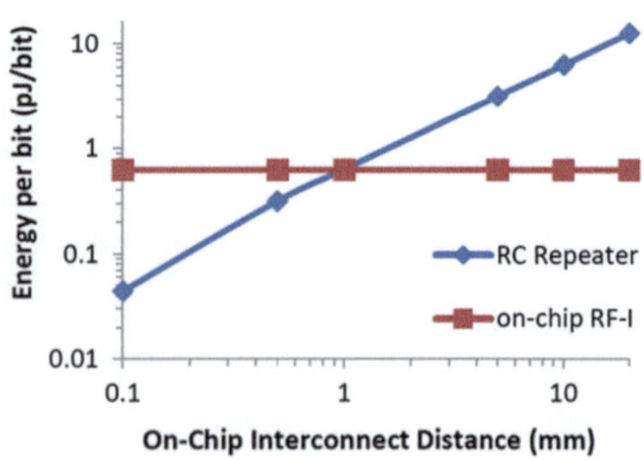

Fig. 15.15 RF-I will crossover the energy efficient curve of the RC repeater and become more energy efficient above a 1 mm interconnect distance at a 16 nm CMOS process [15, 16]

for the global X-Y interconnects. Within some technology parameters, the cross over from RF to Optical could be at over 30 cm [15, 16] (Fig. 15.16).

Wafer availability and cost could have a strong impact upon such choice. It is our assessment that the adoption of the 5G wireless communication standard and the increased use of wafers for RF applications would make RF-I the preferred choice for many applications. Figure 15.17 provides some benchmarks for these interconnect options [17].

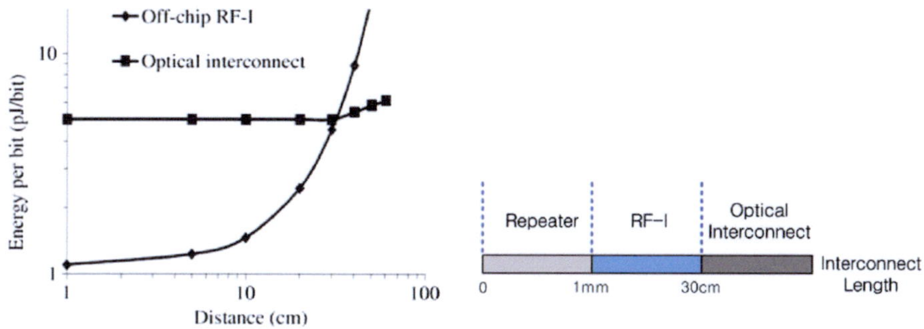

Fig. 15.16 RF-interconnect (RF-I) versus optical interconnect [16]

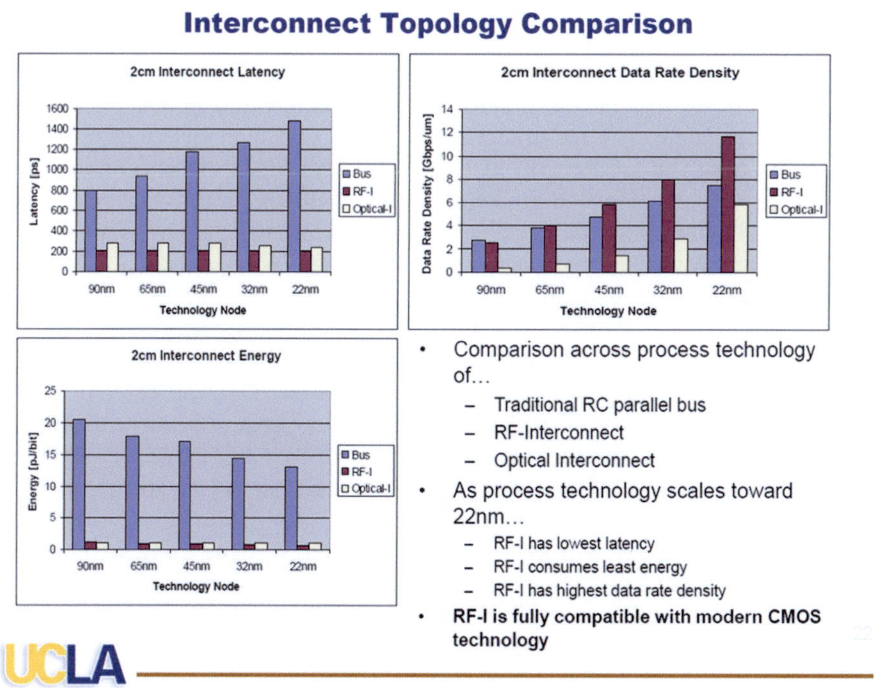

Fig. 15.17 Benchmarks for 2 cm interconnects [17]

An important aspect of the monolithic 3D technologies as presented in Chap. 8 is the enablement of heterogeneous integration, in which one level (wafer) is produced using processes and materials to fabricate logic devices while another level (wafer) is produced using different processes and different materials to fabricate on-chip RF or optical interconnect devices. Furthermore, these levels (wafers) would likely be made in different fabs. Then, using a layer transfer process, one level is transferred over the other enabling fine vertical (3D) integration between the two.

The on-chip RF or optical interconnect level could include more than one sub-level, for example, such as a passive photonic device level(s) for signal routing such as wave guides, photonic crystals, and resonators, and an active device level(s) such as a photo-detectors and a light sources (example e.g., lasers). The photo-detectors and light sources can each reside in its own different levels or they can be in the same level but with the two made with different substrates knitted together side by side. For example, the photodetector may be based on germanium, the light source may be based on a III–V semiconductor, and the passive devices may be based on silicon (core)-silica (cladding) structures.

15.7 Ultra Scale Integration (>1000 mm^2)

The key challenge of large reticle size or wafer level integration is yield. 3D integration may include multiple redundancy structures and repair techniques [13, 18–20] which could be used for robust RF and optical interconnected 3D system. Another alternative is to leverage the fact that RF transmission lines and optical interconnect waveguides are relatively large structures that have a very high yield with today process capabilities. The benchmarks of Fig. 15.17 were based on transmission lines having a 6 μm pitch, compared to advanced semiconductor process having less than 60 nm pitch. Optical waveguides use larger than a micron pitch lines as well. These large structures could be processed at very high yield while the drive electronics could be structured with redundancy for yield robustness (Fig. 15.18).

To allow ultra-scale integration of structures larger than a single reticle, the connectivity structure should extend over more than single reticle (>30 mm). Techniques to use optical lithography to pattern large areas greater than the full reticle field by 'stitching' multiple reticle patterns that had been projected independently are known in the industry, and are currently used for Interposer lithography and other applications. Alternatively some lithography tools are designed to support large area projections [21, 22].

Additionally, some prior work suggests integrating systems using an interposer with optical waveguides [23]. An additional alternative is to pre-test the RF or the optical interconnect components allowing the use of the concept of Known-Good-Die to wafer level die-to-wafer 3D integration by pretesting the RF or the optical interconnect fabric before transfer over to the 3D system. This could be efficiently implemented with the use of a generic RF or optical interconnect which could be

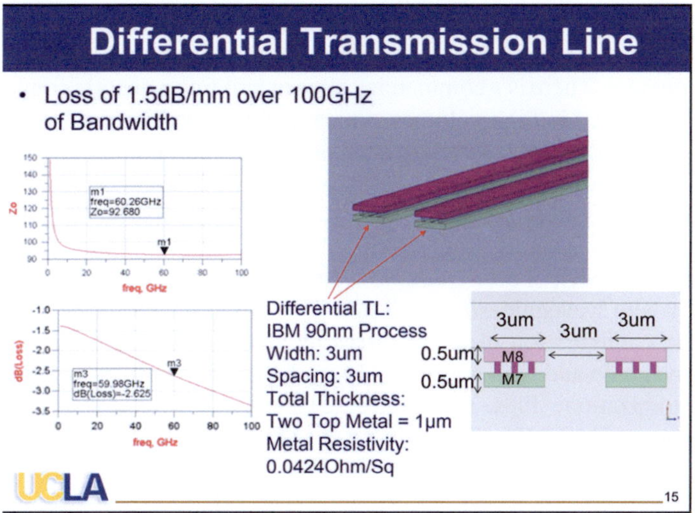

Fig. 15.18 Transmission line example

produced in volume and pretested before use for the specific application. Another option is to avoid the physical interconnects and use wireless interconnects [24, 25].

The use of RF could include the use of differential signaling, which would help reduce the cross talk and interference effects, thus allowing lower supply voltages, and other advantages. The previous concepts for interconnection fabrics could be adapted to use differential transmission lines [26, 27].

Figure 15.19a, b illustrate a 3D system which include X-Y horizontal interconnection fabrics at relatively the upper level of the structure. In general, the horizontal interconnection fabric could be engineered in the middle level of the 3D system or at any other level. Placing it in the center could be advantageous in some systems by having a compute structure on both sides (under it and overlying it) thus allowing shorter vertical paths from the computing structures to the X-Y horizontal interconnection fabric. Figure 15.19a illustrates the structure as a generic continuous array of

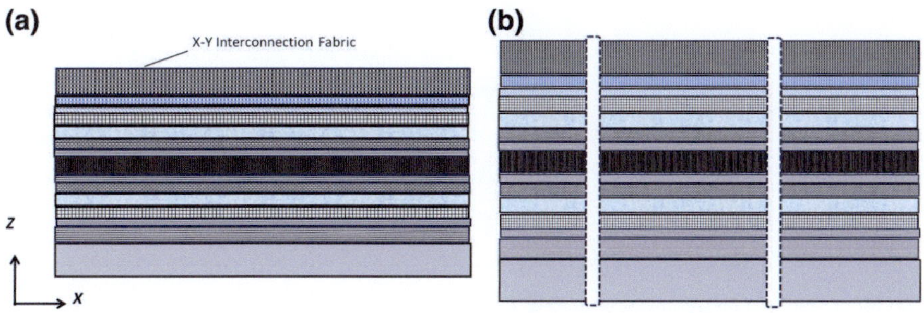

Fig. 15.19 **a** 3D heterogeneous integration. **b** 3D structure diced to smaller devices

cores, each with its own memories on top, and X-Y connectivity structure allowing data transfer between cores. Figure 15.19b illustrates the structure after being diced to smaller devices. There is a commercial value in building a generic computing platform to be produced in high volumes, which could be later used to specific market needs by dicing the generic structure according to the computing power needed for the target application.

A 3D system could include X-Y waveguides or transmission lines with configurable connectivity such as Single Write Multiple Read (SWMR), Multiple Write Single Read (MWSR), or even Multiple Write Multiple Read (MWMR). Connectivity fabrics where waveguides/transmission lines are designed for MWMR [28–30] simplify the configuration of its resources by adapting who gets to 'write' into a specific waveguide and who gets to read based on considerations such as yield and sizing (customization) (Figs. 15.20 and 15.21).

The concept of MWMR allows flexible use of the interconnection fabric in which compute units can sign in and sign out into the system's overall computing fabric.

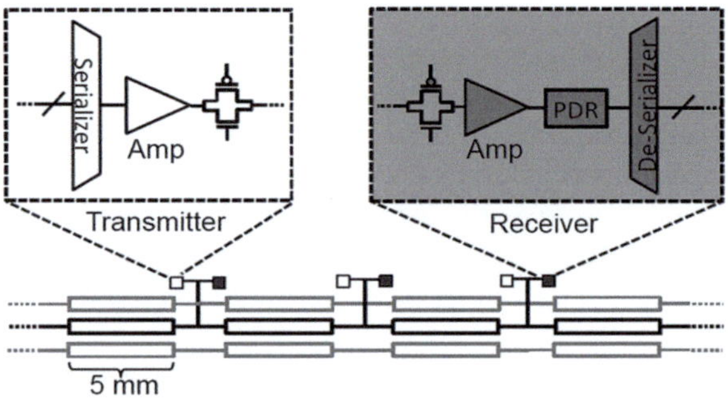

Fig. 15.20 RF interconnect with MWMR

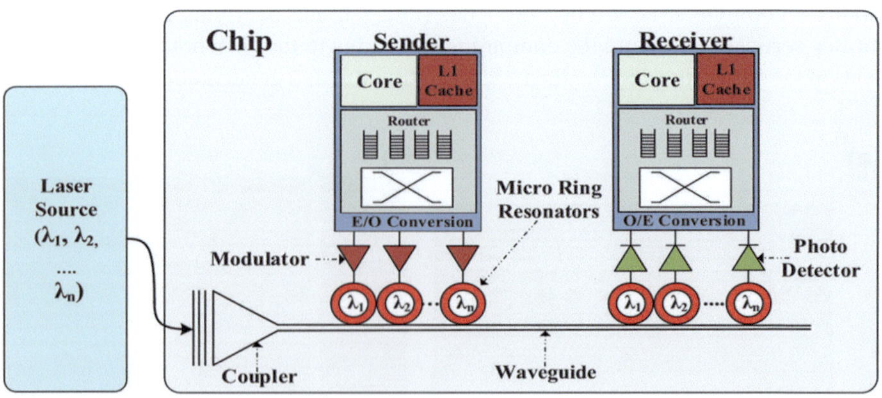

Fig. 15.21 Optical interconnect with MWMR

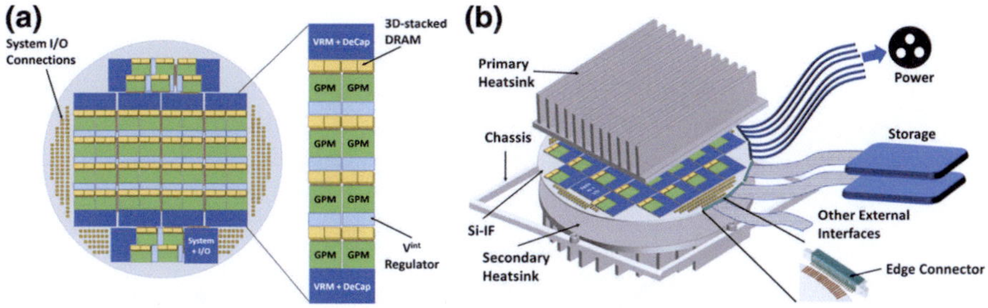

Fig. 15.22 **a** Waferscale GPU with 42 GPM unit (2 redundant). **b** Overall structure [34]

Such an architecture would be very tolerant to yield loss and to system reconfiguration based on yield or field customization.

The concept of wafer scale integration ("WSI") has been considered and explored over many years. It was never adopted due to the challenge of defects and due to the success of scaling. There is more interest these days as conventional scaling has slowed and with the growing interest in Artificial Intelligence (AI) and brain inspired architectures [31–33]. Recent work [34] demonstrated over 100× Energy Delay Product (EDP) for such wafer scale integration of GPU even without the use of EM interconnect. Figure 15.22a, b illustrate such wafer-scale demo.

The concept of leveraging 3D integration for wafer scale integration, or for multi reticles or multi die integration is extending the idea commonly used for memory repair. Memory repair utilizes the availability of redundant similar function memory cells designed with similar access time. Use of EM interconnect with arrays of computing units each with its own memory is similar. The functional units are equivalent and the X-Y EM connectivity is generally dominated by the delay converting a voltage to or from the EM signal, and is far less dependent on the location of the unit within the array. Accordingly redundancy would work well just as it is commonly used for memory repair (Fig. 15.23).

This enables wafer-scale integration and resolves the fundamental limit behind Moore's Law—yield.

It was yield that was driving the cost of integration up beyond some level of integration due to defect density. Once redundancy can be effectively used, defects do not limit the device size, allowing wafer-scale integration with an additional 1000× potential Energy-Delay product advantage.

15.8 Cooling

3D Systems such as those presented herein commonly generate heat while in operation, which must be managed to protect the system from heating up and affecting its operation. Figure 15.22b illustrates air cooling techniques for wafer scale system [34]. The next level of heat removal is the use of Microfluidic Cooling [35–37].

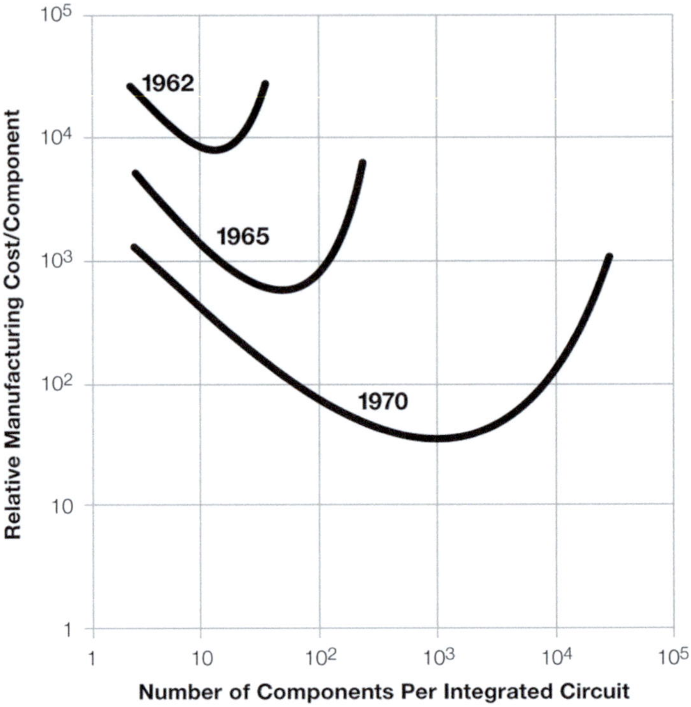

Fig. 15.23 The famous chart resulting in Moore's Law

MC has been proposed and is now used with some 3D devices at the device level (Fig. 15.24).

An additional advantage of the 3D wafer level heterogeneous integration of wafer-scale systems is the option to naturally form a micro-fluid cooling fabric in the substrate. Instead of forming micro-fluidic channels at the individual device level

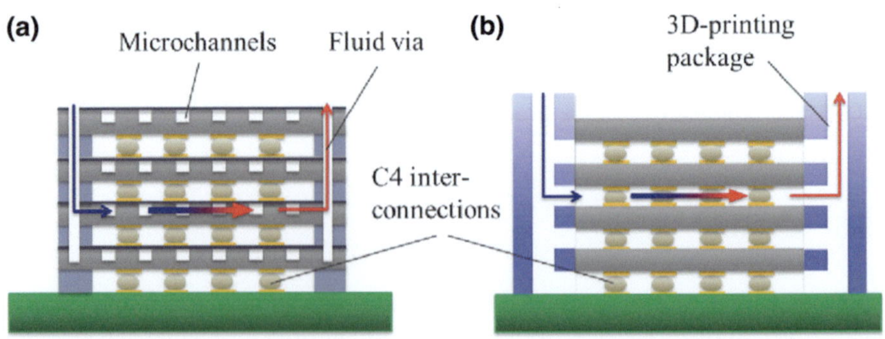

Fig. 15.24 **a** Diagrams of microchannel cooling, **b** fluidic chamber with 3D-printing package

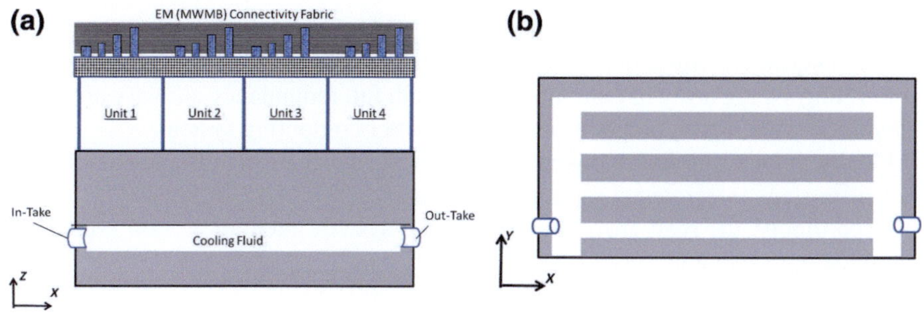

Fig. 15.25 **a** Wafer level microchannel cooling. **b** Horizontal cut view of (**a**)

and connecting them for the system level, the micro-fluidic cooling system could be formed at the wafer substrate and provide effective cooling system to the wafer scale system.

Figure 15.25a illustrates an X-Z cut view of such large scale 3D device integration with a substrate constructed to support fluid cooling. The illustration includes four computing units each with its own memories and connectivity (MWMR) to an EM connectivity fabric. The channeled silicon substrate could include micro-channels designed with fluid in-take and out-take. The substrate could be preprocessed to include the micro-channels at the wafer level, or bonded afterward to a micro-channel structure, for example with silicon to silicon bonding. Figure 15.25b is an X-Y cut-view through a micro-channel structure of the cooled 3D device. The micro-channels could be formed by etching trenches using conventional semiconductor processes into the micro-channel structure and then bonded to the wafer substrate. The micro-channel structure and a thinned wafer substrate could be slightly oxidized to enable a silicon dioxide to silicon dioxide bond if required by engineering and production constraints. Alternatively, the inner surface of the micro-channel may be further protected by silicon nitride or other desired film in order to protect the device from the cooling fluid. The wafer substrate could be thinned down by conventional techniques such as grinding and etch prior to the bonding. Thinning the substrate post device processing down to 50 μm is common in the industry.

15.9 Summary

3D heterogonous Integration with modern high-precision aligners allows the system designer to utilize wafers sourced from different fabs to form a 3D system. Using such integration technology allows constructing a computing system that could be many orders of magnitude better than today's 2D PCB–based integration technology.

By integrating memory on top of the processor logic, the memory wall can be overcome, resulting in a 1000× better computing unit.

Integrating many computing units with X-Y EM connectivity fabric allows overall system integration with orders of magnitude better data flow.

Forming a large array of computing units with full redundancy allows wafer scale integration with, again, orders of magnitude better computing system.

Finally, integrating micro channels at the wafer level allows effective cooling of such ultra-scale integrated computing platforms.

Accordingly, the road ahead for device and system integration with effectively unlimited upside potential is wide open.

References

1. J.L. Hennessy, D.A. Patterson, *Computer Architecture: A Quantitative Approach* (Elsevier, 2011)
2. D. Efnusheva, A. Cholakoska, A. Tentov, A survey of different approaches for overcoming the processor—memory bottleneck. Int. J. Comput. Sci. Inf. Technol. **9**(2), 151–163 (2017)
3. H. Simon, Why we need Exascale and why we won't get there by 2020, in *Optical Interconnects Conference, Santa Fe, New Mexico* (2013)
4. M.M.S. Aly et al., Energy-efficient abundant-data computing: the N3XT 1,000 x. Computer **48**(12), 24–33 (2015)
5. W. Hwang et al., 3D nanosystems enable embedded abundant-data computing: special session paper, in *Proceedings of the Twelfth IEEE/ACM/IFIP International Conference on Hardware/Software Codesign and System Synthesis Companion* (ACM, 2017)
6. S. Mitra, Abundant-data computing: the N3XT 1,000 X, in *2018 International Symposium on VLSI Design, Automation and Test (VLSI-DAT)* (IEEE, 2018)
7. Z. Or-Bach, A 1,000x improvement in computer systems by bridging the processor-memory gap, in *2017 IEEE SOI-3D-Subthreshold Microelectronics Technology Unified Conference (S3S)* (IEEE, 2017)
8. R.J. Gutmann et al., Wafer-level via-first 3D integration with hybrid-bonding of Cu/BCB redistribution layers, in *Proceedings of the International Wafer Level Packaging Congress (IWLPC)* (2005)
9. A. Jourdain et al., Simultaneous Cu-Cu and compliant dielectric bonding for 3D stacking of ICs, in *2007 IEEE International Interconnect Technology Conference* (IEEE, 2007)
10. F. Liu et al., A 300-mm wafer-level three-dimensional integration scheme using tungsten through-silicon via and hybrid Cu-adhesive bonding, in *2008 IEEE International Electron Devices Meeting* (IEEE, 2008)
11. A. Jourdain et al., Electrically yielding collective hybrid bonding for 3D stacking of ICs, in *2009 59th Electronic Components and Technology Conference* (IEEE, 2009)
12. T. Uhrmann, J. Burggraf, M. Eibelhuber, Heterogeneous integration by collective die-to-wafer bonding, in *2018 International Wafer Level Packaging Conference (IWLPC)* (IEEE, 2018)
13. Patent Application WO 2018/071143 in reference to Fig 11F-11H
14. Patent Application WO 2019/060798
15. S.W. Tam, M.-C.F. Chang, RF/wireless-interconnect: the next wave of connectivity. Sci. China Inf. Sci. **54**(5), 1026–1038 (2011)
16. S.-W. Tam, M. Frank Chang, J. Kim, Wireline and wireless RF-interconnect for next generation SoC systems, in *2011 IEEE 54th International Midwest Symposium on Circuits and Systems (MWSCAS)* (IEEE, 2011)
17. M.F. Chang et al., CMP network-on-chip overlaid with multi-band RF-interconnect, in *2008 IEEE 14th International Symposium on High Performance Computer Architecture* (IEEE, 2008)
18. U.S. Patent 7,964,916 (Fig. 41)

19. U.S. Patent 8,395,191 (Fig. 114–132)
20. U.S. Patent 10,014,318
21. W. Flack et al., Large area interposer lithography, in *Electronic Components and Technology Conference (ECTC), 2014 IEEE 64th* (IEEE, 2014)
22. Hao et al., Demonstration of 3–5 μm RDL line lithography on panel-based glass interposers, in *Electronic Components and Technology Conference (ECTC), 2014 IEEE 64th*
23. Y. Arakawa et al., Silicon photonics for next generation system integration platform. IEEE Commun. Mag. **51**(3) (2013)
24. C. Xiao, Z. Huang, D. Li, A tutorial for key problems in the design of hybrid hierarchical noc architectures with wireless/rf. SmartCR **3**(6), 425–436 (2013)
25. S. Deb et al., Wireless NoC as interconnection backbone for multicore chips: promises and challenges. IEEE J. Emerg. Sel. Top. Circuits Syst. **2**(2), 228–239 (2012)
26. S-W. Tam et al., A simultaneous tri-band on-chip RF-interconnect for future network-on-chip, in *2009 Symposium on VLSI Circuits* (IEEE, 2009)
27. B. Sawyer et al., Modeling, design, and demonstration of 2.5 D glass interposers for 16-channel 28 Gbps signaling applications, in *Electronic Components and Technology Conference (ECTC), 2015 IEEE 65th* (IEEE, 2015)
28. A. Brière et al., A dynamically reconfigurable rf noc for many-core, in *Proceedings of the 25th Edition on Great Lakes Symposium on VLSI* (ACM, 2015)
29. A. Carpenter et al., Using transmission lines for global on-chip communication. IEEE J. Emerg. Sel. Top. Circuits Syst. **2**(2), 183–193 (2012)
30. M.R. Yahya et al., Review of photonic and hybrid on chip interconnects for MPSoCs in IoT paradigm, in *2018 21st Saudi Computer Society National Computer Conference (NCC)* (IEEE, 2018)
31. A. Kumar et al., Toward human-scale brain computing using 3D wafer scale integration. ACM J. Emerg. Technol. Comput. Syst. (JETC) **13**(3), 45 (2017)
32. Z. Wan, Three-dimensional wafer scale integration for ultra-large-scale neuromorphic systems. Diss. UCLA (2017)
33. Z. Wan, S.S. Iyer, Three-dimensional wafer scale integration for ultra-large-scale cognitive systems, in *SOI-3D-Subthreshold Microelectronics Technology Unified Conference (S3S), 2017 IEEE* (IEEE, 2017)
34. S. Pal et al., Architecting waferscale processors-a GPU case study, in *2019 IEEE International Symposium on High Performance Computer Architecture (HPCA)* (IEEE, 2019)
35. M.S. Bakir et al., 3D integrated circuits: liquid cooling and power delivery. IETE Tech. Rev. **26**(6), 407–416 (2009)
36. A. Bar-Cohen, J.J. Maurer, D.H. Altman, Gen3 embedded cooling for high power RF components, in *2017 IEEE International Conference on Microwaves, Antennas, Communications and Electronic Systems (COMCAS)* (IEEE, 2017)
37. J.-H. Han et al., Microfluidic cooling for 3D-IC with 3D printing package (IEEE S3S, 2019)
38. A. Karkar et al., A survey of emerging interconnects for on-chip efficient multicast and broadcast in many-cores. IEEE Circuits Syst. Mag. **16**(1), 58–72 (2016)

Chapter 16
High-Performance Computing Trends

Bernd Hoefflinger

16.1 Supercomputers

The supercomputers with complexities of one Million computing cores continue to be the benchmark for technology progress. The performance of the top 500 is recorded by TOP500.org [1] with annual updates. The status of 2014 was summarized and evaluated in [2], and we consider the 5-year development 2014–2019, using the 2019 data in Table 16.1.

The start into the 2010–2020 decade was marked by the 1000-times improvement-per-decade belief in computer performance, supported by the ITRS roadmap. In CHIPS 2020 of 2012 [3], this was corrected already to a 100-times improvement, considering the immanent end of the nanometer roadmap. The rapid introduction of graphics accelerators into supercomputers with hundreds-of-thousands of cores, running at reduced voltages to improve the energy efficiency, helped to report remarkable results in 2014. The projections for 2020 were, that the no. 1 supercomputer would perform 10^{18} Floating-Point Operations per Second (Exa-FLOPS) in 2020 and that the energy per FLOP would improve by a factor of 7× in 5 years.

The 5-years development for the Top-10 supercomputers is summarized in Table 16.2.

In the TOP500 June-2019 Report [1], the no.1 supercomputer achieved 149 PFLOPS, about 5-times higher than the no. 1 of 2014. The Top-10 together achieved a 5× improvement in thruput, ans, what is really significant:

- **From 2014–2019, the Top-10 supercomputers improved their energy efficiency by a factor of 5 so that with 5× higher throughput, their total wall-plug electric power consumption stayed the same**.

As is shown at the bottom of Table 16.1, the latest milestone, reported in August 2019 [4], is that the top US supercomputer to go into operation in 2023, will have

B. Hoefflinger (✉)
Sindelfingen, Baden-Württemberg, Germany
e-mail: bhoefflinger@t-online.de

B. Murmann and B. Hoefflinger (eds.), *NANO-CHIPS 2030*,
The Frontiers Collection, https://doi.org/10.1007/978-3-030-18338-7_16

Table 16.1 The top-10 supercomputers in the June 2019 report of the TOP500 [1]

	Name	Source		Cores thousands	Thruput PFLOPS	Power MW	Efficiency GFLOPS/W (nJ/FLOP)	Thruput FOM PFLOPS/nJ
1	Summit	IBM Power9	USA	2414	149	10.1	14.8 (0.067)	2223 (1)
2	Sierra	IBM Power9	USA	1572	94.6	7.4	13.8 0.073	1295 (2)
3	Sunway Taihu Light	Sunway	China	10,650	93.0	15.4	6.1 (0.160)	581 (3)
4	Tianhe 2A		China	4982	61.4	18.5	3.3 (0.300)	205 (6)
5	Frontera	Dell	USA	448	23.5	?		
6	Piz Daint	Cray	Suisse	388	21.2	2.3	9.1 (0.096)	226 (5)
7	Trinity	Cray	USA	979	20.1	7.6	2.6 (0.384)	53
8	AI bridging	Primary	Japan	302	19.9	1.6	12.9 (0.077)	246 (4)
9	Super MUC	Think	Germany	306	19.5			
10	Lassen	IBM Power9	USA	288	18.2			
	Top 10 Sum			20,100	516.4	63	8.5 (0.172)	295
	No. 1 2023	Cray	USA		1500	30	50 (0.020)	75,000

Thruput in the table = Throughput

Table 16.2 5-years progress no. 1 supercomputer (left) and top 10 sum (right)

	Thruput (PFLOS)	Power (MW)	Efficiency (GFOPS/W)	Thruput (PFLOPS)	Power (MW)	Efficiency (GFLOPS/W)
2014	33.9	17.8	1.9	105	56	2.1
2019	149	10.1	14.8	516.4	63	8.5
	X 4.8	X 0.57	X 7.8	X 5.0	X 1.1	X 4.8

a throughput of 1.5 EFLOPS and an energy efficiency of 50 GFLOPS/W. This will mean a 10× improvement in thruput and 3.3× in energy efficiency. This progress requires sustained investment in

- 3D integration (Chaps. 8–11 and 13)
- Bridging the processor-memory gap (Chaps. 15 and 19)
- New Ultra-low voltage, efficient CMOS logic (Chaps. 6 and 7)
- AI-inspired accelerators (Chaps. 10, 14, 18 and 19).

Table 16.3 Energy efficiency 2014–2030

	2014	2019	2024	2030
CPU	4 GOPS/W	20 GOPS/W	100 GOPS/W	300 GOPS/W
GPU	80 GOPS/W	400 GOPS/W	2 TOPS/W	10 TOPS/W
No.1 supercomputer	2 GFLOPS/W	14.8 GFLOPS/W	50 GFLOPS/W	150 GFLOPS/W
DNN (1 k MAC's)	1.9 TOPS/W	11.5 TOPS/W	100 TOPS/W	1 POPS/W

In such a scenario, a 10× improvement in energy efficiency and a throughput of 10 EFLOPS should be possible by 2030 (Table 16.3).

A special supercomputer ranking has been introduced with the GREEN 500 [5]. Their ranking is governed by their energy efficiency. Here, most of the leaders are much smaller systems, where 10-times fewer cores mean less energy overhead, and voltages = clock-rates = thruput are reduced to reduce energy and cooling requirements. While such systems dominated the GREEN list in 2014 and only one of them (the Swiss Piz Daint) made it into the overall Top 10, it is remarkable that three of the overall TOP 10 in 2019 made it to positions 2, 3 and 7 in the 2019 GREEN list. The almost 8-times improvement of the energy efficiency of the Top-10 2019 vs. the 2014 list was better than the 2014 forecast in [2].

The ultimate Figure-of-Merit (FOM), emphazised in [2], is the

- **Throughput over Energy-per-Operation**

Shown as PFLOPS/nJ in Table 16.1, it is evident, that the 2019 leader reaches an FOM 10-times higher than many others on the Top-10 throughput list. And the projected 2023 leader would offer another 30-times improvement in this ultimate performance measure.

Overall, supercomputers with basically Von-Neumann architectures will continue to benefit from AI-inspired graphics, video and neural-network accelerators, and new benchmarks will evolve to assess their performance.

16.2 Overview Processor Trends Towards 2030

The three-orders-of-magnitude spread in energy efficiency-per-operation was presented in Ref. [2] with its 2014 status, and remarkable developments took place among the just presented supercomputers and the digital neural networks covered in Chaps. 3, 12, 18 and 22.

In Table 16.3, we summarize results since 2014 and projected developments towards 2030. The multi-core CPU's will benefit from new low-voltage CMOS design, new processor-memory communication with 3D integration to enable a 15× improvement of their efficiency over 10 years. This will help supercomputers and servers.

A disruptive development manifested itself in DNN's (Digital Neural Networks), proceeding from 1.9 TOPS/W in 2014 to 11.5 TOPS/Win 2019 with an accelerating global innovation effort, as it is detailed in Chap. 12. A 100× improvement over 10 years in energy efficiency can be quite likely. This level of progress is essential to alleviate the immanent energy crisis in Internet video, autonomous mobility and in the IOT.

16.3 INTERNET Energy Forecast

The exploding INTERNET traffic with annual growth rates of mobile traffic of 48% continues to be the major energy challenge in the information and communication economy and, by now, in the global energy balance. Servers received particular attention in CHIPS 2020 of 2012 [6], and a full overview [7], based on 2014 data, with

- Servers
- Mobile phones and PC
- Infrastructure = networks
- Embodied energy = production, installation, environment control, repair, disposal.

The projection for 2020 for wall-plug electric power was 526 GW with 240 GW for data centers.

A representative article by the magazine NATURE appeared in 2018 [8], and it contains a 2030 forecast of 928 GW, shown in Fig. 16.1.

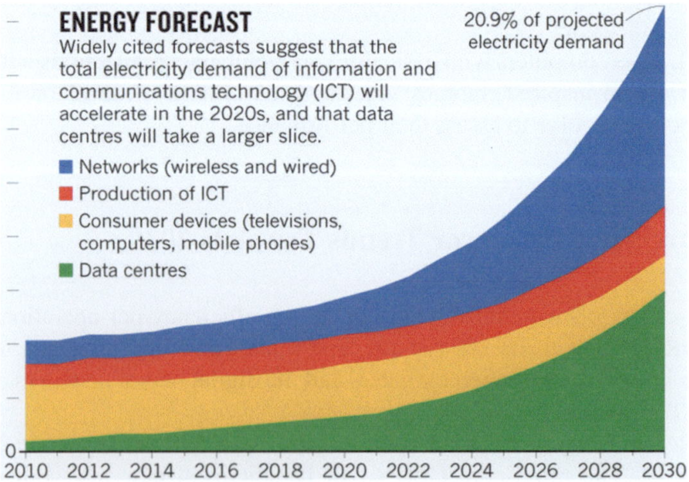

Fig. 16.1 Forecast Electric Energy for Information and Communication. 2030 Total 8 PWh/year equivalent to 928 GW or 20.9% of projected global electric power demand [8]

Table 16.4 Electricity demand (GW) of information and communication

	2020 (GW)	%/10a	2030
CHIPS 2020 [7]	526		
Nature [8]	298	X 3.28	928 GW

The graph in NATURE arrived at a total of 8 Peta-Watt-hours per year (PWh/a) in2030, more than 3-times the demand of 2020 and 20.9% of the projected global electric energy of 40 PWh provided per year in 2030, an increase of 14% over the global electric energy of 2020.

NATURE addresses the total ICT energy. Its non-INTERNET component shows up at 150 GW in 2010, probably decreasing since then due to the INTERNET take-over (e.g. TV and IP-Phone).

A comparison of the NATURE data with those in CHIPS 2020 [7] is shown in Table 16.4.

The 2020 estimate in CHIPS 2020 of 2014 would be reached 5 years later in 2025 according to the graph in NATURE, which continues to leave the progress in energy efficiency as the dominant issue for sustainable growth of the IC economy.

16.4 Conclusion

The decade towards 2030 faces significant challenges and opportunities in high-performance computing. The continuing data explosion of the INTERNET must be managed by making machines learn to provide and to store intelligent data instead of big data and perform such tasks with new architectures. The broad exploration of neural networks may enable a 100× improvement of their energy per operation in the next 10 years, reaching 10^{15} operations/s per Watt by 2030.

References

1. www.top500.org
2. B. Hoefflinger, High-performance computing, Chap. 10 in *CHIPS 2020, Vol. 2—New Vistas in Nanoelectronics* (Springer, 2016). ISBN 978-3-319-22093-2
3. B. Hoefflinger (ed.), *CHIPS 2020—A Guide to the Future of Nanoelectronics* (Springer International). ISBN 978-3-642-23096-20127
4. Cray secures third Exascale Deal, eetasia.com/news, 16 August 2019
5. www.green500.org
6. B. Hoefflinger, *The Energy Crisis*, Chap. 20 in B. Hoefflinger (ed.), *CHIPS 2020—A Guide to the Future of Nanoelectronics* (Springer International, 2012). ISBN 978-3-642-23096-7
7. B. Hoefflinger, Intelligent data instead of big data, Chap. 12 in *CHIPS 2020, Vol. 2—New Vistas in Nanoelectronics* (Springer, 2016). ISBN 978-3-319-22093-2
8. How to stop data centres from gobbling up the world's electricity, *Nature* **561**, 163–166 (2018)

Chapter 17
Analog-to-Information Conversion

Boris Murmann, Marian Verhelst and Yiannos Manoli

17.1 Introduction

Analog-to-digital converters (ADCs) are irreplaceable interface components, residing at the boundary between the physical analog world and the digital backbone of modern electronic systems. As shown in Fig. 17.1, a complete analog-to-digital interface typically involves signal amplification, frequency translation (where needed) and filtering to condition the analog signal. The function of the ADC block is to sample and quantize the conditioned analog signal for subsequent information processing in the digital domain. In many applications, the ADC's attainable speed, resolution and power dissipation have a significant impact on the overall system architecture and its specifications. Thus, it is not surprising that a tremendous amount of effort has been dedicated to improving this building block over the past several decades [1]. However, as we have already pointed out in CHIPS 2020 [2, 3], we are approaching performance asymptotes that will be difficult (if not impossible) to overcome with classical and application agnostic ADC architectures.

As we enter the next decade of innovation, it has become increasingly clear that further improvements must come from a more holistic and application-centric approach in analog-to-digital interface design. Instead of viewing the interface merely as a (nearly) "lossless" mapper between a waveform's continuous and discrete representations, one can exploit knowledge about the signal's information content to

B. Murmann (✉)
Stanford University, Stanford, USA
e-mail: murmann@stanford.edu

M. Verhelst
KU Leuven, Leuven, Belgium
e-mail: marian.verhelst@kuleuven.be

Y. Manoli
Universität Freiburg, Freiburg im Breisgau, Germany
e-mail: yiannos.manoli@imtek.uni-freiburg.de

B. Murmann and B. Hoefflinger (eds.), *NANO-CHIPS 2030*,
The Frontiers Collection, https://doi.org/10.1007/978-3-030-18338-7_17

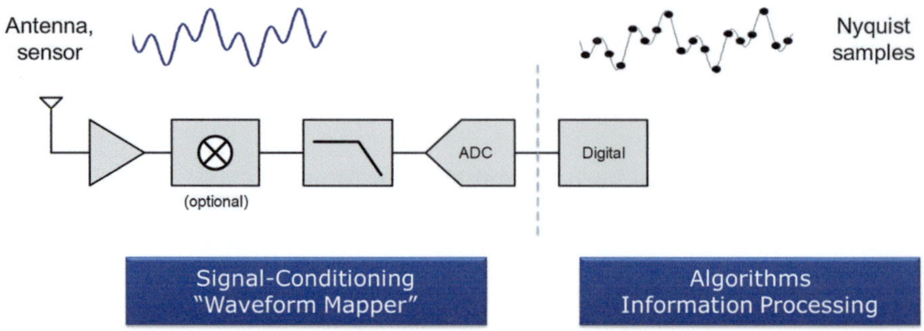

Fig. 17.1 Classical analog-to-digital interface

devise a customized and presumably more efficient "information-centric" digitization approach. For example, it may be possible to sample a signal below its Nyquist rate, or employ a simplified nonlinear quantizer using insight on the requirements for preserving the desired information. We refer to such concepts using the umbrella term of Analog-to-Information (A-to-I) conversion. While this expression was established in 2008 by Dennis Healy in the context of compressive sampling [4], it is now understood to refer to a broader variety of information-centric interface design techniques [5].

The purpose of this chapter is twofold. First, we review recent performance trends of ADCs in Sect. 17.2 as an update to our previous treatments in CHIPS 2020 [2, 3]. Then, in Sects. 17.3 and 17.4, we discuss the A-to-I framework in more detail and provide several state-of-the-art examples. These include an audio signal classifier, a sub-Nyquist observation receiver for digital power amplifier predistortion (DPD), as well as an image sensor for object detection and a machine learning based tactile sensor.

17.2 Performance Trends of Conventional ADCs

As explained in [3], one way to enumerate the progress in ADC design is to track the so-called Schreier Figure of Merit (FoM_S) [6, 7]. This metric considers the power dissipation, signal-to-noise-and-distortion ratio (SNDR) and effective Nyquist sampling rate (f_s) of an ADC to compute a number in dB (or dB/Joule, to be precise) that represents its efficiency. This FoM is constructed such that adding one bit of resolution (6.02 dB increase in SNDR) warrants a power increase of 4×. This is consistent with the fundamental tradeoff in noise-limited analog circuits. The reader is referred to [2, 3] for a more detailed discussion.

Figure 17.2 shows a scatter plot of FoM_S for data converters published between 2000 and 2019 (data from [8]). We can see that ADCs have generally become more efficient over time. Also, for a given efficiency (FoM_S), it has become possible to operate at higher sampling speeds. Given this qualitative observation, it is interesting

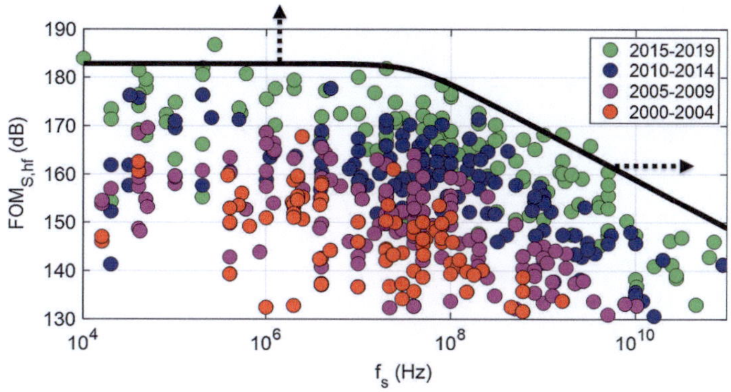

Fig. 17.2 Schreier Figure of Merit (FoM_S) versus ADC sampling rate. The arrows indicate the direction of improvement over time

to extract the "velocities" of the drawn envelope curve in the x and y directions (shown as arrows in Fig. 17.2).

Figure 17.3 plots the progress in the x-direction (using the f_s value at FOM_S = 150 dB for the fitted envelope curve in each year). The rate of improvement has been remarkably stable over the years and corresponds to a doubling every 1.6 years (dashed line). In [3] (published in 2015), we reported doubling every 1.8 years, which implies that the last four years have not led to a significant change in the overall trend.

The ability to build a faster ADC at the same efficiency (FoM_S) is tied to improvements in transistor speed and circuit design. Transistors have become faster (higher transit frequency, f_T), and the design community has found ways to benefit from parallelism (e.g., massively time-interleaved successive approximation register (SAR)

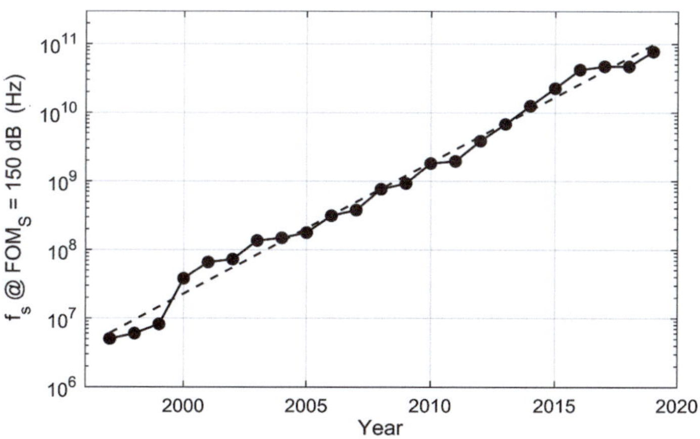

Fig. 17.3 Sampling rate (f_s), at a fixed efficiency of FoM_S = 150 dB, versus time

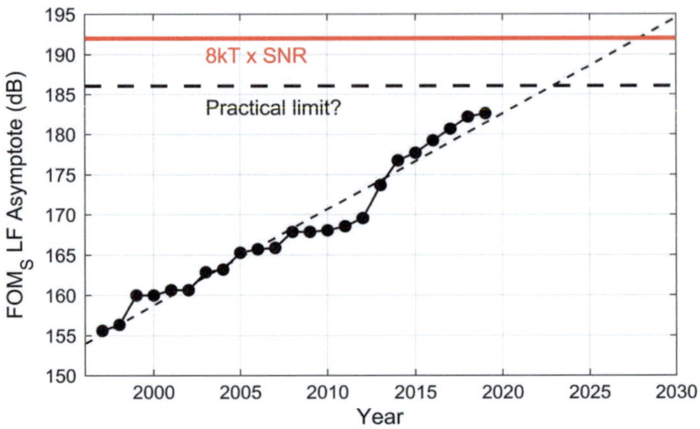

Fig. 17.4 Schreier Figure of Merit (FoM_S) at low sampling rates versus time

ADCs [9]). However, it has become clear that this trend will not continue indefinitely; the temporary "plateau" seen between 2016 and 2018 in Fig. 17.3 could be a first confirmation of this conjecture. First, there has been no major architectural innovation in high-speed ADCs for the past several years. Second, transistor speed improvements have come to an end below the 22 nm CMOS node. FinFET technology at 16 nm and below offers higher integration density but is plagued by extrinsic RC parasitics that significantly load down f_T performance. It will be interesting to track future developments on the shown chart.

Figure 17.4 shows the progress in the y-direction of Fig. 17.2 (FoM_S). Again, the rate of improvement has been remarkably stable at about 1.2 dB increments per year (dashed line). Our 2015 analysis had suggested about 1 dB per year [3]. However, just like for the above-discussed speed improvements, we know that this trend will ultimately come to an end. A relatively hard limit to surpass is the red line drawn at 192 dB, which corresponds to the well-known minimum energy for analog class-B gain stages, given by 8kT $\times$ SNR [10]. As we have already argued in [3], it is in fact quite difficult to surpass values of even 186 dB in conventional ADC realizations due to overheads in biasing, clocking, reference generation, etc. On the other hand, it is actually possible to overcome the stated limit of 192 dB with alternative circuit topologies [11]. Once again, it will be interesting to see how the battle for each extra decibel per year will play out in the future. However, it is clear that order-of-magnitude improvements are not easy to come by anymore, with only incremental improvements in circuit design and transistor technology.

17.3 Basic Considerations for Analog-to-Information Converters

A-to-I conversion interfaces have the potential to sidestep the trend toward diminishing returns discussed in the previous section. To understand this, let's begin by considering the following basic equation for the power dissipation of an ADC:

$$P_{ADC} = \frac{energy}{conversion} \cdot \frac{conversions}{second} = E_{ADC} \cdot f_s \tag{17.1}$$

Conventional ADCs are typically designed for a given sinusoidal signal fidelity (SNDR) and sampling rate (f_S), where the latter is imposed by the Nyquist criterion and/or anti-aliasing considerations. With these two quantities fixed, the task of the ADC designer merely boils down to minimizing the conversion energy (E_{ADC}) for the given rate and fidelity. Most of the designer's degrees of freedom from here on are related to how the circuit is implemented and which process technology is chosen. On the other hand, the designer of A-to-I converters aims to lower the requirements on f_s and SNDR (simpler conversion, less energy) by challenging the naïve waveform mapping paradigm that often underpins the derivation of these specifications. The salient aspect in this task is to understand what constitutes wanted information versus unwanted data in the processed signal.

In modern electronic systems, the signals of interest usually have a complex structure (as opposed to being simple sinusoids) and span a wide frequency band. We can assign a dimensionality to such signals per the Nyquist-Shannon sampling theorem. This theorem states that for a physical bandwidth W over a period of T, the minimum number of samples required for perfect digital reconstruction is 2WT. However, it is often the case that the wanted information after digitization has much lower dimensionality. As an example, consider a device that is designed to detect the sound of a crying baby. One way to perform this detection is to digitize thousands of signal samples per second to obey the Nyquist-Shannon theorem and then run a digital-domain algorithm to produce exactly one bit of information: baby is crying/not crying. This situation is illustrated further in Fig. 17.5. A clear issue with this approach is that we require the ADC to run at a relatively high rate and we expend energy on digitizing many samples. Furthermore, we expend additional energy in the digital domain on what essentially amounts to a dimensionality reduction of the signal. Is there a way to avoid digitizing all this data at high fidelity? In other words, can we perform at least some of the dimensionality reduction before the ADC? The designer of an A-to-I interface attacks this question by assessing the signal structure and typically by devising an ADC pre-processing circuit that exploits this structure for dimensionality reduction.

Many innovative sampling strategies have emerged in the quest for pre-ADC dimensionality reduction. Figure 17.6 contrasts classical Nyquist-rate sampling approaches with approaches that leverage a priori knowledge of the signal structure and the desired information content. Techniques like compressed sensing (Fig. 17.6b)

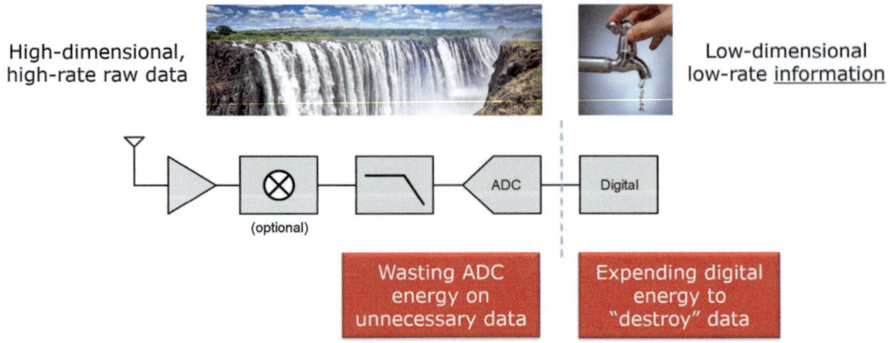

Fig. 17.5 Illustration of the typical dimensionality reduction that occurs after the A/D interface

[12] and innovation rate sampling (Fig. 17.6c) [13] aim to reduce the ADC sampling bandwidth, bringing it as closely as possible to the signal's true information rate. Then, after digitization, the wanted signal is reconstructed using a digital algorithm. In contrast, feature-sampling ADCs (Fig. 17.6d) reduce the dimensionality of the waveform through feature-enhancing filters to retain only decision-relevant signal components. In other words, the goal is to classify the obtained features in a machine-learning setting instead of reconstructing the original waveform. The reader is referred to [5] for an in-depth discussion of these concepts.

Common to all schemes in Fig. 17.6 is that the analog preprocessing blocks are linear. However, there also exists a multitude of opportunities to leverage nonlinear processing for A-to-I interfaces. For instance, the work of [14] uses nonlinearly compressed data for heartbeat detection from muscle noise corrupted ECG signals. It is shown that the nonlinear compression reduces the rms error in heartbeat detection by $2\times$, while also reducing the required digital word-length by 50% for significant power savings the digital backend circuits. Similarly, the feature-extracting image sensor of [15] (see next section) uses logarithmic gradients to reduce its output data by up to $25\times$, enabling significant power savings in the proceeding digital classifier. Another relevant example is the data-compressive, wired-OR array digitizer described in [16], which leverages amplitude sparsity in biological neurons to achieve a $40\times$ data reduction (see Chap. 24 for more details). In addition to highlighting the benefits of nonlinear processing, these examples also show that A-to-I converter design is not only beneficial for alleviating the ADC requirements, but also for reducing the storage and processing burden placed on the digital backend.

Most of the above-discussed schemes require additional analog preprocessing circuitry in front of the ADC. In order to achieve a system-level benefit, these added circuits must be efficient and consume significantly less power than the original combination of a Nyquist ADC and its backend processor. This brings up the age-old question about analog versus digital signal processing efficiency [10, 17]. As shown in Fig. 17.7, digital power tends to scale with the logarithm of SNR, while thermal noise limited analog processing power is approximately linear in SNR. This

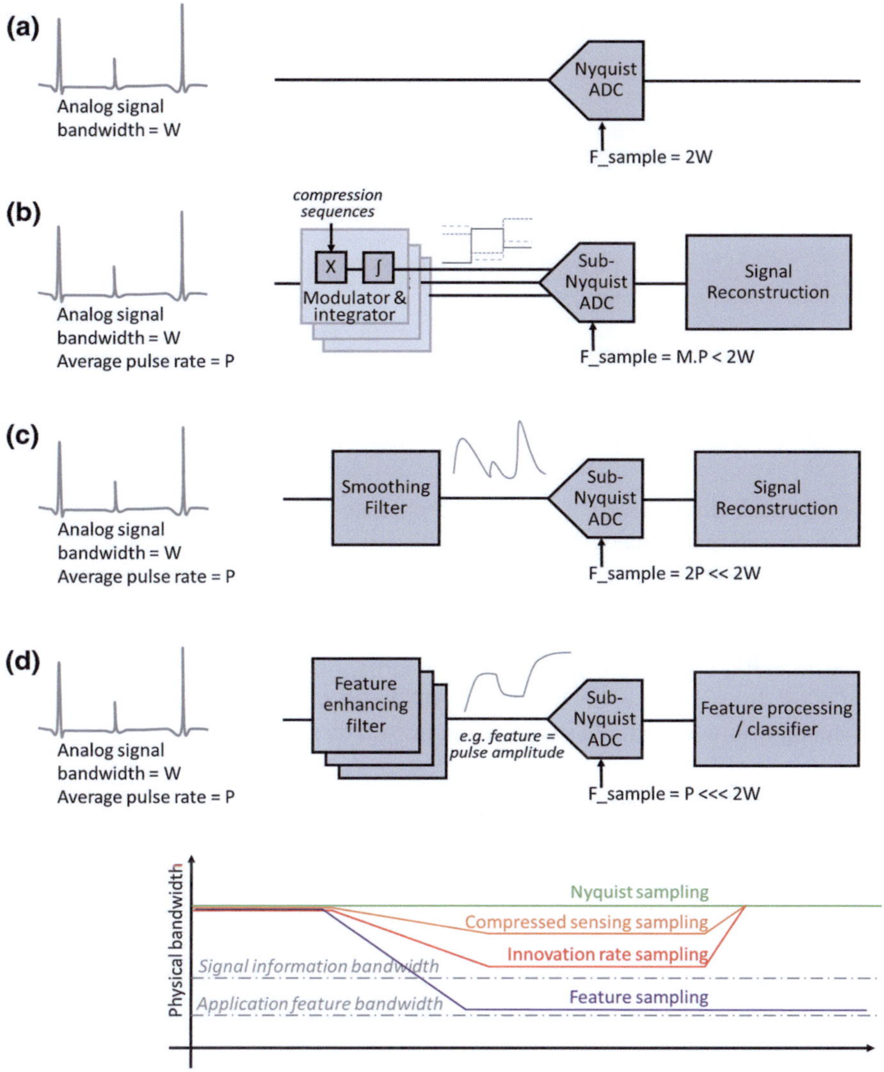

Fig. 17.6 Architecture comparison: **a** Nyquist rate sampling, **b** Compressed sensing sampling, **c** Innovation rate sampling, and **d** Feature sampling using analog analytics. **e** Evolution of the physical bandwidth along the signal chain for the architectures in **a–d** (from [5] © IEEE 2015)

leads to a cross-over point that natively shifts with technology scaling, but typically retains an advantage for analog at low SNR. In addition, since the preprocessing circuitry may benefit from reduced bandwidth and SNR, the analog overhead is usually amortizable.

As A-to-I interface design pushes for a tighter coupling between the analog and digital worlds, it also opens up new opportunities for mitigating circuit imperfections

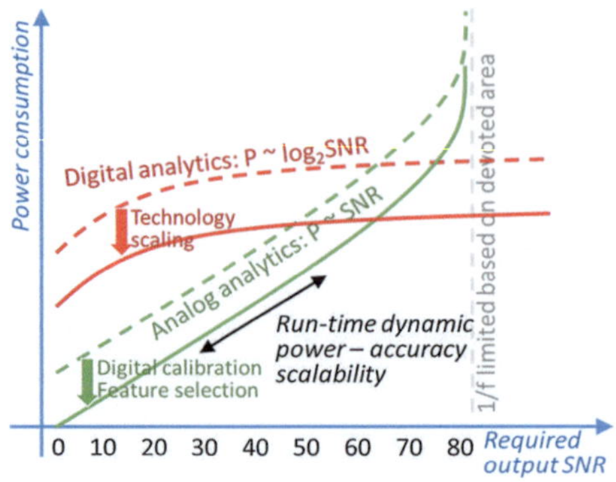

Fig. 17.7 Analog and digital power consumption trends in relation to the required SNR (from [5] © IEEE 2015)

in the spirit of "digital assist" [18]. For example, as shown in Fig. 17.8a, error terms for compensating the analog front-end can be derived from the classification metrics of a machine learning system. These classification error terms can either be obtained based on training sequence inputs or test signal injections. Another option is to take the front-end errors into account during training contained within the digital backend, as illustrated in Fig. 17.8b. Thereby, nonlinearity, offsets, frequency shifts, and as such other distortions can be absorbed in the trained classifier without feedback to the analog front-end and will have limited impact on system performance.

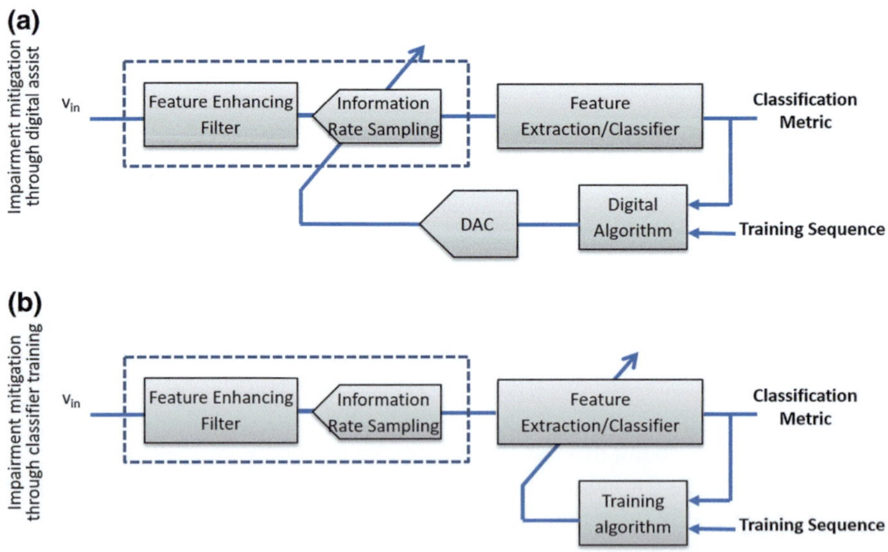

Fig. 17.8 Impairment mitigation with: **a** Digital calibration and **b** Classifier learning (from [5] © IEEE 2015)

In the above context, one should distinguish systematic and relatively constant errors (such as the imperfect N-path filter shape used in [19]) and random errors that may also drift over time. To address random errors, it is typically impractical to perform per-device training during module assembly. However, with the increasing trend toward on-device training, it may ultimately be possible for each chip to "learn" and suppress its own imperfections during normal operation in the field. This would constitute a more sophisticated and system-centric extension to the well-known self-calibration approach that is widely used in conventional analog circuits.

17.4 Examples

The previous section has laid out the general mindset and framework for A-to-I interface design. In this section, we review a number of examples to illustrate the wide range of ideas that are pursued in this field. It is important to note that this survey is not comprehensive, but simply meant to make the case for A-to-I-inspired design more concrete. Many additional examples, dealing for instance with radio transceivers [20] and several other omitted topics can be found in the current literature.

Audio signal classification. Voice control has become a ubiquitous feature in a variety of interactive electronic systems. In most use cases, the employed classification system must be always on and may only have a small battery as its energy supply. To achieve ultra-low power dissipation, such systems can benefit from the analog analytics feature sampling approach shown in Fig. 17.6d. Figure 17.9 shows

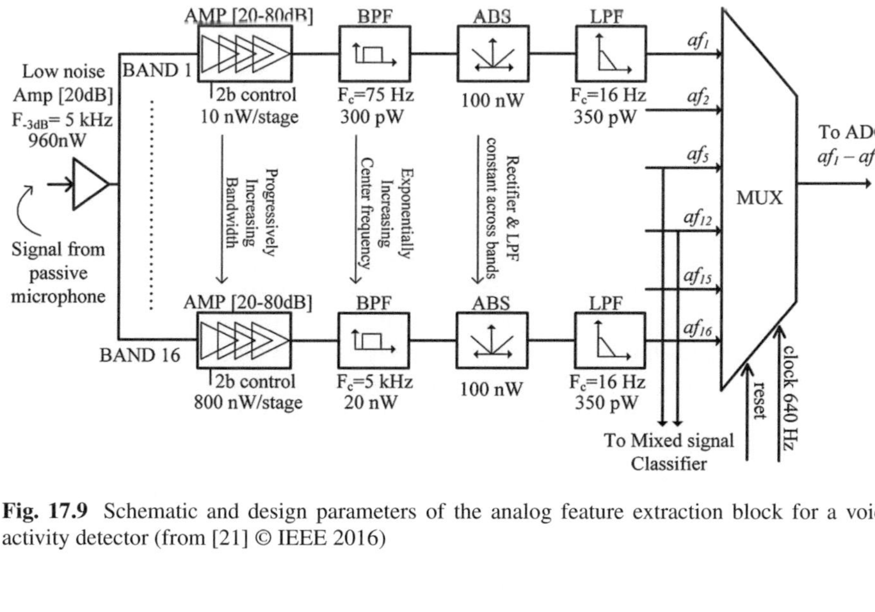

Fig. 17.9 Schematic and design parameters of the analog feature extraction block for a voice activity detector (from [21] © IEEE 2016)

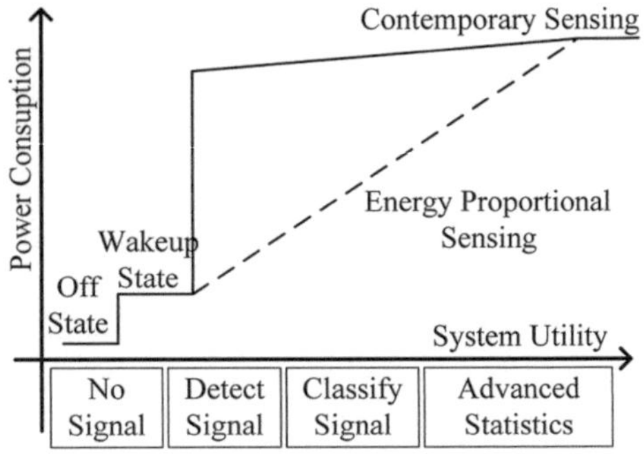

Fig. 17.10 Concept of power-proportional sensing in contrast to state-of-the-art sensing systems (from [21] © IEEE 2016)

an example front-end implementation from the voice activity detector chip detailed in [21]. The ADC that samples the sub-band filters operates at only 640 Hz, as opposed to the much higher rate that would be needed to sample the full audio signal. In this particular work, the filters are implemented with g_m-C circuits. An alternative implementation using switched capacitor N-path filters was recently proposed in [19].

As shown in Fig. 17.10, the overall system of [21] heavily relies on dynamic power gating/scaling and employs cascaded classifiers. The first is a simple always-on wake-up detector that consumes only 700 nW. This detector wakes up a mixed-signal classifier to decide if the signal is speech or non-speech (without using the ADC). If the signal is classified as speech, the system uses the ADC and a microcontroller for more advanced processing. In addition to cascading, the system makes use of the fact that not all computed analog features carry information under all audio contexts, and dynamically disables the computation of features that do not assist in the classification process. In the end result, the acoustic frontend dissipates only 6 μW on average for speech/non-speech classification. This corresponds to a power consumption advantage of about 10× compared to a more conventional system.

Sub-Nyquist observation receiver for DPD systems. System identification is another classical case where the information rate of the involved signal may be much lower than its physical bandwidth. A specific modern instance where this is relevant is digital predistortion for power amplifiers (PAs). As shown in the system diagram of Fig. 17.11, the PA output signal bandwidth can be as large as 7× the input signal bandwidth due to spectral regrowth. For example, with a signal bandwidth of 100 MHz in Long Term Evolution (LTE) systems, the PA's output spectrum spans about 700 MHz. Digitizing the full output spectrum without aliasing requires power-hungry ADCs that are undesired in today's cost-efficient systems. In addition, the aggregate data rate between the ADC and the model extraction block that determines the predistorter coefficients is typically excessively high (20 Gb/s in the shown example).

The A-to-I based approach described in [22] builds on the observation that there are only relatively few degrees of freedom in the desired information, i.e. the predistorter coefficients (100 in the shown example). We can investigate this further by

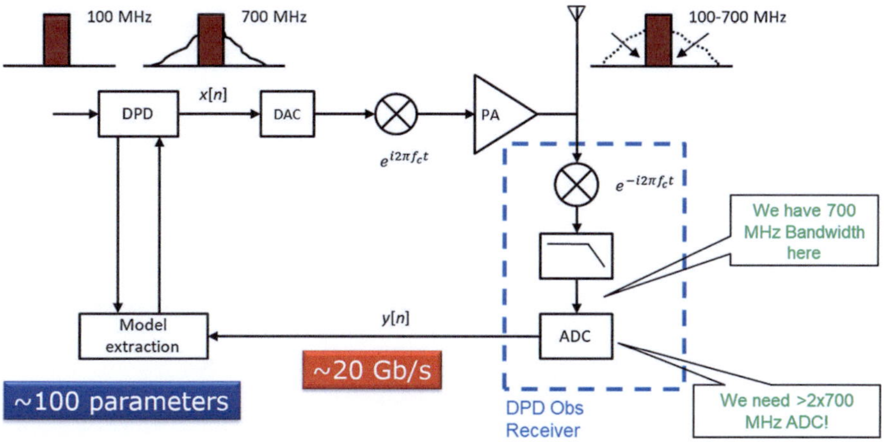

Fig. 17.11 Wireless transmitter system with digital power amplifier predistortion and observation receiver for parameter extraction

considering a basic equation for this setup:

$$y[n] = \sum_{p=1}^{P} \sum_{q=0}^{Q} c_{pq} \cdot x[n-p]^p \tag{17.2}$$

Here, y(n) are the Nyquist-rate ADC output samples, and x(n) is the known signal before the power amplifier. What is unknown are the coefficients c_{pq} that model the PA's nonlinearity. Indeed, as long as we have at least P × Q Nyquist samples, we can solve the system of equations (which is linear in c_{pq}) for the desired coefficients. However, this approach uses high-speed Nyquist samples, which is undesired. The basic idea of [22] is to work with the Fourier transform of (17.2) instead:

$$Y(j\Omega) = \sum_{p=1}^{P} \sum_{q=0}^{Q} c_{pq} \cdot X_p(j\Omega) \cdot e^{-j\Omega T_s} \tag{17.3}$$

Now, what this equation implies is that we only need to measure the signal's Fourier coefficients at P × Q different frequencies to invert the system of equations. Figure 17.12 shows the modified approach that results from this idea. Instead of a fixed-frequency mixer and a Nyquist ADC, it employs a frequency-agile mixer that can move across the output spectrum. The mixer is followed by an integrate-and-dump circuit that measures the Fourier bin of the signal at the mixer frequency. Note that the operation of the integrate and dump circuit corresponds to the exact definition of the Fourier transform for a periodic signal. However, as shown in [22], an extension for practical non-periodic signals merely requires a correction matrix to the applied in the digital domain. The resulting system is shown to lower the

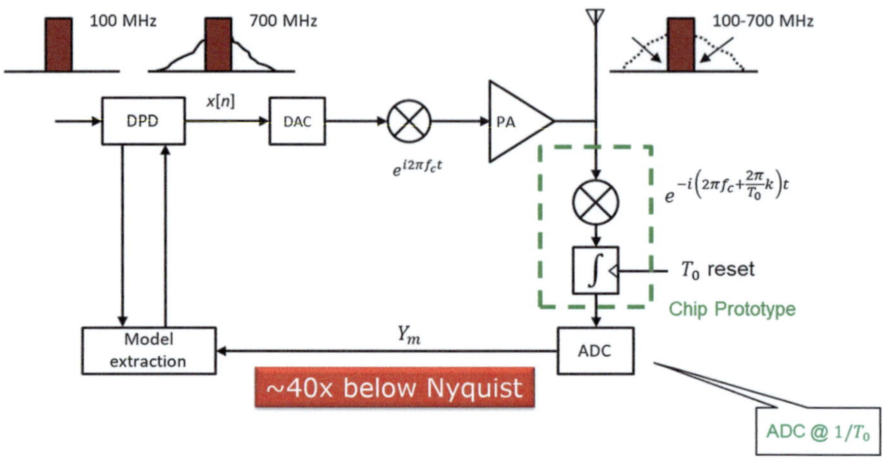

Fig. 17.12 DPD system with sub-Nyquist observation path

ADC sampling rate requirement in an orthogonal frequency-division multiplexing (OFDM) system setting by approximately 30–40× in lab measurements [22].

Imager with analog feature extraction. Machine learning based object detection is a classical application where high-dimensional data (pixels of an image) is processed to produce a relatively low-dimensional output (location of objects belonging to a certain class, e.g. cars). Figure 17.13a shows a typical processing pipeline with a conventional A/D interface, which faithfully digitizes each pixel value and pushes all information processing into the digital domain. The problem with this approach is that a substantial amount of energy is needed for digitization, data movement and storage before even any mathematical operations are applied to the data. An alternative approach was explored in [15], where the imager performs analog gradient feature extraction before A/D conversion (see Fig. 17.13b). The gradients are computed via ratios of pixel values and are logarithmically digitized with 1.5 or 2.75 bits of resolution (see Chap. 3 for a related discussion about logarithmic processing). Working with ratios eliminates unnecessary illumination-related data and allows the features to be compressed by up to 25× relative to a conventional 8-bit readout. As a result, the digital backend-detector, which typically limits system efficiency, incurs less data movement and computation, leading to an estimated 3.3× energy reduction.

Figure 17.14 takes a look at the inner workings of the feature extraction imager. To compute the vertical component of the image gradient (G_V), the scheme requires access to three rows of pixels at a time. Because the pixels are read out in a rolling shutter mode as in a normal image sensor, switched-capacitor-based analog memory cells are employed to circularly store the rows of pixels. Similarly, for the horizontal component of the gradient (G_H), every pixel column needs access to the columns on its left and right, which is provided through an analog multiplexer. The signals are then fed to ratio-to-digital converters (RDCs), which read two voltages sequentially from the cyclic row buffers. While this readout scheme does not achieve lower power

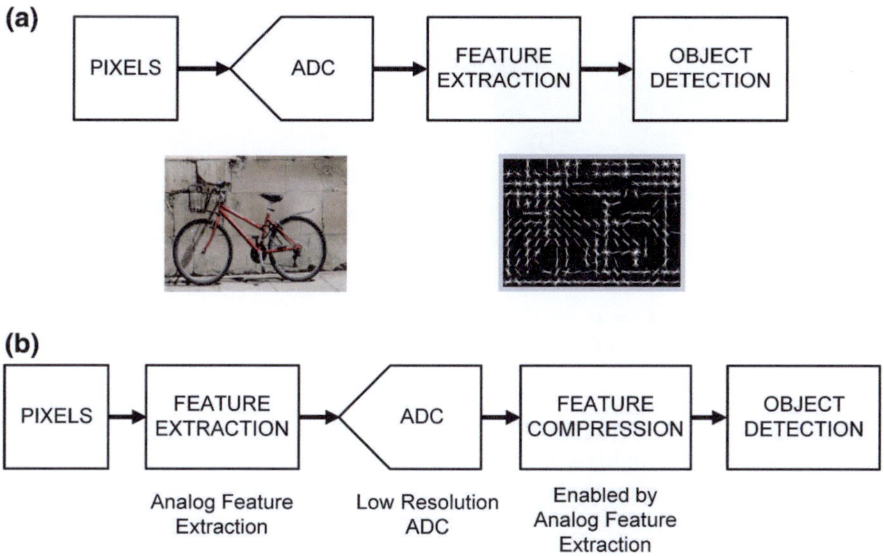

Fig. 17.13 **a** Object detection pipeline producing high-resolution image and extracting low-dimensional features. **b** Alternative approach with analog feature extraction, enabling low-resolution features to be digitized and compressed (from [15] © IEEE 2019)

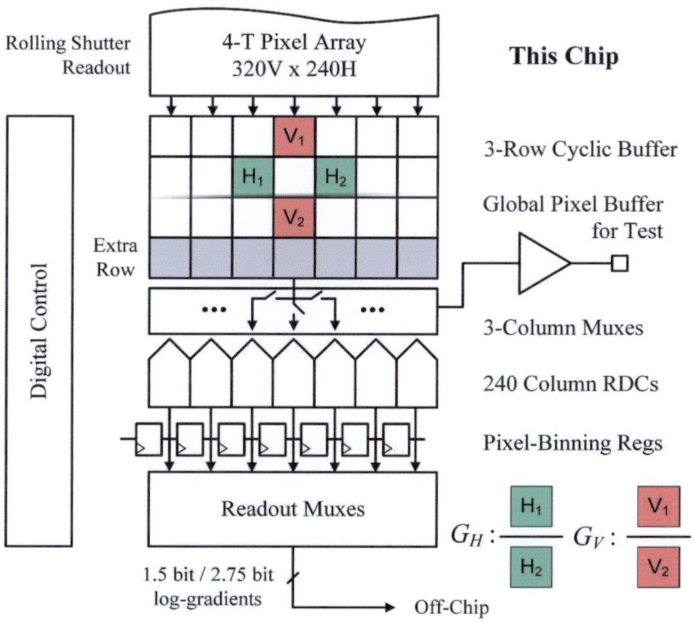

Fig. 17.14 Chip architecture of the imager with log-gradient feature extraction (from [15] © IEEE 2019)

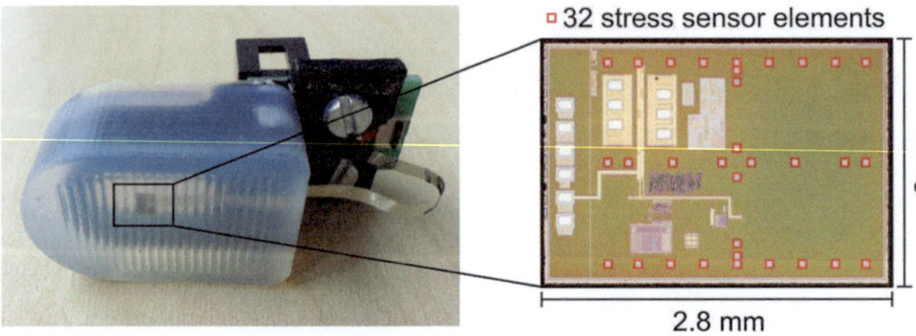

Fig. 17.15 Photograph of the tactile sensor. The sensor chip is visible under the surface of the silicone body (from [25] © IEEE 2019)

dissipation than a standard imager sensor signal chain, the main advantage is that this imager produces highly compressed output data (25×) that lowers the energy consumption of the digital backend processor. At the same time, the log-digitization does not lead to a noticeable performance degradation (compared to a standard 9-bit implementation) for the custom RAW image data set that was used in this study [23].

Tactile sensor. Tactile sensing is widely used for prosthetic and robotic hands to control the grasping force or to detect slip. State-of-the-art sensors achieve a very small form factor (see Fig. 17.15) by employing on-chip, transistor-based transducers [24, 25]. It is thus desirable to simplify the signal processing backend as much as possible to retain this size advantage for a complete platform that communicates only abstract information (e.g., contact/no-contact, slip forward/backward, etc.) to a top-level control unit. In this spirit, the authors of [25] devised a "sensor-to-information processing" approach for tactile sensing.

Conventional signal processing for tactile sensors typically uses frequency-based slip detection using a Fourier transform or frequency-domain filters. The spectral power of a certain frequency range is compared to a threshold to decide if slip is happening. Machine learning is often used in such approaches as opposed to selecting the thresholds by hand. For example, an FFT is computed over a time-series sampled at Nyquist rate and fed into a machine learning algorithm to classify the occurrence of slip. However, this is quite inefficient as it requires a spectral analysis based on a large number of ADC samples. The concept proposed in [25] employs a machine-learning-based processing scheme that detects slip without a spectral analysis and performs one classification per sample from the array of sensor elements. While this scheme does not use A-to-I pre-processing circuits as the general schemes depicted in Fig. 17.6, it uses the selectivity of nonlinear machine learning post-processing to minimize the number of samples that must be taken.

The stress sensing chip used in this work integrates 32 sensor elements that are serially digitized using a single ADC. The digital samples are fed into a supervised machine learning algorithm (random forest classifier). In order to learn how to predict the output for a given sensor array sample, the algorithm takes examples of recorded sensor data, as seen in Fig. 17.16. These sensor examples are labeled according to the abstract information that the data represents. The classifier considers all 32 sensor

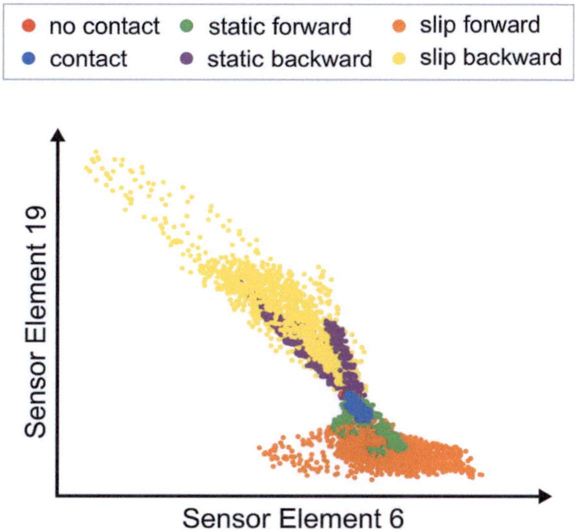

Fig. 17.16 Scatter plot of a slip measurement for two sensor elements. Each point in the scatter plot represents one sensor sample and is color-coded according to the load class (from [25] © IEEE 2019)

elements and finds patterns in the data to reliably classify all loads at a rate of 30 Hz with 99.6% accuracy.

A time-domain measurement with some of the sensor elements is shown in Fig. 17.17, which plots the output of four channels during the following test pattern: pure normal force → static forward force → sliding forward → stop → sliding forward again. The plot also shows the classification result for each sample. The

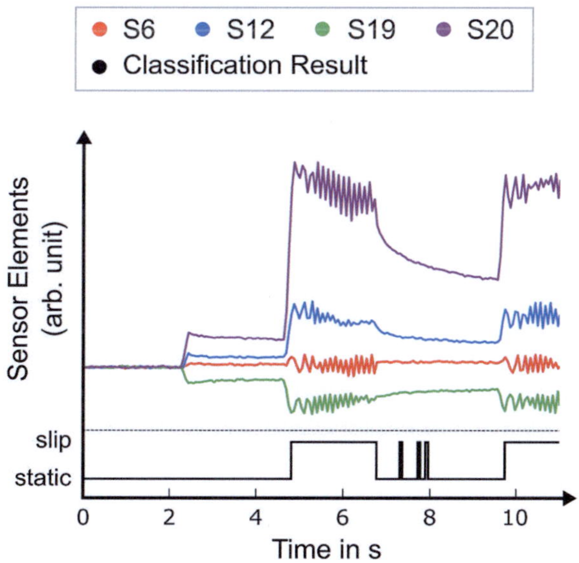

Fig. 17.17 Time behavior of four sensor elements during the slip validation measurement. The tactile sensor is loaded with a pure normal force in the beginning and a forward directed tangential force after 2.5 s. At 4.7 s the finger slips, stops at 6.8 s and slips again at 9.5 s (from [25] © IEEE 2019)

classification is mostly correct and reacts very fast to the changing load. The only misclassification happens during the second static phase, in which the random forest wrongly predicts slip. Such corner cases should be analyzed carefully and included as examples into the training set. Another possible approach is to filter the classification output which is much easier than filtering the sensor outputs.

Overall, just like the discussed audio and image classifier systems, this tactile sensor system advocates for a much tighter coupling and integration of the sensing elements and the classification algorithm. This is expected to become a popular theme in many other applications, and it opens up the opportunity of new interactions between these two domains for sensor and circuit imperfection management, as already discussed above in the context of Fig. 17.8.

17.5 Summary and Conclusions

This chapter presented an overview of the emerging field of analog-to-information processing as a holistic framework for circuit-system co-design. We motivated the need for this new direction by looking at the performance trajectories of conventional A/D converters. First, given that CMOS transistor speed has already saturated, it is unlikely that the future will bring substantially faster ADCs at a given energy efficiency. Second, since we are approaching absolute theoretical efficiency limits even for slow ADCs, the main option that remains is to make the analog-to-digital interface "more intelligent," and "customized to the application." Thus, we should refrain from digitizing data that is deemed irrelevant by a subsequent digital algorithm or system task.

The justification for this departure is not only motivated by process technology limitations, but also by the fact that many modern systems, most notably in the area of machine learning, do not rely on perfect signal reconstruction or exact wave shapes. On the other hand, it is foreseeable that there will remain a good number of applications where signal reconstruction is required, as for instance multimedia applications. These applications will continue to be served by more classical A/D interfaces.

In system scenarios where A-to-I thinking is applicable, it may enable ultralow-power, always-on electronics that can break traditional performance boundaries and fuel new applications. This is what motivates many of the A-to-I demonstrators that we have seen in the literature over the past decade. For most of these emerging concepts, we are now waiting for adoption in mainstream products. In this endeavor, an important question that remains is whether we can generalize these ideas from point solutions toward broadly applicable methodologies.

References

1. B. Murmann, The race for the extra decibel: a brief review of current ADC performance trajectories. IEEE Solid-State Circ. Mag. **7**(3), 58–66 (2015)
2. M. Keller, B. Murmann, Y. Manoli, Analog-digital interfaces, in *2020—A Guide to the Future of Nanoelectronics,* ed. by B. Hoefflinger (Springer, 2012), pp. 95–130
3. M. Keller, B. Murmann, Y. Manoli, Analog-digital interfaces-review and current trends, in *CHIPS 2020 VOL. 2: New Vistas in Nanoelectronics*, ed. by B. Hoefflinger (Springer, 2015)
4. D. Healy, D.J. Brady, Compression at the physical interface. IEEE Sig. Process. Mag. **25**(2), 67–71 (2008)
5. M. Verhelst, A. Bahai, Where analog meets digital: analog-to-information conversion and beyond. IEEE Solid-State Circ. Mag. **7**(3), 67–80 (2015)
6. R. Schreier, G.C. Temes, *Understanding Delta-Sigma Data Converters* (Wiley, New York, 2005)
7. A.M.A. Ali et al., A 16-bit 250-MS/s IF sampling pipelined ADC With background calibration. IEEE J. Solid-State Circ. **45**(12), 2602–2612 (2010)
8. B. Murmann, ADC Performance Survey 1997–2019. [Online]. Available: http://web.stanford.edu/~murmann/adcsurvey.html
9. D. Draxelmayr, A 6b 600 MHz 10mW ADC array in digital 90 nm CMOS, in *ISSCC Dig. Techn. Papers* (2006), pp. 264–265
10. E.A. Vittoz, Future of analog in the VLSI environment, in *Proceedings of ISCAS* (1990), pp. 1372–1375
11. J. Musayev, A. Liscidini, Quantised inverter amplifier. Electron. Lett. **54**(7), 416–418 (2018)
12. E.J. Candes, M.B. Wakin, An introduction to compressive sampling. IEEE Sig. Process. Mag. **25**(2), 21–30 (2008)
13. M. Vetterli, P. Marziliano, T. Blu, Sampling signals with finite rate of innovation. IEEE Trans. Sig. Process. **50**(6), 1417–1428 (2002)
14. K. Badami, J.-C. P. Ramos, S. Lauwereins, M. Verhelst, Mixed-signal programmable non-linear interface for resource-efficient multi-sensor analytics, in *ISSCC Dig. Tech. Papers* (2018), pp. 344–346
15. C. Young, A. Omid-Zohoor, P. Lajevardi, B. Murmann, A data-compressive 1.5/2.75-bit log-gradient QVGA image sensor with multi-scale readout for always-on object detection. IEEE J. Solid-State Circ. 1–15 (2019)
16. D. Muratore, P. Tandon, M. Wootters, E. J. Chichilnisky, S. Mitra, B. Murmann, A data-compressive wired-OR readout for massively parallel neural recording. IEEE Trans. Biomed. Circuits Syst. (2019)
17. R. Sarpeshkar, Analog versus digital: extrapolating from electronics to neurobiology. Neural Comput. **10**(7), 1601–1638 (1998)
18. B. Murmann, Digitally assisted data converter design, in *2013 Proceedings of the ESSCIRC (ESSCIRC)* (2013), pp. 24–31
19. D. Villamizar, D. Battaglino, D.G. Muratore, R. Hoshyar, B. Murmann, Sound classification using summary statistics and N-path filtering, in *2019 IEEE International Symposium on Circuits and Systems (ISCAS)* (2019), pp. 1–5
20. R.T. Yazicigil, T. Haque, M. Kumar, J. Yuan, J. Wright, P.R. Kinget, How to make analog-to-information converters work in dynamic spectrum environments with changing sparsity conditions. IEEE Trans. Circuits Syst. I Regul. Pap. **65**(6), 1775–1784 (2018)
21. K.M.H. Badami, S. Lauwereins, W. Meert, M. Verhelst, A 90 nm CMOS, power-proportional acoustic sensing frontend for voice activity detection. IEEE J. Solid-State Circ. **51**(1), 291–302 (2016)
22. N. Hammler, A. Cathelin, P. Cathelin, B. Murmann, A spectrum-sensing DPD feedback receiver with 30x reduction in ADC acquisition bandwidth and sample rate. IEEE Trans. Circuits Syst. I Regul. Pap. **66**(9), 3340–3351 (2019)
23. A. Omid-Zohoor, D. Ta, B. Murmann, PASCALRAW: raw image database for object detection, in *Stanford Digital Repository*. [Online]. http://purl.stanford.edu/hq050zr7488

24. M. Kuhl et al., A wireless stress mapping system for orthodontic brackets using CMOS integrated sensors. IEEE J. Solid-State Circ. **48**(9), 2191–2202 (2013)
25. J. Kuehn, Y. Manoli, A fingertip-shaped tactile sensor with machine-learning-based sensor-to-information processing, in *International Conference on Solid-State Sensors, Actuators and Microsystems & Eurosensors XXXIII (TRANSDUCERS & EUROSENSORS XXXIII)* (2019), pp. 1811–1814

Chapter 18
Machine Learning at the Edge

Marian Verhelst and Boris Murmann

18.1 The Need for Machine Learning at the Edge

Over the last decade, electronic devices have started to ubiquitously populate our environment. Billions of connected electronic devices such as drones, smart watches, wearable health patches, smart speakers, together form the Internet-of-Things (IoT) [1]. These devices are typically equipped with around a dozen of sensors, to continuously observe the environment and act accordingly. Similarly, also in smartphones the number of integrated sensors keeps rising, to feed the devices with more information about the user and the environmental context [2].

These data collection devices, often denoted as "edge devices," capture raw sensory data streams for further processing. Recent developments resulted in algorithms capable of extracting more accurate information from such sensory data than ever before, through the usage of neural networks and other machine learning models [3]. Yet, this comes at the expense of more computationally complex algorithms, requiring many billions of computations per second, with gigabytes of storage needs [4].

Increasing computational needs are, however, in strong conflict with the limited resource budgets of edge devices: As typically powered by batteries, their energy budget is highly constrained. Furthermore, size and cost constraints limit the amount of affordable memory space and compute power. As a result, until recently, the edge devices were mainly responsible for sensory data capture, with some light preprocessing for data reduction. The compressed data could subsequently be sent to a data center, where ample compute power and memory resources are available. The

M. Verhelst (✉)
KU Leuven, Leuven, Belgium
e-mail: marian.verhelst@kuleuven.be

B. Murmann
Stanford University, Stanford, USA
e-mail: murmann@stanford.edu

B. Murmann and B. Hoefflinger (eds.), *NANO-CHIPS 2030*,
The Frontiers Collection, https://doi.org/10.1007/978-3-030-18338-7_18

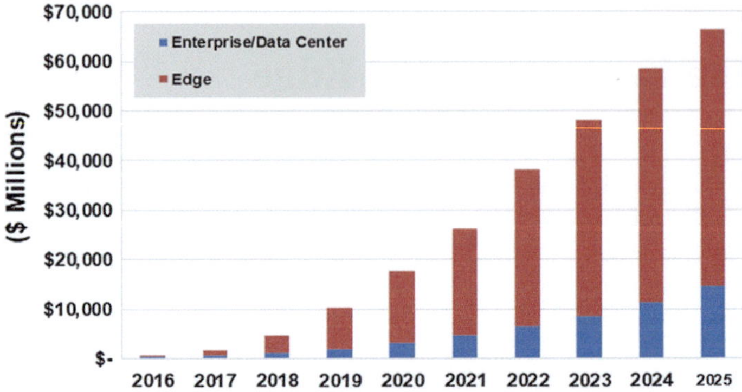

Fig. 18.1 Deep learning chip revenue for edge and data center applications. *Source* Tractica [7]

recent rise of data center activity and investments in machine learning equipment within the data centers is the consequence of this operating scheme [5, 6].

However, increasingly, users and applications shy away from such cloud-centric deployment. The desire to keep sensory data of edge devices private, as well as the energy and latency cost to send all data to the cloud, pushes for device-centric solutions, in which data is kept and processed locally as much as possible [2]. This requires edge devices to become intelligent devices that can autonomously process and interpret data in real time. This emerging operational paradigm will cause a shift in machine learning focus from the data center to the edge. As Tractica predicts (see Fig. 18.1 [7]), edge-based AI chipsets for mobile phones, smart speakers, cars, drones, AR/VR headsets, surveillance cameras and other devices will by 2025 account for more than $50 billion in revenue, or 3.5× larger than in the data center.

To serve this emerging market, heterogeneous compute platforms are required. Special purpose processors help the traditional CPU and GPU compute platforms deployed in the edge towards resource-constrained ML processing of large volumes of sensory data. The market of such machine learning accelerators, ASIPs or ASICs, is hence expected to see the fastest growth (see Fig. 18.2 [7]), with currently already more than 70 specialty AI companies working on some sort of chip-related AI technology [8].

Up until now, this recent evolution has already resulted in a very broad landscape of customized machine learning processors, covering a wide performance space. Figure 18.3 depicts the performance of a range of state-of-the-art neural network processors [9]. State-of-the-art solutions are capable of achieving processing efficiencies of 1–100 TOPS/Watt, enabling processing at several TOPs/second within the edge devices' power budget. Yet, it is important to note that these different state-of-the-art solutions rely on very different algorithmic, architectural and technology assumptions, and cannot be fairly compared purely at the hardware level without considering other system aspects.

In this chapter, we argue and demonstrate the importance of considering the whole stack in a machine learning edge solution (see Fig. 18.4): From algorithm

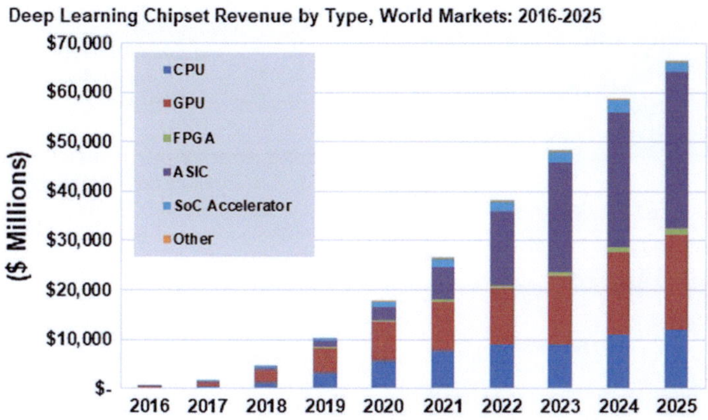

Fig. 18.2 Deep learning chip revenue by type. *Source* Tractica [7]

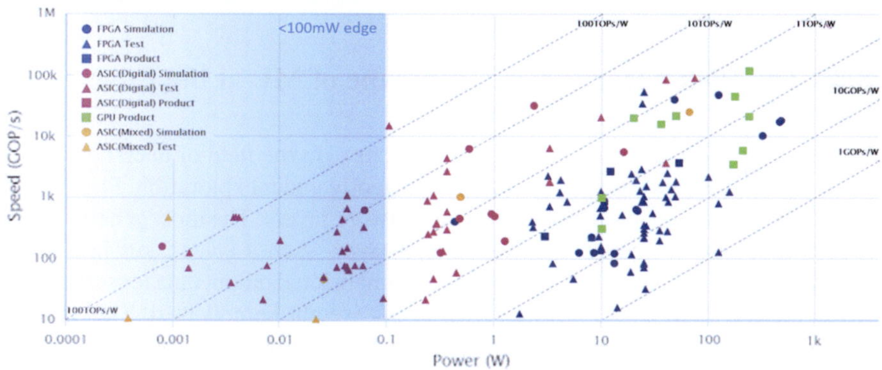

Fig. 18.3 Neural network processor comparison, highlighting the power region <100 mW, interesting for edge devices. Adapted from [9]

and dataflows (Sect. 18.2), over architectures (Sect. 18.3), to circuits and technology options (Sect. 18.4). Only such a vertically integrated approach allows to fairly benchmark different solutions relative to each other (Sect. 18.5) and perform true system optimizations towards efficient deployment of edge intelligence (Sect. 18.6). Throughout these sections, we will mostly focus on neural networks as the main machine learning model. We conclude the chapter with an outlook towards recently emerging trends, such as training at the edge, and newly emerging machine learning models (Sect. 18.7).

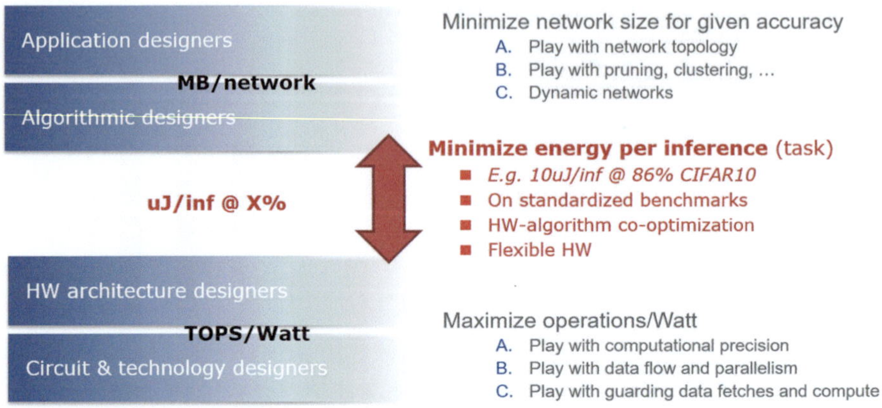

Fig. 18.4 Efficient edge solutions should not be optimized from a sole algorithmic perspective (minimal MB/network), nor from a sole hardware perspective (maximal TOPS/Watt), yet should jointly consider the complete design stack to come to efficient system level solutions

18.2 The Rich Algorithmic Landscape of ML at the Edge

Machine learning models are currently in high flux. In the field of deep neural networks, a wide range of model topologies is currently under exploration. Each model is carefully built out of a sequence of neural network layers. The most generic neural network topology element is a convolutional layer. Such a layer takes in a three-dimensional data tensor and produces a three-dimensional output data tensor through convolving the input tensor with a series of 3D weight kernels [10]. This is illustrated with the relevant data dimensions highlighted in Fig. 18.5. The convolutional operation of one neural network layer can be captured in eight nested for-loops, with a multiply-accumulate operation at the core (see Fig. 18.5). Since in edge devices real time operation requires every input data item to be processed as soon as it comes

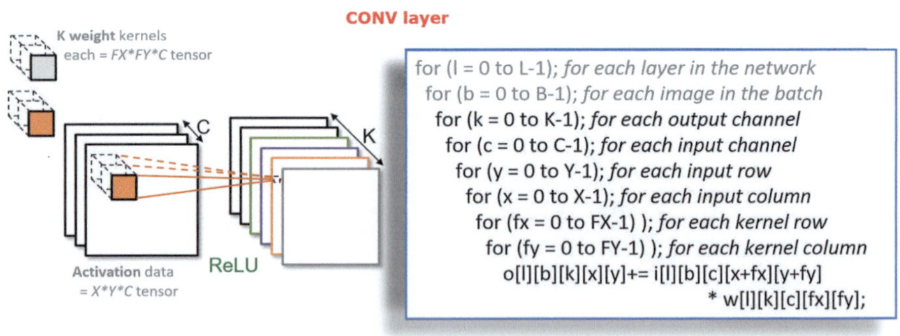

Fig. 18.5 For each item in a batch, each convolutional layer represents six nested for-loops per inference

in, batching is not tolerated and a batch size of one is typically used, making B = 1. As processing efficiency is of such crucial importance in edge devices, research here is focused on algorithmic transformations that impact model size and execution cost without affecting model accuracy. We briefly survey model compaction, model quantization and model pruning techniques, and give an outlook to the future in this area.

Model topology and model compaction: The index ranges of the aforementioned for-loops are determined by the layer and network topology and (as we show later) strongly influence the network's execution efficiency in hardware. Network designers hence use these dimensions in a quest to construct the most compact or efficient models which can fit in small sized embedded memories. Such model compaction research led for instance to the introduction of bottleneck layers [11]. Here, a three-dimensional convolutional layer is replaced by a stack of three layers in which the first and last layer only perform a one-dimensional convolution (FX = FY = 1) to reduce the number of channels (see Fig. 18.6a). Experiments have proven that such structures maintain good modeling capabilities while drastically reducing model computations and coefficients in the three-dimensional convolution (middle layer), thus lowering compute and memory needs. This technique is often combined with the usage of parallel network layers, which are concatenated further in the network, as in the Inception module in Fig. 18.6c [11].

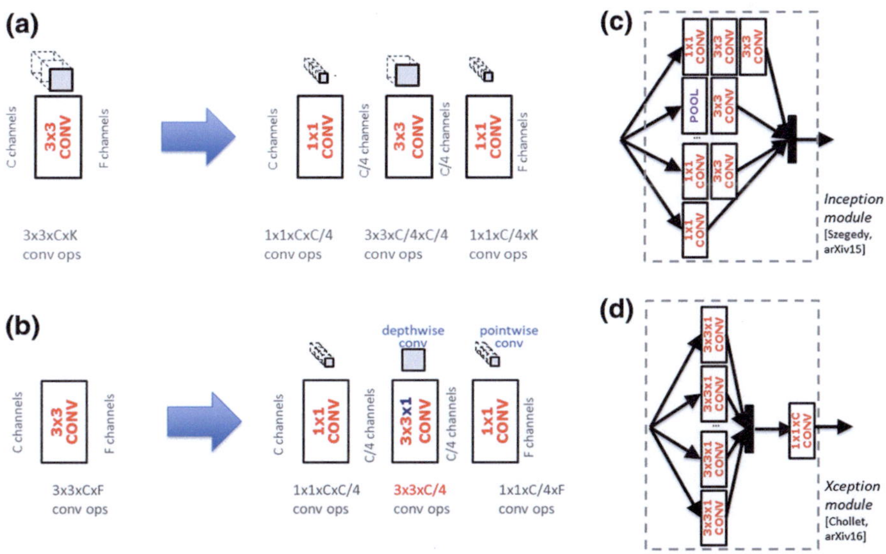

Fig. 18.6 Evolution toward reduced-dimension neural network layers: (**a**) bottleneck layers and (**b**) depth-wise/point-wise layers. These techniques are combined with parallel layers that are subsequently fused, such as in (**c**) the inception and (**d**) Xception modules

To further reduce the computational load of the remaining three-dimensional convolution, the middle layer of this stack was subsequently replaced by a two-dimensional convolution, which only convolves within one channel (removing the for-loop across C, Fig. 18.6b). The resulting "depthwise-pointwise" technique [12] was successfully used in the creation of MobileNet [13], a lightweight network to perform object recognition on mobile phones. In a next generation, MobileNetV2, this technique was further combined with feedforward connections across network layers [14].

Recently, a new paradigm shift emerges: Network topologies are no longer optimized by hand, but are the result of automated neural network search, also denoted by AutoML. Here, reinforcement learning, evolutionary algorithms and/or random sampling strategies are used to find more compact and better performing networks [15–21]. The focus in this field of research is on finding the best performing networks from an accuracy point of view, while minimizing the amount of GPU compute time required for the network search. Very few works [22–24], however, take the neural networks execution efficiency on edge devices into account in the cost function when searching for the most optimal networks. This is discussed further in Sect. 18.6.

From previous discussion, it should be clear that a wide variety of convolutional topologies exists for neural network layers, which are often combined, concatenated and interconnected in many different and irregular ways. When developing hardware architectures, we hence must ensure sufficient flexibility to support the mapping of all these different topologies and dataflows (see also Sect. 18.6).

Beyond convolutional layers, other types of neural network and non-neural network models must also be supported. Yet, interestingly enough, they can often be rewritten in the form of the generic convolutional model in Fig. 18.5. For example, the fully connected neural network layer [10], often found at the end of classification networks, can be rewritten in the same form of the convolutional layer with X = Y = FX = FY = 1, as indicated in Fig. 18.7. Likewise, other machine learning kernels, such as the support vector machine (SVM) [25] and one-class SVMs (often used in

FC layer
(~ support vector computations)

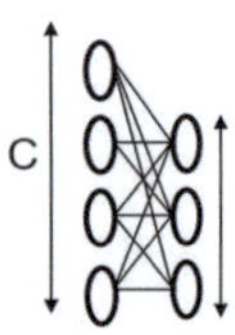


```
for (l = 0 to L-1); for each layer in the network
 for (b = 0 to B-1); for each image in the batch
  for (k = 0 to K-1); for each output channel
   for (c = 0 to C-1); for each input channel
    for (y = 0 to Y-1); for each input row
     for (x = 0 to X-1); for each input column
      for (fx = 0 to FX-1) ); for each kernel row
       for (fy = 0 to FY-1) ); for each kernel column
        o[l][b][k][x][y]+= i[l][b][c][x+fx][y+fy]
                              * w[l][k][c][fx][fy];
```

Fig. 18.7 For each item in a batch, each fully connected neural network layer represents two nested loops. Similarly, each support vector machine evaluation represents two nested loops per inference

anomaly detection), demonstrate similar matrix-vector multiplication kernels, fitting the same framework. This is good news, as it simplifies the development of a generic hardware platform for such machine learning workloads (see Sect. 18.3). Yet, these computational layers have fewer effective nested for-loops, which results in fewer opportunities for efficient hardware mapping, as seen later in this chapter.

Model quantization: Researchers have found that neural network models carry some redundancy, making them to a certain extent robust to perturbations. This enables further model efficiency and storage reduction techniques that exploit sparsity and reduced precision operation. Regarding computational precision, limited-precision fixed point data representations for both weights and activations were shown to be sufficient for nearly all inference tasks, drastically cutting model memory weight storage and MAC complexity [26–28]. Operation down to 8, 4 or even fewer bits has been demonstrated for many machine learning benchmarks, with ternary (−1, 0, 1) and binary (−1, 1) neural networks as the extremes [29]. The best results are achieved when using the dynamic fixed point format [30], and quantizing the network during the training process [31, 32], instead of first training a floating point network and quantizing it afterwards, or smartly unifying the dynamic range of all weights during training [33, 34]. Active research tries to find efficient ways to determine the minimum bit width representation necessary to achieve a target accuracy level for a given task, which at the moment is still relying largely on inefficient exhaustive searches. It is important to realize that this optimum is heavily interwoven with the selected network topology and cannot be looked at in isolation [35].

Model pruning: Instead of just quantizing the weights of a network, one can also remove some weights completely, which is called "pruning the network." Many pruning techniques exist, ranging from after-training techniques that just remove smallest weights of a network [36–38], to during-training regularization techniques that try to force as many coefficients as possible to become approximately zero [39]. This results in sparse neural network models, whose zero values can be exploited to further reduce the model's storage and computational footprint. Several model compression formats have been proposed, such as the Compressed Sparse Column (CSC) format, which encodes the sparse matrices and vectors into fewer words by skipping zero-valued data [36]. The processor must of course be equipped with the corresponding decoding logic to be able to interpret this data [40].

Interdependencies: It is important to realize that all aforementioned optimizations, such as model compaction, model quantization and model pruning are strongly interwoven. It is observed that compact models tend to be less sparse, and less tolerant to quantization [35]. Finding the most efficient model hence requires balancing all three techniques. As this results in an enormous algorithmic search space, current research is strongly invested in exploring this space as efficiently as possible. Breakthroughs have been achieved using automated machine learning (AutoML) techniques exploiting Bayesian optimization, evolutionary algorithms and reinforcement learning [15–21]. Yet, quantization and hardware inference cost has received limited attention in this field.

Processor consequences and outlook: The optimization techniques adopted for networks strongly influence the execution efficiency on the processor hardware. As an example, model compaction techniques typically result in models with smaller (FX, FY) filter kernels or (X, Y) activation sizes, causing a drop in data reuse opportunity [41]. Similarly, pruning breaks the processing regularity that made traditional deep learning processors so efficient. As a result, the smallest model is not necessarily the most efficient one for execution at the edge [35]. This gives rise to new, more hardware-aware algorithmic techniques, such as structured sparsity or dynamic neural networks. To understand this better, let's take a closer look at edge processing architectures.

18.3 The Rich Architectural Landscape of ML at the Edge

From CPU to GPU to NPU: As neural networks are characterized by massively parallel MAC operations, their processing requires widely parallel execution. On traditional Von Neumann CPUs this is achieved by exploiting vector processing instructions for parallel MAC execution [42] (Fig. 18.8). Recently, CPUs have been equipped with additional fused (integer) multiply add (FMA) instructions, which allow to also efficiently accumulate multiplication results. Yet, these processors lack sufficient computational resources to achieve more than 100× parallelization factors, limiting performance to a few hundred GOPs per processing core.

For this reason, GPUs have been extensively used as the main neural network inference platform. They are equipped with many parallel execution units and can achieve 1000× or more parallel MAC operations. Moreover, over the last few years, GPUs have moreover a rapid evolution to serve neural network inference workloads even better. First of all, recent implementations support small word length fixed-point data types instead of only supporting floating point operations. Secondly, traditional GPU architectures did not support efficient spatial and temporal reuse of data across processing elements (see also below). The recent inclusion of tensor cores, which spatially unroll the multiplication of two 4×4 matrices in one timestep, alleviates this issue for certain layer topologies [43]. Still, flexibility and efficiency across kernel sizes and models remains an issue, and (embedded) GPU power consumption exceeds the power budget of many edge solutions.

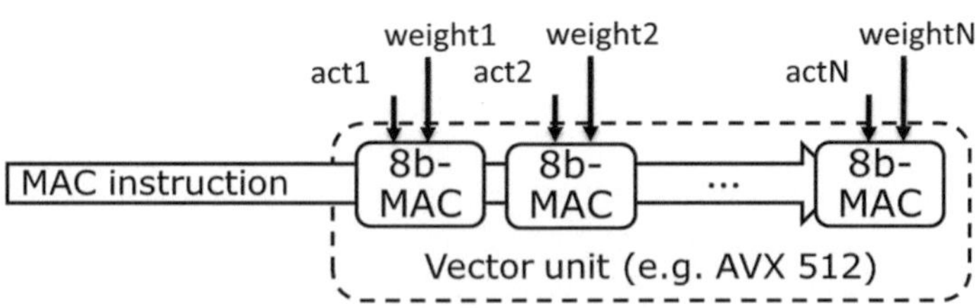

Fig. 18.8 Traditional vector processing units can be used to achieve parallelization for neural network processing. Yet, they are not fully exploiting the neural network data flow properties

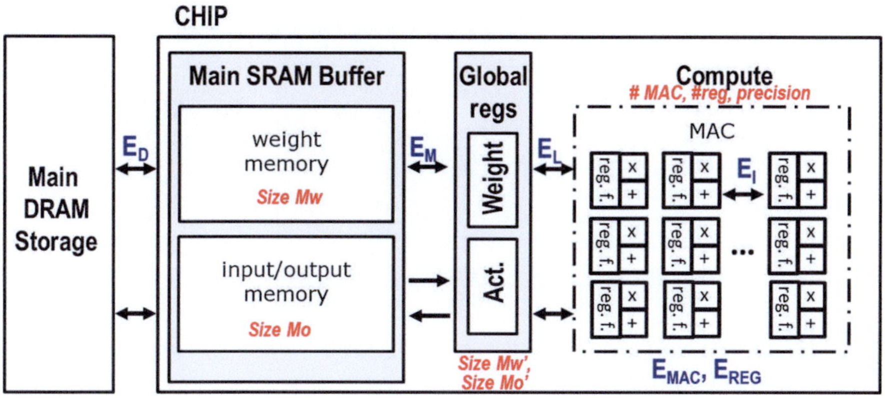

Fig. 18.9 NPU architectural template, which is parametrized across many design dimensions, ranging from the number of parallel MACs and their interconnectivity, to the levels of memory and their sizes and interconnectivity

For this reason, more and more specialized, custom processor cores are appearing, optimized towards neural network inference in resource constrained devices [44]. These class of processors is often denoted as "NPU," or Neural Processing Unit. NPUs consist of a widely parallel datapath equipped with MACs with or without local storage, together with a hierarchy of several optimized memory layers, as shown in Fig. 18.9. Several efficiency techniques are exploited across NPU designs, which we shall discuss in more detail: (1) Spatial and temporal data reuse; (2) hierarchical memories and local storage; (3) sparse or dense processing; (4) reduced precision processing.

Spatial and temporal data reuse: A large fraction of NPU power consumption is spent on data fetches. Good NPU designs therefore try to maximize not only the number of parallel MACs that can be executed in every single clock cycle, but also minimize the average number of data fetches per usefully executed MAC. This can be achieved through spatial or temporal data reuse across different layers of granularity (see Table 18.1). Spatial data reuse exploits the use of multi-dimensional data paths to reuse fetched weights and/or activations across many parallel MAC operations within a processing element (PE) array. Figure 18.10 shows the architecture of the Envision processor [45], in which every weight is multiplied with 16 input activations in parallel, while every input activation is multiplied with 16 weights (of different output channels) in parallel. Also, the number of data stores can be reduced spatially,

Table 18.1 Data reuse opportunities classified across granularity and their spatial/temporal nature

	Intra-PE	Inter-PE (Intra-PE array)	Inter-PE array
Spatial reuse	–	Multi-dimensional datapaths Accumulation trees	Broad-/multi-casting networks
Temporal reuse	Stationarity	Systolic arrays	Systolic/streaming processors

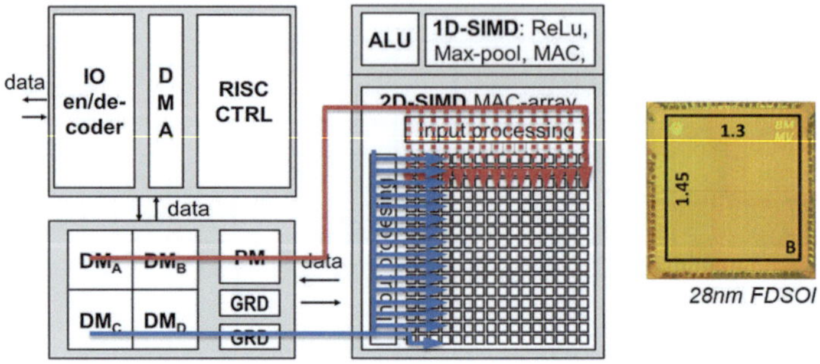

Fig. 18.10 Envision processing architecture, exploiting spatial reuse of input activation data (red) and weight data (blue). Chip photo on the right

by introducing summation trees that accumulate results across PEs before sending them back to memory.

Besides purely reusing data spatially within a single clock cycle, data can also be reused temporally, across clock cycles. Here, a distinction can be made between processing architectures that reuse data across subsequent clock cycles within the same processing element (stationary techniques), and architectures that reuse data across subsequent clock cycles within neighboring processing elements or datapaths (systolic processing architectures). A very common stationarity approach is to keep the MAC output locally and accumulate within a PE in subsequent clock cycles. Envision [45] is an example of such an "output-stationary" approach. Other implementations keep the weights local within a PE across cycles ("weight stationary"), such as the weight stationary TPU processor of Google [46], or the BinarEye processor [47].

Systolic architectures, on the other hand, exploit the fact that it is cheaper to exchange data between neighboring processing elements, instead of sending them to a larger remote memory. Also, systolic principles can be applied at different levels of granularity: At the lowest level, neighboring PEs can pass partial accumulation results and/or weights to each other, as for example done in the aforementioned TPU processor [46] and the Eyeriss chip [48]. But, also across larger clusters, data can be forwarded from processor to processor, with only small streaming buffers in between, avoiding data transfers in and out of large memories. This is done e.g. in [49]. These processor architectures break with the traditional Von Neumann architecture and tightly intertwine processing elements and memory blocks.

Hierarchical memories: To further reduce energy spent on memory fetches and stores, the memory hierarchy is further optimized. Instead of using one large central memory, data is stored as close as possible to the place where it is generated and consumed, while using a memory block that is as small as possible. This results in hierarchical memory structures, while small local memories, complemented with several layers of larger shared memories further up in the hierarchy, as shown in Fig. 18.11. The challenge here is to determine the optimal memory sizes at each level in the hierarchy, not for a single network topology, but across many network topologies (see further discussion below).

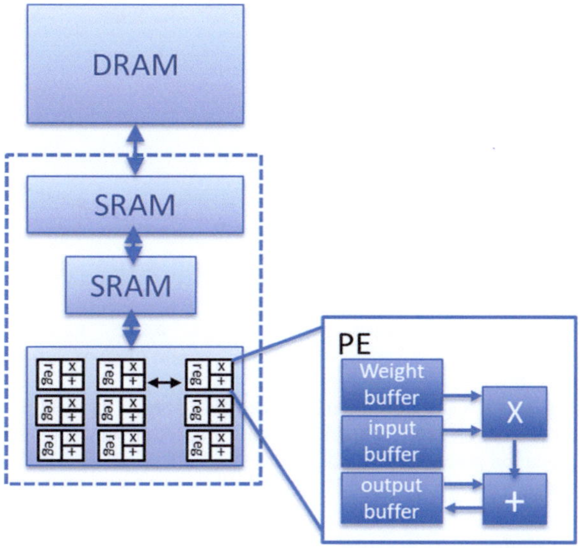

Fig. 18.11 Hierarchical processor memory hierarchy

Sparse or dense: As discussed in Sect. 18.2, neural network models typically exhibit a certain degree of sparsity, which can be exploited in the processing hardware. Indeed, when doing a multiply accumulate operation with one of the multiplication inputs being zero, the accumulation result remains unchanged. The most straightforward way to exploit such sparsity, is to maintain the regular dense processing grid, yet simply clock and data-gate all units that encounter a zero-valued input. Processors such as Envision [45], or Eyeriss [48] support this approach. The operating scheme allows saving power when executing sparse networks, and only comes with very little overhead logic to support the clock and data gating. Yet, the approach only brings (modest) power savings, and does not lead to increased throughput for sparse workloads. Indeed, all idle MAC units are wasting useful processing resources.

This is overcome in sparse NPU processors, which target skipping all zero-valued operations and assign their computational resources only to useful computations. Such an approach allows to automatically speed up processing when the networks are very sparse. Yet, the approach is penalized by large architectural overhead for data decoders, scheduling logic, and irregular data routing. Moreover, data reuse opportunities drop drastically in such processors, often limiting the amount of effective parallel operations that can take place. As a result, such processors prove to be beneficial only when the sparsity is large enough. Parashar et al. [50], have shown this break-even point to lie around 40% sparsity (60% density) for both weights and activations (see Fig. 18.12). While older networks had very high sparsity (e.g., 80% or more for AlexNet), newer networks exhibit different characteristics. The recent network compaction techniques result lower sparsity, ranging between 10 and 70% for networks like GoogleNet, 10–50% for MobileNet and only 10% for MobileNet [51].

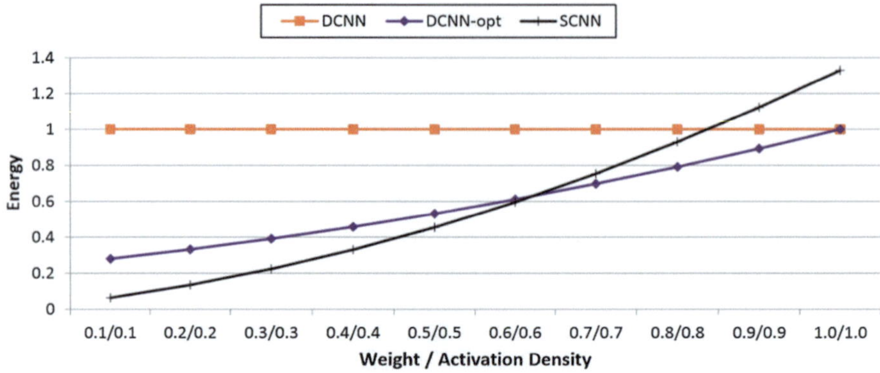

Fig. 18.12 GoogLeNet performance and energy as a function of density for a non-sparsity-aware processor (DCNN), a sparsity-aware processor with datapath gating (DCNN-opt) and a sparse execution processor (SCNN) (from [50], ©IEEE 2017)

Reduced precision: Another algorithmic property that can be exploited at the hardware level is the robustness to reduced computational precision. As discussed in Sect. 18.2, neural networks can be trained to operate with low-resolution fixed-point number representations. Figure 18.13 illustrates this, assuming m-bit integer activation values and n-bit integer weights, drastically reducing the multiplier complexity, area and power consumption. Precision scaling can be done symmetrically (m = n) or asymmetrically (m ≠ n) [52]. Data types used in inference accelerators are often INT8, and more and more frequently also INT4, or even ternary or binary (INT1) values. As can be seen from Fig. 18.14, reduced precision processing typically results in both performance and efficiency boosts.

It is important to realize that different neural networks have different optimal fixed-point word lengths [51]. Even between layers of the same network, optimal quantization values might differ, typically requiring more bits for full resolution input layers for image processing. As a result, a widely deployable NPU processor needs internal MAC units that can operate at different precision settings. The settings should be easily configured, e.g. through a simple processor instruction. Moreover, the overhead of this configurability at the MAC level should be limited to maintain

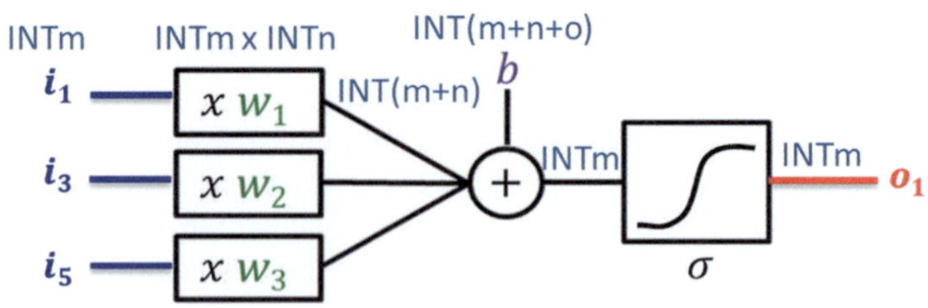

Fig. 18.13 Operating with low-resolution fixed-point number representations

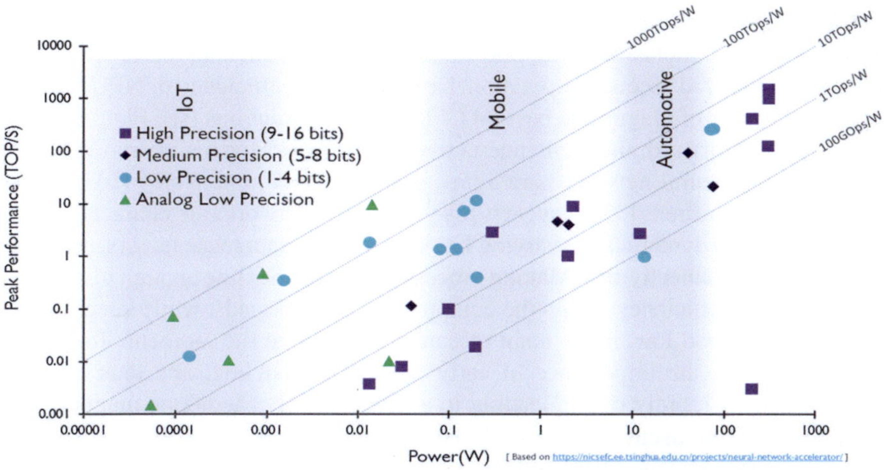

Fig. 18.14 Sample of recent NPU implementations, indicating the precision of the internal MAC units. *Source* [9, 53]

good efficiency across all precision levels. Many precision-scalable MAC designs have been proposed in the literature, each of which coming with their own merits and downsides [52, 54, 55]. Table 18.2 summarizes the main precision scalability architectures in a taxonomy introduced in [55]. 1D scalable designs demonstrate good scalability at weight-only asymmetric scaling, while 2D scalable designs perform well when one wants to scale both activations and weights. 2D scaling can be performed symmetrically across weight and activations, or asymmetrically. This scaling, however, always comes at the expense of increased memory bandwidth in low precision modes, with increased bandwidth pressure on the memory stores when using sum apart techniques, and pressure on the memory loads for the sum together techniques. Across all operating modes, bit serial techniques do not seem to pay off, based on this comparative study. A more elaborate survey can be found in [55].

Table 18.2 Variable precision MAC taxonomy (from [55]) and reported implementations exploiting the various techniques

Architecture types		1D scalable (weight only)	2D asymmetric scalable	2D symmetric scalable
Spatial	Sum apart	[56] (DNPU)		[57] (DVAFS)
	Sum together		[58] (BitFusion)	[54]
Temporal	Serial	[59] (UNPU)	[60] (LOOM)	
	Multi-bit serial	[52]	[60] (LOOM)	

Challenges and outlook: The break from traditional Von Neumann processing architectures, and inclusion of support for multi-dimensional data reuse, sparse processing and reduced precision operation, have pushed the efficiency of NPU processing to 1–2 orders of magnitude beyond CPU and GPU solutions (see Figs. 18.3 and 18.14). Going forward, the challenge is to ensure support for a wide range of new and upcoming neural network paradigms, such as dynamic networks [61], dilated networks [62], shiftnets [63], wavenets [64], etc. These networks are characterized by (sometimes even dynamically) varying kernel sizes, low data reuse factors, and complex layer interconnectivities. Making processors that are flexible enough to maintain good execution efficiency across the complete set of workloads, while keeping configuration overhead low, is the main research challenge at the moment. To achieve these properties, the importance of early processor modeling, and lean dataflow optimizations is rapidly rising, leading to a new class of schedulers, mappers, and compilers that are discussed in Sect. 18.6.

In the future, we'll see these architectures evolve further towards more distributed processing, with small, yet flexible buffers between precision-scalable processing elements. As the memory access remains the main bottleneck, emerging technologies that integrate the memory and computations are rapidly gaining importance and are thus discussed in the next section.

18.4 The Rich Circuit/Technology Landscape of ML at the Edge

The previous sections looked at efficient neural network computation mainly from the perspective of algorithms and architecture, corresponding to the upper three layers of Fig. 18.4. However, a wide range of options are also available at the circuit and technology level, which complicates the search for an optimal implementation even further. In this section, we briefly review the most common innovation vectors.

Analog and mixed-signal computing: There is a rich history of research that promotes the purely analog implementation of neural networks and other machine learning algorithms. This path typically follows neuromorphic principles [65, 66], which build on our (very limited) understanding of the human brain and its "integrate and fire" neurons that are amenable to an analog circuit implementation. While the resulting neurons represent an intriguing and biologically plausible emulation of the units found in the human brain, the networks constructed with them tend to lack scalability. It is fundamentally difficult to array and cascade a large number of analog building blocks and deal with the accumulation of noise and component mismatch. Additionally, and perhaps more significantly, it is challenging to build the required analog memory cells [67]. For this reason, present explorations in neuromorphic design are dominated by digital emulations, such as IBM's TrueNorth processor [68]. A more detailed discussion of such efforts is found in Chap. 22 of this book.

Since purely analog implementations are difficult to scale, could one instead assemble a processor that uses purely digital storage and adds in analog/mixed-signal compute for potential efficiency gains? As shown in [69], mixed-signal computing can indeed be lower energy than digital for low resolutions, typically below 8 bits. The most straightforward way to exploit this would be to embed mixed-signal compute macros into the PE blocks of a mainly digital processor. This was considered in [70, 71], which point to the conclusion that the idea will in practice lead to diminishing returns. In an optimized digital design that conforms to the template of Fig. 18.11, most of the energy is spent on memory access and data movement [72] making even large improvements in the arithmetic units nearly irrelevant. To fully harvest the benefits of mixed-signal processing, one must consider customized architectures. One possible direction is to employ analog and mixed-signal circuits as feature extractors that are placed in front of a digital neural network. This approach is discussed further in Chap. 17 of this book. Another opportunity is to exploit mixed-signal circuits through memory-like processing elements and in-memory computing, which we discuss next.

Memory-like processing elements: Is it possible to re-architect a digital ML processor architecture to benefit more strongly from a mixed-signal compute fabric? This question was the baseline for the research described in [73], which exercises two of the re-use principles stated in Table 18.1 with a mixed-signal mindset: Intra-PE temporal re-use of weights (weight stationarity) and Inter-PE accumulation. The main observation here was that the latter can be done in a particularly efficient way using charge sharing on a wire, instead of a digital accumulation tree. The resulting switched-capacitor PE cell is shown in Fig. 18.15. The overall network that was designed to use this PE is based on the BinaryNet topology from [29], which makes multiplication trivial (XNOR). This enabled a cell size that allowed the on-chip integration of a 64 $\times$ 1024 PE array that computes 64 output activations in one shot. We term this approach "memory-like" since the PE locally stores one bit and otherwise contains only simple add-on-circuits.

Figure 18.16 compares the total neuron energy of a custom digital design with the described mixed-signal approach. The latter shows an improvement of about 4.2×. However, when accounting for other energy consumers (including weight

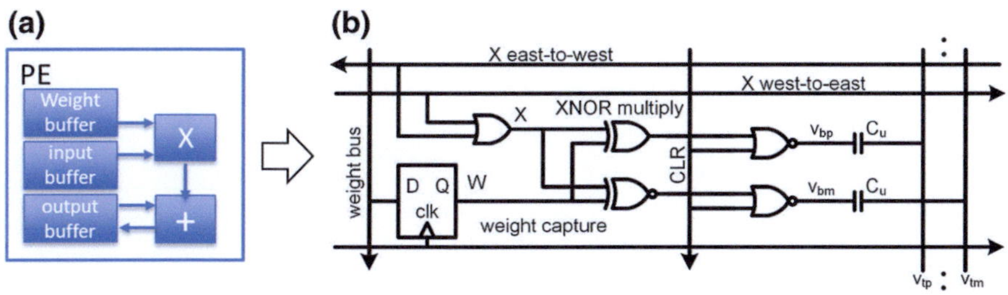

Fig. 18.15 **a** Conventional processing element (PE) versus **b** memory-like mixed-signal PE (single-bit implementation, from [73] ©IEEE 2019)

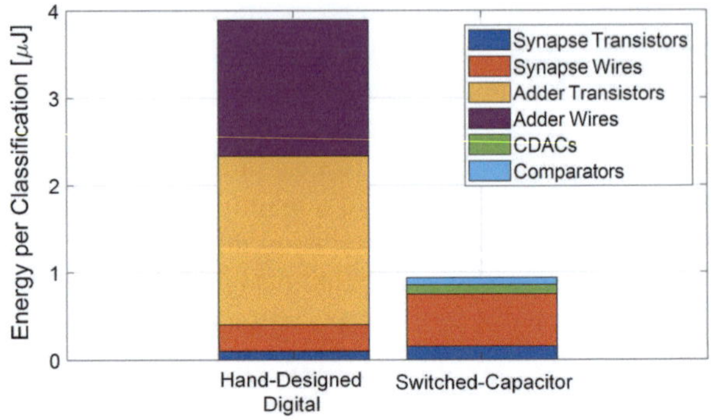

Fig. 18.16 Comparison of total neuron energy (digital versus mixed signal) (from [73] ©IEEE 2019)

and activation memory access), the system-level savings reduce to approximately 1.8×. While this benefit is still significant, this exercise makes it clear that order-of magnitude improvements are hard to come by, unless an even more radical approach is pursued. This brings us to the topic of in-memory computing, which aims to minimize the overhead that diminished the returns from the mixed-signal compute fabric in the example above.

In-Memory Computing: In-memory computing is a relatively old idea [74] that aims to co-integrate memory and compute into a single dense fabric. Conceptually, one could view the memory-like PE in Fig. 18.15 as a compute-in-memory cell. However, its size is relatively large, so that a denser piece of memory is required in its periphery to store the weights and activations of a modern neural network. To overcome this issue, denser cells can be designed as illustrated in Fig. 18.17. The most obvious way to increase density is to handle the memory bit with a standard 6-T SRAM cell (see Fig. 18.17b) as done in [75]. In addition, the logic can be simplified and single-ended signaling can be explored to further reduce the area. While the differential memory-like cell measures 24,000 F^2 (where F is the half pitch the process technology), the SRAM-based cell has an area of only 290 F^2. Further cell size reductions are possible by migrating to emerging memory technologies (see Fig. 18.17c), as discussed in the next sub-section.

At present, SRAM-based in-memory computing is receiving significant attention in the research community [76] and many circuit and network architecture options are being explored. In [77], a complete processor with in-memory compute acceleration

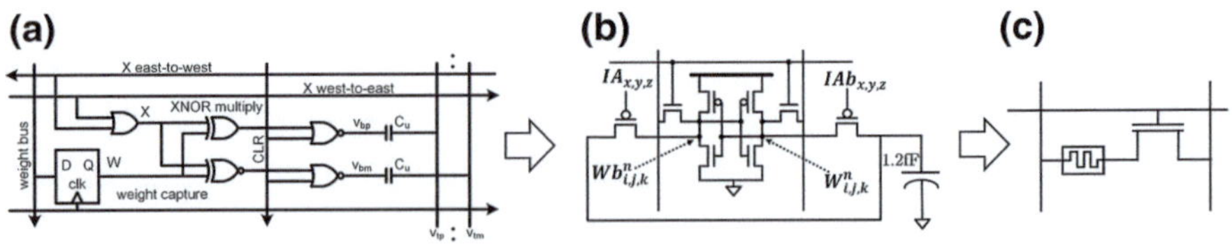

Fig. 18.17 **a** Memory-like PE (from [73] ©IEEE 2019), **b** in-memory computing cell based on SRAM (from [75] ©IEEE 2019), **c** in-memory computing cell based on resistive memory technology

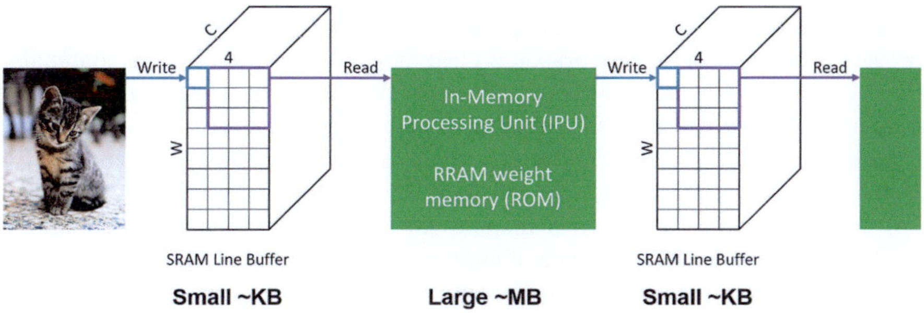

Fig. 18.18 Streaming architecture for neural network processing with emerging memory

is presented. While this design achieves high efficiency within its compute tiles, the overall system efficiency is held back by memory reads from external DRAM, which is typically required for models that exceed several megabytes in size. A promising remedy for this issue lies in embracing emerging memory technologies for in-memory compute.

Emerging Memory: A wide variety of emerging memory technologies are currently under investigation (see Chap. 19 of this book). For instance, Resistive Random Access Memory (RRAM) technology promises to deliver densities that are comparable to DRAM, while being non-volatile and potentially offer multi-level storage. This could open up a future where relatively large machine learning models (>10 MB) can be stored on a single chip to eliminate costly DRAM access. In addition, these memory types are compatible with in-memory-computing by exploiting current summation on the bitlines [78]. While there are many possible ways to incorporate emerging nonvolatile memory into a machine learning processor [79], one attractive option is a streaming topology as shown in Fig. 18.18. Here, large in-memory compute tiles are pipelined between small SRAM line buffers that hold only the current input working set [80]. This scheme can thereby avoid the energy penalty of reading from large SRAMs, which represents a significant energy overhead in the above-discussed processor with memory-like PEs.

Presently, the art of designing of machine learning processors using emerging memory is still in its infancy. Key issues include access to process technology as well as challenges with the relatively poor retention and endurance of emerging memory technologies (see e.g., [81]). Consequently, most existing demonstrators are only sub-systems and use relatively small arrays (see e.g., [82]). However, one important aspect that has already become clear from these investigations is that the D/A and A/D interfaces required at the array boundaries can be a significant showstopper. For example, a state-of-the-art ADC consumes about 1 pJ per conversion at approximately 4–8 bits of resolution [83]. If amortized across 100 memory rows, the energy overhead is 10 fJ per MAC operation, a number that is close to a relatively straightforward digital MAC implementation in 16 nm CMOS [84]. The solution is to work with taller arrays and to push for innovations in the interface and array circuit design (see e.g. [85]), which can lead us to single-digit fJ per MAC.

3D Integration: Given the above-discussed problems of data movement and memory access in large neural networks, it is clear that 3D integration has the potential to play a major role in making NPUs significantly more efficient. The reader is referred to an in-depth discussion of this subject in Chaps. 9 and 10.

Challenges and outlook: While using analog and mixed-signal computing in neural networks is attractive in principle, it is not straightforward to realize large performance gains (e.g., order of magnitude) at the system level. This is simply because a complete NPU has many components and improving only a subset leads to diminishing returns. At present, the most promising option is to pursue mixed-signal processing within in-memory compute tiles and to rely on standard digital processing on the outside. Future work must assess how flexible and programmable such a processor can be, and how much efficiency it may lose due to data sparsity, which can presumably be better managed with a fully digital fabric. Just as with fully digital NPUs, the research on alternative architectures must be guided by a solid system-level benchmarking strategy that will systematically uncover such efficiency losses during the conception of the architecture. The next section therefore looks at this particular aspect.

18.5 Evaluating ML Processors

As discussed in the previous sections, over the past decades, hundreds of custom NPU processing schemes, architectures and technological enhancements have been proposed. It is good practice to benchmark the different solutions relative to one another, and identify which innovations bring actual value. Yet, the main challenge is to determine the right benchmarking metrics.

System-level benchmarks: The only metrics that really matter to an edge device user are: (1) the energy per inference; (2) the latency or throughput per inference; and (3) the cost per inference (determined by chip area and external memory size). To be able to compare different systems, these must be compared on a known standardized benchmarking task, achieving a given target accuracy. Recently, there has been a lot of effort from the MLperf community [86] to pull off such benchmarking. While the current focus is mostly on training tasks in the cloud, it is expanding towards inference benchmarks, also for the edge.

These system-level benchmarks can be improved through different algorithmic, architecture and circuit level techniques. Designers working at these levels tend to use benchmarking metrics focusing at lower level aspects, for instance:

- The number of MACs/inference or model coefficients at *algorithmic* level
- The number of MACs/second or the number of MACs/Watt at *architectural* level
- The number picojoules per memory fetch of per MAC at the *circuit/technology* level.

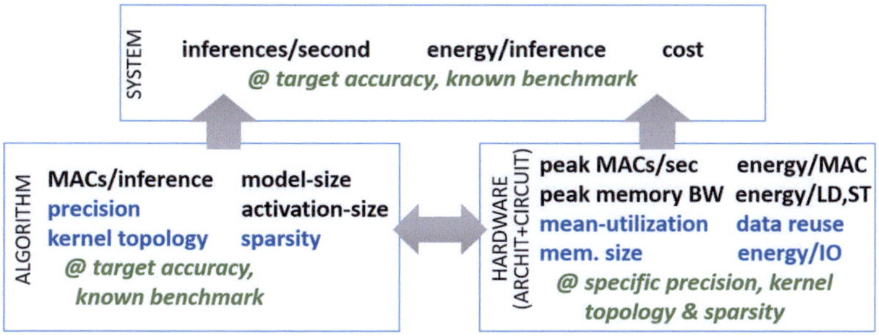

Fig. 18.19 Typical benchmarking metrics at system level, algorithmic level and hardware level (in black), complemented with performance-influencing metrics (blue) and constraints (green) that are often forgotten

Figure 18.19 summarizes some frequently used benchmarks (in black) at these different levels. The figure also highlights important parameters (in blue), and constraints (in green) that are often forgotten at these different levels. It is of crucial importance to see that benchmarks at different levels strongly depend on each other and are often conflicting. For example, one can achieve a very low number of model weights by going to high precision, highly sparse model kernels. Yet, at the hardware level, this will result in very low MAC utilization and high energy per MAC. Similarly, good hardware benchmarks can be achieved by going to very low precision, and highly regular large in-memory compute arrays. Yet, this will result in models that are requiring more MACs and larger model sizes to achieve the same benchmarking accuracy [87].

CIFAR10 example of cross-layer implications: To illustrate this, we compare different solutions for the CIFAR10 benchmark. Table 18.3 shows benchmarking performance across different design levels for three different solutions:

- A high accuracy, 4-bit algorithm running on the Envision chip [45]
- A medium accuracy 4-bit model running on the Envision chip [45]
- A medium accuracy 1-bit model running on the BinarEye chip [88].

It is interesting to observe that at the hardware level, the BinarEye chip [88] seems to beat all performance metrics, showing highest peak performance, at best energy efficiency and with most embedded memory available. At algorithmic level, however, the network capable of execution on the Envision platform show to require less MACs and exhibit more sparsity. However, as their topology cannot be perfectly mapped to the flexible Envision datapath, it cannot achieve maximum utilization of the processor. The network trained for BinarEye on the other hand, was matched to the datapath to achieve 100% utilization. The result of this trade-off shows that for equal accuracy, two solutions consume roughly the same amount of energy to run one CIFAR10 inference. At the system level, also taking external memory accesses into account, BinarEye wins due to the larger embedded memory of BinarEye and the smaller model size of the mapped CIFAR10 model. The table also clearly shows that large energy savings can be achieved if one wants to give in a bit of task accuracy, e.g.

Table 18.3 Benchmarking performance of CIFAR10 task across platforms. Numbers extrapolated from measurements

Platform and task	System level			Algorithm level					Hardware level			
	Inference/s	System energy/inf (uJ)	Chip energy/inf (uJ)	Kernel topology	Precision (bit)	MAC/inference	Sparsity (%)	Model size (MB)	TMAC/Watt	Peak GMAC/s	Mean utilization (%)	Mem size (kB)
Envision, CIFAR10, 90%	90	400	230	$3 \times 3 \times C$	4	6.00E + 08	30–60	12	2.6	102	53	144
Envision, CIFAR10, 86%	1350	25	15	$3 \times 3 \times C$	4	4.00E + 07	30–60	1.2	2.6	102	53	144
BinarEye, CIFAR10, 86%	120	14	13	$2 \times 2 \times C$	1	2.00E + 09	0	0.259	115	2800	100	328

comparing the system level benchmarks for CIFAR10 90% and 86% in Table 18.3. It is hence of crucial importance to always compare data points achieving similar accuracies on known benchmarks to be able to make a fair comparison.

Challenges and outlook: The previous example should make it clear that it is impossible to judge hardware platforms, resp. algorithmic innovations based on hardware-centric, resp. algorithm-centric performance metrics. There is a very strong influence between design decision across different layers. The challenge is hence on being able to report system-level benchmarking improvements for newly proposed algorithmic or hardware innovations, without having to go through the complete optimization across all layers every time. This requires a new set of cross-layer tools and frameworks, as discussed in Sect. 18.6.

18.6 Cross Domain Optimizations, Mapping and Deployment Frameworks

Neural network mapping: When one wants to assess the performance of a specific neural network model on a specific hardware topology, it is necessary to schedule the model's execution on consecutive processing cycles using a mapping supported by the platform. Only the detailed schedule reveals how many data transfers are needed to execute the algorithm, and which layers of the memory hierarchy are involved. For a specific neural network layer, such scheduling starts from the layer's six nested loops, shown in Fig. 18.5. These nested for-loops can be manipulated using loop splitting and loop reordering, denoted as dataflow transformations [89, 90]. Finally, each resulting for-loop should be characterized as a spatial or temporal enrolled loop (in line with what the hardware supported), and its internal data variables should be allocated to a specific level in the hardware's memory hierarchy. Figure 18.20 illustrates this operation for an algorithm mapped on the Envision processor, which supports two-dimensional parallelism along the X and K dimension.

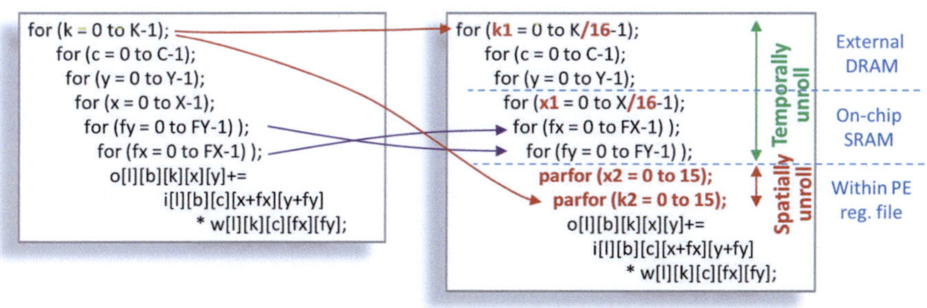

Fig. 18.20 Data transformation and hardware mapping example, targeting the Envision hardware configuration

Yet, this specific set of dataflow transformations is not the only possible option. Many possible loop orderings, loop splitting and loop unrolling options could have been exercised to map the specific network layer on the hardware platform. For realistic networks, there can easily be millions of different valid solutions. While all these mappings would be functionally identical, their resulting system level performance and efficiency benchmark won't be. The challenge is hence to try all possible dataflow transformation, quickly assess their impact at the system level, and pick the best one. It is needless to say that this cannot be done by hand, and automated frameworks are required to support such mapping.

At the moment, several frameworks start to emerge to automate such explorations [90–93]. As shown in Fig. 18.21 (top), these frameworks typically take in a neural network layer representation, together with the constraints imposed by the hardware platform. Based on this information, they are capable of efficiently finding all functionally equivalent data transformations supported by the hardware, and computing the resulting number of compute cycles, and memory accesses required within the platform. This information can then be fed to a high-level processor performance model to find the resulting system level performance of the selected mapping. By repeating this for all possible mappings, the framework can derive a Pareto-optimal set of algorithmic mappings or find the best mapping subject to an application level constraint, such as maximum latency. The selected mapping can subsequently be compiled into micro code to be executed on the platform, as for example integrated in the TVM framework [92].

Challenges and outlook: While these frameworks start to emerge, they are still immature, and many challenges remain. Next, the three most critical challenges are discussed: (1) cross-layer mappings; (2) model-HW co-optimization; (3) exploration space bounding.

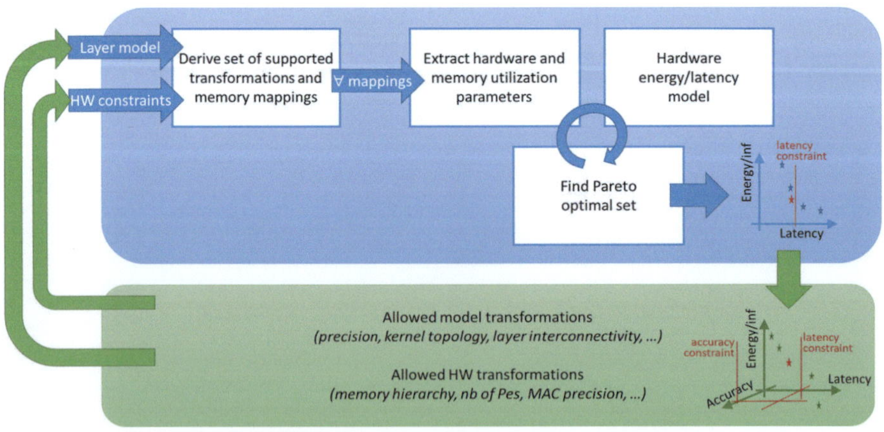

Fig. 18.21 Automated mapping and performance estimation framework assuming a given network model and HW configuration (top), which can be extended with automated neural network model search and optimal hardware topology search (bottom)

1. *Cross-layer mappings*: Current frameworks focus on mapping and scheduling a single neural network layer. This, however, limits the degrees of freedom the mapper has, and excludes interesting solutions such as depth-first network execution, which iterates across layers before executing all tiles of a specific layer [94]. Yet, it is hard to include this into the exploration framework, as it blows up the exploration space.
2. *Model-HW co-optimization*: The framework discussed earlier (Fig. 18.21 (top)) assumes a given network topology, and hardware constellation. Yet, during the design phase, the designer can modify the neural network model, and its computational precision. As shown earlier, many different neural networks can be constructed for the same task achieving the same task accuracy, yet with widely varying hardware mapping consequences. Exploring neural network models, with the hardware mapping tradeoffs in the loop allows to find the optimal neural network topology given the system level benchmarks, instead of just the best algorithmic level benchmarks. This is partially pursued in studies such as MNASnet [23], and the minimum energy QNN study [87], yet still with very crude energy models. Truly integrating this with more realistic hardware models will undoubtedly bring more breakthroughs in the near future.

   When the hardware platform is not decided yet, or the target chip has not yet been taped out, also the hardware configuration can be modified in this iterative exploration loop. As such, the best hardware-model-mapping combination can be found to serve a given task within its application constraints. This is pursued in the Maestro framework [93] and the EyerissV2 studies [41]. Of course, these additional exploration options again increase the search space drastically.
3. *Exploration space bounding*: All aforementioned improvements of the automated exploration, mapping and compilation framework result in yet another increase of the possible exploration space. When only looking at hardware configuration modifications, while keeping the model fixed, the Maestro framework already has to assess millions of design points. On the other hand, also many millions of options have to be searched when only assessing model transformations without considering hardware modifications. It is clear that exhaustively searching this complete design space is simply infeasible. Research towards smart sampling techniques, exploiting Bayesian optimization, or reinforcement learning have been successfully applied to model explorations. It is expected that in the near future they will also start to be successfully applied on joint hardware-model-mapping optimizations. This will undoubtedly give rise to an even more interesting interplay in which novel processor architectures fuel these new dataflow mappings and models, which in turn lead to new processing paradigms.

18.7 Outlook: Towards True Autonomous Intelligence

Looking further out into the future, edge devices will increasingly evolve into truly autonomous intelligent devices: Devices which can not only execute a pre-trained inference model, but can also increase their own knowledge, can reason, and synergistically collaborate with other devices.

Learning at the edge: Several application use cases envision the edge devices to be more than a pure inference engine. The next step is to make the edge NPU also capable of performing update learning on the deployed network model. This capability would allow the edge device to learn for example a user-customized speech interface that works better and better the more it is being used by a specific person. Or, an anomaly detector would be able to use this online training capability to better distinguish anomalies within its specific environment. Many challenges are related to online, in-device learning:

- *At the algorithmic level*, researchers are exploring learning methods that prevent the network to forget previously acquired knowledge [95]. Moreover, researchers are actively exploring whether learning can also be done without the need for full floating-point data types and compute intensive backpropagation steps, e.g. using techniques such as direct feedback alignment [96, 97].
- *At the hardware level*, the support for edge training will require the additional support for higher precision data types within the NPU, and higher precision weight storage. Since the weight matrices have to be read out in transposed form during back-propagation, several recent designs are experimenting with transpose memories, which can be efficiently read out in in a column-parallel manner as well as in a row-parallel scheme [98, 99].
- *At the circuit level*, researchers are looking at ways to embrace emerging resistive memory cells for in-device learning [100]. One direction is to perform standard memory R/W access and to minimize writes to overcome hard endurance limits [101]. Another approach that makes more direct use of the device's physics and treat each device as a "nanokernel" with local feedback during training [102].

Reasoning: Neural networks have shown excellent results in pattern matching and regression tasks, yet they are insufficient towards achieving all intelligence needs of our envisioned future autonomous edge devices. Their main shortcomings are their lack of explainability, their difficulty to integrate expert knowledge or constraints and their inability to support probabilistic reasoning tasks. Other machine learning models, such as Bayesian reasoning, logic reasoning and probabilistic graphical models (PGM, [103]) do possess these features, but come with their own shortcomings, such as their high dataflow irregularity, their inability to efficiently deal with raw data and long training times. Yet, more and more it becomes clear that these two machine learning formalisms form an interesting tandem, in which neural networks can be used as pattern matching layers operating on raw sensor data. The network outputs are then forwarded to reasoning layers on top, which based on these observations make complex decisions in a transparent way. On the algorithmic side, researchers have started to actively explore this using for example Logic Tensor Network models [104], Bayesian Deep Learning models [105] and frameworks such as deep problog [106]. On the hardware side, more challenges are also coming, as the reasoning models are characterized by very different dataflow patterns compared to neural networks, which do not execute efficiently on an NPU, nor CPU or GPU. A new type of processor might yet again have to be invented [107].

Synergistic collaboration: Finally, edge devices are equipped with wireless connections, and hence do not have to operate in isolation. They can exchange data and models among each other, and as such smartly collaborate to perform training and inference on the most suited device at that moment. This will again increase the mapping exploration space discussed in Sect. 18.6 and will now also require incorporating latency and energy complications of sharing data between devices into account into the system cost models. Interestingly, the optimal assignment can change dynamically over time depending on each device's energy availability and current workload, giving rise to real-time scheduling and optimization opportunities. From the hardware side, this will spark an exciting integration of machine learning processors and security hardware, as all models and data that will be exchanged are privacy- and authentication-sensitive.

18.8 Conclusions

Innovations towards more efficient processing of machine learning workloads in edge devices are arriving at a high pace, mostly focused around neural network-based inference. Breakthroughs are realized at the algorithmic level, hardware level and circuit/technology level. Yet, it also becomes increasingly clear that innovations at one level have significant implications at the other levels. As a result, benchmarking initiatives push for system level benchmarks, which jointly consider all levels in an integrated way. To further optimize the complete system stack, integrated frameworks enable to find the most efficient mapping of a neural network model on a given hardware platform. Even one step further, these frameworks can be used to actively explore the algorithmic and hardware design space towards optimal algorithm-hardware co-design. Moreover, new emerging technology options will give rise to very different processor and memory configuration options, and hence new classes of optimal model topologies.

Many challenges remain to effectively enable such cross-layer optimization that covers the complete exploration space, and integrate this in an automated model development, model mapping, and compilation framework. Moreover, workloads will in the future no longer be limited to plain neural network inference but will be expanded with on device learning and the integration with logic and probabilistic reasoning. This will undoubtedly give rise to many more exciting innovations at the algorithmic, architecture and circuit levels.

References

1. A. Al-Fuqaha, M. Guizani, M. Mohammadi, M. Aledhari, M. Ayyash, Internet of things: a survey on enabling technologies, protocols, and applications. IEEE Commun. Surv. Tutorials **17**(4), 2347–2376 (2015)
2. M. Satyanarayanan, P. Simoens, Y. Xiao, P. Pillai, Z. Chen, K. Ha, W. Hu, B. Amos, Edge analytics in the internet of things. IEEE Pervasive Comput. **14**(2), 24–31 (2015)
3. H. Li, K. Ota, M. Dong, Learning IoT in edge: deep learning for the Internet of Things with edge computing. IEEE Netw. **32**(1), 96–101 (2018)
4. A. Canziani, A. Paszke, E. Culurciello, An Analysis of Deep Neural Network Models for Practical Applications. arXiv preprint arXiv:1605.07
5. Caulfield, A.M., Chung, E.S., Putnam, A., Angepat, H., Fowers, J., Haselman, M., Lo, D. et al., A cloud-scale acceleration architecture, in *The 49th Annual IEEE/ACM International Symposium on Microarchitecture* (IEEE Press, 2016), p. 7
6. N. Strom, Scalable distributed DNN training using commodity GPU cloud computing, in *Sixteenth Annual Conference of the International Speech Communication Association* (2015)
7. Tractica report, *Deep Learning Chipsets* (2018). https://www.tractica.com/research/deep-learning-chipsets/
8. Semiconductor Engineering, *AI Chip Architectures Race To The Edge* (2018). https://semiengineering.com/ai-chip-architectures-race-to-the-edge/
9. K. Guo, W. Li, K. Zhong, Z. Zhu, S. Zeng, S. Han, Y. Xie, P. Debacker, M. Verhelst, Y. Wang, Neural Network Accelerator Comparison. [Online]. https://nicsefc.ee.tsinghua.edu.cn/projects/neural-network-accelerator/
10. I. Goodfellow, Y. Bengio, A. Courville, *Deep Learning* (MIT press, Cambridge, 2016)
11. C. Szegedy, V. Vanhoucke, S. Ioffe, J. Shlens, Z. Wojna, Rethinking the inception architecture for computer vision (2015). arXiv preprint arXiv:1512.00567
12. F. Chollet, *Xception: Deep Learning with Depthwise Separable Convolutions* (2016). arXiv preprint arXiv:1610.02357
13. A. Howard et al., MobileNets: Efficient Convolutional Neural Networks for Mobile Vision Applications (2017). arXiv:1704.04861
14. M. Sandler et al., *MobileNetV2: Inverted Residuals and Linear Bottlenecks*. arXiv:1801.04381
15. C. Liu, B. Zoph, M. Neumann, J. Shlens, W. Hua, L.-J. Li, L. Fei-Fei, A. Yuille, J. Huang, K. Murphy, Progressive neural architecture search, in *ECCV2018*
16. E. Real, A. Aggarwal, Y. Huang, Q.V. Le, Regularized evolution for image classifier architecture search, in *The Thirty-Third AAAI Conference on Artificial Intelligence* (2019)
17. S. Xie, A. Kirillov, R. Girshick, K. He, *Exploring Randomly Wired Neural Networks for Image Recognition* (2019). arXiv:1904.01569
18. X. Chu, B. Zhang, J. Li, Q. Li, R. Xu, ScarletNAS: Bridging the Gap Between Scalability and Fairness in Neural Architecture Search (2019). arXiv:1908.06022
19. X. Zhang, Z. Li, C. Change Loy, D. Lin, *PolyNet: A Pursuit of Structural Diversity in Very Deep Networks* (2019). arXiv:1611.05725
20. Google's AutoML, https://research.googleblog.com/2017/11/automl-for-large-scale-image.html?m=1
21. Q. Yao et al., Taking the Human out of Learning Applications: A Survey on Automated Machine Learning. arXiv: 1810.13306
22. Y. He, J. Lin, Z. Liu, H. Wang, L.J. Li, S. Han, Amc: Automl for model compression and acceleration on mobile devices, in *Proceedings of the European Conference on Computer Vision (ECCV)* (2018), pp. 784–800
23. M. Tan, B. Chen, R. Pang, V. Vasudevan, Q.V. Le, Mnasnet: Platform-aware neural architecture search for mobile (2018). arXiv preprint arXiv:1807.11626
24. T.-J. Yang, et al., Netadapt: platform-aware neural network adaptation for mobile applications, in *ECCV* (2018)
25. J.A. Suykens, J. Vandewalle, Least squares support vector machine classifiers. Neural Process. Lett. **9**(3), 293–300 (1999)

26. S. Gupta, A. Agrawal, K. Gopalakrishnan, P. Narayanan, Deep learning with limited numerical precision, in *CoRR*, vol. abs/1502.02551 (2015)
27. M. Courbariaux, Y. Bengio, J.-P. David, Training deep neural networks with low precision multiplications (2014). arXiv preprint arXiv:1412.7024
28. R. Krishnamoorthi, Quantizing deep convolutional networks for efficient inference: a whitepaper. arXiv:1806.08342
29. M. Courbariaux, I. Hubara, D. Soudry, R. El-Yaniv, Y. Bengio, Binarized neural networks: training deep neural networks with weights and activations constrained to $+1$ or -1 (2016). arXiv preprint arXiv:1602.02830
30. N. Mellempudi, A. Kundu, D. Das, D. Mudigere, B. Kaul, Mixed low-precision deep learning inference using dynamic fixed point (2017). arXiv preprint arXiv:1701.08978
31. I. Hubara et al., Quantized neural networks: training neural networks with low precision weights and activations. ArXiv1609.07061
32. B. Jacob et al., Quantization and training of neural networks for efficient integer-arithmetic-only inference, in *CVPR* (2018)
33. M. Nagel, M. van Baalen, T. Blankevoort, M. Welling, Data-free quantization (DFQ) through weight equalization and bias correction (2019). arXiv:1906.04721v1
34. E. Meller, A. Finkelstein, U. Almog, M. Grobman, Same, same but different—recovering neural network quantization error through weight factorization (2019). arxiv:1902.01917
35. B. Moons, K. Goetschalckx, N. Van Berckelaer, M. Verhelst, Minimum energy quantized neural networks (2017). arXiv preprint arXiv:1711.00215
36. S. Han, J. Pool, J. Tran, W. Dally, Learning both weights and connections for efficient neural network, in *Advances in Neural Information Processing Systems* (2015), pp. 1135–1143
37. J. Xue, J. Li, Y. Gong, Restructuring of deep neural network acoustic models with singular value decomposition, in *INTERSPEECH* (2013)
38. T.-J. Yang, Y.-H. Chen, V. Sze, Designing energy-efficient convolutional neural networks using energy-aware pruning, in *The IEEE Conference on Computer Vision and Pattern Recognition (CVPR)* (2017)
39. W. Wei, Learning structured sparsity in deep neural networks, in *NIPS2016*
40. S. Han, X. Liu, H. Mao, J. Pu, A. Pedram, M.A. Horowitz, W.J. Dally, EIE: efficient inference engine on compressed deep neural network (2016). arXiv preprint arXiv:1602.01528
41. Y.-H. Chen et al., Eyeriss v2: a flexible accelerator for emerging deep neural networks on mobile devices, in *JETCAS* (2019)
42. Y. Liu, Y. Wang, R. Yu, M. Li, V. Sharma, Y. Wang, Optimizing CNN model inference on CPUs (2018). arXiv: 1809.02697
43. S. Markidis, S.W. Der Chien, E. Laure, I.B. Peng, J.S. Vetter, Nvidia tensor core programmability, performance & precision, in *2018 IEEE International Parallel and Distributed Processing Symposium Workshops (IPDPSW)* (IEEE, May 2018), pp. 522–531
44. B. Moons, D. Bankman, M. Verhelst, *Embedded Deep Learning: Algorithms, Architectures and Circuits for Always-on Neural Network Processing* (Springer, 2019). ISBN 978-3-319-99223-5
45. B. Moons, R. Uytterhoeven, W. Dehaene, M. Verhelst, Envision: A 0.26-to-10tops/w subword-parallel dynamic-voltage-accuracy-frequency-scalable convolutional neural network processor in 28 nm fdsoi, in *2017 IEEE International Solid-State Circuits Conference (ISSCC)* (IEEE, 2017), pp. 246–247
46. N.P. Jouppi, C. Young, N. Patil, D. Patterson, G. Agrawal, R. Bajwa, S. Bates, S. Bhatia, N. Boden, A. Borchers, R. Boyle, In-datacenter performance analysis of a tensor processing unit, in *2017 ACM/IEEE 44th Annual International Symposium on Computer Architecture (ISCA)* (IEEE), June 2017, pp. 1–12
47. B. Moons, D. Bankman, L. Yang, B. Murmann, M. Verhelst, BinarEye: an always-on energy-accuracy-scalable binary CNN processor with all memory on chip in 28 nm CMOS, in *IEEE Custom Integrated Circuits Conference (CICC)* (2018), pp. 1–4
48. Y.H. Chen, T. Krishna, J.S. Emer, V. Sze, Eyeriss: an energy-efficient reconfigurable accelerator for deep convolutional neural networks. IEEE J. Solid-State Circ. **52**(1), 127–138 (2016)

49. G. Desoli, N. Chawla, T. Boesch, S. Singh, E. Guidetti, F. De Ambroggi, T. Majo, P. Zambotti, M. Ayodhyawasi, H. Singh, N. Aggarwal, A 2.9 TOPS/W deep convolutional neural network SoC in FD-SOI 28 nm for intelligent embedded systems
50. A. Parashar, M. Rhu, A. Mukkara, A. Puglielli, R. Venkatesan, B. Khailany, J. Emer, S.W. Keckler, W.J. Dally, SCNN: an accelerator for compressed-sparse convolutional neural networks, in *Proceedings of ISCA '17*, Toronto, ON, Canada, 24–28 June 2017
51. M. Nikolić, M. Mahmoud, A. Moshovos, Y. Zhao, R. Mullins, Characterizing sources of ineffectual computations in deep learning networks, in *2019 IEEE International Symposium on Performance Analysis of Systems and Software (ISPASS)* (IEEE, 2019), pp. 165–176
52. V. Camus, C. Enz, M. Verhelst, Survey of precision-scalable multiply-accumulate units for neural-network processing, in *2019 IEEE 1st International Conference on Artificial Intelligence Circuits and Systems (AICAS)*, Mar 2019
53. S. Cosemans, Advanced memory, logic and 3D technologies for in-memory computing and machine learning, in *ISSCC2019 Forum Talk*
54. L. Mei, M. Dandekar, D. Rodopoulos, J. Constantin, P. Debacker, R. Lauwereins, M. Verhelst, Sub-word parallel precision-scalable MAC engines for efficient embedded DNN inference, in *2019 IEEE International Conference on Artificial Intelligence Circuits and Systems (AICAS)* (IEEE, 2019), pp. 6–10
55. L. Mei, V. Camus, C. Enz, M. Verhelst, Review and benchmarking of precision-scalable multiply-accumulate unit architectures for embedded neural-network processing, in *2020 IEEE Journal on Emerging and Selected Topics in Circuits and Systems* (2020)
56. D. Shin, J. Lee, J. Lee, H.-J. Yoo, *DNPU: An 8.1 TOPS/W Reconfigurable CNN-RNN Processor for General-Purpose Deep Neural Networks*
57. B. Moons, R. Uytterhoeven, W. Dehaene, M. Verhelst, DVAFS: trading computational accuracy for energy through dynamic-voltage-accuracy-frequency-scaling, in *Design, Automation & Test in Europe Conference & Exhibition (DATE)* (IEEE, 2017), pp. 488–493
58. Sharma et al., BitFusion: bit-level dynamically composable architecture for accelerating deep neural networks, in *ISCA18*
59. J. Lee, C. Kim, S. Kang, D. Shin, S. Kim, H.J. Yoo, UNPU: A 50.6TOPS/W unified deep neural network accelerator with 1b-to-16b fully-variable weight bit-precision, in *2018 IEEE International Solid-State Circuits Conference (ISSCC)* (2018), pp. 218–220
60. S. Sharifymoghaddam et al., Loom: exploiting weight and activation precisions to accelerate convolutional neural networks, in *DAC Conference* (2018)
61. L. Liu, J. Deng, *Dynamic Deep Neural Networks: Optimizing Accuracy-Efficiency Trade-offs by Selective Execution* (2017). arXiv preprint arXiv:1701.00299
62. A. Coucke, M. Chlieh, T. Gisselbrecht, D. Leroy, M. Poumeyrol, T. Lavril, Efficient keyword spotting using dilated convolutions and gating, in *ICASSP 2019–2019 IEEE International Conference on Acoustics, Speech and Signal Processing (ICASSP)* (IEEE, 2019), pp. 6351–6355
63. Z. Yan, X. Li, M. Li, W. Zuo, S. Shan, Shift-net: image inpainting via deep feature rearrangement, in *Proceedings of the European Conference on Computer Vision (ECCV)* (2018), pp. 1–17
64. A.V.D. Oord, S. Dieleman, H. Zen, K. Simonyan, O. Vinyals, A. Graves, N. Kalchbrenner, A. Senior, K. Kavukcuoglu, Wavenet: a generative model for raw audio (2016). arXiv preprint arXiv:1609.03499
65. R.A. Nawrocki, R.M. Voyles, S.E. Shaheen, A mini review of neuromorphic architectures and implementations. IEEE Trans. Electron Devices **63**(10), 3819–3829 (2016)
66. C. Mead, Neuromorphic electronic systems. Proc. IEEE **78**(10), 1629–1636 (1990)
67. E.A. Vittoz, Future of analog in the VLSI environment, in *IEEE International Symposium on Circuits and Systems* (1990), pp. 1372–1375
68. P.A. Merolla, J.V. Arthur, R. Alvarez-Icaza, A.S. Cassidy, J. Sawada, F. Akopyan, B.L. Jackson, N. Imam, C. Guo, Y. Nakamura, B. Brezzo, I. Vo, S.K. Esser, R. Appuswamy, B. Taba, A. Amir, M.D. Flickner, W.P. Risk, R. Manohar, D.S. Modha, A million spiking-neuron integrated circuit with a scalable communication network and interface. Science **345**(6197), 668–673 (2014)

69. B. Murmann, D. Bankman, E. Chai, D. Miyashita, L. Yang, Mixed-signal circuits for embedded machine-learning applications, in *Asilomar Conference on Signals, Systems and Computers* (Nov 2015), Asilomar, CA
70. D. Bankman, B. Murmann, An 8-bit, 16 input, 3.2 pJ/op switched-capacitor dot product circuit in 28-nm FDSOI CMOS, in *Proceedings of IEEE Asian Solid-State Circuits Conference* (Nov 2016), Toyama, Japan, pp. 21–24
71. A.S. Rekhi, B. Zimmer, N. Nedovic, N. Liu, R. Venkatesan, M. Wang, B. Khailany, W.J. Dally, C.T. Gray, Analog/mixed-signal hardware error modeling for deep learning inference, in *Proceedings of Design Automation Conference* (2019), pp. 1–6
72. V. Sze, Y. Chen, J. Emer, A. Suleiman, Z. Zhang, Hardware for machine learning: challenges and opportunities,in *IEEE Custom Integrated Circuits Conference (CICC)* (2017), Austin, TX, pp. 1–8
73. D. Bankman, L. Yang, B. Moons, M. Verhelst, B. Murmann, An always-on 3.8 uJ/86% CIFAR-10 mixed-signal binary CNN processor with all memory on chip in 28-nm CMOS. IEEE J. Solid-State Circ. **54**(1), 158–172 (2019)
74. W.H. Kautz, Cellular logic-in-memory arrays. IEEE Trans. Comput. **C-18**(8), 719–727 (1969)
75. H. Valavi, P.J. Ramadge, E. Nestler, N. Verma, A 64-tile 2.4-Mb in-memory-computing CNN accelerator employing charge-domain compute. IEEE J. Solid-State Circ. **54**(6), 1789–1799 (2019)
76. N. Verma et al., In-memory computing: advances and prospects. IEEE Solid-State Circ. Mag. **11**(3), 43–55 (2019)
77. H. Jia, Y. Tang, H. Valavi, J. Zhang, N. Verma, A microprocessor implemented in 65 nm CMOS with configurable and bit-scalable accelerator for programmable in-memory computing (2018). arXiv preprint, arXiv:1811.04047
78. H. Tsai, S. Ambrogio, P. Narayanan, R.M. Shelby, G.W. Burr, Recent progress in analog memory-based accelerators for deep learning. J. Phys. D Appl. Phys. **51**(28), 283001 (2018)
79. S. Mittal, A survey of ReRAM-based architectures for processing-in-memory and neural networks, in *Machine Learning & Knowledge Extraction* (2018)
80. M. Dazzi, A. Sebastian, P.A. Francese, T. Parnell, L. Benini, E. Eleftheriou, 5 parallel prism: a topology for pipelined implementations of convolutional neural networks using computational memory (2019). arXiv preprint, arXiv:1906.03474
81. Y. Lin et al., Performance impacts of analog ReRAM Non-ideality on neuromorphic computing. IEEE Trans. Electron Devices **66**(3), 1289–1295 (2019)
82. S. Yin, X. Sun, S. Yu, J.S. Seo, High-throughput in-memory computing for binary deep neural networks with monolithically integrated RRAM and 90 nm CMOS (2019). arXiv preprint arXiv:1909.07514
83. B. Murmann, ADC performance survey 1997–2019, [Online]. http://web.stanford.edu/~murmann/adcsurvey.html
84. W.J. Dally et al., Hardware-enabled artificial intelligence, in *Symposium on VLSI Circuits* (2018), pp. 1–2
85. D. Bankman, J. Messner, A. Gural, B. Murmann, RRAM-based in-memory computing for embedded deep neural networks, in *Asilomar Conference on Signals, Systems and Computers*, Asilomar, CA, Nov 2019
86. https://mlperf.org/
87. B. Moons, K. Goetschalckx, N. Van Berckelaer, M. Verhelst, Minimum energy quantized neural networks. arXiv preprint arXiv:1711.00215
88. B. Moons, D. Bankman, L. Yang, B. Murmann, M. Verhelst, BinarEye: an always-on energy-accuracy-scalable binary CNN processor with all memory on chip in 28 nm CMOS, in *Custom Integrated Circuits Conference (CICC)* (IEEE, 2018), pp. 1–4
89. A. Stoutchinin, F. Conti, L. Benini, Optimally scheduling CNN convolutions for efficient memory access (2019). arXiv preprint arXiv:1902.01492
90. X. Yang, M. Gao, J. Pu, A. Nayak, A. Liu, S.E. Bell, J.O. Setter, K. Cao, H. Ha, C. Kozyrakis, M. Horowitz, DNN dataflow choice is overrated (2018). arXiv preprint arXiv:1809.04070

91. A. Parashar, P. Raina, Y.S. Shao, Y.H. Chen, V.A. Ying, A. Mukkara, R. Venkatesan, B. Khailany, S.W. Keckler, J. Emer, Timeloop: a systematic approach to DNN accelerator evaluation, in *2019 IEEE International Symposium on Performance Analysis of Systems and Software (ISPASS)* (IEEE, 2019), pp. 304–315
92. https://tvm.ai/
93. H. Kwon, M. Pellauer, T. Krishna, Maestro: an open-source infrastructure for modeling dataflows within deep learning accelerators (2018). arXiv preprint arXiv:1805.02566
94. K. Goetschalckx, M. Verhelst, Breaking high resolution CNN bandwidth barriers with enhanced depth-first execution, in *JETCAS 2019*
95. R. Aljundi, F. Babiloni, M. Elhoseiny, M. Rohrbach, T. Tuytelaars, Memory aware synapses: learning what (not) to forget, in *Proceedings of the European Conference on Computer Vision (ECCV)* (2018), pp. 139–154
96. A. Nøkland, Direct feedback alignment provides learning in deep neural networks, in *Advances in Neural Information Processing Systems* (2016), pp. 1037–1045
97. C. Frenkel, M. Lefebvre, D. Bol, Learning without feedback: direct random target projection as a feedback-alignment algorithm with layerwise feedforward training (2019). arXiv preprint arXiv:1909.01311
98. J. Yue, R. Liu, W. Sun, Z. Yuan, Z. Wang, Y.N. Tu, Y.-J. Chen, A. Ren, Y. Wang, M.-F. Chang, X. Li, H. Yang, Y. Liu, 7.5 A 65 nm 0.39-to-140.3 TOPS/W 1-to-12b unified neural network processor using block-circulant-enabled transpose-domain acceleration with 8.1 × Higher TOPS/mm 2 and 6T HBST-TRAM-based 2D data-reuse architecture, in *2019 IEEE International Solid-State Circuits Conference-(ISSCC)* (IEEE, 2019), pp. 138–140
99. D. Han, J. Lee, J. Lee, H.J. Yoo, A low-power deep neural network online learning processor for real-time object tracking application. IEEE Trans. Circuits Syst. I Regul. Pap. **66**(5), 1794–1804 (2018)
100. S. Yu, Neuro-inspired computing with emerging nonvolatile memory. Proc. IEEE **106**, 260–285 (2018)
101. A. Gural et al., Low-rank training of deep neural networks for emerging memory technology, unpublished work
102. H. Li, P. Raina, H.-S. P. Wong, Neuro-inspired computing with emerging memories: where device physics meets learning algorithms, in *Proceedings of SPIE 11090, Spintronics XII, 110903L*, Sep 2019
103. L.E. Sucar, Probabilistic graphical models, in *Advances in Computer Vision and Pattern Recognition* (Springer London, London, 2015)
104. L. Serafini, A.D.A. Garcez, Logic tensor networks: deep learning and logical reasoning from data and knowledge (2016). arXiv preprint arXiv:1606.04422
105. H. Wang, D.Y. Yeung, Towards Bayesian deep learning: a framework and some existing methods. IEEE Trans. Knowl. Data Eng. **28**(12), 3395–3408 (2016)
106. R. Manhaeve, S. Dumancic, A. Kimmig, T. Demeester, L. De Raedt, Deepproblog: neural probabilistic logic programming, in *Advances in Neural Information Processing Systems* (2018), pp. 3749–3759
107. N. Shah, L. Galindez, W. Meert, M. Verhelst, Acceleration of probabilistic reasoning through custom processor architecture and compiler, in *Design and Test Conference Europe (DATE)* (2020)

Chapter 19
The Memory Challenge in Ultra-Low Power Deep Learning

Francesco Conti, Manuele Rusci and Luca Benini

19.1 Introduction

Starting from circa 2012, the "viral" revolution of *Deep Learning* [1] has impacted an ever-growing number of fields. *Deep Neural Networks* (DNNs), in particular, have emerged as an almost universal algorithmic "Swiss-Army knife" for tasks related to data analytics, artificial intelligence, and in general, where cognition-like functionality is sought. DNNs and derivative algorithms have been applied successfully to vision [2], which is their de facto benchmark task; speech recognition [3]; big data analytics and financial forecasts [4, 5]; medicine and biomedical engineering [6]; robot control [7]; autonomous driving just to name a few prominent applications.

Arguably, one of the most influential contributions to the development of Deep Learning techniques has been the availability of GPUs capable of high-throughput single-precision computation and with fast and generously sized off chip DDR memories, as well as a vast amount of on-chip SRAM. While GPU architectures have been created to satisfy the requirements of gaming and computer graphics, their architecture is well-suited for the kind of regular computation pattern embodied by DNNs, dominated by linear algebra. In the last few years, the architecture of GPUs and the development of DNN topologies have significantly influenced one another, as GPU vendors consider Deep Learning one of the key target applications for their products. Not only server- and desktop-class GPUs have been influenced by this

F. Conti (✉) · M. Rusci · L. Benini
University of Bologna, viale Pepoli 3/140126, Bologna, Italy
e-mail: f.conti@unibo.it

M. Rusci
e-mail: manuele.rusci@unibo.it

L. Benini
e-mail: luca.benini@unibo.it

F. Conti · L. Benini
ETH Zürich, Zürich, Switzerland

B. Murmann and B. Hoefflinger (eds.), *NANO-CHIPS 2030*,
The Frontiers Collection, https://doi.org/10.1007/978-3-030-18338-7_19

trend, which has recently started also touching mobile GPUs. On the other hand, many dedicated accelerators have been designed to provide higher performance than GPUs in DNNs. These typically rely on some form of numerical approximation, such as using integer numbers instead of floating-point ones, coupled with architectural specialization techniques to minimize the energy spent for executing the dominant multiply-accumulate operation. However even specialized accelerators have in common a strong dependence on off- and on-chip memory.

In a highly constrained embedded system, accesses to off-chip memory are, relatively speaking, extremely expensive and can potentially void the throughput, energy, or cost advantages of an accelerator. One way to alleviate this issue is to use large on-chip SRAM buffers (up to a few MBytes may be used) to capture the locality in the feature map and filter weight accesses) to eliminate main memory traffic [8–10].

However, there are a lot of application targets for which the availability of large amounts of memory, both on- and off-chip, cannot be taken for granted as a commodity: intelligent implantable biomedical devices [6], completely autonomous nano-vehicles [7, 11] for surveillance and search and rescue, cheap controllers that can be "forgotten" in environments such as buildings [12], roads, and fields. These applications are characterized by very stringent constraints in terms of power, area, cost, and durability: they have to work on battery-supplied systems with a peak power envelope of less than 100 mW (and sometimes much less!) while guaranteeing a lifetime of months or years, and they have to cost no more than a few dollars at most. Reducing the memory footprint of Deep Neural Networks, or making memory less expensive, could make the difference between being able to use Deep Learning at in these fields. We call these challenges "the Deep Learning Memory Wall".

Furthermore, enabling emerging artificial intelligence applications at the Very Edge of the Internet-of-Things (IoT), such as time-series analysis and online learning of new capabilities, means coping with more sophisticated and higher-footprint algorithms. In the next 10 years, the combined action of stricter constraints and stronger performance and energy requirements from these applications will require SoC architects to make intelligent and focused usage of emerging technologies to overcome the Deep Learning Memory Wall. In this chapter, we delve into the problem by analyzing the Deep Learning Memory Wall in detail, quantifying how it impacts the design of an embedded system running at the Very Edge of the IoT. Then, we discuss current trends in tailoring the architecture of embedded SoC's to relax the memory footprint of DNNs and how the algorithms can be on turn finely tuned to maximize effectiveness on a given platform. Finally, we analyze how emerging technologies, techniques and algorithms can further help overcome the DL memory wall in the next 10 years.

19.2 The Deep Learning Memory Wall

19.2.1 Memory Footprint of Deep Neural Networks

The memory footprint of modern neural networks dedicated to vision (which is currently their leading application domain) is staggering, considering the capacity of on-chip memories that can be deployed in a low-power chip. The most efficient among recent high-accuracy image classification networks, EfficientNet [13], achieves 84.4% top-1 ImageNet but requires 66 million parameters. This network is explicitly designed to reduce memory footprint (ResNeXt, with similar performance, requires ~10 more parameters [14]); however, considering 8-bit weights, it is still two orders of magnitude beyond anything that can be stored on-chip on a low-power, low-cost system-on-chip. Recent research has shown that while many networks are over-designed, there is a limit to the amount of memory footprint reduction that can be performed without dropping precision significantly. In Table 19.1, we showcase this situation for two kinds of vision-based benchmarks (classification and object detection). We compare them with the memory available in a target current-generation low-power IoT end-node, assuming that it can have at most 1 MB of on-chip SRAM [15, 16], or up to 64 MB of off-chip DRAM [17].

Naturally, not all tasks have such major constraints, meaning there are already applications where Deep Neural Networks fit entirely on-chip. For example, autonomous navigation of a drone can be performed with a residual network using

Table 19.1 Memory footprint of representative DNNs

Task	Algorithm	Performance	Memory Footprint @ 8b (MB)	On-chip (1 MB)	Off-chip (64 MB)
Image classification/ImageNet	ResNeXt-101 32 × 32d [14]	85.1% top-1	466	X	X
Image classification/ImageNet	EfficientNet-B7 [13]	84.4% top-1	66	X	Nearly
Image classification/ImageNet	EfficientNet-B1 [13]	78.8% top-1	7.8	X	V
Image classification/ImageNet	1.0-MobileNet-224 [18]	70.6% top-1	4.2	X	V
Image classification/ImageNet	0.5-MobileNet-224 [18]	63.7% top-1	1.3	Nearly	V
Image classification/ImageNet	0.25-MobileNet-224 [18]	50.6% top-1	0.5	V	V
Detection/COCO	YOLOv3 [19]	0.606 mAP	59	X	V
Detection/COCO	YOLOv3-tiny [19]	0.331 mAP	8.5	X	V
Detection/COCO	Tiny YOLOv2 [20]	0.237 mAP	15	X	V

as little as 300 kB, assuming an 8-bit representation [11]. Also, the analytics of time-series data, such as ExG, can be performed within a reasonable memory budget for ultra-low-power systems. Moreover, some networks, such as MobileNets, are explicitly designed to trade off part of their accuracy in exchange for a much smaller footprint, and reduced versions can be effectively deployed on-chip (e.g., the 0.25-MobileNet-224).

However, looking "ahead of the curve," it is clear that the development of new AI algorithms is trending towards generally bigger architectures to perform more complex and sophisticated tasks, also relying on automatic machine learning (autoML) strategies such as Differentiable Neural Architecture Search [21]. Moreover, if the network has to be re-trained on the field, current training strategies based on back-propagation require keeping track of intermediate tensors in the neural network, further increasing its footprint– possibly by more than one order of magnitude. Therefore, it is clear that to support the expected growth in the next few years in the complexity and functionality of AI-oriented embedded nodes, off-chip DRAM will still play a crucial role.

19.2.2 Performance and Energy Cost of Memory Traffic

The previous discussion could lead to thinking that memory size is the only limiting factor in the deployment of next-generation Deep Learning algorithms. However, memory bandwidth constraints can play just as important a factor. To demonstrate this, we analyzed one of the networks that appear in Table 19.1 (1.0-MobileNet-224) to understand its memory access patterns and required bandwidth. We used the model proposed by Stoutchinin et al. [22] to count the number of data accesses to each vector touched within a typical Deep Neural Network layer (Fig. 19.1).

Using this memory hierarchy model, we assume that on-chip SRAM memory is accessible with unlimited bandwidth (i.e., the data path is designed to maximize its access efficiency to local memory) and it hosts a local copy of a portion (*tile*) of each tensor touched by a DNN layer: *weights*, *input* activations, *partial sums,* and

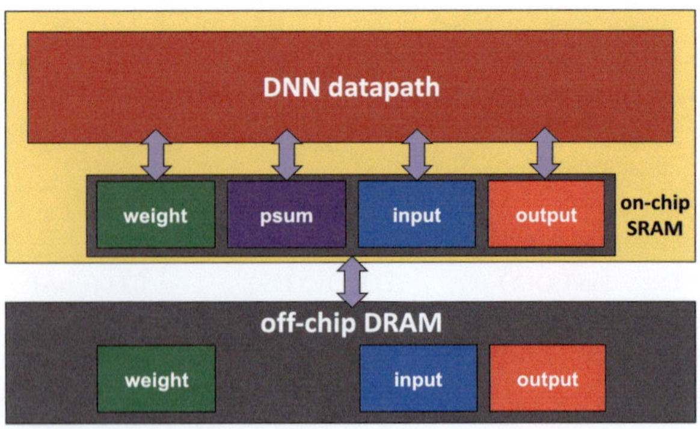

Fig. 19.1 Memory hierarchy model as proposed by Stoutchinin et al. [22]

```
LOF: for m in range(0, K_out):
LIF:  for n in range(0, K_in):
LSY:    for i in range(0, H_out):
LSX:      for j in range(0, W_out):
            psum = b[m]
LFY:        for ui in range(0, F):
LFX:          for uj in range(0, F):
                psum += w[m,n,ui,uj] * x[n,i+ui,j+uj]
            y[m,i,j] = act(psum)
```

Fig. 19.2 Canonical loops of a DNN layer (in pseudo-Python code)

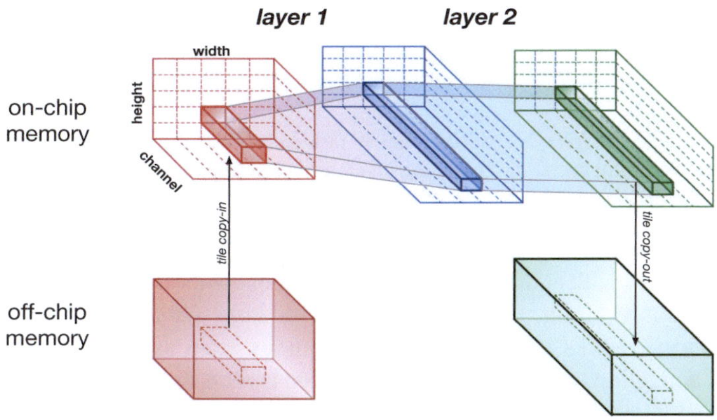

Fig. 19.3 Example of two layers executed in DF fashion; execution proceeds through all layers for a single chain of tiles, then the process is repeated for the following chain of tiles. Tiles are purely spatial and their dimensions are tied to each other

output activations. The off-chip DRAM, on the other hand, has a full copy of all DNN weights, and if necessary, inputs and outputs of each layer.[1]

As a prototype DNN layer, we consider the canonical six-loops of a convolutional layer, which can also represent a fully-connected layer by removing two loops (Fig. 19.2).

In theory, execution of the DNN could optimally happen in a *depth-first* (DF) fashion, i.e., executing sequentially all layers without storing any intermediate data off-chip. In this case, DRAM traffic is limited exclusively to weights. Figure 19.3 visualizes the DF execution pattern for a DNN with two layers, focusing on activation tensors (off-chip memory traffic for weights is not shown). Unfortunately, this model of execution is often not convenient, because it imposes strong constraints in terms of tiling: in most convolutional and linear layers, activations cannot be tiled at all input channels are needed to compute each output channel; moreover, the spatial receptive field of each activation element in the original input of the DNN usually increases along the network, constraining spatial tiling so that the chain of tiles

[1]For simplicity, we do not model the weights as resident in a separate off-chip Flash or non-volatile memory.

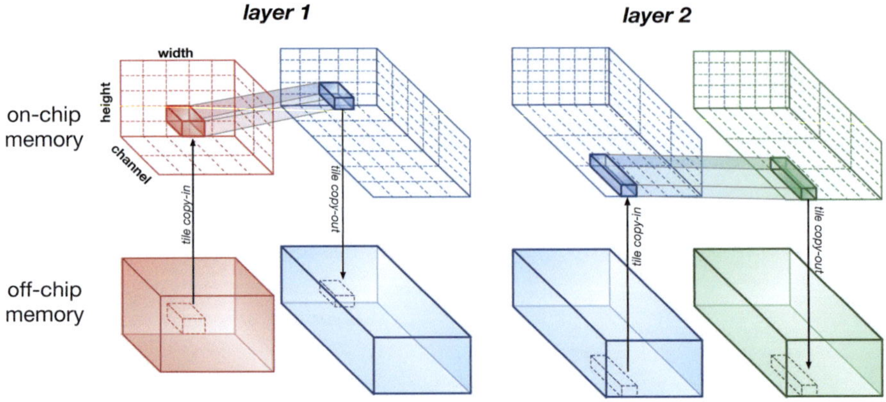

Fig. 19.4 Example of two layers executed in LW fashion; execution proceeds through all tiles for a given layer, then switches to the following layer. Tile dimensions across different layers are not necessarily correlated

always terminates with at least one activation element at the end of the network. With such limitations to tiling, DF execution requires large on-chip buffering capacity. Moreover, in networks with complicated topology such as EfficientNet [13], the lifetime of activation tensors produced by a given layer is not limited to the following layer, but is longer as tensors are utilized multiple times as inputs of layers far apart from each other. This leads to a further increase in buffering requirements.

The alternative execution pattern is *layer-wise* (LW) execution, which is visualized in Fig. 19.4: each layer consumes a full input tensor from off-chip DRAM and produces a full output tensor into the same DRAM. Since the execution of each layer is entirely decoupled from the perspective of on-chip memory, there is no need for consecutive layers to use the same tile shapes: therefore, tiling can be applied much more effectively than in the DF case. In LW execution, fully computed activations constitute an essential part of the traffic to DRAM, but in exchange it is possible to greatly relax on-chip memory requirements, fitting into a much tighter constraint.

The overall amount of DRAM traffic under a given on-chip memory constraint depends chiefly on three factors: (1) the ordering of the six nested loops of Fig. 19.2; (2) at the level of which loop each independent tensor (weights, inputs, outputs) is tiled; (3) the size of each tile. To show how DRAM traffic impacts the maximum performance of a low-power embedded accelerator for DNNs, we performed an exploration in terms of tile size (from 8 to 512, in power-of-two increments) and variety of buffered tiles (checking 180 different loop permutations—with the other combinations equivalent to the ones tested, e.g., with swapped x/y spatial loops). We targeted an on-chip buffer of size 64 or 512 kB to showcase the case of a very small buffer and that of a typical-size one.

Figure 19.5 shows our results for 1.0-MobileNet-224, in terms of minimum DRAM bandwidth to achieve a certain target throughput. Out of our exploration space, we selected the best result for each configuration; each layer uses a different

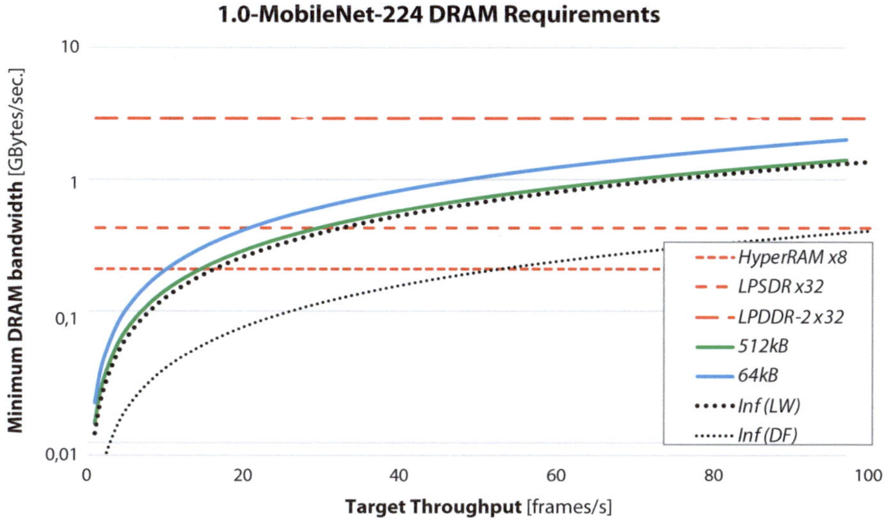

Fig. 19.5 1.0-MobileNet-224 required DRAM bandwidth

configuration and tile size. We also report ideal results obtained while considering layer-wise (LW) and depth-first (DF) execution with unlimited on-chip SRAM (i.e., without tiling). We compare the minimum required DRAM bandwidth at a given target throughput with the available bandwidth of representative DRAM interfaces for the low-power embedded space: Cypress HyperRAM [17], LPSDR DRAM provided by Winbond [23], and a Micron LPDDR2 chip for the automotive market [24]. First of all, we note that the traffic cost of layer-wise execution compared against depth-first is high (~3.3×) when the amount of on-chip SRAM is unlimited. On the other hand, our practical results show that under the two explored constraint settings, no depth-first solution can be run optimally. The best configuration achieved under the 512 kB is practically equivalent to the ideal layer-wise solution, while under a 64 kB constraint, an additional 42% of traffic is required due to tiling. In both cases, however, a low-power HyperRAM is sufficient to achieve ~10 frames per second, enough to be considered real-time for inference, e.g., on a small robotic device. As this network involves ~569 million multiply-accumulate (MAC) operations, 10 fps are equivalent to a workload of 5.7 GMAC/s: a *state-of-the-art ultra-low-power DNN accelerators targeting 20-100 GMAC/s* [8] *would risk being bottlenecked by construction* when used in a real-world DNN such as this, unless they can leverage more complicated—and power-hungry—DRAM controllers.

The same dependence is shown when looking at power: to evaluate this, we used power analysis results from a state-of-the-art convolution accelerator embedded in a 22 nm chip [25], indicating an estimated average expense of 2 pJ/MAC. For the HyperRAM, we assume consumption of 411 pJ/B from its datasheet, neglecting any further (and probably significant) expense due to I/O. The result, shown in Fig. 19.6,

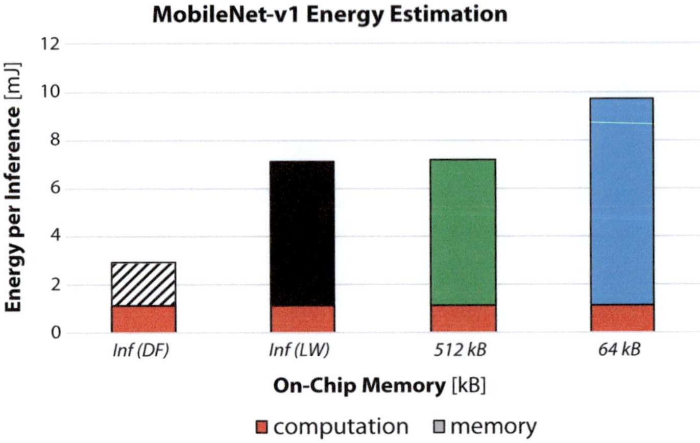

Fig. 19.6 Estimation of energy per inference of 1.0-MobileNet-224

clearly shows that inference energy, even for a relatively small network such as 1.0-MobileNet-224, is dominated by DRAM transfers.

While we have focused on a "reasonably implementable" network for today for this implementation, the issue is exacerbated on bigger, higher-accuracy networks. Taking, for example, EfficientNet-B7, designing a low-power DRAM capable of hosting 66 MB of weights is not unreasonable even in the very near future. On the other hand, its execution at 10 fps would require at least 1.54 GB/s of bandwidth, which looks significantly more challenging to achieve within a low power budget. Finally, even assuming the same energy cost of 411 pJ/B of the previous example, execution at 10 fps would require 633 mW, and the majority of the power (87%) would be spent on the interface. It is therefore clear that technological, architectural, and algorithmic techniques concerning the usage of memory are necessary to enable the next generation of DNNs to run within an ultra-low-power budget.

19.3 Mitigating the Wall

19.3.1 Tiling Techniques for Deployment in IoT-Dedicated Architectures

Section 19.2 shows in which way memory limits the maximum theoretical performance achievable by a DNN-dedicated ultra-low-power node: but can we actually deploy such a network in practice so that its performance is optimal or near-optimal? To answer this question, we focus on a real-world class of devices requiring ultra-low power consumption as well as Deep Learning acceleration capabilities, that of IoT-dedicated end-nodes. Many systems in this class couple a small and fast L1 scratchpad memory, meant to be directly accessed at very high bandwidth by the DNN compute units, with higher capacity and a lower bandwidth L2 background memory, with both levels of the memory hierarchy resident on-chip. Systems of this kind lack a coherent hardware cache to save energy at the cost of labor-intensive explicit memory management, making it practically challenging to achieve high bandwidth utilization rates. Therefore, before even targeting the reduction of the Deep Learning memory wall, it is necessary to verify whether the available memory and bandwidth in real-life systems such as these can be fully exploited.

One of the template architectures for this class of devices is the open-source *PULP*[2] architecture, constituted of a *fabric controller* single-core subsystem, an *I/O subsystem* directly connected to it to enable smart and independent control of external devices, and a *cluster* subsystem acting as a multi-core programmable accelerator. Figure 19.7 shows the (simplified) architecture of a typical PULP-based system with an L3 DRAM based on the 8-bit HyperBus protocol, a relatively large on-chip L2 SRAM of 512 kB and a smaller L1 of 64 kB designed to provide highly parallel access to 8 RISC-V DSP-augmented cores [26]. Such a system is provided with libraries (PULP-NN [27]) achieving more than 1 GMAC/s at 170 MHz on an 8-bit DNN workload consuming ~65 mW; as we have seen in Sect. 19.2.2, this is well-matched with the bandwidth with which the external HyperRAM can be accessed.

In many cases, ultra-low-power nodes with Deep Learning acceleration capabilities couple a small and fast L1 scratchpad memory, meant to be directly accessed at very high bandwidth by the DNN compute units, with higher capacity and lower bandwidth L2 background memory, with both levels of the memory hierarchy resident on-chip. These systems typically lack a coherent hardware cache to save energy at the cost of labor-intensive explicit memory management, making it practically challenging to achieve high bandwidth utilization rates. Before even reducing the Deep Learning memory wall, therefore, the first challenge is to maximize the usage of available on-chip memory, computing resources, and—crucially—off-chip DRAM and bandwidth to achieve the kind of optimal results described in Sect. 19.2.2.

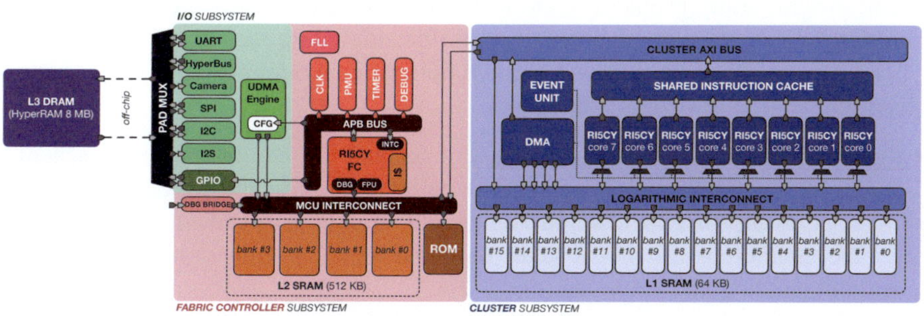

Fig. 19.7 PULP system with 64 KB of L1, 512 KB of L2 (on-chip) and 8 MB of L3 DRAM (off-chip)

---

[2]https://github.com/pulp-platform and https://pulp-platform.org.

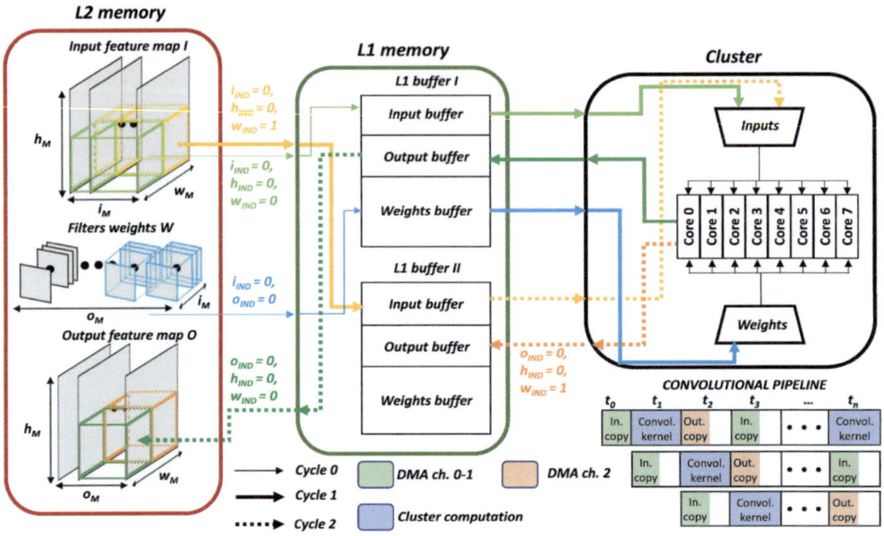

Fig. 19.8 L2/L1 tiling and data movement scheme [29] ©IEEE 2019

As detailed in the previous Section, however, actually delivering this level of performance in practice requires removing memory-related bottlenecks by performing two levels of tiling: from L3 to L2 via the off-chip connection, and from L2 to L1. Tiling can be automatized by exploiting the regular graph-based structure of DNNs to optimize tile sizes (minimizing traffic while keeping within given memory constraints) and automatically generating code to move data between the various layers of the hierarchy [11, 28, 29]. Figure 19.8 shows the L2/L1 data movement scheme targeted by *DORY* (Deployment ORiented to memorY) [29], a tool we developed to perform this operation; a similar scheme operates between L3 and L2.

The optimization of tile sizes for each layer can be abstracted as an integer Constraint Programming problem. DORY receives as input a list of layers and targets minimization of the overall layer-wise traffic, including that generated by overlapped parts of tiles due to the receptive field of convolutional filters. We subject this minimization to several constraints:

- the combined size of all tiles, taking into account also double buffering schemes if present, must be smaller than a given budget (e.g., 64 kB for the L1);
- the relationships between weight, input, and output tile dimensions are mandated by the characteristics of the layer (convolutional vs. fully-connected, with or without padding and stride, etc.);
- the tiles should be sized in such a way to provide a well-parallelizable input to the backend PULP-NN library, maximizing its efficiency.

DORY uses the open-source OR-tools constraint solver from Google AI[3] to derive a solution (in terms of tile sizes) that is compatible with all constraints; then, it directly generates the C code of the DNN running on PULP, including data movement and double buffering, according to these tile sizes.

Figure 19.9 shows the L2/L1 tiling efficiency of the DNN code produced by DORY in the case of a small (142 kB) network run on the PULP-based GAP8 chip. Diamonds represent the optimal solutions chosen by DORY for each layer, while crosses represent sub-optimal solutions. At the worst case, an L2/L1 tiling solution achieves the same performance as operating directly on L2 data, incurring in a latency penalty for every access; at the best case, conversely, it is equivalent to execution directly in L1 without buffer size constraints, whose performance we can gather by means of architectural simulation of the PULP platform. It can be seen from Fig. 19.9 that the DORY scheme can be used as a very specialized software cache to effectively hide the fact that execution happens on L1 instead of L2 for convolutional layers. In fully-connected layers, whose arithmetic intensity is 100× lower, the "caching" mechanism is less efficient—however, performance is still ~2× that achieved with direct execution on the L2 memory. Overall, execution of this small network—which does not fit the L1 of PULP—can run either directly on L2 or with the DORY tiling scheme; in the latter case, it consumes 3.2× less time and 1.9× less energy than the former.

To verify the applicability of this scheme for L3/L2 transfers, we implemented a similar tiling loop between an external DRAM realized with a Cypress 8 MB HyperRAM and the L2 on-chip memory; we can achieve a measured bandwidth of 180 MB/s, around 70% of the ideal one in the same operating point. Most of this loss is due to nonidealities in the DRAM access patterns (e.g., shorter transfers pay a higher penalty), while the loss directly due to the tiling loop overhead is below 1%.

Can such a system scale up to larger networks? The overhead of the tiling scheme is due to the small, but not necessarily negligible, computational overhead introduced by the tiling loops themselves. We measured this second effect to be less than 4% of

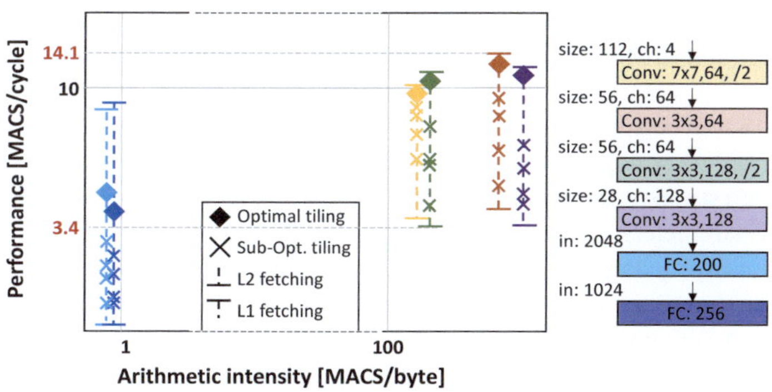

Fig. 19.9 DORY L2-L1 tiling efficiency [29] ©IEEE 2019

[3]https://developers.google.com/optimization.

the overall computation, even for tiny tiles: this leads to the conclusion that a tiling scheme of this kind can be effectively used to unlock execution of relatively large DNNs without incurring in the energy penalty of a data cache, while still hiding the complexity and latency of the memory hierarchy.

19.3.2 Deep Neural Network Adaptation for Deeply Embedded Targets

In addition to the previous strategy, the memory wall can be tackled by reducing the memory footprint of a given DNN model. The amount of data transferred throughout the different memory levels can be reduced by compressing either weights or activations tensors. High compression ratios may eventually lead to dismiss off-chip costly memories for a given DNN workload.

Two main strategies are typically adopted for DNN compression:

- Pruning, to cut *less significant* or *redundant* neural connections within a DNN topology, hence discarding part of the weight parameters.
- Quantization, aiming at lowering the bit precision of either weight parameters and activation maps with respect to the 32-bit size used on the server-side.

Both strategies apply, also in combination, at the cost of an accuracy penalty when incrementing the compression ratio. Pruning strategies consists of cutting edge-connections on a network graph based on importance metrics, typically the absolute value of the weight parameter associated to edge. After pruning, a retraining step results beneficial to learn the final values of the remaining sparse connections [30]. To reduce the high computational complexity introduced by sparsity, a more structured pruning has been explored by enforcing channel-level or filter-level sparsity, also including retraining [31].

Despite reducing the memory footprint and in contrast to pruning, quantization brings faster inference by enabling a high-degree of instruction parallelism thanks to vectorized Single-Instruction-Multiple-Data (SIMD) operations that can be exploited when operating with low-bitwidth (8/16 bits) data formats. Several quantization techniques have been presented—but still it is topic under investigation—to guarantee a minimal accuracy degradation with respect to full precision models [32]. Besides the employed strategy, the accuracy drop depends on the model size, expressed by the number of parameters, i.e. the model capacity: larger model can be quantized with more ‘aggressive’ compression scaling factor, even using only few bits to represent a parameter, due to the over-parametrization.

When considering a large-scale problem such as an image classification among 1000 classes (ImageNet dataset), an 8-bit quantization demonstrates nearly zero accuracy drop even if applied to models already optimized for low number of parameters (e.g. MobileNet), even without retraining steps [38]. However, a quantization-aware retraining of pretrained full-precision models is performed to recover a high-accuracy level [39] in case of sub-byte compression. Table 19.2 reports the accuracy

Table 19.2 Accuracy on Imagenet of low-bitwidth ResNet-18 quantized models

Method	Act bits	Weights bits	Top1/Top5
Full-precision	32	32	69.6/89.2
ABC-Net [33]	5	5	65.0/85.9
PACT [34]	4	4	69.2/89.0
LQ-nets [35]	4	4	69.3/88.8
ABC-Net [33]	3	3	61.0/83.2
PACT [34]	3	3	68.1/88.2
LQ-nets [35]	3	3	68.2/87.9
PACT [34]	2	2	64.4/85.6
LQ-nets [35]	2	2	64.9/85.9
HWGQ [36]	2	1	59.6/82.2
PACT [34]	2	1	62.9/84.7
LQ-nets [35]	2	1	62.6/84.3
XNOR-Net [37]	1	1	51.2/73.2

of a low-bitwidth sub-byte ResNet-18 networks on ImageNet. The PACT approach [34] demonstrated a negligible loss with 4-bit weights and activations to 4-bits by learning the dynamic range through backpropagation. As an extreme corner case, both weights and activations arrays can be compressed to 1 bit, i.e. taking 0 or 1 as values. Doing so, convolutions reduce to bitwise logical operations. Unfortunately, accuracy drops significantly with respect to full-precision (XNOR-NET [37]). To recover accuracy, WRPN [40] proposed to increase the number of filter maps while ABC-Net [33] approximates a full-precision values into a linear combination of binary bases, hence decomposing a convolution of M-bit weight and N-bit activations into N $\times$ M binary convolutions. This latest approach demonstrated the lowest accuracy degradation among the ones involving binary convolutions, but featuring multi-bit memory requirements. Recently, LQ-Nets [35] showed superior accuracy for sub-byte quantization by learning the quantization rule via backpropagation.

We investigated the impact of sub-byte quantization by running a quantized deep learning workload on a STM32H7 microcontroller device, featuring an ARM Cortex M7 core [41]. Figure 19.10 reports latency and energy consumption of convolution kernels featuring weights and activation inputs and outputs compressed down to 8, 4, 2 and 1 bit (denoted as INT-8/4/2/1). All the parameters are stored in the internal second-level RAM (L2). If not enabling any Data Cache, the execution time reduces when decreasing the number of bits from 8 to 2 (green bars), because of the reduced bandwidth—i.e. computation is memory-bounded. On the energy side, the energy consumption slightly increases because of the higher average power costs related to the higher memory access density. When enabling data cache (orange bars), memory hierarchy effects are reduced and computation dominates: INT2 and INT4 kernels result between 30 and 40% slower than INT8 due to unpack operations needed to cast operands to feed the INT16 MAC vector unit. At the same time, power consumption

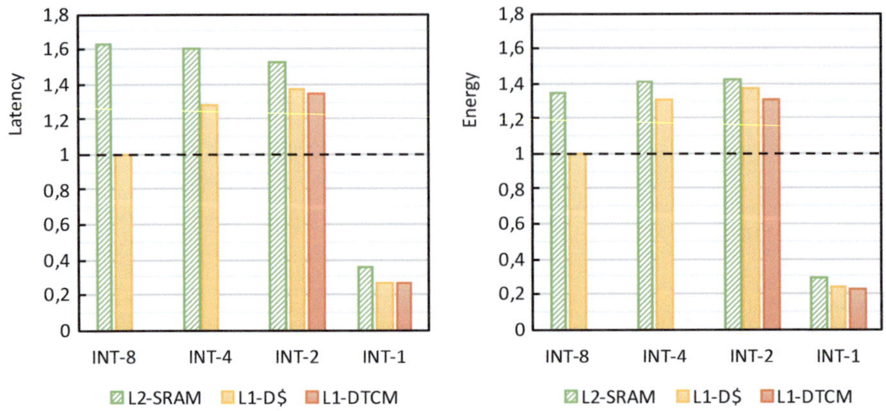

Fig. 19.10 Latency and energy consumption of compressed CONV kernels on ARM Cortex-M7 [41] ©IEEE 2019

increases after enabling the data cache. Hence, the energy gain is lower than the latency gain: the INT8 kernel presents −35% energy consumption with D-cache enabled, while the INT2+D-cache faster execution than the case without D-cache is compensated by the higher power consumption. However, when reducing to 2-bit precision, memory requirements can now fit the smaller L1 memory (red bars). This reflects into an energy reduction of 5% by disabling the D-cache. The INT1 case shows highest latency and energy efficiency, due to the inherent bitwise support of the ISA for binary convolutions. Indeed, running binary kernels on the targeted ARM Cortex M7 cores results 3.7× faster than INT8 convolutions and 4.3× less energy demanding. Our analysis demonstrates that to achieve it full energy-efficiency boosting potential, aggressive bit-width reduction in DNN computations required coupling of data-path and memory hierarchy optimization.

19.3.3 Staged Inference and Heterogeneous on-Chip Memory

Apart from numerical precision scaling, another relevant trend in the algorithmic space of DNNs is to move towards smaller topologies that are as accurate as bigger ones (possibly tolerating a small accuracy loss), with a much lower computational burden and memory footprint. For example, SqueezeNet [42] and ShuffleNet [43] obtain relatively high accuracy on the ImageNet dataset while being tens-to-hundreds of times smaller than first large-but-accurate models, making it clear that relatively small DNNs can be trained to achieve competitive precision. At the same time, many artificial intelligence and data analytics tasks act in a cascaded fashion [29]—most of the data they receive is irrelevant, and their first need is to filter this out, focusing on more interesting information. These techniques can be exploited to enable inference at the edge at a much lower average power budget than would otherwise be possible, by cascading a series of inference techniques and DNNs, each triggering a more expensive, more accurate one—a *staged inference* paradigm.

In Fig. 19.11, we show a system architecture with three stages of inference deployed on a PULP chip. In stage *A*, the chip is actually off, and only a smart always-on sensor is active, acting as a triggering mechanism. These sensors typically produce pre-filtered data using always-on mixed-signal or digital front-end stages [44], operating within a power budget ranging from a few hundred μW for a fully digital vision sensor [45] down to a few μW for mixed-signal biosignal sensor [46]. The smart camera system [47] is based on contrast-based binary visual sensor with address-event readout that wakes up a PULP processing systems once motion is detected on the camera field of view (Stage A), while just paying less than 20mW of average power cost. The second stage B consists on a lightweight classifier filter (e.g. a small DNN model) running on the fabric controller subsystem (i.e. using the on-chip L2 SRAM). The second-stage trainable classifier aims at activating the more computationally intensive cluster sub-system for deep inference tasks after the detection of relevant events. In our camera system, such a classifier is based on a

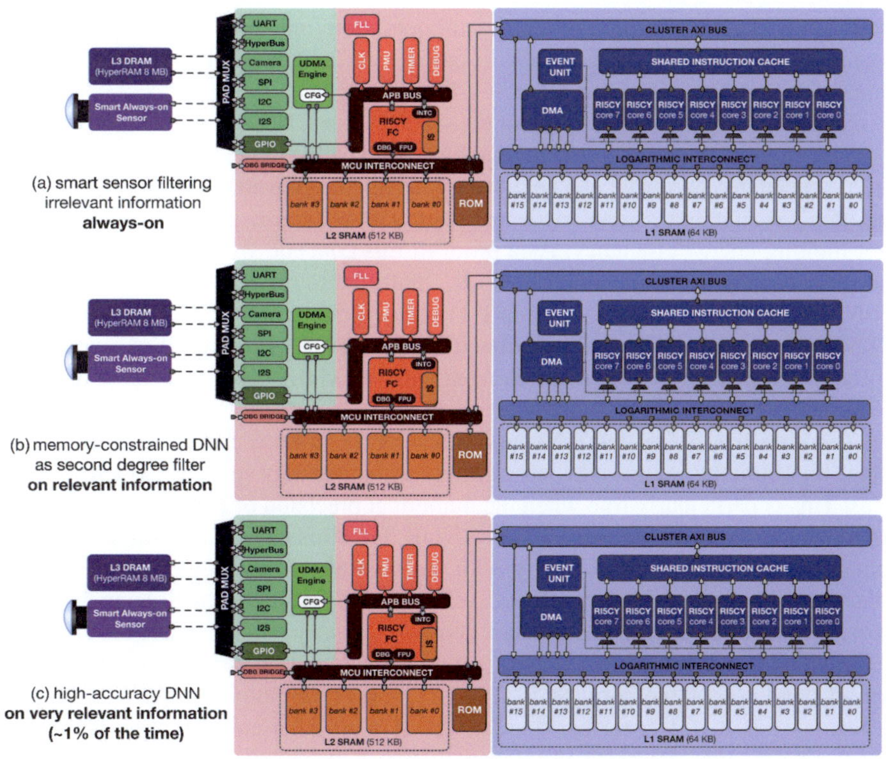

Fig. 19.11 Example of staged inference scheme. Blocks greyed out and blurred are powered down

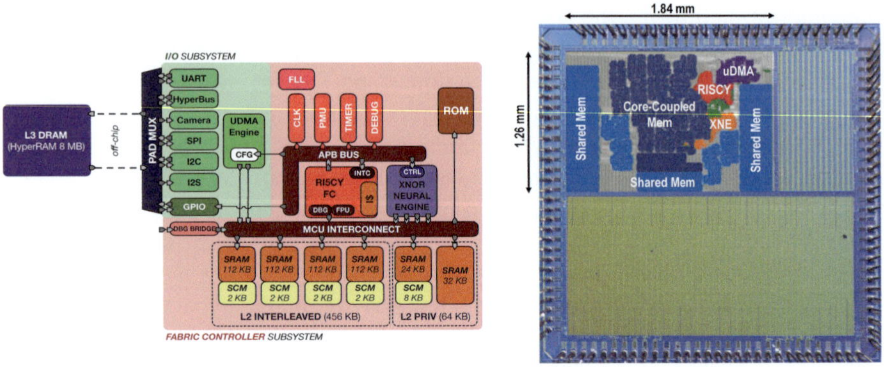

Fig. 19.12 Quentin SoC architecture (left) and die microphotograph (right, [50] ©IEEE 2018)

clustering-based object detector, which wakes-up the cluster when tracked objects enter a virtual loop of the image, corresponding to a relevant region within the external environment (e.g. a door or a street). Finally, the cluster runs an inference task based on binary model to identify the nature of object which causes the alert generation. Thanks to the hierarchical computational model—where the full system is active only during a fraction of the overall operation time—the average power consumption can be as low as 300 μW in case of unfrequent events, 8× lower than a camera-based system running frame-by-frame analytics.

Naturally, the memory footprint of the second stage DNN classifier can be reduced both by topological transformations and by down-scaling numerical precision. Moreover, specialized hardware can be more effectively employed when running entirely on-chip, i.e., in an operating condition where DRAM traffic is absent, and thus it is not limiting energy efficiency. As a proof of this concept, we designed a hardware accelerator mapped inside a PULP Fabric Controller to run Binary Neural Networks (BNNs) [48] directly on the on-chip L2 memory, called the XNOR Neural Engine (XNE). Figure 19.12 shows the architecture of *Quentin*, a prototype SoC constituted by a cluster-less PULP system accelerated with the XNE that was taped out in 22 nm FDX technology.

The ultra-low-power requirements of the staged inference scheme in its second stage highlight the need to take into account also the energy spent for *on-chip memories*. This is particularly true when the numerical precision of DNNs is scaled down so much that products become simple XNOR operations: power consumed by computation scales down superlinearly with the number of bits, while that devoted to memory transfers scales down linearly. To further boost energy efficiency, therefore, the Quentin SoC features a heterogeneous memory scheme where the L2 banks are divided in a large SRAM fraction, and a much smaller part realized with latch-based standard-cell memory (SCM) [49], more parsimonious in terms of energy by a factor of 10×. L2 is further divided into a word-interleaved section accessible by the XNE (456 kB) and a privileged section with priority access from the core

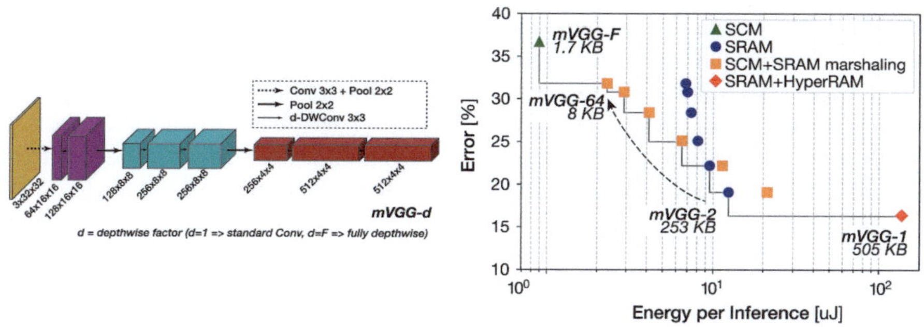

Fig. 19.13 mVGG-d BNN topology (left) and energy versus accuracy estimation on Quentin SoC (right) [48] ©IEEE 2018

(64 kB). Figure 19.12 shows the architecture of the Quentin SoC and an annotated microphotograph of the fabricated die in a multi-project chip.

Utilizing post-layout experiments, we characterized the total energy per inference on a BNN called *mVGG-d* targeting the CIFAR-10 dataset. By modifying the depthwise nature (i.e., the number of groups *d*) used within each convolutional layer, the BNN can be downscaled to fit entirely within SCM execution or to require SCM + SRAM, or even to not fit within the on-chip memory. As the XNE has an intensive access pattern over weight data, we also modeled the case in which this data is marshaled from SRAM to SCM dynamically to have the XNE operate on the more energy-efficient SCM memories. Figure 19.13 shows the results of this elaboration in terms of accuracy vs energy, visualizing clearly that (unsurprisingly) on-chip execution is an order of magnitude more efficient than off-chip execution, while using SCMs can lead to further savings (up to almost another order of magnitude) but with significant accuracy loss due to the extreme memory constraints. The pure-SCM network is significantly more efficient also thanks to the possibility of choosing a lower operating voltage than that at which SRAMs are working (0.4 V down from 0.6 V), achieving higher overall energy efficiency.

19.4 Tearing Down the Wall—Perspectives for the Next 10 Years

19.4.1 Voltage Overscaling: Working with Unreliable on-Chip Memory

As argued in Sect. 19.3.3, one of the key strategies to overcome the Deep Learning memory wall in the next 10 years will actually be to use the memory *less* and organize computation so that for most of the time off-chip memory access can be avoided, driving down the average power consumption to try and meet ultra-low power constraints. However, we have seen in the same section that on-chip SRAMs constitute a significant constraint against running DNNs within an ultra-low power budget. In particular, aggressive voltage scaling techniques [51], which bring quadratic dynamic power savings, cannot be directly applied to conventional six-transistor SRAMs (6T-SRAMs) that behave unreliably at low voltage. Alternative SRAM cells (e.g., [52]) and other memories with better low voltage operation characteristics have not been used widely in the design of System-on-Chips up to now due to the penalty they inflict to density and speed, metrics that are typically considered even more important than energy efficiency.

Enabling aggressive voltage scaling of conventional 6T-SRAMs means letting them work out of a safe operating region, i.e., having them work in a region where run-time errors in both read and write operations are a probable event. In such a region, design margins have to be guaranteed from the application side instead of the hardware itself. Interestingly, there is a class of DNNs that potentially maps well to operation with over-scaled memories: BNNs, where all bits are (potentially) equally vulnerable, and information is spread so that only a high error rate can produce an accuracy drop [53].

We used the fabricated prototype in 22 nm FDX technology of the Quentin SoC discussed in Sect. 19.3.3 to test the resilience of small BNNs such as the *mVGG-d* topology shown in Fig. 19.13 against increased Bit Error Rate (BER) values [54]. We first measured how the BER changes with operating voltage; the nominal voltage in 22FDX is 0.8 V. The measurement was performed by 1000 pseudo-random write/read iterations over the full SRAM arrays, performed directly by the embedded RISC-V core with a small program running exclusively on the error-free SCM memory: thanks to this setup, it was possible to measure the BER down to 0.375 V (the lowest "safe" voltage is 0.6 V—this is the lowest voltage at which no SRAM error can be detected). To map the BER measurement to an overall accuracy level, we trained three BNNs on CIFAR-10: the one proposed by Yang et al. [53]; a slightly reduced version of mVGG-1 entirely fitting in L2; and the network proposed by Hubara et al. [55], which does not fit in the L2 memory. The effect of the varying BER on both weights and activations was simulated in the training framework (PyTorch) to be able to collect relevant statistics (100 experiments).

The results of this exploration, shown in Fig. 19.14, validate the initial assumptions regarding the resilience of BNNs compared to a highly unreliable memory (as long as vulnerable data, such as partial results, is protected in a safe memory—SCMs in this case). Virtually no accuracy loss is detectable with a BER as high as 10^{-4}, and the drop becomes significant only with a BER around 10^{-2}, allowing voltage over-scaling of SRAMs of ~200 mV.

Figure 19.15 shows how voltage over-scaling leads to much better scalability in terms of performance and power. The maximum efficiency point lies at 0.49 V and is accessible only by aggressive voltage scaling, achieving 14.4 Top/s/W. It is worth to note that even with aggressive over-scaling, power in this region is dominated by memory (with around 60% of the overall consumption) due to increased importance of leakage. Without the innate error resilience of BNNs, it would be necessary to

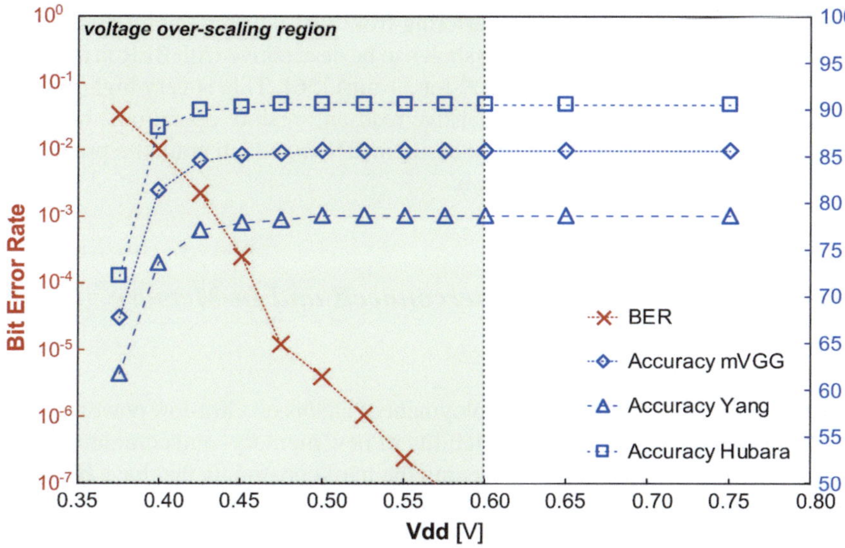

Fig. 19.14 Bit error rate and accuracy of sample BNNs against Vdd in Quentin SoC

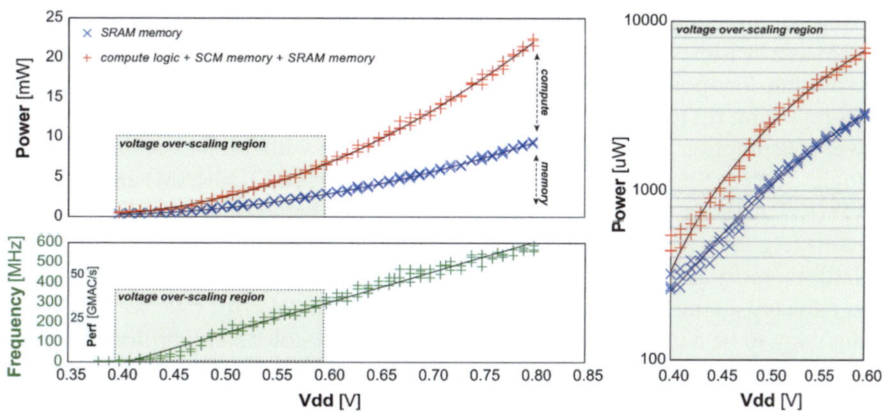

Fig. 19.15 Performance and power consumption of Quentin SoC on BNNs (left), focus on voltage over-scaling region (right)

spend even more power in memory, to keep SRAMs at a higher voltage, implement error correction mechanisms, or use wide voltage range SRAMs.

We believe that increased reliance on error resilience in DNNs will play a significant role in enabling the next 10 years of ultra-low power deep learning. First of all, it reduces the cost in terms of the power of on-chip memory, enabling real ultra-low power (<1 mW active power) inference of small, but non-trivial DNNs. At the same time, error resilience will also be essential in the actual exploitation of novel and emerging memory technologies suffering from real-world reliability issues: for example, STT-MRAM reads have been shown to be destructive with BER in the order of 10^{-6} for 22 nm (projected up to 10^{-4} for 11 nm) [56]. This is very high for conventional computing and requires expensive read-and-restore operations; however, it is in the well-tolerable range of error-resilient BNNs, which could be practically implemented without such a mechanism.

19.4.2 Emerging Memory, Interconnects and In-Memory Computing

A significant amount of help for the deployment of DNNs on ultra-low power embedded systems could come from the availability of new memory- and communication-related technologies, removing or softening the hard constraints that have been discussed in the previous Sections. In this Section, we selected a set of techniques that have already been shown to be potentially feasible and that we believe will achieve maturity within the next 10 years; our target is the deployment of DNNs from the future state-of-the-art on an inference system consuming 1 mW average power, or less.

Non-volatile memory (NVM) has been proposed as a replacement for on-chip SRAM as a last-level cache, as well as for use as off-chip memory. From the perspective of ultra-low-power DL, one of the main advantages of these techniques would be in replacing (in part) on-chip SRAM, exploiting non-volatility to guarantee zero power consumption in the off state in staged inference with full data retention. Technologies such as Spin-Transfer Torque Magnetic RAM (STT-MRAM) and Resistive RAM (ReRAM) are nearing maturity and have started being commercialized in some markets (e.g., as a replacement for embedded Flash memory or as a cache for SSDs). Volatile memory alternatives for SRAMs, such as embedded DRAM (eDRAM), have also attracted attention. Due to their tunability in terms of error rate, we expect this technology to be a useful base technology to run error-tolerant algorithms.

Table 19.3 reports results of recent work on some of these memory technologies compared with a conventional 6T-SRAM (on-chip, including our results from the previous section) and order-of-magnitude parameters of a DRAM and a NOR Flash for embedded applications (off-chip). What appears attractive is that non-volatile memories sit in the position of providing an alternative to SRAM in terms of density, but not in terms of energy/access. Conversely, they do not seem to be overly competitive with DRAMs for what concerns density, but they achieve better energy/access, and, naturally, they consume no power to retain data. eDRAM, on the other hand, could guarantee incremental improvements in density and power compared to conventional SRAM, particularly if used in nominal conditions. Most of the flavors of emerging technology have been shown to be CMOS compatible or to require relatively small changes in terms of process; on the other hand, SRAM and eDRAM are entirely CMOS compatible.

Table 19.3 Current-gen and emerging memory technologies

Technology operating condition	Density (Mbit/mm^2)	Energy per access (pJ/bit)	Data retention power (nW/bit)	Bit error rate	CMOS?
DRAM (LP-DDR2) [24]	~100[a]	>5	>0.003	0	No
NOR Flash [57]	n/a	>50	0	0	No
6T-SRAM 22 nm [54]	<5	>0.13 (@0.8 V) ~ 0.06 (@0.5 V)	~100 (@0.8 V)[b] ~ 30 (@0.5 V)[b]	0 (@0.8 V) $<10^{-5}$ (@0.5 V)	Yes
STT-MRAM 28 nm [58]	4.67	0.7 (read) 4.5 (write)	0	0	In part
Re-RAM 22 nm FinFET [59]	10.1	< 1 (read) n/a (write)	0	$< 10^{-5}$	In part
GC-eDRAM 28 nm [60]	~ 7	n/a	55 (lossless)[b] 10 (lossy)[b]	0 (lossless) $< 10^{-2}$ (lossy)	Yes

[a]Data estimated from https://www.eetimes.com/author.asp?section_id=36&doc_id=1333289
[b]At 85C

A strictly related emerging technology is *computing in-memory* (CIM), i.e., performing part of the computation directly on the memory array, avoiding all cost due to data movement from/to memory. CIM prototypes have been shown both building upon conventional SRAM [61] as well as upon emerging memory technologies such as ReRAM [62]. While architectures targeting larger-scale CIM on DRAMs have also been proposed [63], no practical implementation of this idea has been shown yet. Therefore, the current applicability of CIM is mainly related to (1) replacing on-chip LOAD-MAC-STORE loops with operations running directly on memory, therefore reducing data movement cost, (2) performing multiple computations in a highly parallel fashion with high granularity, potentially achieving very high energy efficiency on computation. However, this kind of CIM does not fundamentally affect the memory footprint constraints: if a DNN does not fit on on-chip memory, external DRAM is still required.

Technologies reducing the cost of storage or making it more efficient to apply a staged inference scheme do not tackle the limitations in terms of memory traffic, which still stand at least in the non-negligible cases in which it is not feasible to fit a full network on-chip. Technologies such as full monolithic 3D integration have been proposed as a solution to the cost of communicating with external memory [64]; however, their real-world feasibility is—for the time being—still unproven. More practically applicable ideas on how to scale the off-die communication capabilities of chips have been shown in the form of *wafer-scale integration* or *multi-chip modules* integrated on-package or over a silicon interposer. Zimmer et al. [65], for example, have recently demonstrated a 16 nm multi-die prototype with chip-to-chip communication at a bandwidth of 11–25 Gbit/s/pin consuming 0.82–1.75 pJ/bit. Technology such as this would enable not only systolic chip-to-chip communication but also high speed and efficiency connection with off-chip memory, "tearing down" the memory wall. For example, in Fig. 19.16, we go back to the 1.0-MobileNet-224 network analyzed in Sect. 19.2.2, updating the results considering a multi-chip-module (MCM) with an in-package L3 memory connected with this technology.

The plot shows that, even without considering any of the other improvements discussed in the chapter, the adoption of this technology alone would greatly relieve the memory wall and significantly reduce the energy cost of off-chip memory, so much that @ 10fps the overall power consumption would be in the order of ~20 mW.

To put it all together, in Fig. 19.17, we try to derive the scheme of a possible future system based on the PULP paradigm dedicated to ULP DNN inference, making extensive usage of the techniques discussed in this chapter to overcome the Deep

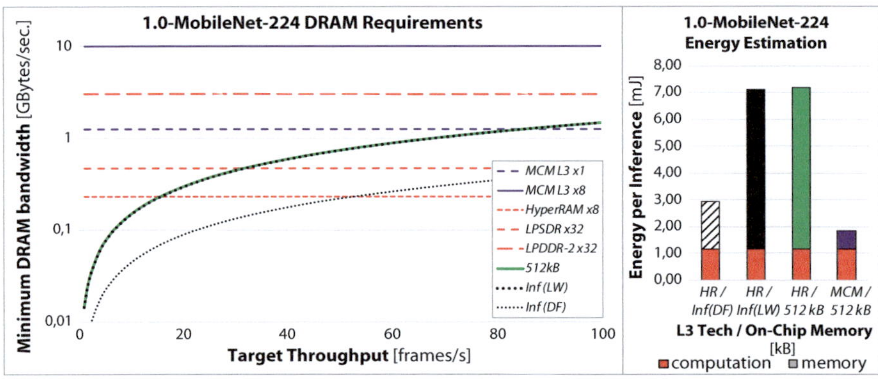

Fig. 19.16 1.0-MobileNet-224 bandwidth requirements and energy estimation with MCM L3

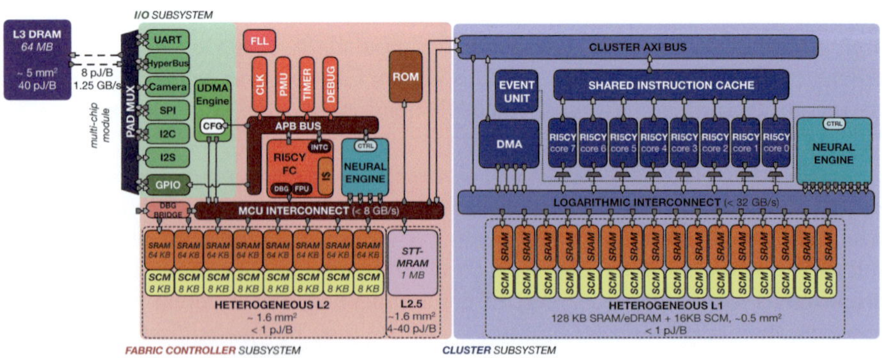

Fig. 19.17 A possible future PULP system for ultra-low power DNN inference

Learning Memory Wall. We assumed the numbers from Table 19.3 and the capability to integrate multiple small chips on an MCM so that the connection can guarantee 10 Gbit/s at 1 pJ/bit. We did not make assumptions on technology scaling, targeting a current-generation process (22 nm). The memory hierarchy is organized into four levels:

- **L1**: 128 KB of SRAM or eDRAM, plus 16 KB of SCM for "safe" data. Access to this memory would happen at high bandwidth (up to 32 GB/s) consuming less than 1 pJ/B. It would be used similarly to the current L1 shown in Fig. 19.7, possibly taking advantage of application-level error tolerance to reduce power. A further improvement to this scheme could be realized, with more architectural changes, by replacing the compound L1 + compute units with a CIM array, potentially saving the cost of communication and achieving higher efficiency on the compute side.
- **L2**: 512 kB of SRAM or eDRAM, plus 64 kB of SCM. Access to this memory would happen at up to 8 GB/s consuming less than 1 pJ/B.
- **L2.5**: 1 MB of STT-MRAM, accessible at up to 8 GB/s and consuming 4 pJ/B in reads and 40 pJ/B in writes. This would be used as on-chip storage for relatively small/precision-reduced neural network weights. Thanks to its non-volatility, it could be used in a staged inference scheme where the PULP chip is fully power-gated, with no need to reload weights.
- **L3**: 64 MB of off-chip DRAM integrated in the same fashion as Zimmer et al. [65]. This would require the design of new DRAM chips dedicated to embedded applications, similar to the HyperRAM and SPI DRAM used in current-generation systems; however, using tight integration the available bandwidth would be 1.25 GB/s for a single pin, paying 48 pJ/B or less (~60 mW at full bandwidth usage).

Putting together these small assumptions, it is clear that the access barrier for ultra-low power deep learning would lower very significantly. High-bandwidth access to L3 would, alone, guarantee a 10–20× improvement in the capability to support memory traffic related to weights and activations. Hardware-aware quantization could achieve further 2–4× improvement. The other changes, notably the addition of 1 MB of L2.5 non-volatile memory, could lead to a dramatic improvement in the capability to use staged inference schemes: at 100 MHz, STT-MRAM bursts would consume ~7 mW and, using voltage-overscaling, SRAM/eDRAM access would consume 10× less power. This means that even a staged-inference scheme characterized by a 90–10% duty cycle between fully-on-chip and off-chip operation will consume, on average, less than 1.3 mW for memory.

19.5 Conclusion

As discussed within the chapter, the memory footprint and traffic generated by deep learning algorithms is proving a significant challenge for the design of ultra-low-power systems—but, thanks to many orthogonal techniques, it is a challenge that can be overcome to provide the next generation of on-chip intelligence. Specifically, we believe that to support future DNN workloads in systems consuming ~1 mW on average, it will be necessary to deploy a combination of:

- staged inference (supported by emerging non-volatile memory technology)
- memory-aware quantization and hardware-aware DNNs
- emerging chip-to-chip interconnect technology
- memory voltage overscaling and algorithmic error tolerance.

We estimate that the combination of these techniques could lead to an improvement of two orders of magnitude in terms of complexity of supported models and the related behavior, without any specific assumption on technological scaling in the post-Moore era.

Moreover, as the memory footprint and traffic become less and less of a bottleneck, emerging applications such as on-device DNN training at the extreme edge become achievable, opening the road for artificial intelligence adaptable on the field.

References

1. Y. LeCun, Y. Bengio, G. Hinton, Deep learning. Nature **521**(7553), 436–444 (2015)
2. C. Szegedy, V. Vanhoucke, S. Ioffe, J. Shlens, Z. Wojna, Rethinking the inception architecture for computer vision, Dec 2015. arXiv: 151200567 Cs
3. Y. Zhang, M. Pezeshki, P. Brakel, S. Zhang, C. L. Y. Bengio, A. Courville, Towards End-to-End Speech Recognition with Deep Convolutional Neural Networks, Jan 2018. arXiv: 170102720 Cs Stat
4. X.W. Chen, X. Lin, Big data deep learning: challenges and perspectives. IEEE Access **2**, 514–525 (2014)
5. M. Dixon, D. Klabjan, J.H. Bang, Implementing deep neural networks for financial market prediction on the Intel Xeon Phi, in *Proceedings of the 8th Workshop on High Performance Computational Finance*, New York, NY, USA, 2015, pp. 6:1–6:6
6. H. Greenspan, B. van Ginneken, R.M. Summers, Guest editorial deep learning in medical imaging: overview and future promise of an exciting new technique. IEEE Trans. Med. Imaging **35**(5), 1153–1159 (2016)
7. A. Loquercio, A.I. Maqueda, C.R. del-Blanco, D. Scaramuzza, DroNet: learning to fly by driving. IEEE Robot. Autom. Lett. **3**(2), 1088–1095 (2018)
8. Y.H. Chen, T. Krishna, J.S. Emer, V. Sze, Eyeriss: an energy-efficient reconfigurable accelerator for deep convolutional neural networks. IEEE J. Solid-State Circ. **52**(1), 127–138 (2017)
9. Z. Du et al., ShiDianNao: shifting vision processing closer to the sensor, in *Proceedings of the 42nd Annual International Symposium on Computer Architecture*, New York, NY, USA, 2015, pp. 92–104
10. V. Sze, Y.-H. Chen, T.-J. Yang, J. Emer, Efficient processing of deep neural networks: a tutorial and survey, Mar 2017. arXiv: 170309039 Cs
11. D. Palossi, A. Loquercio, F. Conti, E. Flamand, L. Benini, A 64mW DNN-based visual navigation engine for autonomous nano-drones. IEEE Internet Things J. **6**(5), 8357–8371 (2019)
12. M. Manic, K. Amarasinghe, J.J. Rodriguez-Andina, C. Rieger, Intelligent buildings of the future: cyber aware, deep learning powered, and human interacting. IEEE Ind. Electron. Mag. **10**(4), 32–49 (2016)

13. M. Tan, Q.V. Le, EfficientNet: Rethinking model scaling for convolutional neural networks, May 2019. arXiv: 190511946 Cs Stat
14. D. Mahajan et al., Exploring the Limits of Weakly Supervised Pretraining, May 2018. arXiv: 180500932 Cs
15. L. Lai, N. Suda, V. Chandra, CMSIS-NN: efficient neural network kernels for arm Cortex-M CPUs, Jan 2018. arXiv: 180106601 Cs
16. E. Flamand et al., GAP-8: A RISC-V SoC for AI at the edge of the IoT, in *2018 IEEE 29th International Conference on Application-specific Systems, Architectures and Processors (ASAP)*, Milano, Italy, 2018, pp. 1–4
17. Cypress 64Mbit—128Mbit HyperRAM Self-Refresh DRAM
18. A.G. Howard et al., MobileNets: Efficient Convolutional Neural Networks for Mobile Vision Applications, Apr 2017. arXiv: 170404861 Cs
19. [1804.02767] YOLOv3: An Incremental Improvement. [Online]. https://arxiv.org/abs/1804.02767. Accessed 15 Oct 2019
20. J. Redmon, A. Farhadi, YOLO9000: Better, Faster, Stronger, Dec 2016. arXiv: 161208242 Cs
21. B. Wu et al., FBNet: hardware-aware efficient ConvNet design via differentiable neural architecture search, presented at the *Proceedings of the IEEE Conference on Computer Vision and Pattern Recognition*, 2019, pp. 10734–10742
22. A. Stoutchinin, F. Conti, L. Benini, Optimally Scheduling CNN Convolutions for Efficient Memory Access, Feb 2019
23. Winbond W989D6KB 512 Mb Mobile LPSDR (2013), p. 66
24. Micron EDB5432BEBH 512 Mb x32 Automotive Mobile LPDDR2 SDRAM (2015), p. 135
25. F. Conti, L. Benini, A ultra-low-energy convolution engine for fast brain-inspired vision in multicore clusters, in *Proceedings of the 2015 Design, Automation & Test in Europe Conference & Exhibition*, San Jose, CA, USA, 2015, pp. 683–688
26. M. Gautschi et al., A near-threshold RISC-V core with DSP extensions for scalable IoT endpoint devices, Aug 2016. arXiv: 160808376 Cs
27. A. Garofalo, M. Rusci, F. Conti, D. Rossi, L. Benini, PULP-NN: accelerating quantized neural networks on parallel ultra-low-power RISC-V processors, Aug 2019. arXiv: 190811263 Cs
28. L. Cecconi, S. Smets, L. Benini, M. Verhelst, Optimal tiling strategy for memory bandwidth reduction for CNNs, in *Advanced Concepts for Intelligent Vision Systems* (2017), pp. 89–100
29. A. Burrello, F. Conti, A. Garofalo, D. Rossi, L. Benini, Work-in-progress: DORY: lightweight memory hierarchy management for deep NN inference on IoT endnodes, in *2019 International Conference on Hardware/Software Codesign and System Synthesis (CODES + ISSS)* (2019), pp. 1–2
30. S. Han, H. Mao, W.J. Dally, Deep compression: compressing deep neural networks with pruning, trained quantization and huffman coding, Feb 2016. arXiv: 151000149 Cs
31. Z. Liu, J. Li, Z. Shen, G. Huang, S. Yan, C. Zhang, Learning efficient convolutional networks through network slimming, Aug 2017. arXiv: 170806519 Cs
32. Y. Guo, A survey on methods and theories of quantized neural networks, Dec 2018. arXiv: 180804752 Cs Stat
33. X. Lin, C. Zhao, W. Pan, Towards accurate binary convolutional neural network, in *Advances in Neural Information Processing Systems 30*, 2017, pp. 345–353
34. J. Choi, Z. Wang, S. Venkataramani, P.I.-J. Chuang, V. Srinivasan, K. Gopalakrishnan, PACT: parameterized clipping activation for quantized neural networks, May 2018. arXiv: 180506085 Cs
35. D. Zhang, J. Yang, D. Ye, G. Hua, LQ-nets: learned quantization for highly accurate and compact deep neural networks, July 2018. arXiv: 180710029 Cs
36. Z. Cai, X. He, J. Sun, N. Vasconcelos, Deep learning with low precision by half-wave Gaussian quantization, Feb 2017. arXiv: 170200953 Cs
37. M. Rastegari, V. Ordonez, J. Redmon, A. Farhadi, XNOR-Net: ImageNet classification using binary convolutional neural networks, in *Computer Vision—ECCV 2016*, 2016, pp. 525–542
38. M. Nagel, M. van Baalen, T. Blankevoort, M. Welling, Data-free quantization through weight equalization and bias correction, Sep 2019. arXiv: 190604721 Cs Stat

39. B. Jacob et al., Quantization and training of neural networks for efficient integer-arithmetic-only inference, Dec 2017. arXiv: 171205877 Cs Stat
40. A. Mishra, E. Nurvitadhi, J.J. Cook, D. Marr, WRPN: wide reduced-precision networks (2018), p. 11
41. M. Rusci, A. Capotondi, F. Conti, L. Benini, Work-in-progress: quantized NNs as the definitive solution for inference on low-power ARM MCUs?, in *2018 International Conference on Hardware/Software Codesign and System Synthesis (CODES + ISSS)*, 2018, pp. 1–2
42. F.N. Iandola, S. Han, M.W. Moskewicz, K. Ashraf, W.J. Dally, K. Keutzer, SqueezeNet: AlexNet-level accuracy with 50x fewer parameters and <0.5 MB model size, Feb 2016. arXiv: 160207360 Cs
43. X. Zhang, X. Zhou, M. Lin, J. Sun, ShuffleNet: an extremely efficient convolutional neural network for mobile devices, July 2017. arXiv: 170701083 Cs
44. F. Glaser et al., Towards a mobile health platform with parallel processing and multi-sensor capabilities. *Rev.* (2017)
45. M. Rusci, D. Rossi, E. Flamand, M. Gottardi, E. Farella, L. Benini, Always-ON visual node with a hardware-software event-based binarized neural network inference engine, in *Proceedings of ACM Computing Frontiers 2018 (to appear)*
46. G. Rovere, S. Fateh, L. Benini, A 2.1 μW event-driven wake-up circuit based on a level-crossing ADC for pattern recognition in healthcare, in *2017 IEEE Biomedical Circuits and Systems Conference (BioCAS)* (2017), pp. 1–4
47. M. Rusci, D. Rossi, E. Farella, L. Benini, A sub-mW IoT-endnode for always-on visual monitoring and smart triggering. IEEE Internet Things J. **4**(5), 1284–1295 (2017)
48. F. Conti, P.D. Schiavone, L. Benini, XNOR neural engine: a hardware accelerator IP for 21.6-fJ/op binary neural network inference. IEEE Trans. Comput.-Aided Des. Integr. Circ. Syst. **37**(11), 2940–2951 (2018)
49. A. Teman, D. Rossi, P. Meinerzhagen, L. Benini, A. Burg, Power, area, and performance optimization of standard cell memory arrays through controlled placement. ACM Trans. Autom. Electron Syst. **21**(4), 59:1–59:25 (2016)
50. P.D. Schiavone, D. Rossi, A. Pullini, A.D. Mauro, F. Conti, L. Benini, Quentin: an ultra-low-power PULPissimo SoC in 22 nm FDX, in *2018 IEEE SOI-3D-Subthreshold Microelectronics Technology Unified Conference (S3S)*, (2018), pp. 1–3
51. A.D. Mauro, D. Rossi, A. Pullini, P. Flatresse, L. Benini, Independent body-biasing of P-N transistors in an 28 nm UTBB FD-SOI ULP near-threshold multi-core cluster, in *2018 IEEE SOI-3D-Subthreshold Microelectronics Technology Unified Conference (S3S)* (2018), pp. 1–3
52. B.H. Calhoun, A.P. Chandrakasan, A 256-kb 65-nm sub-threshold SRAM design for ultra-low-voltage operation. IEEE J. Solid-State Circ. **42**(3), 680–688 (2007)
53. L. Yang, D. Bankman, B. Moons, M. Verhelst, B. Murmann, Bit error tolerance of a CIFAR-10 binarized convolutional neural network processor, in *2018 IEEE International Symposium on Circuits and Systems (ISCAS)* (2018), pp. 1–5
54. A. Di Mauro, F. Conti, P.D. Schiavone, D. Rossi, L. Benini, Pushing on-chip memories beyond reliability boundaries in micropower machine learning applications, in *IEEE International Electron Devices Meeting, 2019. IEDM Technical Digest (in press)*
55. I. Hubara, M. Courbariaux, D. Soudry, R. El-Yaniv, Y. Bengio, Quantized neural networks: training neural networks with low precision weights and activations, Sep 2016. arXiv: 160907061 Cs
56. R. Wang, L. Jiang, Y. Zhang, L. Wang, J. Yang, Selective restore: an energy efficient read disturbance mitigation scheme for future STT-MRAM, in *2015 52nd ACM/EDAC/IEEE Design Automation Conference (DAC)* (2015), pp. 1–6
57. Micron MT25TL01GBBB 1 Gb, Twin-Quad I/O Serial Flash Memory (2015), p. 94
58. Q. Dong et al., A 1 Mb 28 nm STT-MRAM with 2.8 ns read access time at 1.2 V VDD using single-cap offset-cancelled sense amplifier and in-situ self-write-termination, in *2018 IEEE International Solid-State Circuits Conference—(ISSCC)* (2018), pp. 480–482
59. P. Jain et al., 13.2 A 3.6 Mb 10.1 Mb/mm2 embedded non-volatile ReRAM macro in 22 nm FinFET technology with adaptive forming/set/reset schemes yielding down to 0.5 V with

sensing time of 5 ns at 0.7 V, in *2019 IEEE International Solid-State Circuits Conference—(ISSCC)* (2019), pp. 212–214
60. R. Giterman, A. Fish, N. Geuli, E. Mentovich, A. Burg, A. Teman, An 800-MHz Mixed-$V_\textT$4T IFGC embedded DRAM in 28-nm CMOS bulk process for approximate storage applications. IEEE J. Solid-State Circ. **53**(7), 2136–2148 (2018)
61. W.-S. Khwa et al., A 65 nm 4 Kb algorithm-dependent computing-in-memory SRAM unit-macro with 2.3 ns and 55.8TOPS/W fully parallel product-sum operation for binary DNN edge processors, in *2018 IEEE International Solid-State Circuits Conference—(ISSCC)* (2018), pp. 496–498
62. W.-H. Chen et al., A 65 nm 1 Mb nonvolatile computing-in-memory ReRAM macro with sub-16 ns multiply-and-accumulate for binary DNN AI edge processors, in *2018 IEEE International Solid-State Circuits Conference—(ISSCC)* (2018), pp. 494–496
63. L. Jiang, M. Kim, W. Wen, D. Wang, XNOR-POP: a processing-in-memory architecture for binary convolutional neural networks in wide-IO2 DRAMs, in *2017 IEEE/ACM International Symposium on Low Power Electronics and Design (ISLPED)* (2017), pp. 1–6
64. M.M.S. Aly et al., The N3XT approach to energy-efficient abundant-data computing. Proc. IEEE **107**(1), 19–48 (2019)
65. B. Zimmer et al., A 0.11 pJ/Op, 0.32–128 TOPS, scalable multi-chip-module-based deep neural network accelerator with ground-reference signaling in 16 nm, in *2019 Symposium on VLSI Circuits*, Kyoto, Japan (2019), pp. C300–C301

Chapter 20
Multi-sensor Scenarios for Intelligent SOCs

Bernd Hoefflinger

20.1 Introduction

All real-world electronics systems are only as intelligent as their capability of sensing the real world. With the accelerating growth of intelligent Silicon processing, the demand on relevant and possibly Silicon-compatible world sensing increases rapidly.

Silicon sensing, particularly microelectromechanical systems (MEMS), has been a specialty since 1980, promoted to a More-than-MOORE Strategy, and described in Chap. 14 in CHIPS 2020 and updated in Chap. 15 of CHIPS 2020, Volume 2 [1]. The largest monolith effort and market grew out of photo-sensors, treated extra in [1, 2] and again in the present book with the special focus on optimal visual-system image acquisition in Chap. 21. MEMS have been a special challenge for monolithic and heterogeneous 3D integration, now with a gigantic interest because of the broad parallel sensing inputs to deep-learning neural networks with real-time results.

20.2 Compatible Silicon MEMS Systems

Silicon-compatible sensors for acceleration and Coriolis forces with on-site conversion and coding, and cost-effectiveness for automotive were the launch of MEMS mass-production [2], which enabled the introduction into mobile consumer applications. This has pushed the MEMS roadmap forward to a scaling strategy including 3D integration and packaging [1], with a broad spectrum of innovative applications [1], see Fig. 20.1, and meeting the requirements of the IOT (Internet of Things).

MEMS gyroscope systems for consumer devices like smartphones or wearables are often implemented using micromachined sensors with a small area and open loop

B. Hoefflinger (✉)
Leonberger Strasse 5, Sindelfingen, Baden-Württemberg, Germany
e-mail: bhoefflinger@t-online.de

B. Murmann and B. Hoefflinger (eds.), *NANO-CHIPS 2030*,
The Frontiers Collection, https://doi.org/10.1007/978-3-030-18338-7_20

Fig. 20.1 Applications of multi-degrees-of-freedom silicon MEMS (courtesy BOSCH)

readout circuits. This results in low current consumptions of less than 1 mA for three gyroscope axes, while moderate angular rate noise densities (7 mdps/sqrt(Hz)) as well as bias stabilities (10°/h) can be achieved [3, 4].

Automotive or inertial navigation applications require a higher accuracy with lower noise performance. This can be achieved by employing closed loop, mode-matched sensors and by using more sophisticated additional error compensation circuits like active quadrature compensation [5]. Thus, a noise density of 3.9 mdps/sqrt(Hz) in combination with 1.2°/h bias instability was realized [5]. Also larger sensor elements produced using silicon-on-insulator technologies have shown to improve the sensitivity and achieve noise performances down to 0.18 mdps/sqrt(Hz) with a bias instability of 0.08°/h [6]. However, the power consumption for high performance systems is typically strongly increased, e.g., up to more than 2.5 mA for a single gyroscope axis in [6] and 8.8 mA for a triaxial gyroscope in [5].

Gyroscope systems implementing various techniques for active error compensation, low noise performance and at the same time low power consumption are reported in [7, 8]. With a current consumption of about 0.5 mA for one gyroscope axis a noise density of 2 mdps/sqrt(Hz) and a bias instability of 0.9°/h are achieved [7].

Besides mechanical quantities, many others have to be identified like gas, fluids, chemicals, magnetism, and radiation. A general overview will be found in [9].

20.3 CMOS Image Sensors

Silicon image sensors started in 1970, when photons generated electrons in the potential bucket of a silicon pixel, which were then transferred into series-coupled MOS charged-coupled devices. As MOS integrated circuits were scaled down, it became feasible in the late 1980s, to put an MOS source-follower transistor and a select transistor into each pixel for reading out the photo-charge bucket and building "active pixels" MOS image sensors. This evolution perpetuated the linear-response characteristic

- pixel output voltage ~ photons collected.

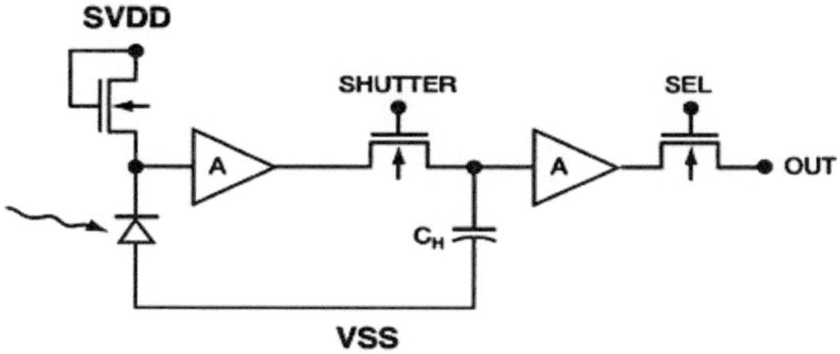

Fig. 20.2 Circuit diagram of a global-shutter HDRC® pixel for high-speed, random-access with a dynamic range of >1.000.000:1

With practical limits on photodiode area and on integration time, this linear-response has limited CCD and MOS sensors to a dynamic range from Dark charge to "white saturation" to <1.000:1. The Human Visual System (HVS) has an instantaneous dynamic range of >200.000:1, with adaptation to 1.000.000:1 (see Fig. 3.1, in Chap. 3 on Real-World Electronics). The very insufficient Low-Dynamic Range (LDR) of CCD and standard CMOS sensors has led to a waste of multiple exposures or multiple sub-pixels per pixel, described in the introduction to the following Chap. 21 on High-Dynamic-Range (HDR) Video.

It was the invention of a very minor change of wiring the source-follower transistor in the standard CMOS pixel, which led to the "Silicon Wonder" of

- a sensor characteristic with the logarithmic characteristic of the Human Visual System (HVS) and exceeding it in speed, dynamic range (dark and white) and in spectral range, including IR and UV.

As shown on the left in the circuit diagram of Fig. 20.2, the transistor is diode-connected, $V_{DS} = V_{GS}$. Here, the drain-current I_D = photodiode-current flows in the sub-threshold region of the MOS transistor, so that the source voltage is ~log I_D over more than seven orders-of-magnitude of the ühoto-current.

First results of this HDRC® sensor [10] were published as a 64 × 64 pixels video sensor in 1993 [11]. The sensor is at the core of a book on HDR Vision [12], and its more recent mega-pixel version with 1.296 parallel 12b-output was covered in Chap. 13 of [1]. Its HVS features like log response, offering color constancy, are the basis of the following Chap. 21.

20.4 LIDAR Arrays

In order to identify objects and their distance, pulses of ultra-sound, light, lasers and microwaves are being transmitted, and their return signals have to be sensed. These techniques of acquiring information on our 3D world, not only in real-time but with updates in micro- and milliseconds, and in sufficient spatial resolution, have become

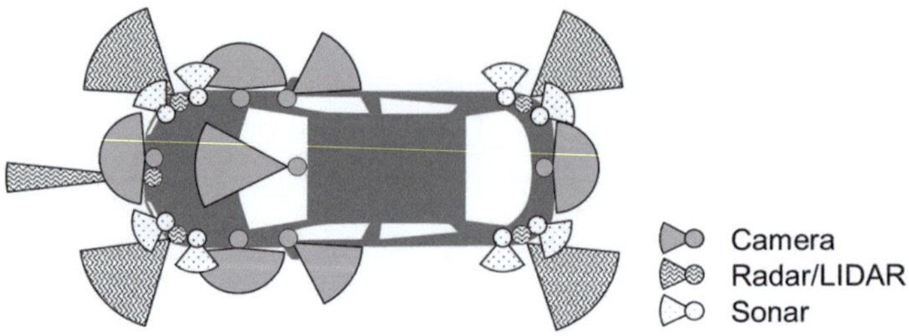

Fig. 20.3 Schematic set-up of sensors for an autonomous vehicle (from Chap. 29). RADAR and LIDAR (laser detection and range-finding) are phased arrays of sensors with 3D detection capability

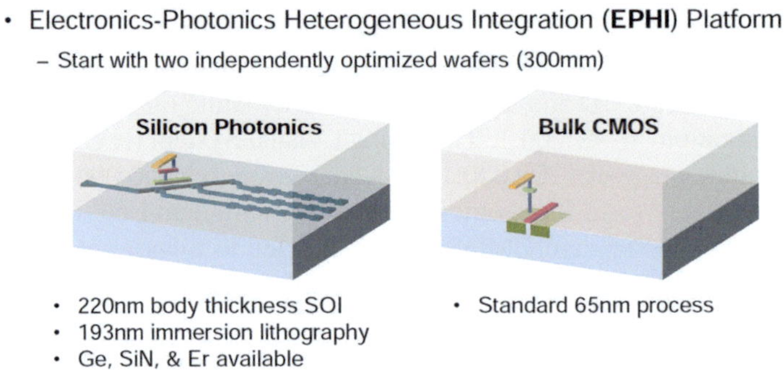

Fig. 20.4 Photonics-electronics wafer-level integration [13]. © IEEE 2019

essential, particularly for autonomous vehicles and robots. A representative set-up for an autonomous vehicle is shown in Fig. 20.3.

Energy-efficient, high-speed, high-resolution, and long-range phased-arrays of sensors need 3D integration with high-performance CMOS steering and processing planes. A leading 3D-integrated SOC realization is shown in Fig. 20.4 [13].

It is the first single-chip Optical Phased Array (OPA) [13]. It has 32×32 transmit elements enabling a $0.15°$ resolution with a steering range of $18.5°$ and a remarkable energy efficiency. Relevant data is shown in Table 20.1. The data means a resolution of $0.25 \text{ m} \times 1 \text{ m}$ at a distance of 100 m, where the lateral steering width would be 30 m. A micrograph is shown in Fig. 20.5.

This single-chip phased-array LIDAR demonstrates the potential for intelligent, Silicon-supported mobility.

Table 20.1 Data on a single-chip 3D-integrated optical phased-array. From Kim et al. [13]

Technology	Silicon photonics + 65 nm CMOS	
Steering dim.	2D (λ/θ)	
Beamwidth (Φ/ϴ)	0.15°/0.6°	0.15°/0.25°
Aperture	500 μm × 130 μm	500 μm × 500 μm
Steering range (Φ/ϴ)	18.5°/16°	18.5°/–
Sidelobe suppression	8.5 dB	7.4 dB
# of elements/independent ctrls	32/32	125/125
Phase shifter	Thermal	
Efficiency	20 mW/π	

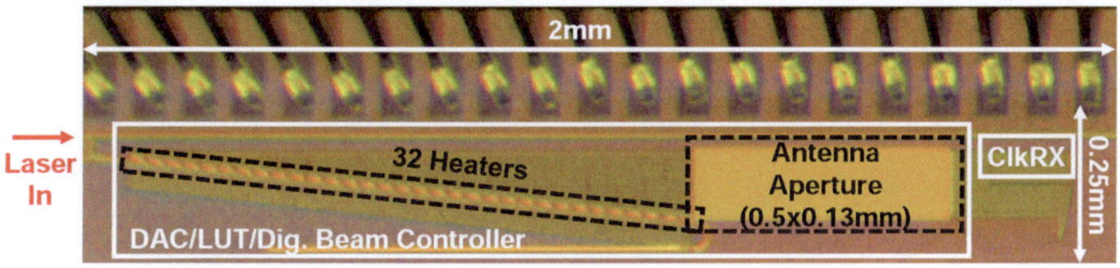

Fig. 20.5 Micrograph of the phased-array LIDAR SOC [13]. © IEEE 2019

20.5 Mobile-System-to-System Communication

Multi-vehicle scenarios like cars, robots, drones et al., need installed-infrastructure control, infrastructure and environment recognition, and high-speed, energy-efficient vehicle-to-vehicle communication, as shown in Fig. 20.6.

Gb/s high-sensitivity, low-energy communication is a major challenge. Optical communication with high-fidelity bandwidth would be a candidate, and GaN

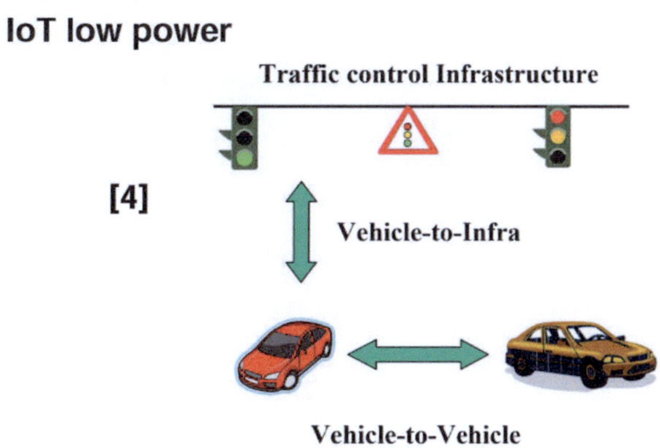

Fig. 20.6 Multi-vehicle control and communication scenario. Efficient vehicle-to-vehicle communication serves the Internet-of-Things (IOT) challenge [14]. ©IEEE 2019

laser diodes with >5 GHz bandwidth at, e.g., 450 nm are such transmitter candidates. CMOS receiver chips use special diodes or diode arrays: Photodiodes (PD), Avalanche Photodiodes (APD) or Single-Photon Avalanche Diodes (SPAD), and the leading realization is a receiver chip with 64 × 64 SPAD elements in a 21 μm pitch in a 130 nm process [14]. Its micrograph is shown in Fig. 20.7 with a chip-size of 2.6 mm × 3.6 mm. It achieves a sensitivity of −46 dBm (0.25 μW) at a rate of 500 Mb/s [13]. This data is listed in a comparison with other reported chips in Table 20.2. The most informative performance diagram shows the sensitivities versus the bit rates in Fig. 20.8. This figure includes the line for the quantum-limit. It shows that the best are about a factor of 40 away from this limit. The table also shows the energy per bit. It shows that the high-sensitivity chips need about 240 pJ/b, which means that high-speed wireless communication, even for a short distance, is not cheap energy-wise.

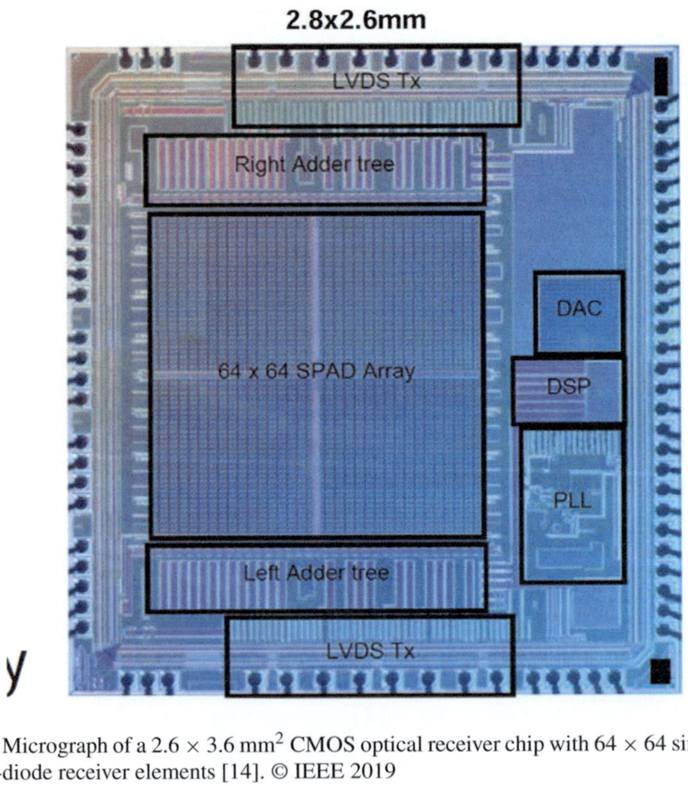

Fig. 20.7 Micrograph of a 2.6 × 3.6 mm^2 CMOS optical receiver chip with 64 × 64 single-photon avalanche-diode receiver elements [14]. © IEEE 2019

Table 20.2 Performance comparison of laser-light receiver-chips [14]. ©IEEE 2019

Ref.	Fahs	O'Brien	Jukic	Fisher	Zhang	Steindl	This work
Type	PD	APD	APD	SPAD	SPAD	SPAD	SPAD
Technology	0.35 μm	N/A	0.35 μm	130 nm	180 nm	0.35 μm	130 nm
Elements	1	1	1	1024	60	4	4096
Fill factor	N/A	N/A	N/A	2.42%	3.2%	48%	43%
Sensitivity (dBm)	−23	−38	−34.6	−31.7	−53	−43.8	−46.1
Modulation type	OOK	OOK	OOK	OOK	OOK	RZ-OOK	4PAM, OFDM
Bit rate	2.5 Gb/s	280 Mb/s	1 Gb/s	100 Mb/s	100 Mb/s	200 Mb/s	500 Mb/s, 350 Mb/s
BER	10^{-3}	10^{-9}	10^{-9}	10^{-9}	10^{-3}	6.5×10^{-3}	2×10^{-3}
Consumption (pJ/bit)	86	N/A	244	800	N/A	248	230

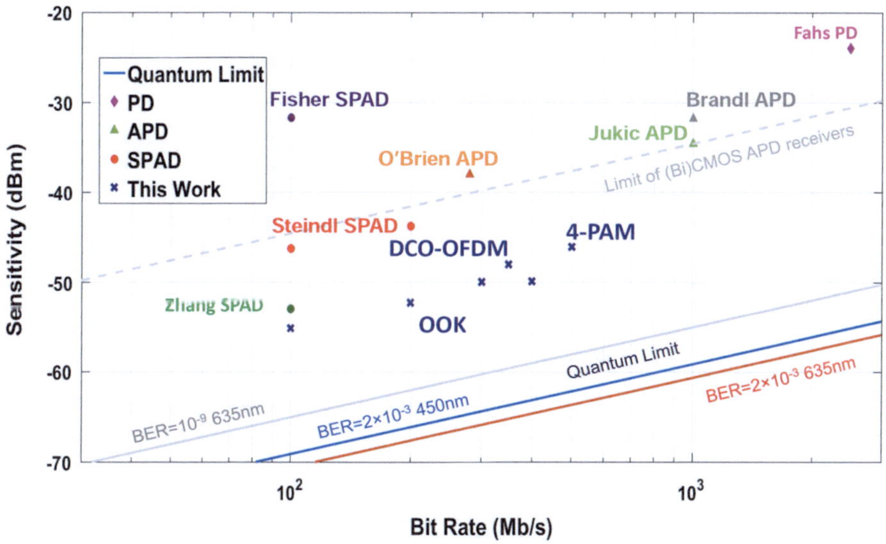

Fig. 20.8 Sensitivity and bit-rate of laser-light receiver chips [14]. See Reference [14] for authors and modulation details. The leader is far down on the far right. © IEEE 2019

20.6 Conclusion

Nano-scale miniaturization, large-scale- and 3D-integration have enabled multi-sensor arrays with gigantic data rates. Their processing towards essential information and inference is a major challenge for deep-learning neural-networks. The focus on optical sensing and receiving in this chapter points out its data-rate challenge. Sensing is an issue in anything, physical or chemical, which makes it a handbook issue as referenced in [9].

References

1. B. Hoefflinger (ed.), *CHIPS 2020 vol. 2—New Vistas in Nanoelectronics* (Springer Science and Business Media, 2016). ISBN 078-3-319-22092-5
2. B. Hoefflinger (ed.), *CHIPS 2020—A Guide to the Future of Microelectronics* (Springer Science and Business Media, 2012). ISBN 978-3-642-22399-0
3. C. Ezekwe, W. Geiger, T. Ohms, A 3-axis open-loop gyroscope with demodulation phase error correction, in *IEEE International Solid-State Circuits Conference Digest of Technical Papers*, 25 Feb 2015, pp. 478–479
4. B. Sensortec, Low noise, low power triaxial gyroscope, *BMG250 datasheet*, 2019
5. G.K. Balachandran, V.P. Petkov, T. Mayer, T. Balslink, A 3-axis gyroscope for electronic stability control with continuous self-test. IEEE J. Solid-State Circuits **51**(1), 177–186 (2016)
6. Y. Zhao et al., A sub-0.1/h bias-instability split-mode MEMS gyroscope with CMOS readout circuit. IEEE J. Solid-State Circuits **53**(9), 2636–2650 (2018)
7. M. Marx, D. De Dorigo, S. Nessler, S. Rombach, Y. Manoli, A 27 μW 0.06 mm^2 background resonance frequency tuning circuit based on noise observation for a 1.71 mW CT-$\Delta\Sigma$ MEMS gyroscope readout system with 0.9/h bias instability. IEEE J. Solid-State Circuits **53**(1), 174–186 (2018)
8. M. Marx, X. Cuignet, S. Nessler, D. De Dorigo, Y. Manoli, An automatic MEMS gyroscope mode matching circuit based on noise observation. IEEE Trans. Circuits Syst. II Express Briefs **66**(5), 743–747 (2019)
9. B. Vigna, Chapter Sensors, in *Springer Handbook of Semiconductor Devices*. To be published
10. B. Hoefflinger, U. Seger, M.E. Landgraf, Image Cell for an Image Recorder Chip. US Patent 5608204, filed 03.23.1993, issued 03.04.1997
11. U. Seger, H.G. Graf, M.E. Landgraf, Vision assistance in scenes with extreme contrast. IEEE Micro **13**, 50–56 (1993)
12. B. Hoefflinger (ed.), *High-Dynamic-Range (HDR) Vision* (Springer, Berlin, 2007). ISBN 978-3-540-44432-9
13. T. Kim, et al., A single-chip optical phased array in a 3D-integrated silicon photonics/65 nm CMOS technology, in *International Solid-State Circuits Conference, 2019, Digest of Technical Papers*, paper 29.5, pp. 454–455, Feb 2019
14. J. Kosman, et al.: A 500 Mb/s −46.1 dBm CMOS SPAD receiver for laser-diode visible light communications, in *International Solid-State Circuits Conference, 2019, Digest of Technical Papers*, paper 29.7, pp. 468–469, Feb 2019

Chapter 21
High-Dynamic-Range and Wide Color Gamut Video

Zhichun Lei, Xin Yu and Markus Strobel

21.1 Introduction

The still or video camera nowadays usually provides Low Dynamic Range (LDR) images. The LDR images contain maximally 256 ($2^8 - 1$) linear levels, in practical terms even less because some levels are for instance reserved for the synchronization purpose in case of video transmission. As a result, sometimes the images appear either too dark or white-saturated.

Illumination levels of real-world scenes can vary from 0.001 lx at clear night sky to 100,000 lx at direct sunlight [1] comprising a ratio of eight decades (1:10^8 or 160 dB). This poses a tough challenge on the dynamic range of cameras and image sensors attempting to acquire the scene information as video sequences. Although this change from very dark to very bright illumination happens gradually over the day (inter-scene), outdoor day- or night-time scenes can exhibit different illumination levels up to four or five decades (80–100 dB intra-scene) easily. The intensities projected by the lens onto the sensor pixels result as the product of the illumination and the object reflectance. The typical range of object reflectance between 5 and 95% (1:19) extends the sensed dynamic range by more than an additional decade (+26 dB). Therefore image sensor technologies capable of acquiring an intra-scene High Dynamic Range (HDR) of at least 100–120 dB are mandatory. The latter value is needed when bright active light sources are in the field of view, visible directly or reflected e.g. on metallic surfaces, for a robust image acquisition.

There are numerous publications pointing out the necessity of HDR images, e.g. [1, 2]. In the European New Car Assessment Program (Euro NCAP) new tests in

Z. Lei (✉) · X. Yu
Tianjin University, School of Microelectronics, 92 Weijin Road, Tianjin 300072, People's Republic of China
e-mail: zclei@tju.edu.cn

M. Strobel
IMS CHIPS, Allmandring 30a, 70569 Stuttgart, Germany

B. Murmann and B. Hoefflinger (eds.), *NANO-CHIPS 2030*,
The Frontiers Collection, https://doi.org/10.1007/978-3-030-18338-7_21

the car-safety rating are included for vehicles to avoid collisions with bicycles and pedestrians and will be expanded for testing the night-time performance. Other applications are traffic monitoring, security and surveillance, industrial vision, robotics as well as welding inspection having the highest demand on sensor dynamic range. While Ultra HD displays become more prominent, HDR is a key element [3] besides high pixel resolution and wide color gamut. For HDR-enabled flat-panel displays, video content acquired with HDR cameras is needed to make full use of its feature.

HDR is also important for the WCG purpose, because less pixel gray levels lead to less color combination possibility. Furthermore, the display color gamut is strongly influenced by the brightness. With the increasing display brightness, the color gamut will shrink. At the maximum display brightness, only white color points are displayed. This problem can be well avoided by HDR because the HDR display is usually of much higher peak brightness than the LDR display. However, HDR display of high brightness often faces an overheating problem. Four-primary-color WCG can reduce display's power consumption, and in turn mitigate the overheating problem of HDR display.

WCG is currently another most acute topic of video technology development. WCG development is driven by the high color-fidelity application requirements, e.g. telemedicine, e-commerce. However, the popular video system using red (R), green (G) and blue (B) primaries, specified by ITU BT.709 standard, only cover 33.24% of the visual locus on the CIE 1931 chromaticity diagram [4]. ITU BT.2020 specified pure RGB three-primary colors, which can cover 63.3% of the visual locus. However, 63.3% coverage will certainly not be the ultimate goal of the WCG technology development. Besides, BT.2020 is so demanding with respect to the purity of the RGB three primaries that until now there is no imaging technique, which can meet its requirement, even though there are standards supporting it, e.g. HDR10, Dolby Vision, HDMI2.0, Display Port1.4, H.265/HEVC. More than RGB three-primary colors can mitigate the above problems.

The remainder of this chapter is organized as follows. Section 21.2 addresses the HDR and WCG imaging techniques with emphasis on the logarithmic HDR imaging technique, which matches the human visual system. Its extension to the WCG imaging is straightforward. Section 21.3 deals with the display techniques of HDR and WCG contents. Section 21.4 talks about the heat dissipation problem of HDR display and the display power saving by the WCG technique. Section 21.5 discusses the delivering of HDR and WCG video contents. Since the delivering technique for HDR video contents is well-established, this section emphasizes the delivery of four-primary-color WCG video contents by means of the state-of-the-art data compression methods and the available YUV bandwidth. It will state that it is possible to deliver four-primary-color WCG video by means of the state-of-the-art image/video coding standards and the available YUV bandwidth. Finally, the authors will summarize this chapter.

21.2 HDR and WCG Imaging Techniques

This section will at first give an introduction to principles achieving HDR video acquisition with linear and non-linear opto-electronic conversion functions (OECF) and show state-of-the-art implementations with signal-to-noise and data processing properties. The HDR techniques for CMOS imagers should fulfill the following properties in order to be useful for the above mentioned applications:

- Dynamic Range >100–120 dB at video speed, i.e. frame rates >30 fps
- Balanced Signal-to-Noise Ratio (SNR) and contrast resolution over the dynamic range with low-light sensitivity
- No or less motion artifacts of moving objects due to HDR processing
- HDR image data processing on imager chip (if needed)
- Possible use of standard image post-processing (image processing pipeline, color, etc.)
- CMOS imager sensor (CIS) implementation in reasonable small pixel size for possible high resolution (Full HD, 4 K)
- Possibility to employ global shutter (snap shot) instead of rolling shutter
- Low imager and system complexity/cost.

These properties can be fulfilled by IMS CHIPS's logarithmic image sensor, which will be described in detail in this section.

After describing the HDR imaging techniques, the WCG imaging techniques are addressed, and the authors will focus on the multi-primary-color imaging techniques.

21.2.1 HDR Imaging

In Sect. 21.2.1, one at first introduces the HDR imaging methods which are suitable for video application. Then, the logarithmic image sensor is compared to other HDR imaging methods.

21.2.1.1 HDR Video Imaging

To obtain an HDR image, one can capture several LDR images under different exposures. These LDR images are then fused into a single HDR image. Because scenes to be imaged can move, e.g. cars move in a motorway, acquiring HDR video by means of sequentially capturing multiple LDR images faces the motion problem. There are many approaches to bypass the motion problem. In the following, several methods will be selected and briefly introduced.

For real-time HDR imaging, one can apply beam splitters to duplicate the optical image of a scene in question. These duplicated optical images are simultaneously detected by image sensors with different exposures [5, 6]. Since all the LDR images

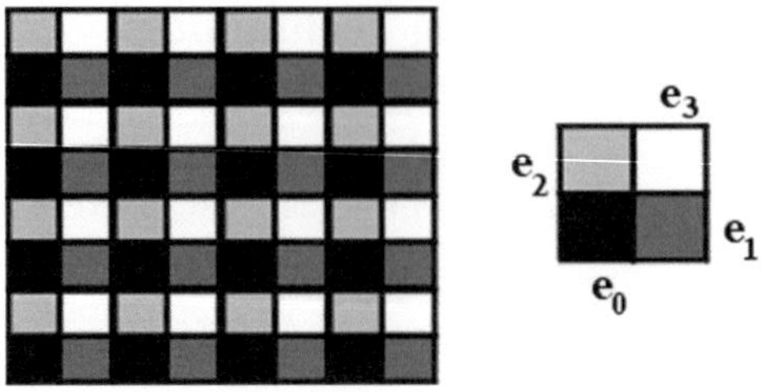

Fig. 21.1 Spatially varying pixel exposure pattern [8]

are detected simultaneously, there is no movement among the LDR images. However, the HDR imaging system is bulky and expensive.

In order to capture videos in real-time, Guthier et al. capture the LDR images with partial re-exposures to save the acquisition time [7]. Nayar et al. use an optical mask with different transparencies in different CMOS sensor areas [8]. As depicted by Fig. 21.1, every four pixels with different transparency are used to form an HDR pixel. The different transparency plays the role of different exposures. Because all LDR pixels are captured at the same time, the motion blur problem is avoided. However, the resolution of each resulting HDR image is reduced by a factor of four. This method is also disadvantageous with respect to color HDR image capturing.

Nayar et al. use the radiance value of the corresponding scene point to adapt the exposure of each pixel on the image detector [9]. The pixel brightness of the image captured before is used to perform the exposure adaptation. The captured image and the exposure adaptation amount are together used to compute the HDR image. Because the image captured before is used to control the exposure of the current scene, it may encounter problem in case of fast object moving or scene change.

Hoefflinger et al. make use of the "leakage" or "parasitic" part of the MOS transistor's characteristic curve for developing an HDR CMOS image sensor, which is branded as HDRC [1]. Figure 21.2 gives a typical characteristic curve between the

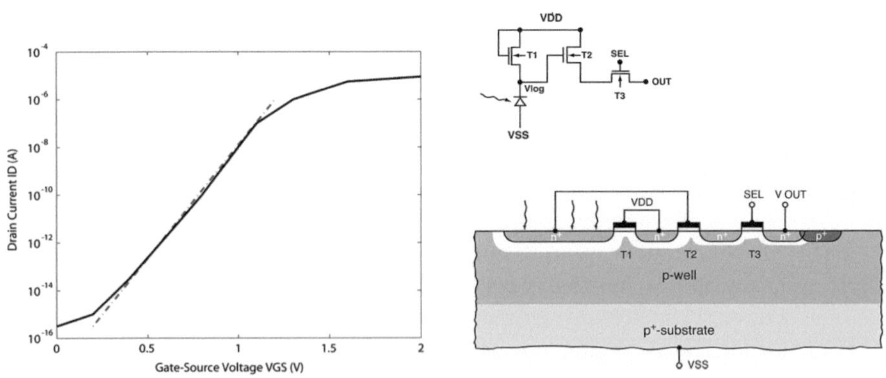

Fig. 21.2 Typical characteristic curve of NMOS transistor and a HDRC imaging cell [1]

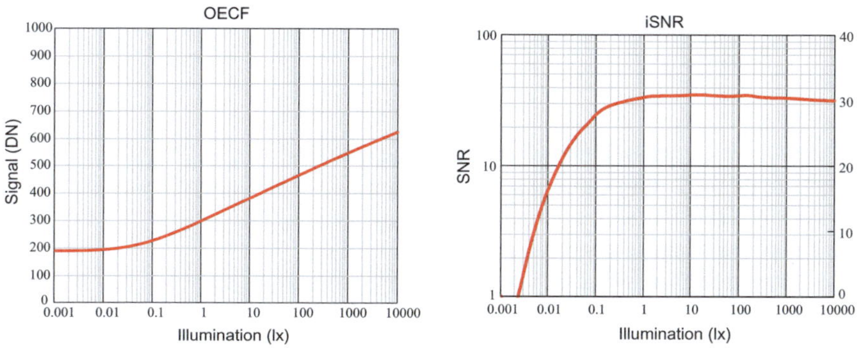

Fig. 21.3 (Left) Digital output of the logarithmic HDRC image sensor versus illumination; (Right) Corresponding Signal-to-Noise ratio of the HDRC image sensor OECF

drain current ID and the gate-source voltage VGS of NMOS transistor in case of VGS = VDS and VBS = 3.0 V. It is clear that the drain-source voltage VDS (=VGS) is the precise logarithm of the drain current ID at least over eight orders of magnitude [1]. Therefore, one can realize a logarithmic imaging and achieve a dynamic range of more than $20 \cdot \log 10^8 = 160$ dB in principle. The right drawing of Fig. 21.2 illustrates a HDRC imaging cell.

The HDRC image sensor output versus illumination intensity has been measured and is reportedly shown in Fig. 21.3 [10] employing the subthreshold transistor characteristic of Fig. 21.2. In addition, [11, 12] include optoelectronic conversion functions (OECF) of HDRC imagers with global shutter feature. In [13], a logarithmic imager using the solar cell mode of the photodiode is reported. In the OECF of Fig. 21.3 (left), one can observe the clear relationship between the output in digits converted with a 10 Bit analog-to-digital converter (ADC) and the incident light level in Lux. Because the abscissa of this figure is logarithmically arranged, whereas the ordinate is plotted linearly, Fig. 21.3 (left) clearly illustrates the logarithmic imaging feature of the HDRC image sensor.

An important figure of merit of the performance of an image sensor or system is the signal-to-noise ratio (SNR). Figure 21.3 (right) shows the measured SNR versus illumination intensity corresponding to the HDRC imager OECF of Fig. 21.3 (left). It increases from a SNR of 1 at 2 mlx low light sensitivity to reach a nearly constant maximum SNR level around 35 (31 dB) above 0.1 lx. The dynamic range of the logarithmic HDRC image sensor is greater than 134 dB ranging from approximatively 2 mlx (SNR of 1) up to more than 10,000 lx (limited by the light source of the measurement setup).

Due to the logarithmic nature of the OECF, the SNR is given as the incremental signal-to-noise ratio (iSNR) as defined by the ISO 15739:2017 standard [14]. ISO 15739:2017 specifies methods for measuring and reporting noise versus signal level and dynamic range of cameras including non-linear characteristics whereas EMVA 1288 standard [15] for machine vision applications uses a linear sensor model but will be extended for HDR cameras in a future release.

21.2.1.2 Logarithmic Image Sensor in Comparison to Other Image Sensors

Comparison with other kinds of HDR image sensors
To compare the logarithmic image sensor with other concepts introduced in the above paragraph A, *HDR Video Imaging*, the basic underlying principle of the dynamic-range extension of those concepts is briefly described. To obtain an HDR response from a saturation-limited LDR pixel, multi exposure LDR captures can be fused together resulting in a linear or piecewise linear OECF with HDR. It is legitimately assumed, that the pixels of these concepts follow the camera model according to EMVA 1288 [15] with an output signals S linear to the exposure H, $S \propto H$. The exposure is proportional to the total quantum efficiency η, the irradiance E and the exposure (integration) time T, which results in $S \propto H \propto \eta \cdot E \cdot T$. In the regime of LDR fusion (light intensities above first saturation having already a reasonable SNR), a further assumption is that the temporal random noise N (N denotes the standard deviation of S) is dominated by the shot noise of the photo generated electrons. Therefore N is proportional to the square root of the exposure and signal [15], $N \propto \sqrt{H} \propto \sqrt{S}$, resulting in a SNR also proportional to the square root of the exposure, $SNR = S/N \propto \sqrt{S} \propto \sqrt{H}$.

Different exposures H_i ($i = 1, 2, 3, \ldots$) of a pixel or a pixel cluster are achieved by (a) multiple integration times T_i ($H_i \propto \eta \cdot E \cdot T_i$) or (b) using beams splitter, optical masks (Fig. 21.1) or different sized photodiodes effecting η_i ($H_i \propto \eta_i \cdot E \cdot T$). Both have equal effects on the resulting total OECF in respect to the SNR behavior. As described in paragraph A for omitting motion artifacts method (b) with constant time T is preferred. A special but similar case is represented by dual conversion gain sensors employing a lateral over-flow integration capacitor (LOFIC), e.g. [16].

In the following example, three different exposures are obtained by three subsequent exposure times T_1 (longest, for low intensity), T_2 and T_3 (shortest, for high intensity) during which the photo-generated electrons are integrated in the pixel photodiode. This results in different slopes of the three partial OECF curves (blue, green and yellow) in Fig. 21.4, each digitized by a e.g. 12 Bit ADC, spanning an extended range of the light intensity axis. The dynamic range extension (DRE) to the sensor's base dynamic range is then given by the ratio $T_1/T_3 = T_1/T_2 \cdot T_2/T_3$. For a given light intensity, the highest digital signal level of a non-saturated partial OECF is taken since it exhibits the highest SNR.

Following this to obtain a complete OECF, the partial OECF curves of T_2 and T_3 are multiplied digitally in post-processing (either on chip or on system) by a constant factor to exhibit the same slope. Therefore, they can be continually fitted together in a linear HDR signal as depicted in Fig. 21.5 with the dashed lines. The linearization step is important for color image processing to prevent false color reproduction at the transition points. After that a piecewise compression can be applied to reduce the necessary wide bit depth of the linear HDR signal, e.g. 20 Bit, for the output data bus. In a rough sense, this approximates a logarithmic characteristic as can be seen with the fitted solid lines in Fig. 21.5.

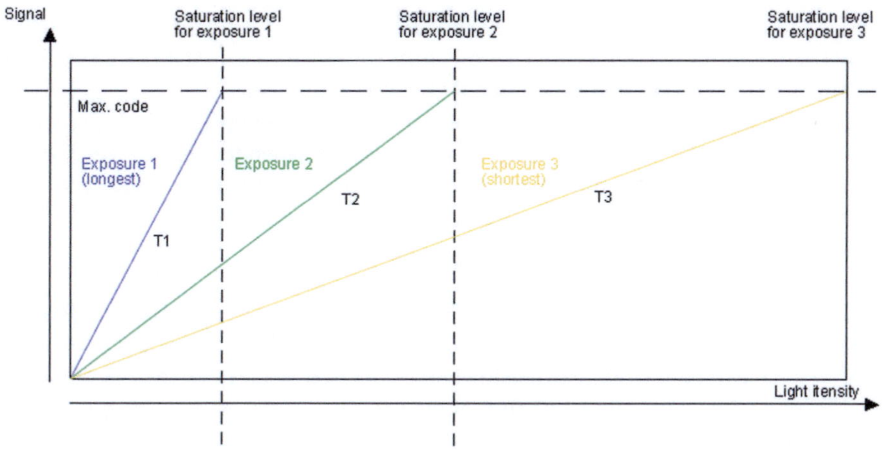

Fig. 21.4 Multiple exposure captures using three different exposure times [17]

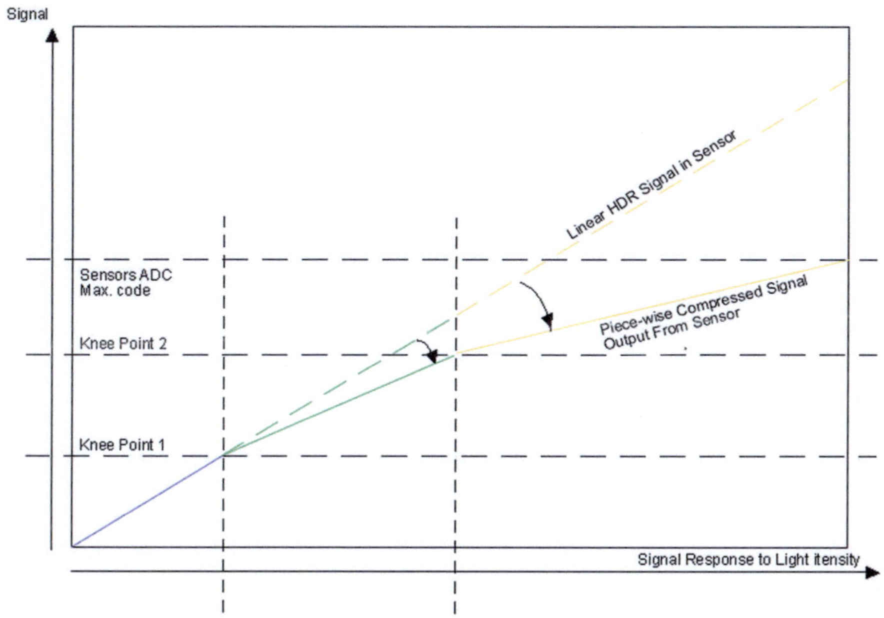

Fig. 21.5 Resulting linear internal HDR signal (dashed) as well as compressed piecewise linear output signal (solid) derived from the 3 captures with different exposure times [17]

If the ratios T_2/T_1 and T_3/T_2 of the exposure times are 2^n a power of two, typically $2^4 = 16$, the multiplication is done by bit shifts of the digitized ADC value by n and 2n Bit for the OEFC T_2 and T_3, respectively.

The wish for a high DRE with high ratios of T_1/T_3, T_1/T_2 and T_2/T_3 respectively, is compromised by the SNR drop at the transition points when the partial OECF curve with the longer exposure time saturates. The factor by which the SNR decreases, is proportional to the square root of the time ratios giving $SNR_{drop} = \sqrt{T_1/T_2} = \sqrt{T_2/T_3}$ since SNR is assumed proportional to the square root of the exposures in the regime of LDR fusion.

To give an example: With a sensor base dynamic range of 12 Bit (72 dB) and exposure times of $T_1 = 32$ ms, $T_2 = 2$ ms and $T_3 = 0.125$ ms, ratios $T_1/T_2 = T_2/T_3 = 2^4 = 16$, the dynamic range extension equals to DRE $= T_1/T_3 = 256$ (48 dB) and a total HDR $= (72 + 48)$ dB $= 120$ dB is achieved. This comes with the disadvantages of performing three times ADC conversion plus memory for storing intermediate results and a 20 Bit $(12 + 2 \cdot 4)$ wide linear data bus for image post processing or at least 14 Bit piecewise compressed for chip-to-chip data transmission. Also the SNR decreases at the transition points by a factor of $\sqrt{16} = 4$ (-12 dB) which will be shown by measurements in the next paragraph.

Triple Exposure HDR sensor compared with logarithmic HDRC image sensor With the understanding of the previous passage, the measured data of the HDRC imager (left and right figure of Fig. 21.3) will be compared with a Triple Exposure HDR sensor [18] with a focus on the SNR performance.

The total (fused) OECF of the triple exposure sensor is plotted in Fig. 21.6 (red curve) having a linear signal range of 1–300,000 digital numbers (DN) in the ordinate. Since the scale for the abscissa is given as exposure values (lx.s) in [18] the illumination (lx) is calculated assuming a repetitive framerate of 40 fps (video speed) with a reciprocal exposure time of 25 ms. This equals a factor of 40 lx/(lx.s) for the illumination scale in Figs. 21.6 and 21.7.

The measured SNR behavior of the Triple Exposure sensor is shown in Fig. 21.7. The drops of SNR with increasing illumination at the transition points due to the switch of exposure time are clearly seen in the SNR curve. The SNR drop is around -13 dB from 40 dB peak SNR down to 27 dB. The ratio T_2/T_1 can be read by the illumination levels at max. SNR peaks to be a factor of 16, which means a theoretical SNR drop of -12 dB. The ratio T_3/T_2 should be 16 too, since it has the same SNR drop. The Dynamic Range Extension (DRE) is up to 48 dB if the SNR curve is extrapolated to a SNR_{max} of 40 dB at maximum illumination (saturation level). With a base DR of around 62 dB the overall DR of the linear Triple Exposure HDR sensor is 110 dB ranging from approx. 10 mlx (SNR of 1) to 3000 lx.

In Table 21.1 the measurement results in respect to OECF and SNR behavior of the logarithmic HDRC imager are compared to the Triple Exposure HDR sensor.

The Triple Exposure sensor serves as a representative of other kinds of HDR sensor concepts showing similar characteristics due to the same underlying principle of extending the dynamic range by multiple exposures. An exposure ratio of $H_i : H_{i+1} = 2^4 = 16$ is a practical compromise achieving a DRE of +24 dB per additional

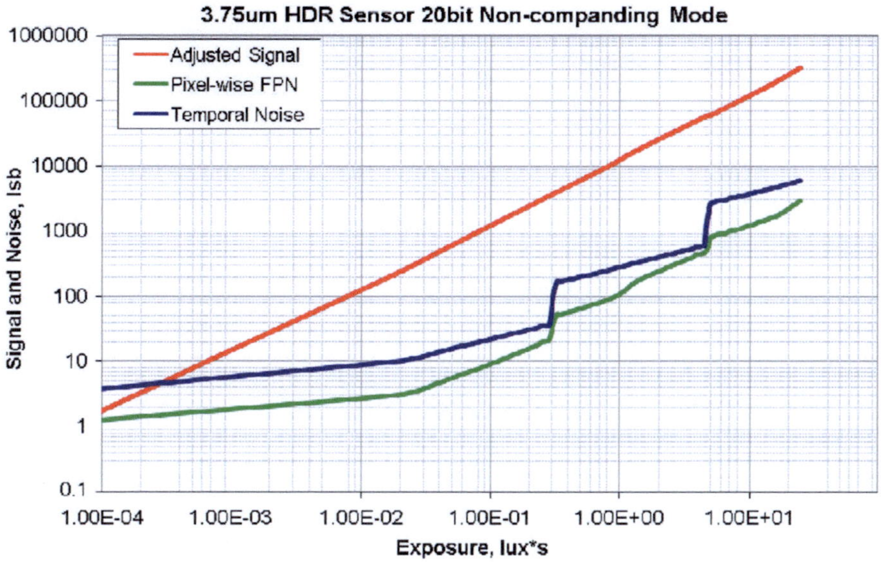

Fig. 21.6 Digital output (red curve) of the linear Triple Exposure HDR sensor versus exposure [18]

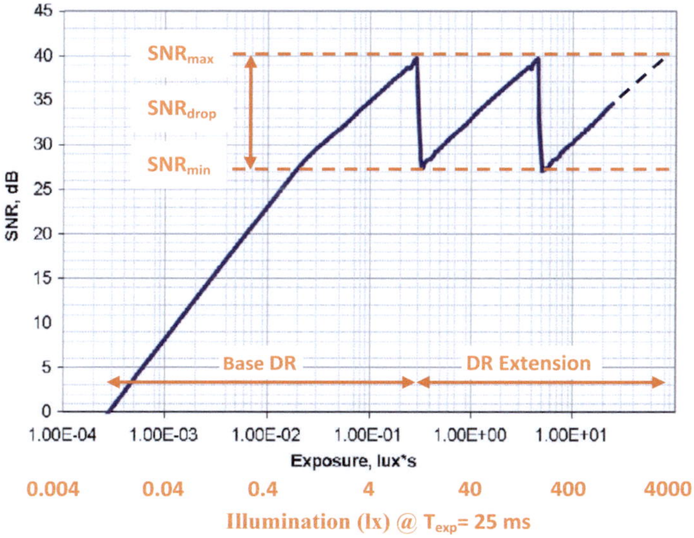

Fig. 21.7 Signal-to-noise ratio of the linear Triple Exposure HDR sensor versus exposure corresponding to Fig. 21.6 [18]

Table 21.1 Comparison of logarithmic HDRC sensor to Triple Exposure HDR sensor

Parameter	Unit	HDRC	Triple Exposure	Remark
OECF type	–	Logarithmic	Linear	
Exposures times	–	–	$T1 = 16 \cdot T2 = 256 \cdot T3$	
ADC resolution	Bit	10	12	
OECF bit depth	Bit	10	20 (14)	(Compressed pw-linear)
Dynamic range	dB	≥134	110	
Max. illumination	lx	≥10,000	≈3000	Compared at 40 fps
Sensitivity (SNR = 1)	mlx	2	10	Compared at 40 fps
SNR behavior of DR	–	Balanced	Exhibiting drops	
SNR_{max}	dB	31	40	
SNR_{drop}	dB	–	−13	At transition points
SNR_{min}	dB	–	27	In LDR fusion regime

exposure with a limited SNR_{drop} of −12 dB maintaining a reasonable minimum SNR_{min} in the regime of LDR fusion. This means that, with a sensor base DR up to 72 dB (12 Bit), for an HDR up to 96 dB (16 Bit) dual exposures and for up to 120 dB (20 Bit) triple exposures are required. Exposure ratios of a power of two permit data fusion by bit shifting which can be realized exactly using digitally generated exposure times T_i. Whereas changes of technological parameter affecting the total quantum efficiency η_i ($H_i \propto \eta_i \cdot E \cdot T$) can suffer from process variations.

As a summary, one can say, that the HDRC sensor achieves its very high dynamic range of more than 134 dB by a logarithmic OECF requiring only a 10 Bit digital output. This is already addressed in Chap. 3 *Real-Word Electronics* in terms of efficient electronic processing of real-world information. It has a well-balanced nearly constant SNR over the illumination range above 0.1 lx where the photo current dominates over photodiode dark current. Other kinds of HDR sensors with a linear OECF, as introduced in previous paragraph A, need a high bit depth of approx. 20 Bit for on-chip or on-system image processing when reaching a dynamic range up to 120 dB. As mentioned they have disadvantages related to circuit complexity, like necessary multiple read operations with ADC conversions per pixel including storage of intermediate values or reduction of spatial resolution and a SNR characteristic with drops reducing their potentially higher peak SNR.

The comparison demonstrates the necessity of HDR acquisition, as the introduction part of Chap. 21 already stated. There are many applications requiring HDR imagers [1, 19] especially in automotive imaging systems including autonomous driving as presented in the road scene in Fig. 21.8 having direct sun in the camera's view (right image). The images were taken with the logarithmic HDRC image sensor clearly demonstrating the capability of robust HDR acquisition and color rendering under such tough conditions.

Fig. 21.8 Two images of a video sequence taken with the HDRC sensor of a road scene demonstrating the need for high dynamic range acquisition with direct sun in view (right)

21.2.2 WCG Imaging

Although the RGB three image components of a color image can be captured by three image sensors covered by red, green and blue color filters, most of the color image sensors consist of only a single image sensor covered by a color filter array (CFA). Each sub-pixel can be captured by the HDRC imaging cell described above. Color imaging with a single image sensor is advantageous for e.g. low cost, low power consumption and compactness. Thus, Chap. 21 only deals with the single image sensor case. In Sect. 21.2.2, the authors first discuss imaging colors of standard color gamut. Then, the WCG imaging approaches will be addresses, which include both the three-, four- and six-primary-color cases.

21.2.2.1 Imaging of Standard Color Gamut

The CFA applied to a single color image sensor usually adopts the Bayern pattern, which arranges 50% green, 25% red and 25% blue color filters on a square grid of photo sensors. Although the saturation of the color filters has a strong influence on the purity of the RGB primaries and in turn on the color gamut, the resulting color gamut is usually limited, it is classified as standard color gamut, i.e. the color gamut specified by BT.709, also called sRGB. The high-definition television video signal has the BT.709 color gamut. As already mentioned, only 33.24% of the visible colors can be represented by the RGB primaries. All other colors, in particular colors with high saturation, lose fidelity.

International organizations have made specifications to extend the standard color gamut. One of the efforts is to exploit the quantization levels reserved to allow for some undershoot and overshoot (taught as Gipp's phenomenon) in the image processing chain without necessitating undesirable clipping. These quantization levels are used to transmit the negative intensity RGB colors of the CIE 1931 RGB color mixture curve, referring to Fig. 21.9. ITU-R BT.1361 [20] takes the Pointer gamut [21], the

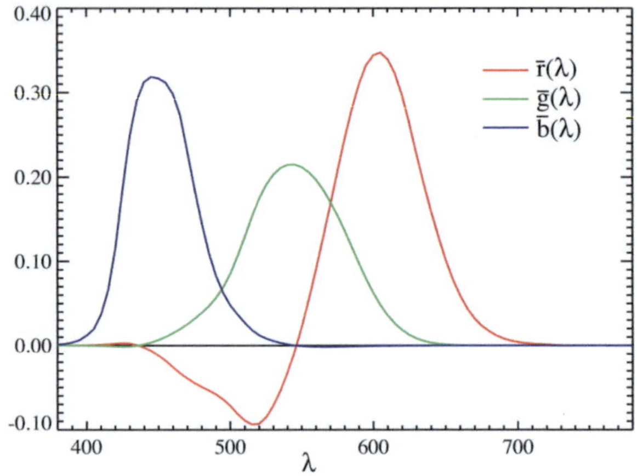

Fig. 21.9 CIE 1931 RGB color matching functions (λ in nm)

gamut of real surface colors, as the target gamut and adopts a non-centrosymmetric gamma correction curve to impose a larger compression on the negative signal part. It should be mentioned that the recommendation ITU-R BT.1361 was suppressed in 2015 [22] and there has been a standard IEC 61966-2-4 [23] on using negative intensity RGB colors to extend color gamut, i.e. the xvYCC color space. However, IEC 61966-2-4 needs to compress the negative color signals. The compression affects the signal fidelity and causes color reproduction errors, in particular for the negative intensity red color because its absolute amplitude is much larger than those of green and blue colors as reportedly shown in Fig. 21.9. Besides, the color gamut extension by means of additional quantization levels is quite limited. The coverage ratio of color gamut is limited to 37.15% [24].

21.2.2.2 Imaging of Wide Color Gamut

ITU-R BT.2020 [25] specifies a wide color gamut to be reached by pure RGB primaries and provides a reasonably good coverage of the Pointer gamut. However, the current imaging technique is not able to meet its requirement and, as a result, no BT.2020 contents can be delivered, at least at the time of this writing. With the application of quantum dot technology to imaging devices [26, 27], in future, one may capture much purer primaries than the CFA, in addition to an efficiency much higher than silicon.

Three-primary colors only form a triangle area on the CIE chromaticity diagram, which represents all the visible colors of the human visual system and is of horseshoe-shape. Four-primary colors form a quadrilateral and can more closely approximate the horseshoe shape than a triangle. The application of four-primary colors is a practicable way to achieve WCG. There are different four-primary-color image CFAs, as illustrated in Fig. 21.10.

Fig. 21.10 Two examples of four-primary-color CFA compared to Bayern CFA

In Fig. 21.10, the left figure gives the popular Bayern CFA pattern using RGB color filters, whereas the middle figure illustrates the RGBE (Red, Green, Blue, and Emerald) CFA pattern developed by the company Sony [28], and the right figure shows the CYGM (Cyan, Yellow, Green, Magenta) CFA pattern [29]. CMOS image sensors of RGBE four-primary colors was commercialized as early as in 2003.

The E color is the complementary color of the R color, whose negative intensity is the largest among the RGB negative intensities as depicted in Fig. 21.9 and whose compression after IEC 61966-2-4 affects the signal fidelity the most, as mentioned before. According to [28], the characteristics of the RGBE color filter come much closer to those of the human visual system, achieving a dramatic reduction in the color reproduction error. The E image component here actually stands for cyan. There are many occasions that require high saturated cyan, e.g. it is the color of shallow water over a sandy beach and the color of clouds of methane gas in the planet Uranus's atmosphere, it is widely used in architecture in Turkey and Central Asia, and surgeons and nurses in some countries often wear gowns colored cyan, and operating rooms are often painted in this color to reduce the emotional response to blood red. Therefore, cyan is used in various aspects such as nature, aviation, architecture and medical care etc. However, the current color gamut standards cannot effectively cover the cyan region in the spectral locus representing the visible gamut of the human vision. RGBE four-primary colors can well cover this region.

The RGB color filters have a narrow-band and thus are able to keep color fidelity. However, the narrow-band RGB color filters will result in low SNR, which one wants to avoid particularly in case of insufficient illumination scenes. On the contrary, the CMY color filters, which are widely applied in the printing industry, are of broad-band and thus are able to achieve a high SNR, which is particularly desired in case of low-illumination scenes, for instance at night. Because in a low-illumination case, the color does not play any important role in the human visual system, e.g. the human eye cannot perceive any color in darkness, the color distortion caused by the broad-band CMY color filters is irrelevant. In case of insufficient illumination scenes, a high SNR is much more important than color fidelity. For this reason, Sajadi et al. presented shiftable layers of CFAs. With them, one camera can capture sets of color primaries, namely RGB, CMY and RGBCMY [30]. It works in the following way: Without shifting the top CFA, one gets the CFA of the CMY pattern; Shifting the top CFA by one tile in the horizontal direction, it results in the CFA of the popular RGB pattern; Shifting the top CFA by one tile in the vertical direction, the RGBCMY pattern CFA is formed.

21.2.2.3 Color Gamut Comparison

Figure 21.11 depicts three color gamuts reached by RGB three primaries and the Pointer gamut. Whereas the Pointer gamut is of irregular shape, whose color gamut is depicted as the blue color curve on the uniform chromaticity diagram, dubbed the CIE 1976 UCS (Uniform Chromaticity Scale) diagram, the color gamuts achievable by using three-primary colors are of a triangle shape. The BT.709 color space, depicted as the cyan triangle on the CIE 1976 UCS, is quite limited as mentioned in the introduction section of this chapter. The BT.2020 color space, depicted as the yellow triangle in Fig. 21.11, becomes much larger than that of BT.709. However, BT.2020 requires pure RGB primaries, and this requirement challenges the current CFA color imaging method.

The BT.2020 color space is not the largest color gamut that is achievable by means of three-primary colors. From Fig. 21.11 one can clearly see that the red

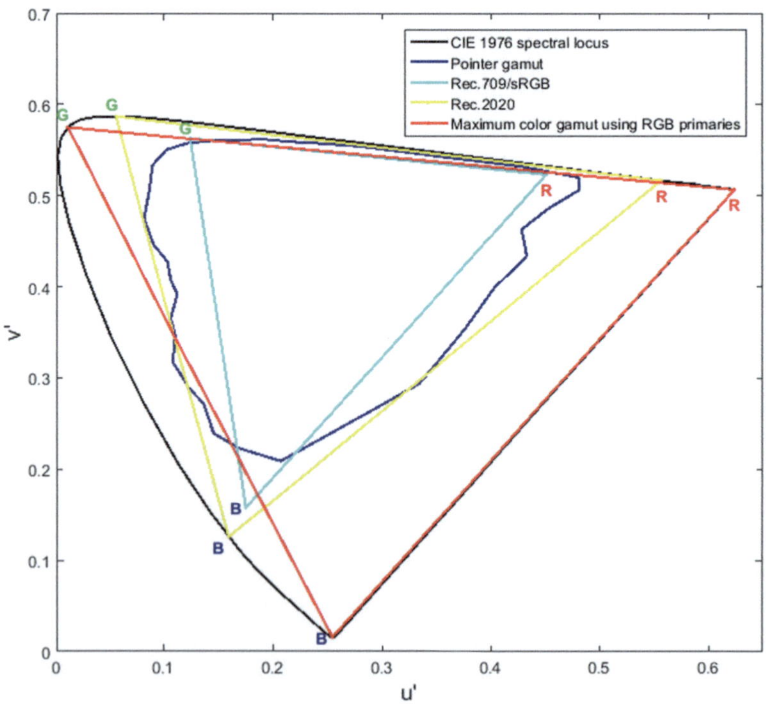

Fig. 21.11 Color gamut of RGB primaries

triangle, which stands for the maximum gamut achievable by three primaries, covers a much larger color space than BT.2020 on the CIE 1976 UCS. In consideration of the representation of the yellow color, however, one cannot adopt the RGB primaries, which can achieve the maximal color gamut. As is well-known, the human visual system can be modelled more precisely by the opponent channel model than the RGB color model [31], e.g. because the popular RGB color system theory cannot explain the color-vision deficiency phenomenon of human being. The opponent channels consist of the black-white (luminance) channel, the red-green channel and the yellow-blue channel. Red and green are two opponent colors and their mixing cannot generate the yellow color, which has its own peculiarity, for instance, even monochromatic yellow/gold color can also be very bright in nature. On the contrary, other colors of high saturation can only exhibit a very low brightness level [32]. In order to represent the yellow color in the RGB color system, one had to replace such red primary by a smaller wavelength color and such green primary by the yellowish green color. Bangert shares a similar view [33]. Moreover, the human eyes are insensitive to the brightness of the red color, in particular the red color with a long wavelength, referring to the photopic luminosity efficiency function curve, which will be discussed later in this chapter. Therefore, the color reproduction using long-wavelength red color will need much power.

The problems discussed above can be well solved by multi-primary-color imaging techniques. Figure 21.12 illustrates the gamuts achieved by four-, five- and six-primary colors. The left figure gives the gamuts achieved by three-, four-, five- and six-primary colors, whose RGB colors are specified by BT.709. One can see that, under the BT.709 color space, even the color gamut of six primaries cannot fully cover the Pointer gamut. The right figure illustrates the gamuts achieved by three-, four-, five- and six-primary colors, whose RGB colors are specified by BT.2020. The last case can cover beyond the Pointer gamut. Pure primary colors are important for the WCG purpose.

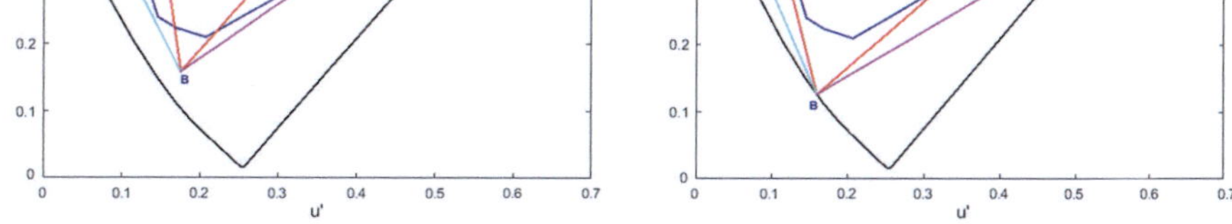

Fig. 21.12 Three-, four-, five- and six-primary-color gamuts and Pointer gamut

21.3 HDR and WCG Display Techniques

The display dynamic range is specified as the ratio between the maximal brightness and the minimal brightness that the display in question can reach, i.e. the ratio between the brightest white value and the darkest black value. One usually applies the ANSI contrast as the display dynamic range. For the measurement, a pattern of 16 alternating black and white rectangles is displayed. The ANSI contrast equals the ratio between the average light output from the white rectangles and the average light output of the black rectangles.

The minimal brightness states the display's performance of black reproduction, i.e. how well a pixel can be turned off. OLED exhibits good black performance and can switch each pixel completely off. As a result, OLED can reach the HDR purpose. For instance, OLED BVM-X300 from Sony is often used as HDR display. However, nowadays the lifetime of OLED operating at high brightness will decrease significantly [34]. OLEDs on the market usually have a low brightness. Ambient light, e.g. room lighting, sunshine through a window, may affect its application. Sometimes, one cannot reduce the ambient light level too much, for instance, in case of hospital operations the physicians need sufficient lighting. Therefore, a high display peak brightness is desired. According to the experiments done by Dolby, a dynamic range of 0:10,000 cd/m^2 can satisfy 90% of the viewers [35]. Thus, the HDR standard Dolby Vision specifies a maximum brightness of 10,000 cd/m^2. To measure the maximum display brightness, usually the screen should display a white rectangle pattern, e.g. at the top-left corner of the screen, with an area of 2% of the whole screen area, whereas the rest of the screen area should be left black. Then, the display is set to its maximal brightness. Using a luminance meter, one can measures the display peak brightness.

If only the reachable maximal brightness is large, but the reachable minimal brightness is not small enough, the display dynamic range will be very limited. This explains why one cannot achieve HDR performance only by increasing the backlight illuminance value of the LCD, although the LCD can reach a much higher peak brightness than the OLED. There exists light leakage from the liquid crystals. Alone increasing the illuminance value of the LCD's backlight, usually LEDs, would wash out the dark tones. If the screen should render dark scenes, only grey scenes occur. Consequently, such a display is unable to precisely reproduce black. In order to increase the dynamic range of the LCD, local dimming is used. Both the LC and the local dimming contribute to the total dynamic range, which is the product of these two subsystem dynamic ranges [36]. For local dimming, the direct-lit backlighting technique is used. The backlight LEDs are divided into many segments, and the brightness of the LEDs in each segment is controlled by the corresponding image-region luminance value. In case of dark image scenes, for instance, the corresponding LEDs are turned off so that the light leakage of the LCD can be strongly reduced. HDR LCD Display HDR47ES6MB from the company SIM2 uses 2202 white LEDs for local dimming, each is individually controlled. It is of more than 6000 cd/m^2 peak brightness and exhibits a dynamic range of up to 1:17.5 f/stops or about 1:1.85 $\times$ 10^5 (=2$^{17.5}$).

There are researchers that use monochromatic RGB LEDs for local dimming [37]. Besides HDR, the monochromatic backlighting technique is advantageous with respect to WCG purpose, because the colors generated by RGB LEDs are much purer or much more saturated than that generated by CFA filtering white light. Nowadays, monochromatic backlighting technique is only used for high-end displays, e.g. Qualia of Sony and DreamColor of HP. The emerging microLED display also uses monochromatic RGB LEDs to emit lights. The monochromatic backlighting technique is also beneficial to reduce the HDR display power consumption. If, for example, only green color scenes are to be reproduced, the red and blue LEDs can be turned off, and thus energy can be saved. This is not the case for white LED, because the white LED has to remain on so that the CFA generates the green light, whereas the blue and red light generated by the CFA are useless. Monochromatic backlighting technique can save display energy. This is also due to the Helmholtz–Kohlrausch effect, stating that a display with a more saturated color is perceived to be brighter [38].

Laser techniques can be used to realize an HDR and in particular WCG display, because the laser color is pure, and the laser technique can realize the BT.2020 color space. However, the energy consumption and cost of laser display may pose a challenge to its wide spreading. Many studies point out that local dimming can significantly reduce display power consumption, e.g. [32]. Local dimming needs many backlight units, which can be inexpensively realized by LEDs, but could be very expensive in case of laser realization. On the contrary, the emerging microLED display technique may provide a good alternative. MicroLED can generate much purer colors than the current widespread RGB color generation, in which CFA is used to filter the white light. Moreover, the microLED display technique does not need liquid crystals any more, which cause the most power of a display. Companies, like Sony and Samsung, have already exhibited microLED displays [39, 40].

21.4 Heat Dissipation of HDR Display and Power Saving by Four-Primary-Color WCG

An HDR display of high peak brightness consumes much electricity and is confronted with problem of heat dissipation. The world's first HDR display DR37-P from BrightSide Technologies, called Sunnybrook Technologies back then and taken over by Dolby, needs water-cooling system [41]! HDR displays often face overheating problems. The overheating problem limits the widespread deployment of HDR display. On the market 2019, there is no HDR display capable of generating peak brightness of 10,000 cd/m^2, which is required by the Dolby Vision HDR standard, although at IBC 2016 the company SIM2 exhibited the world's first HDR display of 10,000 cd/m^2 peak brightness [42]. Among all the display components, the panel makes up the largest proportion of the total display power consumption, e.g. for living room TVs ($\geq$80 cm/32″) more than 85% of the consumed power is caused by the LCD backlight [32].

An HDR display not only faces the technical problem of heat dissipation, but also faces the legislative regulation. According to the United States Environmental Protection Agency, any display manufactured as of January 28, 2020, must meet Version 8.0 requirements to bear the ENERGY STAR mark. Display power consumption reduction becomes more and more prominent and urgent.

Luckily, the four-primary-color WCG, which is essential for the color reproduction of high fidelity, can significantly save the display panel power. RGBW (W: White) four-primary-color pixel format OLED was reported by several studies, e.g. [43, 44]. Besides RGB subpixels, a white subpixel is used so that white light can pass unfiltered through, i.e. with high efficiency. Because white color is the most important color and almost all the composite colors contain white color, high efficiency white color generation will lead to energy saving of RGBW OLED. Since white pixel does not change the hue, strictly saying the RGBW pixel pattern is a three-primary-color format.

In addition to RGBW subpixel format, the RGBY subpixel structure has attracted attention. The Japanese Semiconductor Energy Laboratory developed an OLED with a microcavity structure combined with a blue/yellow tandem structure [45]. The tandem OLED with red, green, blue and yellow subpixels can significantly reduce OLED's power consumption [45]. AU Optronics Corporation developed worldwide the first RGBY format high-definition OLED [46]. It can reduce OLED's power consumption compared to the conventional three-primary-color OLED panel, since the human visual system is very sensitive to the yellow color and for the same brightness lower energy suffices. Thanks to four-primary colors, the color gamut can be widened too, which is essential for high fidelity color reproduction. The color mixing is unique under three-primary-color system. There are myriad color mixing possibilities under an RGBY four-primary-color system. To solve the ambiguous color mixing problem, Yoshiyama suggested to maximize the luminance function resulted from the RGBY four-primary colors [47]. In the course of the luminance function maximization, however, the human visual characteristics is not taken into account. The human eye exhibits a very different sensitivity to the different wavelength light. For instance, the human eye cannot perceive the existence of the infrared light, no matter how strong it is. The human eye is also insensitive to the red and blue light, as reportedly shown by the luminosity efficiency function curves in Fig. 21.13 (left). In fact, there exists the scotopic luminosity efficiency function curve. The scotopic luminosity efficiency function is irrelevant for HDR display of high peak brightness and irrelevant to WCG as well, because the human eye cannot perceive color in case of scotopic vision. To avoid possible misdirecting from the main message of this section, the scotopic luminosity efficiency function curve is removed from Fig. 21.13 (left). The photopic luminosity function curves include the CIE 1931 standard data (solid), the modified data by Judd–Vos (dashed), and the Sharpe, Stockman, Jagla & Jägle data (dotted) [48].

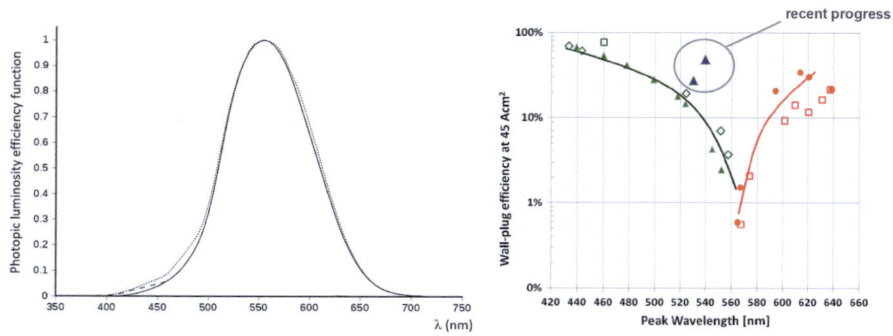

Fig. 21.13 Luminosity efficiency function curves [48] (left) and WPE [49] (right)

Making use of the human visual characteristics, one can significantly reduce the display panel power consumption [50]. The display panel power can be saved by diminishing the usage of that color, to whose brightness the human eyes are not sensitive. From Fig. 21.13 (left), one can see that the human eyes are insensitive to the brightness of the blue and red color. Although for the reproduction of a composite color three primaries are needed, one uses the fourth color, i.e. yellow, of the RGBY four-primary-color OLED WCG display to diminish the usage of the blue or red primary, as illustrated by the left drawing in Fig. 21.14. From this figure, it is obvious that one can diminish the usage of the red color, because colors falling within the ΔYGB triangle can be mixed by YGB three primaries.

Nowadays, the display market is dominated by LCDs, which use LEDs as backlight. As mentioned before, monochromatic LED backlighting technique is widely used in high-end LCDs. After Sony introduced the world's first commercial full-array

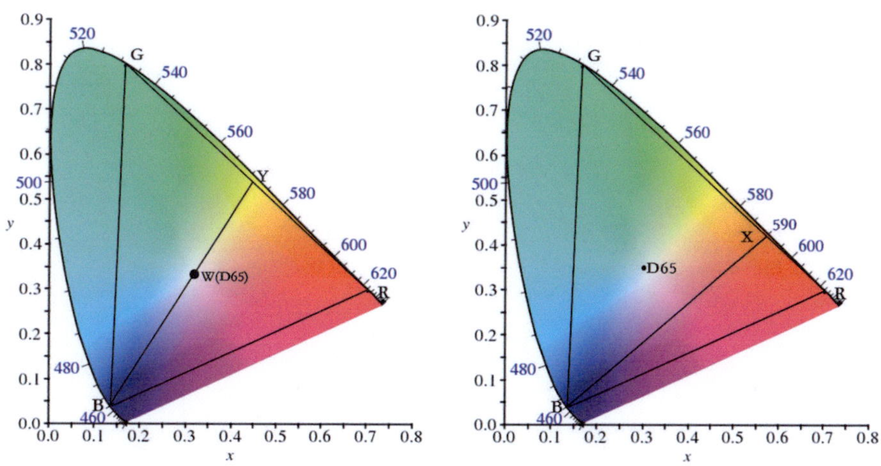

Fig. 21.14 Possible color mixing by YGB (left) and by XGB (right)

monochromatic RGB LED-backlit LCD TV Qualia 005 in 2004, many companies have adopted the monochromatic backlighting technique in their high-end products, e.g. HP, Dell, Eizo, Hazro, LG and NEC [4]. Some Sony's LCDs even use four-primary-color monochromatic LEDs, but one red, one blue and two green LEDs [51]. The emerging microLED display technique uses monochromatic LEDs to generate red, green and blue lights. Energy saving means cost reduction during the operation time. In addition to energy saving, RGBY four-primary-colors WCG can improve the color reproduction performance. Therefore, the authors hope that this chapter can contribute to the spread of the RGBY four-primary-color WCG technique.

The blue LED has a much higher wall-plug efficiency (WPE) than the red LED, as shown by the right drawing in Fig. 21.13. Therefore, one should diminish the usage of red color instead of blue color, although the human eyes are even less sensitive to the blue light brightness than to the red light, in particular to the red color specified by BT.2020, whose wavelength is larger than that of the red color specified by BT.709.

The LED's so-called green gap phenomenon precisely should be called yellow gap, because the WPE value of yellow LEDs is low, which is obvious from the right drawing in Fig. 21.13. Although researchers have made a breakthrough in increasing the efficiency of the yellow LEDs [52], efficient yellow LEDs are not yet available on the market. Therefore, instead of yellow LED, the orange LED is used here. Then, one divides the ΔRGB triangle into ΔRXB and ΔXGB triangles to diminish the usage of red color, as illustrated by the right drawing in Fig. 21.14, where X stands for orange currently, and for yellow in the future. Colors falling within ΔXGB will be mixed by XGB. This way, the usage of the red color is diminished.

Experiments have been conducted to compare the energy consumption by means of the conventional RGB color mixing scheme and the RGBX color mixing scheme described above. To exclude the influence of the blue color, six colors on the $\overline{GX}$ line are used to conduct the comparison, as depicted in Fig. 21.15 (left). The power consumption for the RGB and XGB case is given in Fig. 21.15 (right). It is clear that

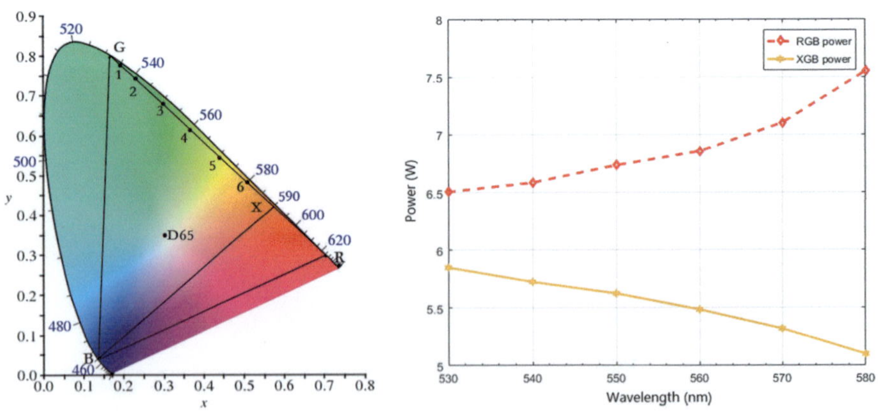

Fig. 21.15 Colors on $\overline{GX}$ line (left) and their energy consumption comparison (right)

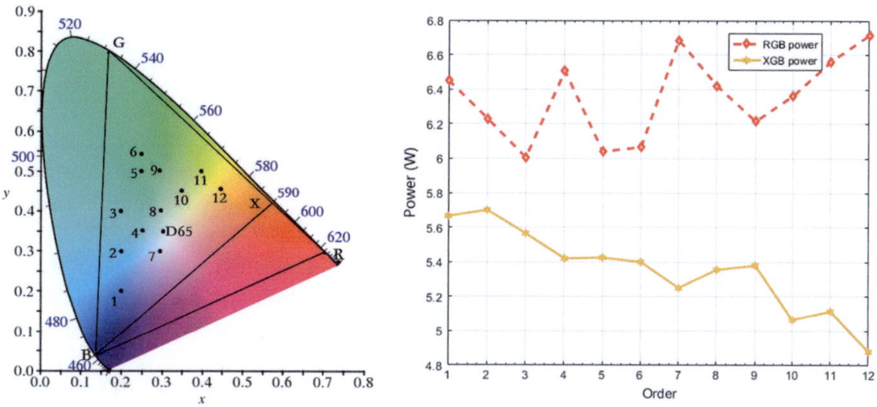

Fig. 21.16 Random colors (left) and their energy consumption comparison (right)

the XGB color-mixing scheme can significantly save energy in comparison to the RGB color-mixing method.

To avoid possible coincidence, 12 other colors are randomly chosen for further power-consumption comparison experiment, as depicted in Fig. 21.16 (left). For mixing the 12 composite colors, the blue color is needed. The power consumed by the RGB color mixing and the XGB color mixing is given in Fig. 21.16 (right). This experiment again proved that the OGB color mixing can significantly reduce the power consumption, compared to the conventional RGB color-mixing method.

Evaluating the results given by Figs. 21.15 and 21.16, on average more than 20% display panel power consumption can be saved. The panel power consumption method can mitigate the overheating problem of HDR displays, on the premise of a wide color gamut.

If one takes into account the fact that most of the pixels of natural images fall within the ΔOGB (O for orange) triangle, the energy-saving effect achieved by using the 4th color is significant. Li et al. have evaluated the ratio of pixels falling within the ΔOGB under BT.709 and BT.2020 color space, respectively [50]. The ratio is beyond 80% for 19 of the 20 test images. The exception image also has a ratio of more than 60%. Image pixels will undergo gamma-correction before display. Li et al. also evaluated the ratio after gamma-correction. The ratio after gamma-correction becomes even higher.

21.5 Delivering Four-Primary-Color Video Using the State-of-the-Art Data Compression Method and YUV Bandwidth

In this chapter, the authors only discuss the delivery issue related to the four-primary-color video signal. Methods to deliver HDR video signals have been addressed by other researchers, e.g. Mantiuk and Myszkowski in Chap. 14 of [11].

The introduction part of this chapter mentioned that the current RGB color system can only cover a part of the visual locus. Multi-primary-colors are an efficient way to widen the color gamut. The emergence of RGBY four-primary-color displays will inevitably demand the corresponding contents, i.e. the delivery of RGBY four-primary-color videos. Besides Semiconductor Energy Laboratory and AU Optronics Corporation, which developed RGBY four-primary-color OLEDs, as mentioned in Sect. 21.4, there are also companies producing four-primary-color LCDs. Sharp has produced several models of RGBY four-primary-color LCD TV, branded as Quattron, for instance LCD-80XU35A. Actually, Quattron is not the first four-primary-color technique, because as early as in the 1970s, Panasonic developed the Quatrecolor display technique with yellow as the 4th color and put it on the market. It is expected that there will be more manufactures producing four-primary-color displays, because in comparison to the conventional RGB display, the power consumption of a display panel can be reduced by adopting the yellow color as the 4th color. This is due to the fact that the human eye is very sensitive to the yellow color light, as the luminosity efficiency function states. RGBY four-primary-color displays need the corresponding four-primary-color contents to be delivered.

There is another important reason necessitating the delivery of RGBY four-primary-color contents. The human visual system can be more precisely modelled by the opponent channels as mentioned before, for which the yellow color in addition to RGB colors is essential. By delivering the yellow color besides RGB one can realize the theory of opponent channels.

The above reasons state that RGBY four-primary-color videos should be delivered. However, four-primary-color video will cause much more data than the current RGB video, and due to the valuable bandwidth, there is no affordable means to deliver RGB + Yellow video contents until now. This section will solve the dilemma between the demand on four-primary-color video contents and the available YUV bandwidth.

This section proposes a four-primary-color video coding scheme that is compatible with the current three-primary-color video system. It aims at enabling the delivery of red, green blue and yellow four-primary-color video by means of the state-of-the-art data compression method and the available YUV bandwidth. As a result, it can efficiently widen the color gamut and match the opponent-channel theory, that more precisely models the human visual system than the popular RGB color space.

The proposed technique will encode the image colors beyond the RGB color gamut in a different way to the image colors within the color gamut covered by the RGB three primaries. More precisely, the image regions, whose colors are within the

RGB color gamut, will be further encoded as the popular YCbCr format, whereas the image regions, whose color gamuts are outside the RGB color gamut, will be encoded differently, i.e. as a pseudo YCbCr format, which can be distinguished from the real YCbCr format at the decoder side without any metadata. If a pixel that is encoded as the pseudo YCbCr format at the deliver side is decoded by means of the real YCbCr format, negative YCbCr values will occur. In this way, the decoder can blindly identify the source of YCbCr signals, i.e. from which three-primary-color the color pixel in question is represented.

The proposed coding scheme does not need to modify the state-of-the-art data compression methods. It also does not need more bandwidth than the available YUV one. One only needs to specify a scheme to encode image contents whose colors are beyond the RGB color gamut so that the receiver can decode them correctly.

The four-primary-color video delivery method is not limited to RGBY four-primary colors. Because Yu et al. addressed the delivery of RGBY image components by means of the state-of-the-art data compression method and the available YUV bandwidth [53], furthermore, until now, RGBY image sensors and cameras are commercially not yet available, this section only discusses the delivery of RGBE image components without necessitating modifying the state-of-the-art coding standards and increasing the bandwidth. Because the 4th primary color can be yellow or emerald, it is denoted as X in the following. In this section, the authors will at first discuss the decomposition of the quadrilateral connecting the red, green, blue and emerald colors into two triangles. Then, the coding scheme for colors falling within the ΔXGB triangle is described that enables the decoder to blindly differentiate between the YCbCr signal encoded from RGB and the pseudo YCbCr signal encoded from XGB.

21.5.1 Decomposing Quadrilateral into Two Triangles

RGB and the fourth primary color ‘E’ constitute a quadri-lateral. As mentioned above, the 4th primary color that can be transmitted by the method herein is not only emerald, it can be yellow. In fact, the 4th color can be even another color that lies outside the ΔRGB triangle on the CIE chromaticiy diagram. Taking the emerald color as the 4th primary color, denoted as ‘X’, ‘X’ is located to the left side of the straight line $\overline{BG}$ as illustrated in Fig. 21.17.

For the current three-primary color image system, the representation of a color covered by the ΔRGB triangle is unique. In case of four-primary colors, the representation of a color is ambiguous, like the composite color C_1, C_2, C_3 and C_4 of Fig. 21.17. Such composite colors can be mixed from R, G, B and E. However, the mixing ratios of R, G, B and X are not unique. In particular, colors represented by four colors will challenge the image transmission due to the high data amount. If one wants to reduce the data amount, the data compression of the emerald color image component will be not efficient, because it is not correlated to the RGB three color image components.

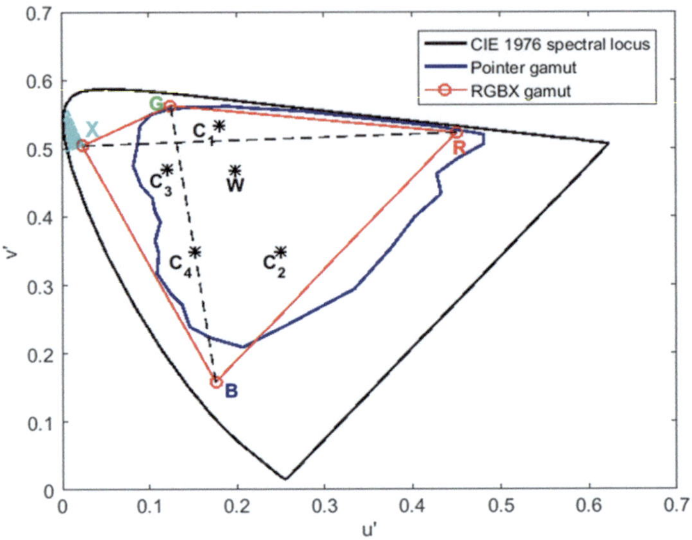

Fig. 21.17 Quadrilateral constituted by R, G, B, X—RGB is BT.709 primaries

To solve the above problems, one decomposes the quadrilateral constructed by R, G, B and X into two triangles ΔRGB and ΔXGB, as schematically shown in Fig. 21.17. In particular, when 'X' is within the cyan area of Fig. 21.17, the RGBX four-primary colors can significantly extend the color gamut.

Colors like color C_1 in Fig. 21.17 is inside the ΔRGB triangle and can be computed from red, green and blue color. Such colors are within the color gamut of conventional displays and can be reproduced correctly by the conventional display devices. Therefore, YUV signals representing colors like color C_1 will be computed from R, G and B. Although colors like color C_1 is also inside triangle ΔRGX and can be computed from R, G and X, its transmission is not compatible with the current RGB image system and one cannot fully make use of the conventional display devices. For the same reason, colors like color C_2 in Fig. 21.17, which lie within both the ΔRGB triangle and the ΔRXB triangle, shall be represented by R, G and B.

On the contrary, colors like color C_3 do not lie within the triangle ΔRGB and cannot be mixed by RGB three colors. Conventional display devices cannot correctly render colors like color C_3. However, colors like color C_3 lies within the ΔXGB triangle and thus can be mixed using X, G and B colors. For transmitting composite colors like color C_3, one must also build Y, U and V components to make use of the available YUV delivery means, as in the RGB case. However, their meanings are different to those of the RGB case.

Colors like C_4 lie on the segment $\overline{BG}$ can be either represented by RGB or XGB, and will not cause a reproduction problem, because in both cases R and X are zero, only G and B colors contribute to mixing them.

21.5.2 Encoding Scheme for Blind Identification

Once the triangle is selected, the Y and the two color difference signals U and V are calculated for delivery purpose. The remaining problem is how to encode the YCbCr signal mixed by XGB primaries so that the decoder can blindly differentiate between the pseudo YCbCr triplex mixed by XGB and the real YCbCr triplex mixed by the conventional RGB. One method to realize the blind identification is to change the luminance signal Y in case of XGB mixing [53]. The nominal range of the Y signal in case of RGB mixing is between 0 and 1, i.e. $Y_{RGB} \in [0, 1]$, and there is no offset. If one changes the nominal range of Y in case of XGB mixing to −1 and 0, i.e. $Y_{XGB} \in [-1\ 0]$, the corresponding YCbCr signals from RGB and XGB can be differentiated. To get digital format like YCbCr format an offset has to be added to Y_{XGB}, which will lead to change in the formula of U_{XGB} and $U_{XGB,}$ too. Besides, one reserves some quantization levels at the upper and at the lower end to deal with the Gipp's phenomenon, as already mentioned in Sect. 21.2.2, as well as the quantization level 0 and 255 for synchronization purpose, which are specified for n = 8 bit digital image signals. Taking all these into account, through maximizing the cost function representing the probability of negative YCbCr values that occur with all the possible combinations of the n bit image signals, one gets the formula for digital image signal Y'_{XGB}, U'_{XGB} and V'_{XGB}, which stand for the quantized Y_{XGB}, U_{XGB} and V_{XGB} signals respectively:

$$\begin{bmatrix} Y'_{XGB} \\ U'_{XGB} \\ V'_{XGB} \end{bmatrix} = \begin{bmatrix} -0.1063 & -0.3576 & -0.0361 \\ -0.0539 & -0.1813 & 0.2352 \\ 0.1347 & -0.1223 & -0.0124 \end{bmatrix} \begin{bmatrix} X' \\ G' \\ B' \end{bmatrix} + \begin{bmatrix} 134 \times 2^{n-8} \\ 67 \times 2^{n-8} \\ 45 \times 2^{n-8} \end{bmatrix} \tag{21.1}$$

where X′, G′ and B′ respectively represent the quantized XGB signals, and n denotes the bit depth of digital image signals and n = 8 in case of LDR images of BT.709 color gamut. For other kind of images, n may take another value, e.g. n = 10 for HDR10 format HDR images.

At the decoder side, one gets:

$$\begin{bmatrix} X' \\ G' \\ B' \end{bmatrix} = \begin{bmatrix} -2 & 0 & 5.8456 \\ -2 & -0.3983 & -1.7377 \\ -2 & 3.9454 & 0 \end{bmatrix} \begin{bmatrix} Y'_{XGB} - 134 \times 2^{n-8} \\ U'_{XGB} - 67 \times 2^{n-8} \\ V'_{XGB} - 45 \times 2^{n-8} \end{bmatrix}. \tag{21.2}$$

By inverse operation of the reserved quantization levels, the XGB three-primary-color signals are recovered by:

$$\begin{bmatrix} X' \\ G' \\ B' \end{bmatrix} = \begin{bmatrix} -2.3288 & 0 & 6.8065 \\ -2.3288 & -0.4638 & -2.0233 \\ -2.3288 & 4.5940 & 0 \end{bmatrix} \begin{bmatrix} Y'_{XGB} - 126 \times 2^{n-8} \\ U'_{XGB} - 67 \times 2^{n-8} \\ V'_{XGB} - 45 \times 2^{n-8} \end{bmatrix}. \tag{21.3}$$

Yu et al. have tested the encoding scheme for the XGB case using the state-of-the-art image coding standards without the necessity of any modification. The state-of-the-art coding methods can compress more efficiently the XGB format image data than the popular RGB format image data, which means that four-primary-color images can be delivered without additional bandwidth to the currently available YUV transmission bandwidth.

21.6 Conclusion

This section discussed the HDR and WCG imaging, display and delivery techniques. They are related to each other. With respect to imaging, the HDRC image sensor can be applied to capture the four-primary-color WCG image video, only a different CFA is needed. A CFA different to the popular Bayern pattern can for instance contribute to improve the SNR of HDR images at dark scenes. The HDR and WCG display techniques benefit from each other as well. Not only HDR display is important to WCG, four-primary-color WCG display can mitigate the heat dissipation problem encountered by the HDR display nowadays. A four-primary-color image coding scheme is described, and it enables to deliver videos of four-primary colors without demanding additional bandwidth.

References

1. B. Hoefflinger (ed.), *High-Dynamic-Range (HDR) Vision* (Springer, Berlin, 2007). ISBN-13 978-3-540-44432-9
2. F. Dufaux, P. Le Callet, R.K. Mantiuk, M. Mrak (eds.), *High Dynamic Range Video* (Elsevier, Amsterdam, 2016). ISBN 978-0-08-100412-8
3. Deutsche TV-Plattform e.V.: ULTRA HD EXPLAINED, 3rd revised edn. (2018) (Online). Available: https://tv-plattform.de/images/stories/download/2018/DTVP-Ultra-HD-explained-3rd-Edition-2018.pdf
4. K. Jansen, The Pointer's Gamut, the coverage of real surface colors by RGB color spaces and wide gamut displays (2014) (Online). Available: https://www.tftcentral.co.uk/articles/pointers_gamut.htm
5. Y. Hara, Y. Kenbo, M. Shiba, Image sensor. Japan Patent 08-223491, Aug 1986
6. K. Saito, Electronic image pickup device. Japan Patent 07-254965, Feb 1995
7. B. Guthier, S. Kopf, W. Effelsberg, Algorithms for a real-time HDR video system. Pattern Recogn. Lett. **34**, 25–33 (2013)
8. S.K. Nayar, T. Mitsunaga, High dynamic range imaging: spatially varying pixel exposures. Proc. IEEE Conf. Comput. Vis. Pattern Recognit. (CVPR) **1**, 472–479 (2000)
9. S.K. Nayar, V. Branzoi, Adaptive dynamic range imaging: optical control of pixel exposures over space and time, in *Proceedings of the 9th IEEE International Conference on Computer Vision (ICCV)* (2003)
10. M. Strobel, Hochdynamische CMOS Bildsensoren – Übersicht der Konzepte, Kontrastauflösung und Rauscheigenschaften, Vortragsunterlagen 46. Heidelberger Bildverarbeitungsforum (2011)

11. B. Hoefflinger (ed.), *CHIPS 2020—New Vistas in Nanoelectronics*, vol. 2 (Springer, Berlin, 2016)
12. M. Strobel, D. Döttling, High dynamic range CMOS (HDRC) imagers for safety systems, in *Advanced Optical Technologies*, vol. 2 (De Gruyter, Berlin, 2013), pp. 147–157
13. Y. Ni, 1280 $\times$ 1024 logarithmic snapshot image sensor with photodiode in solar cell mode. International Image Sensor Workshop (IISW) (2013)
14. International Standard ISO 15739:2017: Photography—Electronic still-picture imaging—Noise measurements. International Organization for Standardization (2017)
15. EMVA 1288 Standard release 3.1: Standard for Characterization of Image Sensors and Cameras, European Machine Vision Association (2016)
16. M. Oh, S. Nicholes, M. Suryadevara, L. Lin, H.-C. Chang, D. Tekleab, M. Guidash, S. Amanullah, S. Velichko, M. Innocent, S. Johnson, 3.0 μm Backside illuminated, lateral overflow, high dynamic range, LED flicker mitigation image sensor. International Image Sensor Workshop (IISW) (2019)
17. mvBlueFOX3 Technical Documentation: Working with HDR (High Dynamic Range Control), MATRIX VISION (2019)
18. J. Solhusvik, S. Yaghmai, A. Kimmels, C. Stephansen, A. Storm, J. Olsson, A. Rosnes, T. Martinussen, T. Willassen, P.O. Pahr, S. Eikedal, S. Shaw, R. Bhamra, S. Velichko, D. Pates, S. Datar, S. Smith, L. Jiang, D. Wing, A. Chilumula, A 1280 $\times$ 960 3.75 um pixel CMOS imager with Triple Exposure HDR. International Image Sensor Workshop (IISW) (2009)
19. J.N. Burghartz, H.-G. Graf, C. Harendt, W. Klingler, H. Richter, M. Strobel, HDR CMOS imagers and their applications, in *Proceedings of 8th International Conference on Solid-State and Integrated Circuit Technology, ICSICT* (2006), pp. 528–531
20. Worldwide Unified Colorimetry and Related Characteristics of Future Television and Image Systems (1998)
21. M.R. Pointer, The gamut of real surface colors. Color Res. Appl. **5**, 145–155 (1980)
22. Worldwide unified colorimetry and related characteristics of future television and imaging systems, ITU-R BT.1361, 15 Feb 2015 (Online). Available: https://www.itu.int/rec/R-REC-BT.1361/en
23. Multimedia systems and equipment—color measurement and management, Part 2–4: colour management—extended-gamut YCC colour space for video applications—xvYCC., IEC 61966-2-4 (2006)
24. Y. Xu, Y. Li, G. Li, Two kinds of methods to extend the color gamut of DTV system with extended quantization levels, in *Proceedings of 2nd International Congress on Image and Signal Processing* (2009), pp. 1–4
25. Parameter values for ultra-high definition television systems for production and international programme exchange, ITU-R BT.2020 (2012)
26. D.V. Volkinburg, A. Chernyakov, Imaging and decoding device with quantum dot imager, US Patent 8,537,245, 17 Sept 2013
27. E. Mandelli, Quantum dot based imagers for multispectral cameras and sensors, in *Proceedings of IEEE Hot Chips 28 Symposium (HCS)* (2016), pp. 1–30
28. Realization of natural color reproduction in Digital Still Cameras, closer to the natural sight perception of the human eye, Sony, 16 July 2003 (Online). Available: https://www.sony.net/SonyInfo/News/Press_Archive/200307/03-029E/
29. G. Sharma (ed.), *Digital Color Imaging Handbook* (CRC Press, Inc., Boca Raton, 2013), pp. 742–743
30. B. Sajadi, A. Majumder, K. Hiwada, A. Maki, R. Raskar, Switchable primaries using shiftable layers of color filter arrays. ACM Trans. Graph. **30**, 1–10 (2011)
31. R.S. Berns, *Billmeyer and Saltzman's Principles of Color Technology* (Wiley-Interscience, New York, 2000)
32. H.V. Mourik, Power saving in LCD panels, Philips Consumer Lifestyle, Advanced Technology Lab. Advanced Research & Technology for Embedded Intelligence and Systems (ARTEMIS), 13 July 2019 (Online). Available: https://artemis-ia.eu/publication/download/288-atc-scalopes-power-savings-in-lcd-panels-mour

33. T. Bangert, An Analysis of Quattron, Queen Mary University of London, 13 July 2019 (Online). Available: http://www.eecs.qmul.ac.uk/~tb300/pub/Appendix_Quattron.pdf
34. N. Miller, F. Leon, OLED Lighting Products: Capabilities, Challenges, Potential, Pacific Northwest National Laboratory, May 2016 (Online). Available: https://www.energy.gov/sites/prod/files/2016/06/f33/ssl_oled-products_2016.pdf
35. Dolby Vision (Online). Available: https://www.dolby.com/us/en/technologies/dolby-vision/dolby-vision-white-paper.pdf
36. H. Seetzen, W. Heidrich, W. Stuerzlinger, G. Ward, L. Whitehead, M. Trentacoste, A. Ghosh, A. Vorozcovs, High dynamic range display systems. ACM Trans. Graph. **23**(3), 760–768 (2004)
37. G. Wang, F. Lin, Y. Huang, Delta-color adjustment (DCA) for spatial modulated color backlight algorithm on high dynamic range LCD TVs. J. Disp. Technol. **6**(6), 215–220 (2010)
38. H. Chen, J. He, S. Wu, Recent advances on quantum-dot-enhanced liquid-crystal displays. IEEE J. Sel. Top. Quantum Electron. **23**(5), 1–11 (2017)
39. The Future of Display: A First Look at Samsung's 146-inch Modular TV, 'The Wall', Samsung, 8 Jan 2018 (Online). Available: https://news.samsung.com/global/the-future-of-display-a-first-look-at-samsungs-146-inch-modular-tv-the-wall
40. Crystal LED Display System, Sony (Online). Available: https://pro.sony/s3/cms-static-content/file/90/1237495157190.pdf
41. G. Richards, BrightSide DR37-P HDR display, 4 Oct 2005 (Online). Available: https://bit-tech.net/reviews/tech/brightside_hdr_edr/5/
42. A. Chalmers, K. Debattista, HDR video past, present and future: a perspective. Sig. Process. Image Commun. **54**, 49–55 (2017)
43. A.D. Arnold, P.E. Castro, T.K. Hatwar, M.V. Hettel, P.J. Kane, J.E. Ludwicki, M.E. Miller, M.J. Murdoch, J.P. Spindler, S.A. Van Slyke, K. Mameno, R. Nishikawa, T. Omura, S. Matsumoto, Full-color AMOLED with RGBW pixel pattern. J. Soc. Inf. Display **13**(6), 525–535 (2005)
44. J.P. Spindler, T.K. Hatwar, M.E. Miller, A.D. Arnold, M.J. Murdoch, P.J. Kane, J.E. Ludwicki, S.A. Van Slyke, Lifetime-and power-enhanced RGBW displays based on white OLEDs. SID Symp. Dig. Tech. Pap. **36**(1), 36–39 (2005)
45. R. Yamaoka, T. Sasaki, R. Kataishi, N. Miyairi, K. Kusunoki, M. Kaneyasu, H. Miyake, N. Ohsawa, S. Seo, Y. Hirakata, S. Yamazaki, K. Ono, T. Cho, H. Mori, High-resolution OLED display with remarkably low power consumption using blue/yellow tandem structure and RGBY subpixels. J. Soc. Inf. Disp. **23**, 451–456 (2015)
46. C.-C. Chen, M.-T. Lee, S.-F. Wu, H.-Y. Yang, S.-M. Shen, Y.-H. Lin, Low power consumption and wide color gamut AMOLED display with four primary colors. SID **46**(1), 1035–1038 (2015)
47. K. Yoshiyama, M. Teragawa, A. Yoshida, K. Tomizawa, K. Nakamura, Y. Yoshida, Power-saving: a new advantage of multi-primary color displays derived by numerical analysis. SID Symp. Dig. Tech. Pap. **41**(1), 416–419 (2010)
48. L. T. Sharpe, A. Stockman, W. Jagla, H. Jägle, A luminous efficiency function for daylight adaptation, J. Vis. **5**, 948–968 (2005)
49. B. Hahn, Closing the green efficiency gap, status and recent approaches. DOE Workshop Raleigh (2016)
50. H. Li, H. Gao, G. Kirca, Z. Lei, Energy-saving display by color pixel re-representation. IEEE Trans. Circuits Syst. Video Technol., https://doi.org/10.1109/TCSVT.2020.2966785, 2020
51. S. Ookubo, RGB LED Backlight Holds Key to Picture Quality of Bravia TVs, Nikkei Electronics, Nikkei Business Publications, Inc., 1 Sept 2008 (Online). Available: https://tech.nikkeibp.co.jp/dm/english/NEWS_EN/20080901/157224/
52. F. Jiang, J. Zhang, L. Xu, J. Ding, G. Wang, X. Wu, X. Wang, C. Mo, Z. Quan, X. Guo, C. Zheng, S. Pan, J. Liu, Efficient InGaN-based yellow-light-emitting diodes. Photonics Res. **7**(2), 144–148 (2019)
53. X. Yu, H. Li, A. Pare, Z. Lei, An image coding scheme to match the opponent channel model of human visual system. Submitted to IEEE Trans. Image Process.

Chapter 22
Update on Brain-Inspired Systems

Ulrich Rueckert

22.1 Introduction

Advances in technology have successively increased our ability to emulate artificial neural networks (ANNs) with speed and accuracy. At the same time, our understanding of neurons in the brain has increased substantially, with improved imaging methods and sophisticated microprobes contributing significantly to our understanding of neural physiology. These advances in both technology and neuroscience stimulated international research projects with the ultimate goal to emulate entire (human) brains. These new approaches are more brain-inspired than the ANN hardware from the nineties. They emulate neural networks on the basis of spiking integrate-and-fire neurons [1] with differences in emphasis. Some approaches aim at a more-detailed and, hence, more computationally-expensive model of neural behaviour, while others use simpler models of neurons but larger networks. In the following, we will consider projects intended to scale up towards millions of neurons, fabricated and tested with currently available technologies.

The majority of larger more bio-realistic simulations of brain areas are done on High Performance Supercomputer (HPS). For example, the Blue Brain Project [2] at EPFL in Switzerland deploys just from the beginning HPSs for digital reconstruction and simulations of the mammalian brain. The Blue Brain project is unusual in its goal to simulate the ion channels and processes of neurons at this fine-grained compartmental level. These models attempt to account for the 3D morphology of the neurons and cortical column, using about 1 billion triangular compartments for the mesh of 10,000 neurons using Hodgkin-Huxley equations [3], resulting in gigabytes of data for each neuron, and presumably a high level of bio-realism based on floating-point

U. Rueckert (✉)
Cluster of Excellence Cognitive Interaction Technology, Bielefeld, Germany
e-mail: rueckert@cit-ec.uni-bielefeld.de

B. Murmann and B. Hoefflinger (eds.), *NANO-CHIPS 2030*,
The Frontiers Collection, https://doi.org/10.1007/978-3-030-18338-7_22

arithmetic. The time needed to simulate brain areas is about two orders of magnitude larger than biological time scales. Based on a simpler (point) neuron model, the simulation could have orders of magnitude lower computational workload.

The network size and the used neuron model mainly determine the computational complexity of the simulation. The implementation cost of the Hodgkin-Huxley model [2] with a high biological plausibility (22 parameters) is orders of magnitude higher (about 1200 FLOPS) than the cost for a simple integrate-and-fire-model ([1], 3 parameters, about 5 FLOPS) (simulation of 1 ms biological time) [4]. At present, the most powerful commercially available HPS on the TOP500 supercomputer list (June 2019) is the Summit IBM Power System AC922 [5]. It has about 2.5 million IBM Power 9 processors (3 GHz), about 2 Peta Byte (PB, 10^{15} Bytes) of memory, 10 Mega Watt (MW) power consumption, and a theoretical peak performance of about 200 PFLOPS (20 GFLOPS/Watt). In principle, this system is able to store all synapse values (if restricted to one byte) of a human brain. The update rate could be 100 Hz, comparable to biology, and energy per synaptic operation of about 50 Pico Joule (pJ, 10^{-12}), orders of magnitude higher compared to biology (about 4 fJ). Though this is a naïve view, it gives an upper limit of what can be theoretically achieved with current HPS technology.

Software emulation of Spiking Neural Networks (SNNs) on HPS is widely used in computational neuroscience laboratories worldwide [6]. Applying HPS for abstract brain simulations is a convenient way but incurs some disadvantages. It does only support batch processing, interactions during simulation are restricted, simulations are in practice by far slower than biological real-time, and getting access to the whole machine for one user is difficult as well. Dedicated brain simulation machines (Neurocomputer) try to overcome these disadvantages.

22.2 Digital Neuromorphic Hardware Systems

The **SpiNNaker** (Spiking Neural Network Architecture) project at Manchester University [7] aims at a massively parallel multi-core computing system. The basic computing node of the current version has one SpiNNaker multi-core chip with 18 low-power ARM 968 processor cores (200 MHz), each with 96 KB of tightly-coupled local on-chip-memory for instructions and data, and a 128 MB SDRAM chip used to store synaptic weights and other information shared by all 18 cores. 16-bit fixed-point arithmetic is used for most of the computation, to avoid the need for a floating-point unit and to reduce energy, computational costs and chip area. A single SpiNNaker chip is able to simulate 16 K neurons with 1000 synapses each within a power budget of 1 W (energy per synaptic event 10 nJ) [7]. The SpiNNaker chip was fabricated in a 130 nm CMOS technology. Both chips are integrated as a System-in-Package (SiP) with the SDRAM wire-bounded on top of the SpiNNaker chip (3D packaging). 48 of these nodes are mounted on a PCB, which can be scaled up to 1200 boards for a full SpiNNaker system with more than 1 million of ARM9 cores and 7.2 TB of distributed RAM. This full system is in operation since 2018 and consumes at most 90 kW of electrical power [7].

The goal of the SpiNNaker system was to simulate SNNs with up to a billion neurons in biological real time (1 ms). A growing number of users worldwide are now using the system for various tasks, including Computational Neuroscience, Neurorobotics and general parallel computing tasks [8]. The software suite SpiNNTools supports users in mapping their computational problem on the massively parallel system [8]. One of the largest SNNs simulated so far is a scalable cortical-like network of micro-columns with about 3.5 million neurons and 380 million synapses. This SNN runs on 98 boards, 75,264 cores, and 74.3 GB of host memory. It utilizes less than 10% of system resources. Data generation, network creation, and data loading sums up to about 7 h [8].

Within the European Human Brain Project [9] the SpiNNaker2 system is under development. It aims at enhancing brain size simulation in biological real-time at 10 × better efficiency. The SpiNNaker2 chip is designed for a 22 nm FDX CMOS technology (GLOBALFOUNDRIES), integrating 144 ARM M4F cores (500 MHz, 1 W) with 128 KB local SRAM, floating-point support, improved power management (dynamic voltage and frequency scaling, adaptive body biasing), energy efficient inter-chip links, and external 8 GB shared memory. Furthermore, the chip provides a dedicated pseudo random number generator, an exponential function accelerator and a Multiply-Accumulate (MAC) array (16 × 48 Bit multiplier) with Direct Memory Access (DMA) for rate based ANN computation. The various processing elements are inter-connected via a Network-on-Chip (NoC) and the SpiNNaker2 router handles on-chip and off-chip spike communication. Chip architecture and a microphotograph of a test chip are shown in Fig. 22.1. The tape-out for the full SpiNNaker2 chip is

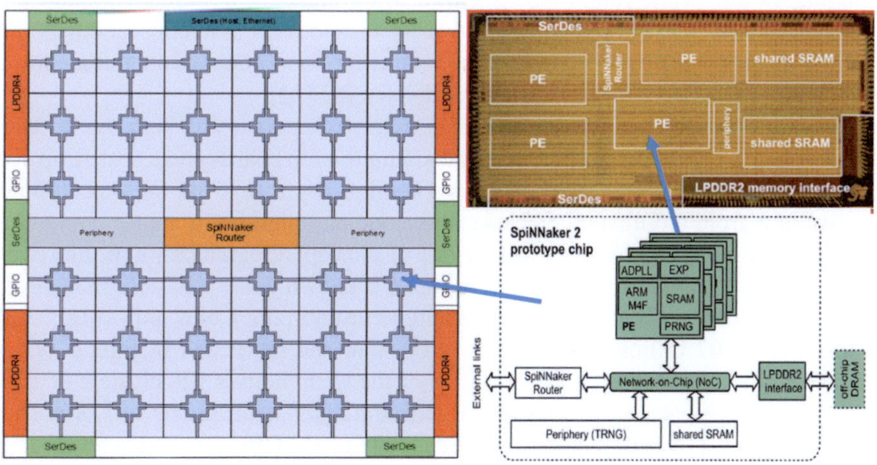

Fig. 22.1 SpiNNaker2 architecture (left), block architecture of one PE (bottom right) and chip photograph of the Santos28 test chip (28 nm CMOS, 4 ARM M4F cores) [11]

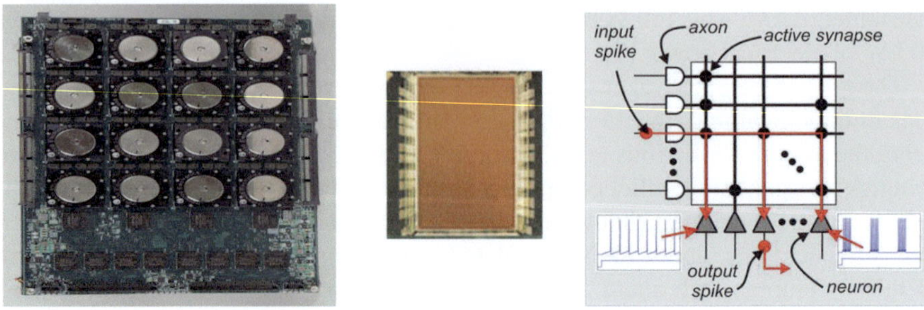

Fig. 22.2 A multi-chip board (left) of 16 IBM TrueNorth chips (die in the middle) integrating 64 × 64 digital neurosynaptic cores. Each core implements 256 neurons with 1024 spike-inputs (right) [12]. ©IEEE 2014

scheduled for 2020. The full 10 million core machine is expected to have 5 PFLOPS and an energy per synaptic spike-based update of 300 pJ (rate-based 300 fJ) [10].

The **IBM TrueNorth chip** (Fig. 22.2) integrates a two-dimensional on-chip network of 4096 digital application-specific cores (64 × 64) and over 400 Mio. bits of local on-chip memory (~100 KB SRAM per core) to store synapses and neuron parameters as well as 256 Mio. individually programmable synapses on-chip [12]. One million individually programmable neurons can be simulated time-multiplexed per chip, sixteen-times more than the current largest neuromorphic chip. The chip with about 5.4 billion transistors is fabricated in a 28 nm CMOS process (4.3 cm^2 die size, 240 μm × 390 μm per core). By device count, TrueNorth is the largest IBM chip ever fabricated and the second largest (CMOS) chip in the world. The routing network extends across chip boundaries through peripheral merge- and split-blocks. The total power, while running a typical recurrent network in biological real-time, is about 70 mW resulting in a power density of about 20 mW/cm^2 (about 26 pJ) which is in turn comparable to the cortex but three to four orders-of magnitude lower compared to 50–100 W/cm^2 for a conventional CPU [13].

IBM laid out an ecosystem for TrueNorth user support which is in use at many universities and government/corporate labs. Single-chip boards and a scaled-up system of tightly integrated 16 chips in a 4 × 4 configuration are in use for a spectrum of applications from mobile and embedded to cloud and high performance computing [14, 15]. However, TrueNorth is still a proof-of-concept research prototype and plans for future generations of such processors are not known yet.

In 2017, **Intel** presented its self-learning neuromorphic **Loihi** chip (Fig. 22.3). It integrates 2.07 billion transistors in a 60 mm^2 die fabricated in Intel's 14 nm CMOS FinFET process. The first iteration of the Loihi houses 128 clusters of 1024 artificial neurons each for a total of 131,072 simulated neurons, up to 128 million (1-bit) synapses (16 MB), three Lakefield (Intel Quark) CPU cores, and an off-chip communication network [16]. An asynchronous NoC manages the communication of packetized messages between clusters. Intel supplies chips in a one- or two-chip USB stick called Kapoho Bay, as well as on printed circuit boards (with four up to 64 chips per boards).

Fig. 22.3 Loihi chip micrograph [17]. ©INTEL 2017

Intel's Loihi is completely digital employing asynchronous processing and supporting inference and learning (Fig. 22.4). It runs SNNs without external memory. Each core has a 256 KB local memory to store configuration and state variables. Asynchronous processing is coordinated by distributed barrier-synchronization. Loihi's most distinctive feature is its ability to learn. A set of learning rules can be programmed through 4-bit microcode operations for modifying synapses with a weight precision between 1 and 9 bits. Learning is local to a single neuron considering pre- and post-synaptic activity from spike trains using long and short time constants. Loihi's performance measurements (pre-silicon) yield a minimal energy per synaptic spike of 23.6 pJ, energy per synaptic update of 120 pJ, maximum time per synaptic update of 6.1 ns, energy per neuron update of 81 pJ (active)/52 pJ (inactive), and time per neuron update of 8.4 ns (active)/5.3 ns (inactive).

Loihi is not a product, but available for research purposes among academic research groups organized in the Intel Neuromorphic Research Community (INRC). Within the community a steadily increasing number of applications for the Loihi system are implemented and benchmarked [17]. In theory, Loihi can scale all the way up to 4096 on-chip cores and 16,384 chips, though Intel has no plans to commercialize a design this large yet.

22.3 Mixed-Signal Neuromorphic Hardware Systems

The European funded research project **BrainScaleS** (Brain-inspired multiscale computation in neuromorphic hybrid systems) aimed at understanding and emulating

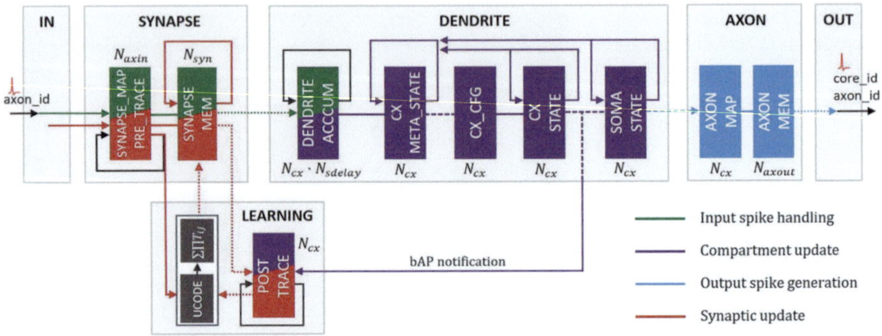

Fig. 22.4 Loihi computing core top-level microarchitecture [16]. ©IEEE 2014

Fig. 22.5 Photograph of the BrainScales wafer (left) and view of the BrainScales system with 20 wafer modules (right) [20]. (Creative Commons Attribution-NoDerivatives CC BY-ND 4.0)

functions and interactions of multiple spatial and temporal scales in brain-information processing [18]. Within this project the basis for the BrainScales hardware was laid and further developed in the HBP. The aim is a neuromorphic hardware of a very-large-scale, mixed-signal implementation of a highly connected, adaptive network of analog neurons. The basic element is the HICANN (High Input Count Analog Neural Network) chip hosting one analog network core (ANC) and necessary support circuitry for communication as well as controlling. The ANC was implemented in a 180 nm CMOS technology and has a total of 112 K synapses and 512 neuron circuits. The area of the analog neuron circuit is 1500 μm^2. The synapse weight is stored in a 4-bit SRAM and is represented as a current generated by a 4-bit multiplying DAC. The synapse area is 150 μm^2. Two synapse columns of the ANC can be combined to realize a weight resolution of 8 bit at the expense of bisecting the number of available synapses for the ANC neuron circuits [19]. A special feature of the BrainScaleS hardware system is wafer-scale integration of the HICANN chips. A total of 384 HICANN chips can be interconnected on an 8-inch silicon wafer (Fig. 22.5), implementing 196,608 neurons and 44 Mio. synapses. One key target of the BrainScaleS hardware is a 10^4-fold speed-up of the natural neuron-firing rate of 10 Hz.

Fig. 22.6 Chip micrograph of the HICANN-DLS test chip [21] (left) and the full size prototype chip (right) [22]

The HICANN analog neurons communicate with each other digitally. The backbone of the communication on the wafer is a grid of horizontal and vertical buses which are placed on top of the manufactured wafer. The wafer is organized into 384 chips with a maximum of 196,608 neurons with 224 inputs (synapses) resulting in about 2×10^9 events/s and 64 Gb/s per wafer.

The second generation of the BrainScales hardware is under development within the HBP. The new mixed-signal neuromorphic computing core integrates a custom Single Instruction Multiple Data (SIMD) processor (32-bit, 128-bit wide vectors) with an Analog Network Core (ANC). The **HICANN-DLS** (High Input Count Analog Neural Network with Digital Learning System) chip is designed for a 65 nm CMOS technology aiming at a better precision of the neuron-(10 bit resolution) and synapse-circuits (6-bit SRAM) and an improved communication system [21]. The chip is operating 1000 times faster than biological real-time. Each analog neuron is configured by 14 current and 4 voltage biases. Plasticity is implemented software-controlled on the embedded processor. A first prototype chip with 32 neurons (200 μm × 11.76 μm per neuron) has been successfully fabricated (Fig. 22.6) and tested [21]. The full size-prototype chip with 128 K synapses, 512 neural compartments, two SIMD plasticity processing units and 1024 ADC channels for plasticity input variables is in fabrication (65 nm CMOS). The estimated power consumption is about 10 pJ/synaptic event.

The **Neurogrid** project at Stanford University uses programmable analog "neurocore" chips [23]. Each $12 \times 14\ \text{mm}^2$ CMOS chip (180 nm CMOS) can emulate over 65,000 neurons, and 16 chips are assembled on a circuit board to emulate over a million neurons. The entire 1 M-neuron system consumes about 3.1 W. The Neurogrid neuron circuit consists of about 300 transistors modelling the components of the cell, with a total of 61 graded and 18 binary programmable parameters. Neurogrid uses local analog wiring to minimize the need for digitization for on-chip communication. Like the other systems, Neurogrid uses an AER packet network to communicate spikes between chips. Like SpiNNaker, the Neurogrid neuron array is designed to run in biological real-time.

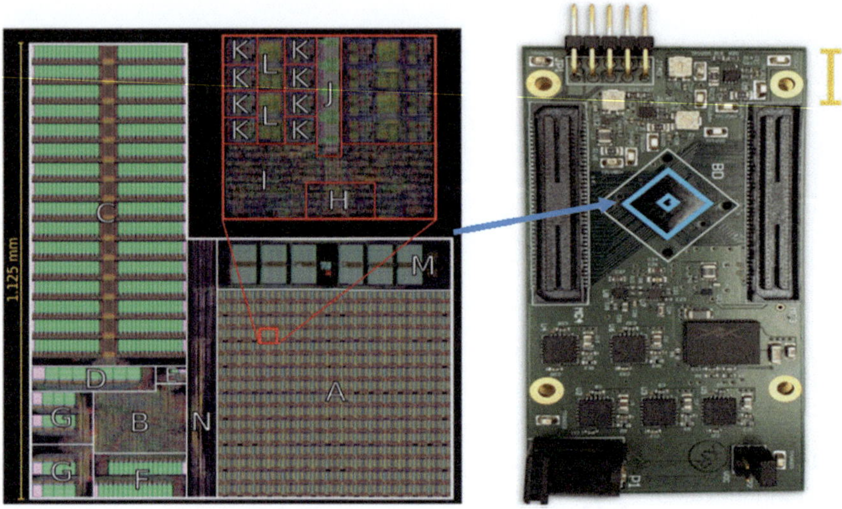

Fig. 22.7 Layout of the Braindrop core (left): inset shows the detail of a 16-neuron tile (red outline). A: 4096-neuron array, B: digital datapath, C: weight memory, D: activation memory, E: pool action table, F: FIFO, G: tag action table, H: AER tree logic, I: AER leaf logic, J: CM, K: neuron, L: synaptic filter, M: 12 DACs and two ADCs, and N: routing between neuron array, data path, and IO pads; Test board with one Braindrop core for embedding into the Nengo framework (right) [24]. ©IEEE 2018

The informal successor of the Neurogrid project is the Braindrop mixed-signal neuromorphic system [24]. The neuromorphic core is fabricated in a 28-nm FDSOI process, integrating 4096 neurons and 64 K synapses (8-bit) in 0.65 mm^2 (Fig. 22.7). Braindrop's computations are specified as coupled nonlinear dynamical systems implemented by subthreshold analog circuits as dynamic computational primitives. Special care is given to device mismatch and temperature sensitivity compensation at the network level. Different techniques are applied to achieve robustness, e.g. reverse body bias, sparse encoding, and deep subthreshold operation. The energy per synaptic operation comes down to 381 fJ for typical network configurations [24]. The Braindrop processor is connected to the neural Engineering framework Nengo [25] for mapping high-level network abstractions to the neuromorphic chip.

The **DYNAP-SEL** is a novel multi-core neuromorphic processor from the lab of Giacomo Indiveri (University of Zurich, Switzerland) [26]. The processors is fabricated in a standard 0.18 μm CMOS process and an advanced 28 nm Fully-Depleted Silicon on Insulator (FDSOI) process [27]. The chip has four neural processing cores, each with 16 × 16 analog neurons and 64 4-bit programmable synapses per neuron, and a fifth core with 1 × 64 analog neuron circuits, 64 × 128 plastic synapses with on-chip learning circuits, and 64 × 64 programmable synapses (Fig. 22.8). All synaptic inputs of all cores are triggered by incoming address events, which are routed among cores and across chips by asynchronous Address-Event Representation (AER) digital

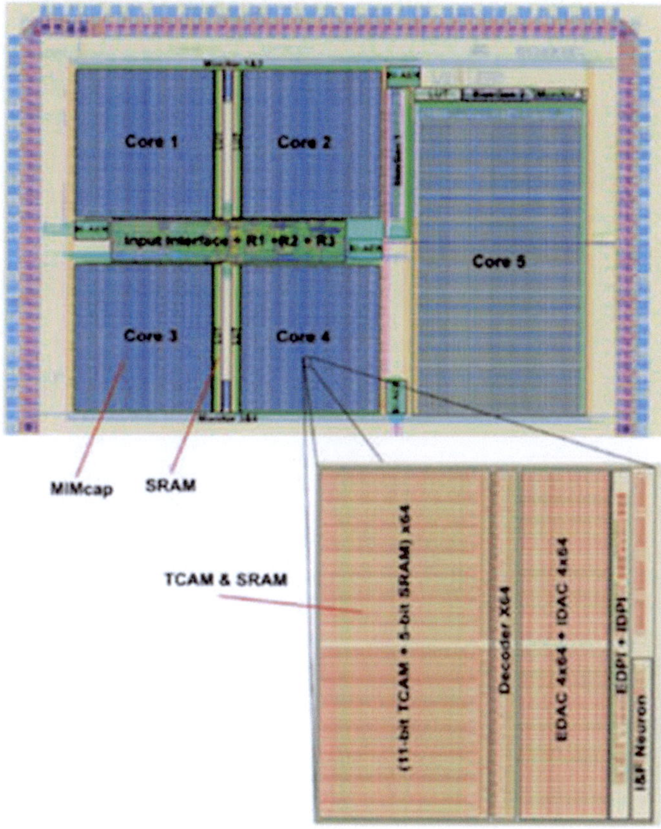

Fig. 22.8 Dynap-SEL chip with four non-plastic cores and one plastic core fabricated using a 28 nm FDSOI process [28]

router circuits. The DYNAP-SEL routing architecture is optimized for AER communication and composed of a hierarchy of routers at three different levels that use both source-address and destination-address routing. Most of the silicon area is occupied by digital SRAM and special content addressable memory cells. The neuromorphic processors support 1088 integrate-and-fire neurons with 78,080 synapses. The chip area is 7.28 mm^2, the area of a neuron is 20 μm^2, and the energy per synaptic event is 2.8 pJ [28].

22.4 Other SNN Accelerator Approaches

The number of digital SNN implementations is constantly increasing, but not as fast as hardware accelerators for non-spiking ANNs. Most of them are research

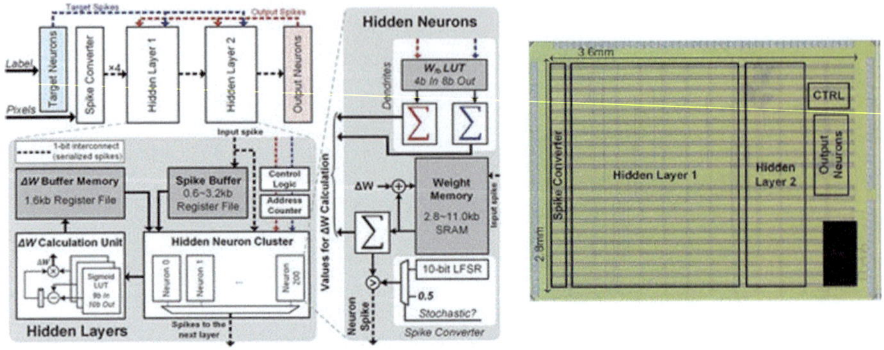

Fig. 22.9 Architecture (left) and die photo (right) of the SNU digital neuromorphic processor [29]

prototypes, only few with a clear application in mind. The Seoul National University (SNU, Korea) designed an on-chip trainable neuromorphic SNN processor for fast pattern classification [29]. The SNU processor is fabricated in 65 nm LP CMOS, has a core area of about 10 mm^2, consumes about 24 mW at 0.8 V, and operates at 20 MHz. It has a fixed structure of two hidden layers and an output layer with 10 target neurons (Fig. 22.9).

Beijing's Tsinghua University Center for Brain Inspired Computing Research designed a hybrid architecture that can concurrently run Convolutional Deep Neural Networks (CDNNs), Recurrent Neural Networks (RNNs), as well as biologically inspired SNNs [30]. The chip, called Tianjic, is manufactured in a 28 nm CMOS technology with a die area of 14.4 mm^2 (Fig. 22.10). Running at 300 MHz, the total chip power is less than 1.0 W. Tianjic implements about 40,000 neurons (256 neurons per core $\times$ 156 Fcores) and 10 million synapses. Each Fcore (unified functional core) includes 16 single-cycle 8-bit multipliers and 24-bit accumulators. "Because each neuron has 256 synapses, the 24-bit accumulators allow up to 256 sequential MAC operations per neuron without overflow. To model 256 neurons per Fcore, the chip repeats that process 16 times. Each Fcore has a 22 KB memory, yielding a total of 3.4 MB for the entire Tianjic chip. At its 300 MHz clock speed, Tianjic's peak performance is 1.5 trillion operations per second (TOPS), but the effective performance is 1.2 TOPS owing to the extra processing time for axon-input organization and transformation, inter-neuron communications, and soma activities" [30]. The Tianjic chip is used for an embedded autonomous-bicycle experiment where it handles balance control, decision-making, object detection, obstacle avoidance, tracking, and voice-command recognition.

Several SNN implementations on FPGAs have been presented as well. Examples are NeuroFlow (Virtex 6, different neuron models [31]), the Neuromorphic Cortex Simulator [32], or the AKIDA accelerator [33]. The company BrainChip sells the AKIDA's core architecture as a configurable intellectual property (IP) core and announced a chip version in 28 nm CMOS for 2020 [34]. FPGA implementations

Fig. 22.10 Tianjic hybrid neural-network processor chip with 156 programmable cores, which can run CDNNs as well as biologically inspired SNNs [30]

offer flexibility on custom devices fabricated with state-of-the-art digital technologies. They suffer from the memory bottle-neck and the reduced computation precision (1–8-bit in general).

22.5 Comparison

The projects focused on in this chapter follow different technological approaches for the implementation of SNNs. The Blue Brain project and the SpiNNaker system simulate ANNs in software on general-purpose processors. Whereas the Blue Brain Project employs high-performance computers (HPC) without bio-inspired architectural hardware adaptations, SpiNNaker relies on embedded low-power processor cores from the mobile world, distributed private memory per core, and a communication network optimized for transmitting "spikes" asynchronously utilizing the address-event-representation (AER). Both approaches make use of the concept of the virtualization that many "neurons" can be simulated time-multiplexed on the same digital core. TrueNorth and Loihi are based on a digital application-specific core, local on-chip-memory for the synapses, and a specialized routing network extending across chip boundaries. TrueNorth and Loihi can employ time multiplexing of several neurons per core. BrainScaleS, DYNAP-SEL and Braindrop use a "neuromorphic" approach, with dedicated, adjustable analog circuitry for every neuron in the ANN, adaptive on-chip synapses, and a configurable interconnection network.

The approaches have their specific pros and cons. The neuromorphic ASICs avoid the substantial computational overhead of software simulation and may produce a more biologically-accurate result in less time. On the other hand, for digital implementations, there is no A/D conversion and the cost of the network routing logic is amortized over 1000 emulated neurons per CPU (virtualization). All approaches face the problem of spike networking. Routing AER packets [35] in real-time from tens of billions of neurons is a challenge. The logic circuitry required for decoding and routing may be much larger than the neuron emulation circuit itself. Another issue with AER networking is the timing of spikes. Neurons adapt to premature and delayed signals over time, and some signal-timing tuning is performed by the axons. According to the routing network, the timing of spikes originating from the same neuron varies (jitter) in the proposed network implementations. Proper synchronization can be achieved by inserting delays or reserving communication bandwidth, as proposed in [36].

Synaptic plasticity and learning present the biggest challenges to artificial brain projects. On the one hand, our knowledge about plasticity, learning, and memory is incomplete [37]. On the other hand, our technologies are far less plastic and compact than neural tissue. Experimental evidence for some basic synaptic plasticity mechanisms exist. There is also evidence for neurons growing new dendrites and synapses to create new connections (structural plasticity) as well as changing the "weight" of existing synapses by increasing or decreasing the number of neurotransmitter vesicles or receptors for the neurotransmitters. Today, the efficient implementation of a writable and non-volatile synapse weight is a hot research topic. Within the discussed projects, the synapses are implemented digitally: 1 bit (TrueNorth), 1–9 bit (Loihi), 4 bit (DYNAP-SEL), 6 bit (HICANN-DLS), 8 bit (Braindrop), 13 bit shared (Neurogrid, off-chip), and 16 Bit (SpiNNaker, off-chip). Whereas TrueNorth and Neurogrid do not support learning, the HICANN chips (hardware-based learning) and SpiNNaker (software) include learning.

Independent of the technological approach, the projects differ in the level of bio-realism and computational sophistication in their emulation of neurons and synapses. SpiNNaker and TrueNorth mainly work with a point-neuron model, as recommended by Izhikevich [4]. A multi-compartment analog model such as Neurogrid's two-compartment circuits, or the BrainScaleS HICANN-DLS chip's separate dendritic membrane circuits, allows more sophisticated neural emulations, depending on the complexity of the compartment emulations. The most bio-realistic approach among the projects is the fully compartmentalized model of the neuron of the Blue Brain Project, representing a biological neuron as hundreds of independent compartments, each producing an output based on adjacent ion-channels and regions, and using the computationally expensive Hodgkin-Huxley equations [37] to compute the potential bio-realistically in each compartment.

Table 22.1 Comparison of neuro-ASICs

Neuro-ASICS	Feature size (nm)	Die size	Neurons	Synapses	Bit/synapse	ESE
SpiNNaker I	130	1.02 cm^2	1600	[a]128×10^6	8	10^{-8} J
SpiNNaker II	22	N/A			8–16	10^{-10} J
IBM TrueNorth	28	4.30 cm^2	10^6	256×10^6	1	10^{-11} J
Intel Loihi	14	60 mm^2	131,072	128×10^6	1–9	10^{-11} J
HICANN	180	0.50 cm^2	8-512	114,688	4–8	10^{-10} J
HICANN-DLS	65	32 mm^2	512	131,072	6	10^{-11} J
Neurogrid	180	1.68 cm^2	65,536	[a]16×10^6	[b]13	10^{-10} J
Braindrop	28	0.65 mm^2	4096	[a]16×10^6	8	10^{-13} J
DYNAP-SEL	28	7.28 mm^2	1088	78,080	4	10^{-12}
Numbers per chip				[a]Off-chip	[b]Shared	ESE: energy/synaptic event

With the increasing number of neuromorphic hardware systems for SNNs there is a demand for objective platform comparison and performance estimation of such systems. At present only few approaches for systematic benchmarking are published [38, 39]. Hence, an objective comparison is not possible yet. Table 22.1 summarizes chip characteristics of the basic building block (Neuro-ASIC) of the discussed approaches as published in the literature. For the energy demand per synaptic event (ESE) only a rough estimation of the expected magnitude is given. The exact determination of the ESE is difficult and not standardized yet.

There is clearly room for scaling for all projects, and it will be interesting to follow the digital-versus-analog strategy, considering the alternative of digital 8b × 8b multipliers with 1 fJ and 100 μm^2 per multiplication [40]. With respect to system scaling, the power efficiency of chip/wafer-level interconnects is relevant. With the optimum efficiency of 1 mW/(Gb/s) [40], the resulting 74 W could be handled. The more likely level at a less advanced technology node would be 1 kW [40]. Scaling this up to mega-neurons clearly shows that power efficiency is the number one concern for these complex systems.

3D integration is the further challenge for any silicon brain. The memory silicon layer on top of the mixed-signal silicon layer can be achieved with a through-silicon-via (TSV) technology [40], and it is only one level in the task of building the whole system, because the global programmable digital control could be added on top. In the BrainScaleS and Neurogrid architecture, the digital control is implemented with FPGA (field-programmable gate array) chips on a printed-circuit board.

22.6 Outlook

The building blocks for ICs and for the Brain are the same at nanoscale level: electrons, atoms, and molecules, but their evolutions have been radically different. The fact that reliability, low-power, reconfigurability, as well as asynchronicity have been brought up so many times in recent conferences and articles makes it compelling that the Brain should be an inspiration (at many different levels), suggesting that future nano-architectures could be neural-inspired. The fascination associated with an electronic replication of the human brain has grown with the persistent exponential progress of chip technology. The last decade 2010–2020 has also made the electronic implementation more feasible, because electronic circuits now perform synaptic operations such as multiplication and signal communication at energy levels of 10 fJ, comparable to biological synapses. Nevertheless, an all-out assembly of 10^{14} synapses will remain a matter of a few exploratory systems for the next two decades because of several challenges.

Currently, it is almost impossible to determine the best way to perform SNN calculations for any given application. This is one reason for the variety of approaches to SNN hardware implementation known in literature. For digital ICs, we can call on efficient software tools for a fast, reliable and even complex design. We can use many process-lines to manufacture the chips down to structure sizes of 7 nm. Digital concepts can use standard technologies with the highest density in devices. On the contrary, the design of analog circuits demands much more design-time, good theoretical knowledge about transistor physics, and a heuristic experience of layout design. Only a few process-lines are characterized by analog circuits. In their favour, we point out that, with integrated analog circuits, some neuron functions are quite simple to implement. For example, summation of the dendritic input signals as a current summing is a fairly convenient electronic analog circuit operation and smarter than with common digital accumulators, or a two-quadrant multiplier demands only five transistors. Nonlinearity or parasitic effects of the devices allow us to realize complex functions, like an exponential or a square-root function [41]. Note however, that analog circuits are not as densely integrated as it may seem at first glance. They demand large-area transistors to assure an acceptable precision and to provide good matching of functional transistor pairs, as used in current-mirrors or differential stages. It is very unclear whether analog implementations provide any power dissipation advantages over digital, and current evidence seems to point in the opposite direction. The problem of benchmarking and an adequate metric for performance evaluation is still open, too.

The systems presented in this chapter are advocating architectures with spiking neurons. SNNs represent an attempt to mimic aspects of the brain's architecture and dynamics with the aim of replicating some functional capabilities in terms of computational power, robust learning and energy efficiency. In some cases an advantage of neuromorphic computation can be shown [42]. In general, the performance of spiking neuron circuits seems considerably inferior to that of traditional digital architectures for realistic convolutional deep neural networks [43]. Current learning algorithms for SNNs do not take advantage of the peculiarities of spiking networks, and no spiking-neuron learning algorithm has been shown to come close to the accuracy of the backpropagation learning algorithm with continuous representations.

In 2009, the US DARPA launched the SyNAPSE program: Systems of Neuromorphic Adaptive Plastic Scalable Electronics [44]. It says in its description: "As compared to biological systems…., today's programmable machines are less efficient by a factor of 1 million to 1 billion in complex, real-world environments". And it continues: "The vision … is the enabling of electronic neuromorphic machine technology that is scalable to biological levels." SyNAPSE is a program with explicit specifications. It requires an existing system-simulation background (like the Blue Brain Project [1]) to assess the likelihood of realizing specified milestones. The last one for 2018 was specified as "Fabricate a multi-chip neural system of ~10^8 neurons and instantiate into a robotic platform performing at "cat" level (hunting a "mouse")". Today, we have to realize that this ambitious milestone has not been reached. Nevertheless, we do have much to learn from brains from the computational standpoint and about the implementation of resource-efficient technical systems. The hardware realization of neural networks should not aim for an exact reproduction of nervous systems, but simply for an efficient use of available technologies for solving practical problems.

References

1. A.N. Burkitt, A review of the integrate-and-fire neuron model: I. Homogeneous synaptic input. Biol. Cybern. **95**, 1–19 (2006)
2. H. Markram, The blue brain project. Nat. Rev. **7**, 153–160 (2006). http://bluebrain.epfl.ch
3. A.L. Hodgkin, A.F. Huxley, A quantitative description of membrane current and its application to conduction and excitation in nerves. J. Physiol. **117**, 500–544 (1952)
4. M. Izhikevich, Which model to use for cortical spiking neurons? IEEE Trans. Neural Netw. **15**(5), 1063–1070 (2004)
5. https://www.top500.org/list/2019/06/. Retrieved 31.10.2019
6. M. Djurfeldt et al., Brain-scale simulation of the neocortex on the IBM Blue Gene/L supercomputer. IBM J. Res. Dev. **52**(1/2), 31–41 (2008)
7. S. Furber et al., Overview of the SpiNNaker system architecture. IEEE Trans. Comput. **62**(12), 2454–2467 (2013)
8. A.G.D. Rowley et al., SpiNNTools: the execution engine for the SpiNNaker platform. Front. Neurosci. **13**, Article 13 (2019)
9. www.humanbrainproject.eu. Retrieved 31.10.2019
10. Y. Yan et al., Efficient reward-based structural plasticity on a SpiNNaker 2 prototype. IEEE Trans. Biomed. Circuits Syst. **13**(3), 579–591 (2019)
11. S. Höppner, C. Mayr, SpiNNaker 2—towards extremely efficient digital neuromorphics and multi-scale brain emulation, in *Proceedings of Neuro Inspired Computing Elements Workshop*. http://niceworkshop.org/wp-content/uploads/2018/05/2-27-SHoppner-SpiNNaker2.pdf (2018)
12. P.A. Merolla et al., A million spiking-neuron integrated circuit with a scalable communication network and interface. Science **345**, 668–673 (2014)

13. A.S. Cassidy et al., Real-time scalable cortical computing at 46 giga-synaptic OPS/Watt with ~100× speedup in time-to-solution and ~100,000× reduction in energy-to-solution, in *Proceedings of the International Conference for High Performance Computing, Networking, Storage and Analysis* (2014), pp. 27–38
14. S.K. Esser et al., Convolutional networks for fast, energy-efficient neuromorphic computing. arXiv:1603.08270v2 (2016)
15. A. Andreopoulo et al., Low power, high throughput, fully event-based stereo system, in *IEEE Conference on Computer Vision and Pattern Recognition* (2018)
16. M. Davies et al., Loihi: A neuromorphic manycore processor with on-chip learning. IEEE Micro **1**, 82–99 (2018)
17. M. Davies, Advancing neuromorphic computing from promise to competitive technology, in *Proceedings of Neuro Inspired Computing Elements Workshop*. https://niceworkshop.org/wp-content/uploads/2019/04/NICE-2019-DAY-2a-Mike-Davies.pdf (2019)
18. http://brainscales.kip.uni-heidelberg.de
19. J. Schemmel et al., A wafer-scale neuromorphic hardware system for large-scale neuron modeling, in *Proceedings of the IEEE International Symposium on Circuits and Systems* (2010)
20. S. Schmitt, Experiments on BrainScaleS, in *Proceedings of Neuro Inspired Computing Elements Workshop*. https://niceworkshop.org/wp-content/uploads/2018/05/3-01-SSchmitt-Experiments-on-BrainScaleS.pdf (2018)
21. S.A. Aamir et al., An accelerated LIF neuronal network array for a large scale mixed-signa-neuromorphic architecture, arXiv:1804.01906v3 (2018)
22. J. Schemmel, Turing or Non-Turing? That is the question, in *Proceedings of Neuro Inspired Computing Elements Workshop*. https://niceworkshop.org/wp-content/uploads/2019/04/NICE-2019-Day-1m-Johannes-Schemmel.pdf (2019)
23. B.V. Benjamin et al., Neurogrid: a mixed-analog-digital multichip system for large-scale neural simulation. Proc. IEEE **102**(5), 699–716 (2014)
24. A. Neckar et al., Braindrop: a mixed-signal neuromorphic architecture with a dynamical system-based programming model. IEEE Proc. **107**(1), 144–164 (2019)
25. C. Eliasmith, C.H. Anderson, *Neural Engineering: Computation, Representation, and Dynamics in Neurobiological Systems* (MIT Press, Cambridge, 2003)
26. S. Moradi et al., A scalable multicore architecture with heterogeneous memory structures for dynamic neuromorphic asynchronous processors (DYNAPs). IEEE Trans. Biomed. Circuits Syst. **12**, 106–122 (2018)
27. N. Qiao, G. Indiveri, Scaling mixed-signal neuromorphic processors to 28 nm FD-SOI technologies, in *IEEE Biomedical Circuits and Systems Conference (BioCAS)* (2016), pp. 552–555
28. C.S. Thakur et al., Large-scale neuromorphic spiking array processors: a quest to mimic the brain. Front. Neurosci. **12**, Article 891 (2018)
29. J. Park et al., A 65 nm 236.5 nJ/classification neuromorphic processor with 7.5% energy overhead on-chip learning using direct spike-only feedback, in *IEEE International Solid-State Circuits Conference* (2019), pp. 140–141
30. M. Demler, Tsinghua pedals hybrid AI processor: Tianjic runs convolutional, recurrent, and spiking neural networks, Microprocessor Report, Sep 2019
31. K. Cheung, NeuroFlow: a general purpose spiking neural network simulation platform using customizable processors. Front. Neurosci. **9**, Article 516 (2016)
32. R.M. Wang et al., An FPGA-based massively parallel neuromorphic cortex simulator. Front. Neurosci. **12**, Article 213 (2018)
33. M. Demler, BrainChip aims to spike neural nets, Microprocessor Report, May 2018
34. M. Demler, BrainChip AKIDA is a fast learner, Microprocessor Report, Oct 2019
35. M. Mahowald, VLSI analogs of neural visual processing: a synthesis of form and function, Ph.D. thesis, California Institute of Technology (1992)
36. S. Philipp et al., Interconnecting VLSI spiking neural networks using isochronous connections, in *Proceedings of 99th International Work-Conference on Artificial Neural Networks*, LNCS 4507 (Springer, Berlin, 2007), pp. 471–478

37. N. Ziv, Principles of glutamatergic synapse formation: seeing the forest for the trees. Curr. Opin. Neurobiol. **11**, 536–543 (2001)
38. C. Ostrau et al., Benchmarking and characterization of event-based neuromorphic hardware, *International Workshop on Performance Analysis of Machine Learning Systems (FastPath)* (2019)
39. M. Davies, Benchmarks for progress in neuromorphic computing. Nat. Mach. Intell. **1**(9), 386–388 (2019)
40. B. Höfflinger, *Chips 2020*, vol. 1, Chap. 18 (Springer, Berlin, 2012)
41. C. Mead, M. Ismail (eds.), *Analog VLSI Implementation of Neural Systems* (Springer, Berlin, 1989). ISBN 978-0-7923-9040-4
42. T. Wunderlich et al., Demonstrating advantages of neuromorphic computation: a pilot study. Front. Neurosci. **13**, Article 260 (2019)
43. C. Farabet et al., Comparison between frame-constrained fix-pixel-value and frame-free spiking-dynamic-pixel ConvNets for visual processing. Front. Neurosci. **6**, 32 (2012)
44. http://www.darpa.mil/Our_Work/DSO/Programs/Systems_of_Neuromorphic_Adaptive_Plastic_Scalable_Electronics_%28SYNAPSE%29.aspx

Chapter 23
Energy-Harvesting Applications and Efficient Power Processing

Thorsten Hehn, Alexander Bleitner, Jacob Goeppert, Daniel Hoffmann, Daniel Schillinger, Daniel A. Sanchez and Yiannos Manoli

23.1 Systems and Applications

The field of energy harvesting has drawn a lot of attention in recent years and research groups across the globe are working in this field. The extraction of energy from the surrounding environment can be beneficial to a large variety of applications ranging from industrial condition monitoring systems all the way to consumer products in everyday life.

23.1.1 Wearables

The ever-increasing number of portable devices in our modern society face one major issue, which is battery lifetime. Energy harvesting technologies offer the opportunity to prolong the device lifetime or replace batteries altogether. The operation of body-worn or textile-integrated systems in particular becomes significantly more comfortable for the end user when the need for battery replacement or recharging and maintenance is minimized or even eliminated. When considering the human body as an energy source, body heat [1, 2] and body motion first come to mind. In terms of human motion, kinetic energy harvesting devices working as knee braces [3, 4], backpacks [5, 6] and devices to be mounted at the ankle [7] or within the shoe [8–11] have been presented in literature. The focus in the next section is on energy harvesting devices for integration into shoes.

T. Hehn (✉) · J. Goeppert · D. Hoffmann · D. Schillinger · D. A. Sanchez · Y. Manoli
Hahn-Schickard-Gesellschaft für angewandte Forschung e.V., Wilhelm-Schickard-Straße 10, 78052 Villingen-Schwenningen, Germany
e-mail: Thorsten.Hehn@Hahn-Schickard.de

A. Bleitner · D. Schillinger · Y. Manoli
Fritz Huettinger Chair of Microelectronics, Department of Microsystems Engineering – IMTEK, University of Freiburg, Georges-Koehler-Allee 102, 79110 Freiburg, Germany

B. Murmann and B. Hoefflinger (eds.), *NANO-CHIPS 2030*,
The Frontiers Collection, https://doi.org/10.1007/978-3-030-18338-7_23

23.1.1.1 Swing-Motion Energy Harvesting

The acceleration of the foot due to the leg swing during walking is one of the major kinetic energy sources of the human gait. With increasing motion speed the acceleration of the foot increases from about 3 g (slow walking) to 15 g and more (fast jogging). In contrast, the step frequency of a leg varies only between 0.8 and 1.2 Hz [12]. Due to the low excitation frequency, a potential harvesting device cannot be continuously operated in resonance mode as it is typically possible with machine vibrations of higher frequencies. Therefore, non-resonant approaches seem to be a promising solution. In this respect a number of linear devices have been developed for integration into shoes [11–13]. Carroll et al. [11] presented a device based on a magnet-in-channel structure. The average output power of a device with optimized parameters was 14 mW at a walking speed of two steps per second. Another magnet-in-channel structure was presented by Wang et al. [13]. The device incorporates a multi-pole magnet and several coils arranged along the channel. The induced voltage output was up to 1 V.

In [12] Ylli et al. introduced six different energy harvesting architectures (HAC1–HAC6), each occupying an active volume V_A of 71 mm by 37.5 mm by 12.5 mm. Numerical analysis revealed architecture HAC5 (Fig. 23.1a) to be the most promising structure with the highest power output. Architecture HAC5 incorporates a movable closed magnetic circuit with an air gap of 2.4 mm and a linear coil array. The magnetic circuit consists of four magnets with alternating polarity and two back iron parts. The coil array is made of 5 coils arranged in a linear manner and connected in series. The magnetic structure is placed within a 3D-printed housing (Fig. 23.1b). The outer device size is 77 mm in length, 41.5 mm in width and 15.75 mm in height, which allows the device to be integrated into a shoe sole. Rubber end stops effectively reduce the motion range to 69 mm. Experimental characterization was carried out on a treadmill. A 390 Ω load resistance equal to the total internal coil resistance was connected to the coil terminal.

The voltage at the load resistance reaches up to 30 V at a motion speed of 4 km/h (Fig. 23.2a). The corresponding power peaks reach 2.5 W (Fig. 23.2b). An average

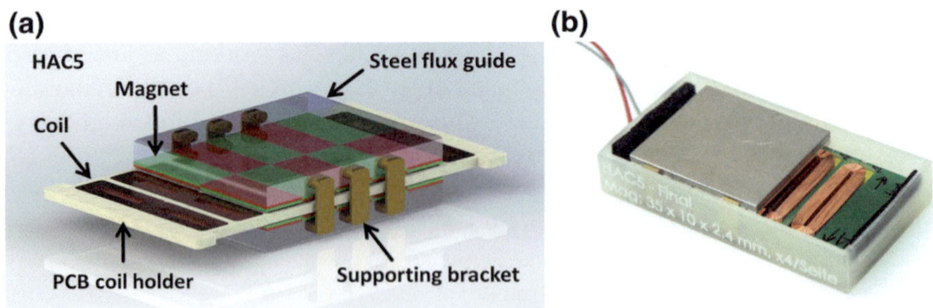

Fig. 23.1 Swing-motion energy harvester for powering wearable devices. **a** Schematic diagram of architecture HAC5 [14], **b** prototype device with optimized system parameters

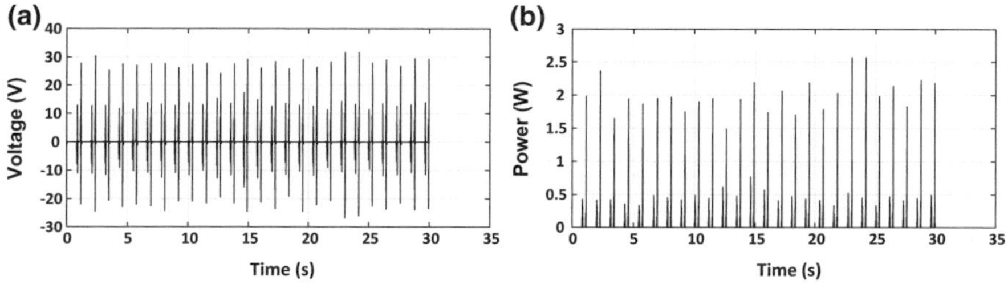

Fig. 23.2 Output parameters for HAC5 measured on a treadmill at a motion speed of 4 km/h. **a** Voltage output across an ohmic load of 390 Ω, **b** calculated power output

Table 23.1 Power output of HAC5 at different walking speeds, the power was measured at a load resistor of 390 Ω connected to the coil terminal

	Walking speed	4 km/h	6 km/h	8 km/h	10 km/h
Average power (mW)	Simulation	21.42	–	–	–
	Experiment person 1	26.02	40.62	18.72	42.1
	Experiment person 2	20.84	29.62	11.87	29.59

power output of 26 mW is obtained at a speed of 4 km/h. The experimental results were compared with simulations using the system models presented in [12]. The results are in good agreement with the simulations, in particularly at low motion speeds.

In Table 23.1 the power output of the prototype device is summarized for two test persons walking at different walking speeds. The highest power output was measured at 6 and 10 km/h. This result corresponds to the higher acceleration values to be found at these walking speeds [9]. In conclusion, at walking speeds as low as 4 km/h an average power output of up to 23 mW can be expected for powering smart wearable devices.

23.1.2 Condition Monitoring

Modern technical assets such as production facilities and construction machines become more and more complex. Therefore, the risk of breakdown of a complete system due to the possible failure of crucial components increases. Therefore, a smart maintenance approach with intelligent devices is essential in order to obtain maximum economic efficiency and safety of the technical system.

The process of continuous or periodic monitoring of system components provides vital information about the physical condition of the system enabling early detection of developing failures. The knowledge about prospective failures allows to turn away from the concept of preventive maintenance towards the smarter approach of predictive maintenance. In this manner subsequent damage and a total breakdown of the system can be avoided leading to a maximum of system availability and safety. However, the installation and operation of condition monitoring systems involves some investment. For instance, the cabling for power supply and signal transmission appears to be a cost-intensive factor, in particular if technical systems are retrofitted. The use of batteries may be an attractive alternative. However, regular replacement of the batteries may implicate high maintenance costs, in particular for technical systems with limited access. Condition monitoring systems become more practical and acceptable if these systems are easy to install and free of maintenance. A key technology for facilitating a self-sustaining operation is based on the process of energy harvesting in which specific devices convert ambient energy into electrical energy.

23.1.2.1 Frequency-Tunable Devices

In most industrial environments kinetic energy is available in form of mechanical vibrations and rotations providing a usable energy source [15–17]. Physical reasons of mechanical vibrations are rotational motion of components (e.g. drive shaft, gear wheel, clutch), contact between parts (e.g. gearing, bearing), machining processes (e.g. milling, turning, grinding, drilling) as well as cavitation (e.g. pumps, compressor, pipe system) [17]. Vibrations are usually predominant at the complete structure of a technical system and exhibit one or more dominant frequencies (Fig. 23.3a). These circumstances make them suitable for energy harvesting purposes. However, the position of dominant frequencies in the vibration spectrum is dependent on the operational state of the technical system and thus changes over time (Fig. 23.3a). As a result, the Eigen-frequency of the vibration energy harvester is not always matched to a dominant frequency and the effectiveness of the power conversion declines significantly (Fig. 23.3b). Therefore, the development of devices with active frequency-adaptation methods is of high interest. The conducted research on these devices aims to increase the effectiveness (maximum power output at minimum size) and to broaden the frequency bandwidth considerably.

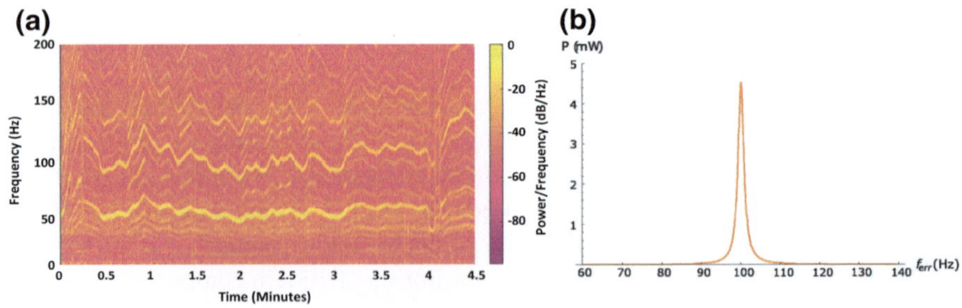

Fig. 23.3 Vibration energy harvesting. **a** Vibration spectrum of a drive component: dominant frequencies (yellow lines) vary over time. **b** Power output of a conventional vibration energy harvester as a function of excitation frequency: the bandwidth of effective power generation is rather narrow

Table 23.2 Mechanical methods for altering the Eigen-frequency of a cantilever structure

Inducing external forces		Variation of cantilever geometry		
Axial	Radial	Length	Width	Height
68–78 Hz [18]	13–18 Hz [21]	23–32 Hz [23]	–	–
4.7–9 Hz [19]		85–149 Hz [24]		
25–50 Hz [20]		21–48 Hz [25]		
150–190 Hz [22]				

In general, there are two main concepts possible for altering the Eigen-frequency of cantilever-based energy harvester (Table 23.2): (i) inducing external forces and (ii) varying the geometry of the cantilever. External forces, axial or radial, applied to a cantilever will change the overall stiffness and thus the Eigen-frequency of the system. A preferred method for inducing axial or radial forces is the application of magnetic fields [18–21]. In case of axial forces, a coupling magnet is attached to the free end of the cantilever. A second magnet, the tuning magnet, is mounted on a movable structure. Translational or rotational motion of the tuning magnet changes the magnetic field interaction and thus the resulting force between the two magnets. Ayala-Garcia et al. [18] and Aboulfotoh et al. [19] demonstrated a system with a translational moving tuning magnet and achieved a frequency bandwidth of 10 Hz and 4.3 Hz, respectively. Hoffmann et al. [20] presented a system with a cylindrical tuning magnet based on a rotary motion. Within a rotation angle of only 180° attractive and repulsive coupling modes between the two magnets can be utilized resulting in a broader frequency bandwidth of 25 Hz. A system with radial force coupling using two coupling magnets and two tuning magnets was demonstrated by Challa et al. [21]. A frequency bandwidth of 9 Hz was achieved. Beside magnetic fields, axial forces can also be applied by direct force coupling using a linear actuator and a mechanical structure. Eichhorn et al. [22] used a piezoelectric actuator, in order to induce axial forces onto two parallel arranged cantilevers. He demonstrated a tunable frequency bandwidth of 40 Hz.

Based on the Euler-Bernoulli beam theory, the Eigen-frequency of a cantilever can be altered by the variation of the cantilever geometry (length, width and height). A favored method is to vary the free length of a cantilever structure. Lee et al. [23] demonstrated a system with a planar coil spring. By rotation of the coil spring the anchor points shift leading to a change in length. A frequency bandwidth of 9 Hz was achieved. A further method was presented by Huang et al. [24]. He used a movable anchor to alter the length of a cantilever beam. Using a total displacement of 30 mm for the movable anchor, a frequency bandwidth of 64 Hz was obtained. Esch et al. [25] followed a similar approach using a U-shaped spring element and a gear spindle (Fig. 23.4a). Both beams of the U-shaped spring element have an effective beam width of 20 mm. To achieve a minimum of mechanical friction between the spring element and the anchor, the material PPS HPV from the company techtron was used. The beam length was varied between 5 and 30 mm resulting in 25 mm displacement range. By careful design of the system parameters a frequency resolution of less

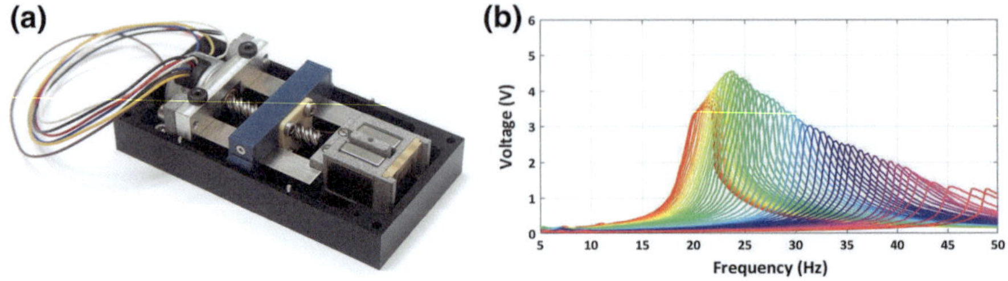

Fig. 23.4 Frequency adaptive energy harvester [25]. **a** Motorized demonstrator with a movable anchor based on a sliding block clamping mechanism. The upper housing and circuit board is removed. **b** Frequency response of the demonstrator with a cantilever beam thickness of 0.4 mm

than 1 Hz was achieved. The frequency response of the demonstrator is shown in Fig. 23.4b. The tuning range of the demonstrator with a beam thickness of 0.4 mm is 21–48 Hz.

23.1.2.2 Self-adaptive Devices with Self-sufficiency

The practical purpose of an energy harvesting system is to ensure the supply of electrical power to a specific electronic system. Therefore, the power demand for frequency adaption should be kept at a minimum. In order to keep the power consumption of a self-adaptive energy harvester low in the first place, it is first necessary to choose a tuning mechanism, which only requires energy at the time of tuning. Second, the energy effort for tuning should be as low as possible requiring a minimum of actuation power. Hoffmann et al. [26] presented a self-adaptive energy harvesting system based on an axial pre-stressed cantilever. A circular tuning magnet, attached to a stepper motor, was used to induce compressive or tensile forces onto a cantilever with a coupling magnet (Fig. 23.5a). The tuning magnet is required to rotate only a half turn (180°) in order to cover the whole tuning bandwidth. A maximum of 12 steps

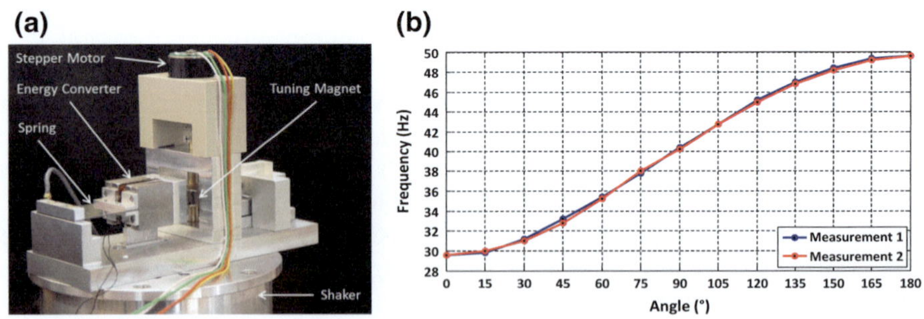

Fig. 23.5 Self-adaptive energy harvesting system [26]. **a** Mechanical structure of the self-tunable energy harvester. **b** Eigen-frequency as a function of the angular position of the tuning magnet

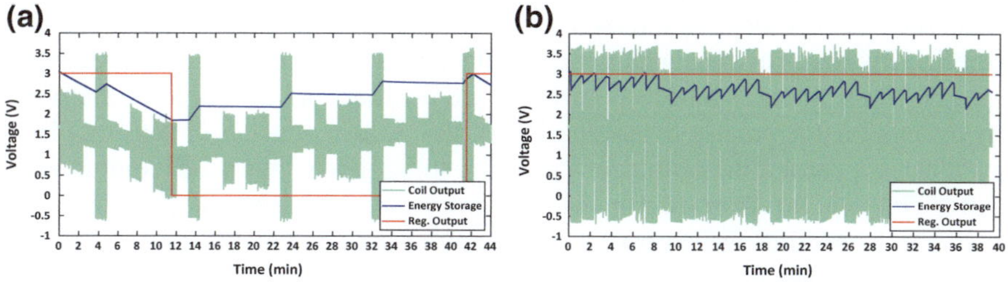

Fig. 23.6 Voltage progression at the energy storage and regulated output port [27]. A 4500 Ω resistor is connected to the regulated output port. The adaption interval is 70 s. **a** Without frequency adaption. **b** With frequency adaption

is required to tune the Eigen-frequency of the energy harvesting system from about 30 to 50 Hz (Fig. 23.5b). The self-adaptive energy harvesting system also included a power management circuit with energy storage in order to power a consumer load and the stepper motor. The execution of one motor step requires an energy portion of 124 mJ. A 4500 Ω load resistance was connected to the regulated output port (3.3 V) of the power management withdrawing a continuous power of 2 mW. On the basis of a specific vibration profile it is demonstrated, that the self-adaptive energy harvesting system is capable of self-sufficient operation while providing a continuous power output of 2 mW for an application.

Figure 23.6 shows the voltage progression at the energy storage and the regulated output port for two different experiments. The first experiment was carried out without frequency adaption while in the second experiment the Eigen-frequency was adapted to the excitation frequency every 70 s. In case of no tuning (Fig. 23.6a) the voltage at the energy storage decreases rapidly due to the power draw at the regulated output port. Although, the voltage level increases temporarily during phase 4 (during phase 4 the energy harvester operates in resonance), the voltage falls below 1.9 V after less than 12 min and the regulated output port is disabled. From there on it requires 30 min to recharge the energy storage to a level of 2.9 V at which the regulated output port is enabled again. During that time, there is no energy available at the output port. If the process of frequency adaption is utilized (Fig. 23.6b), the regulated output port is always in an enabled state considering the same period of time. By adapting the Eigen-frequency of the energy harvester with a rate of 1/70 Hz sufficient energy is generated for both, frequency tuning and application. Certainly, in order to increase the adaption frequency, the power demand for frequency tuning must be further reduced and the system efficiency needs to be improved by implementing smart decision-making algorithms for the adaption process.

23.1.2.3 Linear Devices

In many production environments pneumatic pistons are used as actuators for inducing linear motion. Currently, pneumatic pistons have only a proximity switch to detect the end position after a forward or backward motion. The proximity switch is triggered by means of a magnet, which is attached at the piston rod inside the piston case. However, there is a need to measure the position and the speed of the piston during motion for precise determination of the system state. Hall sensors placed outside at the case of the piston are a possible solution to capture the required information (Fig. 23.7a). For powering such a sensor system, Esch et al. [27] developed an energy harvesting system based on an array of flat and flexible coils. Due to the moving magnet inside the piston (Fig. 23.7b), a voltage is induced in each of the coils, allowing to charge an energy storage.

The energy harvesting system includes 10 coil structures mounted at the outside of the piston (Fig. 23.8b). The folded coil structure is fabricated from a flexible substrate with two copper layers, one at the bottom and one top, each 28 μm in thickness. The flexible substrate contains up to 10 coil elements. By folding the substrate, a multilayer coil with up to 20 coil layers is generated (Fig. 23.8a). The coil elements are designed with a diameter of 18 mm. The linewidth of copper and the spacing between the copper lines is varied. The original magnet (NdFeB N35) in the pneumatic piston was replaced by a slightly larger magnet with a higher magnetization (NdFeB N50).

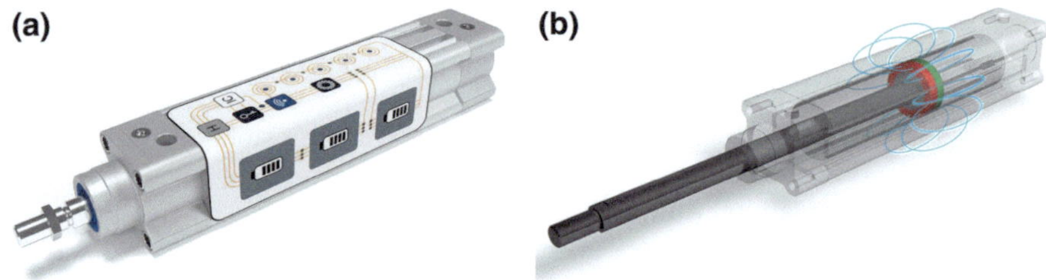

Fig. 23.7 Schematic diagrams of a pneumatic piston (© Festo SE & Co. KG). **a** System-on-a-foil including sensor system and energy harvesting system. **b** Magnetic flux lines from the magnet inside the piston

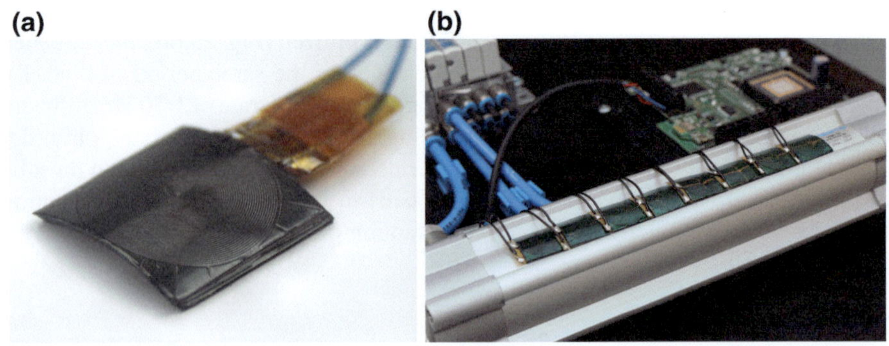

Fig. 23.8 Prototype device of energy harvesting system. **a** Photograph of a folded coil structure [28]. **b** Photograph of the pneumatic piston with mounted coil structures and power management

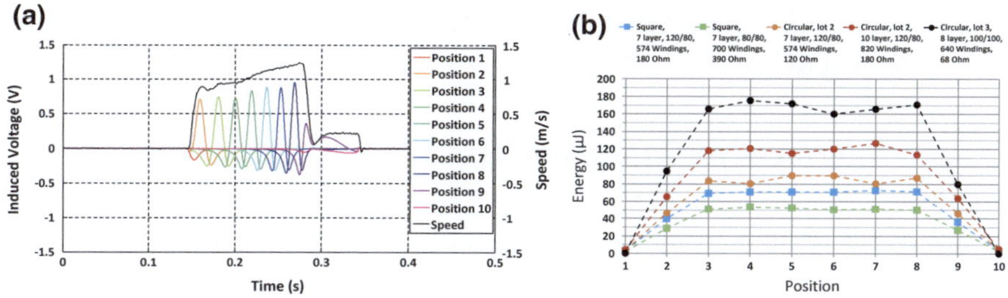

Fig. 23.9 Output parameters of the linear energy harvester. **a** Induced voltage of each coil for one piston motion with the modified magnet and 10-layer coil structure with circular design [28]. **b** Energy comparison for different coil structures including circular and rectangular designs (120/80 means 120 μm copper line width and 80 μm spacing between copper lines)

For characterization, a load resistor was connected to each coil structure, which is equal to the respective coil resistance. The voltage at each load resistor was measured simultaneously. An example is shown in Fig. 23.9a for a circular coil design with 10 folded layers. The induced voltage in the first and the last coil is very low because the magnet starts and stops moving at these positions with a low motion speed. When comparing the data of voltage and speed measurement, a direct correlation between the magnitude of the voltage peaks and the moving speed is evident.

The generated energy for each coil and one motion cycle (one back-and-forth motion) is shown in Fig. 23.9b. The energy is very low for coils at the end positions (position 1, 2, 9 and 10). This is due to the reduced motion speed of the piston when approaching the end position. Figure 23.9b also shows a comparison between square and circular coil designs and different copper line widths. The square designs show larger internal resistance values caused by the fabrication process, where the copper line became unexpectedly narrow in the corners of each winding. When comparing the circular coil designs, experimental data indicates, that a small coil resistance combined with a large number of windings leads to more power output. Therefore, the quality of the fabrication process has a large impact on the power output. The total energy that can be harvested within a complete motion cycle is about 1.18 mJ for a circular coil design with 8 folded layers and a copper line width of 100 μm. This corresponds to 1.18 mW when considering a cycle time of 1 s. This power output is sufficient to power a small sensor system with wireless communication (e.g. BLE).

23.2 Circuit Components for Energy Harvesting Applications

23.2.1 Interfaces for Vibration-Based Energy Harvesting

Vibrations can occur in many environments like industrial environments [28, 29] or railroad tracks [30]. The most popular vibration-based energy harvesting approaches that have been documented are magnetic [31], piezoelectric [32, 33], and capacitive [34].

Piezoelectric energy harvesters (PEH) are popular because of their high power density [35], ease of scaling, and their relative high output voltage [36]. They convert vibration induced mechanical strain into electrical charge by means of the direct piezoelectric effect [37]. Commonly PEHs are cantilever based, in which one or multiple layers of piezoelectric material are mounted on a beam carrier made of e.g. glass fiber, steel, etc. A deflection at the tip of the cantilever, as shown in Fig. 23.10, produces mechanical strain at the top and bottom surfaces, thus the PEH generates charge that can be extracted to power applications or store energy.

The energy extraction is optimized mechanically when the PEH is excited continuously in resonance. However, this is rarely achieved using ambient vibrations, where changes in excitation frequencies and magnitudes are common, or shock excitations occur [9, 26, 28, 33, 39].

Cantilever-beam-based PEHs can be modeled by a spring–mass–damper system (Fig. 23.10), in which the interaction with the piezoelectric layer is also considered by means of the constitutive equations for a linear piezoelectric material [36]. The resulting system of equations can be conveniently used to model both the mechanical and the electrical portions of the PEH as circuit elements. With the electromechanical coupling Γ included in the mechanical parameters, the complete system can be modeled as shown in Fig. 23.11. The resistor $R_M = d/\Gamma^2$ accounts for the damping,

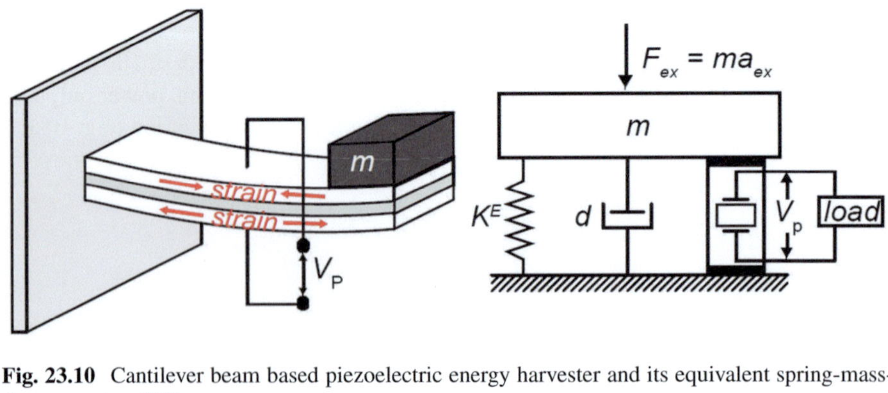

Fig. 23.10 Cantilever beam based piezoelectric energy harvester and its equivalent spring-mass-damper system [38]

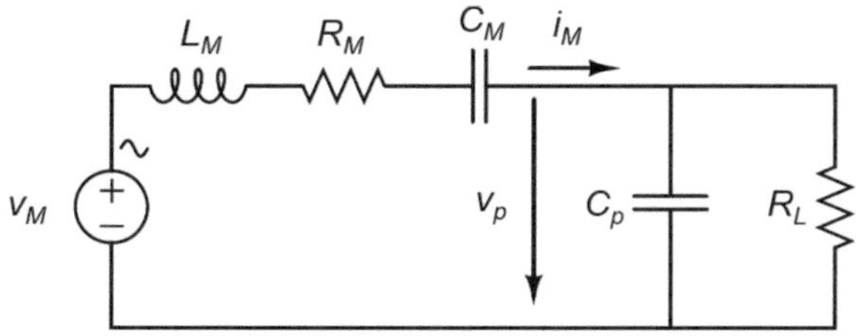

Fig. 23.11 Electrical equivalent circuit model of a piezoelectric harvester with a resistor load R_L

the inductor $L_M = m/\Gamma^2$ is associated with the effective mass, the capacitor CM $= \Gamma^2/K^E$ includes the reciprocal of the spring stiffness, the voltage source $V_M = m_{aex}/\Gamma$ relates to the excitation force, and C_p is the piezoelectric capacitance.

23.2.1.1 Enhancement Schemes for Piezoelectric Generators

Full wave rectifiers are suited as interface circuits for piezoelectric harvesters. However, their performance is dependent on the rectified-output voltage as well as limited by the piezoelectric capacitance for low-coupled piezoelectric energy harvesters [40]. In order to overcome the output voltage dependence, Shim et al. [41] presented a full-wave rectifier circuit in combination with a maximum power point tracking to control the output voltage V_{buf} allowing for optimal output power for a full wave rectifier. It has a speed of only 9.09 ms/V for tracking the maximum power point when the input voltage of the switching converter is changed from 3.4 to 1.2 V.

To overcome the limitations of the piezoelectric capacitance, Sanchez et. al. [40] proposed a circuit implementation based on a parallel-Synchronized Switch Harvesting on Inductor (SSHI). To avoid the charge loss caused by the discharge-charge phase typical in full wave rectifiers (Fig. 23.12), the SSHI scheme flips the piezoelectric voltage when the piezoelectric current is null, every half-wave cycle. This is done by briefly connecting an inductor in parallel to the output terminals of the piezoelectric harvester. This creates an LC circuit, which by properly sizing the inductor, the piezoelectric voltage V_p can be rapidly inverted. This increases the conduction angle, and thus the energy harvesting is significantly increased.

However, the voltage inversion is not perfect due to parasitic losses in the flip switch and the inductor as well as intrinsic losses in the piezoelectric generator. Therefore, the piezoelectric capacitor has to be charged, but for a significant lower time when compared to the full-wave rectifier. The quality of inversion is defined as the flipping efficiency η_F, and a quasi-exponential relation was demonstrated between the flipping efficiency and output power (Fig. 23.13).

The complete circuit implementation, shown in Fig. 23.14, includes an input-voltage control DC/DC converter, which sets the rectified voltage to a configurable value, which in turn can be configured so that it matches the maximum power point (which can be set externally). Additionally, the circuit implementation includes an

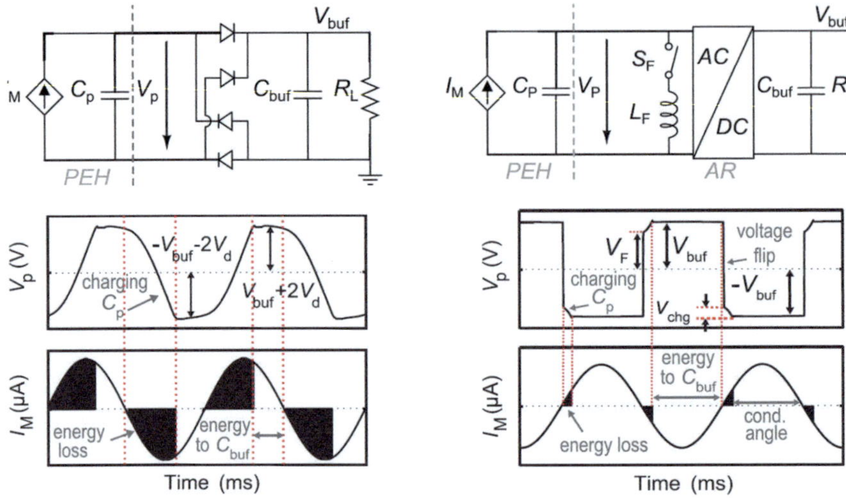

Fig. 23.12 Full-wave rectifier and parallel-SSHI schematics, as well as his operation waveforms [38]

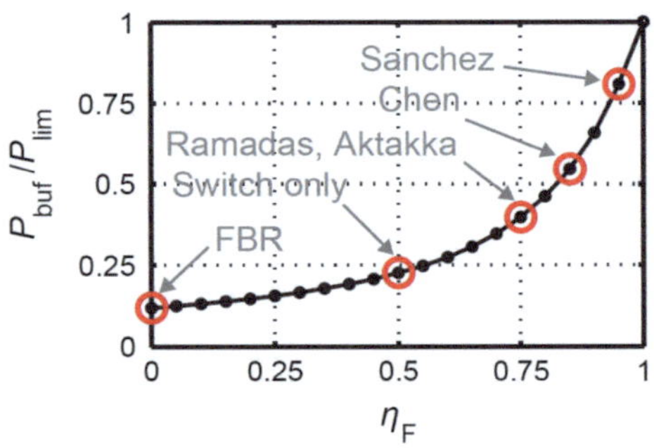

Fig. 23.13 Parameterized calculated harvested power with respect to the flipping efficiency (modified based on [38])

inductor sharing circuit to allow for a single inductor use. A low-dropout regulator as well as over voltage protection are also included, so that the circuit can directly power a device.

The piezoelectric energy harvesting circuit is implemented in a 0.35-μm CMOS technology. Figure 23.15 shows the die micrograph. The total active area is 1.17 mm^2.

The measurement results showed that the implemented chip was able to obtain a flipping efficiency of 0.94 and the complete system is able to harvest up to 6.8 times more energy from a piezoelectric harvester, compared to an ideal full-wave rectifier at its maximum power point. Furthermore, it has the capability to work for both periodic and shock excitations. The chip operates autonomously with high efficiency, powered directly by the harvested energy. It is able to achieve cold startup when the storage capacitors are empty, even for input voltages as low as 670 mV. It

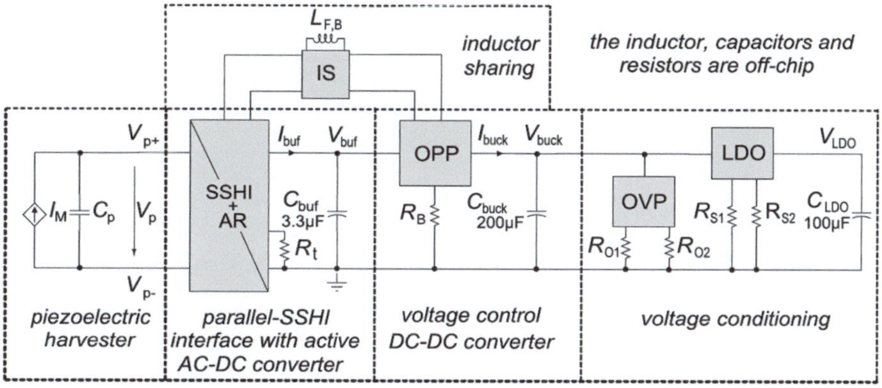

Fig. 23.14 Block diagram of the presented interface [38]. The inductor, capacitors, and resistors are off-chip, whereas all the shaded blocks are on-chip

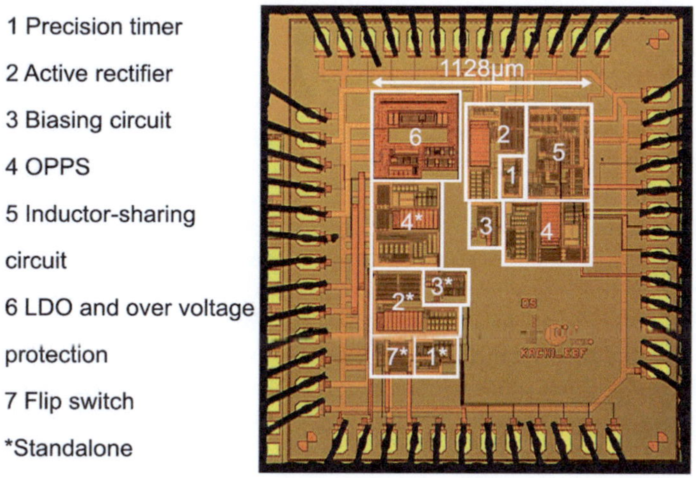

Fig. 23.15 Die micrograph [38]

achieved chip efficiencies up to 95.4% and can harvest from a few μW up to 1 mW of energy (Fig. 23.16).

23.2.1.2 A Piezoelectric Energy-Harvesting Interface Circuit with Fully Autonomous Conjugate Impedance Matching

Piezoelectric generators can be used to convert kinetic energy into electrical energy. Due to their resonant characteristics, piezoelectric generators generate an AC voltage when they are excited by a vibration. The highest output power can be achieved

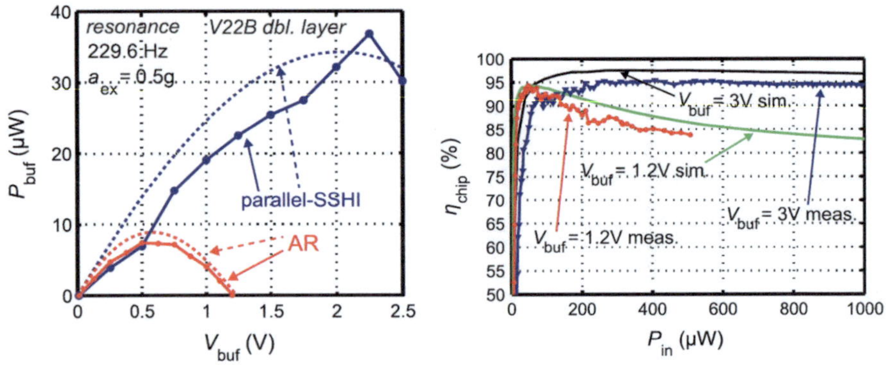

Fig. 23.16 Performance measurement results [38]

when the excitation frequency matches the generator's resonant frequency. In order to extract the energy and store it in a battery or a buffer capacitor, an interface circuit is required. There are mainly two categories of interface circuits: Whereas simple AC/DC rectifiers are considered as passive since they do not require a separate power supply, active interface circuit concepts such as the Synchronous Electric Charge Extraction (SECE) technique are able to significantly increase the harvested energy [38]. This concept periodically extracts the energy stored in the piezoelectric generator within a very short amount of time when its voltage has reached a peak.

In [42], a piezoelectric energy-harvesting interface circuit with fully autonomous conjugate impedance matching and extended bandwidth is proposed. The conjugate impedance matching is achieved by introducing time-delays into the SECE technique. Figure 23.17a shows the block diagram of the proposed circuit, and Fig. 23.17b explains its operating principle: The peak detector senses when the output voltage of the transducer (V_{PEH}) reaches its maximum. Whereas SECE would start the energy extraction process immediately (upper waveform), the proposed technique waits a

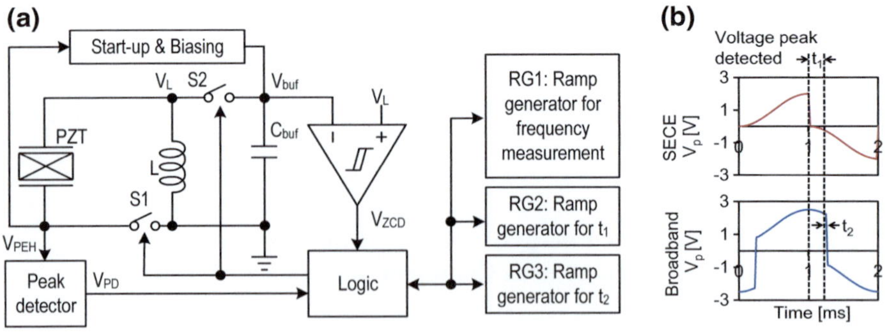

Fig. 23.17 **a** Block diagram of the proposed circuit (modified based on [43]) with **b** corresponding waveform (only for positive half-cycle) [44]

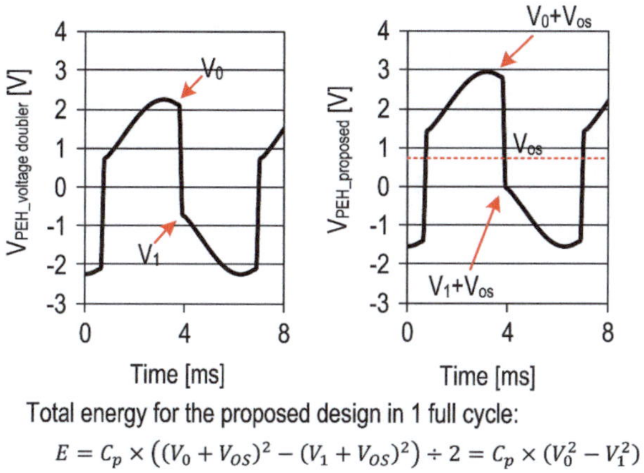

Fig. 23.18 Principle of unbalanced-switching [44] (modified based on [43])

time interval t_1 after having detected the voltage peak before extracting the energy stored in the piezoelectric generator into an inductor L by closing switch S1 for a time interval t_2 (lower waveform). The duration of t_2 depends on the polarity of V_{PEH}: If $V_{PEH} < 0$ before S1 is closed (negative half-cycle), then t_2 is chosen such that all the extracted energy is sent back to the transducer, just switching the polarity of V_{PEH} in order to increase the damping force. If $V_{PEH} > 0$ (positive half-cycle), then t_2 is shorter so that a part of the energy stored in L is transferred to the buffer capacitor C_{buf}.

By choosing different values of t_2 for the negative half-cycle and the positive half-cycle, a DC offset V_{OS} is introduced in the piezoelectric generator's output voltage waveform (see Fig. 23.18). It can be shown mathematically that the offset does not affect the extracted energy. However, by using this method of unbalanced switching, there is no need for an AC/DC rectifier or a voltage doubler, in contrast to other implementations, e.g. [44]. This reduces cost, complexity and energy consumption.

In practical implementations, it is difficult to automatically and accurately generate the optimal t_1/t_2 for maximum power output. Hence, [42] proposes a technique which reduces the number of different t_1/t_2 settings to 6 for a frequency range 84.5–95.5 Hz. A custom-designed ramp generator automatically measures the excitation frequency and chooses the appropriate t_1/t_2 setting for extracting a large part of the power compared to the power which can be extracted by manual tuning. Details about the automatic tuning mechanism can be found in [42].

This interface circuit has been manufactured in a 0.35 μm CMOS technology and tested with a MIDE V21 piezoelectric transducer mounted to a shaker, with its resonant frequency tuned to 90.5 Hz by adding a tip mass. As shown in Fig. 23.19a, the proposed system extends the 3 dB bandwidth by 110% over the conventional SECE or 156% over the natural bandwidth (3.2 Hz) of the unloaded transducer. Regarding the extracted power, there is an improvement of 29% at the resonant frequency and even 96% considering the average power from 85 to 96 Hz. Hence,

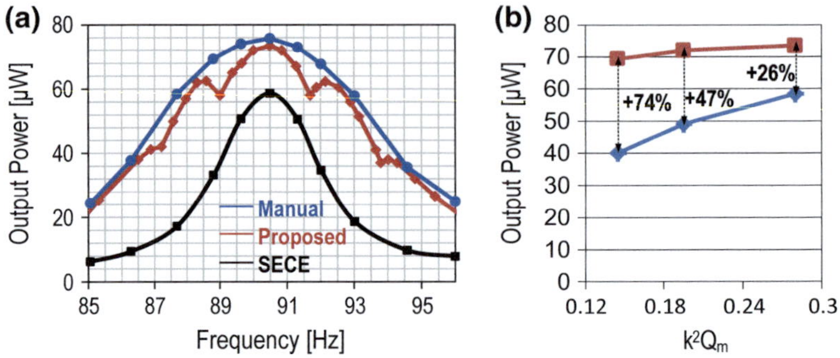

Fig. 23.19 **a** Measured output power over frequency (modified based on [43]) and **b** measured output power at resonant frequency over k^2Q_m (upper waveform: proposed circuit, lower waveform: SECE) [43]

the proposed circuit can be very useful to compensate for frequency drifts in real applications. SECE only reaches its maximum efficiency when the electromechanical coupling coefficient k^2Q_m of the transducer equals $\pi/4$. For a lower k^2Q_m, which is common for transducers with smaller sizes, this gap becomes even larger, as shown in Fig. 23.19b.

The conventional SECE/SSHI methods [38, 45] show no or little bandwidth improvement over the natural bandwidth as shown in Table 23.3. Compared to similar techniques [44, 46], the presented chip is able to run fully autonomously with better figure of merit (FOM), as well as lower chip area and power consumption.

The chip micrograph is shown in Fig. 23.20.

23.2.1.3 Extraction Circuit for Non-periodic Waveforms

While most state of the art circuits work with a sinusoidal input of fixed or slow changing amplitude, this circuit is designed to hold the piezoelectric harvester in the maximum power point at any arbitrary input waveform [47].

Best efficiencies in state of the art are actually reached with circuits based on synchronized switch harvesting. If the energy transfer from generator to storage capacitor is obtained by an inductor, the circuit topology is called SSHI, while using a capacitor array it is called SSHC.

These circuits can nearly extract the maximum possible power which would be extracted with a conjugate complex matching. The drawback of this circuit type is that near maximum power can only be extracted if the voltage of the internal storage capacitor is in a certain relation to the amplitude of the excitation signal [48]. If the amplitude of the excitation signal is continuously changing, also the voltage of the storage capacitor would need to be continuously changed. This is not feasible, due to the high storage capacitance making this circuit type not suited for arbitrary waveforms at the harvester.

Table 23.3 Comparison to other interface circuits (modified based on [43])

	This work	Hsieh, ITPE 2015	Cai, ESSCIRC 2017	Hehn, JSSC 2012	Sanchez, ISSCC 2016
Process technology	0.35 μm	Discrete components	0.35 μm	0.35 μm	0.35 μm
Transducer	V21B	V22B	V22B	V22B	V22B
Scheme	Broadband SECE	Broadband SSHI	Broadband SECE	SECE	SSHI
Load-independent	Yes	No	Yes	Yes	No
Extra mechanical components	No	Accelerometer	No	No	No
Fully-autonomous	Yes	No	No	Yes	No
Power consumption (μW)	0.38	NA	0.85	4.4	NA
Resonant frequency (Hz)	90.5	425.5	175.4	174	224
FOM[a]	+156%	+23%	+71%	<+40%**	<0**
Chip area (mm^2)	0.8/0.27 (active)	NA	3.57	1.25	1.17 (active)

[a]Bandwidth of the output power over the natural bandwidth of the transducer
[b]Calculated from the paper

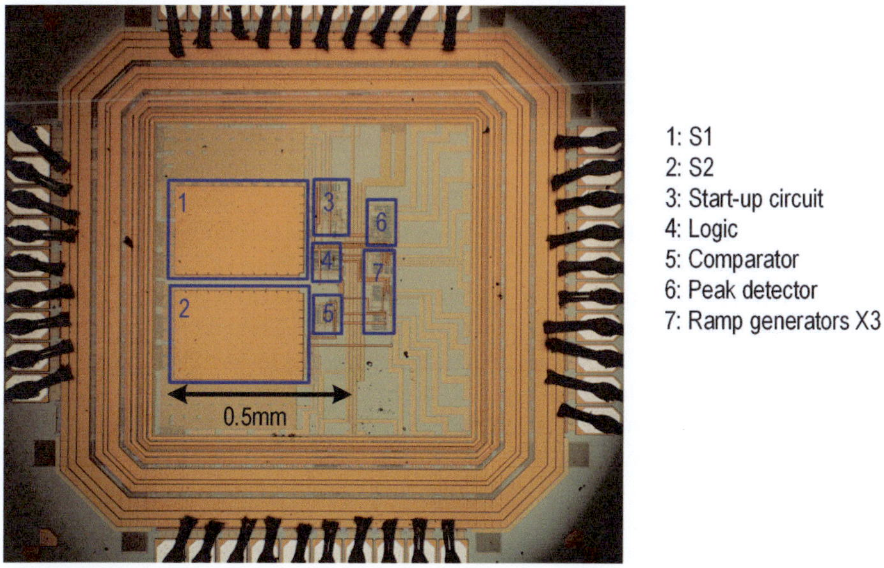

Fig. 23.20 Chip micrograph (modified based on [43])

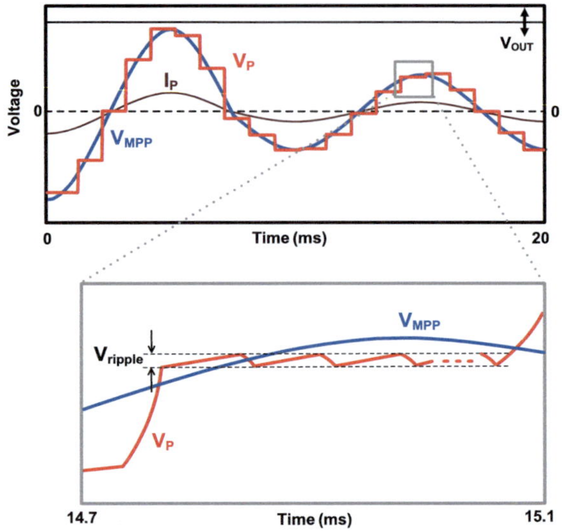

Fig. 23.21 Upper part: the generator voltage V_p following the maximum power point. The generator current I_p is in phase with the maximum power point voltage. Lower part: zoom in the generator voltage V_p. The ripple is a result of a phase wise activation of the switch-converter [48]

Other circuit types like energy pile-up resonant circuits [49] or circuits based on synchronous electric charge extraction (SECE) [50] are in principle better suited to deal with arbitrary waveforms, but even in theory these circuits are not able to extract the maximum possible power.

To overcome these limitations, the circuit described in [48] includes a processing unit which is able to calculate the maximum power point from the generators voltage-curve shape. The circuit adjusts the generator's output voltage stepwise to the Maximum Power Point (V_{MPP}), as can be seen in Fig. 23.21 and is thus in theory able to extract the maximum possible power at any arbitrary waveform.

Each step has a sampling phase and a set phase. In the sampling phase the voltage shape of the generator is detected to calculate the actual value of V_{MPP}. More specifically, sampling the voltage-curve shape means measuring the generator voltage at two different time points that differ by a fixed time difference Δt.

In the set phase a bidirectional switch-converter is used to adjust the generators output voltage to the calculated V_{MPP}. If the generator voltage is higher than V_{MPP} the switch-converter extracts energy from the generator and transfers it to a storage capacitor. If the generator's voltage is lower than V_{MPP} the switch-converter delivers energy from the storage capacitor back to the generator in order to load the generator's intrinsic capacitor to V_{MPP}.

The circuit's behaviour is equal to a conjugate complex matching if exited with a sinosoidal input.

The drawback of the system is its efficiency which is usually below 50% due to the high losses of the switch-converter. Like in a conjugate complex matched

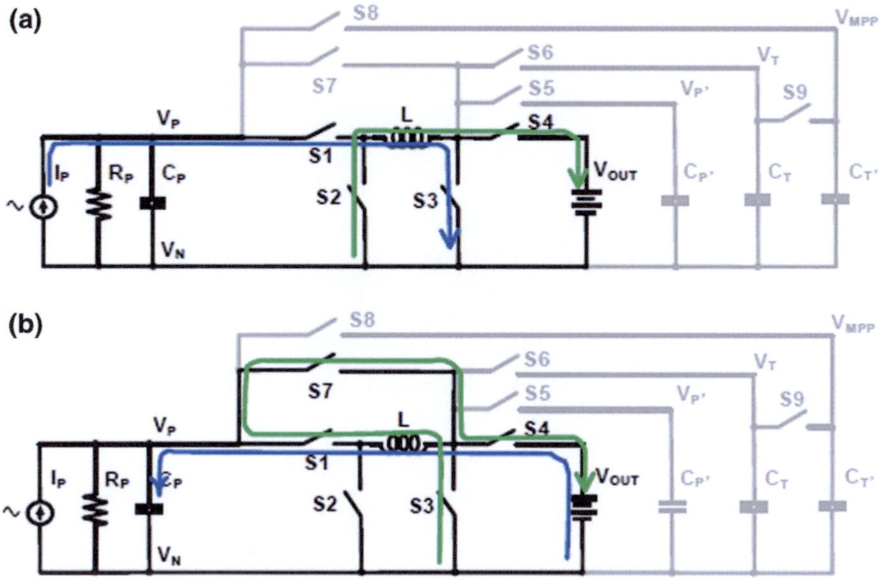

Fig. 23.22 Bidirectional switch converter [48]. **a** Shows the energy transfer the generator to the storage capacitor. **b** Shows the energy transfer from the storage capacitor back to the generator

circuit where currents oscillate between the reactive parts, high currents oscillate here between the generator's intrinsic capacitor and the storage capacitor of the switch-converter as can be seen in Fig. 23.22. These high currents lead to high resistive losses in the switch-converter.

23.2.2 *Thermoelectric Energy Harvesting*

Thermoelectric generators have proven to be one of the most productive and applicable energy harvesting sources to date due to the nearly ubiquitous availability of temperature gradients. Moreover, their nature as a perfect solid-state device enables very compact and flexible implementations where the device dimensions are basically defined only by the power requirements of the application and the corresponding thermal flow and thermal insulation of the device.

As the bandwidth of any thermal signal is extremely limited and orders of magnitude below the internal mechanisms of a solid-state device like a thermoelectric generator, the device essentially acts as a DC voltage source. As such, it can be modelled as a Thevenin equivalent circuit consisting of an ideal DC voltage source and an internal resistor, and load-matching and maximum power point tracking (MPPT) can be comparatively easily achieved by controlling the output voltage to ½ V_{OC}.

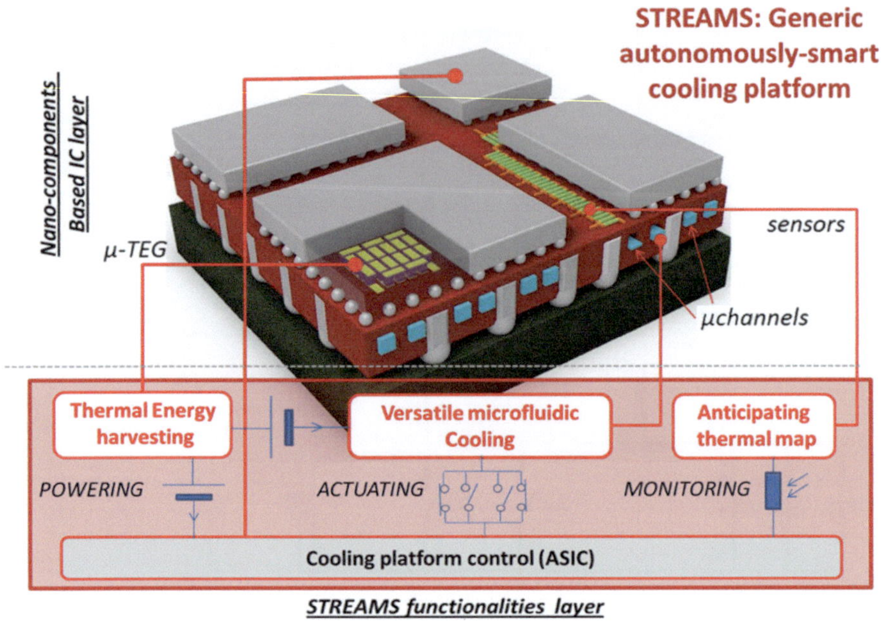

Fig. 23.23 STREAMS proof of concept [52]

The following section explores two interesting use cases of thermoelectric generators: Extremely low temperature gradients, as encountered for example when human body heat is used as an energy source, and multi generator systems.

23.2.2.1 STREAMS—Multi Power Source System

The STREAMS project [51]. was a collaboration of several research institutions, namely CEA in Grenoble, France, the University of Lleida, Spain, the University of Sherbrooke, Canada, Hahn-Schickard, Villingen-Schwenningen, Germany, and the University of Freiburg, Germany and received funding from the European Union's Horizons 2020 research and innovation programme under grant agreement No. 688564. The main goals were the implementation of liquid cooling directly into silicon via microfluidic techniques and to include thermoelectric functionality for both energy extraction and temperature measurement.

System Overview and Key Challenges

Figure 23.23 shows the system that serves as proof-of-concept for the project: Four heater-chips that emulate the power dissipation of active devices like data processors in an actual application are put on a silicon interposer that provides electrical

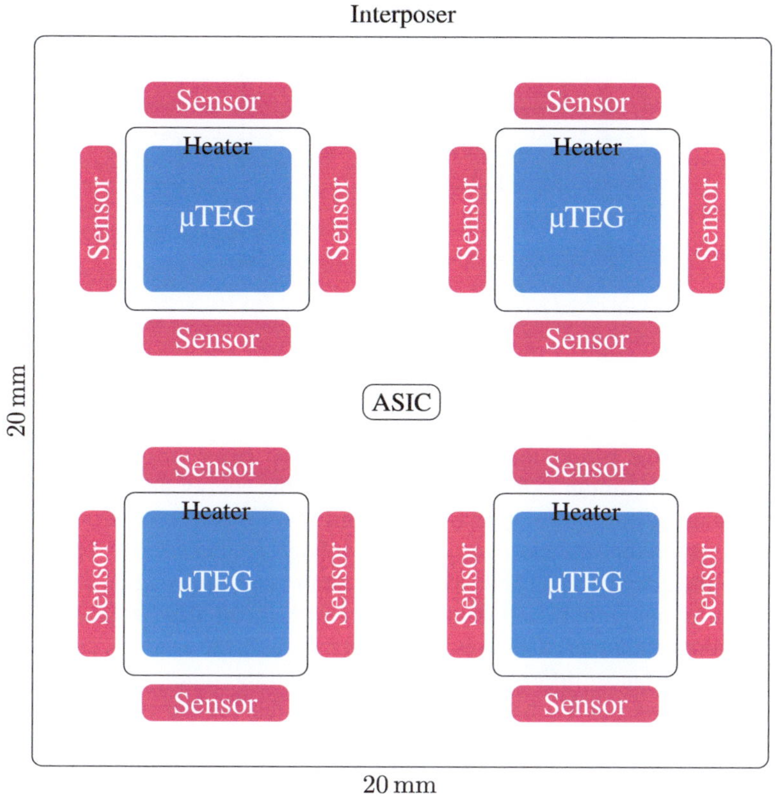

Fig. 23.24 STREAMS system abstraction

interconnectivity and the microfluidic and thermoelectric functionality. Figure 23.24 depicts a schematical representation of the system and highlights the electrically relevant properties and components. The four heat sources are encircled by thermal flow sensors to determine the thermal state of the system, while thermoelectric generators between the microfluidic heat sink and the heaters extract a percentage of the thermal power flow as electrical power. All these devices interact only with a central control chip, whose purpose is to autonomously read out the sensors and transmit the gathered data using only the power extracted from the four generators.

The cooling solution is designed to allow a maximum thermal gradient of 100 K across the generators at the intended heat source activity. In turn, a thermoelectric device with a Seebeck coefficient that leads to an open circuit voltage of 3–5 V at this excitation has an internal resistance of 2–5 kΩ and can provide an output power of 1.5 mW. The maximum input power budget is thus limited to 6 mW, while the sensor interface representing the load is designed for a power demand of 1.5 mW at a stable supply voltage of 3.3 V. In contrast, both the output voltage and power levels of the generators will fluctuate with the actual temperature gradient across the

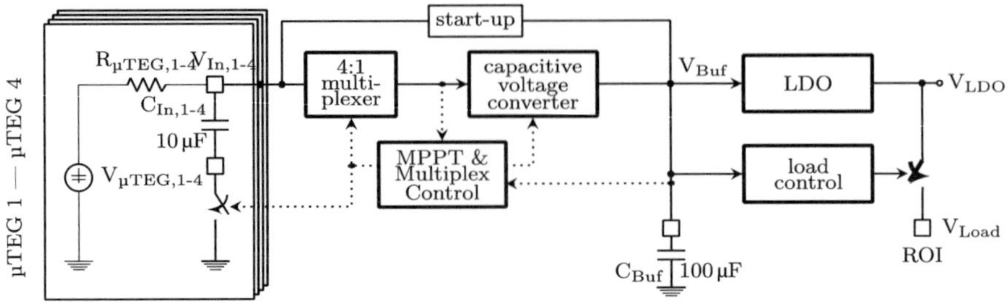

Fig. 23.25 Architecture of the power management unit [52]

respective devices. The primary task of the designed power management unit lies thus in arbitrating between these needs.

Maximizing the power that is actually extracted from the generators towards the theoretical maximum is the key challenge in this energy harvesting system. In this particular case, a power management unit (PMU) has to control individually stimulated generators independently, as the voltage gradient across each device is defined by the activity of its heat source, while the heat sink can be assumed to be at a similar temperature for all devices. Moreover, it can also be assumed that the characteristic parameters of the four generators, namely the Seebeck coefficient and the internal resistance, differ from one generator to the other. The final challenge of this application lies in the extremely limited area for the complete system of roughly 4 by 4 mm^2. This includes the control ASIC consisting of both the power management unit and the sensor read-out system as well as any external component. The power management unit discussed in this section was published at the 2019 European Solid-State Circuits Conference [53].

Power Management Unit Architecture

As previously discussed, in order to maximize the output power of a DC source, its output voltage has to be driven to ½ its actual open circuit voltage. The nominal voltage levels of the PMU's four input nodes are thus well defined at any given moment, as is the output voltage level of 3.3 V. To solve these contradictory demands, the PMU features a two-converter architecture depicted in Fig. 23.25. In this particular system, a primary capacitive voltage converter controls the four generator voltage nodes to their individual maximum power point (MPP) and transfers the extracted energy to an intermediate voltage node connected to a buffer capacitor, which functions as an energy reservoir to arbitrate between differing input- and output power levels.

From this voltage, a secondary converter implemented as a low dropout (LDO) regulator generates the stable 3.3 V output voltage that is also used to supply the internal components of the PMU itself. These are a control unit for the primary converter that implements an MPP-tracking scheme, a voltage monitoring circuit that provides the required information regarding the current state of the system and

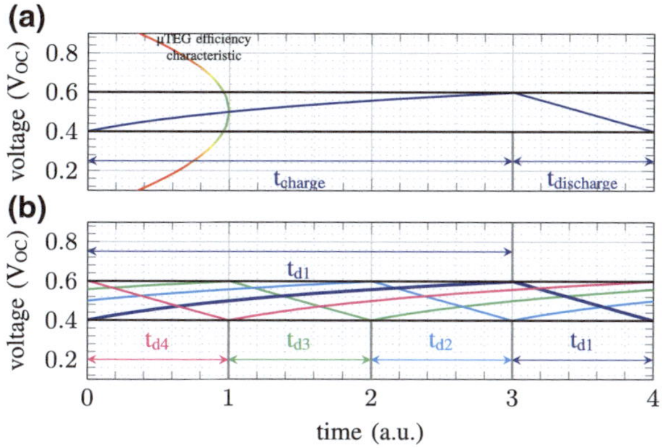

Fig. 23.26 Hysteretic control [52]: **a** single input, **b** four inputs

an output control unit that protects the operation of the PMU against excessive load power levels. A capacitive converter and an LDO were chosen primarily to meet the tight area constraints, as both of these devices can be completely integrated on the ASIC without the need for additional external components, while larger alternative solutions like an inductive voltage converter may have offered superior performance. However, four input filter capacitors and the aforementioned buffer capacitor cannot be eliminated.

The two primary challenges of the MPP-tracking and thus the generator voltage control lies in controlling multiple input nodes simultaneously. A very popular control scheme in the energy-harvesting field is a hysteretic control (Fig. 23.26) because of its low complexity and inherent stability.

In this case, the primary voltage converter has to be able to extract more power from the input node than the generator in question provides to reduce the input node voltage. Once a certain lower hysteresis voltage has been reached, the converter is deactivated and the input power can push the input voltage to an upper hysteresis level. Due to the inherent dead time of this approach, it is perfectly suited for controlling multiple nodes: While the primary converter is discharging one node, the remaining nodes have time to reach the upper voltage level and the converter simply cycles through the different input nodes.

Only the mean value of the controlled node lies at the required voltage level using this control scheme. The losses caused by the deviating waveform as a function of the hysteresis amplitude are depicted in Fig. 23.27. As shown, a hysteresis level of $\pm 10\%$ V_{OC} leads to a tracking loss of 1.34% and has thus been chosen for this system. The hysteretic control itself can also be easily implemented: The open circuit voltage has to be sampled at some point in time. In this particular implementation this is done once at the beginning of observing the charging phase and additionally at the beginning of the discharging phase to compensate for potential leakage from the sampling capacitors. Actually stored in this case are the resulting upper and

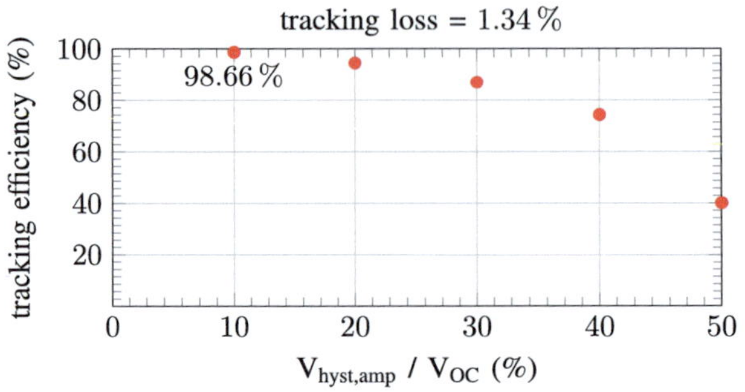

Fig. 23.27 Tracking loss due to the hysteresis amplitude

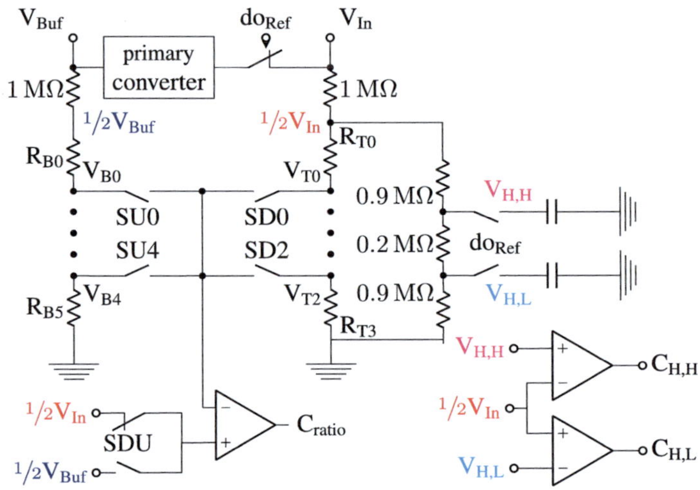

Fig. 23.28 Voltage monitoring circuit [52]

lower hysteresis thresholds generated by a resistive divider (Fig. 23.28), and a single comparator is sufficient to implement the entire MPP control scheme.

Moreover, the constraints imposed on the waveform by the control scheme also simplify controlling the capacitive voltage converter. Such a converter can only implement discrete voltage conversion ratios, and consequently the optimal setting remains constant over a certain range. Since the input voltage of the converter is monotonically falling during the discharging phase, the voltage ratio from input to output will monotonically rise when the converter is active. It is thus sufficient to switch only to an adjacent setting once the ratio crosses a certain threshold. This can be implemented via a single comparator, two sets of resistive voltage dividers, and a set of switches to select the appropriate voltages for the currently prevailing voltage state (Fig. 23.28). In turn, the traditionally relatively complex and demanding analog

Table 23.4 Performance at different input power scenarios [52]

Power scenario	μTEG setup		P_{in} (mW)	Tracking loss (%)	η_{ete} (%)
Min. input	3.7 V	10 kΩ	1.33	1.50	11.2
Max. input	4 V	3.5 kΩ	4.44	2.88	37.7
0.56 Max. input	3.2 V	4 kΩ	2.50	2.30	21.3
0.59 Max. input	$\begin{Bmatrix} 2.8\text{ V to }4\text{ V} \\ 4\text{ k}\Omega\text{ to }6\text{ k}\Omega \end{Bmatrix}$		2.61	2.71	24.1

control circuitry can be reduced to a couple of low power voltage dividers and comparators, while the control itself can be implemented via a low speed and thus low power digital state machine. In turn, the efficiency of the system is defined by the performance of the LDO and the capacitive primary converter. The characteristics of the primary converter, however, are defined by the available chip area, which limits the total switching capacitance to roughly 1.6 pF. This in turn defines the required switching frequency of 6 MHz and in turn the achievable efficiency of about 70%.

System Performance

As discussed previously, the average primary converter's efficiency over the operating range of roughly 70% together with the quiescent power of the LDO and its drop out losses define the performance of the PMU. Table 23.4 summarizes the achieved efficiencies and tracking losses for several TEG excitation scenarios. Here, the highest efficiency can be achieved at the maximum input power scenario, as the constant quiescent losses have a minimal impact in this case, while the dynamic losses become dominant, as they scale with the input power level. Moreover, the low tracking loss of 2.5% for heavily deviating TEG properties confirm the efficacy of the control scheme.

23.2.2.2 Human Body Heat Harvesting—Low Voltage Challenge

In contrast, to the relatively complex system discussed in the last section with high temperature gradients, the very popular application of harvesting power from the human body poses a very different challenge: The human body offers only a very low temperature gradient from its surface to the environment in the single digit Kelvin range [52]. From these low gradients, only specialized harvesters are able to generate a somewhat usable output voltage on the range of tens of millivolts. In turn, such low voltage levels require specific circuit techniques and architectures to provide power to electrical systems. Moreover, such systems need to be comfortable to wear in order to find acceptance by any user, and are thus heavily restricted in terms of physical size.

Table 23.5 State of the art regarding low voltage DC-DC converters

	[55]	[56]	[57]	[58]	[54]
Process	130 nm	65 nm	65 nm	350 nm	PCB
Output voltage	1.25 V	1 V	1.2 V	1.8 V	Unregulated
Start-up voltage	70 mV	95 mV	50 mV	35 mV	69 mV
Peak efficiency @ start-up voltage	58%	NA	65%	58%	NA
Auxiliary power source	No	No	No	Mechanical vibration	No
PCB Inductors	1	1	3	3	Piezoelectric transformer

Low Voltage Start-Up Strategies

Table 23.5 documents the state of the art regarding low-voltage DC converter operation. One approach to overcome the low voltage issue are start-up oscillator circuits based on relatively bulky transformers [54]. This solution runs contradictory to the aforementioned requirement of being easy and comfortable to wear. Moreover, it attempts to achieve both low voltage start-up and optimal operation under nominal conditions, and the achieved efficiency is quite low as a result.

In contrast, two stage voltage converter architectures [55, 56, 58, 57] are able to distribute these contradictory requirements and solve them with different system components, where each can be heavily optimized for its assigned task. The achieved minimum start-up voltage, the efficiency during nominal operation, and the required system complexity can classify these solutions. As can be seen, the minimum start-up voltage can be improved by more complex systems, either in terms of number of off-chip components [57] or even auxiliary energy sources like mechanical vibrations. In contrast, systems that are integrated to the highest degree with the minimum external complexity of a single inductor only achieve slightly higher start-up voltage levels at a comparable nominal efficiency. The following section will thus take a closer look at the best performing fully integrated start-up solution [55].

System Architecture

This system (Fig. 23.29) is an example of a two-stage architecture. During nominal operation, a primary boost converter generates a 1.25 V output voltage from the low voltage input node. With a minimum open circuit startup voltage of 70 mV, the maximum power point lies at an input voltage level of 35 mV and the converter has to implement a voltage conversion ratio of 35. This value is only achievable using inductive converter techniques. In this state, the primary converter supplies itself via its own output voltage, which is connected to an output filter capacitor with a relatively high capacitance of 100 μF. As long as this voltage level is kept above

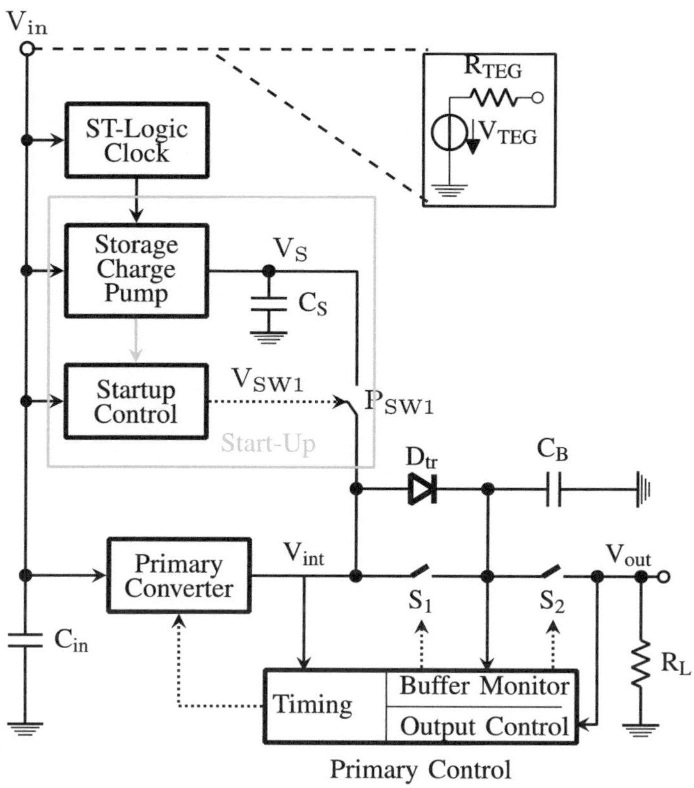

Fig. 23.29 Low voltage start-up architecture [55]

a certain value of about 600 mV, the system will remain operational. The major challenge thus lies in charging this large output capacitor to this voltage level from an energy source with extremely low output power.

This source is implemented as a variation of a low voltage linear charge pump that can generate an output power in the picowatt range. This value is limited by the low voltage operation itself limiting this secondary converters efficiency, and the need not to draw any appreciable power from the source during the start-up phase. Doing so would cause a significant voltage drop across the internal resistance of the generator and in turn increase the minimum start-up open circuit voltage. The low output power in turn is used to accumulate energy on a small fully integrated storage capacitor with a total capacitance of roughly 1 nF. In turn, switching from operation on only the low capacitance storage device to the high capacitance output filter requires an additional transfer phase in order to retain the accumulated voltage level (Fig. 23.30). During this phase, the primary inductive boost converter operates only on the storage capacitor, while the buffer output capacitor gets charged via a diode to transfer any excessive power not needed for operation. Once a sufficient voltage level has been reached on the high value capacitor, the actual power switch connecting it

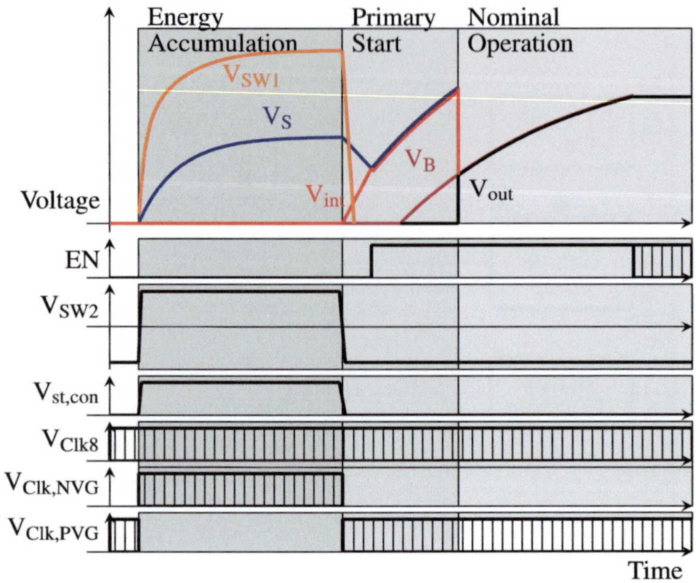

Fig. 23.30 Start-up sequence [55]

to the primary converter output is closed and nominal operation starts. The circuit techniques that enable the low voltage operation of the secondary converter itself and the control of the start-up sequence are discussed in the following.

Ultra Low Voltage Circuits

The basic technique that allows extremely low voltage operation are Schmitt-Trigger (ST) logic gates and libraries discussed in Sect. 23.2.3.1 that can operate at supply voltage levels as low as 62 mV. For the performant functionality of a Dickson style charge pump, however, driving capacitive loads is critical in terms of speed, output power and efficiency. The feedback structure of ST logic (see Fig. 23.33 in Sect. 23.2.3) inhibits this capability by steering the needed active current out of the signal path at the beginning of a signal transition. A feed-forward structure (Fig. 23.31) that disables the feedback and thus the current steering directly at the beginning of a transition can improve the dynamic characteristics of deep subthreshold ST-logic and enables capacitor driving chains with fan-out values greater than 1 per stage. This is the most critical improvement for operating a charge pump at ultra-low input voltage levels.

Controlling start-up sequence comprises to distinct challenges, namely to determine the end of the initial energy accumulation phase on the storage capacitor, and in turn the transition to nominal operation. Both these functionalities have to be implemented in the ultra-low voltage input domain and have to draw minimal power from

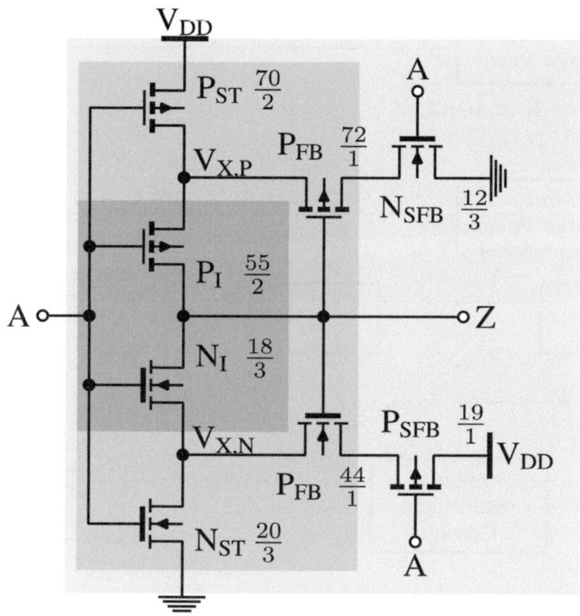

Fig. 23.31 Modified Schmitt-Trigger inverter [55]

the storage node in order to not significantly increase the start-up time. The simplest way to achieve this goal is by using a PMOS device with a positive gate source voltage to pinch off the leakage current in the off state (Fig. 23.32). The control voltage can be generated by an additional charge pump operating from the storage voltage. As this circuit has to supply only the low leakage of a gate current, it poses only a negligible load.

Up to now, the gate voltage only achieves to pinch off the leakage current, but is not actually controllable from the low input voltage domain, in which any controller determining the end of the accumulation phase has to operate. An additional NMOS device pulls the gate voltage of the actual power switch to ground. This auxiliary switch needs to offer a sufficiently high ratio of on-to-off current to both not impede the current pinching during the accumulation phase and to force its gate voltage actually to ground during the transfer phase. This is achieved via an additional set of charge pumps whose activity can be controlled in the low voltage domain and that generate a control voltage of ± 125 mV during the accumulation and the transfer phase respectively.

Regarding the controller, not a precise determination but only an estimation of the end of the accumulation phase is feasible. A simple digital counter can achieve this functionality, as the total charge transferred through a charge pump and thus the resulting voltage on the storage capacitor is dependent on the total number of clock cycles. Counting clocks thus allows for a rough estimation of the voltage level on the storage capacitor. Once the counter reaches a value that indicates that the primary

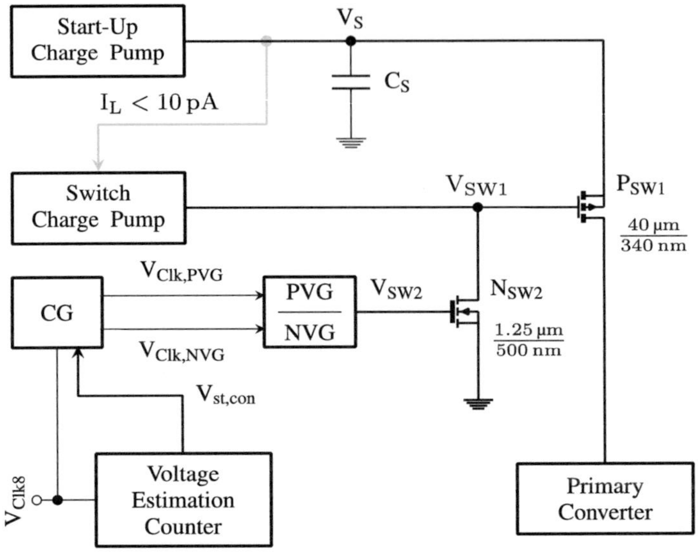

Fig. 23.32 Low voltage low leakage switch [55]

converter should be operational at the accumulated voltage level, the controller terminates the accumulation phase and initiates the transfer phase by controlling the activity of the complementary charge pumps.

23.2.3 Digital Ultra-Low Voltage (ULV) and Ultra-Low Power (ULP) Circuits

Designing startup and control circuits for harvesting interfaces as well as downstream voltage converters constitutes a substantial challenge [59]. The strict limits on the available energy and the ultra-low output voltage of some harvesting principles make it difficult to implement working and reliable circuits. Especially thermoelectric generators running on body heat supply only a few tens of millivolts [59, 60]. This entails the need to design for feasibility, functionality, and absolute power consumption rather than throughput and efficiency.

Given the ultra-low supply voltages, transistors operate in the deep subthreshold region with the following equation for the drain-source current I_D:

$$I_D = I_0 e^{(V_{GS}-V_{th})/nV_T}\left(1 - e^{-V_{DS}/V_T}\right) \tag{23.2.3.1}$$

With the constant

$$I_0 = W/L\mu_{p,n}C_{OX}(n-1)V_T^2 \tag{23.2.3.2}$$

consisting of device geometry W/L, charge carrier mobility $\mu_{p,n}$, unit area oxide capacitance C_{OX}, subthreshold factor n, and the thermal voltage $V_T = kT/q$ which is about 26 mV at room temperature. The threshold voltage is given by

$$V_{th} = V_{th0} - \lambda_{DS}V_{DS} - \lambda_{BS}V_{BS} \tag{23.2.3.3}$$

where V_{th0} is the zero-bias threshold voltage, λ_{DS} the drain induced barrier lowering coefficient and λ_{BS} the body-effect coefficient.

Bringing the supply voltage down to its minimum leads to some major drawbacks for electronic circuits: Due to the exponential dependence of I_D on the threshold voltage V_{th}, device variation increases significantly. Especially with decreasing physical dimensions in the race to ever shrinking process nodes, effects like random-dopant-fluctuation [61] and reverse-short-channel-effect—due to non-uniform doping profiles [62, 63]—increase variability in threshold voltage. While advanced process technologies like FD-SOI have inherent advantages over bulk processes due to the absence of doping atoms in the channel [61, 64] further challenges need to be addressed. As the signal swing of V_{GS} scales with the supply voltage, the on current I_{ON} of transistors approach the same order of magnitude as the leakage current I_{OFF} in the off-state of the devices in the deep subthreshold region [65]. This results in poor digital signal output levels and gain for logic gates as well as very slow operation frequencies. Moreover, P/N mismatch as well as small-channel effects like DIBL [61] have a negative influence on the transfer function of transistors and circuits. It is therefore vital to have circuit techniques that allow building systems that can reliably generate clocks and control the harvesting interface.

23.2.3.1 Schmitt Trigger (ST) Logic

One way to tackle variation and signal level degradation is to use Schmitt Trigger structures for the implementation of digital logic gates. The general structure of a Schmitt Trigger inverter is shown in Fig. 23.33. As demonstrated in [66] the ST structure can easily be extended to NAND, NOR and other types of logic gates. The state of the transistors in Fig. 23.33 is depicted for the example input value of '0' and corresponding output of '1'. Devices N_1, N_2 and P_3 are off while P_1, P_2 and N_3 are conducting. The main idea behind this technique for ULV applications is to increase the gain or steepness and the output levels of the voltage-transfer-curve by reducing the off-current in the currently non-active branch (NMOS in this case) of the gate. This is achieved by splitting each of the transistors of a regular inverter into a stack of two (N_1, N_2) and introducing the feedback transistor N_3 which acts as a source follower. It pulls node L_N up which reduces V_{DS} and more importantly introduces a negative V_{GS} in N_2. The channel therefore exhibits greatly reduced leakage currents and the on-to-off current ratio is significantly improved [66]. Additionally the feedback structure mitigates P/N mismatch [66, 67] while the higher gain reduces the impact of threshold voltage variation [65].

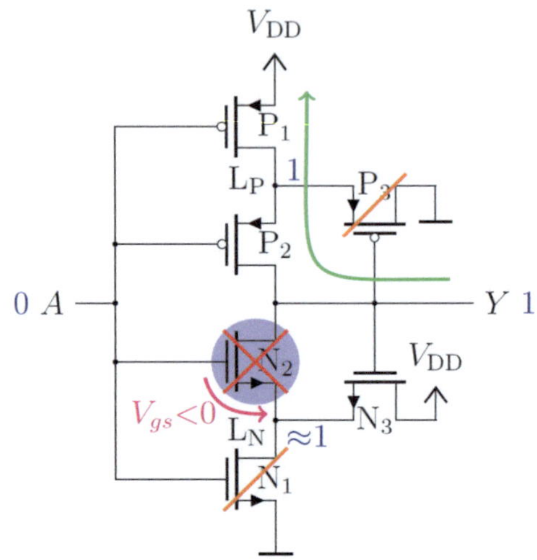

Fig. 23.33 Schematic of ST inverter in input low, output high configuration

It has been shown that ST gates achieve lower minimal supply voltage than regular CMOS implementations both theoretically (31.5 mV vs. 36 mV for an ideal inverter) [68] and practically by reaching $V_{DDmin} = 40$ mV for single inverters [68] and 62 mV for an 8 × 8 bit multiplier consisting of NAND, NOR and Inverter gates [66]. This leads to lower leakage currents and power consumption [65, 66] or to less area and higher efficiency [67] than comparable CMOS gates. Consequently, ST circuits are better suited for always-on blocks or wake-up circuits for IoT and for startup circuits in ULV and ULP harvester interfaces. As an example, the charge-pump based controller presented in the previous section [59] starts at $V_{DDmin} = 70$ mV. There, the dynamic power consumption of the ST is reduced by switching the feedback path through N_3 or P_3 off by an additional switch.

It is noteworthy that not only digital but also analog Schmitt Trigger circuits have been proposed recently [68] paving a way to the implementation of analog amplifiers and control loops in the sub-100 mV range. While conventional analog subthreshold circuits are becoming more common they require a few hundred millivolts to operate in (subthreshold-)saturation and to achieve sufficient gain [69, 70] rendering them non-functional for ULV startup controllers.

Operation at ultra-low supply voltages—as is feasible with ST logic—may be necessary for startup of an interface and a DC-DC converter and might result in the lowest power. However, a more efficient operating point for a particular computation load might exist at supply voltages closer to the threshold voltage.

23.2.3.2 Body Biasing

A different interesting low power technique is body biasing. There a voltage is applied to the bulk terminal of transistors. The bulk-source voltage V_{BS} is then used to tune the threshold voltage of the transistors according to Eq. (23.2.3.3). This directly corresponds to a change in the transistor currents and circuit speed.

There exist different approaches to set the bias voltage. The simplest method is to statically apply a fixed voltage e.g. a positive/forward voltage to lower the threshold voltage. A more advanced measure is regulating the bias voltage according to the given situation. This adaptive body biasing (ABB) can be used to adjust the frequency and power consumption of a circuit to the available power from the harvester, to mitigate variation in order to lower the minimum supply voltage of logic gates [62] and SRAM cells [71], or to extend the lifetime by reacting to aging effects [72].

Since the load capacitance of the logic gates stays constant despite the increase in current drive capability with forward biasing, the power-delay-product reduces and circuit efficiency can be increased. This effect can be utilized in duty-cycled systems where stored harvested energy is used to efficiently process data within a short timeframe and otherwise leakage current is reduced by reverse biasing and/or power-gating [73, 74]. In memory cells, where power-gating is not an option, reverse body bias can save energy during the idle period of a duty-cycle [71, 74, 75] or prevent data loss when insufficient excitation of the harvester diminishes the power budget. This method allows the 6T SRAM cells in [76] or the 8T cells in [71] to achieve a frequency of 200 MHz with an access energy of 20.4 fJ/bit while in sleep the power consumption is only 7.4 pW/bit, which is 5.5× lower than in the active state. Various biasing options for the transistors in a 6T SRAM are analyzed and compared in [76] resulting in 1/5th the read power compared to a zero-bias cell. The microprocessor in [77] is fitted with a purely digitally controlled ABB generator allowing the chip to run at 0.35 V with minimal area and power overhead due to analog circuitry.

Another more granular option called dynamic threshold MOS (DTMOS) or also back-gating uses the bulk terminal as an additional logic-controlled gate for small sub-circuits or individual transistors. By altering the transistors threshold voltage depending on the state of logic signals, the switching speed of a gate or its power consumption in a specific state can be improved. Some prominent examples for this approach are improving static noise margin and energy efficiency in SRAM [76, 75, 78] and reducing on-resistance and leakage of transistors in the critical path [79] or for power-gating [73]. Using this in precharge-evaluate logic results in a 16–19% faster adder with 10–15% less active energy per cycle [79]. In [80], the back gate is used for temporary current boosting during the switching operation gaining a super-linear speed-up of 10%. However, it should be noted that fine-grained body-biasing will introduce a noticeable area overhead due to additional wells/tubs and their corresponding contacts. For a modified 6T SRAM cell, where the bulk of the PMOS devices is switched, 78% more circuit area is occupied [78].

Using body biasing is especially interesting for SOI processes where the drain and source terminals are isolated from the bulk silicon so there is no parasitic diode between them. In bulk processes, this would result in significant leakage currents for forward biased transistors [61]. Further benefits are that no triple-well process is required for biasing individual NMOS devices [76] and the much higher body effect coefficient λ_{BS} (85 mV/V in FD-SOI vs. 25 mV/V in Bulk) [61]. With specialized deeply-depleted channel processes, even $\lambda_{BS} = 375\,\text{mV/V}$ can be achieved [81].

23.3 Conclusion

In this chapter, the possible application scenarios making use of Energy Harvesting have been presented. It has been shown that in the area of wearable devices and condition monitoring, significant progress has been achieved in the past, pushing this topic towards practical applications in the industrial and consumer environment. After optimization efforts towards higher power generation from swing harvesters, it is foreseeable that the power harvested from human walking could supply a wide variety of wearable devices, such as used for health monitoring as well as wearable consumer electronics. Condition monitoring and predictive maintenance are possible applications where batteries can be omitted or complemented by energy harvesting devices, resulting in reduced maintenance effort.

In the past few years, significant attention has been paid to the electronic circuit interfacing the energy harvesting generators. Recent publications have been able to push the energy harvested out of piezoelectric generators close to the theoretical limit by using enhancement techniques like the parallel-SSHI. Furthermore, advancements in enhancement techniques, like the broadband SECE, allows to expand the energy extraction capabilities, when the generators excitation frequency is not exactly the resonance frequency. Circuits tracking the maximum power point for generators with continuously changing excitation conditions have been proposed recently, but this principle seems to be impractical due to a low efficiency.

Thermoelectric generators have proven to be one of the most productive and applicable energy harvesting sources to date due to the nearly ubiquitous availability of temperature gradients. Moreover, their nature as a perfect solid-state device enables very compact and flexible implementations where the device dimensions are basically defined only by the power requirements of the application and the corresponding thermal flow and thermal insulation of the device. Particularly in applications with low temperature gradient, these devices require ultra- low voltage and ultra-low power circuits in order to ensure robust and flawless operation.

References

1. V. Leonov, Thermoelectric energy harvester on the heated human machine. J. Micromech. Microeng. **21**(12), 125013 (2011)
2. M.-K. Kim, M.-S. Kim, S. Lee, C. Kim, Y.-J. Kim, Wearable thermoelectric generator for harvesting human body heat energy. Smart Mater. Struct. **23**(10), 105002 (2014)

3. J.M. Donelan et al., Biomechanical energy harvesting: generating electricity during walking with minimal user effort. Science **319**(5864), 807–810, 18258914 (2008)
4. Q. Li, V. Naing, J.M. Donelan, Development of a biomechanical energy harvester. J. Neuroeng. Rehabil. **6**, 22, 19549313 (2009)
5. Z. Yang, A. Khaligh, A flat linear generator with axial magnetized permanent magnets with reduced accelerative force for backpack energy harvesting
6. Q. Zhang, Y. Wang, E.S. Kim, Power generation from human body motion through magnet and coil arrays with magnetic spring. J. Appl. Phys. **115**(6), 064908 (2014)
7. K. Ylli, D. Hoffmann, A. Willmann, B. Folkmer, Y. Manoli, Investigation of pendulu structures for rotational energy harvesting from human motion. J. Phys.: Conf. Ser. **660**, 012053 (2015)
8. P. Niu, P. Chapman, R. Riemer, X. Zhang, Evaluation of motions and actuation methods for biomechanical energy harvesting, in *2004 IEEE 35th Annual Power Electronics Specialists Conference, PESC 04* (2004), pp. 2100–2106
9. K. Ylli, D. Hoffmann, A. Willmann, P. Becker, B. Folkmer, Y. Manoli, Energy harvesting from human motion: exploiting swing and shock excitations. Smart Mater. Struct. **24**, 025029 (2015)
10. D. Hoffmann, B. Folkmer, Y. Manoli, Human motion energy harvester for biometric data monitoring. J. Phys.: Conf. Ser. **476**, 012103 (2013)
11. D. Carroll, M. Duffy, Modelling, design, and testing of an electromagnetic power generator optimized for integration into shoes. Proc. Inst. Mech. Eng., Part I: J. Syst. Control. Eng. **226**(2), 256–270 (2012)
12. K. Ylli, D. Hoffmann, A. Willmann, B. Folkmer, Y. Manoli, Human motion energy harvesting: numerical analysis of electromagnetic swing-excited structures. Smart Mater. Struct. **25**, 095014 (2016)
13. J.-X. Shen et al., A shoe-equipped linear generator for energy harvesting. IEEE Trans. Ind. Appl. **49**(2), 990–996 (2013)
14. K. Ylli, Energy harvesting from the swing motion of the human leg, Ph.D dissertation, Albert-Ludwigs-Universität Freiburg, 2019. Accessed 28 March 2020. [Online]. Available: https://freidok.uni-freiburg.de/data/17369, https://doi.org/10.6094/UNIFR/17369
15. P. Mitcheson, E. Yeatman, G.K. Rao, A. Holmes, T. Green, Energy harvesting from human and machine motion for wireless electronic devices. Proc. IEEE **96**(9), 1457–1486 (2008)
16. S. Roundy, E.S. Leland, J. Baker, E. Carleton, E. Reilly, E. Lai, B. Otis, J.M. Rabaey, V. Sundararajan, P.K. Wright, Improving power output for vibration-based energy scavengers. IEEE Pervasive Comput. **4**(1), 28–36 (2005)
17. A. Zahid Kausar, A.W. Reza, M.U. Saleh, H. Ramiah, Energizing wireless sensor networks by energy harvesting systems: scopes, challenges and approaches. Renew. Sustain. Energy Rev. **38**, 973–989 (2014)
18. I.N. Ayala-Garcia, D. Zhu, M.J. Tudor, S.P. Beeby, A tunable kinetic energy harvester with dynamic over range protection. Smart Mater. Struct. **19**(11) (2010)
19. N.A. Aboulfotoh, M.H. Arafa, S.M. Megahed, A self-tuning resonator for vibration energy harvesting. Sens. Actuators, A **201**, 328–334 (2013)
20. D. Hoffmann, A. Willmann, B. Folkmer, Y. Manoli, Tunable vibration energy harvester for condition monitoring of maritime gearboxes. J. Phys.: Conf. Ser **557** (2014)
21. V.R. Challa, M.G. Prasad, F.T. Fisher, Towards an autonomous self-tuning vibration energy harvesting device for wireless sensor network applications. Smart Mater. Struct. **20**(2), 025004 (2011)
22. C. Eichhorn, R. Tchagsim, N. Wilhelm, P. Woias, A smart and self-sufficient frequency tunable vibration energy harvester. J. Micromech. Microeng. **21**(10) (2011)
23. B.-C. Lee, G.-S. Chung, Frequency tuning design for vibration-driven electromagnetic energy harvester. IET Renew. Power Gener. **9**(7), 801–808 (2015)
24. S.-C. Huang, K.-A. Lin, A novel design of a map-tuning piezoelectric vibration energy harvester. Smart Mater. Struct. **21**(8), 085014 (2012)
25. J. Esch, D. Hoffmann, D. Stojakov, Y. Manoli, Self-tunable vibration energy harvester, in *Proceedings of PowerMEMS 2018*, Daytona Beach, USA (2018)

26. D. Hoffmann, A. Willmann, T. Hehn, B. Folkmer, Y. Manoli, A self-adaptive energy harvesting system. Smart Mater. Struct. **25**(3), 035013 (2016)
27. J. Esch, K. Ylli, D. Stojakov, A. Willmann, D. Hoffmann, Y. Manoli, Energy harvesting flex-coil system for pneumatic pistons, in *Proceedings of PowerMEMS 2017*, Kanazava, Japan, *Journal of Physics: Conference Series*, vol. 1052 (2018)
28. D. Hoffmann, A. Willmann, B. Folkmer, Y. Manoli, Energy harvesting for high-speed sensor telemetry, in *PowerMEMS* (2012)
29. D. Hoffmann et al., Tunable vibration energy harvester for condition monitoring of maritime gearboxes. J. Phys.: Conf. Ser. **557**(1), 012099 (2014)
30. J. Wang, G. Penamalli, L. Zuo, Electromagnetic energy harvesting from train induced railway track vibrations, in *2012 IEEE/ASME International Conference on Mechatronics and Embedded Systems and Applications (MESA)*, July 2012
31. J. Leicht, M. Amayreh, C. Moranz, D. Maurath, T. Hehn, Y. Marioli, Electromagnetic vibration energy harvester interface IC with conduction angle-controlled maximum-power-point tracking and harvesting efficiencies of up to 90%, in *International Solid State Circuits*, Feb 2015, pp. 1–3
32. Y. Ramadass, A. Chandrakasan, An efficient piezoelectric energy harvesting interface circuit using a bias-flip rectifier and shared inductor. IEEE J. Solid-State Circuits **45**(1), 189–204 (2010)
33. D. Kwon, G. Rincon-Mora, A single-inductor 0.35 mm cmos energy-investing piezoelectric harvester. IEEE J. Solid-State Circuits **49**(10), 2277–2291 (2014)
34. S. Stanzione, C. van Liempd, M. Nabeto, F.R. Yazicioglu, C.V. Hoof, A 500 nW batteryless integrated electrostatic energy harvester interface based on a DC-DC converter with 60 V maximum input voltage and operating from 1 mW available power, including MPPT and cold start, in *2015 IEEE International Solid-State Circuits Conference—(ISSCC)*, Feb 2015, pp. 1–3
35. Y. Manoli, Energy harvesting—from devices to systems, in *2010 Proceedings of the ESSCIRC*, Sept 2010, pp. 27–36
36. S. Roundy, P.K. Wright, A piezoelectric vibration based generator for wireless electronics. Smart Mater. Struct. **13**(5), 1131 (2004)
37. G. Gautschi, *Piezoelectric Sensorics: Force, Strain, Pressure, Acceleration and Acoustic Emission Sensors, Materials and Amplifiers* (Springer, Berlin, 2002)
38. T. Hehn et al., A fully autonomous integrated interface circuit for piezoelectric harvesters. IEEE JSSC **47**(9), 2185–2198 (2012)
39. A. Frey, J. Seidel, M. Schreiter, I. Kuehne, Piezoelectric MEMS energy harvesting module based on non-resonant excitation, in *2011 16th International Solid-State Sensors, Actuators and Microsystems Conference (TRANSDUCERS)*, June 2011, pp. 683–686
40. D.A. Sanchez, J. Leicht, E. Jodka, E. Fazel, Y. Manoli, A parallel-SSHI rectifier for piezoelectric energy harvesting of periodic and shock excitations. J. Solid State Circuits **51**(12), 2867–2879 (2016)
41. M. Shim, J. Kim, J. Jeong, S. Park, C. Kim, Self-powered 30 μW to 10 mW piezoelectric energy harvesting system with 9.09 ms/V maximum power point tracking time. IEEE J. Solid-State Circuits **50**(10), 2367–2379 (2015)
42. Y. Cai, Y. Manoli, A piezoelectric energy-harvesting interface circuit with fully autonomous conjugate matching, 156% extended bandwidth, and 0.38 μW power consumption, ISSCC 2018 paper
43. Y. Cai, Y. Manoli, A piezoelectric energy-harvesting interface circuit with fully autonomous conjugate matching, 156% extended bandwidth, and 0.38 μW power consumption, ISSCC 2018 visuals
44. Y. Cai, Y. Manoli, A piezoelectric energy harvester interface circuit with adaptive conjugate impedance matching, self-startup and 71% broader bandwidth, ESSCIRC 2017
45. D. Sanchez et al., A 4 μW-to-1 mW parallel-SSHI-rectifier with inductor sharing, cold start-up and up to 681% power extraction improvement, ISSCC, Feb 2016, pp. 366–367
46. P. Hsieh et al., Improving the scavenged power of nonlinear piezoelectric energy harvesting interface at off-resonance by introducing switching delay. IEEE Trans. Power Electron. **30**(6), 3142–3155 (2015)

47. Y. Peng et al., An efficient piezoelectric energy harvesting interface circuit using a sense-and-set rectifier. IEEE J. Solid-State Circuits. https://doi.org/10.1109/jssc.2019.2945262
48. D.A. Sanchez, J. Leicht, F. Hagedorn, E. Jodka, E. Fazel, Y. Manoli, A parallel-SSHI rectifier for piezoelectric energy harvesting of periodic and shock excitations. IEEE J. Solid-State Circuits **51**(12), 2867–2879 (2016). https://doi.org/10.1109/jssc.2016.2615008
49. Y. Yuk et al., 23.5 An energy pile-up resonance circuit extracting maximum 422% energy from piezoelectric material in a dual-source energy-harvesting interface, in *2014 IEEE International Solid-State Circuits Conference Digest of Technical Papers (ISSCC)*, San Francisco, CA, 2014, pp. 402–403. https://doi.org/10.1109/isscc.2014.6757488
50. A. Quelen, A. Morel, P. Gasnier, R. Grézaud, S. Monfray, G. Pillonnet, A 30 nA quiescent 80 nW-to-14 mW power-range shock-optimized SECE-based piezoelectric harvesting interface with 420% harvested-energy improvement, in *2018 IEEE International Solid-State Circuits Conference—(ISSCC)*, San Francisco, CA, 2018, pp. 150–152. https://doi.org/10.1109/isscc.2018.8310228
51. J. Colonna, G. Savelli, A. Royer, P. Coudrain, M. Keller, D. Wendler, Y. Manoli, L. Fréchette, L.-M. Collin, S. Billat, J. Barrau, H2020 European project STREAMS: general overview, in *24th International Workshop on Thermal Investigations of ICs and Systems*, Stockholm, 2018
52. V. Leonov, Thermoelectric energy harvesting of human body heat for wearable sensors. IEEE Sens. J. **13**(6), 2284–2291 (2013)
53. J. Goeppert, S. Braun, D. Pellhammer, M. Amayreh, J. Leicht, M. Keller, Y. Manoli, Area constrained multi-source power management for thermoelectric energy harvesting, in *Proceedings of the European Solid-State Circuits Conference*, Krakow, 2019
54. A. Camarda, A. Romani, E. Macrelli, M. Tartagni, A 32 mV/69 mV input voltage booster based on a piezoelectric transformer for energy harvesting applications. Sens. Actuators A: Phys. 341–352 (2015)
55. J. Goeppert, Y. Manoli, Fully integrated startup at 70 mV of boost converters for thermoelectric energy harvesting. IEEE J. Solid-State Circuits **7**, 1716–1726 (2016)
56. P.-H. Chen, X. Zhang, K. Ishida, Y. Okuma, Y. Ryu, M. Takamiya, T. Sakurai, An 80 mV startup dual-mode boost converter by charge-pumped pulse generator and threshold voltage tuned oscillator with hot carrier injection. IEEE J. Solid-State Circuits **11**(47), 2554–2562 (2012)
57. P.-S. Weng, H.-Y. Tang, P.-C. Ku, L.-H. Lu, 50 mV-input batteryless boost converter for thermal energy harvesting. IEEE J. Solid-State Circuits **48**(4), 1031–1041 (2013)
58. Y.K. Ramadass, A.P. Chandrakasan, A batteryless thermoelectric energy-harvesting interface circuit with 35 mV startup voltage. IEEE J. Solid-State Circuits **46**(1), 333–341 (2011)
59. J. Goeppert, Y. Manoli, Fully integrated start-up at 70 mV of boost converters for thermoelectric energy harvesting, in *ESSCIRC Conference 2015—41st European Solid-State Circuits Conference (ESSCIRC)*, Graz, 2015, pp. 233–236
60. Y. Shi, Y. Wang, D. Mei, Z. Chen, Wearable thermoelectric generator with copper foam as the heat sink for body heat harvesting. IEEE Access **6**, 43602–43611 (2018)
61. D. Jacquet et al., A 3 GHz dual core processor ARM cortex TM -A9 in 28 nm UTBB FD-SOI CMOS with ultra-wide voltage range and energy efficiency optimization. IEEE J. Solid-State Circuits **49**(4), 812–826 (2014)
62. M. Hwang, Supply-voltage scaling close to the fundamental limit under process variations in nanometer technologies. IEEE Trans. Electron Devices **58**(8), 2808–2813 (2011)
63. T. Kim, J. Liu, C.H. Kim, A voltage scalable 0.26 V, 64 kb 8T SRAM with V_{min} lowering techniques and deep sleep mode. IEEE J. Solid-State Circuits **44**(6), 1785–1795 (2009)
64. N. Planes et al., 28 nm FDSOI technology platform for high-speed low-voltage digital applications, in *2012 Symposium on VLSI Technology (VLSIT)*, Honolulu, HI, 2012, pp. 133–134
65. A. Bleitner, J. Goeppert, N. Lotze, M. Keller, Y. Manoli, Comparison and optimization of the minimum supply voltage of Schmitt Trigger gates versus CMOS gates under process variations, in *ANALOG 2018, 16th GMM/ITG-Symposium*, Munich/Neubiberg, Germany, Sept 2018, pp. 111–116

66. N. Lotze, Y. Manoli, A 62 mV 0.13 μm CMOS standard-cell-based design technique using Schmitt-Trigger logic. IEEE J. Solid-State Circuits **47**(1), 47–60 (2012)
67. N. Lotze, Y. Manoli, Ultra-sub-threshold operation of always-on digital circuits for IoT applications by use of Schmitt Trigger gates. IEEE Trans. Circuits Syst. I Regul. Pap. **64**(11), 2920–2933 (2017)
68. L.A. Pasini Melek, M.C. Schneider, C. Galup-Montoro, Operation of the classical CMOS Schmitt Trigger as an ultra-low-voltage amplifier. IEEE TCAS II **65**(9), 1239–1243 (2018)
69. F.M. Yaul, A.P. Chandrakasan, A noise-efficient 36 nV/$\sqrt{}$ Hz chopper amplifier using an inverter-based 0.2-V supply input stage. IEEE J. Solid-State Circuits **52**(11), 3032–3042 (2017)
70. S. Orguc, H.S. Khurana, H. Lee, A.P. Chandrakasan, 0.3 V ultra-low power sensor interface for EMG, in *ESSCIRC 2017—43rd IEEE European Solid State Circuits Conference*, Leuven, 2017, pp. 219–222
71. T. Haine, D. Flandre, D. Bol, 8-T ULV SRAM macro in 28 nm FDSOI with 7.4 pW/bit retention power and back-biased-scalable speed/energy trade-off, in *2018 IEEE SOI-3D-Subthreshold Microelectronics Technology Unified Conference (S3S)*, Burlingame, CA, USA, 2018, pp. 1–3
72. A.P. Shah, N. Yadav, A. Beohar, S.K. Vishvakarma, On-chip adaptive body bias for reducing the impact of NBTI on 6T SRAM cells. IEEE Trans. Semicond. Manuf. **31**(2), 242–249 (2018)
73. J. Le Coz, B. Pelloux-Prayer, B. Giraud, F. Giner, P. Flatresse, DTMOS power switch in 28 nm UTBB FD-SOI technology, in *2013 IEEE SOI-3D-Subthreshold Microelectronics Technology Unified Conference (S3S)*, Monterey, CA, 2013, pp. 1–2
74. E. Ashenafi, M.H. Chowdhury, A new power gating circuit design approach using double-gate FDSOI. IEEE Trans. Circuits Syst. II Express Briefs **65**(8), 1074–1078 (2018)
75. M.U. Mohammed, A. Nizam, M.H. Chowdhury, Double-gate FDSOI based SRAM Bitcell circuit designs with different back-gate biasing configurations, in *2018 IEEE Nanotechnology Symposium (ANTS)*, Albany, NY, USA, 2018, pp. 1–4
76. A. Biswas, A.P. Chandrakasan, A 0.36 V 128 Kb 6T SRAM with energy-efficient dynamic body-biasing and output data prediction in 28 nm FDSOI, in *ESSCIRC Conference 2016: 42nd European Solid-State Circuits Conference*, Lausanne, 2016, pp. 433–436
77. H. Okuhara, A.B. Ahmed, H. Amano, Digitally assisted on-chip body bias tuning scheme for ultra low-power VLSI systems. IEEE Trans. Circuits Syst. I Regul. Pap. **65**(10), 3241–3254 (2018)
78. M.-E. Hwang, K. Roy, A 135 mV 0.13 μW process tolerant 6T sub-threshold DTMOS SRAM in 90 nm technology, in *2008 IEEE Custom Integrated Circuits Conference*, Sept 2008, pp. 419–422
79. S. Jayapal, J. Stuijt, J. Huisken, Y. Manoli, Energy efficient computation with self-adaptive single-ended body bias, in *23rd IEEE International SOC Conference*, Las Vegas, NV, 2010, pp. 326–329
80. H. Koike, T. Sekigawa, The missing XDXMOS found!—A SOTB circuit acceleration technique using front and back gate interaction, in *2015 IEEE SOI-3D-Subthreshold Microelectronics Technology Unified Conference (S3S)*, Rohnert Park, CA, 2015, pp. 1–2
81. M. Pons et al., A 0.5 V 2.5 μW/MHz microcontroller with analog-assisted adaptive body bias PVT compensation with 3.13 nW/kB SRAM retention in 55 nm deeply-depleted channel CMOS, in *2019 IEEE Custom Integrated Circuits Conference (CICC)*, Austin, TX, USA, 2019, pp. 1–4

Chapter 24
Artificial Retina: A Future Cellular-Resolution Brain-Machine Interface

Dante G. Muratore and E. J. Chichilnisky

24.1 Brain-Machine Interfaces of the Future

A *brain-machine interface* (BMI) is a device capable of providing a direct communication path between the nervous system and an external device. BMIs can be used in research to better understand the brain, and are increasingly intended for clinical applications, including treating hearing and vision loss, paralysis, and other consequences of degeneration and injury [1, 2]. In the future, BMIs will likely be used to augment human capabilities, including sensory acuity, control of complex devices, memory, attention and more. However, to realize this futuristic promise requires major advances in the design of circuits and systems for interfacing to the brain.

A BMI usually consists of a neural interface capable of sensing and/or eliciting neural activity, and a computing device that controls its operation. The neural interface can operate in any of several modalities—e.g. optical, electrical, magnetic—each with advantages and disadvantages [3]. In this chapter, we will focus on electrical neural interfaces; typically, arrays of electrodes for stimulating and recording neural activity. Thus, the performance of the interface is determined by the channel count and density, the signal-to-noise ratio (SNR) of recording and stimulation, and the bandwidth of wireless transmission of data to and from the device. However, the specifications for a neural interface that attempts to approach or exceed the capability of the neural circuitry can pose major challenges in terms of size and power consumption for an implanted device.

D. G. Muratore (✉)
Department of Electrical Engineering, Stanford University, Stanford, CA 94305, USA
e-mail: dantegmuratore@gmail.com

Wu Tsai Neurosciences Institute, Stanford University, Stanford, CA 94305, USA

E. J. Chichilnisky
Department of Neurosurgery, Stanford University, Stanford, CA 94305, USA
e-mail: ej@stanford.edu

Department of Ophthalmology, Stanford University, Stanford, CA 94305, USA

B. Murmann and B. Hoefflinger (eds.), *NANO-CHIPS 2030*,
The Frontiers Collection, https://doi.org/10.1007/978-3-030-18338-7_24

What exactly are the specifications for an effective neural interface? Ideally, one would like to independently access each neuron in a region of the nervous system. Obviously, this is often not feasible: systems that communicate with many neurons usually do not achieve single-cell resolution, while high-resolution systems can only record from a limited number of neurons. The optimal trade-off of resolution and scale may depend on the specific application; thus, the design of future BMIs should be guided by a deep understanding of how information is encoded in the targeted part of the nervous system.

For example, accumulating evidence in intracortical motor BMIs shows that, for current devices, *spike sorting* (distinguishing electrical impulses, or spikes, recorded from different neurons) does not produce a substantial increase in decoding performance [4–8]. Instead, if only threshold detection is performed on the recorded data, the specifications on the neural interface can be relaxed and the power consumption per channel can be drastically decreased, allowing more channels to be implanted [9]. However, as a better understanding of the neural circuitry in the motor cortex develops and recording devices with greater capabilities become available, systems that are capable of resolving individual cells could be valuable.

This evolution of design tradeoffs is already evident in the retina, likely because its function is better understood than other parts of the nervous system. Decades of research on the retina and visual system indicate that an effective interface that can replace retinal function lost to disease should reproduce the natural pattern of activation of the cells that transmit visual signals to the brain [10]. Because diverse cell types are intermixed in the retinal circuitry, such a device will need to sort spikes coming from different cells and sort the recorded cells into different cell types. This in turn will permit the device to stimulate each cell and cell type in a way that matches natural function. No BMI has ever been developed that can achieve these goals; however, advances in circuit design as well as in our understanding of the retina now bring this goal within reach.

In this chapter, we describe how the retina transduces light into a neural code that then is processed by the brain, and why this architecture implies that a high-fidelity retinal implant should operate at cellular and cell-type resolution. We then discuss the limited performance of first-generation retinal prostheses, which were not designed with this goal in mind. Finally, we describe a project at Stanford University aimed at developing an *artificial retina* that takes into consideration cell-type specificity of retinal signals and aims to reproduce the neural code at its natural resolution. Finally, we comment briefly on the possible implications for future BMIs.

24.2 Neuroscience of the Retina and Vision

The retina is a multi-layer structure at the rear of the eye containing neural circuitry that transduces the visual image into electrical signals, processes those signals, then transmits the results to the brain so that a visual image of the external world can be

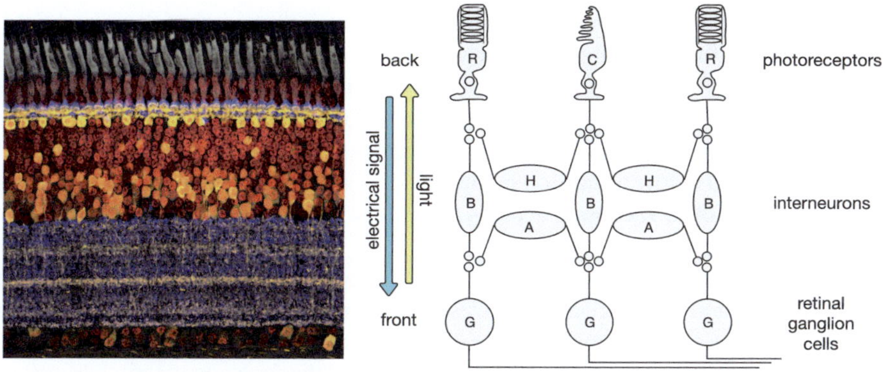

Fig. 24.1 Left: retina cross-section (from [12], © Abrams 2010). Right: schematic of retina layers and cell classes: (R)od and (C)one photoreceptors; (H)orizontal, (B)ipolar, and (A)macrine interneurons; (G)anglion cells

formed [11]. A cross-section of the retina highlights three different layers: *photoreceptors*, *interneurons*, and *retinal ganglion cells* (RGCs) (Fig. 24.1). The incoming light passes through the retina, which is mostly transparent, and arrives at the photoreceptors, where it is transduced into electrical signals. The signals coming from photoreceptors are integrated by interneurons, which in turn synapse onto RGCs, the output cells of the retina. As with most neurons, RGCs consist of a cell body (or *soma*), and a long nerve fiber (or *axon*) that transmits its all-or-none electrical impulses (or *spikes*) to target neurons in the brain. The axons of RGCs form the optic nerve, which routes visual information to many different parts of the brain for subsequent processing. Vision is the result of this fascinating distributed biological system.

The retina has several gross structural features that are important for neural interface design. In an *en face* view of the primate retina (Fig. 24.2; [13]) bundles of axons are visible, traveling from RGCs towards the optic nerve. Interestingly, axon bundles avoid passing through an area in the central retina called the *fovea*, presumably to

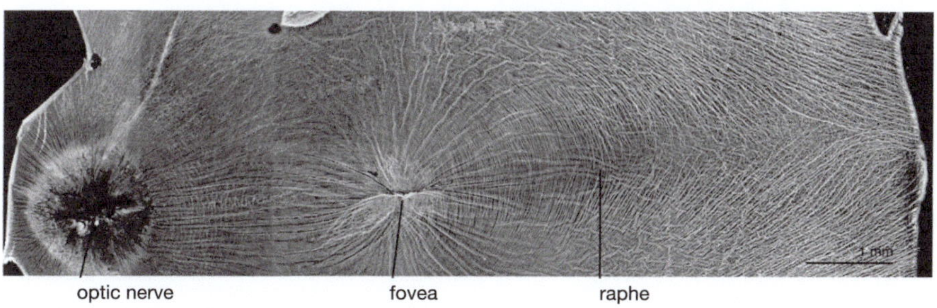

Fig. 24.2 Photograph of a flattened macaque retina, with landmarks and locations relevant for epiretinal implants shown (from [13], © APS 2017)

preserve optical clarity in this high-resolution region at the center of the visual field. In the center of the fovea, interneurons and RGCs are displaced laterally, providing a direct light path to the photoreceptors. As a result, the RGC layer is several cells thick, as opposed to the rest of the retina, where RGCs form a single layer. The high-resolution fovea occupies a small part of the retina, while the remainder of the visual scene is captured at low resolution by the peripheral retina. In fact, at approximately 15° outside the fovea, the visual acuity of a healthy person is equal to that of a legally blind patient.

The reason we can look at the Colosseum in Rome or the Golden Gate Bridge in San Francisco and reconstruct a high-resolution image, in spite of the small size of the fovea, is that our eyes constantly explore the scene in a series of active fixations connected by rapid eye movements. The eye focuses the fovea on certain features of the visual image, and moves quickly between them (Fig. 24.3; [14]). These different views are then integrated by the brain to produce a coherent representation of the visual scene.

Certain kinds of vision loss, such as macular degeneration and retinitis pigmentosa, arise from the loss of photoreceptors which normally transduce light. However, many other neurons, notably the RGCs, remain in large numbers. Therefore, a potential way to treat this kind of blindness is a device that uses an implanted neural interface to electrically stimulate RGCs (Fig. 24.4), causing them to fire spikes that are transmitted to the brain. In addition to a neural interface, such a device requires a camera to capture the image, and a processing unit to determine the appropriate patterns of electrical stimulation. First-generation devices with this general design have been shown to produce visual perception in profoundly blind patients, though performance has been limited (see Sect. 24.4).

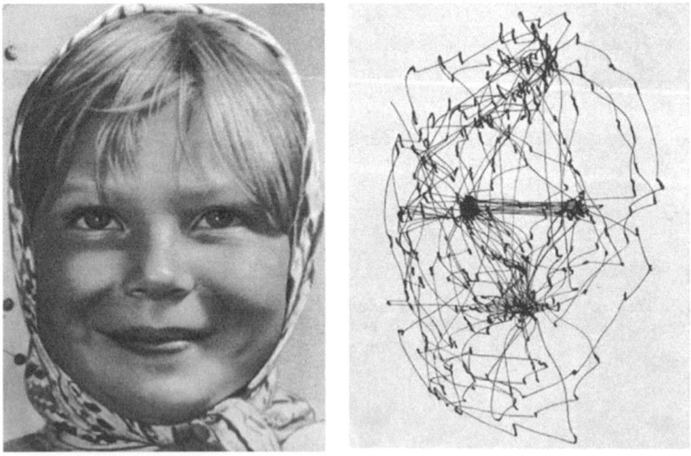

Fig. 24.3 Pattern of eye movements over an image during 3 minutes of free viewing by a human subject (from [14], © Springer 1967)

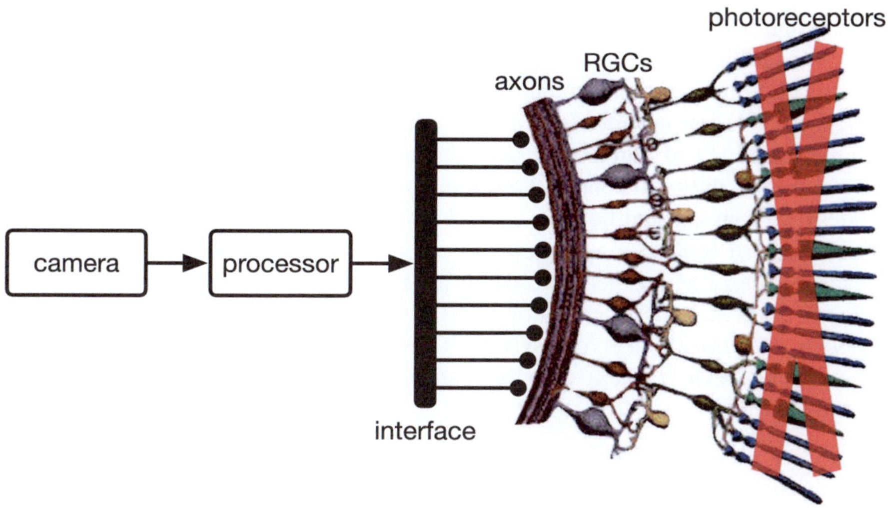

Fig. 24.4 Schematic of retina with degenerated photoreceptors, and epi-retinal prosthesis. The prosthesis consists of a camera, a processing unit, and a neural interface to electrically stimulate RGCs

A major obstacle to restoring high-fidelity vision with such devices, and arguably a key reason for the limited performance of first generation devices, is that there are many types of RGCs in the retina that deliver distinct visual signals to different targets in the brain (see [15–18]). For example, simultaneous activation of so-called ON and OFF type cells at a given location sends conflicting "messages" to the brain, indicating both a light increase, and a light decrease, at the same retinal location at the same time. Present-day retinal prostheses make no attempt to distinguish distinct cells or cell types, and therefore produce a non-specific, and thus profoundly scrambled, retinal signal of exactly this kind (see Sect. 24.4). These considerations suggest that to re-create the naturalistic neural code, and to produce high-fidelity artificial vision, requires that the distinct cell types be addressed independently. Because the different cell types are intermixed in the neural circuitry of the retina, this requires cellular resolution interfaces, with high channel count to recreate complex patterns in the neural network. Neural interfaces that operate this way—at the natural resolution of the neural circuitry—do not currently exist.

Note that, in principle, highly plastic neural circuits in the brain could adjust over time to accommodate scrambled retinal signals. Although a thorough treatment of visual plasticity and learning is out of the scope of this chapter, two major considerations argue against a major role for plasticity in artificial vision. First, plasticity is costly and complex to implement and regulate in any circuit, be it neural or electronic, and it is unlikely that evolutionary pressures would favor a visual brain with the ability to adjust itself to the highly non-biological stimulation provided by current prostheses. Second, a substantial rewiring of the distinct projections of different RGC types into the brain after prosthesis implantation is highly unlikely in adults, and existing studies suggest that visual perception with prostheses changes little with experience [19].

Assuming that a retinal implant must reproduce the neural code at cellular and cell-type resolution in order to provide high-fidelity artificial vision, what are the specific scientific and engineering requirements? From an engineering point of view, we require the ability to control the spiking activity of each cell and cell type independently. From a scientific point of view, we must understand the natural pattern of activation of RGCs for a given image in order to reproduce it faithfully. Several decades of basic neuroscience research have accumulated a great deal of knowledge about the pattern of activation of RGCs; indeed, the retina is one of the best understood parts of the nervous system. Thus, at the moment, the limiting factor is engineering a device capable of cellular and cell type resolution over a large area of the retina.

24.3 Neurophysiology and Electrical Stimulation of Neurons

To understand how such a device might work and the state of the art, we will describe extracellular recording and stimulation of neural activity. Information in most neurons (including RGCs) is encoded in spikes, which are brief electrical impulses, i.e. perturbations in the voltage across the cell membrane. Because spikes are stereotyped all-or-none signals, information is conveyed only by the temporal pattern of spikes, not by the spike shape.

Understanding how spikes are generated in the cell sets the context for thinking about how electrical recording and stimulation can be used to monitor and control neural activity. The cell's membrane potential is controlled by a wide variety of *ion channels*, proteins that control the flow of ions [predominantly sodium (Na+) and potassium (K+)] across the cell membrane by opening and closing in response to voltage changes and binding of neurotransmitters [20]. Under resting conditions, the cell potential is ≈ -70 mV with respect to the extracellular medium, due to the different ionic concentrations inside and outside of the cell that are maintained by cellular pumps. If a signal from another neuron, or an external stimulus, depolarizes the cell above a certain threshold level (≈ -50 mV), then positive feedback from voltage-gated channels will cause the neuron to generate a spike—a stereotyped depolarization to about $+30$ mV that lasts about 1 ms. During the spike, the Na+ permeability initially increases rapidly, bringing in positive charge and further increasing the membrane potential (positive feedback). Then, the slower K+ channels are activated, allowing positive charges to flow out of the cell, and returning the intracellular potential to its resting state (negative feedback). Note that because ion channels open in response to voltage changes, spikes can be generated artificially by applying electric fields outside the cell. Furthermore, because these ionic currents are large, the polarization during a spike can be monitored outside the cell by recording the extracellular potential using a microelectrode. The dispersive medium between the cell and the microelectrode introduces a signal attenuation, so that, typical extracellularly recorded spikes are $\approx$100 μV in amplitude, as opposed to the $\approx$100 mV change in the membrane potential.

Thus, neurons are de facto electrical processing units: they signal information with electrical impulses, and these impulses can be recorded and elicited by an electrical device outside the cell. Hence, the electrical domain is a natural one for the development of neural interfaces.

The way neurons produce spikes is crucial for identifying the origin of spikes recorded extracellularly (*spike sorting*; [21–23]) and for achieving single-cell resolution. The extracellularly-recorded spike waveform for any given cell is consistent over time and reflects the distribution of ion channels on the cell and spatial relationship between the cell and the recording electrode. Thus, if an electrode records spikes from two cells, their waveforms will typically be different, making it possible to distinguish the signals from the two cells. If an array of electrodes is used, the same cell can be recorded on multiple electrodes, increasing the spatial information for spike sorting.

The way neurons produce spikes can also inform the design of electrical stimulation for neural interfaces. An electrical model of RGCs can be used to simulate the membrane potential as a response to an external stimulus [24]. This simulation can inform the specifications on the stimulation pulse (i.e. duration and amplitude)—for example, results in [25] present numerical and analytical models of strength-duration curves for eliciting a spike.

In summary, interfacing with the nervous system in the electrical domain is a natural choice, given that neurons communicate with electrical signals. In particular, neurons encode information in spikes, which are comparatively easy to record and elicit without manipulating the internals of the cell.

24.4 First-Generation Epiretinal Prostheses

The first generation of retinal prostheses used electrical stimulation to elicit neural activity and artificial vision in blind human patients [10]. This technology development produced striking advances and an exciting proof of concept, but also fell short of the goal of restoring high-quality vision. Here, we focus on epiretinal devices, which are implanted on the surface of the retina to directly stimulate RGCs (see [10, 26–30] for a more complete review). The Argus devices developed by Second Sight constituted the first and largest commercial effort. Argus I was a first-generation prototype approved for a clinical trial aimed at establishing safety. Positive results of this trial on 6 patients motivated the development of Argus II, the only epiretinal device approved (in 2013) by the U.S. Food and Drug Administration (FDA) for clinical use. Argus II was initially implanted on 30 patients between 2007 and 2009, and in more than 150 additional patients since FDA approval [31]. However, in 2019, Second Sight stopped selling the device.

First-generation epiretinal prostheses provided a wealth of useful information. Most notably, stimulation of RGCs in profoundly blind patients with retinal degeneration elicits artificial visual perception (*phosphenes*). Phosphenes have been described by patients as being large, elongated and irregular [32, 33]. Their brightness and size were modulated with variations in the stimulation parameters (amplitude and frequency) [34]. Stimulation thresholds for light perception were well below the safety limits for electroporation [35, 36] and charge injection for most common electrode materials [37–39]. Notably, stimulation thresholds were highly dependent on the distance between the implant and the retinal surface, making obvious the need for keeping the array in close proximity to the retina [40].

However, the current state of the art of epiretinal prostheses, including the Argus II, can be summed up as such: no blind patient would trade their cane or guide dog for one. The artificial vision achieved with these devices is coarse, irregular, and difficult to relate to the objects in the visual scene [10]. A feature of first-generation devices that may play a role in their limited function is the small number of channels (e.g., 60 electrodes in Argus II), which is arguably insufficient to recreate a high-resolution image. One way to gain intuition about the required number of channels is to examine an image rendered with different numbers of pixels (Fig. 24.5; [41]). In the example reported below, the content of the image becomes clear when using a few thousand pixels. Because retinal neurons are not mere pixel detectors, the number of channels in a retinal implant and the effective number of pixels in perception are not directly comparable quantities. However, it seems clear that high-resolution visual restoration will require more independent stimulating channels.

Another issue with existing epiretinal devices is that they simultaneously activate many ganglion cell types due to the size of the stimulating electrodes (typically 50–500 μm in diameter). As discussed above, different cell types encode very different types of information, and failing to respect this specificity will likely lead to poor visual restoration. Increasing the spatial resolution of electrodes will also help with another problem faced by epiretinal devices: activation of unwanted axon bundles, which reside between the target ganglion cells and the microelectrode array and carry spikes from distant RGCs to the brain (Fig. 24.6). Activating axon bundles is almost

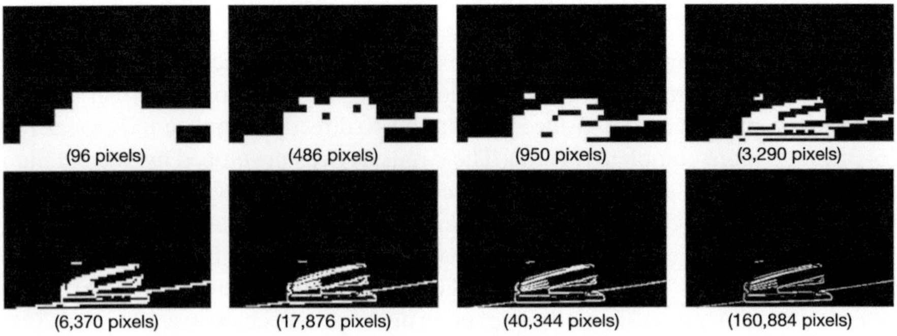

Fig. 24.5 Image of a stapler rendered with different numbers of pixels (from [41], © Elsevier 2015)

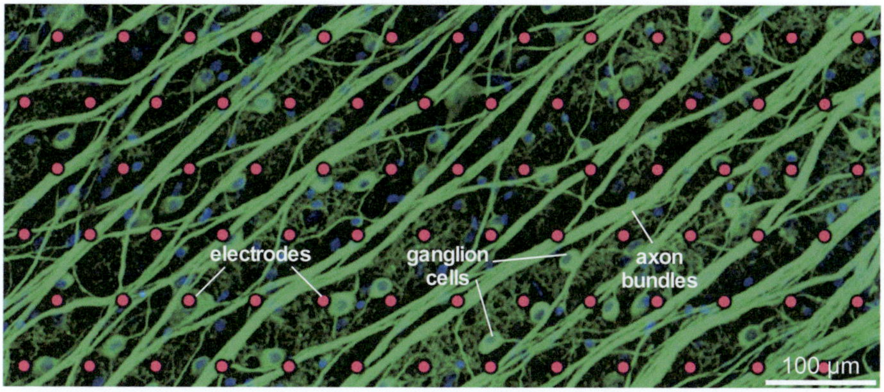

Fig. 24.6 Primate retinal ganglion cells, axon bundles, and overlying electrode array [13]

certain to degrade the image, because the originating cells and thus the location and nature of the visual signal they convey are varied and unknown. High-density arrays of small electrodes can increase the probability of being able to stimulate neurons without also activating axon bundles [13].

Axon activation can also be avoided by *subretinal* prostheses, which activate retinal interneurons rather than RGCs ([42, 43]; see [10]). These devices can be relatively simple and modular and show greater promise for vision restoration in the near future. However, the technical advantages of axon avoidance and interfacing to neurons at an earlier level of retinal processing are accompanied by severe limitations: limited scientific information about the neural code of the interneuron cell types, and the inability to record and precisely control their activity—both limitations arising because retinal interneurons are among the few cell groups in the central nervous system that signal with graded voltages rather than spikes. Hence, subretinal implants have little promise for accurately reproducing the neural code of the retina with cellular and cell-type resolution, and no path for extension to brain interfaces, and will not be discussed further.

In summary, first-generation prostheses demonstrate the possibility of vision restoration, but lack the ability to stimulate many cells with cellular and cell-type resolution. As a result, the neural code that is transmitted to the brain is severely distorted, limiting sight restoration.

24.5 The Stanford Artificial Retina

To improve on the state of the art, we propose a novel architecture for retinal implants—an *artificial retina* designed to adapt itself to the particular cells and cell types in the host circuitry, in order to faithfully reproduce the naturalistic neural code. Specifically, the device will have larger and denser electrode arrays to interface with many cells at their native spatial resolution, and the ability to record spikes in order to calibrate the device to the underlying biological circuit. We make a distinction between retinal prostheses of the past, and artificial retinas of the future, to highlight that the goal of an artificial retina is to actually replace the natural function of the neural circuitry of the retina.

Our approach is based on an extensive set of neurophysiological experiments in isolated retina, which serve as a laboratory prototype for a future clinical device. This work builds on a long history of experiments using various technologies and animal models; here, we focus specifically on high-density large-scale electrical recording and stimulation in the monkey retina, the most relevant animal model for human vision. The results reveal that it is possible to reproduce the natural retinal signal with high fidelity [13, 44–46]. In the experiments, electrical activity in the peripheral macaque retina is recorded on a 512-electrode array with 30–60 μm pitch, and spike sorting is performed to identify distinct RGCs [47, 48]. To distinguish different cell types, the spatial and temporal light response properties of each cell are measured by correlating the images focused on the retina with the spiking activity of the cell, yielding functionally distinct clusters of cells [47, 48]. The accuracy of this cell type classification is confirmed by the fact that the spatial sensitivity profiles, or receptive fields, of each cell type form a mosaic covering the region of retina recorded [48–50] (Fig. 24.7). Electrical stimulation is then calibrated by passing varying amounts of current through each electrode on the array, while recording the elicited activity, exploiting the fact that the spike waveforms of different cells have already been learned during recording [13, 44–46]. Finally, electrical stimulation is tailored to the recorded cells and cell types in a manner that most accurately reproduces the natural

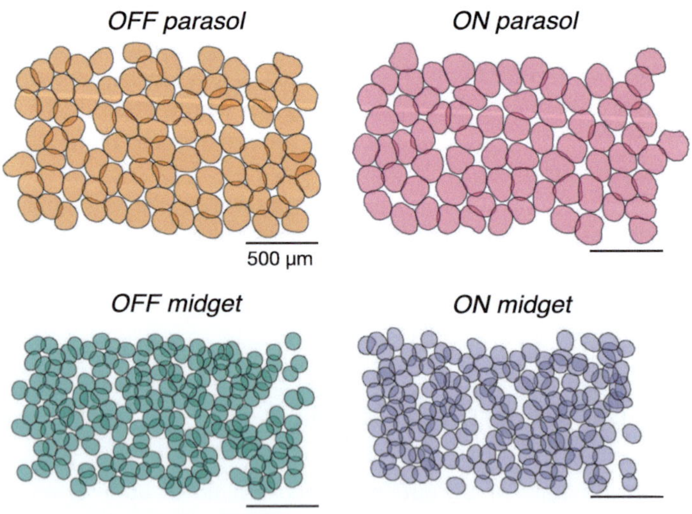

Fig. 24.7 Mosaics of receptive fields of the four major primate RGC types, obtained in a single multi-electrode recording from an isolated retina. The different RGC types overlap, but here, have been displaced for visibility (adapted from [53], © Cell Press 2019)

retinal activity (Fig. 24.8; [13, 44]). Spatial patterns of electrical stimulation through multiple electrodes simultaneously can also be used to fine-tune the activation of RGCs [51, 52].

The experimental approach described above informs the design of the future clinical device, and serves as a rapid prototyping platform for circuit-algorithm co-design and optimization. The proposed device [54] will operate bi-directionally, in three modes: *cell calibration* (recording only), *dictionary calibration* (stimulating and recording), and *runtime* (stimulating only) (Fig. 24.9). During cell calibration, the interface identifies the cells and cell types in close proximity to the electrode array, by recording spontaneous neural activity and sorting spikes from different RGCs. During dictionary calibration, the device determines how the different electrodes activate these cells, by recording and stimulating simultaneously. During runtime,

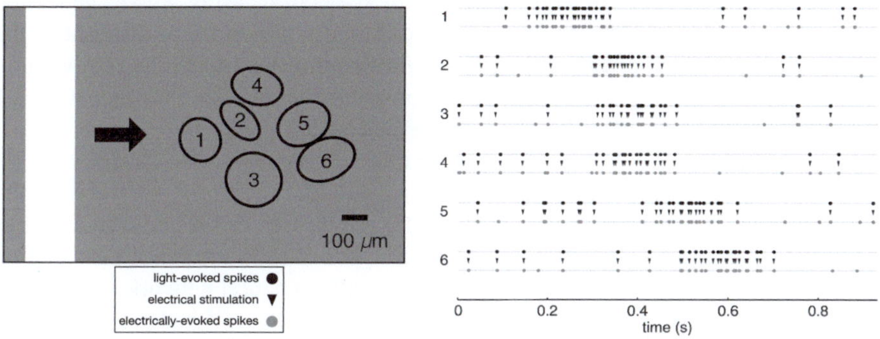

Fig. 24.8 Spatiotemporal visual and electrical activation of a local population of retinal ganglion cells. Left: A moving bar light stimulus is shown while recording from a population of six RGCs, with receptive fields indicated by ellipses, to determine the natural visual signal. Right: For each numbered cell, times are indicated for: spikes recorded during the moving bar light stimulus (black dots), applied current pulses during electrical stimulation (black triangles), and spikes evoked by electrical stimulation (grey dots). The alignment indicates that natural responses to light can be replicated by electrical stimulation with sub-millisecond temporal precision (from [44], © Cell Press 2014)

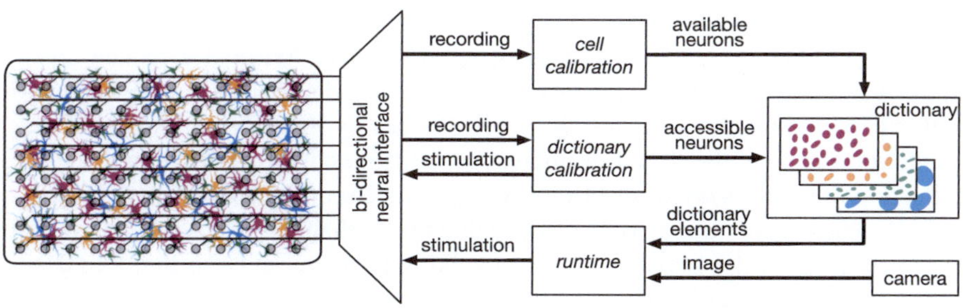

Fig. 24.9 System-level diagram of the proposed bi-directional neural interface with three operating modes

the interface stimulates the available cells to best approximate the correct neural signal based on the visual image captured by a camera. To produce meaningful visual signals (Fig. 24.5), the electrode array will have $>10^4$ channels, covering a relatively small part of the visual field, but one that is sufficient for useful vision. To interface at a resolution that roughly matches the neural circuitry, the electrode pitch will be ~30 μm or less. To faithfully encode a wide variety of visual images, the IC will have the ability to record and stimulate every channel independently.

24.5.1 Cell Calibration

Cell calibration (Fig. 24.10a) involves recording spontaneous activity to characterize the location and cell type of RGCs near the microelectrode array. Spikes generated by neurons are recorded by the interface and digitally processed to distinguish the spikes originating in different cells (*spike sorting*), requiring substantial computation to make sense of the large amount of data captured at the interface. We then leverage the fact that cell type classification can be performed solely based on electrical features of neurons [55]. A critical feature for this classification is the *electrical image* (EI) [47], which is the average spatiotemporal voltage waveform produced by the spike on the electrode array, a unique signature for each recorded cell (Fig. 24.11a). Another feature that helps distinguish different types of RGCs is the autocorrelation function of their spike trains (Fig. 24.11b). Typically, a neural network is trained to perform cell classification using these features (*classification*). Finally, the receptive field of each cell can be inferred from the EI, and its normal light response properties inferred from existing data.

Challenges: Although electrical features of RGCs can be very useful in classification, it is uncertain how completely classification can be accomplished, particularly given the variability between retinas, and the potential effect of retinal degeneration on the electrical properties of cells. Current work focuses on the use of very large data sets to improve classification. Also, to perform classification, the interface must record from many or all channels simultaneously in order to track the spatiotemporal evolution of spikes, resulting in data rates that are prohibitive in terms of energy and transmission bandwidth: e.g., 10,000 channels with 10-bit resolution at 20,000 samples per second generate 2 Gbps of recorded data. Without a radical change in the way neural interfaces are designed, the power dissipation during calibration will exceed the target for clinically viable devices by more than an order of magnitude. Fortunately, the signal of interest requires much less bandwidth, because a collection of neurons on the electrode array produce temporally and spatially sparse signals—thus, a continuous voltage recording on all channels is not needed. Such an observation about the statistics of neural spiking provides an opportunity to design a very efficient neural interface. This data explosion problem relates to most BMI applications and researchers have investigated a wide range of options to address it, such as on-chip spike sorting [56], on-chip compression [57–59], compressive sensing [60], and active analog multiplexing [61–63].

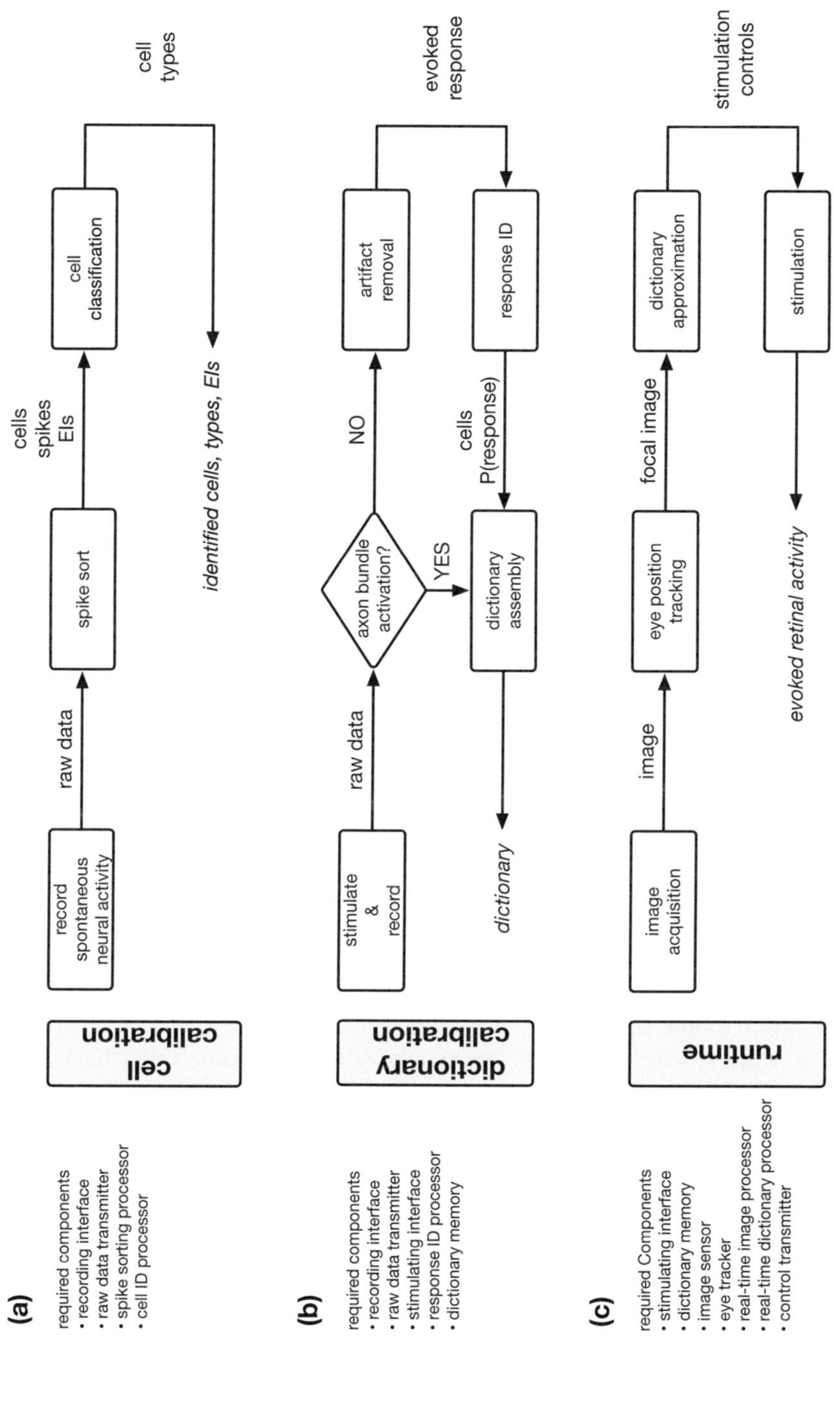

Fig. 24.10 Block diagram of each mode of operation for the proposed interface: **a** cell calibration, **b** dictionary calibration, and **c** runtime. Required components for each mode of operation are shown on the left

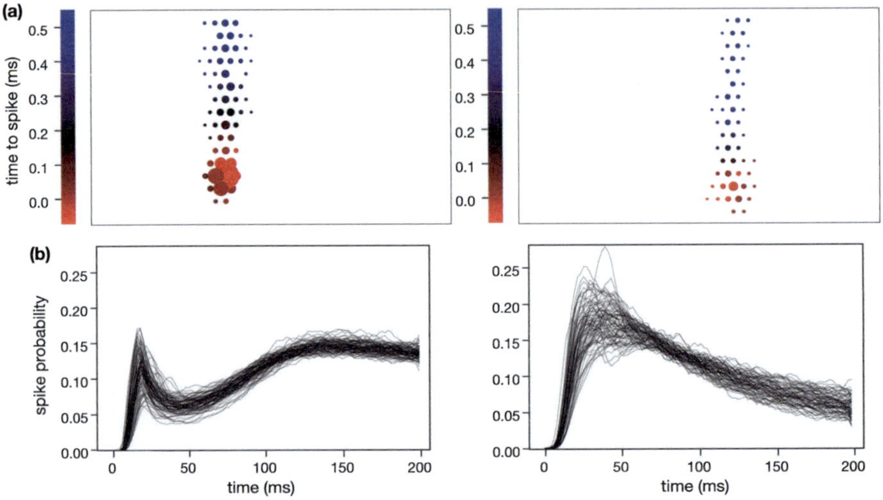

Fig. 24.11 Electrical features of neurons to be used for cell-type classification in blind retina. **a** Electrical image (EI) for an ON parasol and an ON midget cell from a primate retina. Each circle represents an electrode in the array, with radius indicating the average amplitude of the recorded spike, and color indicating its arrival time. Parasol EIs are usually larger and exhibit faster spike propagation than midget EIs. **b** Autocorrelation function of the spike train for a population of ON parasol cells and a population of OFF parasol cells in the primate retina

The electrical recording approach we propose performs lossy compression in the mixed-signal domain (i.e. before full digitization), exploiting two principles [64]. First, spikes are sparse in space and time, therefore we need only record spikes, and not voltage samples between spikes. Second, it is necessary only to identify and distinguish the spikes produced by different cells, therefore, spike waveforms need not be perfectly recorded. A scheme that exploits these facts is one in which the digitized voltage on a given electrode is retained only if it is different from the values on other electrodes. This can be accomplished efficiently with a *ramp analog-to-digital converter* (*ADC*) coupled with a *wired-OR readout* and a *unique-signal decoder*. During each sample, the ramp ADC indicates the input voltage with a brief pulse at a discrete time step proportional to the quantized voltage (the number of distinct time steps sets the ADC resolution and the ramp voltage range sets the ADC full-scale range). This is achieved by comparing the input signal to a ramp voltage that steps through the entire input range (Fig. 24.12a). The time of the pulse is captured using a counter that keeps track of the ramp steps. This is an efficient algorithm for digitizing many channels in an array because the ramp and the counter can be shared between all channels. Then, instead of reading the output pulses from each channel individually, channels are combined with an OR logic, across the rows and across the columns (Fig. 24.12b), to achieve the desired compression. Consequently, if only a single channel produces a pulse at a given time step (i.e., it is the only channel with a quantized voltage corresponding to the time step), then the channel location

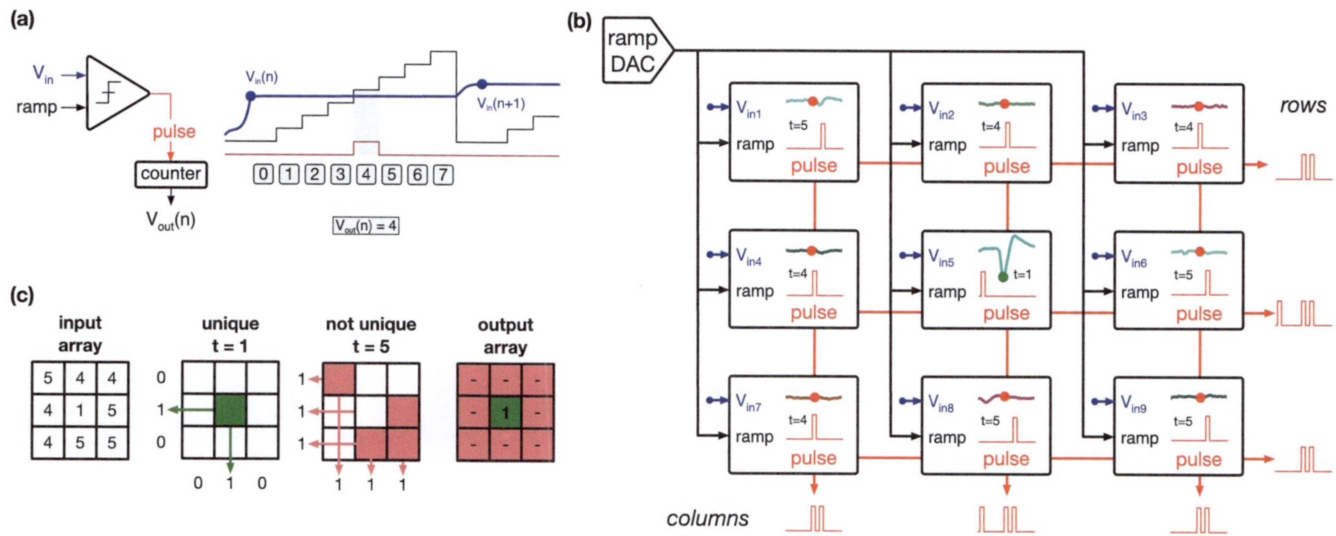

Fig. 24.12 Compressive readout strategy. **a** *Ramp ADC*: schematic and operation. **b** *Wired-OR readout*: 3 × 3 array example. **c** *Unique-signal decoder*: only the channel(s) presenting a unique digitized voltage is recorded in the output. Examples of unique and non-unique activated locations at two time steps are also shown

is indicated by a uniquely decoded row and column (pulse at t = 1 in Fig. 24.12c). On the other hand, if multiple pulses from different channels occur at the same time step (i.e., the quantized voltages on many channels are equal) multiple rows and/or columns are activated, and no uniquely decoded channel is indicated (pulse at t = 5 in Fig. 24.12c). Only the uniquely decoded samples are stored, leading to substantial compression (output in Fig. 24.12c).

Direct measurements from large collections of RGCs indicate that this compression approach is effective for reconstructing real neural spikes. The probability distribution of the input signal reveals that spikes primarily inhabit the tails (Fig. 24.13a), which implies that voltages associated with spikes tend to be unique. Consequently, spike samples are typically retained while other samples are typically discarded (Fig. 24.13b). The result is an accurate reconstruction of spike waveforms (Fig. 24.13c) accompanied by substantial compression (~40x). Most importantly, these approximately reconstructed waveforms are sufficient to distinguish spikes

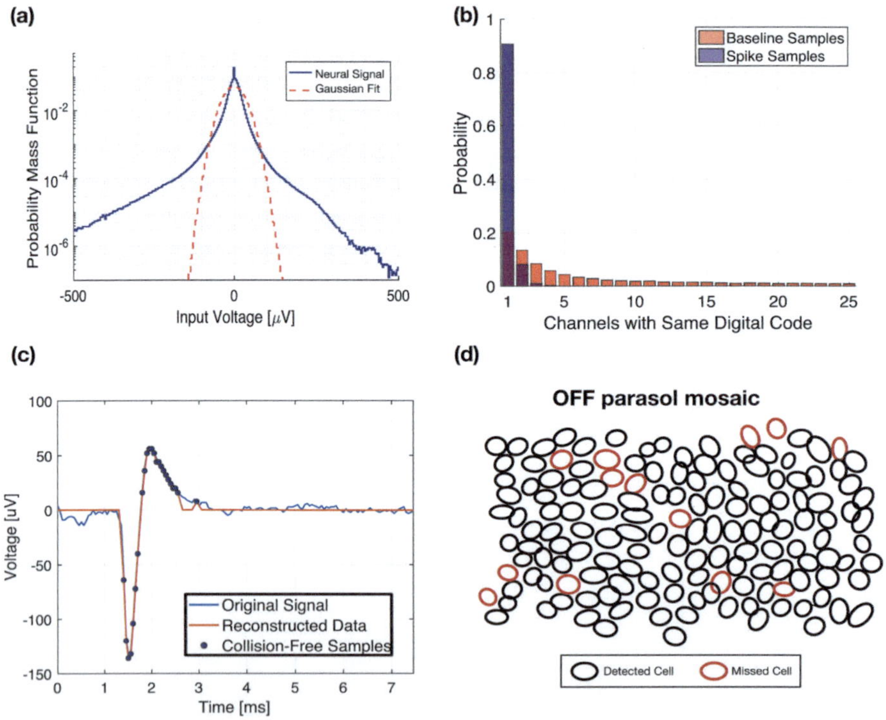

Fig. 24.13 Neural signal characteristics and compressive readout results. **a** Probability mass function of 100,000 samples from 512-electrode recording, after offset removal. **b** Probability of channels having the same digitized voltage for spike and baseline samples—here, baseline samples refer to samples that are within $\pm 4\sigma_n$ (standard deviation of noise in the recording channel) and spike samples refer to samples outside the $\pm 4\sigma_n$ band. Results from data recorded in isolated primate retina. **c**, **d** Example of a reconstructed spike (**c**) and a reconstructed OFF parasol mosaic (**d**) in the primate retina using the described readout strategy

from different cells, allowing recordings from nearly complete collections of neurons (~95%; Fig. 24.13d and [64]).

24.5.2 Dictionary Calibration

Dictionary calibration involves learning how current passed through the stimulating electrodes activates RGCs (Fig. 24.10b). The dictionary consists of elements that indicate the probability of generating a spike in each of the recorded RGCs, given a particular set of stimulating electrode(s) and current(s) [65]. Typically, for each dictionary entry, one or a few electrodes are used and a small number of cells are activated (but more complex dictionary entries are also possible). The generation of this dictionary is critical to obtain single-cell resolution. For each possible electrode and current level on that electrode, stimulation is applied, and the neural response is recorded on the entire array. The evoked response is compared against all the recorded EIs to identify which cell(s), if any, were activated (*response ID*). This step is performed repeatedly to estimate the probability of response of each RGC. The combined information about electrical stimulation and cellular activation probability constitutes one element in the dictionary. Note that dictionary elements which include axon bundle activation are typically not used (see Sect. 24.4 and [13]).
Challenges: A serious technical challenge for the above calibration is removing the *stimulation artifact* resulted from injecting a current into the high electrode impedance. Stimulation artifacts are large recorded waveforms that can obscure the neural response of interest, and are thus a severe problem for bi-directional neural interfaces. Typically, a combination of front-end mitigation techniques and back-end cancellation methods have to be implemented to overcome this issue [66]. Another challenge is the size of the dictionary: it is impractical to characterize the responses of all cells to all possible patterns of stimulation through a large electrode array. Approaches to this problem currently being explored include adaptive methods for developing models of electrically-evoked response [67], the use of prior information obtained from recording to predict the results of stimulation [67], and modeling interactions between electrodes in producing electrical stimulation [52].

24.5.3 Runtime

Runtime operation involves stimulating the available cells based on the incoming image, using the dictionary calibration (Fig. 24.10c). An external camera captures the visual scene, and the eye position is used to extract a focal image, i.e. the region of visual space that the interface should be encoding. Given the focal image, the goal is to optimize the stimulation pattern in real time such that the elicited cell responses lead to a faithful perception of the focal image (*dictionary approximation*). An obvious approach is to approximately mimic the normal RGC responses that would be produced by the focal image, using existing understanding of natural image coding by the various RGC types, and the available dictionaryelements. A different approach is to assume a model for how the brain interprets the visual information transmitted by the retina, and tune the evoked RGC responses accordingly to optimize perception. These two approaches have benefits and drawbacks that depend on both the richness of the dictionary and the degree of our understanding of how the brain interprets visual signals. In either case, the calibrated dictionary is leveraged to efficiently determine the desired electrical activation of RGCs.

Challenges: In general, a large collection of RGCs must be activated to produce useful vision. This usually requires passing current through many electrodes. However, as described above, a dictionary that specifies RGC responses to all possible patterns of current through many electrodes would be prohibitively large to create and store on an implantable chip. To avoid this problem, temporal dithering of simpler (e.g. single-electrode) stimulation patterns has been proposed [65], with stimulation patterns from the dictionary interleaved at sub-millisecond time resolution—faster than the integration time of the brain. Results from ex vivo experiments show that greedily selecting the stimulation sequence from a simple dictionary to most rapidly reduce the expected error between the target image and a linear image reconstruction from the neural responses (as a surrogate for perception) leads to an efficient encoding of the image. However, future algorithms could improve upon the strong assumptions of this approach—such as integration times, independence of dictionary elements, linearity of perception, measures of perceptual error, the role of eye movements—while allowing for efficient runtime implementation in portable hardware.

24.5.4 System Architecture

To accomplish the above goals, we envision a system with a minimalist device implanted at the back of the eye that maximizes the number of channels within the tight power budget, while the more complex system components are implemented elsewhere (Fig. 24.14). The interface communicates wirelessly with a relay on the outside of the eye, which in turn communicates with a digital signal processing (DSP) chip outside the body. Finally, a camera and an eye tracker provide the desired image to the DSP. In this system, the data for calibration can be sent out to be processed by experts in the clinic, while the patient keeps using the device. During runtime operation, the DSP processes the data from the camera and the eye sensor and transmits the stimulation control signals appropriate for the focal image to the interface, via the relay.

Challenges: Novel circuits and systems are required to implement the bi-directional neural interface, the data-power relay, and the runtime DSP chip. The interface needs to maximize the number of channels, while maintaining low power, data bandwidth, and area for implantation. The implanted relay chip will optimize the wireless power and data link for efficiency, helping to overcome the difficulties introduced by eye

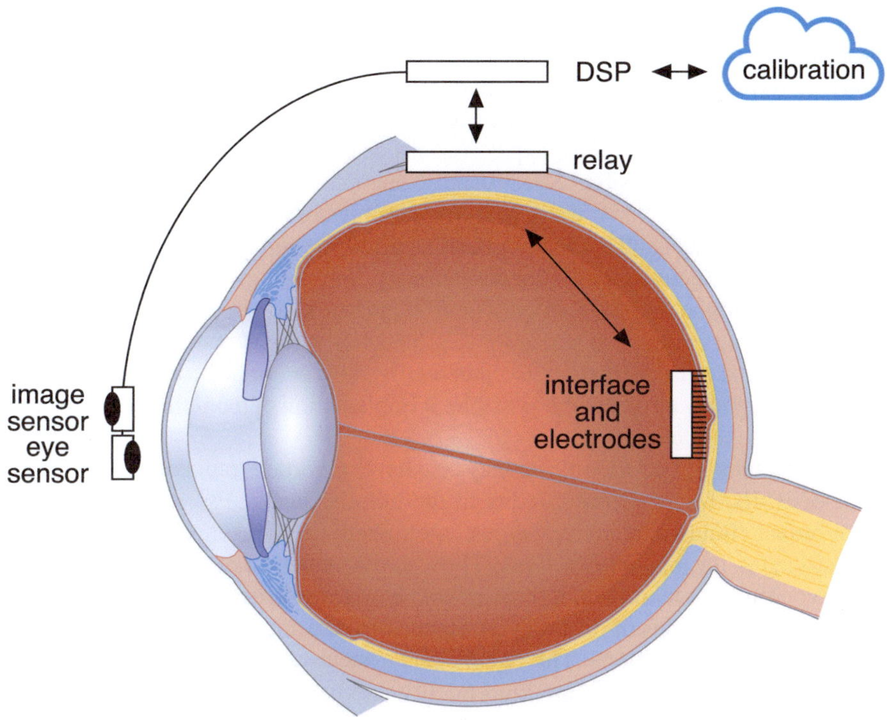

Fig. 24.14 Conceptual sketch of the Stanford Artificial Retina

movements and the different wireless communication environments inside and outside the eye. While the DSP chip will not be implanted and thus will have more power available, it will also have to implement compute-intensive algorithms that operate in real time with short latencies. Finally, other system challenges need to be faced, such as high precision eye movement tracking, encapsulation of active devices for long term implantation, CMOS-MEA integration, and surgical procedures. All these challenges will promote innovation in IC design and system integration.

24.6 Conclusion

Future retinal implants for restoring vision will need to achieve cell and cell-type resolution over a large area of the central retina to overcome the limitations of current devices. To do so requires novel circuits and systems coupled to bi-directional neural interfaces that interact efficiently with the neural circuitry via large and dense electrode arrays. The retina is an ideal target for pioneering this kind of high-fidelity, adaptive neuroengineering, as it is relatively accessible and well-understood. The Stanford

Artificial Retina is the first attempt at such an architecture, exploiting data-driven algorithm-circuit co-design for high-fidelity artificial vision.

In addition to treating incurable blindness, such a device may open other doors as well. Once the device is developed and able to interface specifically and configurably to many distinct RGC types, various scientific, medical, and commercial applications become possible. Scientifically, one may be able to modify the neural code of the retina in diverse ways to test how subtle and targeted alterations of the natural visual signal, such as changes in spike timing or cell type signaling diversity, influence visual acuity in experimental animals, and perhaps in humans that have already been implanted. Medically, it may be possible to encode visual scenes in abstracted or augmented ways in order to provide an artificial visual signal that is of even greater utility to the patient. Commercially, diverse applications of augmented artificial vision may also become possible, if implantation of the retina becomes safe and routine.

These technology developments may also have important implications for a wide range of BMIs. Many neural circuits in the brain share the essential architecture of the retinal circuitry: many cell types, intermixed, transmitting distinct signals to distinct targets for distinct functions. Thus, as our understanding of other areas of the nervous system increases, interfaces capable of reproducing the natural pattern of activation of neurons of different types may provide much higher performance in a range of BMI applications. The circuits and systems we propose for the artificial retina may be adaptable to future high-fidelity interfaces to the brain.

Acknowledgements The authors would like to thank Marty Breidenbach, Ruwan Silva, Stephen Weinreich, Matthias Kuhl, Daniel Palanker, Nishal Shah and the Stanford Artificial Retina Group [54] for useful discussions and comments, and Peter Li, Chris Sekirnjak, and Nora Brackbill for figure contributions. DM, EJC and the Stanford Artificial Retina Project are funded by the Wu Tsai Neurosciences Institute. EJC is funded by NEI grant EY021271 and a Stein Innovation Award from Research to Prevent Blindness.

References

1. M.W. Slutzky, Brain-machine interfaces: powerful tools for clinical treatment and neuroscientific investigations. Neuroscientist **25**, 139–154 (2019)
2. M.A. Lebedev, M.A.L. Nicolelis, Brain-machine interfaces: from basic science to neuroprostheses and neurorehabilitation. Physiol. Rev. **97**, 767–837 (2017)
3. A.H. Marblestone, B.M. Zamft, Y.G. Maguire, M.G. Shapiro, T.R. Cybulski, J.I. Glaser, D. Amodei, P.B. Stranges, R. Kalhor, D.A. Dalrymple, D. Seo, E. Alon, M.M. Maharbiz, J.M. Carmena, J.M. Rabaey, E.S. Boyden, G.M. Church, K.P. Kording, Physical principles for scalable neural recording. Front. Comput. Neurosci. **7**, 137 (2013)
4. G.W. Fraser, S.M. Chase, A. Whitford, A.B. Schwartz, Control of a brain–computer interface without spike sorting. J. Neural Eng. **6**, 055004 (2009)
5. C.A. Chestek, V. Gilja, P. Nuyujukian, J.D. Foster, J.M. Fan, M.T. Kaufman, M.M. Churchland, Z. Rivera-Alvidrez, J.P. Cunningham, S.I. Ryu, K.V. Shenoy, Long-term stability of neural prosthetic control signals from silicon cortical arrays in rhesus macaque motor cortex. J. Neural Eng. **8**, 045005 (2011)

6. B.P. Christie, D.M. Tat, Z.T. Irwin, V. Gilja, P. Nuyujukian, J.D. Foster, S.I. Ryu, K.V. Shenoy, D.E. Thompson, C.A. Chestek, Comparison of spike sorting and thresholding of voltage waveforms for intracortical brain-machine interface performance. J. Neural Eng. **12**, 016009 (2015)
7. J. Li, Z. Li, Sums of spike waveform features for motor decoding. Front. Neurosci. **11**, 406 (2017)
8. E.M. Trautmann, S.D. Stavisky, S. Lahiri, K.C. Ames, M.T. Kaufman, D.J. O'Shea, S. Vyas, X. Sun, S.I. Ryu, S. Ganguli, K.V. Shenoy, Accurate estimation of neural population dynamics without spike sorting. Neuron **103**, 292–308.e4 (2019)
9. N. Even-Chen, D.G. Muratore, S.D. Stavisky, L.R. Hochberg, J.M. Henderson, B. Murmann, K.V. Shenoy, Motor intracortical interface design opportunities for an order of magnitude power saving. Nat. Biomed. Eng. (2020) (In Press)
10. G.A. Goetz, D.V. Palanker, Electronic approaches to restoration of sight. Rep. Prog. Phys. **79**, 096701 (2016)
11. E.R. Kandel, J.H. Schwartz, T.M. Jessell, *Principles of Neural Science* (McGraw-Hill, New York, 2012)
12. C.E. Schoonover, *Portraits of the Mind: Visualizing the Brain from Antiquity to the 21st Century*. Abrams (2010)
13. L.E. Grosberg, K. Ganesan, G.A. Goetz, S.S. Madugula, N. Bhaskhar, V. Fan, P. Li, P. Hottowy, W. Dabrowski, A. Sher, A.M. Litke, S. Mitra, E.J. Chichilnisky, Activation of ganglion cells and axon bundles using epiretinal electrical stimulation. J. Neurophysiol. **118**, 1457–1471 (2017)
14. A.L. Yarbus, *Eye Movements and Vision* (Plenum, New York, 1967)
15. R.W. Rodieck, *The First Steps in Seeing* (Sinauer, Sunderland, 1998)
16. B.A. Wandell, *Foundations of Vision* (Sinauer, Sunderland, 1995)
17. B. Roska, M. Meister, The retina dissects the visual scene into distinct features, in *The New Visual Neurosciences* (2014), pp. 163–182
18. T. Gollisch, M. Meister, Eye smarter than scientists believed: neural computations in circuits of the retina. Neuron **65**, 150–164 (2010)
19. M. Beyeler, A. Rokem, G.M. Boynton, I. Fine, Learning to see again: biological constraints on cortical plasticity and the implications for sight restoration technologies. J. Neural Eng. **14**, 051003 (2017)
20. P. Dayan, L.F. Abbott, *Theoretical Neuroscience: Computational and Mathematical Modeling of Neural Systems* (MIT Press, 2001)
21. H.G. Rey, C. Pedreira, R. Quian Quiroga, Past, present and future of spike sorting techniques. Brain Res. Bull. **119**, 106–117 (2015)
22. J.E. Chung, J.F. Magland, A.H. Barnett, V.M. Tolosa, A.C. Tooker, K.Y. Lee, K.G. Shah, S.H. Felix, L.M. Frank, L.F. Greengard, A fully automated approach to spike sorting. Neuron **95**, 1381–1394.e6 (2017)
23. D. Carlson, L. Carin, Continuing progress of spike sorting in the era of big data. Curr. Opin. Neurobiol. **55**, 90–96 (2019)
24. J.F. Fohlmeister, P.A. Coleman, R.F. Miller, Modeling the repetitive firing of retinal ganglion cells. Brain Res. **510**, 343–345 (1990)
25. D. Boinagrov, J. Loudin, D. Palanker, Strength-duration relationship for extracellular neural stimulation: numerical and analytical models. J. Neurophysiol. **104**, 2236–2248 (2010)
26. G.J. Chader, J. Weiland, M.S. Humayun, Artificial vision: needs, functioning, and testing of a retinal electronic prosthesis. Prog. Brain Res. **175**, 317–332 (2009)
27. E. Zrenner, Fighting blindness with microelectronics. Sci. Transl. Med. **5**, 210ps16 (2013)
28. A.T. Chuang, C.E. Margo, P.B. Greenberg, Retinal implants: a systematic review. Br. J. Ophthalmol. **98**, 852–856 (2014)
29. Y.H.-L. Luo, L. da Cruz, A review and update on the current status of retinal prostheses (bionic eye). Br. Med. Bull. **109**, 31–44 (2014)
30. K. Grifantini, Aiding the Eye, Watching the Brain: James Weiland, IEEE Fellow, explores the unique challenges of retinal prostheses. IEEE Pulse **8**, 39–41 (2017)

31. M.S. Humayun, E. de Juan, G. Dagnelie Jr., The bionic eye: a quarter century of retinal prosthesis research and development. Ophthalmology **123**, S89–S97 (2016)
32. D. Nanduri, M.S. Humayun, R.J. Greenberg, M.J. McMahon, J.D. Weiland, Retinal prosthesis phosphene shape analysis, in *2008 30th Annual International Conference of the IEEE Engineering in Medicine and Biology Society* (2008), pp. 1785–1788
33. C. de Balthasar, S. Patel, A. Roy, R. Freda, S. Greenwald, A. Horsager, M. Mahadevappa, D. Yanai, M.J. McMahon, M.S. Humayun, R.J. Greenberg, J.D. Weiland, I. Fine, Factors affecting perceptual thresholds in epiretinal prostheses. Invest. Ophthalmol. Vis. Sci. **49**, 2303–2314 (2008)
34. A. Caspi, J.D. Dorn, K.H. McClure, M.S. Humayun, R.J. Greenberg, M.J. McMahon, Feasibility study of a retinal prosthesis: spatial vision with a 16-electrode implant. Arch. Ophthalmol. **127**, 398–401 (2009)
35. R.J. Jensen, J.F. Rizzo 3rd, O.R. Ziv, A. Grumet, J. Wyatt, Thresholds for activation of rabbit retinal ganglion cells with an ultrafine, extracellular microelectrode. Invest. Ophthalmol. Vis. Sci. **44**, 3533–3543 (2003)
36. A. Butterwick, A. Vankov, P. Huie, Y. Freyvert, D. Palanker, Tissue damage by pulsed electrical stimulation. IEEE Trans. Biomed. Eng. **54**, 2261–2267 (2007)
37. S.F. Cogan, Neural stimulation and recording electrodes. Annu. Rev. Biomed. Eng. **10**, 275–309 (2008)
38. D.R. Merrill, M. Bikson, J.G.R. Jefferys, Electrical stimulation of excitable tissue: design of efficacious and safe protocols. J. Neurosci. Methods **141**, 171–198 (2005)
39. L.S. Robblee, T.L. Rose, Electrochemical guidelines for selection of protocols and electrode materials for neural stimulation, in *Neural Prostheses: Fundamental Studies* (1990), pp. 25–66
40. A. Horsager, S.H. Greenwald, J.D. Weiland, M.S. Humayun, R.J. Greenberg, M.J. McMahon, G.M. Boynton, I. Fine, Predicting visual sensitivity in retinal prosthesis patients. Invest. Ophthalmol. Vis. Sci. **50**, 1483–1491 (2009)
41. J.-H. Jung, D. Aloni, Y. Yitzhaky, E. Peli, Active confocal imaging for visual prostheses. Vis. Res. **111**, 182–196 (2015)
42. E. Zrenner, K.U. Bartz-Schmidt, H. Benav, D. Besch, A. Bruckmann, V.-P. Gabel, F. Gekeler, U. Greppmaier, A. Harscher, S. Kibbel, J. Koch, A. Kusnyerik, T. Peters, K. Stingl, H. Sachs, A. Stett, P. Szurman, B. Wilhelm, R. Wilke, Subretinal electronic chips allow blind patients to read letters and combine them to words. Proc. R. Soc. B Biol. Sci. **278**, 1489–1497 (2011)
43. H. Lorach, G. Goetz, R. Smith, X. Lei, Y. Mandel, T. Kamins, K. Mathieson, P. Huie, J. Harris, A. Sher, D. Palanker, Photovoltaic restoration of sight with high visual acuity. Nat. Med. **21**, 476–482 (2015)
44. L.H. Jepson, P. Hottowy, G.A. Weiner, W. Dabrowski, A.M. Litke, E.J. Chichilnisky, High-fidelity reproduction of spatiotemporal visual signals for retinal prosthesis. Neuron **83**, 87–92 (2014)
45. C. Sekirnjak, P. Hottowy, A. Sher, W. Dabrowski, A.M. Litke, E.J. Chichilnisky, High-resolution electrical stimulation of primate retina for epiretinal implant design. J. Neurosci. **28**, 4446–4456 (2008)
46. L.H. Jepson, P. Hottowy, K. Mathieson, D.E. Gunning, W. Dabrowski, A.M. Litke, E.J. Chichilnisky, Focal electrical stimulation of major ganglion cell types in the primate retina for the design of visual prostheses. J. Neurosci. **33**, 7194–7205 (2013)
47. A.M. Litke, N. Bezayiff, E.J. Chichilnisky, W. Cunningham, W. Dabrowski, A.A. Grillo, M. Grivich, P. Grybos, P. Hottowy, S. Kachiguine, R.S. Kalmar, K. Mathieson, D. Petrusca, M. Rahman, A. Sher, What does the eye tell the brain?: Development of a system for the large-scale recording of retinal output activity. IEEE Trans. Nucl. Sci. **51**, 1434–1440 (2004)
48. E.J. Chichilnisky, R.S. Kalmar, Functional asymmetries in ON and OFF ganglion cells of primate retina. J. Neurosci. **22**, 2737–2747 (2002)
49. E.S. Frechette, A. Sher, M.I. Grivich, D. Petrusca, A.M. Litke, E.J. Chichilnisky, Fidelity of the ensemble code for visual motion in primate retina. J. Neurophysiol. **94**, 119–135 (2005)
50. G.D. Field, J.L. Gauthier, A. Sher, M. Greschner, T.A. Machado, L.H. Jepson, J. Shlens, D.E. Gunning, K. Mathieson, W. Dabrowski, L. Paninski, A.M. Litke, E.J. Chichilnisky, Functional connectivity in the retina at the resolution of photoreceptors. Nature **467**, 673–677 (2010)

51. V.H. Fan, L.E. Grosberg, S.S. Madugula, P. Hottowy, W. Dabrowski, A. Sher, A.M. Litke, E.J. Chichilnisky, Epiretinal stimulation with local returns enhances selectivity at cellular resolution. J. Neural Eng. **16**, 025001 (2019)
52. L.H. Jepson, P. Hottowy, K. Mathieson, D.E. Gunning, W. Dąbrowski, A.M. Litke, E.J. Chichilnisky, Spatially patterned electrical stimulation to enhance resolution of retinal prostheses. J. Neurosci. **34**, 4871–4881 (2014)
53. C.E. Rhoades, N.P. Shah, M.B. Manookin, N. Brackbill, A. Kling, G. Goetz, A. Sher, A.M. Litke, E.J. Chichilnisky, Unusual physiological properties of smooth monostratified ganglion cell types in primate retina. Neuron **103**, 658–672.e6 (2019)
54. Stanford Artificial Retina Project. http://med.stanford.edu/artificial-retina.html
55. E. Richard, G.A. Goetz, E.J. Chichilnisky, Recognizing retinal ganglion cells in the dark, in *Advances in Neural Information Processing Systems* (2015), pp. 2476–2484
56. V. Karkare, S. Gibson, D. Marković, A 75-μW, 16-channel neural spike-sorting processor with unsupervised clustering. IEEE J. Solid-State Circuits **48**, 2230–2238 (2013)
57. M. Pagin, M. Ortmanns, A neural data lossless compression scheme based on spatial and temporal prediction, in *2017 IEEE Biomedical Circuits and Systems Conference (BioCAS)* (2017), pp. 1–4
58. C. Aprile, K. Ture, L. Baldassarre, M. Shoaran, G. Yilmaz, F. Maloberti, C. Dehollain, Y. Leblebici, V. Cevher, Adaptive learning-based compressive sampling for low-power wireless implants. IEEE Trans. Circuits Syst. I Regul. Pap. **65**, 3929–3941 (2018)
59. T. Wu, W. Zhao, E. Keefer, Z. Yang, Deep compressive autoencoder for action potential compression in large-scale neural recording. J. Neural Eng. **15**, 066019 (2018)
60. T. Okazawa, I. Akita, A time-domain analog spatial compressed sensing encoder for multi-channel neural recording. Sensors **18**, 184 (2018). https://doi.org/10.3390/s18010184
61. V. Majidzadeh, A. Schmid, Y. Leblebici, A 16-channel, 359 μW, parallel neural recording system using Walsh-Hadamard coding, in *Proceedings of the IEEE 2013 Custom Integrated Circuits Conference* (2013), pp. 1–4
62. D. Tsai, R. Yuste, K.L. Shepard, Statistically reconstructed multiplexing for very dense, high-channel-count acquisition systems. IEEE Trans. Biomed. Circuits Syst. **12**, 13–23 (2018)
63. M. Sharma, A.T. Gardner, H.J. Strathman, D.J. Warren, J. Silver, R.M. Walker, Acquisition of neural action potentials using rapid multiplexing directly at the electrodes. Micromachines (2018). https://doi.org/10.3390/mi9100477
64. D.G. Muratore, P. Tandon, M. Wootters, E.J. Chichilnisky, S. Mitra, B. Murmann, A data-compressive wired-or readout for massively parallel neural recording. IEEE Trans. Biomed. Circuits Syst. **13**, 1128–1140 (2019)
65. N.P. Shah, S. Madugula, L. Grosberg, G. Mena, P. Tandon, P. Hottowy, A. Sher, A. Litke, S. Mitra, E.J. Chichilnisky, Optimization of electrical stimulation for a high-fidelity artificial retina, in *2019 9th International IEEE/EMBS Conference on Neural Engineering (NER)* (2019)
66. A. Zhou, B.C. Johnson, R. Muller, Toward true closed-loop neuromodulation: artifact-free recording during stimulation. Curr. Opin. Neurobiol. **50**, 119–127 (2018)
67. N.P. Shah, S. Madugula, P. Hottowy, A. Sher, A. Litke, L. Paninski, E.J. Chichilnisky, Efficient characterization of electrically evoked responses for neural interfaces, in *Neural Information Processing Systems (NeurIPS)* (2019)

Chapter 25
Augmented and Virtual Reality

Gordon Wetzstein

Immersive computer graphics systems, such as virtual and augmented reality (VR/AR) displays, aim at synthesizing perceptually realistic user experiences. To achieve this goal, several components are required: interactive, photorealistic rendering; a high-resolution, low-persistence, stereoscopic display; and low-latency head tracking. Modern VR/AR systems provide all of these capabilities and create experiences that support many, but not all, of the depth cues of the human visual system. They fall short of passing a "visual Turing test for displays". Imagine a person using a wearable computing system and that system delivering user experiences that are indistinguishable from the real world. That is, the user would not be able to tell whether an image is computer generated or real. While the field of computer graphics has been developing algorithms to generate photorealistic images, to pass the visual Turing test for displays, a VR/AR system must deliver perceptually realistic experiences. This challenge requires displays with high resolution, color fidelity, dynamic range, and adequate support of all the depth cues of human vision. Moreover, for such a system to be practical, device form factor, weight, power, heat, battery life, limited compute power, and bandwidth have to optimized as well and set physical constraints on the capabilities of a wearable computing system.

A second, equally important goal of AR/VR systems is that they are wearable, operating at low power to provide sufficiently long battery life, and that the thermal management works well enough to prevent overheating. After all, users will only adopt wearable computing systems if they are comfortable, functional, and in many but not all cases also fashionable. Application-specific integrated circuits dedicated to solving the computationally demanding tasks of these systems, including rendering, display, and tracking, are therefore crucial to the success of AR/VR systems.

Although significant research and engineering efforts have focused on reducing the size, weight, and power (SWaP) characteristics and also the user experiences of AR/VR displays, we are far from being able to deliver experiences that pass the

G. Wetzstein (✉)
Stanford University, Stanford, CA, USA
e-mail: gordon.wetzstein@stanford.edu

B. Murmann and B. Hoefflinger (eds.), *NANO-CHIPS 2030*,
The Frontiers Collection, https://doi.org/10.1007/978-3-030-18338-7_25

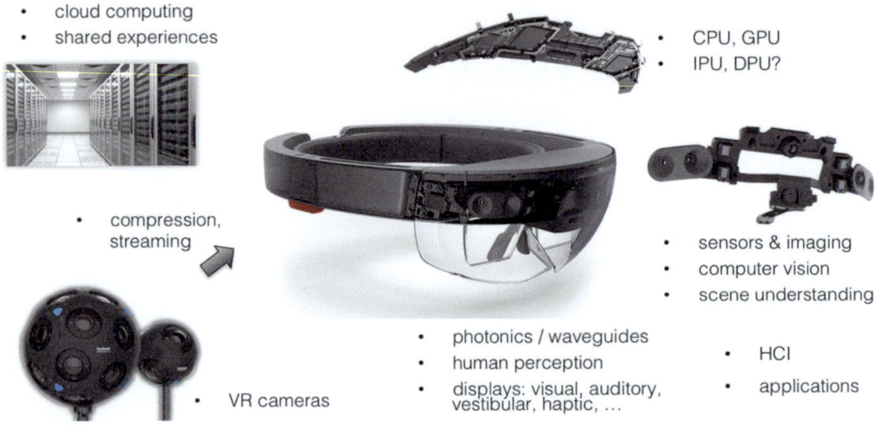

Fig. 25.1 Overview of components of AR/VR systems. From dedicated, application-specific processors to microdisplays, photonic waveguides, 360° cameras, novel cloud services to new approaches to computer vision and human-computer interaction, AR/VR system design includes a plethora of aspects that all together define the user experience. Parts of this image are reproduced from Microsoft and Facebook promotional material

aforementioned visual Turing test for displays with wearable form factors. In the following sections, we outline challenges and solutions for rendering and perception, near-eye displays, tracking, and capturing and editing cinematic VR content. Figure 25.1 shows an illustration outlining the many components that need to be considered for engineering a VR/AR system.

25.1 Near-Eye Displays

25.1.1 Foveated Rendering and Display

Each of our eyes has a field of vision of about 150 $\times$ 135 degrees horizontal and vertical, respectively. Designing a display that provides retinal resolution, i.e. about 60 pixelsper degree of visual angle, would thus require about 9000 $\times$ 8100 pixels per eye to achieve 20/20 vision. Rendering this massive amount of visual data at 90–120 frames per second is a major challenge for any graphics processing unit (GPU). Moreover, this data also has to be transmitted to the head-mounted display and shown there. To address some of the challenges associated with high-resolution rendering in AR/VR, we can exploit some of the limitations of the human visual system (HVS). We know that the visual acuity of the HVS is higher in the fovea than in the periphery of the visual field. Using eye tracking technology, we can easily determine where the user is fixating and adaptively rendering images of varying resolution, a technique know as foveated rendering. Due to the large field-of-view

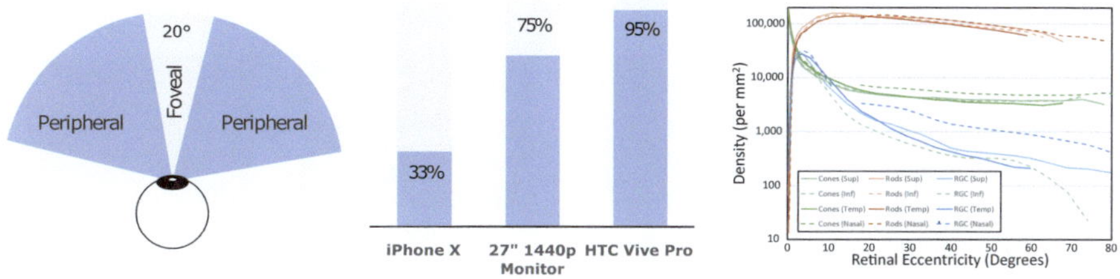

Fig. 25.2 (Left) While viewing most displays, a fraction of pixels lie in our foveal vision, while the remaining lie in our peripheral vision. For displays like smartphones, foveal pixels dominate, and for computer monitors peripheral pixels are a majority. However, for near-eye displays like contemporary VR, almost all pixels are peripheral. (Right) Density of photoreceptors and retinal ganglion cells (RGCs) varying with eccentricity. There is a strong preference for central vision as compared to peripheral vision. Data measured by Curcio and Allen [28] and Curcio et al. [29]

of near-eye displays, we observe a large majority of VR/AR pixels through our peripheral vision (see Fig. 25.2, left). Combined with the fact that visual acuity of peripheral vision is significantly lower than foveal or central vision, adaptively and dynamically distributing image quality and detail across the visual field is an important class of perceptual optimizations for near-eye displays. In this section we discuss the physiological and perceptual bases for foveation, as well as the relevant rendering and display technologies proposed in recent literature.

Human visual perception starts at the optical components (the lens, pupil, etc.), followed by retinal cells like rods, cones, and ganglion cells, and finally by higher level neural processing. Each of these optical, retinal, and neural components exhibit a strong preference for the central area of the visual field. On the retina this region is also called the fovea, and is marked by high density of retinal cells (see Fig. 25.2, right). As a consequence of the variation in processing density across the pathway, our foveal vision has a much higher acuity than our peripheral vision. Hence, it is better for near-eye displays to provide more detail in the foveal region than in the peripheral region.

The degradation in visual acuity from foveal to peripheral vision is also known to be highly non-uniform [189]. For instance, while we cannot perceive fine details in images through our peripheral vision, we are extremely sensitive to moving and flickering images. Researchers have identified several such non-uniformities in peripheral visual acuity, e.g. in color perception [54, 147], in existence of a peripheral aliasing zone [199], and the anisotropy of peripheral perception [167]. While designing foveated rendering and display applications, we should be aware of this peculiarity. On the other hand, these effects can create additional opportunities to improve rendering performance or image quality.

Many researchers have proposed foveated rendering techniques to improve rendering performance for gaze-contingent displays. The most prominent class of techniques work by reallocating image pixels such that the density is highest at the fovea, and lowest in the periphery. There are can be done in two main ways (also see Fig. 25.3):

Fig. 25.3 Illustrations of two prominent techniques for foveated rendering. Left: We can render multiple views of a scene with varying resolution, and blend the resulting buffers to obtain the final foveated image. Right: We can render the scene into a distorted buffer that prioritizes foveal pixels, and after rendering undistort it into the final foveated image

- By rendering the fovea, periphery, and zero or more intermediate regions into different framebuffers of varying size and resolution, and blending them together to produce the final foveated image [52].
- By rendering the image into a distorted framebuffer that oversamples the fovea, but undersamples the periphery [26, 27, 45, 133].

Other techniques for foveated rendering work by reducing expensive computations like pixel shading operations [133, 155, 186, 202] and geometric evaluation [194].

While foveated rendering solutions seek to improve performance by reducing pixel computations in the periphery, a recent class of techniques moves the foveal-peripheral adaptivity directly to the display. Such novel display system designs match the nature of human vision. One example in VR is to expand the 2D foveated to 4D light field display [192]. The system is shown to offer both foveation (performance) and accommodation (comfort). More recently, the idea of foveating display has been advanced to augmented reality as well [89]. Whereas foveated rendering algorithms reduce the rendering and bandwidth requirements, foveated displays use clever combinations of optically foveated (i.e. steered) microdisplays to create a perceived resolution of a display that exceeds the actually pixel count.

Overall, foveated rendering and display techniques can help address the computational challenges of rendering and displaying computer-generated images at retinal resolution over a wide field of view. However, accurate and low-latency eye tracking are required for this technique as are GPUs, rendering algorithms, and microdisplay backplanes that support these display modes.

25.1.2 Enabling Focus Cues in VR/AR

Human depth perception relies on a variety of cues [68, 152]. Many of these cues are pictorial and can be synthesized using photorealistic rendering techniques, including occlusions, perspective foreshortening, texture and shading gradients, as well as

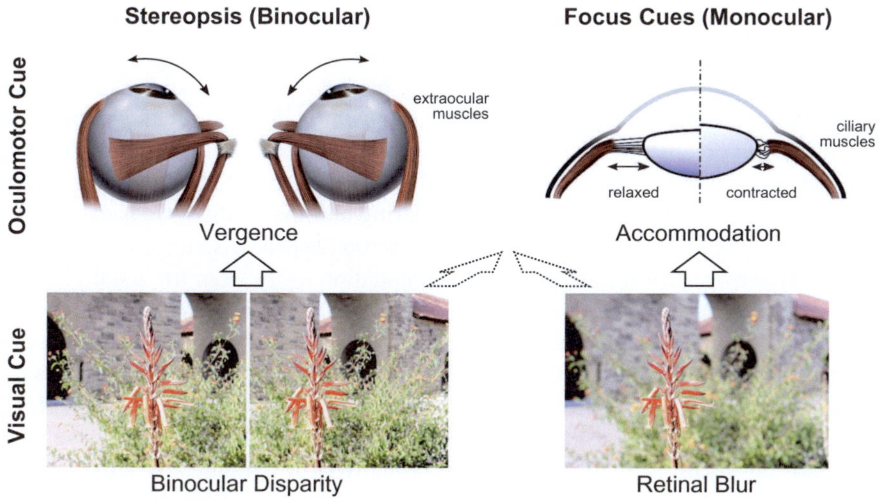

Fig. 25.4 Overview of several depth cues that are important for near-eye displays. Vergence and accommodation are oculomotor cues whereas binocular disparity and retinal blur are visual cues. In normal viewing conditions, disparity drives vergence and blur drives accommodation. However, these cues are cross-coupled, so there are conditions under which blur-driven vergence or disparity-driven accommodation occur

relative and familiar object size. Compared with conventional 2D displays, head-mounted displays (HMDs) use stereoscopic displays and head tracking and can thus support several additional depth cues: binocular disparity, motion parallax, and vergence (see Fig. 25.4) as well as ocular parallax [93]. All of these cues are important for human depth perception to varying degrees, depending on the fixation distance [31]. In this section, we review an area of active research and development: emerging near-eye displays that support focus cues, i.e. retinal blur, accommodation, and chromatic aberrations. For a more detailed survey of 3D displays and perceptual related issues, the interested reader is referred to [6].

Current near-eye displays cannot reproduce the changes in focus that accompany natural vision, and they cannot support users with uncorrected refractive errors. For users with normal vision, this asymmetry creates an unnatural condition known as the vergence–accommodation conflict [97, 102]. Symptoms associated with this conflict include double vision (diplopia), compromised visual clarity, visual discomfort, and fatigue [97, 177]. Moreover, a lack of accurate focus also removes a cue that is important for depth perception [31, 63, 66, 206]. Note that adequate reproduction of focus cues in VR/AR is most important for younger users, while older users tend to be presbyopic, i.e. they lost the ability to accommodate their eyes [149].

In the following, we outline several approaches to enabling focus cues in VR/AR and to mitigating the vergence–accommodation conflict. For a more comprehensive review of this topic, we refer the interested reader to the survey papers by Kramida [101] and Hua [71].

Gaze-contingent focus cue rendering is a software-only approach that tries to mimic focus cues by changing the way 2D images are being rendered. This includes gaze-contingent depth-of-field rendering [40, 64, 84, 123, 131] and rendering the chromatic aberrations of the human eye [24]. Although depth-of-field rendering may improve perceived realism, several recent studies have demonstrated that this approach alone does not drive accommodation [83, 94], therefore it does not reduce the vergence–accommodation conflict. Rendering chromatic aberrations can drive a user's accommodation in a monocular display setup [24] and also improve perceived realism. However, driving the user's accommodation away from the focal plane of the display may result in degradation of perceived image sharpness.
Varifocal displays present a single image plane to the observer, the focus distance of which can be dynamically adjusted. This is typically done by estimating the distance of the user's fixation point with eye tracking and then optically driving the display's focal plane to this distance. Two approaches for optical focus adjustment have been proposed: physically actuating the screen [149, 191] or dynamically adjusting the focal length of the lens via focus-tunable optics (programmable liquid lenses or reflectors) [41, 56, 83, 94, 100, 117, 149, 187, 188]. Several such systems have been incorporated into the form factor of a near-eye display [94, 117, 149]. Figure 25.5 shows both benchtop and wearable varifocal display prototypes along with data measured for users of all ages demonstrating that varifocal displays effectively drive accommodation for non-presbyopic users.
Multifocal displaysare probably the most common approach to focus-supporting displays by approximate the 3D space in front of the eye using a few virtual planes that are generated by beam splitters [2, 38, 134], time-multiplexed focus-tunable optics [20, 70, 117, 118, 121, 142, 160, 166, 204, 219], or phase-modulating spatial light modulators [130]. Naïve implementations with beam splitters seem impractical for wearable displays because they compromise the device form factor, but this concept is promising, especially for see-through AR displays, when implemented with stacked diffractive optical elements [110] or waveguides, such as in the Magic Leap ML1. One of the biggest challenges with time-multiplexed multi-plane displays is that they require high-speed displays and may thus introduce perceived flicker. Specifically, an N-plane display requires a refresh rate of $N\times$ 60–120 Hz. Digital micromirror devices (DMDs) are of the fastest available microdisplay technologies and seem particularly promising for this direction, as also realized by recent research [20, 160] as well as Avegant's commercial AR Video Headset. Content-adaptive multifocal displays [130, 219] seem particularly interesting, because they have the capability of minimizing the number of required focal planes based on the saliency of the content. However, optically generating non-planar or adaptive focal planes is challenging.
Near-eye light field displays aim to synthesize the full 4D light field in front of each eye [105, 107, 126, 213, 214]. Conceptually, this approach allows for parallax over the entire eyebox to be accurately reproduced, including monocular occlusions, specular highlights, ocular parallax, and other effects that cannot be reproduced by varifocal or multifocal displays. However, current-generation near-eye light field displays provide limited resolution due to the spatio-angular resolution tradeoff of

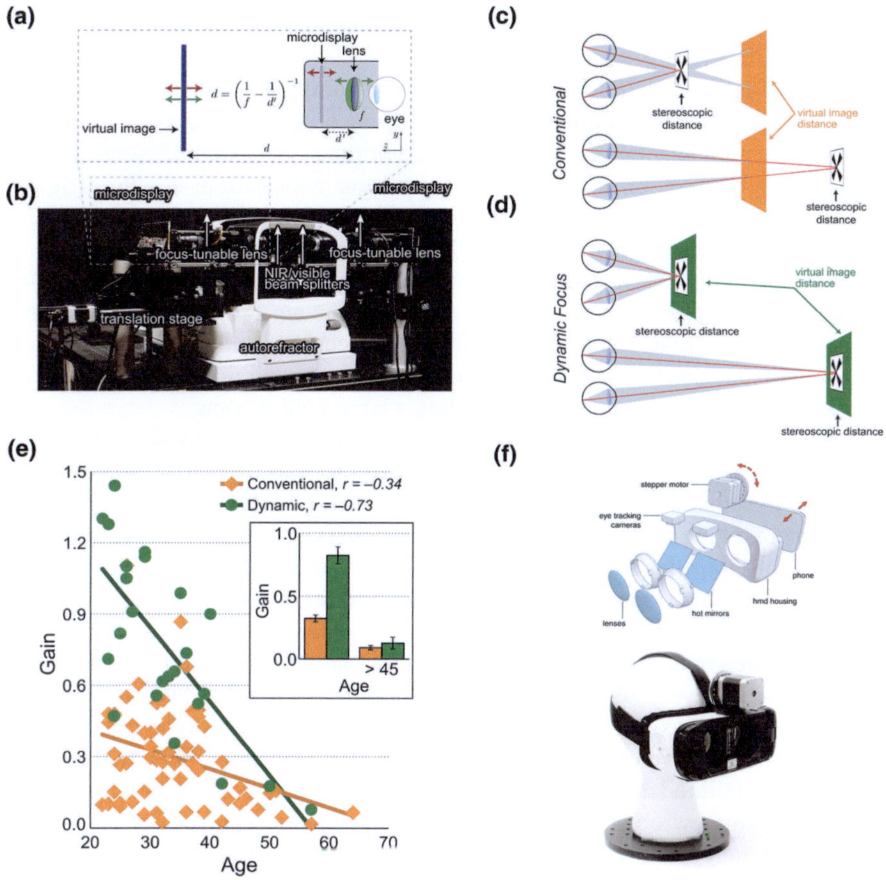

Fig. 25.5 Varifocal display prototypes and user experiments. **a** A typical near eye display uses a fixed-focus lens to show a magnified virtual image of a microdisplay to each eye. The focal length of the lens, f, and the distance to the microdisplay, d′, determine the distance of the virtual image, d. Dynamic focus can be implemented using either a focus-tunable lens (green arrows) or a fixed-focus lens and a mechanically actuated display (red arrows), so that the virtual image can be moved to different distances. **b** A benchtop setup designed to incorporate dynamic focus via focus-tunable lenses, and an autorefractor to record accommodation. **c** The use of a fixed-focus lens in conventional near-eye displays means that the magnified virtual image appears at a constant distance (orange planes). However, by presenting different images to the two eyes, objects can be simulated at arbitrary stereoscopic distances. To experience clear and single vision in VR, the user's eyes have to rotate to verge at the correct stereoscopic distance (red lines), but the eyes must maintain accommodation at the virtual image distance (gray areas). **d** In a dynamic focus display, the virtual image distance (green planes) is constantly updated to match the stereoscopic distance of the target. Thus, the vergenceand accommodation distances can be matched. **e** These accommodative gains plotted against the user's age show a clear downward trend with age, and a higher response in dynamic. Inset shows mean and standard error of the gains for users grouped into younger and older cohorts relative to forty-five years old. **f** A wearable varifocal prototype using a conventional near-eye display (Samsung Gear VR) that is augmented by a gaze tracker and a motor that is capable of adjusting the physical distance between screen and lenses. Figures reproduced from [149]

microlens-based systems [72, 106] or the diffraction limit of dual layer liquid crystal displays (LCDs) [73].

Renewed interest in **holographic near-eye displays** for applications in virtual and augmented reality has emerged. Recently, much progress has been made both on hardware implementations and efficient algorithms. For example, several recent near-eye displays combine a holographic projector with various see-through eyepieces in innovative ways: holographic optical elements [113], waveguides [221], and lenses with beamsplitters [23, 50, 138]. Moreover, algorithms for computer-generated holography have significantly advances at the same time [125, 151, 176]. Although holographic near-eye displays are one of the most promising directions of near-eye display research, they also face significant challenges. Holographic displays may suffer from speckle and have extreme requirements on pixel sizes that are not afforded by near-eye displays also providing a large field of view.

Near-eye display systems that remove the accommodation-dependent change in retinal blur, also known as **Maxwellian-view displays** [96, 101, 209], allow accommodation to remain coupled to the vergence distance of the eyes, and thus allow for accommodating freely in a scene and mitigating the vergence–accommodation conflict. Conceptually, the idea of accommodation invariance can be illustrated by imagining that a user views a display through pinholes—the depth of focus becomes effectively infinite and the eyes see a sharp image no matter where they accommodate.

Vision is one of the primary modes of interaction with which humans understand and navigate the everyday world, so it is natural to ask whether **vision-correcting displays** could also correct for visual aberrations like myopia or hyperopia. In addition to these refractive errors, the aging process is accompanied by a hardening of the eye's crystalline lens; the end result is that by their late 40s or 50s, most people struggle to view objects that are within arm's reach in sharp focus [39]. This reduction in range of accommodation, known as presbyopia, affects more than a billion people [67] and will become more prevalent as the population ages. While several types of eyeglasses and contacts exist to correct myopia, hyperopia, and also presbyopia, corrective eyewear could be directly integrated into AR/VR displays. For example, Padmanaban et al. studied age-related effects of accommodation in VR/AR and showed that varifocal displays drive accommodation in a natural way for non-presbyopes [149]; they also demonstrated vision-correcting capabilities for myopia and hyperopia. Varifocal display technology can also correct for presbyopia in see-through AR systems [18] or, integrated into electronic eyeglasses, for presbyopes viewing the real world [55, 114, 150]. Finally, light field display technology has been demonstrated to enable vision-correction for myopia, hyperopia, and higher order aberrations [75, 153].

25.1.3 Occlusion-Capable Optical See-Through AR Displays

Optical see-through augmented reality (AR) systems are a next-generation computing platform that offer unprecedented user experiences by seamlessly combining physical and digital content. Many of the traditional challenges of these displays have been significantly improved over the last few years, but AR experiences offered by today's systems are far from seamless and perceptually realistic. One of the most important image characteristics for improved seamlessness between digital and physical content is mutually consistent occlusions between physical and digital content in optical see-through augmented reality. When digital content is located in front of physical objects, the former usually appear semi-transparent and unrealistic (Fig. 25.6). To adequately render these objects, the light reflected off of the physical object toward the user has to be blocked by the display before impinging on their retina. This occlusion mechanism needs to be programmable to support dynamic scenes and it needs to be perceptually realistic to be effective. The latter implies that occlusion layers are correctly rendered at the distances of the physical objects, allowing for pixel-precise, or hard-edge, control of the transmitted light rays. In the following, we discuss several recent approaches to enabling mutually consistent occlusions in AR.

Projection-based lighting can be used to control the lighting of a scene in a spatially varying manner. Using such controlled illumination, mutually consistent occlusions, shading effects, and shadows in projector-based AR systems can be synthesized [4, 11, 12, 127]. However, it may not always be possible to use a projector to control the illumination of a physical scene, for example due to device form factor constraints or in the presence of strong ambient illumination.

Commercial AR displays (e.g., Microsoft HoloLens, Magic Leap) often use a neutral density filter placed on the outside of the display module to reduce ambient light uniformly across the entire field of view. An adaptive version of such a **global**

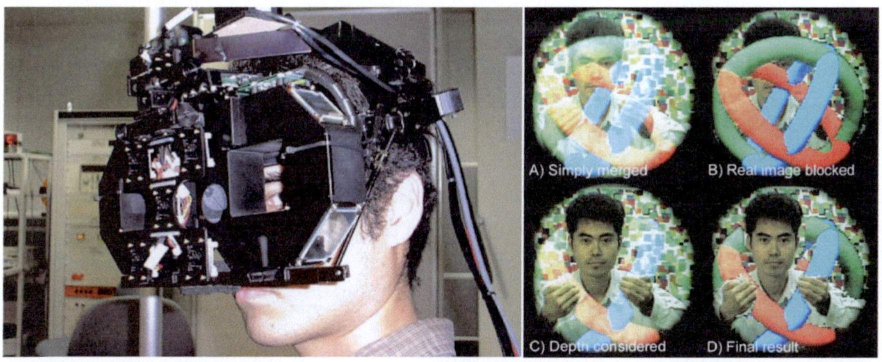

Fig. 25.6 Occlusion-capable optical see-though AR display (left). The display includes relay optics and spatial light modulators that allow for hard-edge per-pixel control of the observed scene before it hits the user's retina. The right panel shows views through the display with **a** no occlusion control, i.e. digital and physical image are simply superimposed, **b** occlusion enabled to block light from the physical scene everywhere where there is digital content, **c** occlusion disabled but depth considered, i.e. physical objects can occlude digital objects but selectively rendering the latter, **d** occlusion enabled and depth considered, i.e. both physical and digital objects can correctly occlude the other one. Figure reproduced from [90] © IEEE 2003

dimming approach was recently proposed by Mori et al. [139], where the amount of dimming is controlled by a single liquid crystal cell and responsive to its physical environment. While these approaches may be useful in some scenarios, they do not provide spatial control of the occlusion layer.

The physical scene can be focused onto an occlusion spatial light modulator (SLM) which selectively blocks its transmission in a spatially varying manner before it reaches the user's eye. Known as **soft-edge** or **hard-edge occlusion**, depending on whether the SLM is in focus or out of focus, this idea was first proposed by the seminal work of Kiyokawa et al. [90–92] (see Fig. 25.6). Improvements of related systems were later demonstrated [16, 17, 48, 49, 53, 69, 80, 124, 161, 211, 216, 228]. Although all of these approaches are successful in providing mutually consistent occlusions in AR systems, miniaturizing these capabilities into a wearable device form factor remains one of the primary challenges of optical see-through AR today.

25.1.4 Optimizing Other Display Characteristics

Spatial AR systems and optical see-through AR display often aim at providing radiometrically consistent, color-corrected or even color-stylized imagery (e.g., [13, 81, 103, 104, 210, 211]). Some of the most important display characteristics that determine how well a digital visual experience could match a physical one are resolution, dynamic range/ brightness, and color. We briefly review computational display strategies to address these display characteristics. A comprehensive survey of these topics can be found in [129, 212].

Examples of **superresolution displays** include optical configurations that combine the contribution of multiple overlapping devices [34] or single devices with either two stacked LCDs [168] or one LCD and a double-lens system [169]. Superresolution display with monitors, as opposed to projectors, can be achieved by fast mechanical motion of the screen [10] or using two stacked LCDs [61, 62]. Finally, Hirsch et al. [65] proposed a light field and HDR projector using stacked spatial light modulators. They used formal optimization to derive optimal pixel states in the display and demonstrate superresolution on a diffuse projection screen rather than a monitor.

High dynamic range displays overcome the limited contrast of LCDs. In their seminal work, Seetzen et al. [173] introduced the concept of dual-layer modulation where a low-resolution LED backlight is modulated by a high-resolution LCD. While the LED array has low resolution, it offers ultra-large dynamic range. An image decomposition algorithm is applied to decompose a target HDR image into the pixel states of the two display layers. This technical approach has become standard practice in industry and is now marketed using the terms "micro dimming" or "local dimming" in consumer products. Extensions to more than two display layers have been discussed [213] and high dynamic range projectors have also been proposed [33]. These typically build on light steering using phase-only spatial light modulators [5], dual layer modulation [65], or adaptive control of the peak brightness over time [19].

A large color gamut can be achieved by **multi-primary displays** [137, 163, 198] that extend the set of reproducible colors by using more than three primaries. Related algorithmic problems include selecting the optimal color primaries [8, 74, 86, 115, 120] as well as gamut mapping (e.g., [7]), where pixels of an image are processed to fit within the fixed gamut provided by a display.

25.2 Tracking Headsets, Controllers, and Hands

One of the core challenges of any AR/VR system is tracking headsets, controllers, and the user's hands. Tracking a headset is crucial if we want to render digital 3D content from the user's perspective and update this perspective in the computer-generated content as they move. While this capability may not be as important for information displays that simply show text, email, or other content that is independent of the physical environment, any virtual or augmented reality experience requires the user's head to be tracked at low latency. Head tracking enables us to move through a virtual environment and always see it from the correct perspective, but it does not necessarily allow us to interact with this environment. For this reason, most commercial VR/AR system offer additional controllers that act as a prop for various digital interfaces in VR/AR. For example, one could imagine overlaying a digital wand, sword, remote control, or another object that the user sees in the virtual environment but that they physically interact with through the controller. Finally, as AR systems become more ubiquitous, controller-free interaction paradigms are desired. The most natural way for humans to interact with their environment is using their hands. Therefore, tracking mechanisms for capturing the precise motion of hands and fingers are a sought-after goal for AR/VR systems.

Tracking a headset or controller is a slightly different problem from tracking hands, because the former are rigid bodies whereas the latter are deformable bodies. And although headset and controller tracking are slightly different problems, technological approaches to both share many similarities. Broadly speaking, headset and controller tracking approaches can be classified as either "outside in" or "inside out". The outside-in approach refers to technologies that use cameras, special light sources, WiFi routers or other infrastructure external to the device for tracking. Imagine an array of cameras spread out on the ceiling or walls of a room, all looking towards the headset or controller. The challenge with this approach is obviously that the area in which tracking is supported is confined to the physical space that is covered by the cameras. The inside-out approach aims to provide a fully integrated system that does not rely on external tracking hardware. A gyroscope, accelerometer, camera, or other sensors are mounted on the headset that track its position and orientation relative to the world around them are examples of inside-out systems. As it is not confined in space, this is certainly the preferred approach, but the camera-based inside-out approach can be computationally costly, power hungry, and introduce additional latency. The next two sections outline a number of outside-in and inside-out tracking technologies, respectively.

What makes rigid body tracking easier than tracking deformable objects, like hands, is that rigid bodies only have six degrees of freedom: position and orientation in 3D space. Together, position and orientation are typically referred to as *pose* in the computer vision community and there are 3-DOF (i.e., either position or orientation) or 6-DOF (i.e., both position and orientation) pose tracking systems that track either three or all six degrees of freedom. A deformable object, such as a hand, can have many more degrees of freedom, for example one may be interested in 6-DOF pose of the hand as well as the joint angles of the fingers. We review approaches to hand tracking in Sect. 25.2.3.

25.2.1 Outside-In Pose Tracking

Pose tracking can be implemented with a variety of technologies. Commercially-available systems include mechanical trackers, magnetic trackers, ultrasonic trackers, and GPS or WiFi-based tracking [132]. The most widely used technology, however, is optical pose tracking. In the following, we briefly review these technological approaches and discuss their relative benefits and disadvantages.

Mechanical tracking is probably the most intuitive tracking approach. Mechanical linkages are physically attached to a headset or controller and encoders read out the joint angles between linkages. Examples of mechanical trackers in VR include the system used in Ivan Sutherland's "Ultimate Display" [193], which is widely recognized as the first head-mounted display, and also Fakespace Labs' BOOM (Binocular Omni-Orientation Monitor). The advantages of mechanical tracking are low latency and high accuracy while disadvantages include the limited range of user motion and device form factors.

Ultra-sonic tracking can be offered by light-wight, small, and inexpensive transducer systems. For example, one or more transmitters can be mounted on the headset or controller and emit pulses that are being triangulated by three or more receivers based on their relative time of flight of the pulses. These types of trackers are susceptible to acoustic interference, they may have low update rates, and they are subject to line-of-sight constraints, i.e. the direct line of sight between transmitter and received should not be blocked. In addition to a mechanical tracker, Ivan Sutherland's "Ultimate Display" for example also used an ultra-sonic tracker [193].

Magnetic tracking has been widely used for pose tracking throughout the last few decades. These systems typically comprise a magnetic field generator along with one or more magnetometer sensors. The generator is usually capable of re-orienting the magnetic poles of the field along the three axis in rapid succession. Synchronizing the oscillating magnetic field with the magnetometers allows the sensors to estimate their own position and orientation within the field. An example of a magnetic tracking system is Polhemus' Fastrak. Generally, magnetic tracking offers reasonably good accuracy and latency for 6-DOF pose tracking, small and low cost sensors, and it does not suffer from line-of-sight constraints. However, the strength of the magnetic field determines the working volume, which is typically somewhat limited and the system is susceptible to distortions of the magnetic field, for example caused by electronic devices or metal. Wireless magnetic trackers exist but the clocks of sensors and magnetic field generator need to be synchronized.

Although magnetic trackers have been largely replaced by optical trackers in many VR/AR applications, they have found a niche in tracking controllers. For example, the Magic Leap ML1 controller uses a miniaturized magnetic field generator with the magnetometer sensors being integrated in the headset [78]. This would allow the controller to be tracked relative to the headset without being subject to line-of-sight constraints, which is crucial when the direct line of sight between controller and headset is blocked, for example by the user's body.

Optical tracking is probably the most widely used tracking system for headsets today and several different flavors of optical tracking are available. In a multi-camera configuration, several calibrated cameras observe the target from different perspectives. After being detected in the individual 2D images, the target is triangulated from several cameras to estimate its pose. This approach is similar to motion capture systems in the visual effects industry, where actors wear suits with retroreflective markers or light emitting diodes (LEDs) that are being tracked by a set of cameras. In a single-camera configuration, multiple markers with a known relative orientation to each other are tracked in the 2D image of a single camera. The inverse problem of estimating the 6-DOF pose of a target from the 2D projections of a set of markers is known as the *perspective-n-point problem*, which can be solved efficiently as long as at least four markers are visible to the camera [85, 162].

Many commercial VR/AR systems use optical tracking with infrared LEDs or actively illuminated retroreflective markers mounted on a VR controller or a headset (e.g., Oculus Rift and Sony's Playstation VR headset; see Fig. 25.7). The system ships with an additional camera that estimates the 6-DOF pose from the measured 2D locations of the LEDs or markers. The arrangement of the markers on the tracked device is usually known from its design or calibrated by the manufacturer. Typically, these systems do not solely rely on optical data but use a sensor fusion approach between the optical data and inertial measurement units.

Fig. 25.7 Examples of optical tracking in VR. Left: near-infrared (NIR) LEDs of the Oculus Rift recorded with a camera that is sensitive to NIR (image reproduced from ifixit.com). Center: HTC Vive headset and controllers with exposed photodiodes (image reproduced from roadtovr.com). Right: disassembled HTC Lighthouse base station showing two rotating drums that create horizontal and vertical laser sweeps as well as several LEDs that emit the sync pulse (image reproduced from roadtovr.com)

The HTC Vive also uses an optical tracking system, but rather than using a camera to observe LEDs on the headset, the Vive uses a slightly different approach where the camera is replaced by a projector and instead of LEDs, photodiodes are mounted on the device. The projector emits structured illumination to help the photodiodes determine their own 2D location in the reference frame of the projector. An early paper on this technology was published by Raskar et al. [159], who used spatially-structured illumination. HTC calls their technology Lighthouse and it uses temporally-structured illumination. Specifically, the Lighthouse projector or base station sweeps horizontal and vertical laser stripes across the room (hence the name Lighthouse). It does that very fast—60 times per second for a full horizontal and vertical sweep with sync pulses in between. The photodiodes are fast enough to time stamp when the laser sweeps hit them relative to the last sync pulse. Using these measurements, one of several optimization techniques can be employed to estimate the 6-DOF pose of the tracked device with respect to the base station. Sensor fusion between an inertial measurement unit (i.e., gyroscopes and accelerometers) and the photodiodes allow for low-latency tracking.

Other tracking approaches include the *global position system (GPS)* or *WiFi positioning*, but these systems are not as common in VR/AR.

25.2.2 Inside-Out Pose Tracking

Inside-out tracking systems represent the ideal-case scenario for VR/AR because they enable a device, such as a headset or controller, to track itself within a physical environment without the need for external tracking infrastructure. The most intuitive inside-out systems are inertial measurement units (IMUs) that typically include gyroscopes and accelerometers, often also magnetometers. IMUs are microelectromechanical systems (MEMS) that offer low cost, extremely low latency (up to thousands of samples per second), they require little power, and they are readily available, for example in all cellphones. Although gyroscopes suffer from drift and accelerometers from noise, sensor fusion between these sensors enables robust orientation tracking with low latency in VR/AR [108]. Unfortunately, positional tracking is not easily possible with IMUs alone, because of two challenges. First, separating the effects of gravity and linear acceleration from potentially noisy accelerometer measurements is very difficult. Second, the double integration required to turn linear acceleration measured by an accelerometer into relative position introduces integration errors and thus drift. Even with high sampling rates, i.e. small time steps, the estimated position will quickly diverge from the true position.

One of the most popular inside-out tracking approaches that does not suffer from these limitations is optical tracking, where the images of a single or multiple moving cameras are analyzed to determine the relative 6-DOF pose of the camera. This is known as simultaneous localization and mapping (SLAM) [42] or visual odometry [146] in the literature. Generally speaking, an image is analyzed to find 2D features that can be tracked in subsequent frames of a video, these features are matched between successive frames, and an inverse problem known as bundle adjustment then determines the relative poses of the cameras in each frame. This is now the standard approach for tracking standalone headsets, like Microsoft Hololens, Magic Leap ML1, or Oculus Quest. All of these systems likely use a hybrid, visual-inertial odometry approach, which uses sensor fusion between camera data and an IMU for 6-DOF tracking. While the benefits of camera-based inside-out tracking are clear, this is a computationally challenging problem that requires substantial computational resources, which in turn creates challenges in thermal management, power consumption, and latency.

25.2.3 Hand Tracking

Early VR/AR systems supporting hand tracking typically used data gloves. A data glove is a wired or wireless electronic glove-like device that the user wears [190]. Conventional outside-in or inside-out tracking systems, as discussed above, can be used to track the 6-DOF pose of the hand. Additional flex sensors also track the amount of finger bending. Commercial flex sensors use fiber optics, conductive ink, or capacitance to estimate the amount of bending. Examples of commercial data gloves include Nintendo's Power Glove, Virtual Technology's CyberGlove, or VPL's DataGlove. Gloves are less common in modern hand tracking systems, which often use camera-based input along with optimization or neural network-based algorithms that fit some parametric representation, for example a skeleton, of a hand to the input data [111, 141, 183, 184, 205]. Some of the most successful commercial products offering hand tracking for VR/AR are Leap Motion and Microsoft's Hololens, which includes an integrated gesture recognition system.

25.3 Cinematic VR

Capturing the real world for rendering in virtual reality [164] is closely related to work on 3D reconstruction [30] and particularly image-based rendering (IBR) [37, 179]. These are active, long-running fields of research that have produced a large variety or techniques and systems towards the goal of capturing the real world in all its visual fidelity. Many of the proposed systems share a similar structure, which is embodied by the *VR capture pipeline*:

Capture → Reconstruction → Represenation → Compression → Rendering

The goal of this section is to look at each stage of this pipeline, and to provide an overview of the range of techniques used by existing VR capture approaches as well as their trade-offs. For any particular approach or system, the most important design choice is the data representation to be used, as this constrains many of the other pipeline stages, in particular reconstruction, compression and rendering.

25.3.1 Capture

Most virtual reality capture approaches rely on one or more color cameras to capture the visual appearance and dynamics of a scene. Sometimes, special cameras are used, such as RGBD cameras which capture a depth maps in addition to color footage, or special attachments like mirrors.

One static camera can only capture a limited view of a larger scene due to its limited field of view. The content captured in this fashion can still be compelling, as demonstrated by Facebook's 3D photos [98], which are captured by dual-lens cameras on commodity mobile phones to provide depth in addition to color. However, wider views require wider camera optics, such as fisheye lenses or catadioptic systems [1] for omnidirectional video.

One moving camera can capture a more complete picture of a scene by sweeping it over time. Traditional panorama stitching approaches [14, 197] assume a camera that rotates on the spot, so that it essentially captures all light rays converging at a single point in space, the center of the panorama. By moving the camera in space, even more light rays can be captured, for instance for omnidirectional stereo [9, 156, 165], layered depth panoramas [225] or 3D photography [57]. Towards the extreme end, a camera can also be moved along the surface of a plane or sphere, to capture a more complete light field [122, 136, 148].

One moving RGBD camera makes it easier to reconstruct the geometry of the scene from the captured depth maps. A pioneer in this category is KinectFusion [143], which reconstructs a global truncated signed distance field representation of a scene from registered input depth maps alone. There are many more recent variants that improve on the scale and robustness of this kind of scene reconstruction [32, 145, 215]. Instant 3D photography [58] aligns multiple RGBD images captured with a dual-lens camera into a consistent, textured 3D panoramic surface.

Multi-camera rigs are required for video capture and to capture multiple viewing directions simultaneously. Consumer 360° cameras are now commercially available as commodity devices that stitch two or more video streams into one 360° video [109, 158, 208]. Stereo cameras capture two viewpoints side by side, and their baseline can be magnified in post [226]. Multiple viewpoints can also be interpolated and manipulated in a post-process after video capture [116]. A ring of video cameras captures sufficient information for compelling omnidirectional stereo video [3, 43, 172], while a rotating camera rig can even capture live omnidirectional stereo video [95] (see Fig. 25.8). Light fields [44, 51, 112] are based on a dense sampling of viewpoints, which requires many co-located cameras. A different camera setup distributes cameras on a dome or around a capture volume, for example to capture objects and people in a light stage [36] or as volumetric video [25].

Fig. 25.8 SpinVR: a rotating camera with two line sensors directly captures omnidirectional stereo panoramas

25.3.2 Reconstruction

Reconstruction is all about interpreting and combining the information contained in the captured imagery and depth maps, if available.

The first step is often camera calibration and structure from motion, i.e., characterizing the imaging devices used, including their lens distortion, and determining which views of a scene they captured. Multiple structure-from-motion implementations are publicly available, including Bundler [182], VisualSFM [218], AliceVision [82, 140], MVE [46], Theia [195] and COLMAP [170], with the latter currently enjoying the widest use. However, general-purpose structure-from-motion tools do not perform well for the kind of inside-out capture commonly used for environment capture [9, 57]. This has led to the development of specifically tailored structure-from-motion solutions that assume camera motion on a spherical surface [196, 203], which is a good match for handheld [9, 57, 165] or spherical [122, 148] capture approaches. One of the outputs of structure from motion is also a sparse 3D point cloud of feature points in the scene, which can be useful for image alignment [109] or view warping [76].

Once the viewpoints are reconstructed, the next step is generally to combine all the captured information into a single model of the scene. In classical panorama stitching, this is achieved by aligning and blending the individual input views on a spherical or cylindrical image surface [14, 197]. While still panoramas can hide alignment artifact to some degree using clever blending approaches [57, 223, 224], this becomes much harder for panoramic videos, as the visual content, and hence any artifacts, keep changing over time. To address this, the stitching needs to vary over time in accordance with the scene [109, 158, 208]. To achieve more complex projections, such as the multi-perspective omnidirectional stereo (ODS) projection [79, 156], requires dense correspondence between input views so that intermediate views can be synthesized [3, 15, 165, 172]. Most approaches use optical flow for this purpose, as it provides useful flexibility in case of calibration errors or scene motion.

The reconstruction of 3D geometry goes beyond the purely image-based approaches discussed before by recovering the 3D structure of a scene or object. Most approaches start by estimating per-view depth maps using multi-view stereo

(MVS) techniques [47, 77, 171, 174], unless depth maps are directly available from RGBD cameras. In theory, these per-view depth maps can be integrated into a global geometry model of the scene [22, 57, 175] if the camera poses and depth maps are estimated sufficiently accurately. Approaches such as KinectFusion [143] and BundleFusion [32] integrate noisy depth maps over time to improve the accuracy of the surface reconstruction. Having a large number of views also leads to a cleaner geometry reconstruction [25]. Hedman et al. [58] introduce a locally varying depth map alignment step to integrate differently normalized depth maps from mobile phones into a globally consistent depth map. However, because of calibration and depth estimation errors, better view synthesis results can often be obtained with per-view geometry [21, 60, 148] that is smoothly blended across the synthesized novel view.

25.3.3 Representation

Over the years, various approaches have been proposed for representing captured scenes or objects. To provide an overview, Shum et al. [178, 179] organized approaches along a continuum according to how much geometry is being used: no geometry, implicit geometry or explicit geometry. 'No geometry' refers to purely image-based approaches, such as panoramas or 360° video. 'Implicit' geometry comprises approaches using posed images and/or relying on 2D image correspondences, such as optical flow. And 'explicit' geometry includes textured meshes with actual 3D geometry. Figure 25.9 contains an updated version of Shum et al.'s continuum of representations.

There is no universally best representation—all have their advantages and disadvantages and provide different trade-offs. There is also often no hard boundary between representations, so there is some overlap and hybrids are possible. In the limit, i.e., with infinite resolution, the representations are theoretically interchangeable. However, any conversion always requires resampling, which is an inherently lossy process that reduces overall accuracy and fidelity. There are usually also practical limits, for example the physical size of cameras that limits the maximum camera density achievable for light field approaches.

Images and Panoramas provide the most basic snapshot of what a scene or object looked like. They represent a photographic likeness that captures visual appearance

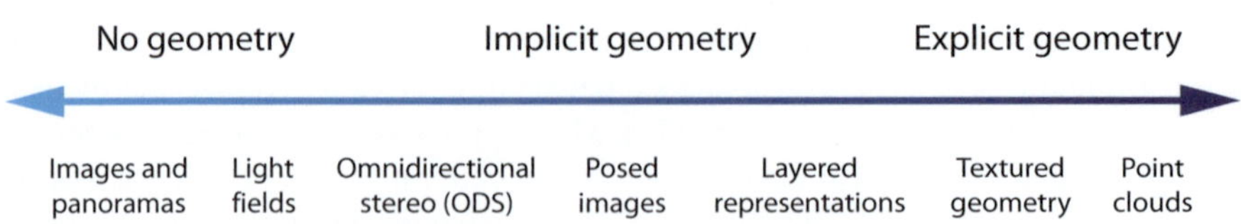

Fig. 25.9 The continuum of image-based rendering representations, inspired by Shum et al. [178, 179]

of a scene or object from a single point of view with a fixed field of view. Panoramas [14, 197] and 360 videos [109, 158] capture a wide or even complete field of view. Images and panoramas enjoy great popularity as they are easy to capture with modern mobile phones and straightforward to share. However, their main limitation is that they only provide information for a single point of view and no depth perception, and thus do not support any transnational change of viewpoint.

Light Fields represent a dense spatio-angular sampling of a scene [112], generally using a regular 2D grid of camera viewpoints. More general camera configurations are supported by the Lumigraph [51], a closely related variant of light fields. As the comprehensive coverage of an object in a scene is challenging to obtain in practice, Davis et al. [35] proposed a guidance approach that helps users in capturing missing viewpoints. Videos captured with a moving camera can also be considered to be a densely light field along the camera path, which can be exploited for particularly accurate scene reconstruction [87, 220].

Omnidirectional Stereo (ODS) is a multi-perspective, circular projection [79, 156] that has become a popular medium for stereoscopic and 360° VR photos and videos [3, 15, 165, 172]. ODS encodes two panoramic view, one for the left eye and one for the right eye. This has the advantage that there is binocular disparity, and hence a feeling of depth, in all viewing directions along the equation. Furthermore, the format is an excellent fit for existing video processing, compression and transmission pipelines, as both views are encoded in a single top-bottom configuration.

Posed Images have known camera geometry (camera position and orientation) in addition to the image data. This enables scene reconstruction in the form of point clouds using multi-view stereo. Even sparse point clouds are sufficient for and an overview of community photo collections as demonstrated by the seminal Photo-Tourism work by Snavely et al. [182]. Correspondences between adjacent viewpoints can be used for interpolating novel views from existing ones. Novel views can be interpolated from existing ones by establishing correspondences between adjacent viewpoints. In practice, optical flow is often used for flow-based blending [9, 122, 165], which significantly reduces blurry ghosting artifacts and produces results with high visual fidelity.

Layered Representations consist of multiple semi-transparent layers that encapsulate the appearance of a scene or object without any explicit geometry. The underlying core idea goes back to Disney's multi-plane camera (1937), in which multiple transparent cel sheets are positioned at different depths from the camera. This allows each cel sheet to be moved independently and creates the effect of motion parallax over time. Early approaches by Wetzstein et al. computed layered representations using custom-tailored optimization frameworks [213, 214].

Textured Geometry makes it easy to render novel views in real time with existing 3D graphics pipelines, even on mobile devices. Mesh geometry is good at modeling hard occlusion boundaries, but it needs to be reconstructed from depth maps. For the highest quality depth maps, many observations from different viewpoints need to be combined, for example for volumetric video [25] or Google's light fields [148]. One consumer-facing example are Facebook's 3D Photos [98], which are based on an image and lower-resolution depth map from an off-the-shelf mobile phone. The final 3D photo can be looked at from different directions by tilting the phone. Several approaches separate foreground and background objects in a scene into multiple textured layers [57, 58, 175, 225], to preserve clean occlusion boundaries. This generally requires some kind of inpainting to fill the areas behind foreground objects. In the real world, the appearance of objects also often depends on the viewing direction, e.g. when objects are shiny. This effect can be modeled using surface light fields [217] or view-dependent blending [59]. In general, modeling and editing favors geometric approaches, as there are better software tools available for textured meshes.

Point Clouds represent a scene as an unordered collection of points, which may or may not have colors and/or surface normals. They are readily obtained from structure-from-motion and multi-view stereo tools, or Lidar scans of a scenes. However, they are inherently sparse, tend to be noisy and non-uniformly distributed, and contain gaps that make them impractical for render high-quality novel views (although this is slowly changing thanks to neural re-rendering [135]). Nevertheless, they are often useful intermediate representation or debugging tool.

Neural Scene Representations have emerged as a new representation that is coupled with machine learning. The idea behind these algorithms is similar to classical approaches: given a set of input views, distill these into an intermediate representation, and then render the scene from novel viewpoints using the intermediate representation. However, a neural representation differs from a classical scene representation in being differentiable with respect to its parameters. In combination with a differentiable renderer that takes the neural scene representation as well as a camera position and orientation, i.e. a pose, as input and computes a 2D image from the camera's perspective, neural scene representations allow for end-to-end optimization of the representation supervised only on the images.

Several different types of neural scene representations have been proposed. For example, building on differentiable proxy geometries and neural textures [88, 200, 227], multiplane image representations [44, 136, 157, 185, 201, 226], voxel grids [119, 144, 180, 207], point clouds [128, 222], or continuous differentiable functions [154, 181].

25.3.4 Compression

Raw scene representations can become very large. This can make it difficult to store them given limited space on disk or in memory, to transmit over networks in a reasonable time, or even to render them in real time. Compression and decompression are therefore indispensable for practical scene capture and rendering systems.

The light fields introduced by Levoy and Hanrahan [112] in 1996 were up to 1.6 GB in size. This would easily fill a large hard drive at the time, but thus would not fit in memory. However, light fields are highly redundant within images and between images, so they are highly compressible. Levoy and Hanrahan designed a custom light-field compression scheme that combines vector quantization of 2D or 4D tiles (24:1 compression) with gzip entropy encoding (another 5:1 compression) for a total compression of 120:1. This scheme allowed fast random-access decompression entirely in software, so that real-time rendering became feasible.

Recently, image compression techniques such as JPEG have become computationally affordable, even in real-time applications. Existing video codecs, such as h.264 and h.265, can also often be used directly for compressing video-based representations, such as 360° video [109, 158] or omnidirectional stereo videos [3, 172].

Collet et al. [25] encode their volumetric free-viewpoint videos in a standard MPEG-DASH file. Thanks to mesh tracking, their geometry has a temporally consistent parameterization. Therefore, the resulting texture atlases are unwrapped consistently and can be compressed effectively using the standard h.264 video codec. The mesh geometry is encoded as a custom unit inside the video stream and compressed using linear motion prediction, 16-bit quantization of vertex positions and UV coordinates, and Golomb coding.

Google's panoramic light fields require 46 GB of image data each [148] and thus also need significant compression. Like for the original light fields [112], fast random access is required for rendering novel views of the light field. Overbeck et al. [148] build on the open-source VP9 codec and encode most light-field images relative to a sparse set of reference views, which are like key frames in standard videos. In practice, they decode all reference images when loading the light field from disk and keep them in memory. They also contribute an extension to VP9 that enables random access to individual image tiles. This allows them to decode any tile from any other image immediately. Most light fields can be compressed at high quality by $40\times$–$200\times$.

25.3.5 Rendering

The final step of the VR capture pipeline is to render the novel views corresponding to the user's location, so that they see the right views of the captured scene as they move. Most rendering approaches adopt the standard graphics pipeline, which has the benefit of efficient hardware implementations across a large range of devices, from mobile to desktop setups. This efficient rendering hardware enables rendering in real time, and even hitting the high frame rates of 80–120 Hz required to feed state-of-the-art VR head-mounted displays [99].

Panoramas and omnidirectional stereo content only require a change of projection, to perspective projection, to be viewed by users. This does not require any explicit geometry and can be implemented in 2D, or, equivalently, using textured spheres viewed from virtual perspective cameras. Many other approaches also use textured geometry directly [32, 57, 58, 143, 175]. Even multi-plane images [44, 136, 185, 226] can be rendered using textured geometry, by texturing the semi-transparent layers on parallel planes that are appropriately spaced, and using alpha compositing in the z-buffer during rendering.

Modern graphics pipelines are also programmable using shaders, which provides an opportunity to influence the rendering more locally depending on the viewing direction, for example. Flow-based blending has been used to interpolate novel views on the fly [122] and per pixel or light ray [165], also in a view-dependent fashion [9]. When many input views are combined to synthesize novel views, they also require spatial blending to ensure smooth transitions [148]. Ultimately, the decision of how to blend multiple observations of a single surface point can even be optimized using a deep neural network [59]. However, evaluating the neural network per frame at run time noticeably impacts the overall frame rate that is achievable with this approach.

25.4 Conclusion

In summary, wearable computing systems, such as virtual and augmented reality displays, have made significant progress over the last few years. By exploiting the specific characteristics of human vision, researchers have developed techniques that overcome many previous challenges for these types of displays. For example, foveated rendering and display minimize computational resources and bandwidth requirements whereas varifocal display modes can support focus cues, improving perceptual realism and visual comfort. However, all of these and also many other techniques rely on eye tracking. To date, accurate and low-latency eye tracking remains out of reach with consumer AR/VR systems. In contrast to the developments on the research side, commercial head-mounted displays often provide only limited resolution and field of view, they do not support focus cues or mutual occlusion in optical see-through AR settings, and they are restricted in many other ways.

To bridge the gap between research on AR/VR and the engineering reality of wearable computing, low-power and low-latency application-specific integrated circuits are necessary for these tasks. Microsoft's Hololens and Magic Leaps ML1 pave the way for dedicated processing with wearable computing, supporting inside-out tracking, real-time rendering, and many other capabilities for untethered headsets. However, eye tracking needs to improve and these accelerators should become easier to re-purpose so that the research community can help push the frontier of VR/AR systems.

References

1. R. Aggarwal, A. Vohra, A.M. Namboodiri, Panoramic stereo videos with a single camera, in *Proceedings of the International Conference on Computer Vision and Pattern Recognition (CVPR)*, June 2016, pp. 3755–3763
2. K. Akeley, S. Watt, A. Girshick, M. Banks, A stereo display prototype with multiple focal distances. ACM Trans. Graph. (SIGGRAPH) **23**(3), 804–813 (2004)
3. R. Anderson, D. Gallup, J.T. Barron, J. Kont-kanen, N. Snavely, C. Hernandez, S. Agarwal, S.M. Seitz, Jump: virtual reality video. ACM Trans. Graph. (Proc. SIGGRAPH Asia) **35**(6),

198:1–13 (2016)
4. G. Avveduto, F. Tecchia, H. Fuchs, Real-world occlusion in optical see-through ar displays, in *Proceedings of the 23rd ACM Symposium on Virtual Reality Software and Technology* (ACM, 2017), p. 29
5. A. Ballestad, R. Boitard, G. Damberg, G. Stojmenovik, Advances in HDR display technology for cinema applications, including light steering projection. Inf. Disp. **35**(3), 16–19 (2019)
6. M.S. Banks, D.M. Hoffman, J. Kim, G. Wetzstein, 3d displays. Annu. Rev. Vis. Sci. **2**(1), 397–435 (2016)
7. F. Banterle, A. Artusi, T.O. Aydin, P. Didyk, E. Eisemann, D. Gutierrez, R. Mantiuk, K. Myszkowski, Multidimensional image retargeting, in *SIGGRAPH Asia 2011 Courses* (ACM, 2011), p. 15
8. M. Ben-Chorin, D. Eliav, Multi-primary design of spectrally accurate displays. J. Soc. Inf. Disp. **15**(9), 667–677 (2007)
9. T. Bertel, N.D.F. Campbell, C. Richardt, MegaParallax: casual 360° panoramas with motion parallax. IEEE Trans. Vis. Comput. Graph. **25**(5), 1828–1835 (2019)
10. F. Berthouzoz, R. Fattal, Resolution enhancement by vibrating displays. ACM Trans. Graph. (TOG) **31**(2), 15 (2012)
11. O. Bimber, B. Fröhlich, Occlusion shadows: Using projected light to generate realistic occlusion effects for view-dependent optical see-through displays, in *Proceedings of IEEE ISMAR* (2002)
12. O. Bimber, A. Grundhöfer, G. Wetzstein, S. Knödel, Consistent illumination within optical see-through augmented environments, in *Proceedings of IEEE ISMAR* (2003), pp. 198–207
13. O. Bimber, D. Iwai, G. Wetzstein, A. Grundhoefer, The visual computing of projector-camera systems, in *Computer Graphics Forum* (2008)
14. M. Brown, D.G. Lowe, Automatic panoramic image stitching using invariant features. Int. J. Comput. Vis. **74**(1), 59–73 (2007)
15. B. Cabral, VR capture: designing and building an open source 3D-360 video camera, in *SIGGRAPH Asia Keynote*, December 2016
16. O. Cakmakci, Y. Ha, J. Rolland, Design of a compact optical see-through head-worn display with mutual occlusion capability, in *Proceedings of SPIE*, vol. 5875 (2005)
17. O. Cakmakci, Y. Ha, J.P. Rolland, A compact optical see-through head-worn display with occlusion support, in *Proceedings of IEEE ISMAR* (2004), pp. 16–25
18. P. Chakravarthula, D. Dunn, K. AkÅŸit, H. Fuchs, Focusar: auto-focus augmented reality eyeglasses for both real world and virtual imagery. IEEE Trans. Vis. Comput. Graph. **24**(11), 2906–2916 (2018)
19. J.-H.R. Chang, B.V.K.V. Kumar, A.C. Sankaranarayanan, 216 shades of gray: high bit-depth projection using light intensity control. Opt. Express **24**(24), 27937–27950 (2016)
20. J.-H.R. Chang, B.V.K.V. Kumar, A.C. Sankaranarayanan, Towards multifocal displays with dense focal stacks. ACM Trans. Graph. (SIGGRAPH Asia) **37**(6), 198:1–198:13 (2018)
21. G. Chaurasia, S. Duchene, O. Sorkine-Hornung, G. Drettakis, Depth synthesis and local warps for plausible image-based navigation. ACM Trans. Graph. **32**(3):30, 1–12 (2013)
22. G. Chaurasia, O. Sorkine-Hornung, G. Drettakis, Silhouette-aware warping for image-based rendering, in *Computer Graphics Forum (Proceedings of Eurographics Symposium on Rendering)*, vol. 30, no. 4, June 2011, pp. 1223–1232
23. J.-S. Chen, D.P. Chu, Improved layer-based method for rapid hologram generation and real-time interactive holographic display applications. Opt. Express **23**(14), 18143–18155 (2015)
24. S.A. Cholewiak, G.D. Love, P.P. Srinivasan, R. Ng, M.S. Banks, Chromablur: rendering chromatic eye aberration improves accommodation and realism. ACM Trans. Graph. (SIGGRAPH Asia) **36**(6), 210:1–210:12 (2017)
25. A. Collet, M. Chuang, P. Sweeney, D. Gillett, D. Evseev, D. Calabrese, H. Hoppe, A. Kirk, S. Sullivan, High-quality streamable free-viewpoint video. ACM Trans. Graph. (Proc. SIGGRAPH) **34**(4), 69:1–13 (2015)
26. N. Corporation. VRWorks—Lens Matched Shading (2016)
27. N. Corporation. VRWorks—Multi-Res Shading (2016)

28. C.A. Curcio, K.A. Allen, Topography of ganglion cells in human retina. J. Comp. Neurol. **300**(1), 5–25 (1990)
29. C.A. Curcio, K.R. Sloan, R.E. Kalina, A.E. Hendrickson, Human photoreceptor topography. J. Comp. Neurol. **292**(4), 497–523 (1990)
30. B. Curless, S. Seitz, J.-Y. Bouguet, P. Debevec, M. Levoy, S.K. Nayar, 3D photography, in *SIGGRAPH Courses* (2000)
31. J. Cutting, P. Vishton, Perceiving layout and knowing distances: the interaction, relative potency, and contextual use of different information about depth, in *Perception of Space and Motion*, Chap. 3, ed. by W. Epstein, S. Rogers (Academic Press, 1995), pp. 69–117
32. A. Dai, M. Nießner, M. Zollhofer, S. Izadi, C. Theobalt, BundleFusion: Real-time globally consistent 3D reconstruction using on-the-fly surface reintegration. ACM Trans. Graph. **36**(3), 24:1–18 (2017)
33. G. Damberg, H. Seetzen, G. Ward, W. Heidrich, L. Whitehead, High dynamic range projection systems, in *SID Symposium Digest of Technical Papers* (2007), pp. 4–7
34. N. Damera-Venkata, N.L. Chang, Display supersampling. ACM Trans. Graph. (TOG) **28**(1), 9 (2009)
35. A. Davis, M. Levoy, F. Durand, Unstructured light fields, in *Computer Graphics Forum (Proceedings of Eurographics)*, vol. 31, no. 2, May 2012, pp. 305–314
36. P. Debevec, The light stages and their applications to photoreal digital actors, in *SIGGRAPH Asia Technical Briefs* (2012)
37. P. Debevec, C. Bregler, M.F. Cohen, L. McMillan, F. Sillion, R. Szeliski, Image-based modeling, rendering, and lighting, in *SIGGRAPH Courses* (2000)
38. E. Dolgoff, Real-depth imaging: a new 3D imaging technology with inexpensive direct-view (no glasses) video and other applications, in *Proceedings of SPIE*, vol. 3012 (1997), pp. 282–288
39. A. Duane, Normal values of the accommodation at all ages. J. Am. Med. Assoc. **59**(12), 1010–1013 (1912)
40. A.T. Duchowski, D.H. House, J. Gestring, R.I. Wang, K. Krejtz, I. Krejtz, R. Mantiuk, B. Bazyluk, Reducing visual discomfort of 3d stereoscopic displays with gaze-contingent depth-of-field, in *Proceedings of the ACM Symposium on Applied Perception* (ACM, 2014), pp. 39–46
41. D. Dunn, C. Tippets, K. Torell, P. Kellnhofer, K. Aksit, P. Didyk, K. Myszkowski, D. Luebke, H. Fuchs, Wide field of view varifocal near-eye display using see-through deformable membrane mirrors. IEEE TVCG **23**(4), 1322–1331 (2017)
42. H. Durrant-Whyte, T. Bailey, Simultaneous localization and mapping: part i. IEEE Robot. Autom. Mag. **13**(2), 99–110 (2006)
43. Facebook, Filming the future with RED and Facebook 360, Sept 2018
44. J. Flynn, M. Broxton, P. Debevec, M. DuVall, G. Fyffe, R. Overbeck, N. Snavely, R. Tucker, DeepView: view synthesis with learned gradient descent, in *Proceedings of the International Conference on Computer Vision and Pattern Recognition (CVPR)*, June 2019, pp. 2367–2376
45. S. Friston, T. Ritschel, A. Steed, Perceptual rasterization for head-mounted display image synthesis. *ACM Trans. Graph. (Proc. SIGGRAPH 2019)* **38**(4), 1–14 (2019)
46. S. Fuhrmann, F. Langguth, M. Goesele, MVE: a multi-view reconstruction environment, in *Proceedings of the Eurographics Workshop on Graphics and Cultural Heritage* (2014), pp. 11–18
47. S. Galliani, K. Lasinger, K. Schindler, Massively parallel multiview stereopsis by surface normal diffusion, in *Proceedings of the International Conference on Computer Vision (ICCV)*, Dec 2015, pp. 873–881
48. C. Gao, Y. Lin, H. Hua, Occlusion capable optical see-through head-mounted display using freeform optics, in *Proceedings of IEEE ISMAR* (2012), pp. 281–282
49. C. Gao, Y. Lin, H. Hua, Optical see-through head-mounted display with occlusion capability, in *Proceedings of SPIE*, vol. 8735 (2013)
50. Q. Gao, J. Liu, J. Han, X. Li, Monocular 3d see-through head-mounted display via complex amplitude modulation. Opt. Express **24**(15), 17372–17383 (2016)

51. S.J. Gortler, R. Grzeszczuk, R. Szeliski, M.F. Cohen, The lumigraph, in *Proceedings of the Annual Conference on Computer Graphics and Interactive Techniques (SIGGRAPH)*, Aug 1996, pp. 43–54
52. B. Guenter, M. Finch, S. Drucker, D. Tan, J. Snyder, Foveated 3d graphics. ACM Trans. Graph. (TOG) **31**(6), 164 (2012)
53. T. Hamasaki, Y. Itoh, Varifocal occlusion for optical see-through head-mounted displays using a slide occlusion mask. IEEE TVCG **25**(5), 1961–1969 (2019)
54. T. Hansen, L. Pracejus, K.R. Gegenfurtner, Color perception in the intermediate periphery of the visual field. J. Vis. **9**(4), 26 (2009)
55. N. Hasan, A. Banerjee, H. Kim, C.H. Mastrangelo, Tunable-focus lens for adaptive eyeglasses. Opt. Express **25**(2), 1221–1233 (2017)
56. A. Hasnain, P.-Y. Laffont, S.B.A. Jalil, K. Buyukburc, P.-Y. Guillemet, S. Wirajaya, L. Khoo, T. Deng, J.-C. Bazin, Piezo-actuated varifocal head-mounted displays for virtual and augmented reality, vol. 10942 (2019)
57. P. Hedman, S. Alsisan, R. Szeliski, J. Kopf, Casual 3D photography. ACM Trans. Graph. (Proc. SIGGRAPH Asia) **36**(6), 234:1–15 (2017)
58. P. Hedman, J. Kopf, Instant 3D photography. ACM Trans. Graph. (Proc. SIGGRAPH) **37**(4), 101:1–12 (2018)
59. P. Hedman, J. Philip, T. Price, J.-M. Frahm, G. Drettakis, Deep blending for free-viewpoint image-based rendering. ACM Trans. Graph. (Proc. SIGGRAPH Asia) **37**(6), 257:1–15 (2018)
60. P. Hedman, T. Ritschel, G. Drettakis, G. Brostow, Scalable inside-out image-based rendering. ACM Trans. Graph. (Proc. SIGGRAPH Asia) **35**(6), 231:1–11 (2016)
61. F. Heide, J. Gregson, G. Wetzstein, R. Raskar, W. Heidrich, Compressive multi-mode superresolution display. Opt. Express **22**(12), 14981–14992 (2014)
62. F. Heide, D. Lanman, D. Reddy, J. Kautz, K. Pulli, D. Luebke, Cascaded displays: spatiotemporal superresolution using offset pixel layers. ACM Trans. Graph. (TOG) **33**(4), 60 (2014)
63. R. Held, E. Cooper, J. O'Brien, M. Banks, Using blur to affect perceived distance and size. ACM Trans. Graph. **29**(2), 1–16 (2010)
64. S. Hillaire, A. Lecuyer, R. Cozot, G. Casiez, Using an eye-tracking system to improve camera motions and depth-of-field blur effects in virtual environments, in *2008 IEEE Virtual Reality Conference* (2008), pp. 47–50
65. M. Hirsch, G. Wetzstein, R. Raskar, A compressive light field projection system. ACM Trans. Graph. (TOG) **33**(4), 58 (2014)
66. D. Hoffman, A. Girshick, K. Akeley, M. Banks, Vergence-accommodation conflicts hinder visual performance and cause visual fatigue. J. Vis. **8**(3) (2008)
67. B.A. Holden, T.R. Fricke, S.M. Ho, R. Wong, G. Schlenther, S. Cronjé, A. Burnett, E. Papas, K.S. Naidoo, K.D. Frick, Global vision impairment due to uncorrected presbyopia. Arch. Ophthalmol. **126**(12), 1731–1739 (2008)
68. I.P. Howard, B.J. Rogers, *Seeing in Depth* (Oxford University Press, New York, 2002)
69. I.D. Howlett, Q. Smithwick, Perspective correct occlusion-capable augmented reality displays using cloaking optics constraints. J. Soc. Inf. Display **25**(3), 185–193 (2017)
70. X. Hu, H. Hua, Design and assessment of a depth-fused multi-focal-plane display prototype. J. Disp. Technol. **10**(4), 308–316 (2014)
71. H. Hua, Enabling focus cues in head-mounted displays. Proc. IEEE **105**(5), 805–824 (2017)
72. H. Hua, B. Javidi, A 3D integral imaging optical see-through head-mounted display. Opt. Express **22**(11), 13484–13491 (2014)
73. F.-C. Huang, K. Chen, G. Wetzstein, The light field stereoscope: immersive computer graphics via factored near-eye light field display with focus cues. ACM Trans. Graph. (SIGGRAPH) **34**(4) (2015)
74. F.-C. Huang, D. Pajak, J. Kim, J. Kautz, D. Luebke, Mixed-primary factorization for dual-frame computational displays. ACM Trans. Graph. (SIGGRAPH) **36**(4), 149–1 (2017)
75. F.-C. Huang, G. Wetzstein, B.A. Barsky, R. Raskar, Eyeglasses-free display: towards correcting visual aberrations with computational light field displays. ACM Trans. Graph. (SIGGRAPH) **33**(4), 59 (2014)

76. J. Huang, Z. Chen, D. Ceylan, H. Jin, 6-DOF VR videos with a single 360-camera, in *Proceedings of IEEE Virtual Reality (VR)*, Mar 2017, pp. 37–44
77. P.-H. Huang, K. Matzen, J. Kopf, N. Ahuja, J.-B. Huang, DeepMVS: learning multi-view stereopsis, in *Proceedings of the International Conference on Computer Vision and Pattern Recognition (CVPR)* (2018)
78. IFIXIT, Magic leap one teardown (2018). https://www.ifixit.com/Teardown/Magic+Leap+One+Teardown/112245
79. H. Ishiguro, M. Yamamoto, S. Tsuji, Omni-directional stereo. IEEE Trans. Pattern Anal. Mach. Intell. **14**(2), 257–262 (1992)
80. Y. Itoh, T. Hamasaki, M. Sugimoto, Occlusion leak compensation for optical see-through displays using a single-layer transmissive spatial light modulator. IEEE TVCG **23**(11), 2463–2473 (2017)
81. Y. Itoh, T. Langlotz, D. Iwai, K. Kiyokawa, T. Amano, Light attenuation display: subtractive see-through near-eye display via spatial color filtering. IEEE TVCG **25**(5), 1951–1960 (2019)
82. M. Jancosek, T. Pajdla, Multi-view reconstruction preserving weakly-supported surfaces, in *Proceedings of the International Conference on Computer Vision and Pattern Recognition (CVPR)*, June 2011, pp. 3121–3128
83. P.V. Johnson, J.A. Parnell, J. Kim, C.D. Saunter, G.D. Love, M.S. Banks, Dynamic lens and monovision 3d displays to improve viewer comfort. OSA Opt. Express **24**(11), 11808–11827 (2016)
84. P.M.S. Julian, P. Brooker, Operator performance evaluation of controlled depth of field in a stereographically displayed virtual environment, vol. 4297 (2001)
85. H. Kato, M. Billinghurst, Marker tracking and HMD calibration for a video-based augmented reality conferencing system, in *Proceedings of International Workshop on Augmented Reality* (1999), pp. 85–94
86. I. Kauvar, S.J. Yang, L. Shi, I. McDowall, G. Wetzstein, Adaptive color display via perceptually-driven factored spectral projection. ACM Trans. Graph. (SIGGRAPH Asia) **34**(6), 165–1 (2015)
87. C. Kim, H. Zimmer, Y. Pritch, A. Sorkine-Hornung, M. Gross, Scene reconstruction from high spatio-angular resolution light fields. ACM Trans. Graph. (Proc. SIGGRAPH) **32**(4), 73:1–12 (2013)
88. H. Kim, P. Garrido, A. Tewari, W. Xu, J. Thies, M. Nießner, P. Pérez, C. Richardt, M. Zollhofer, C. Theobalt, Deep video portraits. ACM Trans. Graph. (Proc. SIGGRAPH) **37**(4), 163:1–14 (2018)
89. J. Kim, Y. Jeong, M. Stengel, K. Akşit, R. Albert, B. Boudaoud, T. Greer, J. Kim, W. Lopes, Z. Majercik, P. Shirley, J. Spjut, M. McGuire, D. Luebke, Foveated AR: dynamically-foveated augmented reality display. ACM Trans. Graph. **38**(4), 99:1–99:15 (2019)
90. K. Kiyokawa, M. Billinghurst, B. Campbell, E. Woods, An occlusion-capable optical see-through head mount display for supporting co-located collaboration, in *Proceedings of IEEE ISMAR* (2003)
91. K. Kiyokawa, Y. Kurata, H. Ohno, An optical see-through display for mutual occlusion of real and virtual environments, in *Proceedings of ISAR* (2000), pp. 60–67
92. K. Kiyokawa, Y. Kurata, H. Ohno, An optical see-through display for mutual occlusion with a real-time stereovision system. Comput. Graph. **25**(5), 765–779 (2001)
93. R. Konrad, A. Angelopoulos, G. Wetzstein, Gaze-contingent ocular parallax rendering for virtual reality, ACM Trans. Graph. **39**(2) (2020)
94. R. Konrad, E.A. Cooper, G. Wetzstein, Novel optical configurations for virtual reality: evaluating user preference and performance with focus-tunable and monovision near-eye displays, in *Proceedings of SIGCHI* (2016)
95. R. Konrad, D.G. Dansereau, A. Masood, G. Wetzstein, SpinVR: towards live-streaming 3D virtual reality video. ACM Trans. Graph. (Proc. SIGGRAPH Asia) **36**(6), 209:1–12 (2017)
96. R. Konrad, N. Padmanaban, K. Molner, E.A. Cooper, G. Wetzstein, Accommodation-invariant computational near-eye displays. ACM Trans. Graph. (SIGGRAPH) **36**(4), 88:1–88:12 (2017)

97. F.L. Kooi, A. Toet, Visual comfort of binocular and 3d displays. Displays **25**(2–3), 99–108 (2004)
98. J. Kopf, S. Alsisan, F. Ge, Y. Chong, K. Matzen, O. Quigley, J. Patterson, J. Tirado, S. Wu, M.F. Cohen, Practical 3D photography, in *Proceedings of CVPR Workshops* (2019)
99. G.A. Koulieris, K. AkÅŸit, M. Stengel, R.K. Mantiuk, K. Mania, C. Richardt, Near-eye display and tracking technologies for virtual and augmented reality. Comput. Graph. Forum **38**(2), 493–519 (2019)
100. G.-A. Koulieris, B. Bui, M.S. Banks, G. Drettakis, Accommodation and comfort in head-mounted displays. ACM Trans. Graph. (SIGGRAPH) **36**(4), 87:1–87:11 (2017)
101. G. Kramida, Resolving the vergence-accommodation conflict in head-mounted displays. IEEE TVCG **22**, 1912–1931 (2015)
102. M. Lambooij, M. Fortuin, I. Heynderickx, W. IJsselsteijn, Visual discomfort and visual fatigue of stereoscopic displays: a review. J. Imaging Sci. Technol. **53**(3):30201–1 (2009)
103. T. Langlotz, M. Cook, H. Regenbrecht, Real-time radiometric compensation for optical see-through head-mounted displays. IEEE TVCG **22**(11), 2385–2394 (2016)
104. T. Langlotz, J. Sutton, S. Zollmann, Y. Itoh, H. Regenbrecht, Chromaglasses: computational glasses for compensating colour blindness, in *Proceedings of SIGCHI* (2018), pp. 390:1–390:12
105. D. Lanman, M. Hirsch, Y. Kim, R. Raskar, Content-adaptive parallax barriers: optimizing dual-layer 3d displays using low-rank light field factorization, in *ACM Transactions on Graphics (SIGGRAPH Asia)*, vol. 29 (ACM, 2010), p. 163
106. D. Lanman, D. Luebke, Near-eye light field displays. ACM Trans. Graph. (SIGGRAPH Asia) **32**(6), 220:1–220:10 (2013)
107. D. Lanman, G. Wetzstein, M. Hirsch, W. Heidrich, R. Raskar, Polarization fields: dynamic light field display using multi-layer LCDs, in *ACM Transactions on Graphics (SIGGRAPH Asia)*, vol. 30, p. 186 (2011)
108. S.M. LaValle, A. Yershova, M. Katsev, M. Antonov, Head tracking for the oculus rift, in *IEEE International Conference on Robotics and Automation (ICRA)* (2014), pp. 187–194
109. J. Lee, B. Kim, K. Kim, Y. Kim, J. Noh, Rich360: Optimized spherical representation from structured panoramic camera arrays. ACM Trans. Graph. (Proc. SIGGRAPH) **35**(4), 63:1–11 (2016)
110. S. Lee, C. Jang, S. Moon, J. Cho, B. Lee, Additive light field displays: realization of augmented reality with holographic optical elements. ACM Trans. Graph. (SIGGRAPH Asia) **35**(4), 60:1–60:13 (2016)
111. T. Lee, T. Hollerer, Multithreaded hybrid feature tracking for markerless augmented reality. IEEE Trans. Vis. Comput. Graph. **15**(3), 355–368 (2009)
112. M. Levoy, P. Hanrahan, Light field rendering, in *Proceedings of the Annual Conference on Computer Graphics and Interactive Techniques (SIGGRAPH)*, Aug 1996, pp. 31–42
113. G. Li, D. Lee, Y. Jeong, J. Cho, B. Lee, Holographic display for see-through augmented reality using mirror-lens holographic optical element. Opt. Lett. **41**(11), 2486–2489 (2016)
114. G. Li, D.L. Mathine, P. Valley, P. Äyräs, J.N. Haddock, M.S. Giridhar, G. Williby, J. Schwiegerling, G.R. Meredith, B. Kippelen, S. Honkanen, N. Peyghambarian, Switchable electro-optic diffractive lens with high efficiency for ophthalmic applications. Proc. Natl. Acad. Sci. **103**(16), 6100–6104 (2006)
115. Y. Li, A. Majumder, D. Lu, M. Gopi, Content-independent multi-spectral display using superimposed projections, in *Computer Graphics Forum*, vol. 34 (Wiley Online Library, 2015), pp. 337–348
116. C. Lipski, C. Linz, K. Berger, A. Sellent, M. Magnor, Virtual video camera: image-based viewpoint navigation through space and time. Comput. Graph. Forum **29**(8), 2555–2568 (2010)
117. S. Liu, D. Cheng, and H. Hua. An optical see-through head mounted display with addressable focal planes. In *Proc. ISMAR*, pages 33–42, 2008
118. P. Llull, N. Bedard, W. Wu, I. Tosic, K. Berkner, N. Balram, Design and optimization of a near-eye multifocal display system for augmented reality, in *OSA Imaging and Applied Optics* (2015)

119. S. Lombardi, T. Simon, J. Saragih, G. Schwartz, A. Lehrmann, Y. Sheikh, Neural volumes: learning dynamic renderable volumes from images. ACM Trans. Graph. (Proc. SIGGRAPH) (2019)
120. D. Long, M.D. Fairchild, Optimizing spectral color reproduction in multiprimary digital projection, in *Color and Imaging Conference*, vol. 2011 (Society for Imaging Science and Technology, 2011), pp. 290–297
121. G.D. Love, D.M. Hoffman, P.J.W. Hands, J. Gao, A.K. Kirby, M.S. Banks, High-speed switchable lens enables the development of a volumetric stereoscopic display. Opt. Express **17**(18), 15716–15725 (2009)
122. B. Luo, F. Xu, C. Richardt, J.-H. Yong, Parallax360: stereoscopic 360° scene representation for head-motion parallax. IEEE Trans. Vis. Comput. Graph. **24**(4), 1545–1553 (2018)
123. G. Maiello, M. Chessa, F. Solari, P.J. Bex, Simulated disparity and peripheral blur interact during binocular fusion. J. Vis. **14**(8), 13 (2014)
124. A. Maimone, H. Fuchs, Computational augmented reality eyeglasses, in *Proceedings of IEEE ISMAR* (2013), pp. 29–38
125. A. Maimone, A. Georgiou, J.S. Kollin, Holographic near-eye displays for virtual and augmented reality. ACM Trans. Graph. (SIGGRAPH) **36**(4), 85:1–85:16 (2017)
126. A. Maimone, G. Wetzstein, M. Hirsch, D. Lanman, R. Raskar, H. Fuchs, Focus 3D: compressive accommodation display. ACM Trans. Graph. **32**(5), 153–1 (2013)
127. A. Maimone, X. Yang, N. Dierk, A. State, M. Dou, H. Fuchs, General-purpose telepresence with head-worn optical see-through displays and projector-based lighting, in *2013 IEEE Virtual Reality (VR)* (IEEE, 2013), pp. 23–26
128. R. Martin-Brualla, R. Pandey, S. Yang, P. Pidlypenskyi, J. Taylor, J. Valentin, S. Khamis, P. Davidson, A. Tkach, P. Lincoln, A. Kowdle, C. Rhemann, D.B. Goldman, C. Keskin, S. Seitz, S. Izadi, S. Fanello, LookinGood: enhancing performance capture with real-time neural re-rendering. ACM Trans. Graph. (Proc. SIGGRAPH Asia) **37**(6), 255:1–14 (2018)
129. B. Masia, G. Wetzstein, P. Didyk, D. Gutierrez, A survey on computational displays: pushing the boundaries of optics, computation, and perception. Comput. Graph. **37**(8), 1012–1038 (2013)
130. N. Matsuda, A. Fix, D. Lanman, Focal surface displays. ACM Trans. Graph. (SIGGRAPH) **36**(4), 86:1–86:14 (2017)
131. M. Mauderer, S. Conte, M.A. Nacenta, D. Vishwanath, Depth perception with gaze-contingent depth of field, in *Proceedings of the SIGCHI Conference on Human Factors in Computing Systems* (ACM, 2014), pp. 217–226
132. T. Mazuryk, M. Gervautz, Virtual reality—history, applications, technology and future, 12 (1999)
133. X. Meng, R. Du, M. Zwicker, A. Varshney, Kernel foveated rendering. Proc. ACM Comput. Graph. Interact. Tech. (I3D) **1**(5), 1–20 (2018)
134. O. Mercier, Y. Sulai, K. Mackenzie, M. Zannoli, J. Hillis, D. Nowrouzezahrai, D. Lanman, Fast gaze-contingent optimal decompositions for multifocal displays. ACM Trans. Graph. (SIGGRAPH Asia) **36**(6) (2017)
135. M. Meshry, D.B. Goldman, S. Khamis, H. Hoppe, R. Pandey, N. Snavely, R. Martin-Brualla, Neural rerendering in the wild, in *Proceedings of the International Conference on Computer Vision and Pattern Recognition (CVPR)* (2019)
136. B. Mildenhall, P.P. Srinivasan, R. Ortiz-Cayon, N.K. Kalantari, R. Ramamoorthi, R. Ng, A. Kar, Local light field fusion: practical view synthesis with prescriptive sampling guidelines. ACM Trans. Graph. (Proc. SIGGRAPH) (2019)
137. A. Mohan, R. Raskar, J. Tumblin, Agile spectrum imaging: programmable wavelength modulation for cameras and projectors, in *Computer Graphics Forum*, vol. 27 (Wiley Online Library, 2008), pp. 709–717
138. E. Moon, M. Kim, J. Roh, H. Kim, J. Hahn, Holographic head-mounted display with RGB light emitting diode light source. Opt. Express **22**(6), 6526–6534 (2014)
139. S. Mori, S. Ikeda, A. Plopski, C. Sandor, Brightview: increasing perceived brightness of optical see-through head-mounted displays through unnoticeable incident light reduction, in *Proceedings of IEEE VR* (2018), pp. 251–258

140. P. Moulon, P. Monasse, R. Marlet, Adaptive structure from motion with a Contrario model estimation, in *Proceedings of the Asian Conference on Computer Vision (ACCV)* (2012), pp. 257–270
141. F. Mueller, F. Bernard, O. Sotnychenko, D. Mehta, S. Sridhar, D. Casas, C. Theobalt, Generated hands for real-time 3d hand tracking from monocular RGB, in *Proceedings of IEEE Conference on Computer Vision and Pattern Recognition (CVPR)* (2018), pp. 49–59
142. R. Narain, R.A. Albert, A. Bulbul, G.J. Ward, M.S. Banks, J.F. O'Brien, Optimal presentation of imagery with focus cues on multi-plane displays. ACM Trans. Graph. (SIGGRAPH) **34**(4) (2015)
143. R.A. Newcombe, A.J. Davison, S. Izadi, P. Kohli, O. Hilliges, J. Shotton, D. Molyneaux, S. Hodges, D. Kim, A. Fitzgibbon, KinectFusion: real-time dense surface mapping and tracking, in *Proceedings of the International Symposium on Mixed and Augmented Reality (ISMAR)*, Oct 2011, pp. 127–136
144. T. Nguyen-Phuoc, C. Li, L. Theis, C. Richardt, Y.-L. Yang, HoloGAN: unsupervised learning of 3D representations from natural images, in *Proceedings of the International Conference on Computer Vision (ICCV)* (2019)
145. M. Nießner, M. Zollhofer, S. Izadi, M. Stamminger, Real-time 3D reconstruction at scale using voxel hashing. ACM Trans. Graph. (Proc. SIGGRAPH Asia) **32**(6), 169:1–11 (2013)
146. D. Nister, O. Naroditsky, J. Bergen, Visual odometry, in *Proceedings of Conference on Computer Vision and Pattern Recognition (CVPR)*, vol. 1 (2004)
147. C. Noorlander, J.J. Koenderink, R.J. Den Olden, B.W. Edens, Sensitivity to spatiotemporal colour contrast in the peripheral visual field. Vis. Res. **23**(1), 1–11 (1983)
148. R.S. Overbeck, D. Erickson, D. Evangelakos, M. Pharr, P. Debevec, A system for acquiring, compressing, and rendering panoramic light field stills for virtual reality. ACM Trans. Graph. (Proc. SIGGRAPH Asia) **37**(6), 197:1–15 (2018)
149. N. Padmanaban, R. Konrad, T. Stramer, E.A. Cooper, G. Wetzstein, Optimizing virtual reality for all users through gaze-contingent and adaptive focus displays. Proc. Natl. Acad. Sci. U.S.A. **114**, 2183–2188 (2017)
150. N. Padmanaban, R. Konrad, G. Wetzstein. Autofocals: evaluating gaze-contingent eyeglasses for presbyopes. Sci. Adv. **5**(6) (2019)
151. N. Padmanaban, Y. Peng, G. Wetzstein, Holographic near-eye displays based on overlap-add stereograms. ACM Trans. Graph. (SIGGRAPH Asia) **38**(6) (2019)
152. S.E. Palmer, *Vision Science—Photons to Phenomenology* (MIT Press, 1999)
153. V.F. Pamplona, M.M. Oliveira, D.G. Aliaga, R. Raskar, Tailored displays to compensate for visual aberrations. ACM Trans. Graph. (SIGGRAPH) **31**(4), 81:1–81:12 (2012)
154. J.J. Park, P. Florence, J. Straub, R. Newcombe, S. Lovegrove, DeepSDF: learning continuous signed distance functions for shape representation, in *Proceedings of the International Conference on Computer Vision and Pattern Recognition (CVPR)* (2019)
155. A. Patney, M. Salvi, J. Kim, A. Kaplanyan, C. Wyman, N. Benty, D. Luebke, A. Lefohn, Towards foveated rendering for gaze-tracked virtual reality. ACM Trans. Graph. (TOG) **35**(6), 179 (2016)
156. S. Peleg, M. Ben-Ezra, Y. Pritch, Omnistereo: panoramic stereo imaging. IEEE Trans. Pattern Anal. Mach. Intell. **23**(3), 279–290 (2001)
157. E. Penner, L. Zhang, Soft 3D reconstruction for view synthesis. ACM Trans. Graph. (Proc. SIGGRAPH Asia) **36**(6), 235:1–11 (2017)
158. F. Perazzi, A. Sorkine-Hornung, H. Zimmer, P. Kaufmann, O. Wang, S. Watson, M. Gross, Panoramic video from unstructured camera arrays. Comput. Graph. Forum (Proc. Eurographics) **34**(2), 57–68 (2015)
159. R. Raskar, H. Nii, B. deDecker, Y. Hashimoto, J. Summet, D. Moore, Y. Zhao, J. Westhues, P. Dietz, J. Barnwell, S. Nayar, M. Inami, P. Bekaert, M. Noland, V. Branzoi, E. Bruns, Prakash: lighting aware motion capture using photosensing markers and multiplexed illuminators. ACM Trans. Graph. (SIGGRAPH) **26**(3) (2007)
160. K. Rathinavel, H. Wang, A. Blate, H. Fuchs, An extended depth-at-field volumetric near-eye augmented reality display. IEEE Trans. Vis. Comput. Graph. **24**(11), 2857–2866 (2018)

161. K. Rathinavel, G. Wetzstein, H. Fuchs, Varifocal occlusion-capable optical see-through augmented reality display based on focus-tunable optics. IEEE TVCG (Proc. ISMAR) (2019)
162. J. Rekimoto. Matrix: a realtime object identification and registration method for augmented reality, in *Proceedings of Asia Pacific Computer Human Interaction* (1998), pp. 63–68
163. J.P. Rice, S.W. Brown, J.E. Neira, R.R. Bousquet, A hyperspectral image projector for hyperspectral imagers, in *Algorithms and Technologies for Multispectral, Hyperspectral, and Ultraspectral Imagery XIII*, vol. 6565 (International Society for Optics and Photonics, 2007), p. 65650C
164. C. Richardt, P. Hedman, R.S. Overbeck, B. Cabral, R. Konrad, S. Sullivan, Capture4VR: from VR photography to VR video, in *SIGGRAPH Courses* (2019)
165. C. Richardt, Y. Pritch, H. Zimmer, A. Sorkine-Hornung, Megastereo: constructing high-resolution stereo panoramas, in *Proceedings of the International Conference on Computer Vision and Pattern Recognition (CVPR)*, June 2013, pp. 1256–1263
166. J.P. Rolland, M.W. Krueger, A. Goon, Multifocal planes head-mounted displays. Appl. Opt. **39**(19), 3209–3215 (2000)
167. J. Rovamo, V. Virsu, P. Laurinen, L. Hyvärinen, Resolution of gratings oriented along and across meridians in peripheral vision. Invest. Ophthalmol. Vis. Sci. **23**(5), 666–670 (1982)
168. B. Sajadi, M. Gopi, A. Majumder, Edge-guided resolution enhancement in projectors via optical pixel sharing. ACM Trans. Graph. (TOG) **31**(4), 79 (2012)
169. B. Sajadi, D. Qoc-Lai, A.H. Ihler, M. Gopi, A. Majumder, Image enhancement in projectors via optical pixel shift and overlay, in *IEEE International Conference on Computational Photography (ICCP)* (IEEE, 2013), pp. 1–10
170. J.L. Schönberger, J.-M. Frahm, Structure-from-motion revisited, in *Proceedings of the International Conference on Computer Vision and Pattern Recognition (CVPR)* (2016), pp. 4104–4113
171. J.L. Schönberger, E. Zheng, J.-M. Frahm, M. Pollefeys, Pixelwise view selection for unstructured multi-view stereo, in *Proceedings of the European Conference on Computer Vision (ECCV)*, ed. by B. Leibe, J. Matas, N. Sebe, M. Welling (2016), pp. 501–518
172. C. Schroers, J.-C. Bazin, A. Sorkine-Hornung, An omnistereoscopic video pipeline for capture and display of real-world VR. ACM Trans. Graph. **37**(3), 37:1–13 (2018)
173. H. Seetzen, W. Heidrich, W. Stuerzlinger, G. Ward, L. Whitehead, M. Trentacoste, A. Ghosh, A. Vorozcovs, High dynamic range display systems. ACM Trans. Graph. **23**(3), 760–768 (2004)
174. S. Seitz, B. Curless, J. Diebel, D. Scharstein, R. Szeliski, A comparison and evaluation of multi-view stereo reconstruction algorithms, in *Proceedings of the International Conference on Computer Vision and Pattern Recognition (CVPR)*, vol. 1, June 2006, pp. 519–528
175. A. Serrano, I. Kim, Z. Chen, S. DiVerdi, D. Gutierrez, A. Hertzmann, B. Masia, Motion parallax for 360° RGBD video. IEEE Trans. Vis. Comput. Graph. **25**(5), 1817–1827 (2019)
176. L. Shi, F.-C. Huang, W. Lopes, W. Matusik, D. Luebke, Near-eye light field holographic rendering with spherical waves for wide field of view interactive 3d computer graphics. ACM Trans. Graph. (SIGGRAPH Asia) **36**(6), 236:1–236:17 (2017)
177. T. Shibata, J. Kim, D.M. Hoffman, M.S. Banks, The zone of comfort: predicting visual discomfort with stereo displays. J. Vis. **11**(8), 11 (2011)
178. H. Shum, S.B. Kang, Review of image-based rendering techniques, in *Visual Communications and Image Processing*, vol. 4067 (2000)
179. H.-Y. Shum, S.-C. Chan, S.B. Kang, *Image-Based Rendering* (Springer, Berlin, 2007)
180. V. Sitzmann, J. Thies, F. Heide, M. Niessner, G. Wetzstein, M. Zollhofer, DeepVoxels: learning persistent 3D feature embeddings, in *Proceedings of the International Conference on Computer Vision and Pattern Recognition (CVPR)* (2019), pp. 2437–2446
181. V. Sitzmann, M. Zollhofer, G. Wetzstein, Scene representation networks: continuous 3D-structure-aware neural scene representations, in Proceedings of the Conference on Neural Information Processing Systems (NeurIPS) (2019). arXiv:1906.01618
182. N. Snavely, S.M. Seitz, R. Szeliski, Photo tourism: exploring photo collections in 3D. ACM Trans. Graph. (Proc. SIGGRAPH) **25**(3), 835–846 (2006)

183. S. Sridhar, F. Mueller, A. Oulasvirta, C. Theobalt, Fast and robust hand tracking using detection-guided optimization, in *Proceedings of the IEEE Conference on Computer Vision and Pattern Recognition* (2015), pp. 3213–3221
184. S. Sridhar, F. Mueller, M. Zollhofer, D. Casas, A. Oulasvirta, C. Theobalt, Real-time joint tracking of a hand manipulating an object from RGB-D input, in *European Conference on Computer Vision* (Springer, Cham, 2016), pp. 294–310
185. P.P. Srinivasan, R. Tucker, J.T. Barron, R. Ramamoorthi, R. Ng, N. Snavely, Pushing the boundaries of view extrapolation with multiplane images, in *Proceedings of the International Conference on Computer Vision and Pattern Recognition (CVPR)*, June 2019, pp. 175–184
186. M. Stengel, S. Grogorick, M. Eisemann, M. Magnor, Adaptive image-space sampling for gaze-contingent real-time rendering, in *Computer Graphics Forum*, vol. 35 (Wiley Online Library, 2016), pp. 129–139
187. R.E. Stevens, T.N. Jacoby, I.Ş. Aricescu, D.P. Rhodes, A review of adjustable lenses for head mounted displays, in *Digital Optical Technologies 2017*, vol. 10335 (International Society for Optics and Photonics, 2017), p. 103350Q
188. R.E. Stevens, D.P. Rhodes, A. Hasnain, P.-Y. Laffont, Varifocal technologies providing prescription and VAC mitigation in HMDs using Alvarez lenses, vol. 10676 (2018)
189. H. Strasburger, I. Rentschler, M. Jüttner, Peripheral vision and pattern recognition: a review. J. Vis. **11**(5), 13 (2011)
190. D.J. Sturman, D. Zeltzer, A survey of glove-based input. IEEE Comput. Graph. Appl. **14**(1), 30–39 (1994)
191. T. Sugihara, T. Miyasato, 32.4: A lightweight 3-D HMD with accommodative compensation. SID Dig. **29**(1):927–930 (1998)
192. Q. Sun, F.-C. Huang, J. Kim, L.-Y. Wei, D. Luebke, A. Kaufman, Perceptually-guided foveation for light field displays. ACM Trans. Graph. **36**(6), 192:1–192:13 (2017)
193. I.E. Sutherland, A head-mounted three dimensional display, in *Proceedings of Fall Joint Computer Conference* (1968), pp. 757–764
194. N.T. Swafford, J.A. Iglesias-Guitian, C. Koniaris, B. Moon, D. Cosker, K. Mitchell, User, metric, and computational evaluation of foveated rendering methods, in *Proceedings of the ACM Symposium on Applied Perception* (ACM, 2016), pp. 7–14
195. C. Sweeney, Theia multiview geometry library: tutorial & reference (2016). http://theia-sfm.org
196. C. Sweeney, A. Holynski, B. Curless, S.M. Seitz, Structure from motion for panorama-style videos (2019). arXiv:1906.03539
197. R. Szeliski, Image alignment and stitching: a tutorial. Found. Trends Comput. Graph. Vis. **2**(1), 1–104 (2006)
198. M. Teragawa, A. Yoshida, K. Yoshiyama, S. Nakagawa, K. Tomizawa, Y. Yoshida, Multi-primary-color displays: the latest technologies and their benefits. J. Soc. Inf. Disp. **20**(1), 1–11 (2012)
199. L.N. Thibos, D.L. Still, A. Bradley, Characterization of spatial aliasing and contrast sensitivity in peripheral vision. Vis. Res. **36**(2), 249–258 (1996)
200. J. Thies, M. Zollhofer, M. Niessner, Deferred neural rendering: Image synthesis using neural textures. ACM Trans. Graph. (Proc. SIGGRAPH) (2019)
201. S. Tulsiani, R. Tucker, N. Snavely, Layer-structured 3D scene inference via view synthesis, in *Proceedings of the European Conference on Computer Vision (ECCV)*, Sept 2018
202. K. Vaidyanathan, M. Salvi, R. Toth, T. Foley, T. Akenine-Möller, J. Nilsson, J. Munkberg, J. Hasselgren, M. Sugihara, P. Clarberg et al., Coarse pixel shading, in *Proceedings of High Performance Graphics* (Eurographics Association, 2014), pp. 9–18
203. J. Ventura, Structure from motion on a sphere, in *Proceedings of the European Conference on Computer Vision (ECCV)*, ed. by B. Leibe, J. Matas, N. Sebe, M. Welling (2016), pp. 53–68
204. M. von Waldkirch, P. Lukowicz, G. Tröster, Multiple imaging technique for extending depth of focus in retinal displays. Opt. Express **12**(25) (2004)
205. R. Wang, S. Paris, J. Popović, 6d hands: markerless hand-tracking for computer aided design, in *Proceedings of ACM Symposium on User Interface Software and Technology (UIST)* (2011)

206. S.J. Watt, K. Akeley, M.O. Ernst, M.S. Banks, Focus cues affect perceived depth. J. Vis. **5**(10), 834–862 (2005)
207. S.-E. Wei, J. Saragih, T. Simon, A.W. Harley, S. Lombardi, M. Perdoch, A. Hypes, D. Wang, H. Badino, Y. Sheikh, VR facial animation via multiview image translation. ACM Trans. Graph. (Proc. SIGGRAPH) **38**(4), 67:1–16 (2019)
208. C. Weissig, O. Schreer, P. Eisert, P. Kauff, The ultimate immersive experience: panoramic 3D video acquisition, in *Advances in Multimedia Modeling (MMM)*, ed. by K. Schoeffmann, B. Merialdo, A.G. Hauptmann, C.-W. Ngo, Y. Andreopoulos, C. Breiteneder, vol. 7131 of *Lecture Notes in Computer Science* (2012), pp. 671–681
209. G. Westheimer, The Maxwellian view. Vis. Res. **6**, 669–682 (1966)
210. G. Wetzstein, O. Bimber, Radiometric compensation through inverse light transport, in *15th Pacific Conference on Computer Graphics and Applications (PG'07)* (2007), pp. 391–399
211. G. Wetzstein, W. Heidrich, D. Luebke, Optical image processing using light modulation displays. Comput. Graph. Forum **29**(6), 1934–1944 (2010)
212. G. Wetzstein, D. Lanman, Factored displays: improving resolution, dynamic range, color reproduction, and light field characteristics with advanced signal processing. IEEE Sig. Process. Mag. **33**(5), 119–129 (2016)
213. G. Wetzstein, D. Lanman, W. Heidrich, R. Raskar, Layered 3d: tomographic image synthesis for attenuation-based light field and high dynamic range displays, in *ACM Transactions on Graphics (SIGGRAPH)*, vol. 30 (2011), p. 95
214. G. Wetzstein, D. Lanman, M. Hirsch, R. Raskar, Tensor displays: compressive light field synthesis using multilayer displays with directional backlighting. ACM Trans. Graph. (SIGGRAPH) **31**(4), 1–11 (2012)
215. T. Whelan, S. Leutenegger, R.F. Salas-Moreno, B. Glocker, A.J. Davison, ElasticFusion: dense SLAM without a pose graph, in *Proceedings of Robotics: Science and Systems (RSS)*, July 2015
216. A. Wilson, H. Hua, Design and prototype of an augmented reality display with per-pixel mutual occlusion capability. OSA Opt. Express **25**(24), 30539–30549 (2017)
217. D.N. Wood, D.I. Azuma, K. Aldinger, B. Curless, T. Duchamp, D.H. Salesin, W. Stuetzle, Surface light fields for 3D photography, in *Proceedings of the Annual Conference on Computer Graphics and Interactive Techniques (SIGGRAPH)* (2000), pp. 287–296
218. C. Wu, VisualSFM: a visual structure from motion system (2011). http://ccwu.me/vsfm/
219. W. Wu, P. Llull, I. Tosic, N. Bedard, K. Berkner, N. Balram, Content-adaptive focus configuration for near-eye multi-focal displays, in *IEEE International Conference on Multimedia and Expo (ICME)* (2016), pp. 1–6
220. K. Yücer, A. Sorkine-Hornung, O. Wang, O. Sorkine-Hornung, Efficient 3D object segmentation from densely sampled light fields with applications to 3D reconstruction. ACM Trans. Graph. **35**(3), 22:1–15 (2016)
221. H.-J. Yeom, H.-J. Kim, S.-B. Kim, H. Zhang, B. Li, Y.-M. Ji, S.-H. Kim, J.-H. Park, 3d holographic head mounted display using holographic optical elements with astigmatism aberration compensation. Opt. Express **23**(25), 32025–32034 (2015)
222. W. Yifan, F. Serena, S. Wu, C. Öztireli, O. Sorkine-Hornung, Differentiable surface splatting for point-based geometry processing (2019). arXiv:1906.04173
223. J. Zaragoza, T.-J. Chin, Q.-H. Tran, M.S. Brown, D. Suter, As-projective-as-possible image stitching with moving DLT. IEEE Trans. Pattern Anal. Mach. Intell. **36**(7), 1285–1298 (2014)
224. F. Zhang, F. Liu, Parallax-tolerant image stitching, in *Proceedings of the International Conference on Computer Vision and Pattern Recognition (CVPR)*, June 2014, pp. 3262–3269
225. K.C. Zheng, S.B. Kang, M.F. Cohen, R. Szeliski, Layered depth panoramas, in *Proceedings of the International Conference on Computer Vision and Pattern Recognition (CVPR)*, June 2007

226. T. Zhou, R. Tucker, J. Flynn, G. Fyffe, N. Snavely, Stereo magnification: Learning view synthesis using multiplane images. ACM Trans. Graph. (Proc. SIGGRAPH) **37**(4), 65:1–12 (2018)
227. M. Zollhöfer, J. Thies, P. Garrido, D. Bradley, T. Beeler, P. Pérez, M. Stamminger, M. Niessner, C. Theobalt, State of the art on monocular 3D face reconstruction, tracking, and applications. Comput. Graph. Forum **37**(2), 523–550 (2018)
228. B. Krajancich, N. Padmanaban, G. Wetzstein, Factored Occlusion: Single Spatial Light Modulator Occlusion-capable Optical See-through Augmented Reality Display, IEEE TVCG (Proc. VR) (2020)

Chapter 26
Cryogenic-CMOS for Quantum Computing

Edoardo Charbon, Fabio Sebastiano, Masoud Babaie and Andrei Vladimirescu

26.1 Introduction to Quantum Processors and Qubits

In the 2010s quantum technologies have emerged as a compelling complement to classical technologies for a number of applications, including quantum sensing, metrology, imaging, communications, security, and computing. In particular, quantum computing is a promising alternative to von Neumann machines and it holds the promise for solving today's intractable problems [1]. Quantum processors, the core of a quantum computer, comprise an array of quantum bits (qubits), the fundamental computational unit. Unlike conventional bits, qubits can take a coherent state ranging from $|0\rangle$ to $|1\rangle$ on a continuous sphere, known as the Bloch sphere (Fig. 26.1).

When in superposition, qubits can take multiple states simultaneously and thus, in principle, multiple computations can be performed at the same time, whereas the number of possible states of a quantum processor is 2^N, N being the number of qubits. The entanglement of qubits is the second important quantum mechanical property in qubit states, so that knowing the state of one would imply knowing the state of the other.

Qubits are however fragile and their fragility arises from the fact that a qubit needs to be coherent at all times. As qubits tend to lose coherency, they need to be constantly monitored and, if necessary, corrected. Figure 26.2 shows a quantum processor and its classical control, with a bidirectional interface that is generally electrical but that could also have optical or opto-electrical components.

E. Charbon (✉)
EPFL, Lausanne, Switzerland
e-mail: edoardo.charbon@epfl.ch

F. Sebastiano · M. Babaie · A. Vladimirescu
Delft University of Technology, Delft, The Netherlands

A. Vladimirescu
University of California, Berkeley, CA, USA

B. Murmann and B. Hoefflinger (eds.), *NANO-CHIPS 2030*,
The Frontiers Collection, https://doi.org/10.1007/978-3-030-18338-7_26

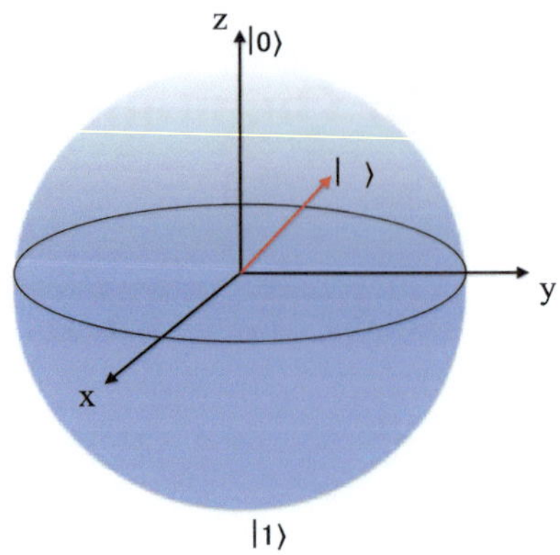

Fig. 26.1 Bloch sphere

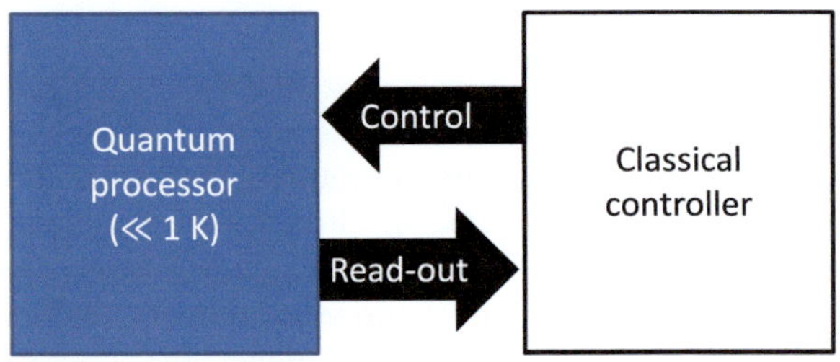

Fig. 26.2 Classical control of a quantum processor. © IEEE 2016

The general operating temperature of qubits is in the milli-Kelvin domain. However, higher temperatures can be achieved, while still retaining the main properties of entanglement and superposition, with reasonably long coherence times. As of today (2019), qubits are still interfaced with a classical controller operating at room temperature. A recent trend however has been to bring part or all of this control to temperatures that are closer than those of qubits for a more compact arrangement, potentially enabling a scalable machine with thousands or millions of qubits operating in a relatively small volume. This trend poses a problem to the electronics, that needs to be designed to match stringent specifications both in terms of noise and frequency. In addition, linearity and, in the case of mixed-signal systems, quantization granularity need to be achieved at a reasonable power, typically in the order of milliwatts and even microwatts per qubit.

This chapter describes the challenges and opportunities encountered in designing the electronic interface for quantum processors. We will specifically focus on the use of standard CMOS technology to design and fabricate integrated circuits operating at cryogenic temperatures. The chapter is organized as follows. In Sect. 26.2, an example of state-of-the-art quantum processor and the electronic interface currently employed for its control are presented, with emphasis on the current limitations. Section 26.3 explains our proposal of a generic classical controller based on cryo-CMOS technology. Cryo-CMOS device behavior is briefly covered in Sect. 26.4, while examples of the circuits necessary to achieve qubit control are presented in Sect. 26.5 and their verification in Sect. 26.6. Finally, a perspective of the proposed approach is outlined in Sect. 26.7, along with concluding remarks in Sect. 26.8.

26.2 Quantum Processors and Their Control Interface

A. *State-of-the-art in spin qubit controller and readout*

Several qubit technologies have been proposed but today there is not yet a clear winner. Qubits based on solid-state fabrication technology are often considered a favorable alternative, as they promise a scaled approach by exploiting fabrication techniques borrowed from 60 years of experience in the semiconductor industry. Figure 26.3 shows an illustration of some solid-state qubit technologies [2–10]. Although many more exist, a complete overview is however beyond the scope of this chapter. What is most common for solid-state qubits is a deep-cryogenic temperature of operation ($\ll$1 K) and the need for continuous monitoring and control for proper operation. In the reminder of the chapter we will focus on spin qubits possibly operated in the high milli-Kelvin or even in the low Kelvin domain [11, 12]. Among solid-state qubits, spin qubits are characterized by relatively long coherence time (above 100 μs) and the extremely low pitch (~100 nm), which would in principle allow the integration of the millions of qubits required in a practical quantum computer on a silicon die of few mm^2. Furthermore, as spin qubits are fabricated on a semiconductor substrate, they can be in principle be co-integrated on the same die with standard microelectronic circuits to be used for control and readout. However, although large efforts are currently devoted to this research field, state-of-the-art spin qubit processors comprise only up to 2 qubits.

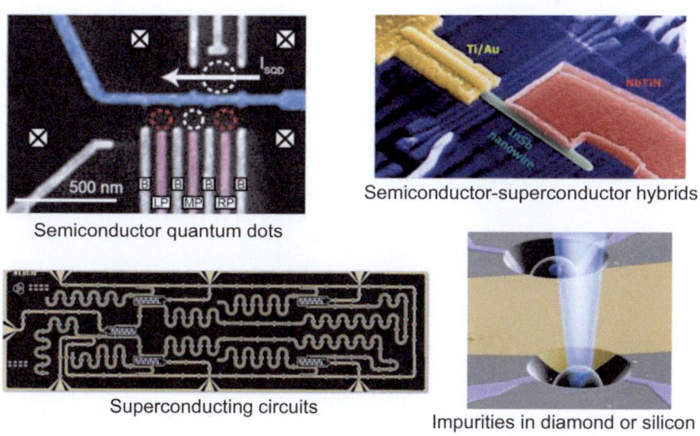

Fig. 26.3 A few candidates of solid-state qubits available today. Clockwise from top, a spin qubit, a topological qubit, NV-centers, and a superconductive qubit. © IEEE 2019

To realize a spin qubit, a single electron is isolated in an extremely small site, i.e. a quantum dot, on the surface of a semiconductor die. A large magnetic field is applied to ensure that the spin-up and spin-down state of the electron corresponds to distinct energy levels. Those two states are then used to encode the qubit quantum states $|1\rangle$ and $|0\rangle$. Figure 26.4 illustrates the block diagram of the state-of-the-art system for readout and control of a two-qubit single-electron spin-qubit chip [13]. The qubit chip is placed into a dilution refrigerator at a base temperature of 20 mK to ensure that the thermal energy is much smaller than the energy scales of the qubits. Room temperature digital-to-analog converters (not shown in Fig. 26.4) are used to provide proper DC voltages to isolate electrons in neighboring quantum dots separated by tunnel barriers.

Single-qubit operations can be achieved by applying a microwave magnetic field at a frequency corresponding to the energy difference between the spin-up and the spin down state $\left[f = \left(E_{spin,up} - E_{spin,down}\right)/h\right]$. Since a magnetic-field gradient is applied on-chip by a micromagnet, a microwave excitation applied to the qubit gate (e.g., P3 terminal for qubit-1 in Fig. 26.4) causes the electron to oscillate in the magnetic gradient, thus applying a magnetic excitation and hence a quantum

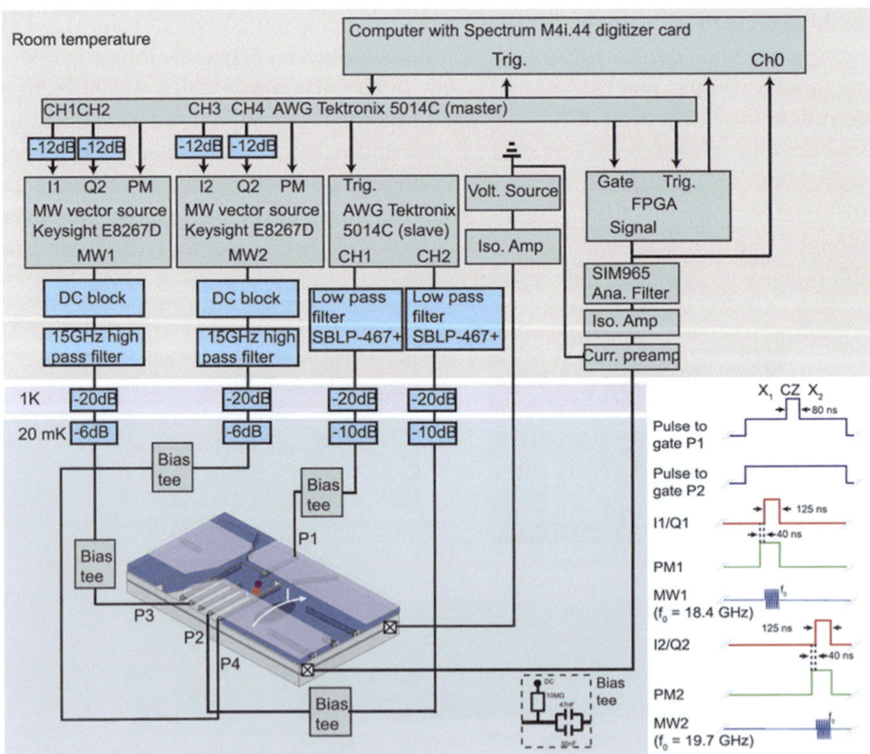

Fig. 26.4 Control and readout circuit for single-electron spin qubits in quantum dots (image taken from [13]). © Nature Publishing Corp. 2018

operation. In the experimental setup of Fig. 26.4, the microwave signals required for the single-qubit gates are generated by two vector source generators (VSG, Keysight E8267D), whose phase, frequency, amplitude, and modulation time are controlled by an arbitrary wave generator (AWG, Tektronix 5014C). The required microwave frequency is typically in the range of 10–40 GHz (e.g., 18.4 GHz for qubit-1, and 19.7 GHz for qubit-2 in Fig. 26.4) with modulation time of 0.1–2 μs for a π-rotation (e.g., 125 nsec in Fig. 26.4). The microwave signals pass through DC blocks, high-pass filters, and attenuators at different stages of the fridge to isolate the qubits from the noise of the room temperature instruments. At the base temperature, the incoming microwave signals added to the required gates DC voltage through bias tees.

For the two-qubit operation, the wave functions of the two electrons encoding the qubits must overlap, so that they can entangle and influence each other state. This is accomplished by tuning the energy of the two qubits and/or by lowering the potential of the tunnel barrier separating the quantum dots hosting the electrons. The required voltage pulses are applied to the proper gates (e.g., P1 and P2 in Fig. 26.4) with a duration of ~100 ns and are generated by an AWG (e.g., Tektronix 5014C in Fig. 26.4) with 1 GHz clock rate connected to the gate via a low-pass filter and a bias-tee. For proper control, the instruments used for both single- and two-qubit operation are synchronized by an external trigger provided by the master (e.g., Tektronix 5014C).

Read-out is typically performed by converting the electron spin information into the position of the electron and sensing its charge. In the figure, the spin-to-charge conversion is performed by spin-selective tunneling to a reservoir [14], and a single-electron transistor (SET) is then used for sensing the charge through measuring its electrical impedance. The SET input resistance changes by a few percent due to the movement of the electron (e.g., ~100 k$\Omega \pm 1\%$). This impedance modulation can be directly read by measuring the device current when biased at a fixed voltage [15, 16]. Before the digitalization, the SET current is converted to a voltage signal by a transimpedance amplifier and then filtered by a 20-kHz low-pass filter. Unfortunately, the bandwidth and thus the speed of the current sensing readout is limited by the parasitic capacitance of the long wire connecting the SET to the amplifiers. To increase the readout bandwidth, RF reflectometry is usually used [17–19]. In this approach, the input resistance of the charge sensor is matched to 50-Ω by an LC matching network closely connected to the SET. Consequently, by sending an RF pulse and measuring the reflected power steered by a directional coupler, SET impedance variations, and thus the qubit state can be monitored.

B. *Need for Cryo-CMOS Control and Readout Circuits*

As mentioned earlier, classical control is generally operated at room temperature. While this is convenient for small qubit numbers, large arrays, typically in the hundreds, may pose a problem of feasibility and reliability due to the large temperature gradient and the complex interconnect between classical and quantum devices. We thus believe ensuring the scalability of future quantum computers will be a necessary condition for the development of this field so as to lead to a practical realization.

To address scalability, in 2016 we have proposed cryogenic electronics, and in particular, cryogenic CMOS, or cryo-CMOS, as the key technology will enable large numbers of qubits [20]. Classical cryo-CMOS control electronics will generate the signals necessary to control qubits at amplitudes that are far smaller than today at room temperature, since no thermalization will be required. Low temperatures will also be advantageous to reduce thermal noise in front-ends, to reach levels far lower than achievable at room temperature. The proximity in space and temperature of cryo-CMOS circuits and qubits will drastically reduce the complexity of the cabling, and possibly enable superconductive interconnect, thus optimizing thermal isolation with virtually zero electrical losses.

Given that cryo-CMOS circuits and systems will operate at temperatures close, ideally equal, to those of the qubits, the main limitation is the power dissipation of classical circuits. This will need to be budgeted to be within the limits of thermal absorption by the refrigeration system used. Up until a few years ago, non-CMOS devices were proposed, such as high-electron-mobility transistors (HEMTs), SiGe heterojunction bipolar transistors (HBTs), GaAs logic, and rapid single-flux quantum (RSFQ) circuits [21–23]. However, none of these technologies, except perhaps HBTs, can take advantage of over 60 years of innovation and optimization at industrial levels, as CMOS. Only CMOS technology can offer the integration of billions of transistors on a single chip, while ensuring low-power consumption and sub-Kelvin functionality, thus representing the ideal choice for integrating complex electronic systems, potentially extending into single-digit Kelvin regimes [24–26].

26.3 Our Control Paradigm and Trade-Offs

As described in the previous section, the electronic interface for a typical spin-qubit quantum processor must be able to provide the following functionalities:

- Generate accurate DC voltages to properly bias the quantum dots
- Generate microwave pulses to perform single qubit operations
- Generate fast baseband pulses to perform two-qubit operations (and activate the read-out sequence)
- Detect (at DC or RF) the signals from the quantum processors to perform qubit read-out.

Similar features are required for quantum processors employing different qubit technologies, such as superconducting qubits and multi-electron spin qubits. While those functionalities are currently implemented using room-temperature equipment, moving to a fully cryo-CMOS implementation is beneficial to facilitate the progress towards future large-scale quantum computers. The architecture of such generic classical control system is shown in Fig. 26.5. For generality, the system also includes an optical link, which may be required for specific types of qubits. The block diagram in Fig. 26.5 closely resembles a radio-frequency transceiver, including a receiving path and a transmitting path, and comprises typical functions found in standard radios, including an RF section (amplifiers, down/up-converters, frequency generators) and a mixed-signal/digital section (analog-to-digital and digital-to-analog converters and

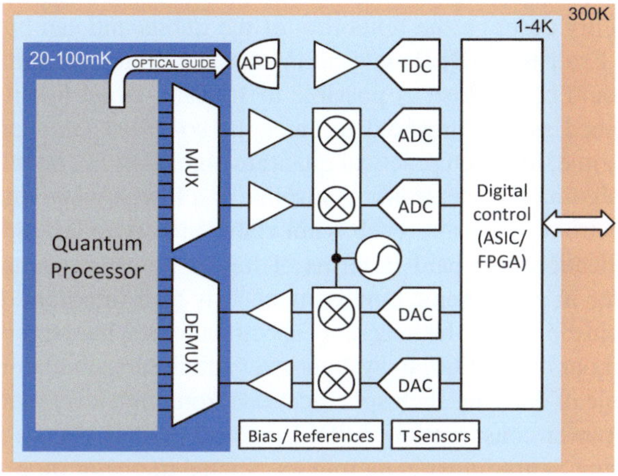

Fig. 26.5 Generic classical control system for quantum processors. The system includes conventional radio-frequency components found in most radio transceivers, along with optical detectors, useful in certain qubits. © IEEE 2016

a digital signal processor). As in state-of-the-art radio, we envision the fabrication of such system in an advanced nanometer CMOS node to take advantage of the high-frequency capabilities of the transistors to efficiently implement both RF and digital circuit blocks. As in standard radio, the power dissipation of such "*quantum transceiver*" must be minimized. While in typical radios such constraint is imposed by the limited capacity of the batteries in portable electronics, the power dissipation electronics in quantum computers is limited by the cooling capabilities of the cryogenic infrastructure. For this reason, most researchers anticipate that the majority of the control system in Fig. 26.5 will operates at the 4-K stage of typically employed cryostats, for which several watts of thermal absorption capability is available. However, some circuitry must be designed to operate in direct contact with qubits, typically housed in another cryogenic stage at sub-1-K temperature, which has much more limited cooling capabilities, well below 1 mW. For instance, multiplexing and demultiplexing (both in time and in frequency) are used to reduce the number of interconnects between qubits and the classical control [20]. Since these circuits require only a few transistors, it is conceivable that they be implemented in cryo-CMOS technology too, whereas the power dissipation could be kept under a few μW, thus ensuring compatibility with milli-Kelvin environments. For those reasons, designing qubits that can operate at a more practical temperature between 1 and 4 K is attracting significant research effort, as it would allow qubits and electronics all operating in close proximity and, eventually, on the same chip.

Unlike a room-temperature RF transceiver used for 4/5G or WiFi, the specifications for the cryo-CMOS control interface are not enumerated in any standard. Today's standard approach in quantum-computing labs is using the top-notch room-temperature equipment to ensure that qubit performance is not limited by the electrical control and read-out. Qubit performance are measured in terms of their *fidelity*.

In practice, fidelity describe the reliability of the qubits and can be used in a similar way as Bit Error Rate (BER) is used in classical digital systems. Although the fidelity that should be as close as possible to 100%, a target fidelity above 99.9% is usually assumed as the threshold to enter the so-called *fault-tolerant quantum computation* regime, in which practical quantum algorithm can be reliably executed. While over-designing the electronic controller is allowed when employing room-temperature electronics, this approach is not viable in a cryo-CMOS implementation as better specifications are paid in terms of higher power consumption, which is a scares resource in a cryogenic environment. It is then important to design cryo-CMOS circuits just meeting the necessary specifications. Thus, significant effort has been devoted in our consortium to system-level simulators, emulators, and verification tools capable of deriving such specifications from high level requirements, such as fidelity and power consumption. SPIN Emulator (SPINE) is one such approach, capable of solving Schrödinger equations associated to one or more single-electron spin qubits and translating qubit fidelity into specifications related to the control signals that must be applied to the qubit(s) [27, 28].

Figure 26.6 show an example of such simulations using SPINE, while in Table 26.1 we report the results of an example study we conducted on the impact of amplitude, phase and timing errors for a microwave pulse used for a single-qubit rotation on a spin qubit. Increasingly, researchers are using these types of specifications to design complex systems for qubit control [29, 30]. The main challenge for those designers is providing those required specifications while achieving high *power efficiency*, so as to be able to drive and read-out the largest possible number of qubits in the allowed power budget.

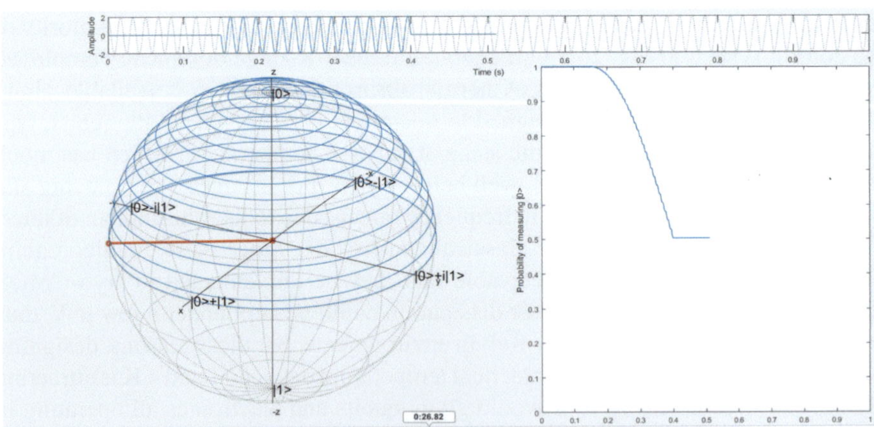

Fig. 26.6 Simulation via the SPINE software platform of a qubit driven by a microwave excitation (top). The qubit evolution on the Bloch sphere (bottom left) shows a transition from $|0\rangle$ (North pole) to $|1\rangle$ (South pole) and through a maximum superposition state (equator). The probability of measuring the state $|0\rangle$ is also shown (bottom right). © IEEE 2019

Table 26.1 Example specifications for a π-rotation on a single-electron spin qubit for a fidelity of 99.9%

Frequency inaccuracy	11 kHz
Phase noise	−106 dBc/Hz at 1 MHz offset, −20 dB/dec slope
Wideband additive noise	7.1 nV/Hz
Phase inaccuracy	0.64°
Amplitude inaccuracy	14 μV on 2.0 mV amplitude
Amplitude noise	SNR = −40 dB
Duration inaccuracy	3.6 ns on 500 ns nominal duration
Timing jitter	3.6 ns_{rms}

26.4 Cryo-CMOS Device Behavior

A supplementary, but not negligible, challenge for cryo-CMOS designers is operating CMOS devices at cryogenic temperatures. Unlike traditional CMOS designers that can rely on standard CMOS models provided by the CMOS foundry, no device models are readily available for temperatures below −55 °C, and even more so for cryogenic temperatures down to 4 K. It is not possible to make use of CMOS model parameters tuned to room temperature and compact models qualified for the standard temperature range, since, although CMOS transistors behave properly even at sub-1-K temperatures, new physical effects come into play and device parameters for both active and passive devices are dramatically different when cooling them from 300 to 4 K. Figure 26.7 shows a sample of measurements of 40-nm CMOS transistor I_d-V_{ds} and I_d-V_{gs} characteristics.

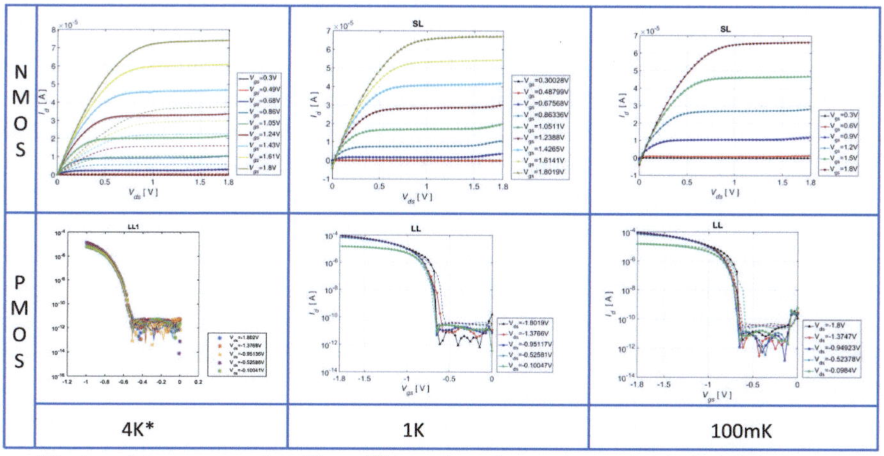

Fig. 26.7 CMOS I_d-V_{ds} and I_d-V_{gs} characteristics at various deep-cryogenic temperatures and W/L combinations. © IEEE 2019

A first noticeable difference is the larger I_d current at 4 K (solid lines) compared to 300 K (dotted lines), see first upper-left plot for a long and wide NMOS device; this is due to the temperature dependence of physical quantities such as the increase in mobility ($\approx 2\times$), which is countered by an increase of threshold voltage ($\approx 30\%$). The former is dominant and is due to an overall decrease in electron scattering, while the latter is due to an increase in required ionization energy. Another important difference at 4 K is the reduction of the velocity saturation leading to I_d curves to be equally spaced compared to those at 300 K where saturation occurs due to pinch-off. The following two I_d-V_{ds} characteristics of a short and wide device show the good matching of simulated characteristics with matched parameters compared to measurements for both 1 K and 100 mK.

The second row of Fig. 26.7 shows the characteristics of a PMOS long and wide device at the same three temperatures. The first row displays the difference between 4 and 300 K, while the other two point to the matching between model and measurement. The improvement in subthreshold slope (*SS*) when cooling down the devices as can be seen from Fig. 26.7, was expected due to its intrinsic dependence on temperature,

$$SS(T) = \left[\frac{\partial \log(I_D)}{\partial V_{GS}}\right]^{-1} = \ln(10)\frac{nkT}{q}.$$

However, the measured *SS* improvement is only around 3.4× and not equal to the ratio of temperatures; this can be explained by the incomplete ionization of impurity atoms at cryogenic temperatures leading to an important increase in the non-ideality factor $n(T)$. Another physical phenomenon observed at 4 K is carrier freeze-out in the substrate. This can lead to a kink effect in some older processes due to the increase of the substrate potential caused by impact ionization generating electron-hole pairs with holes flowing through the substrate ultimately forward biasing the source-substrate junction.

The additional physical effects observed at cryogenic temperatures lead to different behaviors of MOSFETs depending on the process they are fabricated in; thus, FDSOI MOSFETs and FinFETs will show different behavior at 4 K than standard Bulk-CMOS process; however, they all will show correct transistor operation.

From a circuit design point of view the important differences at cryogenic temperatures are the reduction in leakage and the increase in the transconductance efficiency (g_m/I_D) by up to 3.4× in weak inversion [31, 32]. Device mismatch however is generally higher at cryogenic temperatures and, while thermal noise is lower, flicker noise can be significant, thus especially impacting analog and mixed-signal circuits [33, 34]. While physics-based models are emerging [31], model fitting based on experiments has been the technique of choice thus far [32]. Digital modeling follows a similar path, though electronic design automation (EDA) tools have yet to reach the necessary level of maturity to automate the design of large systems in design flows modified and optimized for deep-cryogenic temperatures [35]. Nevertheless, successful attempts have been made, as we will see later, to automate the place-and-route process of digital circuits.

26.5 Components Required and Some Examples

Several cryo-CMOS components have already been successfully designed and test from Fig. 26.5 [36]. After verifying our cryo-CMOS models by means of so-called farms of active and passive components, voltage regulators were built for various scenarios. Bandgap references are generally the solution of choice to create well controlled, stable voltage and current sources. At cryogenic temperatures, these solutions cannot be used, since they are based on traditional BJTs that suffer from a strong increase in base resistance and a strong decrease of current gain at temperatures below 100 K [37]. As an alternative, one can use HBTs fabricated in SiGe technologies or MOS and dynamic-threshold MOS (DTMOS) transistors biased in weak inversion. Figure 26.8 shows a schematic of a conventional bandgap reference compared to a DTMOS reference, along with the voltage stability plot from room temperature down to 4 K. Current research [38] is focused on understanding which device (among NMOS, PMOS and DTMOS) can ensure the best performance both in terms of minimum temperature coefficient and statistical variations due to process spread and mismatch.

Low-dropout (LDO) circuits can be used in combination with voltage references to drive the necessary currents at the wanted voltage levels. Several LDOs have been designed and successfully tested but only a few components have been shown to operate correctly at 4 K. Figure 26.9 shows an example of such design implemented in discrete technology [39].

Next, let us consider the front-end electronics that is in direct contact with the qubits.

Figure 26.10 shows two configurations often used while interfacing with qubits and qubit arrays. The challenge is to build these circuitries in such a way that the power

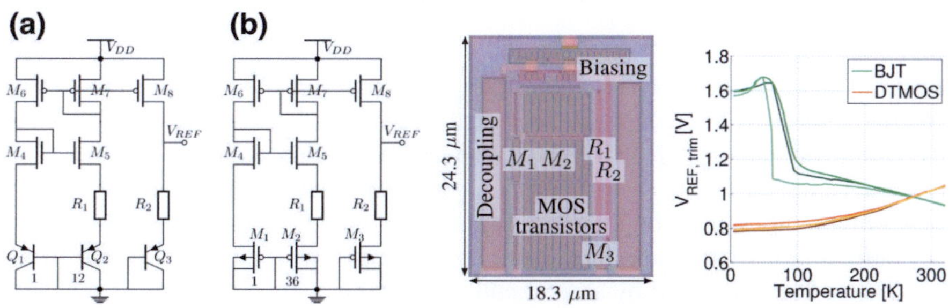

Fig. 26.8 Schematics of a conventional bandgap reference based on BJTs (**a**) and a reference based on DTMOS transistors (**b**); photomicrograph of DTMOS reference and its voltage stability over a wide temperature range after trimming, compared to BJT based bandgaps (from [37]) © IEEE 2018

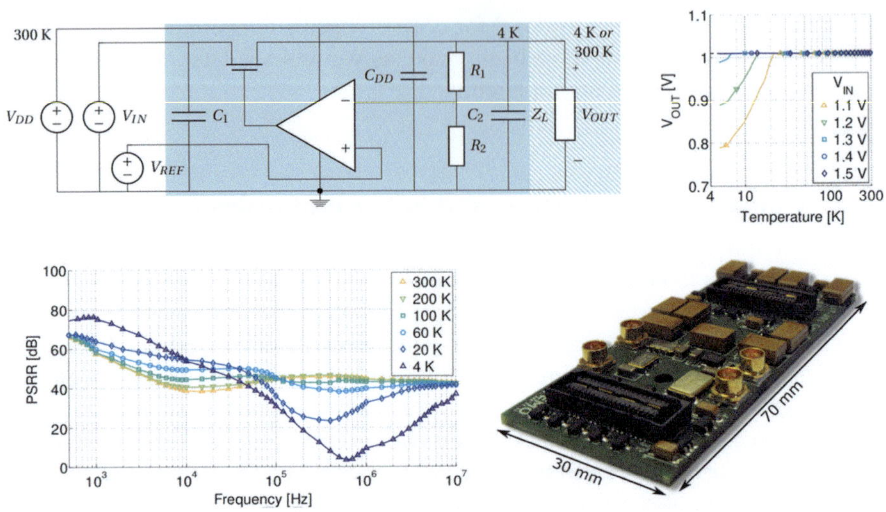

Fig. 26.9 Discrete low top-out (LDO) voltage regulator. Clockwise from top left: schematics of the LDO; I-O behavior as a function of temperature; PSRR measurement [39]; implementation of the LDO in a printed circuit board. © IEEE 2018

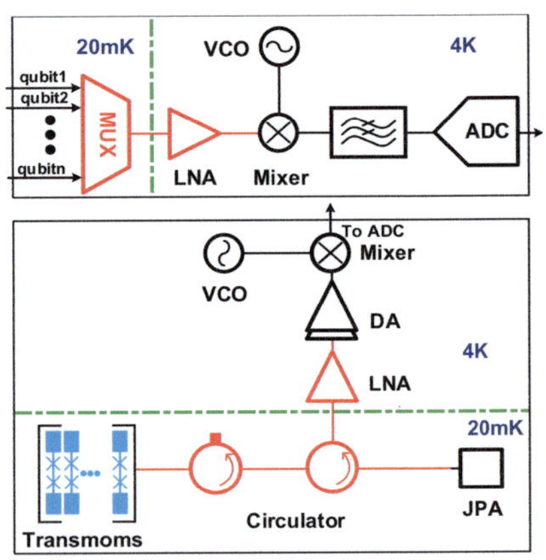

Fig. 26.10 Front-end electronics: multiplexer and low-noise amplifier (top, in red). Circulator and low-noise amplifier (bottom, in red)

consumption is sufficiently low for mixing chambers operating at deep-submilli-Kelvin regimes. A simple multiplexer based on MOS pass-transistors was proposed in [20] and implemented in [40]. With a voltage drop of only a few millivolts and negligible current, this circuit can dissipate only a few microwatts of power, nevertheless scalability to thousands of qubits is questionable.

Configurations for reading out (and controlling) qubit states organized in 2D arrays have been proposed by several authors, notably [40, 41]. In Fig. 26.11 (left), the selection of quantum dot is exposed by means of column and row lines (CL, RL) and qubit lines (QL), thus enabling the transfer of electrons between dots, so

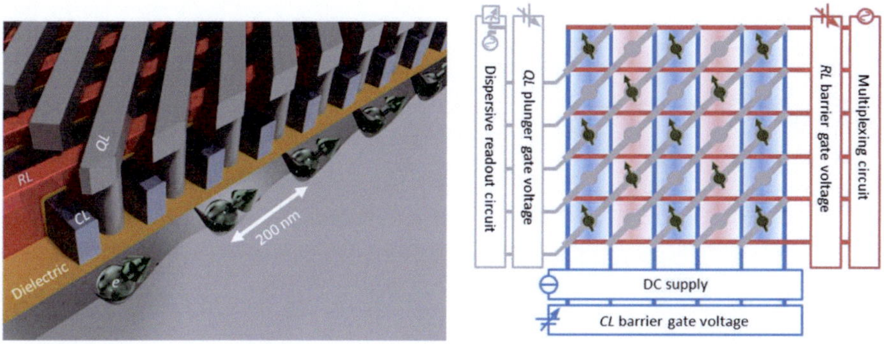

Fig. 26.11 2D crossbar readout configuration proposed in [41]. Left: artist's rendering and 3D construction of the array, dielectrics are not shown. Right: column/row lines (CL, RL) and qubit lines (QL) are used to select the qubit to control or read out. The quantum dots in the array have 50% electron occupancy, whereas the electron spin encodes the state of the qubit. Electron transfer occurs between quantum dots through this means to achieve two-qubit gates (for nearest neighbors) and long-range entanglement for non-adjacent quantum dots

as to enable two-qubit gates for neighboring dots and long-range entanglement for non-adjacent quantum dots. The spin of these electrons, encoding the state of the corresponding qubit, is read out by peripheral electronics, shown in Fig. 26.11 (right) placed to the exterior of the array and organized in $R + C$ channels, with R and C being the number of rows and columns, respectively. This configuration is preferable to using one channel per qubit, which would result in RC channels. A circulator is indispensable in many of these classical controllers. Usually circulators are bulky, often requiring a large footprint. In [42] for instance, a passive cryo-CMOS circulator was proposed, capable of achieving isolation better than 10.5 dB (port 2/3) and 16.5 dB (port 1) and attenuation less than 1.5 dB (port 2/3) and 4.5 dB (port 1) in the 6.5–7 GHz frequency range.

Next, let us consider the front-end used to read out the state of the qubits, implemented as a low-noise amplifier (LNA). The most stringent specifications are input-referred noise (NET < 10 K) and power dissipation (<1mW/qubit). The bandwidth is determined by the number of qubits the LNA will be able to read out. For instance, assuming a power dissipation of 100 mW, the LNA needs to serve 150 qubits. Assuming a FDMA scheme and 3.3 MHz separation per qubit, a 3-dB bandwidth of 500 MHz is required. In our implementation (Fig. 26.12), we could achieve this requirement dissipating a power of 91 mW, thus resulting in 0.61mW/qubits.

To satisfy the noise specification, we chose a noise-canceling architecture, proposed in [43]. The IC was fabricated in 160 nm CMOS technology with a gain (S21) of 57 dB, an input matching (S11) of −8 dB, and a noise figure better than 0.3 dB at 4 K. IIP2 was measured at −5 dBm and IIP3 at −47 dBm. However, given that the input signal is expected to be weaker than −110 dBm, linearity is not of concern in this application.

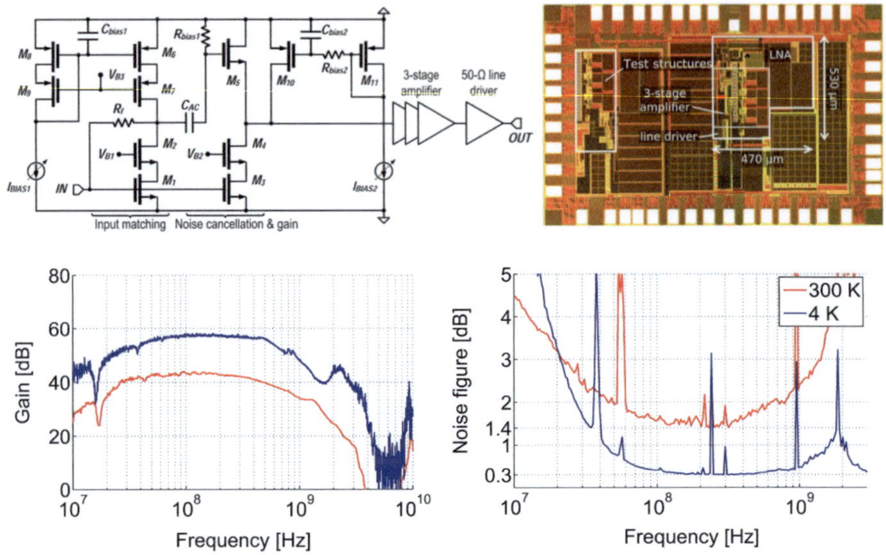

Fig. 26.12 Clockwise from top: low-noise amplifier (LNA). I_{BIAS1} and I_{BIAS2} determine the best possible noise figure at any temperature. Photomicrograph of the IC fabricated in 160-nm CMOS technology. Noise figure, S11, and S21 as a function of frequency. The measured 3 dB bandwidth is 500 MHz. © IEEE 2018

Finally, let us consider logic circuits. The obvious choice is a field-programmable gate-array (FPGA). Virtually all functions of commercially available FPGAs have been successfully tested at 4 K [44, 45]. See Table 26.2 for a list of tested functions.

These tests have enabled us to design all digital components required by the classical control and even A/D converters (ADCs) with up to 2.4 GSa/s conversion rate and 8 bits of resolution [46]. The principle of operation of the ADC is shown in Fig. 26.13. A square wave is generated in the MMCM block and converted to a shark fin wave through external resistor R_{REF} and parasitic capacitor C_{INT}.

A time-to-digital converter (TDC) measures the time from the rising edge of the clock to the crossing point of the input signal to the fin, which is detected by a comparator in a standard LVDS (see Fig. 26.13). Figure 26.13 also shows the spectral purity of the reconstructed signals with two frequencies at 15 K and a sampling rate of 1.2 GSa/s. The TDC's block diagram is shown in Fig. 26.14; it consists of a delay line implemented by a carry chain, so as to enable high uniformity, and thus low INL/DNL, a set of latches, and a thermometer decoder. The IO delay is necessary to use the TDC in parallel with other TDCs, so as to achieve up to 2.4 GSa/s sampling.

C. Cryogenic Frequency Generation

In order not to degrade the fidelity of a single qubit gate, the *integrated* frequency noise (FN) of the control signal should be $<1.9\ kHz_{rms}$ from a 6 GHz carrier frequency [27]. The lower integration bound of the phase noise (PN) profile of phase-locked loops (PLL) is set by the quantum operation cycle (worst case: $1/T^* \approx 8.3$ kHz)

Table 26.2 Functions of a Xilinx Artix-7 FPGA verified at 4 K

Module	Usable	Test	Performance w.r.t. RT
IOs	✓	Output swing versus frequency	Drive strength increases
LVDS	✓	Differential clocks from RT	
LUTs	✓	LUTs connected as oscillator in full columns	Propagation delay decreases <3%, jitter increases <78% (151–270 ps)
CARRY4	✓	Carry chains connected as oscillator in full columns	Propagation delay decreases <3%, jitter increases <16% (18.6–21.5 ps)
DFF	✓	n-bit counters + comparators	Maximum operating frequency reduces
BRAM	✓	Transfers of 8 kB (write and read)	No corruption in 100 test sets of 8 kB
MMCM	✓	100 MHz differential input clock multiplied by 12 and divided by various values, single ended output	Jitter decreases on average 52% (13.6–6.5 ps)
PLL	✓	100 MHz differential input clock multiplied by 12 and divided by various values, single ended output	Jitter decreases on average 54% (13.7–6.4 ps)
IDELAYE2		IDELAYE2 elements connected as tunable oscillator (output frequency variable 13–70 MHz)	Delay decreases <30%, jitter increases <50%
DSP48E1	✓	Random calculations (additions, subtractions and multiplications)	No corruption in 400 calculations
Temperature diode	✓	Operating range 4–300 K	

and the qubit operation speed determines the higher limit (e.g., ~10 MHz) [36]. By considering a PLL bandwidth of 300 kHz, the required FN, and the integration bandwidth, an in-band PN of −115 dBc/Hz and an out-of-band PN of −147 dBc/Hz at a 10 MHz offset from a 6 GHz carrier are required [36]. Since the phase detector and oscillator respectively dominate the in-band and out-of-band PN profile of the PLL, we investigate the performance of those blocks at cryogenic temperatures next.

(1) *Performance of Cryo-CMOS Phase Detectors*

The in-band PN of an all-digital PLL is usually determined by a time resolution of a TDC [47]. To achieve the target in-band PN, the TDC resolution has to be finer than 2 ps even when considering a reference frequency as high as 100 MHz. The TDC core is usually based on a digital delay line, whose time resolution is equal to an inverter propagation delay (10 ps in 40-nm CMOS at RT). Figure 26.15 shows the measured

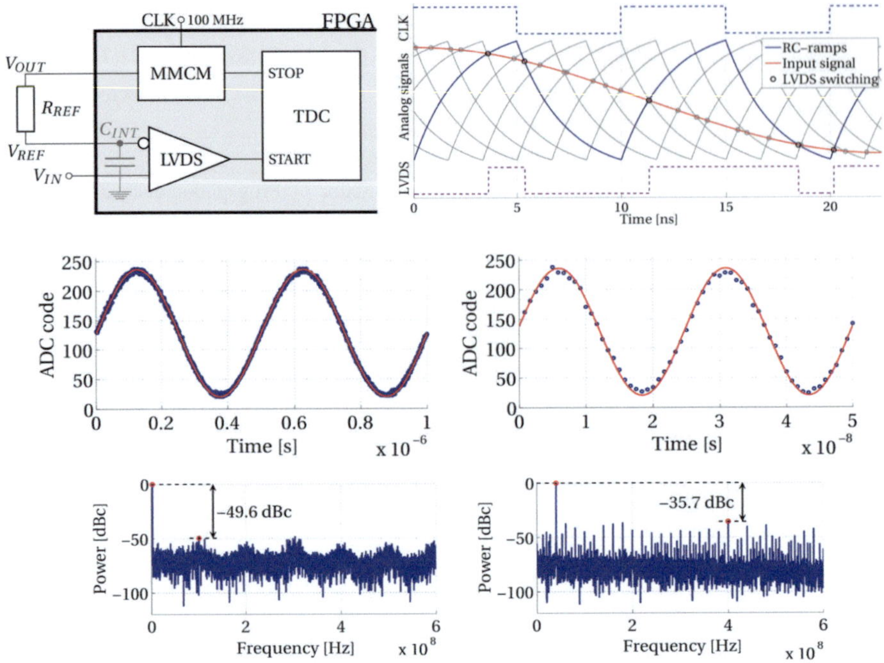

Fig. 26.13 A/D converter implemented with an FPGA. Top: block diagram of the ADC in the FPGA; sinusoidal wave sampled through a sequence of shark fin waves, the 100 MHz clock and LVDS response are also shown. Middle: reconstruction at 15 K of 2 and 40 MHz sinusoidal waves sampled at 1.2 GSa/s. Bottom: FFT of the reconstructed signals at 15 K. The achieved effective number of bit (ENOB) were 8 and 5.6 bits, respectively. © IEEE 2016

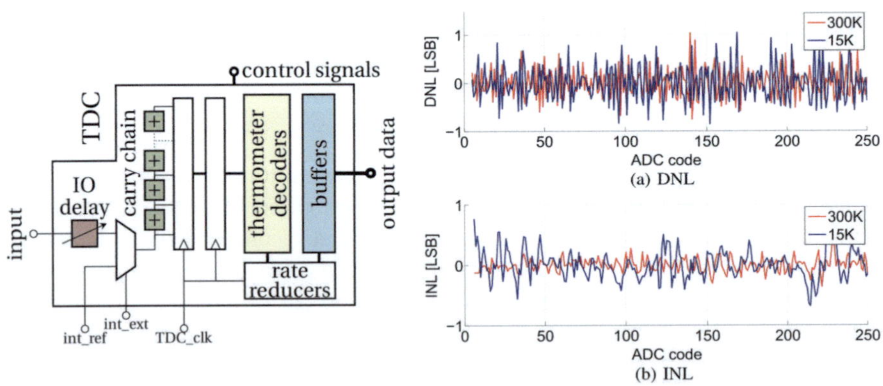

Fig. 26.14 TDC implementation in an Artix-7 FPGA. 6 TDCs operating in parallel with a delay enable an ADC sampling rate of up to 2.4 GSa/s. In the inset: measured differential non-linearity (DNL) and integral non-linearity (INL). © IEEE 2016

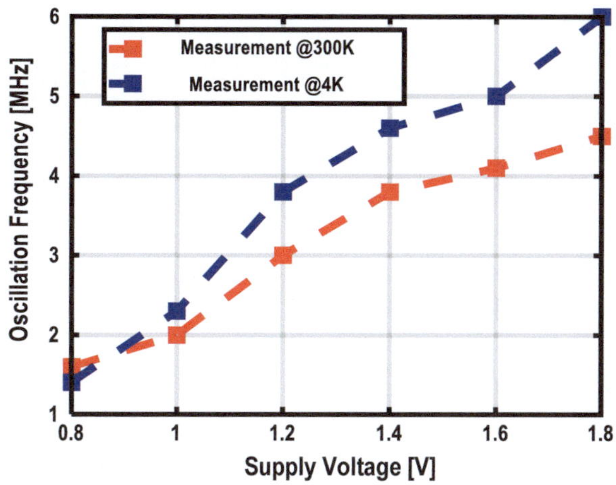

Fig. 26.15 Measured frequency of a ring oscillator versus supply voltage taken from [32]. © IEEE 2018

oscillation frequency of a ring oscillator versus different supply voltages for both 300 and 4 K [32]. The oscillation frequency increases by 36% due to the higher current driving capability of the transistors at 4 K, resulting in 6–7 ps for the inverter delay. Despite this improvement in time resolution, to satisfy the in-band PN requirement, there is a need for sub-gate-delay resolution for TDCs. Furthermore, the current mismatch of CMOS transistors also increases by 20% at cryogenic temperatures, as shown in [33]. As a result, the TDC nonlinearity becomes much severer at 4 K, thus degrading PLL's jitter and spurious tone emissions. Therefore, innovative PLL architectures (i.e., injection-locked structure in [48, 49]) associated with intensive digital calibration techniques are highly desirable at cryogenic temperatures to tackle the challenges mentioned above [50].

(2) *Performance of Cryo-CMOS RF Oscillators*

To investigate the performance of RF oscillators over a wide range of temperature (i.e., 4–300 K), a 1:2 step-up transformer-based oscillator is designed and prototyped in a 40-nm 1P7M CMOS process with an ultra-thick metal layer [36]. The oscillator schematic, and its chip micrograph are shown in Fig. 26.16a, b, respectively. At the bottom of the core transistors ($M_{1,2}$), a 5-bit binary-weighted switchable resistor is implemented to roughly control the oscillator current. The single-ended primary and differential secondary capacitor banks are realized using two 6-bit switchable metal-oxide-metal (MoM) capacitors. Figure 26.16c shows that the oscillation frequency shifts up ~7% at the same tank configuration from 300 to 4 K, mainly because of the reduction of the tank single-ended parasitic capacitance due to carrier freeze-out in the substrate.

The measured PN plot of the oscillator is shown in Fig. 26.16d for different temperatures at the same power consumption [36]. To better understand the measured

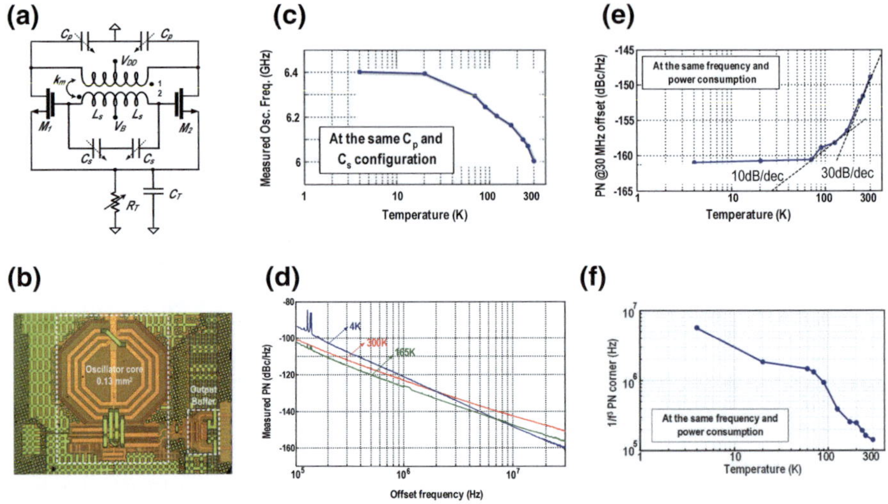

Fig. 26.16 **a** Oscillator schematic and **b** chip micrograph; **c** oscillation frequency versus temperature; **d** measured PN at 6.3 GHz at various temperatures; **e** oscillator PN at 30 MHz and **f** $1/f^3$ PN corner versus temperature for a carrier frequency of 6.3 GHz. © IEEE 2018

PN, the oscillator PN at 30 MHz offset frequency due to white noise upconversion is drawn in Fig. 26.16e. From 300 to 170 K, the PN improves in the white noise region by 30 dB/decade over temperature, out of which 10 dB/decade is attributed to the white noise decrease due to temperature reduction, while the tank's Q-factor enhancement realizes the remaining 20 dB/decade. At room temperature, magnetically induced image currents from the tank flow in the low-resistive substrate, reducing the quality factor. Due to the carrier freeze-out at cryogenic temperatures, the substrate becomes highly resistive, hence lowering substrate losses considerably. Moreover, the inductor's series resistance also decreases at lower temperatures, thus further improving the tank's Q-factor [36]. Interestingly, the oscillator PN improvement reduces to ~10 dB/dec from 170 to 77 K and a negligible PN reduction is observed by further reducing the temperature to 4 K. This phenomenon can be justified by the following reasons. First, the resistivity of the metals, and hence the quality factor of passives does not improve by further reducing the temperature due to the impurities and crystallographic defects in metal layers [51]. Second, the quality factor of an LC tank also depends on the loss of switched capacitors used for frequency tuning. Unfortunately, the transistor's on-resistance and thus the quality factor of switched capacitors just improves by 2× and limits the tank Q-factor below a certain temperature [36]. Third, white noise in nanoscale CMOS devices is limited by temperature-independent shot noise at cryogenic temperatures and just scales 10× from 300 to 4 K [52–54].

The oscillator's $1/f^3$ PN corner increases dramatically at lower temperatures, as shown in Fig. 26.16f, thus indicating in a larger 1/f noise corner for MOS transistors at cryogenic temperatures [55, 56]. At cryogenic temperatures, it appears that the oscillator PN is dominated by the 30 dB/dec region, and the $1/f^3$ corner of RF oscillators can go beyond the PLL bandwidth, significantly degrading PLL's jitter and integrated FN [36]. In order to mitigate the effect of the increase in the $1/f^3$ corner, one needs to resort to oscillator topologies with low flicker noise upconversion. It is well-known that the flicker noise upconversion in an LC oscillator significantly reduces if the common-mode of the circuit also resonates at twice the oscillation frequency [57, 58]. However, it is not trivial to adapt this technique for cryogenic oscillators where the value of single-ended capacitance significantly varies due to carrier freeze-out in the substrate [50]. Consequently, this issue calls for a new digital calibration loop to automatically adjust the common-mode resonance of the oscillator at its optimum point [59].

26.6 Design Verification of the Complete Quantum-CMOS System

Once all the circuit blocks of the CMOS controller are in place a final verification of the performance and fidelity of operating the qubits can be achieved simulating the complete system in SPINE. As an example, a full system containing a controller targeting multiple qubits is simulated. A fidelity of 99.9% is targeted while performing a π-rotation in 50 ns (Table 26.1). Figure 26.17 shows the system under consideration, containing a high-level description of the quantum computer's controller, Verilog-A models of the digital-to-analog converters (DACs) and an analog mixer circuit at transistor-level integrated together with SPINE.

The performance was verified by simulating a small quantum algorithm executed by the controller and consisting of 4 gates: a π-rotation around the X-axis in 50 ns followed by three additional $\pi/2$-rotations in 25 ns around the X- and Y-axis, also shown in Fig. 26.17. It can be seen that in response to the controller, the DACs generate the required in-phase and quadrature-phase signals for the mixer, and the analog mixer circuit performs the required upconversion. In response to the generated RF-signal *Vout*, the qubit performs the expected rotations as evident from the simulated spin-up probability, finally achieving a 99.98% chance of success meeting the required system performance.

26.7 Vision for the Future, Trends

The classical control schemes presented here for qubits require smart interfaces to quantum algorithms. What is usually referred to as quantum stack is the ensemble of architectural components designed to translate such algorithms to quantum circuits and finally to quantum control via assembly languages, like QASM. Such architectures are complex and significant development is ongoing [60].

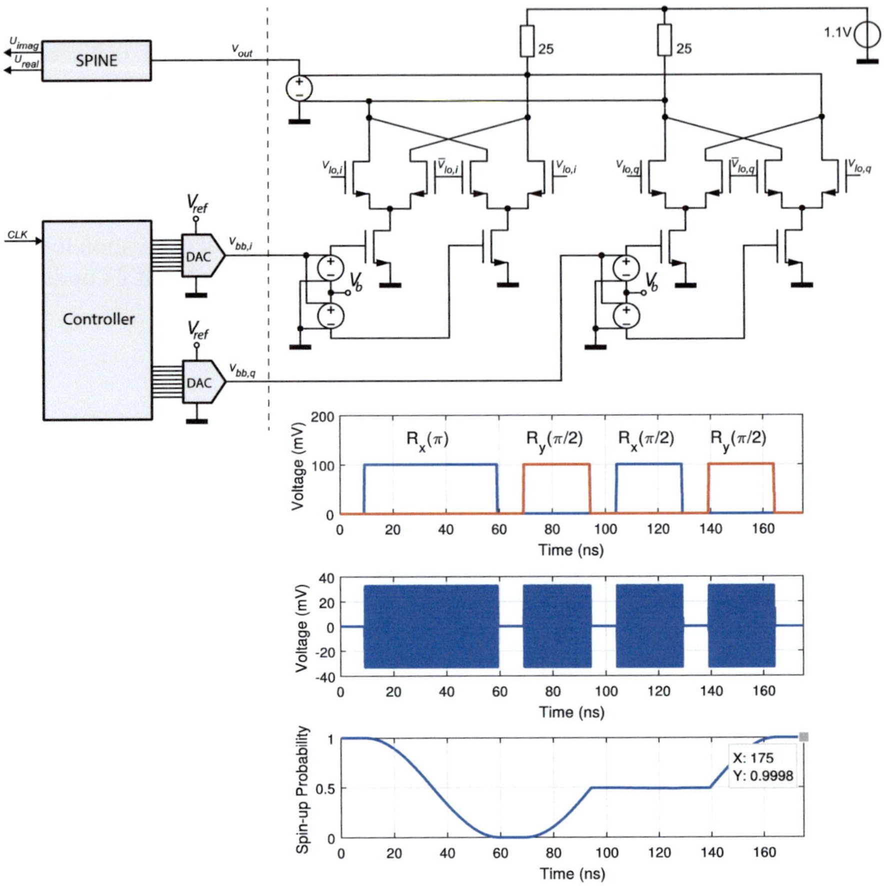

Fig. 26.17 From top to bottom: schematic of a full system containing the controller and the quantum circuit in SPINE; result of the full system simulation, including the voltage at the output of the DACs, the I- and Q-signals driving the mixer and the quantum gate, the voltage at the output of the mixer, driving the qubit, and the qubit spin-up probability. © IEEE 2018

While most solid-state qubits operate around 20 mK today, there has been an important general trend to increase such temperature to one Kelvin or higher [26]. This could enable even more compact setups and, in the short to medium term, fully integrated solutions. Moreover, an operating temperature close to 4 K could enable pure liquid helium (LHe) refrigeration, thus significantly reducing system complexity and cost.

Furthermore, with the overall increase of temperatures, the power budget of cryo-CMOS circuits and systems could drastically increase, thus enabling an even higher acceleration of quantum processor sizes and, with that, the computational power of the machines.

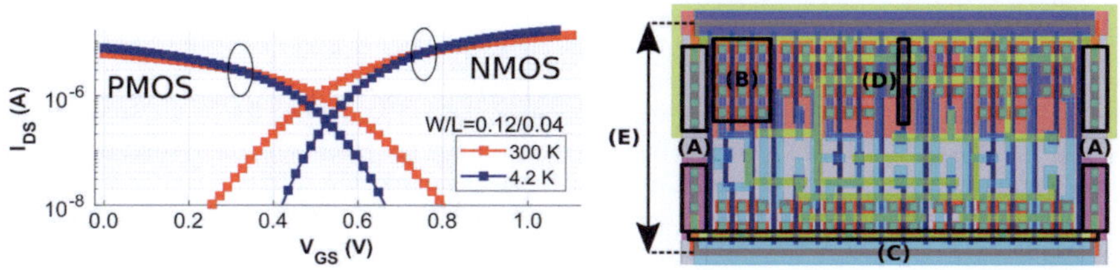

Fig. 26.18 MOS transistor I_{DS}-V_{GS} characteristics for minimum-size devices at room temperature and 4 K (left). Cryo-CMOS library element implemented in 40 nm CMOS technology. The component is designed to minimize the risk of latchup and to reduce threshold voltage increases due to low temperatures. We took several measures (A–E) described in the text. © IEEE 2019

ASIC implementations of logic circuits are preferable, as they don't require overhead circuits as in FPGAs, this results in significant less power, which is critical in cryo-CMOS applications, where every milliwatt counts. For this purpose, we developed a library, 'cooLib', specifically designed to operate at deep-cryogenic temperatures. An example of a cell designed using these principles is shown in Fig. 26.18.

We designed integrated well taps on cell boundary (A) for tight biasing of the wells, so as to prevent latchup at low temperatures. Back-biasing and, when possible, back-gate biasing were used via secondary power rails (C) to reduce voltage threshold increases due to low temperatures. Moreover, inverse-narrow-width effect (B) and increased mismatch (D, E) aware sizing were adopted. Preliminary results show that ultra-low-power cryo-CMOS logic is feasible and easy to design.

26.8 Conclusions

Quantum computing is a completely new paradigm that holds the promise of large speedup over conventional von Neumann machines. However, controlling qubits using classical electronics is a necessary but daunting task. We have proposed to bring classical electronics closer to the qubits in space and temperature, ultimately enabling full integration in a not so distant future. This goal requires new qubits operating at higher temperatures and electronics operating at deep-cryogenic temperatures. We believe that this can be achieved by means of cryo-CMOS circuits and systems implemented in advanced technological nodes with strict power budgets at multi-gigahertz bandwidths. This trend may lead to larger quantum processors, lower power requirements, and ultimately lower cost.

Acknowledgements The authors are grateful to the members of the coolgroup: the analysis tools, circuits, and systems presented here have been designed and tested by them, and to Intel Corp. for funding.

References

1. R.P. Feynman, Simulating physics with computers. Int. J. Theor. Phys. **21**(6), 467–488 (1982)
2. L. Vandersypen, Quantum computing—the next challenge in circuit and system design, in *International Solid-State Circuits Conference*, San Francisco, CA (2017)
3. H. Bluhm, L.R. Schreiber, Semiconductor spin qubits—a scalable platform for quantum computing?, in *IEEE International Symposium on Circuits and Systems*, Sapporo, Japan (2019)
4. M.A. Rol, C.C. Bultink, T.E. O'Brien, S.R. de Jong, L.S. Theis, X. Fu, F. Luthi, R.F.L. Vermeulen, J.C. de Sterke, A. Bruno, D. Deurloo, R.N. Schouten, F.K. Wilhelm, L. DiCarlo, Restless tuneup of high-fidelity qubit gates. Phys. Rev. Appl. **7**(4), 041001 (2017)
5. J. Heinsoo, C.K. Andersen, A. Remm, S. Krinner, T. Walter, Y. Salathé, S. Gasparinetti, J.-C. Besse, A. Potocnik, A. Wallraff, C. Eichler, Rapid high-fidelity multiplexed readout of superconducting qubits. Phys. Rev. Appl. **10**(3), 034040 (2018)
6. M. Veldhorst, J.C.C. Hwang, C.H. Yang, A.W. Leenstra, B. de Ronde, J.P. Dehollain, J.T. Muhonen, F.E. Hudson, K.M. Itoh, A. Morello, A.S. Dzurak, An addressable quantum dot qubit with fault-tolerant control-fidelity. Nat. Nanotechnol. **9**, 981–985 (2014)
7. J.T. Muhonen, J.P. Dehollain, A. Laucht, F.E. Hudson, R. Kalra, T. Sekiguchi, K.M. Itoh, D.N. Jamieson, J.C. Mccallum, A.S. Dzurak, A. Morello, Storing quantum information for 30 seconds in a nanoelectronic device. Nat. Nanotechnol. **9**(12), 986–991 (2014)
8. D.M. Zajac, A.J. Sigillito, M. Russ, F. Borjans, J.M. Taylor, G. Burkard, J.R. Petta, Resonantly driven CNOT gate for electron spins. Science **359**(6374), 439–442 (2018)
9. J. Yoneda, K. Takeda, T. Otsuka, T. Nakajima, M.R. Delbecq, G. Allison, T. Honda, T. Kodera, S. Oda, Y. Hoshi, N. Usami, K.M. Itoh, S. Tarucha, A quantum-dot spin qubit with coherence limited by charge noise and fidelity higher than 99.9%. Nat. Nanotechnol. **13**, 102–106 (2018)
10. J.P. Gaebler, T.R. Tan, Y. Lin, Y. Wan, R. Bowler, A.C. Keith, S. Glancy, K. Coakley, E. Knill, D. Leibfried, D.J. Wineland, High-fidelity universal gate set for $^{9}Be^{+}$ ion qubits. Phys. Rev. Lett. **117**(6–5), 060505 (2016)
11. L. Petit, J.M. Boter, H.G.J. Eenink, G. Droulers, M.L.V. Tagliaferri, R. Li, D.P. Franke, K.J. Singh, J.S. Clarke, R.N. Schouten, V.V. Dobrovitski, L.M.K. Vandersypen, M. Veldhorst, Spin lifetime and charge noise in hot silicon quantum dot qubits. Phys. Rev. Lett. **121**(7–17), 076801 (2018)
12. C.H. Yang, R.C.C. Leon, J.C.C. Hwang, A. Saraiva, T. Tanttu, W. Huang, J. Camirand Lemyre, K.W. Chan, K.Y. Tan, F.E. Hudson, K.M. Itoh, A. Morello, M. Pioro-Ladrière, A. Laucht, A.S. Dzurak, Silicon quantum processor unit cell operation above one Kelvin. arXiv preprint arXiv: 1902.09126
13. T. Watson, S. Philips, E. Kawakami, D. Ward, P. Scarlino, M. Veldhorst, D. Savage, M. Lagally, M. Friesen, S. Coppersmith et al., A programmable two-qubit quantum processor in silicon. Nature **555**(7698), 633 (2018)
14. J. Elzerman, R. Hanson, L.W. Van Beveren, B. Witkamp, L. Vandersypen, L.P. Kouwenhoven, Single-shot read-out of an individual electron spin in a quantum dot. Nature **430**(6998), 431 (2004)
15. L. Vandersypen, J. Elzerman, R. Schouten, L. Willems van Beveren, R. Hanson, L. Kouwenhoven, Real-time detection of single-electron tunneling using a quantum point contact. Appl. Phys. Lett. **85**(19), 4394–4396 (2004)
16. I. Vink, T. Nooitgedagt, R. Schouten, L. Vandersypen, W. Wegscheider, Cryogenic amplifier for fast real-time detection of single-electron tunneling. Appl. Phys. Lett. **91**(12), 123512 (2007)
17. E.A. Laird, J.M. Taylor, D.P. DiVincenzo, C.M. Marcus, M.P. Hanson, A.C. Gossard, Coherent spin manipulation in an exchange-only qubit. Phys. Rev. B **82**(7), 075403 (2010)
18. D. Reilly, C. Marcus, M. Hanson, A. Gossard, Fast single-charge sensing with a RF quantum point contact. Appl. Phys. Lett. **91**(16), 162101 (2007)
19. J. Hornibrook, J. Colless, A. Mahoney, X. Croot, S. Blanvillain, H. Lu, A. Gossard, D. Reilly, Frequency multiplexing for readout of spin qubits. Appl. Phys. Lett. **104**(10), 103108 (2014)

20. E. Charbon, F. Sebastiano, A. Vladimirescu, H. Homulle, S. Visser, L. Song, R.M. Incandela, Cryo-CMOS for quantum computing, in *International Electron Device Meeting*, San Francisco, CA (2016)
21. J.D. Cressler, H.A. Mantooth (eds.), *Extreme environment electronics* (CRC Press, Boca Raton, FL, 2013)
22. J.D. Cressler, J.H. Comfort, E.F. Crabbé, J.M.C. Stork, J.Y.-C. Sun, On the profile design and optimization of epitaxial Si- and SiGe-base bipolar technology for 77 K applications. I. Transistor DC design considerations. IEEE Trans. Electron Devices **40**(3), 525–541 (1993)
23. L. Najafizadeh, J.S. Adams, S.D. Phillips, K.A. Moen, J.D. Cressler, S. Aslam, T.R. Stevenson, R.M. Meloy, Sub-1-K Operation of SiGe Transistors and Circuits. IEEE Electron Device Lett. **30**(5), 508–510 (2009)
24. S.R. Ekanayake, T. Lehmann, A.S. Dzurak, R.G. Clark, A. Brawley, Characterization of SOS-CMOS FETs at low temperatures for the design of integrated circuits for quantum bit control and readout. IEEE Trans. Electron Devices **57**(2), 539–547 (2010)
25. T. Lehmann, Cryogenic support circuits and systems for silicon quantum computers, in *IEEE International Symposium on Circuits and Systems*, Sapporo, Japan (2019)
26. C.H. Yang, R.C.C. Leon, J.C.C. Hwang, A. Saraiva, T. Tanttu, W. Huang, J. Camirand Lemyre, K.W. Chan, K.Y. Tan, F.E. Hudson, K.M. Itoh, A. Morello, M. Pioro-Ladrière, A. Laucht, A.S. Dzurak, Silicon quantum processor unit cell operation above one Kelvin (2019). arXiv:1902.09126
27. J. van Dijk, E. Kawakami, R.N. Schouten, M. Veldhorst, L.M.K. Vandersypen, M. Babaie, E. Charbon, F. Sebastiano, The impact of classical control electronics on qubit fidelity (2019). arXiv:1803.06176
28. J. van Dijk, E. Charbon, F. Sebastiano, The electronic interface for quantum processors. Microprocess. Microsyst. **66**, 90–101 (2019)
29. C. Degenhardt, A. Artanov, L. Geck, C. Grewing, A. Kruth, D. Nielinger, P. Vliex, A. Zambanini, S. van Waasen, Systems engineering of cryogenic CMOS electronics for scalable quantum computers, in *IEEE International Symposium on Circuits and Systems*, Sapporo, Japan (2019)
30. J.C. Bardin, E. Jeffrey, E. Lucero, T. Huang, O. Naaman, R. Barends, T. White, M. Giustina, D. Sank, P. Roushan, K. Arya, B. Chiaro, J. Kelly, J. Chen, B. Burkett, Y. Chen, A. Dunsworth, A. Fowler, B. Foxen, C. Gidney, R. Graff, P. Klimov, J. Mutus, M. McEwen, A. Megrant, M. Neeley, C. Neill, C. Quintana, A. Vainsencher, H. Neven, J. Martinis, A 28 nm Bulk-CMOS 4-to-8 GHz 2mW cryogenic pulse modulator for scalable quantum computing, in *International Solid-State Circuits Conference*, San Francisco, CA (2019)
31. A. Beckers, F. Jazaeri, H. Bohuslavskyi, L. Hutin, S. De Franceschi, C. Enz, Characterization and modeling of 28-nm FDSOI CMOS technology down to cryogenic temperatures. Solid-State Electron. **159**, 106–115 (2019)
32. R.M. Incandela, L. Song, H. Homulle, E. Charbon, A. Vladimirescu, F. Sebastiano, Characterization and compact modeling of nanometer CMOS transistors at deep-cryogenic temperatures. IEEE J. Electron Devices Soc. **6**, 996–1006 (2018)
33. P.A. 't Hart, J. van Dijk, M. Babaie, E. Charbon, A. Vladimirescu, F. Sebastiano, Characterization and model validation of mismatch in nanometer CMOS at cryogenic temperatures, in *IEEE European Solid-State Circuits Conference*, Dresden, Germany (2018)
34. P.A. 't Hart, M. Babaie, E. Charbon, A. Vladimirescu, F. Sebastiano, Subthreshold mismatch in nanometer CMOS at cryogenic temperatures, in *IEEE European Solid-State Circuits Conference*, Krakow, Poland (2019)
35. F. Sebastiano, H. Homulle, B. Patra, R.M. Incandela, J. van Dijk, L. Song, M. Babaie, A. Vladimirescu, E. Charbon, Cryo-CMOS electronic control for scalable quantum computing, in *Design Automation Conference*, Austin, TX (2017)
36. B. Patra, R.M. Incandela, J. van Dijk, H. Homulle, L. Song, M. Shahmohammadi, R.B. Staszewski, A. Vladimirescu, M. Babaie, F. Sebastiano, E. Charbon, Cryo-CMOS circuits and systems for quantum computing applications. IEEE J. Solid-State Circuits **53**(1), 309–321 (2018)

37. H. Homulle, F. Sebastiano, E. Charbon, Deep-cryogenic voltage references in 40-nm CMOS. IEEE Solid-State Circuits Lett. **1**(5), 110–113 (2018)
38. J. van Staveren, C. Garcia Almudever, G. Scappucci, M. Veldhorst, M. Babaie, E. Charbon, F. Sebastiano, Voltage references for the ultra-wide temperature range from 4.2 K to 300 K in 40-nm CMOS, in *Proceedings of ESSCIRC 2019* (2019)
39. H. Homulle, E. Charbon, Cryogenic low-dropout voltage regulators for stable low-temperature electronics. Cryogenics **95**, 11–17 (2018)
40. S. Schaal, A. Rossi, V.N. Ciriano-Tejel, T.-Y. Yang, S. Barraud, J.J.L. Morton, M.F. Gonzalez-Zalba, A CMOS dynamic random access architecture for radio-frequency readout of quantum devices. Nat. Electron. **2**, 236–242 (2019)
41. R. Li, L. Petit, D.P. Franke, J.P. Dehollain, J. Helsen, M. Steudtner, N.K. Thomas, Z.R. Yoscovits, K.J. Singh, S. Wehner, L.M.K. Vandersypen, J.S. Clarke, M. Veldhorst, A crossbar network for silicon quantum dot qubits (2017). arXiv:1711.03807
42. A. Ruffino, Y. Peng, F. Sebastiano, M. Babaie, E. Charbon, A 6.5-GHz cryogenic all-pass filter circulator in 40-nm CMOS for quantum computing applications, in *IEEE RFIC*, Boston, MA (2019)
43. F. Bruccoleri, E.A.M. Klumperink, B. Nauta, Wide-band CMOS low-noise amplifier exploiting thermal noise canceling. IEEE J. Solid-State Circuits **39**(2), 275–282 (2004)
44. H. Homulle, S. Visser, B. Patra, G. Ferrari, E. Prati, F. Sebastiano, E. Charbon, A reconfigurable cryogenic platform for the classical control of scalable quantum computers (2016). arXiv:1602.05786
45. H. Homulle, S. Visser, B. Patra, G. Ferrari, E. Prati, F. Sebastiano, E. Charbon, A reconfigurable cryogenic platform for the classical control of quantum processors. Rev. Sci. Instrum. **88**(4), 045103 (2017)
46. H. Homulle, S. Visser, E. Charbon, A cryogenic 1 GSa/s, soft-core FPGA ADC for quantum computing applications. IEEE Trans. Circuits Syst. I **63**(11), 1854–1865 (2016)
47. R.B. Staszewski et al., All-digital PLL and transmitter for mobile phones. IEEE J. Solid-State Circuits **40**(12), 2469–2482 (2005)
48. A. Elkholy, A. Elmallah, M.G. Ahmed, P.K. Hanumolu, A 6.75–8.25-GHz 250-dB FoM rapid on/off fractional-N injection-locked clock multiplier. IEEE J. Solid-State Circuits **53**(6), 1818–1829 (2018)
49. J. Gong, Y. He, A. Ba, Y.-H. Liu, J. Dijkhuis, S. Traferro, C. Bachmann, K. Philips, M. Babaie, A 1.33 mW, 1.6 psrms-integrated-jitter, 1.8–2.7 GHz ring-oscillator-based fractional-N injection-locked DPLL for internet-of-things applications, in *2018 IEEE Radio Frequency Integrated Circuits Symposium* (RFIC), June 2018, pp. 44–47
50. M. Mehrpoo et al., Benefits and challenges of designing cryogenic CMOS RF circuits for quantum computers, in *2019 IEEE International Symposium on Circuits and Systems (ISCAS)*, Sapporo, Japan (2019), pp. 1–5
51. J. Ekin, *Experimental Techniques for Low-Temperature Measurements: Cryostat Design, Material Properties and Superconductor Critical-Current Testing* (Oxford University Press, 2006)
52. A. Coskun, J. Bardin, Cryogenic small-signal and noise performance of 32 nm SOI CMOS, in *Microwave Symposium (IMS)* (2014), pp. 1–4
53. J. Wang, X.-M. Peng, Z.-J. Liu, L. Wang, Z. Luo, D.-D. Wang, Observation of nonconservation characteristics of radio frequency noise mechanism of 40-nm n-MOSFET. Chin. Phys. B **27**(2), 027201 (2018)
54. X. Chen, C.-H. Chen, R. Lee, Fast evaluation of the high-frequency channel noise in nanoscale MOSFETs. IEEE Trans. Electron Devices **65**(4), 1502–1509 (2018)
55. J. Chang, A. Abidi, C. Viswanathan, Flicker noise in CMOS transistors from subthreshold to strong inversion at various temperatures. IEEE Trans. Electron Devices **41**(11), 1965–1971 (1994)
56. K. Hung, P. Ko, C. Hu, Y. Cheng, Flicker noise characteristics of advanced MOS technologies, in *Electron Devices Meeting, 1988. IEDM'88. Technical Digest., International* (IEEE, 1988), pp. 34–37

57. M. Shahmohammadi, M. Babaie, R.B. Staszewski, A 1/f noise upconversion reduction technique for voltage-biased RF CMOS oscillators. IEEE J. Solid-State Circuits **51**(11), 2610–2624 (2016)
58. D. Murphy, H. Darabi, H. Wu, Implicit common-mode resonance in LC oscillators. IEEE J. Solid-State Circuits **52**(3), 812–821 (2017)
59. J. Gong, Y. Chen, F. Sebastiano, E. Charbon, M. Babaie, A 200 dB FOM 4–5 GHz cryogenic oscillator with an automatic common-mode resonance calibration for quantum computing applications, in *International Solid-State Circuits Conference*, San Francisco, CA (2020) (accepted)
60. L. Riesebos, X. Fu, A.A. Moueddenne, L. Lao, S. Varsamopoulos, I. Ashraf, J. van Someren, N. Khammassi, C.G. Almudever, K. Bertels, Quantum accelerated computer architectures, in *IEEE International Symposium on Circuits and Systems*, Sapporo, Japan (2019); F. Sebastiano, L.J. Breems, K.A.A. Makinwa, S. Drago, D.M.W. Leenaerts, B. Nauta, A 1.2-V 10-μW NPN-based temperature sensor in 65-nm CMOS with an inaccuracy of 0.2 °C (3σ) from −70 °C to 125 °C. IEEE J. Solid-State Circuits **45**(12), 2591–2601 (2010)

Chapter 27
Quantum Computing

Large-Scale Quantum Systems Based on Superconducting Qubits

Albert Frisch, Harry S. Barowski, Markus Brink and Peter Hans Roth

27.1 Introduction

Quantum Computing emerged from the study of how little energy was required to perform a computational operation. It was realized that the laws of quantum mechanics allow for a richer approach to computation that would enable more efficient information processing for certain types of problems. As Richard Feynman noted in 1981: "If you want to make a simulation of nature, you'd better make it quantum mechanical." In other words, quantum systems map straight-forwardly onto other quantum mechanical systems, but mapping them onto a classical system requires significant overhead.

Early theoretical work on quantum algorithms and quantum information science showed that a speed-up in the runtime compared to classical algorithms can be expected for specific problems. Furthermore, fundamental research in atomic and condensed-matter physics has developed the coherent control, manipulation, and readout possibilities of many different quantum mechanical systems to a mature level suitable for quantum computing. Advances in increasing coherence times of quantum systems by many orders of magnitude propelled the field further, and the first small-scale quantum computing devices consisting of only a few quantum bits (qubits) became accessible. Technical developments in system control and improved

A. Frisch (✉) · H. S. Barowski · P. H. Roth
IBM Germany Research and Development, Schönaicher Str. 220, 71032 Böblingen, Germany
e-mail: alfr@de.ibm.com

H. S. Barowski
e-mail: barowski@de.ibm.com

P. H. Roth
e-mail: peharo@de.ibm.com

M. Brink
IBM T.J. Watson Research Center, Yorktown Heights, NY 10598, USA
e-mail: mbrink@us.ibm.com

B. Murmann and B. Hoefflinger (eds.), *NANO-CHIPS 2030*,
The Frontiers Collection, https://doi.org/10.1007/978-3-030-18338-7_27

error rates of quantum gates allowed to grow the number of qubits constantly, and the field of noisy intermediate-scale quantum devices appeared on the horizon [1].

In 2019, IBM introduced IBM Q System One, the world's first integrated universal approximate quantum computing system for commercial use, see Fig. 27.1a, and opened the first quantum computation center [2]. At this point, a total of 13 quantum computing devices have become available for enterprise users, researchers, and developers within the IBM Q Network, which consists of six 5-, one 14-, five 20-, and one 53-qubit device(s), see Fig. 27.1b, c and [3] for an industry-wide overview of quantum processors.

In this chapter, we will first give a general overview on current physical implementations of quantum processors suitable for quantum computing in Sect. 27.2, then discuss superconducting circuits in more detail in Sect. 27.3, introduce methods for performance benchmarking in Sect. 27.4, present open-source software tools available for a complete quantum computing ecosystem and infrastructure in Sect. 27.5, and give an outlook on future near-term developments in quantum computing in Sect. 27.6.

27.2 Quantum Processors

David DiVincenzo introduced a set of five requirements for the implementation of a quantum computer [4]. They are known as the DiVincenzo criteria and provide practical conditions that a quantum system must meet to be useful for quantum computation purposes. In an abbreviated format, these criteria are:

1. **A scalable physical system** with well characterized two-level systems that can act as qubits.
2. **The ability to initialize the state of the qubits** to a simple fiducial state. In practice, this often is the ground state $|000\ldots0\rangle$ of the system.
3. **Long relevant decoherence times of the qubits**. While longer coherence is advantageous, this condition requires coherence times to be much longer than the gate operation time.
4. **A "universal" set of quantum gates**.
5. **A qubit-specific measurement capability**, i.e. the ability to readout the state of each qubit. A quantum non-demolition measurement scheme, which preserves the projected state of the qubit at the end of the readout, is preferable.

Ultimately, a universal quantum computer supports the execution of arbitrary sequences of quantum gates, i.e. a gate-based quantum computer, which is fully quantum error corrected.

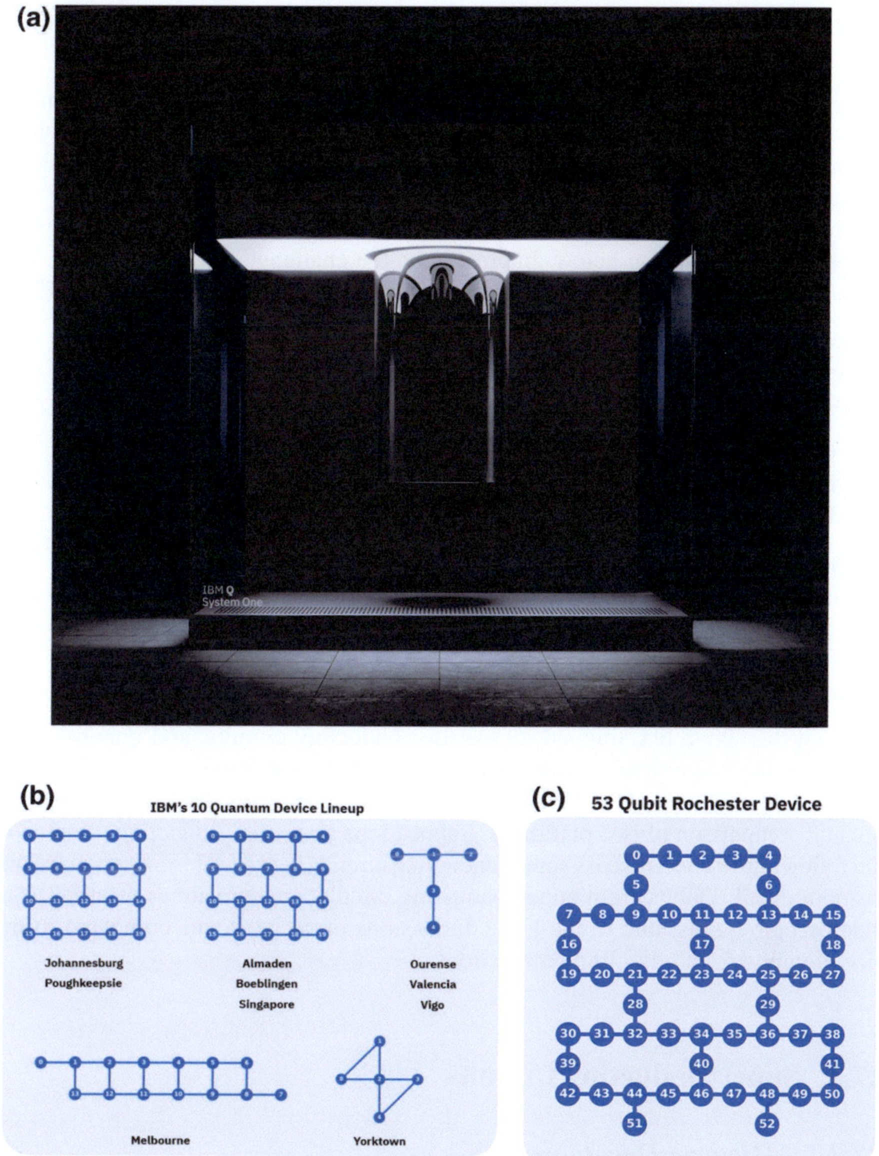

Fig. 27.1 **a** IBM Q System One, the first commercially available quantum computing system from IBM, contains a 20-qubit chip with a layout similar to the Poughkeepsie device. **b** Overview of currently available devices via the IBM Q Experience, see Sect. 27.5. **c** The 53-qubit device Rochester offers a larger lattice of qubits for advanced quantum algorithms. Device layouts from [2]

27.2.1 Physical Implementations

Early implementation of qubits relied on microscopic systems that naturally behaved quantum-mechanically. Examples include single molecules addressed by nuclear-magnetic resonance (NMR) techniques and trapped ions that can be manipulated by lasers.

Today, there are a number of demonstrated implementations of qubits, some of which are touched upon below. Interestingly, the challenges for scaling them to a useful quantum processor are often specific for the particular implementation [5].

- **Trapped Ions** [6, 7]
- **Neutral Atoms** [8, 9]
- **Spin Qubits**
  - Quantum Dots [10–12]
  - Lattice Defects: Nitrogen-Vacancy Centers in Diamond [13, 14]
  - Impurities: Phosphorus donors in isotopically pure Silicon (Si:P) [15, 16]
- **Photonic Computing** [17, 18]
- **Topological Qubits**
  - Majorana Fermions [19, 20], e.g., in Nanowires [21]
- **Superconducting Circuits**

From the above list, trapped ions, superconducting circuits, and quantum dots are, due to their favorable scaling behavior compared to other implementations, most promising for future large-scale quantum computing platforms, see [22] for an experimental comparison of two platforms, trapped ions and superconducting circuits. In the following, we will discuss superconducting circuits based on Josephson junctions in more detail. Qubits from superconducting circuits are sometimes referred to as macroscopic qubits, due to the large dimensions of such circuits compared to the size of qubits from other implementations.

27.3 Superconducting Circuits

27.3.1 Historic Development

Superconducting quantum circuits based on Josephson junctions arrived as a later addition to qubit implementations. Nakamura [23] first demonstrated driven coherent oscillations in a Cooper-pair box, the superconducting analog of a single electron transistor. While the coherence times were short, it proved that more macroscopic, larger structures such as superconducting circuits can exhibit quantum coherence.

Early implementations of superconducting qubits generally fell into three canonical structures, see Fig. 27.2:

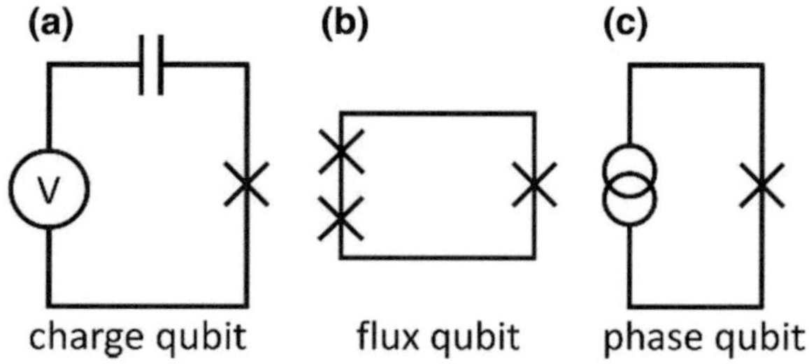

Fig. 27.2 Schematic diagrams for early Josephson-junction-based qubits. **a** Charge qubit based on the Cooper-pair box, **b** flux qubit, and **c** phase qubit

- **charge qubit**: based on the Cooper-pair box. The Cooper-pair box is sensitive to offset charges (charge noise).
- **flux qubit**: consisting of superconducting loops interrupted by three (or more) Josephson junctions [24]. The two qubit states can be thought of as clockwise versus counterclockwise persistent currents in the loop. The energy landscape depends on the flux threading the superconducting loop. The flux qubit is sensitive to flux noise.
- **phase qubit**: current-biased large junction, with a finite bias current less than the junction critical current (zero-voltage state), which shows a washboard energy potential [25]. The energy levels in a local minimum of the washboard potential can be used for computation.

The Quantronium [26] was a hybrid qubit circuit design that showed a significant boost in coherence times. It also demonstrated that clever design of the circuit can have a significant impact on qubit performance.

Another notable development was the invention of cavity quantum electrodynamics (cavity QED) as well as circuit QED and the dispersive readout, where a qubit is coupled to a superconducting microwave resonator, which can assess the state of the qubit [27].

Early experiments in combining superconducting circuits with other physical systems to hybrid quantum circuits have been carried out recently to gain advantages over simple circuits and harness the best properties of different systems, see [28] for an overview.

27.3.2 Current Implementations

A major invention for superconducting qubits was the Transmon qubit [29]: The Transmon realized that the charge dispersion of low energy levels of the Cooper-pair box can be practically eliminated by increasing the ratio of inductive (Josephson) energy to capacitive (charging) energy, E_J/E_C. The only drawback is that the Transmon qubit becomes more harmonic, making it harder to isolate (the lowest) two states

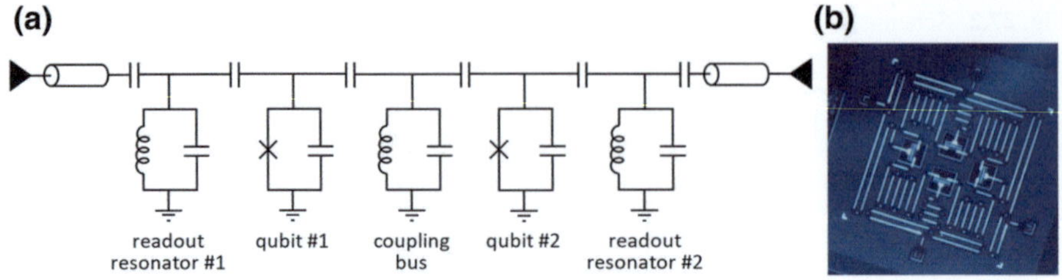

Fig. 27.3 **a** Schematic of two coupled Transmon qubits with coupling bus resonator in the center and readout resonators on the sides. **b** 4-qubit chip layout with resonators as coplanar waveguides and Transmon qubits, which are visible as square-like shapes, each containing a single Josephson junction and a large shunt capacitor, from [30]

for computation. But the charge dispersion of the lowest levels vanishes exponentially as E_J/E_C increases, while the anharmonicity decreases polynomially, opening an operational sweet spot around $E_J/E_C \approx 50$ (Fig. 27.3).

Many flavors of Transmon qubits have been made since, including X-mon, Starmon, Pacmon, to name a few. While the ideas of the Transmon qubit have taken superconducting qubits by storm, there have been several other noteworthy developments since.

The Fluxonium qubit [31] shunts the qubit junction by a large inductor (instead of a large capacitor in a Transmon qubit), which is typically implemented as a chain of Josephson junctions. A simple way of thinking of the inductive shunt is as a short at low frequencies, which eliminates charge noise across the qubit junction, and an open at microwave frequencies, which permits the operation as a qubit. By design, the Fluxonium contains a closed superconducting loop, and a flux threading the loop changes the potential energy landscape of the Fluxonium qubit. Consequently, the Fluxonium is sensitive to flux noise, which limits its coherence times.

Similar to the large shunt capacitor of a Transmon qubit, the capacitively shunted flux qubit (CSFQ) uses a large shunt capacitor in parallel with the superconducting loop of the flux qubit. The CSFQ contains three Josephson junctions of two different sizes in the loop. Both the ratio of Josephson critical currents and the magnetic flux through the superconducting loop determine the energy levels of the CSFQ.

The Quarton qubit arose from a classification of existing Josephson qubits and uses four Josephson junctions in total. The goal of the Quarton design was to increase qubit coherence and anharmonicity.

With relaxation times in the milliseconds, superconducting three-dimensional (3D) cavities have been coupled to a Transmon qubit. As the 3D cavity is strictly harmonic, the Transmon lends its non-linearity to the resulting hybrid qubit, while the cavity provides a significant boost in lifetime. Instead of single photon states, so-called cat states (i.e. superposition of two coherent states with opposite phase) have been used for the computational basis, and their parity for error correction schemes.

Another interesting development for superconducting quantum circuits are topologically protected qubits, such as the $0-\pi$ qubit. The topological protection has to be turned off for computational operations and coupling between such qubits is still under development.

27.3.3 Current Status and Trends

While it is difficult to compare all the flavors of superconducting qubits on an equal footing, it is fair to say that their quality has improved continually over the years, the size of superconducting quantum circuits (as measured by the number of qubits) has scaled significantly, and the performance of quantum processors (as measured by the Quantum Volume, see Sect. 27.4) has likewise increased steadily. As an example, see Fig. 27.4 for the improvement of the controlled NOT gate errors for various IBM devices over time and a full error map for the IBM device named 'Boeblingen'.

Comparing the implementations of larger quantum chips, most rely on a version of the Transmon qubit. A notable distinction between these implementations is fixed-frequency versus tunable versus hybrid fixed and tunable approaches (where some qubits have fixed frequencies and others are tunable). Fixed-frequency qubits offer higher coherence times, but suffer from so-called frequency collisions, as the frequencies of coupled qubits have to obey certain conditions for specific two-qubit gates, such as the all-microwave cross-resonance gate. Tunable qubits on the other hand provide much flexibility, as their resonance frequencies can be tuned on the fly. Thereby their interaction strength can be turned on and off, resulting in fast two-qubit gates. Because magnetic flux threading a loop is typically used as the tuning knob, tunable qubits are susceptible to flux noise.

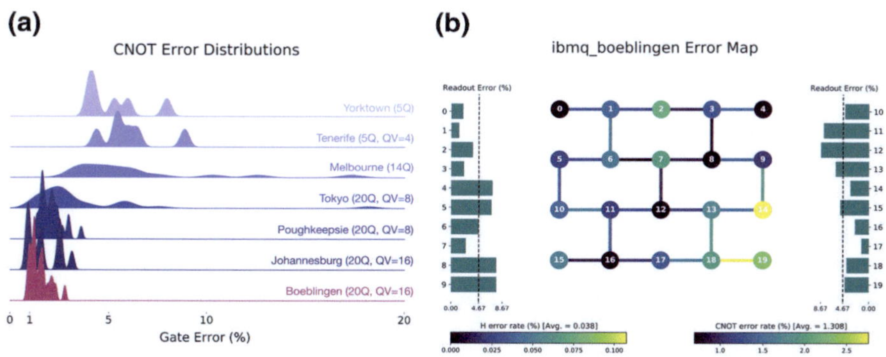

Fig. 27.4 **a** Controlled-NOT (CNOT) gate error distributions for different IBM devices. The average gate errors have been steadily improved over time from top to bottom. The number of qubits (Q) available in the system and the Quantum Volume (QV) of the system is given next to the device name. **b** Error map for 20-qubit IBM device named 'Boeblingen'. Hadamard (H) gate errors and CNOT gate errors are shown using a color code for individual qubits (circles) and the couplings between two qubits (edges), respectively. Additionally, the readout error is given for every qubit, the average readout error is shown by the dashed line. Both figures from [32]

Given the size of existing quantum chips, some useful algorithms and circuit demonstrations have successfully been implemented on currently available (i.e. not error-corrected) systems. These include ab-initio chemistry simulations of small molecules, optimization codes and machine learning, and implementations of classic quantum algorithms [33]. Today, quantum processors stand at the edge of showing quantum supremacy—the point where a quantum processor outperforms any existing classical (super)computer for some computational problem [34, 35].

Error-correcting codes have been demonstrated, where a number of physical qubits combine to form a logical qubit with extended effective coherence times [36]. But one has to consider that, first, improvements of effective coherence times have been small so far and only over a range of operation times [37], and second, the overhead on hardware requirements is high, i.e. a significant number of qubits have to combine to form a logical qubit that can correct higher weight errors. Consequently, the quality of qubits and qubit operations (as measured by their gate fidelity) needs to improve, and quantum processors have to scale much further in size before error correcting codes are adopted for practical applications.

27.4 Performance Benchmarking

For qualifying and measuring the performance of a quantum computer, terms like quantum supremacy and quantum advantage were used in recent years. Quantum supremacy describes the dominant power of a quantum computer assuming that, with a given number of qubits, it can perform tasks which are not feasible with classical computers. More recently, the quantum advantage has been used to describe the point, when a quantum computer outperforms classical computers for a practically relevant (e.g., business) problem, such as an optimization.

27.4.1 Performance Metrics

To classify and characterize the computational power of quantum computers, various metrics have been proposed. Similar to classical computers where benchmarks characterize the computational power in various aspects like LinPack or SpecInt or instructions per second (MIPS), metrics for quantum computers can characterize several quantum gates or focus on quantum algorithms.

Since performance metrics allow to determine the computational performance of a quantum computing system, they also measure how close current quantum processors are to reaching quantum advantage, where quantum computers outperform existing technologies (classical computers and neural networks) and achieve commercial advantage.

The first proposed benchmarks were based on tiny implementations of well-known quantum algorithms like Shor's and Grover's algorithm [38, 39]. Random quantum gate operations directly lead to randomized benchmarking, eventually based on Clifford gates [40]. Randomized benchmarking is a popular technique, because it is easy to implement and gives a single number to characterize the (average) gate performance, but it does not provide much detail of what is limiting the performance.

Quantum tomography requires repeated measurements until tomographical completeness to reconstruct the quantum state. The statistical state is described by a density matrix which allows to calculate the probability of measurement outcomes of a quantum system.

The quantum gate performance is targeted by gate-set tomography, the characteristics of individual qubits by quantum-state tomography. Both give extensive characterization information, but are complex tasks. They can easily be extended to multiple qubits, but the number of parameters to measure scales exponentially and is not feasible for large systems [41].

Alternative approaches are direct fidelity estimation, to determine the gate errors, and robust phase estimation, to target the characterization of qubits.

Furthermore, the preparation of multi-qubit entangled states, so-called GHZ states, can be used as another performance metric by verifying the presence of multipartite entanglement [40].

At present there is no accepted best practice in performance metrics for the field of quantum computing. Each performance metric only allows to compare the performance for certain aspects and not necessarily to determine quantum advantage. For example, gate set tomography targets and emphasizes different properties from randomized benchmarking. IBM's proposal of the Quantum Volume (QV) as a metric aims to assess the overall systems performance. It includes all properties of a quantum computer system like qubit characteristics, qubit topology, quantum errors, and readout errors. However, it allows optimization at all levels of a quantum system and can also leverage the quantum transpiler.

27.4.2 Benchmarking

Randomized benchmarking is a concept to specify the capability of a quantum computing system by executing long sequences of random quantum gate operations that allow to determine the average error rates by measurements. The original theory of randomized benchmarking assumed pseudo-random quantum gates to validate the quantum operators. The result is a fidelity metric, which describes the noisy operations of a quantum system. However, the original concept suffered from practical limitations—there was no well-defined metric—which led to a more efficient version of randomized benchmarking based on uniformly-distributed random Clifford operations.

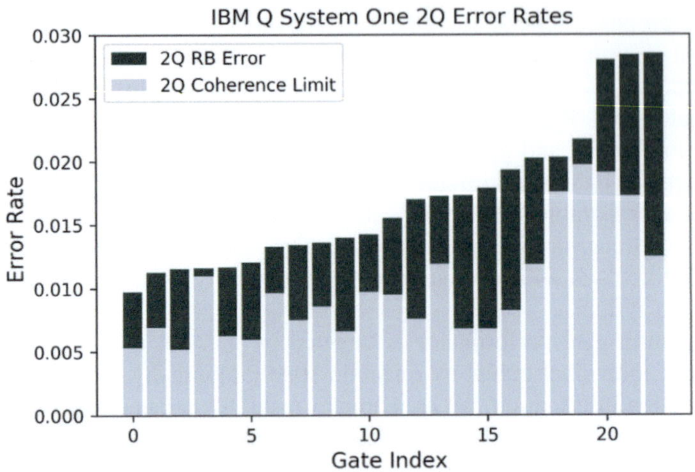

Fig. 27.5 CNOT gate (2Q) error rates (green) measured by randomized benchmarking (RB) and the contribution by the coherence limit (gray) of all possible two-qubit gates in the IBM Q System One, from [42]

Randomized benchmarking allows to identify various features, like type and strength of errors affecting quantum gates, and thus allows to validate and qualify the quantum operations of a quantum system, e.g. see Fig. 27.5.

A challenge in improving a quantum computer is the characterization of the noise affecting the qubits and quantum gates. A full noise characterization would be helpful for noise mitigation or error correction schemes. Full process tomography in principle allows this characterization, but it grows exponentially with the number of qubits and thus requires exponential resources. Thus, this method becomes impractical for large systems.

A proposal for randomized benchmarking over a full n-qubit space constructs gate sequences from the n-qubit Clifford group. If characterizing the n-qubit space still requires too much effort, the n-qubit space is divided into smaller sets of n_i qubits and the randomized benchmarking for the n_i-qubit subdivided space is executed in parallel. Either way, a metric of fidelity of the n-qubit space can be derived.

A proposal [43] for a protocol for Clifford gates overcomes problems for real systems where noise is not independent of the chosen quantum gate or even not understood for multiple-qubit systems. The main idea is to construct a m-length sequence of random n-qubit Clifford gates, and add as last gate the inverse of the sequence, which can be efficiently calculated (according to the Gottesman-Knill theorem).

The protocol involves starting in the initialized state $|0\rangle^{\otimes n}$, applying the sequence of Clifford gates repeated l-times for different random sequences [43]. In the limit for large l the error map is twirled into a depolarization error map. From measurements of the population on $|0\rangle$ versus the sequence length, an average error over the Clifford gates can be determined.

This average gate error determines the average gate fidelity for a set of operations (Clifford group). This can be understood as making stronger assumption of the operations to achieve a more reliable noise characterization of the quantum gates.

A further method, by interleaving random Clifford gates between the gates of interest, allows for a specific quantum gate to characterize its noise individually, by so-called interleaved randomized benchmarking a specific gate fidelity is derived [44].

27.4.3 Quantum Volume

Since the usefulness and power of a quantum computer is not only dependent on the number of qubits, n, but also relies on the connectivity of the qubits, as well as on the supported set of quantum gates, and the error rates of quantum gates, IBM suggested to introduce a new metric called the Quantum Volume (QV) [40].

In this metric, different aspects are included like the number of qubits, which describes the possible width of a quantum circuit, the topology, which describes the interconnectivity and thus possible two-qubit gates, the set of quantum gates supported by a certain quantum architecture, and the quantum gate error rates. A given system with a specific QV will indicate the width and depth of the quantum circuit, which can be executed with sufficiently high fidelity, and thus the QV must be considered by the circuit compiler.

Therefore, QV as a metric is spanned by the maximum width and maximum length of the quantum circuit before decoherence occurs and the quantum state collapses into a classical state. In this sense, IBM's metric of a QV is considered as a general approach to measure the ability of a quantum computer to perform complex algorithms on the passkey to quantum advantage.

The ability of a quantum computer to solve complex quantum algorithms is directly linked to the quantum circuit width, for which the accumulated error rates need to be below a certain acceptable overall error rate for the full quantum algorithm, and the quantum circuit depth, for which the execution time is well within the range of the coherence times of the qubits.

Since the topology and the error rates of a set of quantum gates varies with the architecture of a quantum computer, circuit compilers may optimize the quantum algorithm. The width and depth of the optimized quantum algorithm thus is dependent on the architecture and may be system-specific. Optimization should aim to maximize quantum-gate parallelism to increase the efficiency of quantum computation. Since the QV is derived as the maximum achievable area of number of qubits and circuit depth. It is also highly dependent on the optimization quality of the circuit compiler [40].

The model circuit used to determine the QV is built up by several layers of quantum operations of random permutations and random two-qubit gates, see Fig. 27.6. As stated in [40], any quantum algorithm may be regarded as polynomially-sized circuits of two-qubit unitary gates. Random circuits in general are a base proposal

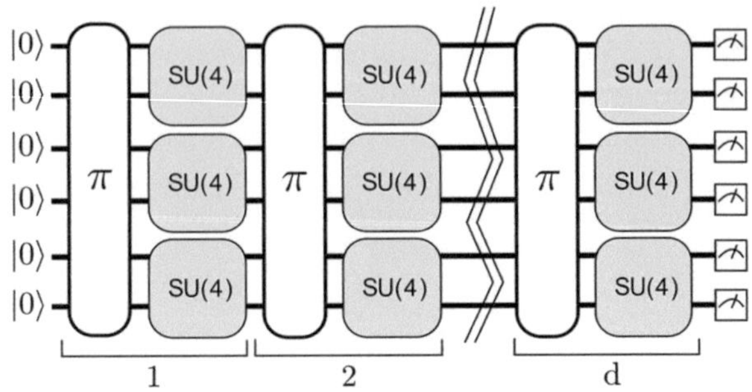

Fig. 27.6 Model circuit to measure QV using *d* iterations of layers consisting of random permutations of qubits followed by random two-qubit gates generated from SU(4) unitary matrices, from [40]

to demonstrate quantum advantage and are the underlying concept of randomized benchmarking.

In detail, the proposed model circuit has a width, *m*, spanned by the number of qubits and a depth, *d*, counting the number of layers of random permutations of qubits and random two-qubit gates generated from SU(4) unitary matrices. Thus, the sequence $U = U^{(d)} \dots U^{(2)} U^{(1)}$ with *d* layers given by

$$U^{(t)} = U^{(t)}_{\pi_t(m'-1),\pi_t(m')} \otimes \cdots \otimes U^{(t)}_{\pi_t(1),\pi_t(2)}$$

and labeled with time *t* is acting on $m' = 2\lfloor n/2 \rfloor$ qubits, which covers the cases of even and odd number of qubits.

The circuit has to be adopted to the specifics of each quantum system. It may depend on the gate set that was implemented and the qubit topology, which could require additional operations like SWAP gates to satisfy the two-qubit operations in each layer, even if the qubit connectivity does not allow a two-qubit gate directly between them. To optimize across all aspects, like qubit placement, the determination of *m*-qubit unitaries, U', may take a great computational effort.

The QV is derived using a concrete protocol to check whether a generated circuit is heavy, see Fig. 27.7, and hence is not tailored to any particular system, but can be implemented on any quantum computer. The only system requirement is the ability to implement a universal set of quantum gates.

Fig. 27.7 Pseudocode for testing whether the output of a random model circuit of width *m* and depth *d* executed on real hardware is heavy, from [40]

Algorithm 1 Check heavy output generation

function ISHEAVY($m, d; n_c \geq 100, n_s$)
  $n_h \leftarrow 0$
  for n_c repetitions **do**
    $U \leftarrow$ random model circuit, width m, depth d
    $H_U \leftarrow$ heavy set of U from classical simulation
    $U' \leftarrow$ compiled U for available hardware
    for n_s repetitions **do**
      $x \leftarrow$ outcome of executing U'
      if $x \in H_U$ **then** $n_h \leftarrow n_h + 1$
  return $\frac{n_h - 2\sqrt{n_h(n_s - n_h/n_c)}}{n_c n_s} > \frac{2}{3}$

Noisy intermediate-scale quantum devices, which currently exhibit non-negligible gate error rates, will need to start with small model circuits and continue to larger model circuits where the confidence of the executed model circuit with a given width m and depth $d(m)$ with $m \in [n]$ for the largest possible d will satisfy $h_d > 2/3$ with $h_1, h_2, \ldots, h_{d(m)} > 2/3$ and $h_{d(m)+1} \leq 2/3$.

Over repeated executions while sweeping the parameters m and d the largest square of d and m allows to derive the QV, V_Q, defined as

$$\log_2 V_Q = \text{argmax}_m \min(m, d(m)).$$

The QV accomplishes the complexity of a classical quantum circuit simulation. This implies a heuristic that the width of a classical tree of random circuits has a width scaling roughly with the depth d. See Fig. 27.8 for an experimental evaluation of the QV of a real quantum computing device.

As stated above, the QV is strongly related to the gate error rates. Thus, the achievable QV scales like $1/\varepsilon$, with ε the two-qubit error rate, as single qubit gates tend to have smaller error rates than two-qubit gates.

The QV for different systems is dependent on the qubit topology, which defines the coupling maps of the qubits, as well as performance parameters like coherence times, calibration errors, crosstalk, fidelity of initialization, gate fidelity, and readout fidelity.

As shown in Fig. 27.9, the achievable QV is expected to double every year, comparable to Moore's law in the semiconductor industry, which predicted doubling of the number of transistors about every two years.

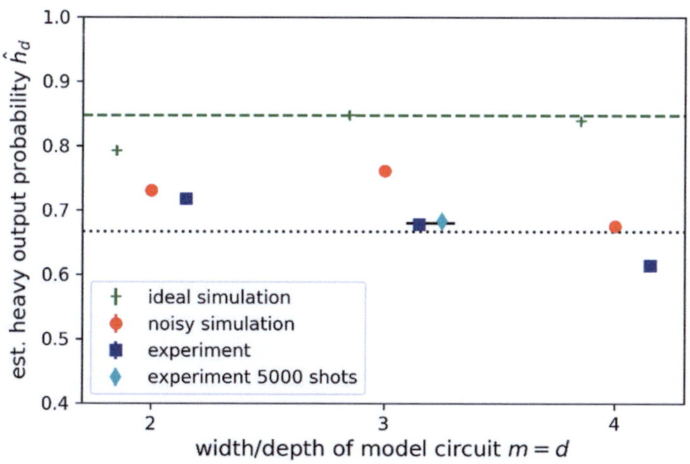

Fig. 27.8 Experimental data and ideal simulation results for the 20-qubit IBM device 'ibmq_tokyo' with the heavy output probability threshold of 2/3 (dotted) and the ideal asymptotic heavy output probability (dashed). For the experimental data up to $m = d = 3$ the heavy output probability is above the threshold and thus the system is being characterized as a system with a QV of 8, from [40]

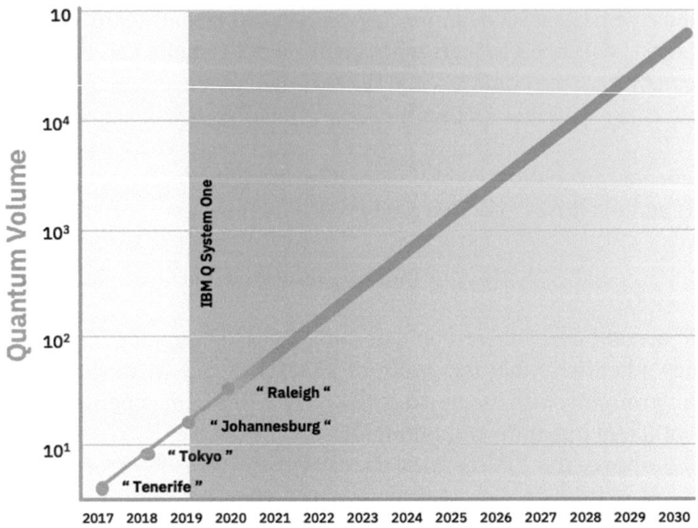

Fig. 27.9 Exponential forecast of the QV depicting an exponential growth of quantum processing power by doubling the QV every year for the next few years, from [45]

27.5 Software Ecosystem

To make quantum computing successful in solving real-world applications with an improvement over current classical computing, the right ecosystem and programming infrastructure is a key requirement. Such an ecosystem spans from integration of the quantum processor into the quantum computing system, over classical electronic hardware for controlling and reading-out qubits without limitations to the quantum circuit, to the software stack that allows rapid development and in-depth optimization of quantum algorithms.

For the development of quantum processors with superconducting circuits, two open-source Python frameworks are publicly available, QuTiP [46] and QuCAT [47]. QuCAT allows the definition of a quantum circuit either via a graphical user interface or programmatically, and it creates a netlist, that can then be used by QuTiP to simulate the dynamics of the quantum system. Important system parameters, like resonance frequencies, loss-rate, anharmonicities, or cross-Kerr coupling, can be easily extracted and analyzed further as design parameters. See [48] for a review on engineering concepts and challenges in superconducting quantum circuits.

For programming quantum systems, several open-source quantum programming platforms have been developed by different organizations in recent years, including Forest, ProjectQ, Q#, Cirq with OpenFermion, and Qiskit, see [49] for an overview. Most platforms offer methods and functions to create and manipulate quantum circuits with gate-level control, and they execute them on real hardware backends or simulators. Translators between different platforms are available in some cases. In the following, we will focus on Qiskit as an example for an actively developed platform that has one of the largest quantum programming communities and is notable for its large number of tutorials on various topics [49].

27.5.1 *Qiskit*

Qiskit [50] is an open-source software framework designed for working with quantum computers at multiple levels. The scope of Qiskit ranges from the level of pulses [51] and open quantum assembly language (QASM) [52], over quantum circuits and quantum algorithms, to applications. Thus, it supports various users with expertise in different topics, from researchers to engineers and developers.

A central goal of Qiskit is to provide a software stack, see Fig. 27.10, that makes it easy for anyone to use quantum computers requiring only basic knowledge. Additionally, Qiskit is also designed to support quantum computing research on today's most important open questions.

Qiskit is divided into four major parts with distinct functions and features, which can be used as a flexible set of tools for quantum programming. Qiskit consists of the following four elements:

- **Terra**: Composing quantum programs at the level of circuits and pulses.

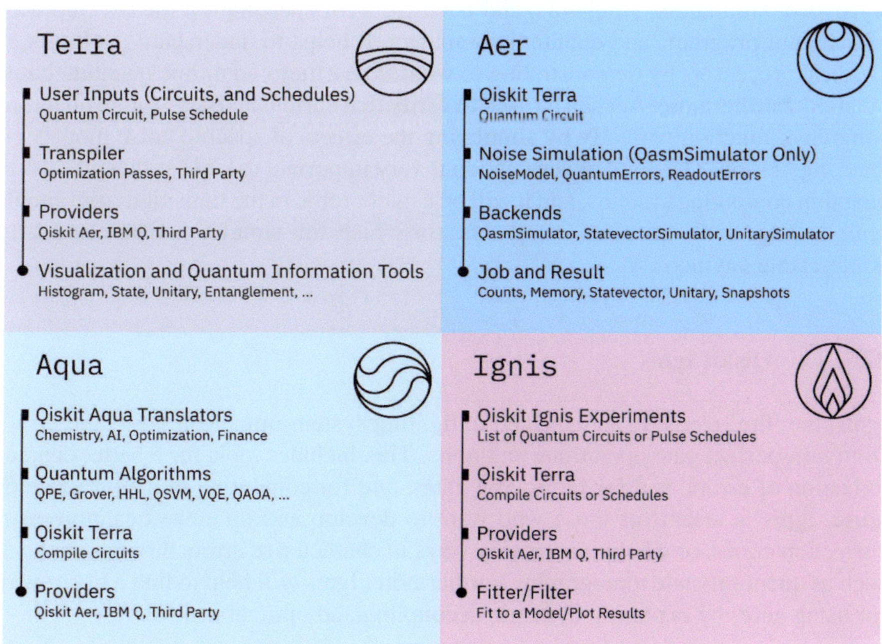

Fig. 27.10 Qiskit software stack for each of its elements, from [50]

- **Aer**: Optimizing quantum programs with simulators, emulators, and debuggers.
- **Ignis**: Characterizing and mitigating noise and errors.
- **Aqua**: Building quantum algorithms and applications.

27.5.1.1 Qiskit Terra

Terra, the ‘earth’ element, is the foundation, which supports the other elements of Qiskit. Terra provides tools for composing quantum programs at the level of circuits and pulses. Circuit optimization, e.g. for the constraints of a particular device and for error rates and decoherence, is carried out after compilation in a dedicated transpilation step, which offers different degrees of optimization. Terra also manages the execution of batches of experiments on remote-access devices, so-called backends, and it communicates via an application programming interface (API) with the provider for a specific backend. Terra defines the interfaces exposed to the end-user. A Python code example of running a simple quantum circuit on a real hardware backend is given in Fig. 27.11.

27.5.1.2 Qiskit Aer

Aer, the ‘air’ element, permeates all other Qiskit elements as a universal tool. It consists of simulators, emulators, and debuggers for speeding-up the development of quantum programs and quantum computers. It helps to understand the limits of classical processors by demonstrating to what extent they can mimic quantum computation. Furthermore, Aer can be used to verify that current and near-future quantum computers function correctly by simulating the effects of specific noise models on the computation. This kind of verification is very important to build sufficient trust in quantum computing systems, and it will be a major topic in the transition from small-scale to large-scale quantum computers, for which full simulations are classically not tractable anymore.

27.5.1.3 Qiskit Ignis

Ignis, the ‘fire’ element, is dedicated to fighting system-immanent noise and errors from non-perfect gate operations on qubits. This includes tools for a better characterization of errors, and for improving gates, and for computing in the presence of noise. Ignis is meant for those who want to develop and optimize quantum error correction codes, or who wish to study ways to characterize errors through methods such as quantum-state tomography. Furthermore, Ignis will help to find a better way for using gates by exploring dynamic decoupling and optimal control.

```
from qiskit import QuantumCircuit, execute, IBMQ

IBMQ.load_account()
provider = IBMQ.get_provider()
simulator = provider.get_backend('ibmq_boeblingen')
circuit = QuantumCircuit(2, 2)

circuit.h(0)
circuit.cx(0, 1)
circuit.measure([0, 1], [0, 1])

job = execute(circuit, simulator, shots=1000)
result = job.result()
counts = result.get_counts(circuit)

print("Total counts are:", counts)
print(circuit)
```

```
Total counts are: {'00': 348, '10': 46, '01': 46, '11': 560}
```

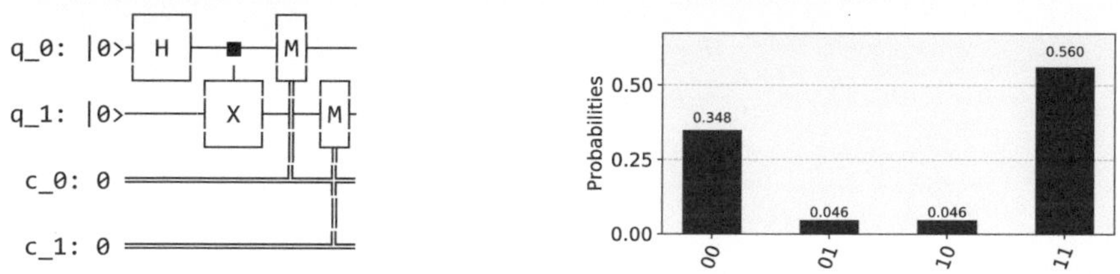

Fig. 27.11 Complete Qiskit Terra software example (dark box with output below) that creates an entanglement of two qubits and runs the quantum circuit on the real hardware backend called 'ibmq_boeblingen' for the IBM device 'Boeblingen'. The result shows the counts for a total number of shots of 1000 and the corresponding quantum circuit visualized as text output. The inset shows the counts converted into measurement probabilities and visualized as a histogram

27.5.1.4 Qiskit Aqua

Aqua, the 'water' element, is the element of life within Qiskit. To make quantum computing live up to its expectations, we need to find real-world applications and solutions for practical problems. Aqua is the place where algorithms for quantum computers are developed. These algorithms can then be used to build applications for quantum computing by connecting the quantum algorithm with the data from the problem to solve. In this sense, Aqua is accessible to domain experts in chemistry, optimization, finance and artificial intelligence, who want to explore the benefits of using quantum computers as accelerators for specific computational tasks. A Python code example, solving a simple linear system of equations, is given in Fig. 27.12.

```
from qiskit.aqua import run_algorithm
from qiskit.aqua.algorithms.classical import ExactLSsolver
import numpy as np

params = {
    'problem': {'name': 'linear_system'},
    'algorithm': {'name': 'HHL'},
    'eigs': {'name': 'EigsQPE'},
    'reciprocal': {'name': 'Lookup'},
    'backend': {
        'provider': 'qiskit.BasicAer',
        'name': 'statevector_simulator'
    }
}

matrix = [[1, 0], [0, 2]]
vector = [1, 4]
params['input'] = {
    'name': 'LinearSystemInput',
    'matrix': matrix,
    'vector': vector
}

result = run_algorithm(params)
print("solution ", np.round(result['solution'], 5))

result_ref = ExactLSsolver(matrix, vector).run()
print("classical solution ", np.round(result_ref['solution'], 5))
```

```
solution  [1.02398+0.j 1.99696+0.j]
classical solution  [1. 2.]
```

Fig. 27.12 Complete Qiskit Aqua software example (dark box with output below) that runs the HHL algorithm [53] to solve a system of linear equations of size 2 on a simulator backend called 'statevector_simulator'. The result from the quantum algorithm is compared with the result from a classical linear algebra solver, which corresponds to an overall fidelity of 99.9897%

27.5.2 IBM Q Experience

The IBM Q Experience [54] is an online platform for cloud-based quantum computing. It provides access to IBM's quantum processors for the public as well as for members within the IBM Q Network. The platform features multiple ways to construct and execute a quantum circuit: Users can employ a circuit composer, which is a graphical user interface for drag-and-drop quantum gates onto a score, a circuit editor, which allows editing QASM code and importing a pre-defined QASM file. Alternatively, they can execute programs via Qiskit by running Jupyter notebooks on the cloud platform or via connecting to IBM Q backends via an API.

Furthermore, the IBM Q Experience lets the user manage the results generated by previous executions of quantum circuits on the backends available via the IBM Q provider. Various run details as well as the result data are stored on the platform and can be accessed at a later time.

Administrators or members within the IBM Q Network use the platform to configure available backends, make system reservations in case of planned high loads, and manage users by group- and project-based access controls.

Since the launch of the IBM Q Experience in May 2016, its primary focus of attention lies on the education and enablement of the quantum computing community. Topics include various guides for beginners with exercises [55], which are closely connected to the circuit composer and real hardware backends, as well as advanced tutorials [56], which are written as Jupyter notebooks and which can be executed online. This infrastructure supports an interactive way to learn the basics of quantum computing and to run simple demonstrators of quantum algorithms for further study.

In addition, in [57], an open-source online textbook is being published called "Learn Quantum Computation Using Qiskit", which guides self-learners and educators to become quantum developers and to teach others. The textbook is maintained by the Qiskit community.

27.6 Outlook

Algorithms and methods for verifying quantum programs will become a very important aspect in moving beyond the point of classically simulatable quantum algorithms, especially for problems where the correctness of an answer cannot easily be verified. This will help in building trust in the operating principles of a quantum computer and its applications. Making quantum algorithms tractable is of key importance when entering the era of quantum advantage.

As one of the major next steps, quantum algorithms will be benchmarked and compared to classical algorithms in great depth regarding runtime, performance, quality of results, and resource requirements. As examples, comparisons are shown in [58, 59] for the results of simple quantum chemistry problems between the quantum method, including error mitigation running on current noisy hardware, and the well-established classical method, regarded as the exact result running on classical computers.

The increase of the QV of future quantum computing systems will give users and researchers the ability to solve larger and more complex problems across different fields and disciplines. It will enable the exploration of new algorithms for quantum chemistry, finance, protein folding, optimization, among others, and bring quantum computing to the next level with prospects on commercial values in solving real-world problems. Major challenges to the quantum hardware will be the growth of the number of qubits in such systems to hundreds or thousands and even more importantly the reduction of error rates.

The opening of quantum computation centers will provide more systems with larger QV via cloud access. Highly integrated quantum computing systems will be the first on-premise systems delivered to and operated by the customer.

Trademarks

The following are trademarks of the International Business Machines Corporation ("IBM") in the United States and/or other countries:

IBM, IBM Q, IBM Q Experience, IBM Q Network, IBM Q System One, Qiskit.

References

1. J. Preskill, *Quantum Computing in the NISQ Era and Beyond* (2018). arXiv:1801.00862
2. https://www.ibm.com/blogs/research/2019/09/quantum-computation-center/
3. https://en.wikipedia.org/wiki/List_of_quantum_processors. Retrieved 26 Sept 2019
4. D. DiVincenzo, The physical implementation of quantum computation. Fortschr. Phys. **48**, 771 (2000)
5. G. Popkin, Quest for qubits. Science **354**, 1090 (2016)
6. R. Blatt, C.F. Roos, Quantum simulations with trapped ions. Nat. Phys. **8**, 277 (2012)
7. S. Debnath et al., Demonstration of a small programmable quantum computer with atomic qubits. Nature **536**, 63 (2016)
8. T. Xia et al., Randomized benchmarking of single-qubit gates in a 2D array of neutral-atom qubits. Phys. Rev. Lett. **114**, 100503 (2015)
9. A. Omran et al., Generation and manipulation of Schrödinger cat states in Rydberg atom arrays. Science **365**, 570 (2019)
10. M. Veldhorst et al., An addressable quantum dot qubit with faulttolerant control-fidelity. Nat. Nanotechnol. **9**, 981 (2014)
11. D.P. Franke et al., Rent's rule and extensibility in quantum computing. Microprocess. Microsyst. **67**, 1 (2019)
12. T.F. Watson et al., A programmable two-qubit quantum processor in silicon. Nature **555**, 633 (2018)
13. F. Dolde et al., Room-temperature entanglement between single defect spins in diamond. Nat. Phys. **9**, 139 (2013)
14. K. Nemoto et al., Photonic architecture for scalable quantum information processing in diamond. Phys. Rev. X **4**, 031022 (2014)
15. Y. Wang et al., Characterizing Si:P quantum dot qubits with spin-resonance techniques. Sci. Rep. **6**, 31830 (2016)
16. Y. He et al., A two-qubit gate between phosphorus donor electrons in silicon. Nature **571**, 371 (2019)
17. J.L. O'Brien, Optical quantum computing. Science **318**, 1567 (2007)
18. A. Aspuru-Guzik, P. Walther, Photonic quantum simulators. Nat. Phys. **8**, 285 (2012)
19. A. Fornieri et al., Evidence of topological superconductivity in planar Josephson junctions. Nature **569**, 89 (2019)
20. V. Mourik et al., Signatures of Majorana fermions in hybrid superconductor-semiconductor nanowire devices. Science **336**, 1003 (2012)
21. S. Nadj-Perge et al., Spin-orbit qubit in a semiconductor nanowire. Nature **468**, 1084 (2010)
22. N.M. Linke et al., Experimental comparison of two quantum computing architectures. Proc. Natl. Acad. Sci. **114**, 3305 (2017)
23. Y. Nakamura, YuA Pashkin, J.S. Tsai, Coherent control of macroscopic quantum states in a single-Cooper-pair box. Nature **398**, 6730 (1999)
24. J.E. Mooij et al., Josephson persistent-current qubit. Science **285**, 5430 (1999)
25. J.M. Martinis et al., Rabi oscillations in a large Josephson-junction qubit. Phys. Rev. Lett. **89**, 117901 (2002)

26. D. Vion et al., Rabi oscillations, Ramsey fringes and spin echoes in an electrical circuit. Fortschr. Phys. **51**, 4 (2003)
27. A. Wallraff et al., Strong coupling of a single photon to a superconducting qubit using circuit quantum electrodynamics. Nature **431**, 162 (2004)
28. Z.-L. Xiang et al., Hybrid quantum circuits: Superconducting circuits interacting with other quantum systems. Rev. Mod. Phys. **85**, 632 (2013)
29. J. Koch et al., Charge-insensitive qubit design derived from the Cooper pair box. Phys. Rev. A **76**, 042319 (2007)
30. A.D. Córcoles et al., Demonstration of a quantum error detection code using a square lattice of four superconducting qubits. Nat. Commun. **6**, 6979 (2015)
31. V.E. Manucharyan, J. Koch, L.I. Glazman, M.H. Devoret, Fluxonium: single cooper-pair circuit free of charge offsets. Science **326**, 5949 (2009)
32. A.D. Córcoles et al., *Challenges and Opportunities of Near-Term Quantum Computing Systems* (2019). arXiv:1910.02894 (2019)
33. M. Amico, Z.H. Saleem, M. Kumph, Experimental study of Shor's factoring algorithm using the IBM Q Experience. Phys. Rev. A **100**, 012305 (2019)
34. F. Arute et al., Quantum supremacy using a programmable superconducting processor. Nature **574**, 505 (2019)
35. E. Pednault et al., *Leveraging Secondary Storage to Simulate Deep 54-qubit Sycamore Circuits* (2019). arXiv:1910.09534
36. J.M. Gambetta, J.M. Chow, M. Steffen, Building logical qubits in a superconducting quantum computing system. npj Quant. Inf. **3**, 2 (2017)
37. N. Ofek et al., Extending the lifetime of a quantum bit with error correction in superconducting circuits. Nature **536**, 441 (2016)
38. L.M.K. Vandersypen et al., Experimental realization of Shor's quantum factoring algorithm using nuclear magnetic resonance. Nature **414**, 883 (2001)
39. L.M.K. Vandersypen et al., Implementation of a three-quantum-bit search algorithm. Appl. Phys. Lett. **76**, 646 (2000)
40. A.W. Cross et al., Validating quantum computers using randomized model circuits. Phys. Rev. A **100**, 032328 (2019)
41. E. Onorati, A.H. Werner, J. Eisert, Randomized benchmarking for individual quantum gates. Phys. Rev. Lett. **123**, 060501 (2019)
42. K.X. Wei et al., *Verifying Multipartite Entangled GHZ States via Multiple Quantum Coherences* (2019). arXiv:1905.05720
43. D.C. McKay et al., Three qubit randomized benchmarking. Phys. Rev. Lett. **122**, 200502 (2019)
44. E. Magesan et al., Efficient measurement of quantum gate error by interleaved randomized benchmarking. Phys. Rev. Lett. **109**, 080505 (2012)
45. https://www.ibm.com/blogs/research/2020/01/quantum-volume-32/
46. J.R. Johansson, P.D. Nation, F. Nori, QuTiP: an opensource Python framework for the dynamics of open quantum systems. Comput. Phys. Commun. **183**, 1760 (2012)
47. M.F. Gely, G.A. Steele, *QuCAT: Quantum Circuit Analyzer Tool in Python* (2019). arXiv:1908.10342
48. P. Krantz et al., A quantum engineer's guide to superconducting qubits. Appl. Phys. Rev. **6**, 021318 (2019)
49. R. LaRose, Overview and comparison of gate level quantum software platforms. Quantum **3**, 130 (2019)
50. https://qiskit.org
51. D.C. McKay et al., *Qiskit Backend Specifications for OpenQASM and OpenPulse Experiments* (2018). arXiv:1809.03452
52. A.W. Cross et al., *Open Quantum Assembly Language* (2017). arXiv:1707.03429
53. A.W. Harrow, A. Hassidim, S. Lloyd, Quantum algorithm for linear systems of equations. Phys. Rev. Lett. **103**, 150502 (2009)
54. https://quantum-computing.ibm.com
55. https://quantum-computing.ibm.com/support

56. https://quantum-computing.ibm.com/jupyter
57. A. Asfaw et al., *Learn Quantum Computation Using Qiskit* (2019). https://community.qiskit.org/textbook
58. A.J. McCaskey et al., Quantum chemistry as a benchmark for near-term quantum computers. npj Quant. Inf. **5**, 99 (2019)
59. A. Kandala et al., Error mitigation extends the computational reach of a noisy quantum processor. Nature **567**, 491 (2019)

Chapter 28
Human-Machine Interaction and Cognitronics

Ulrich Rueckert

28.1 Introduction

Our society is on its way into a truly hybrid society of natural and artificial agents. Digital assistants such as Amazon's Alexa, autonomous cars, drones, mobile robots, and humanoid robots such as Aldebaran's Pepper becoming more and more part of our lives in our homes, at work as well as in public areas. With ever-more smart technological capabilities opportunities exist for advancing the mechanisms that we employ for intuitive human-machine interaction. Interaction will take many different forms from immersive virtual worlds to direct and precise physical interaction with everyday appliances, production machines, or robots. It is imperative that the interaction is safe, smooth, ergonomically well designed, and easy to operate for human users. Future technology should be easy to handle and adapt to human preferences through individualized information, processing, and behaviour. "Cognitive Interaction Technology" [1] is a major step forward in all application domains: at home, at work and in leisure time.

Man-machine interaction requires a resource-efficient Cognitive System Architecture (CSA). Cognitive psychology and Artificial Intelligence (AI) have a long history of building cognitive architectures [2]. Some work focuses on modelling the invariant aspects of human cognition, whereas other efforts view architectures as an effective path to building intelligent agents. Important basic capabilities of CSAs are perception, situation assessment, prediction, planning, decision making, action, communication, and learning (Fig. 28.1). Basic properties are the representation, organization, utilization, acquisition and refinement of knowledge. Evaluation criteria for CSA are for example generality, versatility, rationality, reactivity, persistence, efficiency, improvability, and scalability [2].

U. Rueckert (✉)
Bielefeld University, Bielefeld, Germany
e-mail: rueckert@cit-ec.uni-bielefeld.de

B. Murmann and B. Hoefflinger (eds.), *NANO-CHIPS 2030*,
The Frontiers Collection, https://doi.org/10.1007/978-3-030-18338-7_28

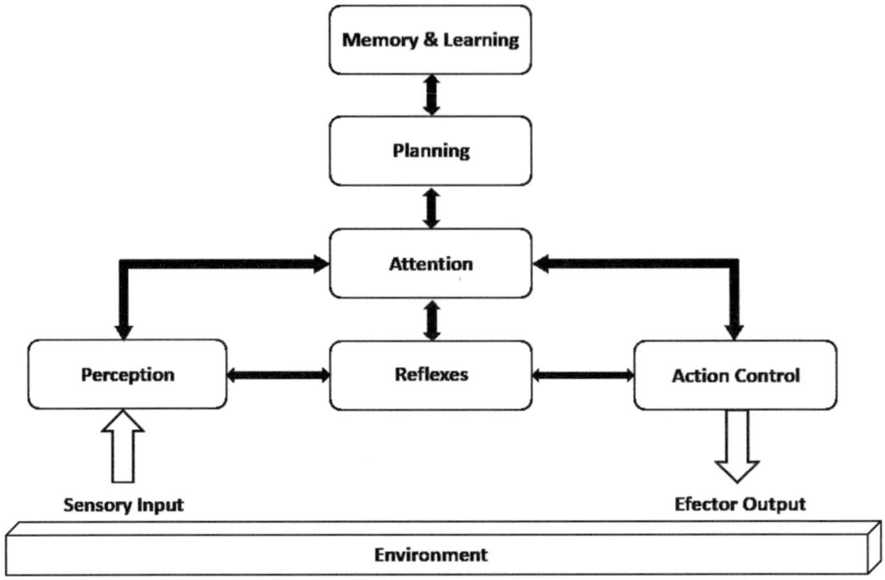

Fig. 28.1 Building blocks of a simplified cognitive system architecture

There is still a steady flow of new research on cognitive architectures, but only few considering system level approaches for integrated systems. Physical agents have limited resources for perceiving the world and affecting it, yet few architectures address this issue. Required are approaches for the management of an agent's resources to selectively focus its perceptual attention, its effectors, and the tasks it pursues. Although many architectures interface with complex environments, they rarely confront the interactions between body and mind that arise with real embodiment. For instance, to examine the manner in which physical embodiment impacts thinking and consider the origin of an agent´s primary goals in terms of internal drives. This demand can be met by an increased focus on system-level architectures that support complex cognitive behaviour and the CSA specifies the underlying infrastructure for an intelligent system. However, with no clear definition and no general theory of cognition, each architecture is based on a different set of premises and assumptions, making comparison and evaluation difficult [2].

In the following, technical issues of embedded CSAs for human-machine interaction are considered. Today, there are three main application areas of embedded CAS: smartphones, robotics, and automotive systems.

28.2 Human-Machine Interaction

Humans interact with machines invading our daily and professional life in many ways. With the increasing application of information technology in almost all areas of life, human-machine interaction has become a key technology of our modern information society. With the advancing integration of information and automation technology, technical objects increasingly become more independent, more flexible and capable of autonomous acting. The enormous functional expansion of the individual devices goes hand in hand with the availability of cheap communication technology, which facilitates the ubiquitous and spontaneous networking of everyday objects of all kinds. The result is a consistent but heterogeneous infrastructure that enables an unforeseen variety of new applications, services, and products.

All kind of technical objects can be enriched with computing and communication power as well as new user interfaces. This can be integrated almost inconspicuously in our professional and private environment for the peoples' benefit. Technical objects of all kinds become active nodes in complex networks and thus turn into a cooperative medium in our environment. The situation-oriented integration of technical products and services into open dynamic systems requires the combination of different aspects, which range from technological product features, questions of spontaneous networking, the development and configuration of services to user interfaces and safety mechanisms [3].

For human-machine interaction real-time operation is crucial. Human reaction and interaction times are between 1 ms and 1 s depending on the involved modalities (sight, hearing, touch, smell, and taste) [4]. The response time of human tactile to visual feedback control is approximately 1 ms. The development of the fifth generation wireless communication systems (5G [5]) promises 1000-fold performance gains with very low latency on the order of 1 ms or less and ultra-high reliability. Potential 5G applications range from industry, robots and drones, and virtual and augmented reality, to healthcare, road traffic, and smart grid. This perspective initiated the emergence of the Tactile Internet enabling real-time control and physical tactile experiences remotely [6]. Obviously, requiring a 1 ms round-trip latency and ultra-reliable as well as ultra-responsive network connectivity are huge challenges effecting all communication layers. The vision of the Tactile Internet and its potential impact on society is expected to add a new dimension to human-machine interaction as it opens up the access to knowledge available in huge databases in the cloud. The progress in wireless networking and the ubiquitous presence of Internet resources are the sources for the emergence of Cloud Robotics [7]. Cloud computing empowers robots by offering them more powerful computational capabilities and higher data storage facilities in the cloud.

The more natural approach is to have the CSA embedded in the mechanical body of the physical agent. All living creatures have their nervous system on board. As an alternative to conventional approaches for mobile robotics the field of Neurorobotics emerged recently [8]. Since brains are so closely coupled to the body, Neurorobots aim at studying neural functions in a holistic fashion. A key feature of brains is their resource-efficient ability to process cross modal information from multisensory input providing a robust perceptual experience and behavioural responses. Hence, the processing and integration of multisensory information streams such as vision, audio, haptics, and proprioception play a crucial role in the development of cognitive robots, yielding a situated interaction with the environment also under conditions of sensory uncertainty. The perception, integration, and segregation of multisensory cues improve the capability to physically interact with objects and persons with higher levels of autonomy in the real world. However, multisensory inputs must be represented and integrated in an appropriate way so that they result in a reliable perceptual experience aimed to trigger adequate behavioural responses. The modelling of cross modal processing in robots is of crucial interest for learning, memory, cognition, and behaviour, and particularly in the case of uncertain and ambiguous or incongruent multisensory input.

At present, CSAs are implemented mainly on standard hardware. The technology push for cognitive technical systems coming from nanotechnology is still impressive. Anticipating this technology means to shape system architectures for an increasing number of processors, memory capacity and embedded sensors. However, having this massive parallelism on board does not mean that we have to use these resources constantly in a massively parallel way. More likely, concurrent as well as sequential processes have to be coordinated on different abstraction levels in order to meet task requirements and resource efficiency. Application specific integrated circuits and intellectual property blocks for embedded intelligence are entering the market of cognitive devices. These chips are improving conventional symbolic AI and bio-inspired algorithms (see Chaps. 12 and 22).

28.3 Cognitronics

The term Cognitronics is a coinage, the combination of Cognition and Electronics. It was created by myself in 1989 after an inspiring discussion with Eduardo Caianiello, an Italian Professor of Theoretical Physics. The topic was "Is there a silicon way to intelligence?". His main conclusion, which anyone can certainly agree with, was: "Until fundamental concepts become better understood, the many advantages of silicon should not blind us to alternative neural network design" [9]. However, up to now we only have this technology available for making systems smarter: either by software or alternative hardware architectures or a combination of both.

The recent shift in computation towards massive parallelism is not a result based on breakthroughs in novel software or architectures for parallelism; instead, this shift is actually a technology push based on the progress in nanoelectronics offering 1000 cores on a chip today. The basic idea is to exploit the massive parallelism of current IC technologies for ultra-low power and fault-tolerant information-processing systems. Aiming at overcoming the big challenges of deep-submicron CMOS technology (power wall, reliability, and design complexity), bio-inspiration offers alternative ways to (embedded) artificial intelligence. The challenge is to understand, design,

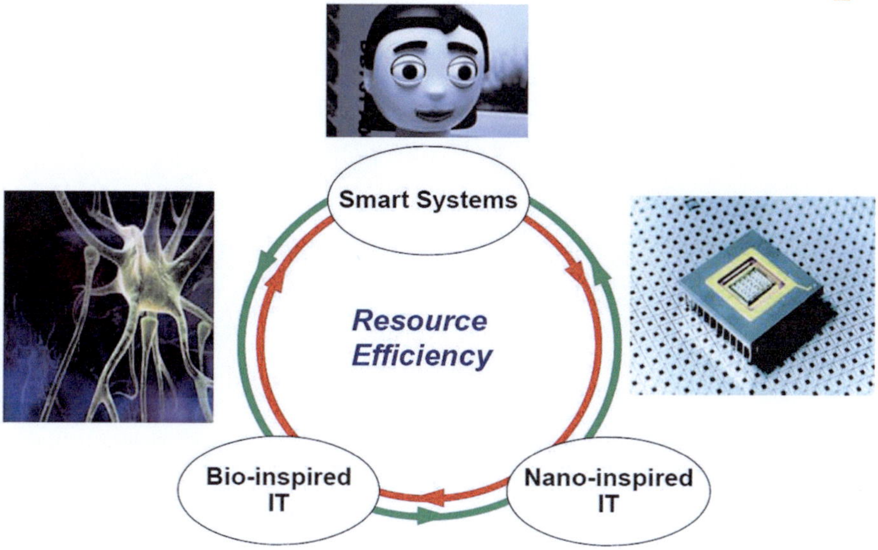

Fig. 28.2 Cognitronics: interplay of bio-inspired and nano-inspired information processing

build, and use new architectures for nanoelectronic systems, which unify the best of brain-inspired information processing concepts, conventional signal processing, and of nanotechnology hardware, including both algorithms and architectures (Fig. 28.2). The goal of Cognitronics is the implementation of such embedded resource-efficient architectures for cognitive technical systems; similar to neuromorphic systems introduced by Carver Mead [10]. The focus is on embedded massively parallel and reconfigurable system architectures, which are characterized by a decentralized organization utilizing architectures that autonomously adapt their system resources to changing task requirements in order to increase resilient and robust performance within complex, dynamic, and uncertain environments.

There are interesting 'engineering' aspects of biological neural networks from the computational standpoint and about the implementation of resource-efficient technical systems. The hundred billion neurons have on the order of 10^{15} connections, each coupling an action potential at a mean rate of not more than a few hertz. This amounts to a total computational rate of about 10^{16} complex operations per second. With structure sizes smaller than 0.1 μm, semiconductor technology starts falling below the level of biological structures forming the brain. However, the brain efficiently uses all three dimensions, whereas nanoelectronics mainly use the two physical dimensions of the silicon die surface and a restricted number of wiring layers. Nevertheless, taking an area of one square-millimetre-roughly the square dimension of a Purkinje cell (a type of neuron) in the cerebellar cortex, shown in Fig. 28.3 (right), we can use 10-nm CMOS technology to implement a digital neuron processor (Fig. 28.3, left) with one million 32-bit weight synapses and a 32-bit microprocessor as a neural processing unit. Weights are the practical implementation (in hardware, software,

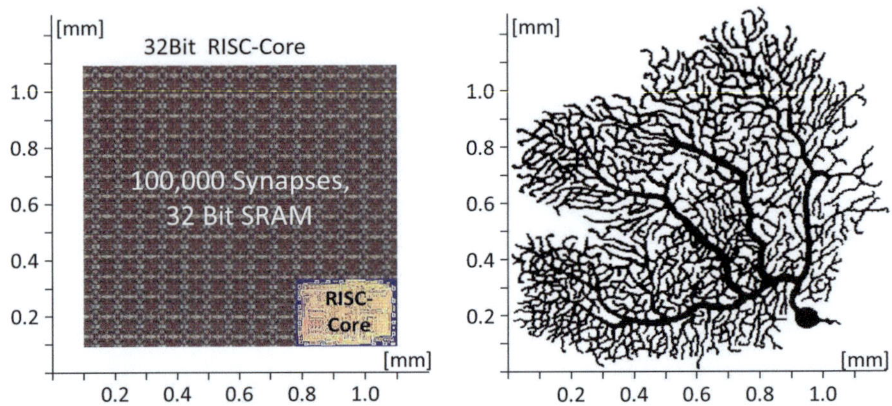

Fig. 28.3 Area comparison of a digital neuron in 40 nm standard CMOS technology (left) and a biological neuron (purkinje cell, right)

theory) of (biological) synapses (contacts between nerve cells). Such a processing unit is in general memory bound and can emulate several hundred artificial neurons.

An even greater challenge is the issue of power efficiency. The power efficiency of neurons (measured as the energy required for a given computation) exceeds that of computer technology, possibly because the neuron itself is a relatively slow component. While computer engineers measure gate speeds in picoseconds, neurons have time constants measured in milliseconds. While computer engineers worry about speed-of-light limitations and the number of clock cycles it takes to get a signal across a chip, neurons communicate directly at a few meters per second. This very relaxed performance at the technology level is, of course, compensated by the very high levels of parallelism and connectivity of the biological system. Finally, neural systems display levels of fault-tolerance and adaptive learning that artificial systems have yet to approach [11].

Out of the empire of acquired knowledge, some biological data are summarized in Table 28.1 as a performance guide to the human brain, which is interesting to compare with technical data from biomorphic silicon brains. The basis of this charge model is a charge density of 2 x 10^{-7} As cm^{-2} per neuron firing, fairly common to biological brains [12] and a geometric mean for axon cross sections, which vary from 0.1 to 100 μm^2.

We are still a long way from fully comprehending the functional mechanisms of the brain, far from any accepted description of the principles of information processing in brains and from the reconstruction of its capabilities. Nevertheless, we do have much to learn from brains from the computational standpoint and about the implementation of resource-efficient technical systems. Despite the revolutionary development of nanotechnology in the last decades, there is still an impressive difference between the resource efficiency of biological and technical systems. Hence, engineers are eager to learn how nature achieves such resource efficient implementations of

Table 28.1 Charge and energy model of the human brain[13]

Parameter	Human brain
Number of neurons	10^{11}
Synapses/neuron	10^3–10^5
Ionic charges per neuron firing	6×10^{-11} As
Mean cross section synaptic gap	30 μm^2
Charge/synaptic gap	6×10^{-14} As $= 4 \times 10^5$ ions
Action potential	70 mV
Energy per synapse operation	4 fJ $= 2.5 \times 10^4$ eV
Energy per neuron firing	4 pJ
Average frequency of neuron firing	10 Hz
Average brain power	$(2 \times 10^{11}) \times (4 \times 10^{-12}) \times 10 = 8$ W

complex and flexible behaviour. For cognitive systems, we must also involve the integration of the innumerable sensors and actuators (motors) enabling living creatures to survive in nature. In this outside-of-electronics domain, that of MEMS (Micro-Electromechanical Systems), remarkable progress has been made and continues with high growth [14].

This leads to the challenging question, of how to generate complex real-time behaviour in the limits of restricted resources. More precisely, how much cognition can be implemented in the limits of 100 Watts, one Terabyte of memory and one Tera-operations per second within a volume of one litre? These are limits relevant for embedding a cognitive architecture into a head of a mobile robot (Fig. 28.1]. For example, in the head of the robot Pepper from Aldebaran Robotics (having a volume of about one litre) we find actually one quad-core processor, 4 GByte of memory, about a dozen integrated sensors, motors and a battery pack of 795 Wh [15], only.

We are still a long way from fully comprehending the functional mechanisms of the brain. Nevertheless, we do have much to learn from brains from the computational standpoint and about the implementation of resource-efficient technical systems. The hardware realization of neural networks should not aim for an exact reproduction of nervous systems, but simply for an efficient use of available technologies for solving practical problems. Furthermore, as the research on CSA models is still ongoing, the system architecture should be flexible to support different approaches. But what are interesting guiding principles for resource-efficient embedded CSA?

We know from systems engineering that there is close interdependence between the main three system views (function, architecture, technology) for the development of resource-efficient technical systems (Fig. 28.4). The very high levels of parallelism and connectivity of brains have already been mentioned. Parallelism serves as a compensation for the slow speed of the processing elements (neurons)

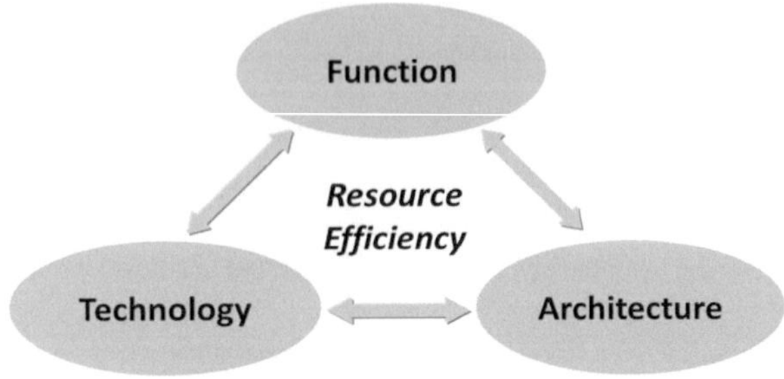

Fig. 28.4 Integrated design view on resource-efficient systems engineering

and as a source for redundancy resulting in low power consumption and higher robustness. Despite the massively parallel organization, not all processing elements are active at the same time. Brains employ sparse data representations in the form of activated cell assemblies [16]. Only a small number of neurons are active at the same time. As a consequence, the input of each neuron is sparse as well. Sparse codes and activities simplify internal computations and external communication of the processing elements, which has a positive effect on power consumption as well. It is assumed that neurons operate asynchronously as long as they are not loosely coupled in cell assemblies. Continuous self-organization based on local rules lead to stable and robust global behaviour. These three bio-inspired mechanisms should be taken as higher-level design guidelines for embedded, massively parallel system architectures for cognitive technical systems.

28.4 The Generic CoreVA-MPSoC

A flexible approach for emulating large CSAs is a Multiprocessor System-on-Chip (MPSoC) with an appropriate trade-off between performance, energy consumption and chip area. MPSoCs are defined by a very high number of small-sized CPU cores, which are able to run applications with high resource efficiency. In our research group, we developed the CoreVA-MPSoC [17] featuring a hierarchical interconnect architecture and scalable number of reconfigurable clustered processing cores (Fig. 28.5). The employed CoreVA CPUs can be easily extended by application specific instructions and dedicated hardware accelerators, as discussed in [18]. Generally, a CoreVA CPU can be used for general purpose applications, so that all kinds of application domains can be addressed. However, currently our CoreVA-MPSoC and especially our automatic partitioning tool focuses on streaming applications, like signal processing.

The CPU core used in our MPSoC is the 32-bit VLIW processor architecture CoreVA, which is designed to provide a high resource efficiency [19]. It features a six-stage pipeline. VLIW-architectures omit complex hardware schedulers and leave the scheduling task to the compiler. The CoreVA architecture allows to configure the number of VLIW slots, their functional units, as well as the number of

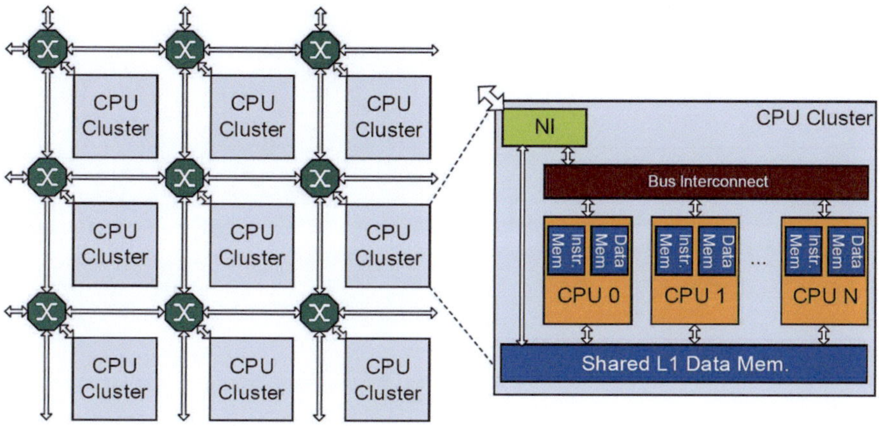

Fig. 28.5 Generic CoreVA-MPSoC architecture [17]

load/store units (LD/ST) at design-time. Functional units are arithmetic-logic-units (ALUs), dedicated units for multiply-accumulate (MAC) and division (DIV). Almost all instructions have a latency of one cycle, excluding memory loads, multiplications and MAC operations, which have a latency of two cycles. Additionally, both, ALU and MAC units, support a 16-bit single instruction multiple data (SIMD) mode. Due to the highly configurable architecture, it is possible to tailor the processors' performance to match the needs of a given application domain. As a typical Harvard architecture, the CPU features separated instruction and data memories. To verify our physical implementation flow, performance and power models, two chip prototypes based on the CoreVA CPU architecture have been manufactured in a 65 nm process using a conventional low power standard-cell library from ST Microelectronics [20]. This chip consumes about 100 mW at an operating clock frequency of 300 MHz. Additionally, an ultra-low power version, the CoreVA-ULP, is build using a custom standard cell library that was designed for sub-threshold operation using a multi-objective approach to optimize noise margins, switching energy, and propagation delay simultaneously (Fig. 28.6) [21, 22]. Operation voltage range from 1.2 V down to 200 mV and frequency from 94 MHz down to 10 kHz. The CPU´s lowest energy consumption per clock cycle of 9.94 pJ is observed at 325 mV and a clock frequency of 133 kHz. At this point the CPU core consumes only 1.3 mW. A performance and power management subsystem provides dynamic voltage and frequency scaling (DVFS) combined with an adaptive supply voltage generation for dynamic process and temperature variation (PVT) compensation.

To couple hundreds or thousands of CPUs, the CoreVA-MPSoC features a hierarchical interconnect architecture with a NoC interconnect that couples several CPU clusters. Each CPU cluster tightly couples several VLIW CPU cores via a bus-based interconnect using a common address space. Different interconnect topologies (shared bus, crossbar) and bus standards (AMBA AXI, OpenCores Wishbone) for use within CPU clusters are compared in [23]. This hierarchical interconnect

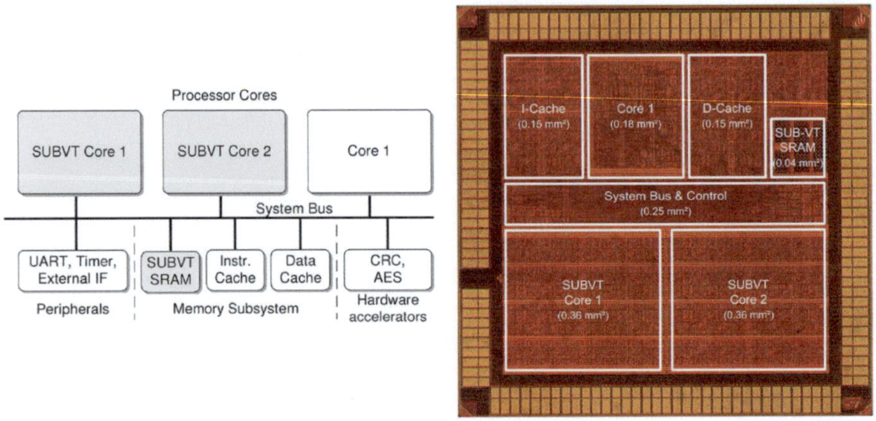

Fig. 28.6 Block diagram (left) and die photograph of the implemented CoreVA test chip in 65 nm CMOS [20]

allows for different memory architectures as discussed in [24, 25]. The efficient coupling of a cluster's shared memory to the network interface for highly efficient NoC communication is discussed in [17]. Physical implementation results utilizing a 28 nm FD-SOI standard cell technology (Fig. 28.7) show only minor differences in area and energy requirements between the use of tightly coupled shared and local data memory architectures.

To allow for large-scale MPSoCs with thousands of CPU cores, a Network-on-Chip (NoC) is used to connect multiple CPU clusters. The NoC features packet switching and wormhole routing. Packets are segmented into small flits each containing 64-bit payload data. Routers forward the flits along a path of network links. Each router has a configurable number of ports to allow for the implementation of different network topologies at design-time. An asynchronous router design, which indicates

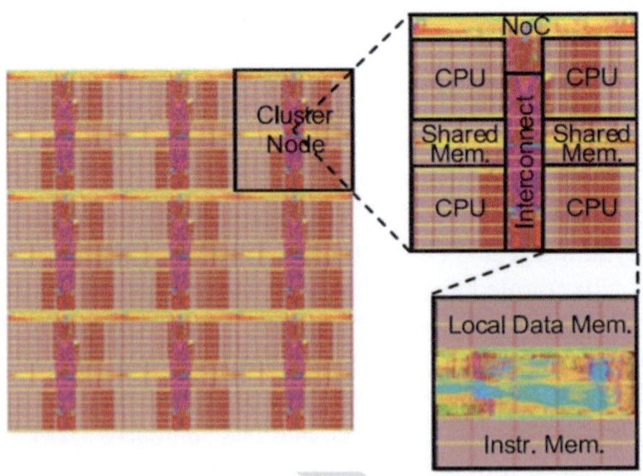

Fig. 28.7 3×3 2D-Mesh MPSoC layout and cluster node with 4 CPU macros (16 KB local data memory each, 64 kB shared data memory 0.817 mm^2, 28 nm FD-SOI CMOS) [17]

a lower area and power consumption compared to the synchronous NoC, is employed [26]. Additionally, this asynchronous NoC allows for a Globally-Asynchronous Locally-Synchronous chip design.

The CoreVA-MPSoC platform particularly targets streaming applications. Streaming applications consist of many different tasks which are connected via a directed data flow graph. An efficient communication model is required to allow communication between the tasks executed on different CPUs. Within the CoreVA-MPSoC a communication model with unidirectional communication channels is used. This approach promises more scalability and efficiency compared to shared memory concepts where the memory access can become the performance bottleneck. The parallelizing CoreVA-MPSoC compiler for streaming applications assists in programming the CoreVA-MPSoC [27, 28].

The CoreVA MPSoC is a scalable architecture for flexible, yet resource-efficient implementation of CSAs. Conventional AI approaches can be combined with ANN and neuromorphic systems. The hierarchical network topology supports sparse data communication with low latency and real-time guarantees. The GALS approach simplifies system design and increases power efficiency. DVFS and PVT enable self-organization on the circuit level. On the system level, self-organization can be achieved by reconfiguration on the architecture and adaptation on the algorithmic level. Fault-tolerance and robustness are achieved by redundancy within the massively parallel system architecture. In conclusion, the generic CoreVA-MPSoC is a versatile platform for design space exploration of embedded CSAs. A prototype is available on our FPGA rapid prototyping system RAPTOR [29].

28.5 Outlook

Modern application areas, like mobile robotics, require high computational power combined with low energy and manufacturing costs in embedded systems. Future technology should be easy to handle and adapt to human preferences through individualized information, processing, and behaviour. A major step forward for human-machine interaction are Cognitive Interaction Technologies [1], which require resource-efficient Cognitive System Architectures. Models of cognitive systems have evolved in nature in the course of biological evolution in large numbers. Therefore, it makes sense to transfer biological information processing principles to technical systems.

Biology has taken its own way through evolution based on its own special technology (real wet tissue). Exact brain emulation in software or implementation in dry solid-state circuitry may guide us the wrong way to artificial machine intelligence as we do not adequately account for the influence of the technology on function (behaviour) and system architecture. Probably there are many technological artefacts in brain measurements, which are irrelevant for systems behaviour and hence for system emulation. For example, the impressive brain simulations on supercomputers with neuron numbers comparable to the brains of a mouse or cat are still not able to perform some simple tasks within a natural environment. In 2009, the US DARPA launched the SyNAPSE program: Systems of Neuromorphic Adaptive Plastic Scalable Electronics [30]. It says in its description: "As compared to biological systems..., today's programmable machines are less efficient by a factor of 1 million to 1 billion in complex, real-world environments". And it continues: "The vision ... is the enabling of electronic neuromorphic machine technology that is scalable to biological levels." SyNAPSE is a program with explicit specifications and five milestones into four phases of ~2 years each. For example, the fourth milestone for 2016, was to fabricate a single-chip neural system of ~10^6 neurons packaged into a fully functional system, and to design and simulate a neural system of ~10^8 neurons and ~10^{12} synapses performing at "cat"-level environment. The last milestone for 2018 was to fabricate a multi-chip neural system of ~10^8 neurons and instantiate into a robotic platform performing at "cat" level (hunting a "mouse"). Today, we have to realize that these ambitious milestones have not been reached. Consequently, we are far from any accepted description of the principles of information processing in brains and from the reconstruction of its capabilities.

Despite the many success stories of DNNs, which have spurred a wave of public and corporate interest in AI, there are still many unsolved issues. One of these is how to make DNN architectures more efficient, especially to be used in embedded devices. The trend in deep learning research is that the models get larger and more complex. Consider e.g. AlphaGo, Google's famous AI powered Go program that beat the world champion Lee Sedol. AlphaGo has a power consumption of 1 MW compared to about 30 W for a human brain. Furthermore, AI is not real intelligence, because it doesn´t have the ability to react comprehensible to unknown situations. The typical AI model uses lots of data and computing power, but it is just a complex black box that can´t explain how it came to a specific recommendation or decision. As AI models are increasingly used to support human decision making, explainability becomes a key feature for future AI systems. Hence, DAPRA started the program Lifelong Learning Machines (L2M) for the "third-wave AI system", which would understand the context and environment in which they operate and, over time, build underlying explanatory models that allow them to characterize real-world phenomena [31]. L2M are developing systems that can learn continuously during execution and become increasingly expert while performing tasks.

Nature offers a fascinating source of inspiration for engineers. These solutions that the slow biological hardware of our brains can implement are by their very nature entirely different from technological solutions. Biological wetware is difficult or impossible to scale up and its networking with other brains has to go through the bottlenecks of language and shared perception. Hence, the technical realization of bio-inspired information processing should not aim for an exact reproduction of the biological model, but simply for an efficient use of available technologies for solving practical problems. Remember that creating human life had fascinated researchers for centuries, and every technological epoch had its view on how to do this (e.g. mechanical automata of the eighteenth century). And yet it is very early in the evolution of CSAs, so that all practical systems will be focused, application-specific solutions, certainly with growing intelligence, but with confined features of biological cognition. An outstanding example of an intelligent vision sensor system based on an efficient combination of classical computer vision and brain-inspired hardware architecture was developed by Ulrich Ramacher from Infineon Technologies in the course of several research projects funded by the German Federal Ministry of Education and Research (BMBF) [32].

Companies and research institutions are now starting to come up with devices for embedded AI that will stream continuous data needing to be processed in real-time from an increasing number of sensors. This technological development progressively driven by commercial interests will be accompanied by unknown impacts on our privacy, our social life, our means of communication and our own self-image. It will take a major, globally consolidated effort involving many disciplines from reliability and ethics to social science to achieve broad acceptance of brain-inspired hardware for demanding applications as human-machine interaction. The advancement of care-bots in various world regions will be a good test-ground for this evolution of cognitive systems. Particularly attractive is the application of CSAs in those domains where, at present, humans outperform any currently available high-performance computer, e.g. in areas like vision, auditory perception, or sensory motor control. Neural information processing is expected to have wide applicability in areas that require a high degree of flexibility and the ability to operate in uncertain environments where information usually is partial, fuzzy, or even contradictory. Nonetheless, the ability of human and animal brains to generalise is unparalleled and new machine learning methods are only just starting to get close in a few domains, such as object recognition based on sensory data.

References

1. https://www.cit-ec.de/en. Retrieved 15 Nov 2019
2. I. Kotseruba, J.K. Tsotsos, 40 years of cognitive architectures: core cognitive abilities and practical applications. Artif. Intell. Rev. (2018). https://doi.org/10.1007/s10462-018-9646-y
3. U. Rückert, MEDIATRONICS—things that communicate and cooperate. in *Proceedings of the International Conference Automatics and Informatics*, Sofia (2003), pp. 9–12
4. J. Johnson, *Designing with the Mind in Mind* (Morgan Kaufman Publishers, 2010)
5. M. Simsek et al., 5G-Enabled tactile internet. IEEE Sel. Areas Commun. **34**(3), 460–473 (2016)
6. P.G. Fettweis, The tactile internet: applications and challenges. IEEE Veh. Technol. Mag. **9**(1), 64–70 (2014)
7. O. Saha, P. Dasgupta, A comprehensive survey of recent trends in cloud robotics architectures and applications. *Robotics* (2018)
8. J. Kirchmer, Neurorobotics—a thriving community and a promising pathway toward intelligent cognitive robots. Front. Neurorobotics **12**, Article 42, (2018)
9. E.R. Caianiello, Is there a silicon way to intelligence? IEEE Micro **9**(6), 75–76 (1989)
10. C. Mead, M. Ismail (eds.) *Analog VLSI Implementation of Neural Systems* (Springer, Berlin 1989). ISBN 978-0-7923-9040-4
11. S. Furber, S. Temple, Neural Systems Engineering. J. R. Soc. Interface **4**(13), 193–206 (2007)
12. B. Sengupta et al., Action potential energy efficiency varies among neuron types in vertebrates and invertebrates. *PLoS Computat. Biol* (2010). https://doi.org/10.1371/journal.pcbi.1000840

13. B. Höfflinger, in *Chips 2020*, vol. 1, Chap. 18 (Springer, 2012)
14. J. Marek et al., MEMS—mirco-electromechanical sensors for the internet of everything. in *Chips 2020*, vol. 2, Chap. 15, (Springer 2012)
15. http://doc.aldebaran.com/2-5/family/pepper_technical/index_pep.html. Retrieved 15 Nov 2019
16. G. Palm et al., Neural associative memories, in *Associative Processing and Processors*, ed. by A. Krikelis, C.C. Weems (IEEE CS Press, Los Alamitos, 1997), pp. 307–326
17. J. Ax et al., The CoreVA-MPSoC: a many-core including tightly coupled shared and local data memories. IEEE Trans. Parallel Distrib. Syst. **29**(5), 1030–1043 (2018)
18. G. Sievers et al., The CoreVA-MPSoC: a multiprocessor platform for software-defined radio (2017). http://dx.doi.org/10.1007/978–3-319-49679-5_3
19. B. Hübener et al., CoreVA: a configurable resource-efficient VLIW processor architecture. in *Proceedings of the International Conference on Embedded and Ubiquitous Computation* (2014), pp. 9–16
20. S. Lütkemeier et al., A 65 nm 32 b subthreshold processor with 9T multi-Vt SRAM and adaptive supply voltage control. IEEE J. Solid-State Circuits **48**(1), 8–19 (2013)
21. G. Sievers et al., Design-space exploration of the configurable 32 bit VLIW processor CoreVA for signal processing applications. in *Proceedings of NORCHIP* (2013), pp. 1–4
22. M. Vohrmann et al., A 65 nm standard cell library for ultra-low-power applications. in *Proceedings of the European Conference on Circuit Theory and Design* (2015), pp. 1–4
23. G. Sievers et al., Evaluation of interconnect fabrics for an embedded MPSoC in 28 nm FD-SOI. in *Proceedings of the IEEE International Symposium on Circuits and Systems* (2015), pp. 1925–1928
24. T. Jungeblut et al., Design space exploration for memory subsystems of VLIW architectures. in *Proceedings of the 5th IEEE International Conference on Networking, Architecture and Storage* (2010), pp. 377–385
25. G. Sievers et al, Comparison of shared and private L1 data memories for an embedded MPSoC in 28 nm FD-SOI. in *Proceedings of the International Symposium on Embedded Multicore/Many-core Systems-on-Chip* (2015), pp. 175–181
26. J. Ax et al., Comparing synchronous, mesochronous and asynchronous NoCs for GALS based MPSoC. in *Proceedings of IEEE 11th International Symposium on Embedded Multicore/Many-core Systems-on-Chip* (2017)
27. W. Kelly et al., A communication model and partitioning algorithm for streaming applications for an embedded MPSoC. in *Proceedings of the International Symposium on System-on-Chip* (2014), pp. 1–6
28. M. Flasskamp et al., Performance estimation of streaming applications for hierarchical MPSoCs. in *Proceedings of the 2016 Workshop on Rapid Simulation and Performance Evaluation*, Article No. 3 (2016)
29. M. Porrmann et al., RAPTOR—a scalable platform for rapid prototyping and FPGA-based cluster computing. in *Parallel Computing: From Multicores and GPU's to Petascale, Advances in Parallel Computing*, (IOS press 2010), pp. 592–599
30. http://www.darpa.mil/Our_Work/DSO/Programs/Systems_of_Neuromorphic_Adaptive_Plastic_Scalable_Electronics_%28SYNAPSE%29.aspx
31. http://lifelongml.org/. Retrieved 15 Nov 2019
32. U. Ramacher, C. von der Marlsburg (eds.), *On the Construction of Artificial Brains* (Springer, Berlin, 2010)

Chapter 29
Efficient System-on-Chip (SOC) for Automated Driving with High Safety

Yutaka Yamada and Katsuyuki Kimura

29.1 Introduction

Advanced Driver Assistance Systems (ADAS) and Automated Driving Systems (ADS) are drawing attention as ways of reducing traffic accidents and improving vehicle usefulness.

As shown in Fig. 29.1, according to the World Health Organization (WHO), traffic accidents rank as the eighth leading cause of deaths worldwide with a total of 1.6 million fatalities in 2016 [1]. Reducing traffic accidents is therefore crucial for the improvement of the public health. According to NHTSA, 94% of serious traffic accidents are caused by human errors [2]. Therefore, to avoid traffic accidents, it is important to reduce such driver errors.

The conventional approach to collision safety was to reduce the damage caused by traffic accidents. An example of this is the airbag feature to reduce damage to passengers in the event of a vehicle collision. Although it may be able to reduce the damage caused by the collision, the airbag cannot prevent the collision itself. Thus, it was not helpful in preventing accidents from happening.

In recent years, an approach called Preventive Safety has been drawing attention as a way to reduce the number of vehicular accidents. To support this approach, vehicles equipped with ADAS features, such as emergency braking, are emerging. In fact, current evaluation standards by car safety assessment agencies such as Euro NCAP now include preventive safety features, not just the conventional collision safety measures [3].

Y. Yamada (✉) · K. Kimura
Toshiba Electronic Devices & Storage Corporation, 5801,
Horikawacho, Saiwaiku, Kawasaki, Kanagawa 2128520, Japan
e-mail: yutaka6.yamada@toshiba.co.jp

B. Murmann and B. Hoefflinger (eds.), *NANO-CHIPS 2030*,
The Frontiers Collection, https://doi.org/10.1007/978-3-030-18338-7_29

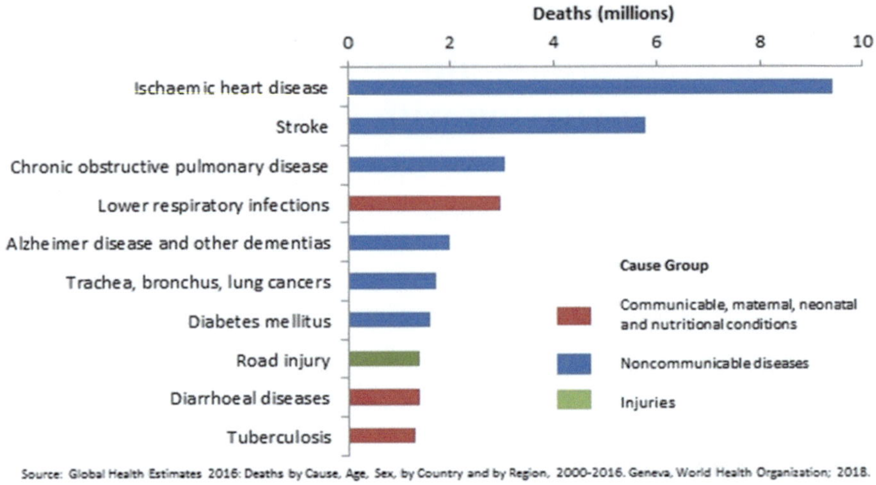

Fig. 29.1 Top 10 global causes of deaths, 2016. *Source* [1]

Vehicles are widely used for transportation all over the world. It is common to encounter problems such as traffic congestion in urban areas, environmental pollution, and transportation issues for some people in rural areas. With an aging population worldwide, travel by vehicle is increasingly important as a means for the elderly to move easily and safely.

Recently, MaaS [4] and other ride-sharing services have become popular as a way to improve vehicle usage efficiency and convenience. These services provide users with suitable and timely transportation by registered transportation services. Currently, public transport, taxis, and registered personal vehicles are the standard modes of transportation. However, the disparity in services between urban and suburban areas is a problem. This gap can be addressed by adding automated driving vehicles to provide those services. Logistically, labor shortage is a problem due to the growing number of e-commerce businesses. Labor shortage can be resolved with the introduction of automated driving into logistics.

Based on these premises, ADAS and ADS can be considered as highly promising technologies for accident reduction and convenience.

29.2 ADAS and ADS

ADAS and ADS are vital technologies for improving the convenience of automobiles.

Vehicles equipped with ADAS are widely used in today's market. The original ADAS capability was limited to simple vehicle controls, such as warning signals during lane departures and speed controls for emergency brakes. In recent years, advances in technologies have led to more complex controls for performing both

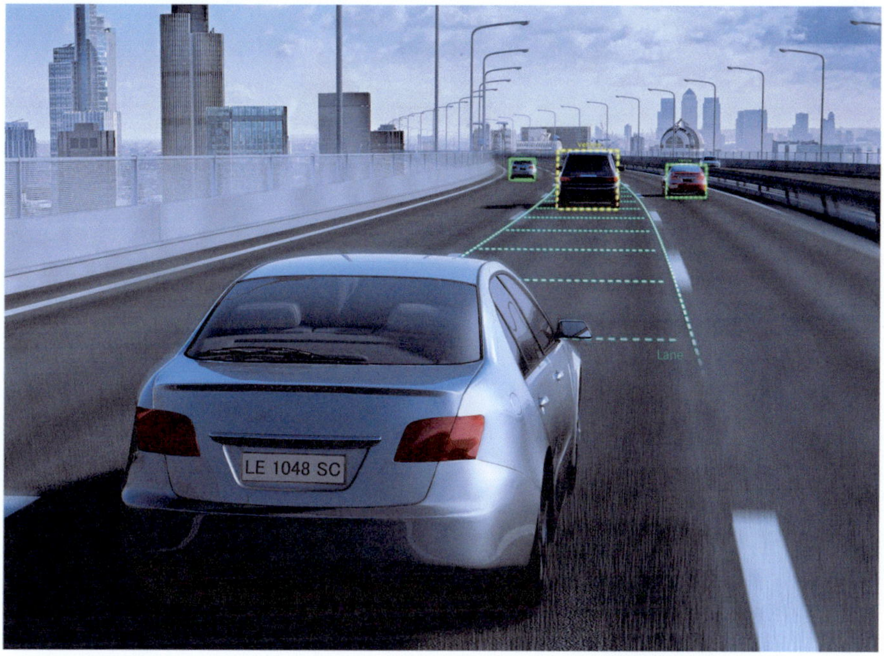

Fig. 29.2 Adaptive cruise control

steering and speed controls, such as Adaptive Cruise Control (Fig. 29.2) and Intelligent Parking Assist. The current operation of these features is restricted to limited environments such as highways and parking lots. However, it is expected that more varied environments will be supported in the future.

Table 29.1 shows the commonly used automated driving levels as defined by SAE [5]. A brief description of each is below. Levels 0, 1, and 2 are classified under ADAS since the driver is responsible for controlling the vehicle. An example for each level is described below.

- LV0 No Automation
  The vehicle does not have driving-support features for vehicle control.
  This includes warnings when an obstacle is detected during parking or the vehicle is exiting a lane.
- LV1 Driver Assistance
  This level supports driving with a single function of either steering or speed control.
  This applies to emergency braking and cruise control only with speed control.
- LV2 Partial Automation
  This level supports driving with both steering and speed control functions.
  An example is cruise control that helps the moving vehicle stay in its lane.

Level 3 and higher levels are automated driving systems in which the system is responsible for controlling the vehicle. In certain conditions when it is difficult to

Table 29.1 Levels of driving automation

Level		Driving	System request to take over driving	Assistance
0	No automation	Human	–	Warnings and momentary assistance
1	Driver assistance	Human	–	Steering or brake/acceleration
2	Partial automation	Human	–	Steering and brake/acceleration
3	Conditional automation	System/Human	Yes	Driving the vehicle under limited conditions
4	High automation	System	No	Driving the vehicle under limited conditions
5	Full automation	System	No	Driving the vehicle under all conditions

drive in Level 3 mode, the driver takes over control of the vehicle. For example, in Level 3 mode, drivers control the vehicle on ordinary roads, and automated driving can only be used on expressways. Levels 4 and up have more advanced automated driving operations and there is no handover for human control. In Level 4, automated driving is only supported in predetermined areas. In Level 5, the areas allowing automated driving are not restricted.

Based on these definitions, most vehicles currently available on the market are classified at Level 2 or below. Various companies and organizations are conducting research and development of advanced automated driving systems for practical use at Level 4 or higher [6]. Some organizations even have open-source frameworks for automated driving and are actively working toward the practical uses of automated driving vehicles [7, 8].

Figure 29.3 shows an example of ADAS and automated driving applications. The outline of each process is explained below.

- Sensing
  Acquire external information using various sensors such as camera, radar, LiDAR, millimeter wave, and sonar.

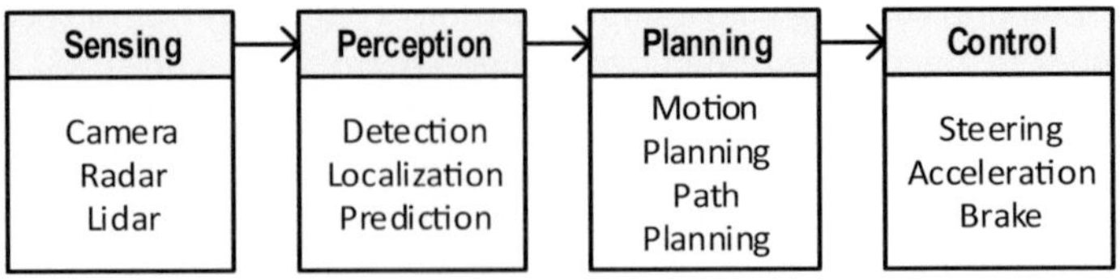

Fig. 29.3 A simple example of ADAS/ADS application

- Perception
  Using acquired information by Sensing, the vehicle recognizes itself and the surrounding environment by detecting and identifying objects, estimating its own position, creating peripheral maps, and predicting the movement of nearby objects.
- Planning
  Determine the route and movement of the vehicle based on the Perception result.
- Control
  Control the vehicle's steering, acceleration, and braking according to the route and operation determined by Planning.

Sensing and Perception processing recognize the surrounding situation, and it is important for both ADAS and ADS. Although, there is a difference between ADAS and ADS as to whether the vehicle is controlled by the driver or the system, the objects to be detected are almost the same.

Figure 29.4 shows an example of the sensor configuration for ADAS and ADS vehicles. Sensing processing obtains the image and object distance information using various sensors such as cameras, radar, LiDAR, and sonars mounted on the front, rear, left, and right sides of the vehicle. Each sensor has different characteristics. Cameras are effective in identifying objects because they can capture high-definition images, but it is difficult to capture images at night or through dense fog. Radar and LiDAR, on the other hand, can detect objects at night or through dense fog. In addition, the object detection range is different for telephoto and wide-angle lenses even when the same camera is used. Therefore, combining multiple sensors can improve detection accuracy and situational performance. Perception detects obstacles such as pedestrians and vehicles through image/signal processing of sensor information.

On the other hand, Planning and Control processing is highly involved in vehicle control, and it is very different between ADAS and ADS. The driver is mainly responsible for data processing in ADAS. In ADS, however, the system is in charge of this processing. In ADS, an overall route to the destination must first be selected. The vehicle can then be controlled according to local road conditions. In dangerous situations, how the vehicle is controlled, is also different between ADAS and ADS.

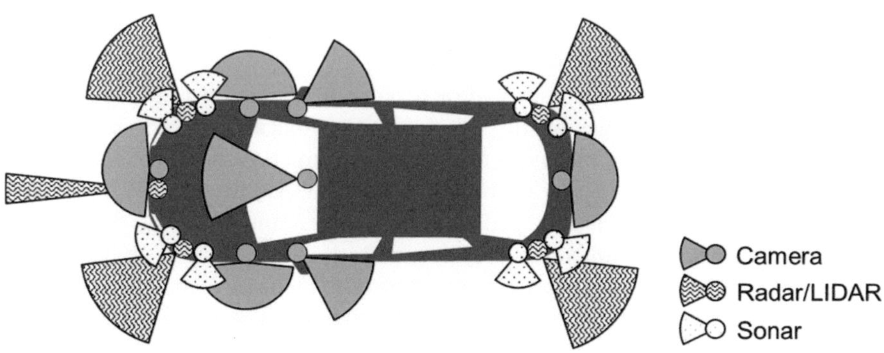

Fig. 29.4 An example of sensors for ADAS/ADS system

Since ADAS is not responsible for vehicle control, it is the driver's responsibility to avoid danger. However, when a dangerous situation is detected, ADAS notifies the driver with a warning and, in some cases, takes over control of the vehicle to do emergency braking. ADS, on the other hand, avoids dangerous situations by taking control of the vehicle and, if possible, operates continuously until the vehicle reaches its destination.

ADAS/ADS safety and reliability are important to avoid serious accidents that could be caused by system malfunctions. Current vehicle systems are becoming more and more complex, so it is crucial to ensure safety on the assumption that malfunctions may occur. Functional safety standards are widely enforced to improve safety and reliability. In 2012, ISO 26262, the international standard for functional safety of automotive electrical and/or electronic systems, was established. A revised version was released in 2018 [9] to include important specific requirements for semiconductors. This trend shows increasing awareness of the need for greater functional safety of automobile systems.

29.3 SoC Requirements for ADAS/ADS

Aside from sensor modules for sensing, an SoC is required to control Perception, Planning, and Control processing to implement ADAS and ADS functions. This section describes these SoC requirements.

29.3.1 High Performance

As explained in Sect. 29.2, Perception processing is important for both ADAS and ADS. To meet higher safety measures, improvements of recognition accuracy, detection and tracking of more objects, and support for higher-definition sensors are all necessary. Figure 29.5 shows the detection and recognition targets in ADAS/ADS. As shown in this figure, perception processing requires optimal signal and recognition processing to detect and identify various objects.

Improving recognition accuracy and increasing the variations of recognition targets are important factors that directly affect safety and that are crucial for automotive systems. By increasing the variations of recognition targets, dangers in more diverse conditions can be detected. For example, in Euro NCAP, the emergency braking target only considers vehicles and pedestrians. However, this target has expanded to other objects such as motorcycles, bicycles, and irregularly shaped objects. In the past, algorithms based on feature descriptor were used in image recognition. In recent years however, deep learning algorithms show higher image recognition accuracy [10–14]. This shows how applications for the automotive field are expanding and several SoCs including deep-learning accelerators are recently developed [15–17].

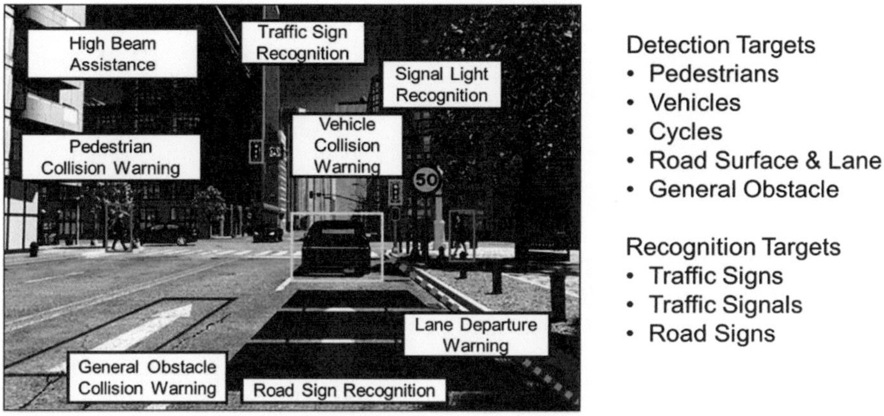

Fig. 29.5 Detection and recognition targets of ADAS/ADS application

By supporting sensors with better performance, it is possible to accurately recognize a greater variety of objects. For example, objects in the distance can be detected in a wider range by increasing camera resolution. Moreover, since the dynamic range is widened, an object can be detected accurately even in varying brightness conditions such as inside and outside a tunnel or at night. Since the image captured by the camera has much more data as a result of a higher resolution and wider dynamic range, higher signal processing performance is required.

ADS also requires processing performance for Planning and Control. Planning selects the optimal route based on the surrounding situation, and it controls the vehicle. For example, it controls the speed in relation to the vehicle's distance from the vehicle in front of it, and it selects a route matching the lane curvature. Furthermore, it controls the operation of the vehicle so that it can move towards the route determined by Control. If more route options are available, Planning can select the most suitable route to take.

29.3.2 High Power-Performance Efficiency

For stable operation in a vehicle, the SoC on vehicle should have low heat generation and low power consumption. Since a vehicle often operates at a high temperature, performance should be stable even at such temperatures. In addition, the ADAS/ADS module is often located behind the front mirror. In this case, providing cooling devices such as a fan may be difficult. Therefore, the module should only generate low heat even in a room temperature environment.

Table 29.2 Target failure metrics of ISO26262

	ASIL-A	ASIL-B	ASIL-C	ASIL-D
Failure rate	<1000FIT	<100FIT	<100FIT	<10FIT
SPFM	–	>90%	>97%	>99%
LFM	–	>60%	>80%	>90%

FIT Failure In Time; *SPFM* Single Point Failure Metrics; *LFM* Latent Failure Metrics

29.3.3 Functional Safety

To avoid serious accidents caused by automated driving-system malfunctions, it is important to comply with functional safety standards at the SoC level. The functional safety approach is to reduce risks by limiting the severity and frequency of failures to an acceptable level. Risk mitigation is implemented by introducing safety mechanisms that reduce fault incidence or its severity of the result of a failure. The main roles of safety mechanisms are malfunction detection, notification of the malfunction to people or systems outside of the malfunctioning system, and the prevention of a malfunction from propagating to other elements.

ISO26262 describes the safety risk level called Automotive Safety Integrity Level (ASIL). The levels are classified as ASIL A to ASIL D according to risk assessment. As shown in Table 29.2, target failure metrics are defined for each ASIL. The SoC must satisfy the target failure metrics of ASIL corresponding to the vehicle system.

29.3.4 Connectivity

The connection with other systems is essential to constructing a complex vehicle system.

An SoC for ADAS/ADS requires sensors such as camera, radar, and GPS/GNSS, storage for map information, log and etc., a monitor providing situation update to the driver, and interfaces that connect to other modules of the vehicle system.

29.4 SoC Architecture for ADAS/ADS Application

This section describes the architecture of the SoC. The following discusses what is required to implement ADAS/ADS functions and the safety mechanisms required for functional safety.

29.4.1 Elements for ADAS/AD Features

Figure 29.6 shows an example of an ADAS/ADS SoC.

The components of the SoC in this example are as follows:

- General Purpose Processors
  In an ADAS/ADS system, processes such as image recognition, signal processing, path determination, and vehicle control operate in parallel. The processor cooperates with other computing units to perform these processes. With a multiprocessor configuration, a number of processes can be processed efficiently. Also, recognition processing of Perception includes a large proportion of signal processing having high parallelism. Therefore, it is possible to offload these processes to DSP, GPU and accelerators. On the other hand, Planning and Control involve complex control processing, requiring processors of higher performance. Thus, the SoC for ADS system needs processors of higher performance.
  In addition, processors perform functional safety processing such as failure diagnosis control and error handling. Higher reliability is required for such processors.
- DSP and/or GPU
  DSP and GPU, having high parallelism, can perform signal processing and image processing more efficiently than processors. As these computing units are programmable, they can adapt to the progress of the algorithm.

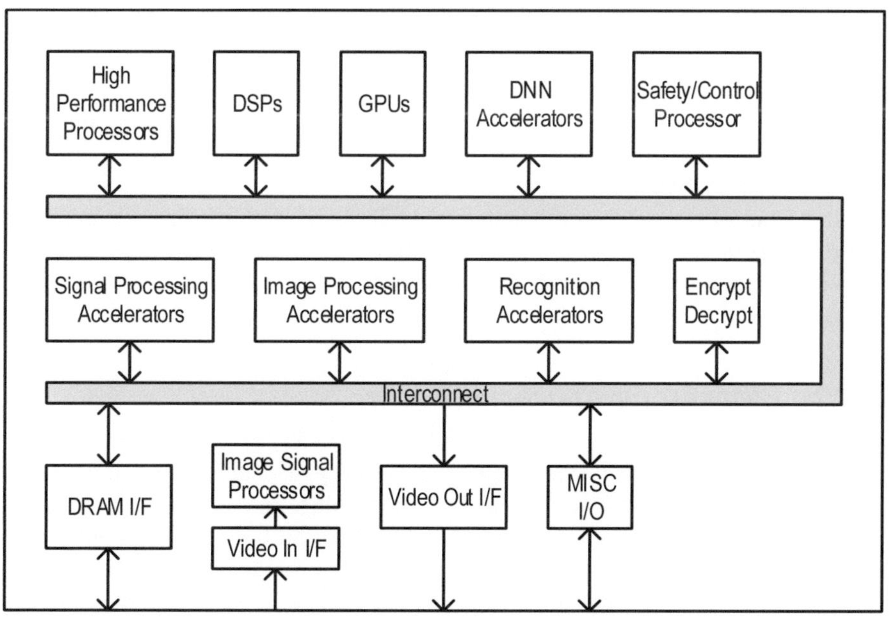

Fig. 29.6 An example of an ADAS/ADS SoC [15]

- Hardware dedicated accelerator
  This is a computing unit that performs the signal processing and image processing as with DSP and GPU. Unlike DSP and GPU, it can perform only certain processes; however it can achieve higher processing performance and power efficiency than DSP and GPU by optimizing parallelism configuration for each algorithm. Thus, the computing unit is suitable for efficient processing of mature algorithms, including signal processing such as filtering, image processing such as geometry transformation, image recognition such as template matching, and standardized algorithms such as codec and encryption/decryption.
- Image Signal Processor
  This is a type of hardware dedicated accelerator that processes images received from the camera module. In general, it has a very deep pipeline structure and can perform image processing such as noise reduction, demosaic processing, and quality enhancement. It has very high parallelism and can perform real-time signal processing of data from a high resolution camera, which is difficult to process with DSP and GPU.
- DNN Accelerator
  This is a hardware dedicated accelerator that specializes in DNN processing. It has a large number of MAC calculators to perform DNN processing efficiently. In general, DNN accelerators can process various neural network operations on a single accelerator and are programmable after the hardware is mounted. In that way, they are similar to processors.
- Various Interfaces
  These are the interfaces to connect with sensors (such as MIPI) and other modules in a vehicle (such as CAN, Flex-ray, and Ethernet). They include extended interfaces such as PCI express to consider the connection with coprocessors.

These computing units are complementary and a different computing unit may be used for the same processing.

The SoC for ADAS mainly supports Perception, so the computing units for Planning and Control may be fewer than that of ADS. Since ADS requires Planning and Control processing, high-performance computing units (processor, DSP, etc.) that support complex control processing are required.

29.4.2 Elements for Functional Safety

To meet the SoC functional safety standards, a safety mechanism should be implemented. If a failure occurs inside the SoC, this failure may lead to more serious problems if left uncontrolled. For example, failure of an ordinary component may affect other more important components, which may eventually lead to an accident. This is why a safety mechanism is an important quality factor in functional safety. It detects failures promptly, issues notifications, and prevents failures from affecting other components.

Table 29.3 An example of safety mechanisms

	Low coverage	Middle coverage	High coverage
Random logic	Software test	Runtime logic BIST	Duplicated logic
Memory	Parity		ECC
Interconnect	Parity	ECC	Duplicated logic
Clock and Voltage		Monitor	Duplicated monitor

Table 29.3 shows an example of a safety mechanism used in SoC. Systems with higher ASIL require a more robust safety mechanism.

29.5 Future Trends in ADAS/ADS

This section describes future ADAS/ADS trends and predictions for the functions required for SoC's.

- Enhanced functionality
  Most ADAS-equipped vehicles currently on the market only monitor the front while driving and left/right/rear side monitoring is limited to parking. In the future, it will be possible to react to left/right and rear monitoring while driving, such as assisting with lane changes. Since camera monitoring on the left/right and rear is also required during high-speed driving, it is expected that camera resolution will continue to increase. In addition, to respond to various environmental conditions, the trend towards combining sensors not only with cameras but with LiDAR and radar will continue as well.
  SoCs are also required to improve the performances of signal and DNN processing so they can process more data.
- Improvement of automated driving levels
  In the future, the level of automated driving is expected to improve so that the usefulness of vehicles can be enhanced. As the level of automated driving increases, the demands for system safety and reliability will also increase. For example, when a failure occurs in a Level 4 automated driving system, the system must not only notify the driver, but it must also also automatically move the vehicle to a safe place.
  To achieve this, SoCs for advanced automated driving systems are required to be highly reliable. Reliability is ensured by providing not just a single SoC but multiple SoCs working together.
- Development of connected cars
  To build a safer vehicle system, a mechanism for communicating with the surrounding environment, such as Vehicle-to-Vehicle (V2V) and Vehicle-to-Infrastructure (V2I), is being studied. For example, sharing information with a vehicle in front can make it easier to check for blind spots. Sharing traffic and accident information via the internet or other means can also improve traffic efficiency.

Consequently, security becomes more important as these vehicles connect via external networks.

29.6 Conclusion

This chapter discussed the trends in ADAS/ADS, SoC requirements, and the corresponding SoC architecture. An introduction to the prospects of ADAS/ADS and SoC advancement are also discussed.

ADAS/ADS need to perform multiple signal processing and image recognition processing simultaneously. To achieve this, SoCs should have both high performance and low power consumption. High safety standards are also required to avoid serious accidents due to malfunctions. An SoC for ADAS/ADS can achieve high performance and low power consumption by combing a variety of computing units such as processors, DSPs, GPUs, and DNN accelerators. Additionally, it can ensure safety by introducing various safety mechanisms.

In the future, ADAS/ADS will require higher safety standards and higher levels of automated driving. Further improvements in computing performance will be needed to achieve this. To effectively manage a variety of environmental conditions, various sensors should be supported. Moving forward, vehicles are expected to be connected externally, via the Internet or other means, which means that the security of vehicle systems will also become increasingly important.

References

1. WHO, The top 10 causes of death. Available: https://www.who.int/news-room/fact-sheets/detail/the-top-10-causes-of-death (Online)
2. NHTSA, Automated vehicles for safety. Available: https://www.nhtsa.gov/technology-innovation/automated-vehicles-safety (Online)
3. Euro NCAP, Euro ncap 2025 roadmap: in pursuit of vision zero, (Leuven, Belgium, 2017). Available: https://cdn.euroncap.com/media/30700/euroncap-roadmap-2025-v4.pdf (Online)
4. S. Pippuri, S. Hietanen, K. Pyyhti, Maas finland. Available: http://maas.global/ (Online)
5. SAE international, Taxonomy and definitions for terms related to driving automation systems for on-road motor vehicles. SAE International, (J3016) (2016). Available: https://www.sae.org/standards/content/j3016201806/ (Online)
6. M. Daily, S. Medasani, R. Behringer, M. Trivedi, Self-driving cars. Computer **50**(12), 18–23 (2017)
7. H. Fan, F. Zhu, C. Liu, L. Zhang, L. Zhuang, D. Li, W. Zhu, J. Hu, H. Li, Q. Kong, Baidu apollo em motion planner (2018), arXiv preprint arXiv:1807.08048. Available: https://arxiv.org/abs/1807.08048 (Online)
8. S. Kato, S. Tokunaga, Maruyama, S. Maeda, M. Hirabayashi, Y. Kitsukawa, A. Monrroy, T. Ando, Y. Fujii, and T. Azumi, Autoware on board: enabling autonomous vehicles with embedded systems. in *Proceedings of 2018 ACM/IEEE 9th International Conference on Cyber-Physical Systems (ICCPS)*, Apr 2018, (IEEE, 2018) pp. 287–296

9. ISO, 26262: Road vehicles-functional safety, International Standard ISO/FDIS, vol. 26262 (2018)
10. A. Krizhevsky, I. Sutskever, G. E. Hinton, Imagenet classification with deep convolutional neural networks. in *Proceedings of Advances in neural information processing systems (NIPS)*, Dec 2012, pp. 1097–1105
11. K. Simonyan, A. Zisserman, Very deep convolutional networks for large-scale image recognition (2014), arXiv preprint arXiv:1409.1556. Available: https://arxiv.org/abs/1409.1556 (Online)
12. R. Girshick, J. Donahue, T. Darrell, J. Malik, Rich feature hierarchies for accurate object detection and semantic segmentation. in *Proceedings of 2014 IEEE Conference on Computer Vision and Pattern Recognition (CVPR)*, June 2014, pp. 580–587
13. J. Long, E. Shelhamer, T. Darrell, Fully convolutional networks for semantic segmentation. in *Proceedings of 2015 IEEE Conference on Computer Vision and Pattern Recognition (CVPR)*, June 2015 (IEEE, 2015), pp. 3431–3440
14. K. He, X. Zhang, S. Ren, J. Sun, Deep residual learning for image recognition. in *Proceedings of the IEEE Conference on Computer Vision and Pattern Recognition (CVPR)* (IEEE, 2016), pp. 770–778
15. Y. Yamada, T. Sano et al., A 20.5TOPS and 217.3GOPS/mm^2 Multicore SoC with DNN accelerator and image signal processor complying with ISO26262 for automotive applications. in *2019 IEEE International Solid-State Circuits Conference-(ISSCC)*. 2018 Digest Tech. Papers, 7.2 (IEEE, 2019), pp. 131–132
16. M. Ditty, A. Karandikar, D. Reed, Nvidia's xavier system-on-chi. in *2018 IEEE Hot Chips 30 Symposium (HCS)*, Aug 2018 (IEEE, 2018)
17. D. Sarma, G. Venkataramanan, Compute and redundancy solution for tesla's full self driving computer. in *2019 IEEE Hot Chips 31 Symposium (HCS)*, Aug 2019 (IEEE, 2019)

Chapter 30
The Thirties

Boris Murmann and Bernd Hoefflinger

As expressed in a famous Danish proverb, "making predictions is difficult, especially about the future." Nonetheless, the preceding 28 chapters have put us in a position to make some predictions about the future of nano chips beyond the horizon of 2030. In the following sections, we extrapolate upon the key points presented by our authors and speculate about the most significant trends and asymptotes that we expect to pass the test of time. We begin with a big-picture overview, followed by an inspection of the following major themes that have coalesced throughout this book:

- The grand challenge: Intelligent machines for man-machine cooperation and navigation
- Artificial intelligence as an application enabler and new driver for the chip industry
- The challenge of flexible and efficient computing at all complexity scales
- The challenge of moving bits at all length scales and 3D integration
- Interfacing with the physical world and the human nervous system.

30.1 The Big Picture

Since the very beginning of the semiconductor era, innovation has relied on application drivers that warrant large investments in process technology and make us dream about future possibilities. However, especially over the last decade, the weight and significance of application pull has dramatically increased and tends to dwarf the strong technology push that marked the big bang of our industry. The statement "data

B. Murmann (✉)
Stanford University, Stanford, CA, USA
e-mail: murmann@stanford.edu

B. Hoefflinger
Sindelfingen, Germany
e-mail: bhoefflinger@t-online.de

B. Murmann and B. Hoefflinger (eds.), *NANO-CHIPS 2030*,
The Frontiers Collection, https://doi.org/10.1007/978-3-030-18338-7_30

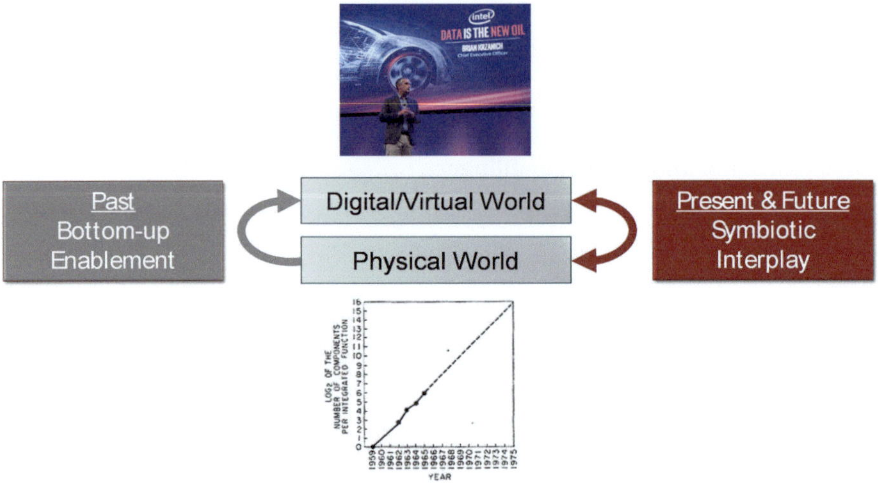

Fig. 30.1 Symbiotic interplay between the physical and virtual worlds. Top photo from [1]. Bottom chart showing Gordon Moore's famous prediction [2]

is the new oil," which was made by Intel's CEO Brian Krzanich in 2016, exemplifies this trend. It is important to understand that our reliance on symbiotic interplay with application developers (who may not be semiconductor experts) will only continue to grow (see Fig. 30.1). Within this context, our "big picture" ecosystem predictions for the 2030s are summarized as follows:

- Differentiation among competitors has transitioned from fab-level to design-level in the last two decades. Differentiation will further rise to the application and domain level.
- An increasing number of large corporations that define and drive new applications will have chip design teams. The onset of this trend is already marked by chip releases from Tesla, Amazon and others.
- The industry will continue to change from vendors following ITRS-like scaling roadmaps, to an ecosystem filled with corporation- and application-specific roadmaps.
- System design has become and will continue to be even more democratized through the Internet. Makers will enjoy increased access to easy-to-use compute platforms, 3D printers and design software. This community will play an important role in designing the future.
- The abundance of learning chips and machines in the data-driven world will fuel a totally new discipline of Industrial Control.

30.2 The Grand Challenge: Intelligent Machines for Man-Machine Cooperation and Navigation

The cross-cutting grand challenge that will continue to drive a significant portion of semiconductor applications lies in improving man-machine interaction and enabling intelligent machines that can perform highly complex tasks without human intervention. In this book, we outlined some of the possibilities in Chaps. 18, 20, 24, 25, 28 and 29. Establishing these capabilities has the potential to revolutionize healthcare, transportation, and entertainment. While improving chips and electronic systems will be a necessary ingredient, many of the most significant breakthroughs are due at higher levels of abstraction. Most significantly, research into what constitutes intelligence and how an intelligent machine should behave is still at its infancy. At the same time, the electronic systems that are built with our current understanding are already immensely complex and difficult to build, debug and operate reliably. Within this context, our predictions for the 2030s are summarized as follows:

- Virtual and augmented reality will be mainstream.
- These systems will be able to pass a "visual Turing test for displays," where a user would not be able to discern digital from physical content.
- The community will define new standards for levels of machine autonomy.
- Managing complexity in electronic systems is an ongoing challenge that could become a significant showstopper in the thirties. System and chip design must be become even more automated, modular, and debug friendly.

30.3 Artificial Intelligence as an Application Enabler and Driver for the Chip Industry

A typical question for the semiconductor community is to ask "what is the next killer application?" Since the introduction of the smartphone, we have been patiently waiting for a new answer. However, from today's perspective, it may very well be that there is no new killer application in form of a concrete new device, but it instead comes in form a of a new capability in our data-driven environment. We believe that this capability is machine learning (ML) and its ultimate manifestation as artificial intelligence (AI). As enumerated in [3], there are currently more than 100 papers per day published on this subject, with a rate of doubling in less than two years. This signifies the immense potential of this technology for a wide range of applications, including the ones highlighted in the previous section. Chaps. 9, 10, 12–14, 18, 19 and 22 of this book have provided a closer look at various aspects of the ML/AI landscape and its foundations. With this background, our predictions for the 2030s are summarized as follows:

- AI and AI-enabled applications will be the key drivers for the semiconductor industry and related new businesses.

- While we differentiate today between "neuromorphic" and more generic deep learning approaches, the thirties will bring a unified perspective that is based on improved understanding in how the physical world is constructed (see e.g. [4]).
- In contrast to the majority of today's systems, which are trained "offline," AI systems of the thirties will learn continuously in the field. This will not only require algorithmic innovations, but also improvements in non-volatile memory technology, real-time self-test and self-repair.

30.4 The Challenge of Flexible and Efficient Computing at All Complexity Scales

Our needs for ubiquitous computing have continued to grow exponentially. The complexity scale of today's computing systems ranges from tiny platforms running on harvested energy, through battery powered handheld devices, desktop computers, all the way to super computers and data centers. The needs across this space have become considerably more heterogenous and have followed a trend of increasing specialization. Specialized computing helps us offset the dramatic slowdown in progress from CMOS scaling alone and has led to new directions in computer architecture [5]. However, while specialization can boost performance, it can adversely affect programmability and compatibility with new and emerging algorithms. This trade-off is felt particularly strongly in hardware design for ML and AI [3] and Chaps. 11, 13, 14, 18, 19 have looked at some of the relevant aspects. At the other end of the spectrum, we find traditional supercomputers (see Chap. 16) and emerging systems for quantum computing (see Chaps. 26 and 27), which come with their own unique challenges. Within this context, our predictions for the 2030s are summarized as follows:

- The rising cost of complex monolithic devices will lead to a wide spectrum of disaggregated programmable devices with multiple dice connected at the package or at wafer level.
- Field programmable and coarse-grained reconfigurable architectures will play a critical role in designing computing systems for the next generation of applications.
- The software experience for field-programmable devices will change to temporal models instead of today's largely spatial approach. Their role moves further from customizing logic to customizing applications and systems.
- As predicted in CHIPS 2020 [6], synaptic operations (multiply & add) in custom chips for ML and AI have already reached the single digit femtojoule range (see Chap. 18). It is unlikely that this number will reduce significantly in the thirties. Instead, the challenge is shifted to moving bits instead of computing (see next section), naturally leading to 3D integration.
- AI processors will be more dynamic and contain processing structures that are normally off, fire up to perform bursts of computation, and go back to "dark" after storing the results in their 3D fabric.

- AI processors will natively support staged inference (supported by emerging non-volatile memory technology), memory-aware quantization, memory voltage over-scaling and algorithmic error tolerance.
- As already discussed in CHIPS 2020 Vol. 2 [7], providing electric power to exponentially growing large-scale computing systems will remain a key challenge in the thirties.
- Quantum computers may help mitigate the energy problem and could become as ubiquitous as conventional microprocessors if we find a way to operate qubits at higher temperatures.
- Algorithms and methods for verifying quantum programs will become a very important aspect in moving beyond the point of classically-simulatable quantum algorithms.

30.5 The Challenge of Moving Bits at All Length Scales and 3D Integration

As already highlighted in CHIPS 2020 Vol. 2 [7], the energy cost of moving a bit across the Internet is relatively large and continues to be a potential showstopper for the future growth of data traffic. Similar concerns hold for moving bits even across much smaller length scales, e.g. from chip-to-chip and within a large chip. The associated energies are summarized in Table 30.1 (from [8]) and are seen to be significantly larger than for typical on-chip compute operations. The recent development of data-intensive AI chips has once again highlighted this fact and provides fuel for the development of 3D chips that minimize the physical distance between

Table 30.1 Energies for Communications and computations. From [8] ©IEEE 2017

Operation	Energy per bit
Wireless data	10–30 μJ
Internet: access	40–80 nJ
Internet: routing	20 nJ
Internet: optical WDM links	3 nJ
Reading DRAM	5 pJ
Communicating off chip	1–20 pJ
Data link multiplexing and timing circuits	~2 pJ
Communicating across chip	600 fJ
Floating point operation	100 fJ
Energy in DRAM cell	10 fJ
Switching CMOS gate	~50 aJ–3 fJ
1 electron at 1 V, or	0.16 aJ
1 photon @ 1 eV	160 zJ

compute and memory functions. This topic was extensively discussed in this book through Chaps. 8–11, 15, 18–20. Extrapolating further, we anticipate the following trends for the adoption of 3D technologies in the 2030s:

- AI chips and other data-intensive applications will embrace monolithic 3D integration in conjunction with in-memory computing across their chip layers.
- Storage class memory (SCM) will use 3D cross point technology.
- High-speed memory (DRAM) will use 3D cube technology.
- Logic foundries will use 3D Stacking with hybrid bonding for heterogeneous integration.
- Ultra-low voltage differential transmission-gate logic in 3D communication with local 3D CMOS SRAM (both at 300 mV) will be used to improve the logic/memory interface.
- 3D-integration will enable new multi-sensor arrays with gigantic data rates.

30.6 Interfacing with the Physical World and the Human Nervous System

The physical world remains stubbornly analog. No matter how sophisticated our compute systems will become, they must ultimately rely on analog interfaces to digitize real-word information. We have closely tracked the progress in A/D interfaces in both CHIPS 2020 [6] and CHIPS 2020 Vol. 2. The update in the present book (see Chap. 17) concludes that energy scaling of generic A/D interfaces will come to an end in the 2020s. Going forward, innovation will be driven by specialized architectures, very similar to the way this is already happening in digital computing. Neural interfaces are an example of a futuristic application that is amenable to application-specific optimizations (see Chap. 24). Generally, a key requirement for further efficiency gains is to consider the proper "domain" in which the information should be interpreted, digitized and presented (see Chaps. 3 and 21). Example domains are logarithmic versus linear, time domain versus frequency domain, etc. In this area, our predictions for the 2030s are summarized as follows:

- Generic A/D interfaces will operate at ~200 fJ per conversion for an effective number of bits of 10 ($FOM_S \approx 186$ dB, see Chap. 17). This energy number scales $4\times$ per bit, and improvements beyond this limit must be realized through application customization.
- HDR vision and display systems that embrace the logarithmic scale will become mainstream.
- Semi-intelligent A/D interfaces that act as feature extractors for ML and AI backends will become mainstream. They will play a critical role in mitigating the data deluge at the physical edge of AI systems.

- The number of electrodes used in interfaces to the human nervous system will continue to grow exponentially (see Chap. 24 and [9]), reaching millions in the thirties.
- We will know how to design electronic circuits that learn how to "talk" to neural circuits, instead of assuming the reverse.

References

1. "Data is the new oil" declares Intel CEO Brian Krzanich. Available: https://hexus.net/ce/news/automotive/99277-data-new-oil-declares-intel-ceo-brian-krzanich/ (Online)
2. G.E. Moore, Cramming more components onto integrated circuits. Electron. **38**(8), 114–117 (1965)
3. J. Dean, The deep learning revolution and its implications for computer architecture and chip design. arXiv preprint, https://arxiv.org/abs/1911.05289
4. H.W. Lin, M. Tegmark, D. Rolnick, Why does deep and cheap learning work so well? arXiv preprint, https://arxiv.org/abs/1608.08225
5. J.L. Hennessy, D.A. Patterson, A new golden age for computer architecture. Commun. ACM **62**(2), 48–60 (2019)
6. B. Hoefflinger (ed.), *CHIPS 2020—A Guide to the Future of Microelectronics*, (Springer Science and Business Media, 2012). ISBN 978-3-642-22399-0
7. B. Hoefflinger (ed.), *CHIPS 2020 Vol. 2—New Vistas in Nanoelectronics*, (Springer Science and Business Media, 2016). ISBN 078-3-319-22092-5
8. D.A.B. Miller, Attojoule optoelectronics for low-energy information processing and communications. J. Light. Technol. **35**(3), 346–396 (2017)
9. C. MoraLopez, Unraveling the brain with high-density CMOS neural probes: tackling the challenges of neural interfacing. IEEE Solid-State Circuits Mag. **11**(4), 43–50 (2019)

Index

B. Murmann and B. Hoefflinger (eds.), *NANO-CHIPS 2030*,
The Frontiers Collection, https://doi.org/10.1007/978-3-030-18338-7

MIX
Papier aus verantwortungsvollen Quellen
Paper from responsible sources
FSC® C105338

If you have any concerns about our products,
you can contact us on
ProductSafety@springernature.com

In case Publisher is established outside the EU,
the EU authorized representative is:
Springer Nature Customer Service Center GmbH
Europaplatz 3, 69115 Heidelberg, Germany

Printed by Libri Plureos GmbH
in Hamburg, Germany